The Philosophy of Biology

History, Philosophy and Theory of the Life Sciences

Volume 1

For further volumes:
http://www.springer.com/series/8916

Kostas Kampourakis
Editor

The Philosophy of Biology

A Companion for Educators

Springer

Editor
Kostas Kampourakis
Secretariat of Educational Research and Development
Geitonas School
Vari Attikis, Greece

ISSN 2211-1948 ISSN 2211-1956 (electronic)
ISBN 978-94-007-6536-8 ISBN 978-94-007-6537-5 (eBook)
DOI 10.1007/978-94-007-6537-5
Springer Dordrecht Heidelberg New York London

Library of Congress Control Number: 2013938263

Printed on acid-free paper

Springer is part of Springer Science+Business Media (www.springer.com)

To my father, Giorgos Kampourakis, who encouraged and supported my intellectual life ever since I was a child

Foreword

There is an ancient Chinese curse: "May you live in interesting times!" My strong suspicion is that the contributors to this splendid volume would think of it as a challenge rather than a curse, which is just as well, for we do indeed live in interesting times. Apart from threats of violence and upheaval, we live in a time when the world is changing rapidly, as societies previously poor and excluded are starting to rise up and demand their places at the table of comfort and satisfaction and meaning. One thinks of the huge changes that have already occurred in China and are now starting to make India and other parts of the world full and functioning societies, with food and health and the possibility of lives of joy and worth.

No one, certainly no one in this volume, would regret this change at all. Everyone is united in hoping that all can share in nature's bounties. But the changes do bring challenges, especially to those in long-established cultures and civilizations. No longer can we automatically take to ourselves the lion's share of the world's energy resources. No longer can we expect that our food supplies will be furnished by others, to the detriment of themselves. No longer can we assume that disease and sickness are the burden of denizens of faraway lands. With change and the rise of others, with electronic devices ever-conquering, with travel becoming so much more commonplace, we live in a world that is integrated but that also makes demands on those already established if they will merely stay still let alone move ahead.

Demands that we must meet and science and technology, as they have been for the centuries since the Scientific Revolution, are going to be in the forefront. And more than this, biological science and technology are going to be vital. Already, for instance, plants are being utilized for new energy sources. Foodstuffs and fuels are being produced by means that depend crucially on knowing the details of life down to the smallest molecules. Sickness and disease are being fought with new medicines and new techniques, again dependent on biological knowledge as never before.

We cannot, we must not, stand still. Biological understanding must be produced and improved and extended and developed, by us and by future generations. And this means education. If worthwhile life on this shared planet is to continue –worthwhile life not just for us humans but for the other inhabitants also – then the training of young minds in our understanding of the nature of the organic world is absolutely crucial. There is no greater task or higher or more challenging calling. For this reason, if for no other, a volume such as this – written and edited with understanding and sensitivity – is to be celebrated.

But why philosophy? What has philosophy to say to biology? What has philosophy to say to biological education? Judging from the comments of many of today's active scientists, including active biologists, not much! The comments by scientists come thick and fast, and they tend not to be complimentary. Philosophy at best is something best done by scientists at the end of a busy day in the laboratory, and even then it is little more than common sense and an extension of the science produced by professionals in the field. Is this a fair judgment? After our initial excitement of finding a book that is going to tackle the all-important matter of biological education, should our enthusiasm gush out as air from a punctured tire when we discover that this is a volume dedicated to the place of philosophy in science education?

Three responses come to mind. First, the scientists are quite wrong about philosophy. It is not irrelevant. It is not just an occupation for tired scientists at the end of the day or – even worse – finished scientists at the end of a career, now going through what is unkindly known as the "philosopause." My inclination is to say that the scientists, the biologists, are wrong. There is more to philosophy than this. However, I cannot but agree that today's philosophers have worked hard to bring on the contempt of the biologists. Analytic philosophy, particularly as it was influenced by Wittgenstein, has always been antipathetic to biology. In a move that only academics could make, it is quite denied that the fact that we are modified monkeys rather than modified mud has any relevance to problems of knowledge, epistemology, or problems of morality, ethics.

Fortunately, although this mindset still stalks the land, as the contributors to this volume show so convincingly, no longer are all philosophers turning their backs on biology. The very opposite is the case. They are taking biology so seriously that they feel the need to grasp the science at such a level of sophistication that they can engage fact to face with the practitioners, which brings me to the second response. Perhaps philosophy must meld with science, with biology, so essentially there is no difference, and a biological education should include a philosophical component because such a component is essentially biology anyway.

Now I do not want to deny that there have been philosophers of biology who have gone so far down the route of biology that it is at times difficult to know how to categorize them. The late David Hull, the founder of modern philosophy of biology, was so engaged in the taxonomic struggles of the 1970s – the old guard versus the cladists – that he was neither one nor the other, but both. Significantly, he was president of both the leading philosophy of science association and of the leading taxonomic association. But generally I don't think this is the way to go, and I think in this I am joined by the editor and contributors to this volume. The cobbler should stick to his last. We are philosophers not biologists. Ultimately we are not trained to do the best cutting-edge science. We should be proud of what we are.

We should be proud of what we are. That is the third response, and that is the key. Philosophy, the discipline of the contributors to this volume, is a subject with a justifiably proud heritage – Plato and Aristotle, Augustine and Aquinas, Descartes and

Kant, Hegel and Russell, and even – dare I say it! – Wittgenstein. Philosophy does deal with understanding, in knowledge about the world and in conduct in this world, and much more. We philosophers have training in these areas, and this is something that we can bring to biology, to biological education. You need to grasp science, biology, for what it is today. But science, biology, thank god never stands still. It is always moving, and the scientists, the biologists, must contribute to this movement or at the least understand and appreciate the movement.

It is here that, as this volume shows, philosophy has a contribution to make. We philosophers think about structure, about evidence, about classification, about the connection between facts and demands, and much more. The biologist-in-training needs to know the Hardy-Weinberg law. The biologist-in-training needs to know the Krebs cycle. The biologist-in-training needs to know the genetic code. But he or she needs also to have the tools, the methods, to move beyond these and to extend our understanding to the next and future generations. It is here that philosophy is not just important. It is fundamental.

The pragmatic virtues of a philosophical component to a biological education are compelling. But I do not want to end on just a pragmatic note. Man does not live by bread alone. To be alive is a privilege. To understand this glorious, threatening, beautiful, dreadful world in which we all live is the end point of being a human being. Don't take my word for it. Socrates told us so! A world seen through a biology informed by philosophy! What a triumph and a joy!

Michael Ruse
Program in the History and Philosophy of Science
Florida State University, Tallahassee, FL, USA
e-mail: mruse@fsu.edu

Contents

Contributors

Denis R. Alexander The Faraday Institute for Science and Religion, St. Edmund's College, University of Cambridge, Cambridge, UK

Robert Arp Independent Scholar, Overland Park, KS, USA

John C. Avise Department of Ecology and Evolutionary Biology, University of California, Irvine, CA, USA

Francisco J. Ayala Department of Ecology and Evolutionary Biology, University of California, Irvine, CA, USA

William Bechtel Department of Philosophy and Center for Chronobiology, University of California, La Jolla, San Diego, CA, USA

Pierrick Bourrat Department of Philosophy, University of Sydney, Sydney, NSW, Australia

Pierre-Alain Braillard UFR de Biologie, Université Lille 1, France

Ingo Brigandt Department of Philosophy, University of Alberta, Edmonton, AB, Canada

Richard M. Burian Department of Philosophy, Virginia Polytechnic Institute and State University, Blacksburg, VA, USA

Carol E. Cleland Department of Philosophy, Center for Astrobiology, University of Colorado, Boulder, CO, USA

Jennifer DeBerardinis Sydney Centre for the Foundations of Science, University of Sydney, Sydney, NSW, Australia

David Depew Project on the Rhetoric of Inquiry, University of Iowa, Iowa City, IA, USA

Michael R. Dietrich Department of Biological Sciences, Dartmouth College, Hanover, NH, USA

Michael J. Duncan Department of Philosophy, University of Sydney, Sydney, NSW, Australia

Patrick Forber Department of Philosophy, Tufts University, Medford, MA, USA

Giuseppe Fusco Department of Biology, University of Padova, Padova, Italy

Lisa Gannett Department of Philosophy, Saint Mary's University, Halifax, NS, Canada

Annie Jamieson School of Philosophy, Religion and History of Science, University of Leeds, Leeds, UK

James Justus Department of Philosophy, History and Philosophy of Science Program, Florida State University, Tallahassee, FL, USA

Kostas Kampourakis Secretariat of Educational Research and Development, Geitonas School, Vari Attikis, Greece

Marc Lange Philosophy Department, University of North Carolina, Chapel Hill, NC, USA

James G. Lennox Department of History and Philosophy of Science, University of Pittsburgh, Pittsburgh, PA, USA

Alan C. Love Department of Philosophy, Minnesota Center for Philosophy of Science, University of Minnesota, Minneapolis, MN, USA

Alfredo Marcos Department of Philosophy, University of Valladolid, Valladolid, Spain

Kevin McCain Department of Philosophy, University of Alabama, Birmingham, AL, USA

Roberta L. Millstein Department of Philosophy, University of California, Davis, CA, USA

Alessandro Minelli Department of Biology, University of Padova, Padova, Italy

David S. Moore Pitzer College, Claremont, CA, USA

Claremont Graduate University, Claremont, CA, USA

Maureen A. O'Malley Department of Philosophy, University of Sydney, Sydney, NSW, Australia

Massimo Pigliucci Philosophy Program, The Graduate Center, City University of New York, New York, NY, USA

Anya Plutynski Department of Philosophy, University of Utah, Salt Lake City, UT, USA

Angela Potochnik Department of Philosophy, University of Cincinnati, Cincinnati, OH, USA

Gregory Radick School of Philosophy, Religion and History of Science, University of Leeds, Leeds, UK

Michael Ruse Program in the History and Philosophy of Science, Florida State University, Tallahassee, FL, USA

Andrew W. Siegel Berman Institute of Bioethics, Johns Hopkins University, Baltimore, MD, USA

Tobias Uller Edward Grey Institute, Department of Zoology, University of Oxford, Oxford, UK

Brad Weslake Department of Philosophy, University of Rochester, Rochester, NY, USA

John S. Wilkins Department of Philosophy, University of Sydney, Sydney, NSW, Australia

Historical and Philosophical Studies, University of Melbourne, Melbourne, Australia

Arno G. Wouters Department of Philosophy, Erasmus University Rotterdam, Rotterdam, The Netherlands

Michael Zerella Department of Philosophy, University of Colorado, Boulder, CO, USA

Philosophy of Biology and Biology Education: An Introduction

Kostas Kampourakis

1 Prolegomena: The Rationale and Aims of this Book

This book presents analyses of philosophical topics of importance to biology education. It is intended foremost for biology educators and teachers, and aims to show how philosophy of science in general, and philosophy of biology in particular, can enrich their intellectual horizons and eventually their classes. The book was initially conceived as a sequel expansion of a thematic issue I have guest edited for the journal *Science & Education* (2013, numbers 1, 2). As soon as the contents of that issue were finalized, I realized that there was much more to be written about what philosophy of biology might contribute to biology education. So I started inviting scholars to contribute to a book on this topic. The outcome was astounding: a book with 30 new chapters covering more topics than I could hope for. This is a book I always wanted to read but so far did not exist.

That philosophy of science has a lot to contribute to science education will not be news to many people. The International History, Philosophy and Science Teaching Group (IHPST) and the journal *Science & Education* have for 20 years focused on contributions of history and philosophy of science to science education. Why then prepare a whole book about philosophical topics which are relevant to biology education? There are two reasons for this. First, context matters. Despite some common features, important differences exist between the different areas of natural science, such as physics, chemistry, biology and geology. Consequently, it is important to discuss and explain what is special in each case, such as the methods and approaches used by biologists which are not shared by those studying the non-living world. The second reason has to do with recent advancements in biology. The active fields of biological research today are quite different from those I encountered as an

K. Kampourakis (✉)
Secretariat of Educational Research and Development, Geitonas School,
Vari Attikis, 16602, Greece
e-mail: konstantinos.kampourakis@gmail.com; kamp@geitonas-school.gr

K. Kampourakis (ed.), *The Philosophy of Biology: A Companion for Educators*, History, Philosophy and Theory of the Life Sciences 1, DOI 10.1007/978-94-007-6537-5_1,

undergraduate student 20 years ago. Nothing like systems biology or evolutionary developmental biology were mentioned in biology education then. Thus, there is a need for updating our current scientific knowledge and understanding. As a result there is also a difference in the focus of the philosophy of biology between now and then.

An important and novel feature of this book is that philosophers of biology and philosophically minded biologists bring their work to bear on biology education. I strongly believe that philosophers could and should make important contributions to biology education, predominantly in two domains: (a) understanding concepts and (b) understanding the nature of science. All chapters included in this book make such contributions. All authors were right from the start perfectly aware that this book should be as readable and as comprehensible as possible in order for people without any formal training in philosophy of science to be able to benefit from it. I believe that this book is characterized both by top-rate philosophical scholarship and by a sense of practical usefulness. There are important implications for biology education in each chapter; most of them are explicit, but I am also confident that science educators will identify the implicit ones, too.

This book looks like an anthology of topics in philosophy of biology and it might indeed be used as one. However, the topics were selected with purely educational criteria in mind. All chapters address issues which are central to biology education, along with some neglected ones. For instance, molecular evolution and non-genetic inheritance are rarely discussed in biology textbooks. Teleology and essentialism are related to important conceptual obstacles to understanding evolution and, yet, the relevant philosophical scholarship is not always taken into account in science education and conceptual development research. Concepts such as genes and dominance are ambiguous and yet these are what we are teaching at schools. Biology textbooks often refer to ethics but without the necessary background from philosophy. And so on.

I should note that I thought a lot about how to arrange the 30 chapters of this book. Because there were many ways in which the chapters could be grouped and because all topics are more or less related to each other anyway, I have decided not to divide the book into parts. I have only tried to arrange chapters in such a way so that consecutive ones are thematically related to each other. The careful reader will notice a flow from evolution and development, to genetics and ethics but there is much more in the book. However, in this introduction I describe the contents of the chapters in groups, thus indicating one way in which the book could have been divided into sections. Quite unintentionally, and despite my great interest in evolution, half of the chapters of this book focus on evolutionary theory. This is no surprise, of course, since topics related to evolutionary theory often predominate in philosophy of biology books. Furthermore, this is something to be expected since evolutionary theory is the central unifying theory of biology. The first three chapters of the book serve as introductory to biological science. The other chapters can be considered to fall under the following main topics: "proximate" phenomena, genetics, and ethics.

The chapters included in this book are intellectual contributions towards what I consider an ideal biology curriculum that would take into account the philosophical issues raised by biological sciences. Although a detailed discussion of these issues is not always possible in the classroom, curriculum developers, science educators and biology teachers need to be aware of them and appropriately introduce them during biology instruction. This could be achieved if a biology curriculum ideally had the following characteristics:

1. *An evolutionary framework:* Evolution is the central unifying theory in biology as it explains both the unity and the diversity of life. Thus, an evolutionary framework could form the basis for teaching about all biological phenomena and the origin of biological forms and functions, for understanding the conclusions of research in microbiology, evolutionary developmental biology and genomics, and for realizing that humans are (at least biologically speaking) a very short branch in the evolutionary network of life.
2. *A developmental perspective:* Biology education cannot focus on DNA and genes and then make a leap to organisms and their phenotypes, overlooking the developmental processes that produce them. There is more in biology than nucleotide sequences, as there is more in language than letter sequences. All cells in an organism contain the same genes, but their expression is differentiated according to their environment. Several effects such as epistatic and pleiotropic ones influence phenotypes. Thus, it is important for biology education to make clear that development is a complex process of which DNA is an important, but not the only, factor.
3. *An integrative approach:* Development notwithstanding, keeping in mind that microbial life is the predominant form of life on Earth is important, too. In addition, the new research fields of evolutionary developmental biology and systems biology suggest an integrative approach to the study of biology, as well as to its teaching. Integration of disciplines is currently a common approach and consequently biology education should make clear that life requires not only DNA but also a complex cellular machinery, that not only phenotypes but also regulatory mechanisms driving their development evolve, that organisms consist of numerous interacting and interdependent parts and that both the evolution and the development of organisms is influenced by the ecological characteristics of their environments.
4. *A socio-ethical dimension:* The life sciences have many direct implications for several aspects of human life. Any discussion of biomedical research on cells, tissues or organs should also include an explicit discussion of their social and ethical implications. There is more to science than the quest for knowledge; research also depends on funding, personal ambitions and the potential benefits, and also has a number of implications for human lives. The social and ethical implications of biological research should be explicitly addressed in biology instruction, and teachers should be prepared for doing that.
5. *A contemporary view:* In the post-genomic era, it is important that biology education is appropriately updated. For example, concepts like genes and phenotypes seem to be somewhat outdated; much contemporary research focuses on genomes and life cycles, instead. Of course, it is not possible for biology education to follow every detail of current developments. However, an updated view is

necessary so that future citizens will be able to understand and make informed decisions about the respective issues.

These five characteristics form the underlying framework of the book, on which the selection of topics was based. All chapters should be seen as contributions to one or more of these characteristics.

One important point to clarify is that the phrase "biology education" as used in this book refers to both secondary school and undergraduate-level university courses. All authors have tried to make their chapters relevant to both. I think that the book can be useful both for future scientists, studying biology at the undergraduate level, and for future citizens studying biology at secondary school level. Most of the topics of the book can be covered in an undergraduate rather than in a secondary school course. However, I believe that secondary teachers should also be aware of the philosophical issues discussed in this book which would enrich their teaching, even if they never actually teach about them.

Furthermore, the philosophical topics discussed in this book can be useful and enlightening for biology professors. Many of them teach undergraduates courses, and it is important that they make their students realize the philosophical issues that arise from biological research, as well as the possible implications of this research for society. Biologists are studying organisms, and the moral, social, and religious implications of their research might be direct and deep. All biologists, no matter if they will be teaching at secondary or tertiary levels or if they are going to get involved in the public communication of science, would be better prepared to engage in addressing socio-ethical issues if they had already gained the required philosophical background from their undergraduate studies.

Thus, this book should be useful to biology educators, undergraduates who are going to become biology teachers, students in biology teacher programs (both pre-service and in-service), curriculum developers, and others who teach biological science or are involved in its communication to the public. In addition, it can be valuable to philosophers of biology as it shows how they can make practical and useful contributions to education, while it might also be used in their undergraduate courses on the philosophy of biology as an introduction to some central topics. The ultimate aim of this book is to initiate a discussion about what the philosophy of biology can contribute to biology education so that more scholars will start working on projects like this, with a view of preparing a new generation of philosophically minded biology teachers and educators, and pedagogically sensitive philosophers of biology.

2 The Science of Life

In the first chapter, titled *What Is Life?*, Carol Cleland and Michael Zerella explain why it is difficult to clearly and accurately define life. This is very important because this may intuitively seem easy to do. Although one may be familiar with various kinds of organisms living on Earth these usually are animals and plants, which are just a minority of all life forms living on Earth (the chapter by Duncan et al. in this volume is illuminating in this respect). Perhaps most importantly, we have no idea what forms of life may exist in other parts of the universe. Cleland and Zerella conclude that it is a mistake to define life in general based on the single example of familiar life on Earth. Without additional examples we cannot discriminate between features which are universal to life and features derived from mere physical and chemical contingencies on Earth. In addition, a narrower definition restricted only to familiar life on Earth is problematic because there are cases, such as viruses, which share some but not all features of familiar life on Earth.

The discussion of whether we can accurately define life is especially important given that biology textbooks often describe, and sometimes define, life in terms of particular characteristics (metabolism, reproduction, development, response to stimuli etc.) and tend to exclude viruses because they lack several of them. The point made by the authors is that we may be able to define cell-based life but this does not mean that we can extrapolate such a definition to life in general. The topic of this chapter is also ideal in order to raise a discussion about how much scientists do not know and the question "What is life?" is one that eventually becomes very difficult to answer. Cleland and Zerella make this clear and they are also explicit about how this topic can be used to teach about important aspects of nature of science.

In the next chapter, titled *Biological Explanation,* Angela Potochnik discusses what constitutes explanation in biology. After describing the different accounts of explanation from philosophy of science and how they apply to biology, Potochnick suggests that an account of biological explanation could be based on the idea that causal information is explanatory. Thus, by showing how certain types of causes tend to bring about certain types of effects, one might reveal causal patterns on which explanations could be based. Such explanations do not include all causal factors but only those which are relevant. This leads to simplifications and idealizations, and eventually to the construction of models with explanatory character. Which model should be used might depend on the particular explanatory aims of scientists. Different explanatory aims drive scientists to focus on different causal patterns and this brings about pluralism in explanation. For example, explanations for micro-evolutionary phenomena are different from explanations for macro-evolutionary phenomena. Potochnick concludes with particular suggestions about how biology education should attend to issues related to scientific explanation.

Understanding the nature of explanations in biology and engaging students in formulating explanations should be an important part of biology education. Textbooks and teachers often serve as transmitters of knowledge. However, this knowledge on its own is not useful if students are not taught how to use it to explain the phenomena of the biological world. This is, in my view, what teachers should guide their students to learn. Not all propositions have an explanatory role and students should be taught what is required in order to construct a genuine and accurate explanation. To achieve this, of course, teaching should go beyond any kind of lecturing to engage students in inquiry activities in which they would not just have to collect raw data, but most importantly use it to formulate explanations or justify why this is not possible. The analysis provided by Potochnick in her chapter is very useful for this purpose.

Traditionally, explanations in science have been presented as law-based. However, this is not what one finds in biology. But if natural laws existed in biology, what kinds of laws would those be? This is the topic of the next chapter, *What Would Natural Laws in the Life Sciences Be?*, by Marc Lange. The author starts by discussing the distinction between laws and accidents. Then he describes various kinds of generalizations used in biology and concludes that they are important for the respective domains of inquiry, no matter if they are laws of nature in the strict sense or not. Having exceptions or being contingent does not make biological generalizations less important than the laws of physics. Lange concludes that although the laws of physics apply to living systems, biological generalizations provide explanations that cannot be replaced, even in principle, by the most fundamental physical explanations. Thus, biological explanations are irreducible to the explanations available at more fundamental levels and so they provide a distinctive kind of understanding, even though organisms are made of physical matter.

This is a very important point because it counters the implicit reductionism of textbooks focusing on DNA and other molecules. Biological systems consist of parts at various levels of organization and one should pay attention to the phenomena characterizing each level of organization under study (cells, organisms, ecosystems - see also the chapter by Braillard in this volume). Moreover, given the widespread misconception that laws are somehow superior to theories, this chapter clarifies that law-likeness is not as important as people may think it to be. What is important, and clearly stated in Lange's chapter, is that biological principles such as natural selection are explanatorily important and predictively accurate even if they are not laws like the ones used in physics.

3 The Nature of Evolutionary Theory

Having concluded that scientific theories are legitimate even if they do not include laws of nature, evolutionary theory is brought to focus. Evolutionary theory is the central unifying theory of biology, with important implications for all biological disciplines and for biology education. Thus, it is no surprise that half of the chapters of this book are directly and explicitly related to it. In the first of these chapters, *The Nature of Evolutionary Biology: At the Borderlands Between Historical and Experimental Science*, Massimo Pigliucci describes and explains the nature of evolutionary biology, which combines elements of both experimental and historical science. Pigliucci describes the criticisms that Darwin's theory received by prominent philosophers such as Whewell and Mill. These criticisms set the stage for important debates that followed on the nature of evolutionary theory such as the one between Fisher and Wright. Important in this context is an understanding of what "drift" is about. The author suggests that we should not think of a drift as antagonistic to selection, but as a deviation of what we might expect from selection. Then

Pigliucci turns to another debate on whether the pattern of evolution is a gradual or a punctuated one as well as on whether developmental constraints limit the power of natural selection, both instigated by Stephen J. Gould's writings. The conclusion from detailed studies and experiments has been that stochastic events have an important role in evolution, along with selection. Thus, evolutionary biology attempts to explain patterns which are the outcome of either selection or stochastic events. Moreover, evolutionary biology has a dual nature: it is both experimental and historical. Pigliucci concludes that evolutionary biology is taught within a narrow perspective and outside the appropriate historical and philosophical contexts.

Indeed, the characteristics of historical science and how they relate to the respective characteristics of experimental science should be in the core of any teaching of evolutionary biology. Evolution is often defined in textbooks as "change in gene frequencies" and no emphasis is put on the historical aspect of such processes. In addition, evolutionary biology is often perceived as a purely descriptive science based on observation, without the manipulative powers of experiments. However, this is not the case and Pigliucci shows that both approaches, the historical and the experimental, are crucial for understanding the history of life on Earth. Over the past 100 years biologists have studied (micro-) evolutionary processes in both natural and experimental populations and have shown how evolution proceeds. These studies have been complemented by detailed studies of historical (macro-) evolutionary patterns which reveal traces of past processes. The result is an account of how natural processes have produced the enormous diversity of extinct and extant life forms from a common starting point.

These studies of micro- and macro-evolutionary phenomena produce knowledge. But what exactly can evolutionary biologists know? This the topic of the next chapter, *Evolutionary Theory and the Epistemology of Science,* by Kevin McCain and Brad Weslake. In this chapter, the authors address various criticisms against evolutionary theory: that it is a "theory" rather than a fact, that it cannot be proven, that it is not falsifiable, that it has been falsified, and that it does not make predictions. By explaining why these criticisms are false or misconceived, teachers can explain to their students how to respond to these criticisms, as well as some basic features of the structure of scientific theories and their evidential support. Knowledge requires evidence and in the case of evolutionary theory there are many kinds of evidence coming from independent sources. The authors also explain that falsifiability is not a good criterion for accepting a theory as scientific because all scientific theories are non-falsifiable. Scientific theories do not make predictions on their own but in conjunction with other (auxiliary) hypotheses. In the case of failed predictions scientists often reject these hypotheses rather than the theory. A much better criterion for accepting a theory as scientific is whether it makes testable predictions. Evolutionary theory has made some predictions which have failed but this does not imply that the theory should be rejected. Evolutionary theory, as all scientific theories, encompasses many different hypotheses, some of which are fundamental whereas others are rather peripheral. Refining such hypotheses is a mark of good, not failed, science and this is the case for evolutionary theory. Finally, the authors describe inference to the best explanation (that the hypothesis that best explains the phenomena compared to other hypotheses is the likeliest to be true) as the principle of scientific inference that can be used to show how evolutionary theory is supported by the evidence.

The common ancestry of organisms is supported by evidence from biogeography, fossils, molecular studies, embryology, and comparative anatomy among else. That natural selection has been a major process driving evolutionary change is supported by evidence from experiments in natural and laboratory contexts, coupled with mathematical modeling and computer simulations. The chapter by McCain and Weslake overall explains that the aim of scientists is not to uncover any absolute truth, if anything like this exists at all. Rather scientists ask questions and/or formulate hypotheses which they test, and try to answer their questions. These answers constitute explanations and provide knowledge. Some of these are well founded and are gradually incorporated into the fundamental part of the theory. Others will remain in the periphery until they are further confirmed or rejected. Evolutionary theory is accepted on the basis of evidence and it is not a matter of belief. Evidence from independent sources supports the core of contemporary evolutionary theory which provides an account for the evolution of life on Earth, even though experts may disagree on the details.

In the following chapter, *Conceptual Change and the Rhetoric of Evolutionary Theory: 'Force Talk' as a Case Study and Challenge for Science Pedagogy*, David Depew discusses the rhetoric of evolutionary theory and in particular the ways used to present it to non-experts. Depew focuses on "force talk" about natural selection, taking into account both the history of evolutionary biology and conclusions from evolution education research. He first explains why the expression "survival of the fittest", initially coined by Herbert Spencer and eventually accommodated by Darwin himself, does not accurately represent natural selection as Darwin had conceived it. Depew carefully distinguishes between Darwin's view of natural selection as a long process of selection of particular variants among others and Spencer's view of the quick elimination of all individuals except from those who happened to have an advantage. Thus, natural selection came to be perceived as an eliminative force, rather than a gradual process of discrimination among variants. However, the creative role that Darwin assigned to natural selection was revived in the middle of the twentieth century by the population genetic theory of natural selection. It was Ronald Fisher who claimed that natural selection adapts populations to environments and his use of mathematical models, similar to those used in physics, made Darwinian theory more respectable that it had ever been before. What is more important, and less obvious in textbooks and the public image of evolutionary theory, is that the Darwinism of the Modern Synthesis was in important respects a different theory than Darwin's Darwinism. Depew notes that contemporary Darwinism begins to look, more like Darwin's Darwinism than like the Darwinism of the Modern Synthesis, but the reasons for this are rhetorical, not conceptual. Depew concludes that biology textbooks should point to the different conceptual frameworks of evolution rather than bringing them into one, cumulative form.

These are very important points. First, as already mentioned, evolution is in many cases presented as "change in gene frequencies", often ignoring the organismal, and especially the developmental, component of evolutionary change. Attention to changes in populations rather than exclusively in individuals is of course important, as students should understand that it is populations and not individual organisms which adapt. However, this should not take the focus of instruction to the other extreme and neglect "proximate" phenomena in individuals. Second, Darwinism and neo-Darwinism are often conflated in textbooks and the view of evolutionary process as consisting of two steps, mutation and natural selection, is often described as Darwinian although Darwin had no idea about the origin of variation. In other words, textbooks often give the impression that Darwin got the whole theory right from the start, something that is not the case. Darwin carefully constructed a theory about the evolution of life on Earth by means of natural selection, but was not able to explain the origin of those variations which were necessary for natural selection to occur. More work was required in order to develop a legitimate theory of heredity, which was a prerequisite for a successful theory of evolution. Depew's chapter makes these points clear and also illustrates a crucial aspect of science: that scientific theories are refined in the light of new (or more) evidence and that, in a sense, scientific theories themselves evolve.

One of the central concepts of evolutionary theory is adaptation, which is also a misconceived one among non-experts and a controversial one among philosophers and biologists. This is the topic of the next chapter, *Debating the Power and Scope of Adaptation,* by Patrick Forber. First, the author describes that the term adaptation has been used to refer to a process, as well as to either a trait that has emerged due to selection (historical definition) or a trait that confers an advantage to its bearers (ahistorical definition). Forber then clarifies the distinction between empirical, methodological and explanatory adaptationism. Empirical questions e.g. about the prevalence of adaptation in the biological world are distinct from methodological questions about how we should investigate the world by looking for adaptations, as well as from the explanations formulated for them based on natural selection. Then Forber discusses issues raised by each kind of adaptationism. Starting with empirical adaptationism, he considers suggestions for testing adaptationist hypotheses with different models and concludes that determining whether natural selection provides a sufficient explanation for a trait is difficult. Methodological adaptationism makes claims about how we should investigate the world. Forber suggests that models tested should not only provide evidence for some hypothesis, but also against rival hypotheses. Finally, he turns to explanatory adaptationism which is about whether we should adopt a particular perspective for explaining life more generally. Forber concludes that the controversy over adaptationism is interesting as a case study of science in action because it raises several questions: empirical questions about the prevalence and power of natural selection; methodological questions about how to test evolutionary hypotheses; and questions about the importance and status of core concepts in evolutionary theory.

Understanding the controversy over adaptationism is important in order to understand wider controversies relevant to evolutionary biology. Consequently, clarifying what adaptations are, how they should be studied and what conclusions should and could be made from such a study is important. There are two kinds of controversies here. The first is one among scientists on the importance of natural selection: is it the only important, the most important or just one factor among others in the evolution of life on Earth? There are different perspectives here, clearly described in the chapter by Forber. The second controversy is between those who argue that adaptations cannot have emerged solely by natural processes, such as natural selection, and who imply that there is something more behind their emergence. Intuitively, it seems more plausible to understand adaptations as the outcome of rational design. This is an old question related to the argument from design, which has religious assumptions and implications. Evolutionary biologists have clearly shown that, given the time available and the details of how evolution proceeds, all kinds of adaptations can have evolved through natural processes. A major controversy exists about whether evolution by natural selection debunks the argument from design, religion or both, and this is the topic of the next three chapters. Before turning into this, it should be made clear that there is no debate among scientists about the importance of natural selection for the emergence of adaptations. What is at stake is whether an adaptational perspective should be the dominant approach or just one approach among others in studying evolution.

4 Evolutionary Theory and Religion

Given these considerations it is useful to examine what the relation between evolutionary biology and religion is, and this is actually the topic of the next two chapters by Francisco Ayala and by Denis Alexander. In his chapter, *Biology and Religion: The Case for Evolution*, Francisco Ayala first describes the "argument from design". This argument consists of two parts; the first part asserts that there is design in the universe, whereas the second part affirms that only an omnipotent and omniscient Creator could account for this design. Ayala explains how this argument has been conceptualized over the centuries, and provides an account of perhaps its more detailed elaboration by William Paley. Then Ayala turns to Charles Darwin. He notes that Darwin's most important accomplishment was not that he accumulated evidence for evolution, but rather that he managed to provide a scientific account for the adaptations of organisms as the outcome of natural processes. Darwin argued that natural selection could be responsible for the production of new species, and most importantly he accumulated evidence indicating that evolution could indeed have occurred. Ayala continues by discussing two important sources of evidence for evolution, namely fossils and molecules. He notes that since Darwin's time more fossil evidence for evolution has been accumulated, including fossils of the so-called missing links such as *Archaeopteryx* and *Tiktaalik*, as well as of hundreds of fossils from individual hominids, i.e. the human lineage after its separation from the ape lineages. The study of molecular data also provides abundant evidence for evolution, and Ayala explains that this is done in two ways: by revealing similarities in the structure and functions of nucleic acids or proteins from different organisms, as well as by making possible the reconstruction of evolutionary relationships. Having argued for the evidence for evolution, Ayala then explains how it helps resolve the problem of evil in the world. Instead of having to explain catastrophes and imperfection in the living world as failed outcomes of God's design, these can all be convincingly explained as the outcomes of natural processes, including evolution. Ayala concludes that science and religion concern different aspects of the human experience.

The next chapter, *The Implications of Evolutionary Biology for Religious Belief* by Denis Alexander, is a complementary contribution to the previous chapter. Alexander first notes that the implications of biology for religious belief vary widely depending on the religion under discussion, and explains his focus on Christianity as biology was mainly related with it, having emerged in Europe over the last few centuries. Starting from Darwin's time, Alexander outlines the historical background of the relationship between religion and evolutionary biology, which frames the contemporary discussions on this topic. Then he goes on to describe four models which are used to frame and explore the complex relationships between science and religion which are: (a) the conflict model (science and religion are in fundamental opposition), (b) the NOMA model (there can be no conflict between science and religion because they address different kinds of questions), (c) the fusion model(s) (the distinction between scientific and religious types of knowledge is blurred altogether, or science is used in order to construct religious systems of thought, or vice versa) and (d) the complementarity model (science and religion are addressing the same reality from different perspectives, providing non-rival but rather complementary explanations). Alexander also distinguishes between methodological reductionism and ontological reductionism, and explains that it is one thing to implement a reductionist approach in order to study a system and another to claim that such an approach is the only legitimate one. Then, he turns to the relationship between evolutionary theory and religion and explains that it is a complicated one as the idea of evolution has been used in support of various, often mutually exclusive, political and social ideas. There are important issues here, some of which are: (a) the changing views on Biblical hermeneutics, the methods by which texts are interpreted, which have had an impact on the growth of US creationism; (b) what *creation* really means (the meaning of this word in theological contexts is different from its everyday meaning); (c) the role of chance (chance events have an effect on evolutionary process but the latter is not a chance process; this is also supported by numerous cases of evolutionary convergence); and (d) the relation between evolution and morality (e.g. whether or not morality has a biological basis). Alexander concludes that no matter what the perceived implications of evolutionary biology for religion may be, there is no reason why biology and religion should be in conflict, especially since religious belief has played an important positive role in the development of the biological sciences.

What the two complementary chapters of Ayala and Alexander suggest is that biology and religion are not necessarily incompatible. It is important to show that research scientists do not perceive science and religion as being in conflict, and clarify not only why they think so but also what kind of compatibility they perceive between them. This is not to deny that evolutionary biology has implications for religious belief. However, having implications for something is very different from being in conflict or incompatible with it. The real conflict would arise if evolutionary theory was used to found a secular religion that would most likely be in conflict with traditional religions, or when unscientific theories are promoted as alternatives to the theory of evolution and time to discuss these is required in biology classes. Teaching about Genesis or Noah's flood in a theology class is perfectly acceptable, but teaching about these in a biology class in order to defend Intelligent Design or creationism as an alternative to evolutionary theory is entirely unacceptable. One should be cautious here. The problem is not about teaching theological topics that address the same questions with science. The problem is about the demand to teach theological topics in biology classes as alternatives to the currently accepted scientific theory.

This is the topic of the chapter titled *Intelligent Design and the Nature of Science: Philosophical and Pedagogical Points* by Ingo Brigandt. Intelligent Design (ID) is actually a modern form of creationism and unquestionably it is religiously founded. Whereas many people would not object to the teaching of any religious doctrine or perspective in religious classes, it would be entirely insensible to include ID as an alternative to the theory of evolution in biology classes given than there is enormous evidence for the former and no evidence for but mostly evidence against the latter. Brigandt first discusses Behe's argument that irreducibly complex systems cannot have emerged through evolution, since a system that becomes non-functional if one of its components is removed cannot have evolved from simpler ones. Brigandt explains that this argument fails because complex systems may have evolved from simpler ones performing entirely different functions, but also from systems with more parts if the latter were redundant and were eventually eliminated during evolution. Brigandt then criticizes machine metaphors used by both scientists and philosophers as inappropriate and suggests that organisms should rather be viewed as flexible developmental systems. Then he turns to the idea, expressed in its fullest form by Dembski, that organisms are so complex that it is extremely improbable for them to have originated through natural evolutionary processes. It is explained that nothing can be inferred about the probable truth or probable falsity of some hypothesis based on its small probability, no matter how small that is. The probability of a series of many individual events will be small anyway and series of events with small probabilities happen all the time in nature. Brigandt suggests that whether such a probability is small makes sense only when it is compared to the probability of another series of events. Finally, Brigandt clarifies the distinction between methodological and metaphysical naturalism. Methodological naturalism admits that science does not study supernatural entities, no matter whether these exist or not, whereas metaphysical naturalism denies the existence of supernatural entities and argues that only natural ones exist. Brigandt concludes that it is important to have students understand and evaluate scientific approaches in terms of their epistemic and social practices, and so evaluate whether ID is a scientific approach or not.

The topics raised in the chapter by Ingo Brigandt are many and have direct implications for biology education. Evolution is not a process that cannot produce complex entities; it certainly can, but the consecutive stages may be functionally different from each other. Evolution is not a process with a low probability of occurring; in contrast, its explanations have higher probability to be accurate than those advanced by ID proponents. Finally, evolution does not rule out God and supernaturalism; as all science, which is empirically based, it only admits that these cannot be its objects of study. This chapter makes clear once again that in the case of Intelligent Design the conflict is not between science and religion(s) but between a well-established scientific theory and an unscientific, religiously founded one. The criticism that ID proponents receive from the majority of philosophers and scientists is not about the religious views they hold (which may shared by the philosophers and scientists themselves) but about their unwillingness to rationally consider the widely accepted scientific conclusions. The evolution/ID controversy may be deliberately portrayed by ID proponents as an instance of the science/religion conflict, but it certainly is not one.

5 Evolution at the Molecular Level

The persistence of adaptation talk, and of the controversy about the grounds on which the adaptations of organisms should be explained, may have made molecular evolution one of the most neglected topics in biology education. Organismal evolution seems to predominate in all relevant discussions and usually little attention is paid, both in textbooks and in the public presentation of evolution, on molecular evolution. Michael Dietrich, in his chapter titled *Molecular Evolution,* explains its importance. He first makes clear that molecular evolution was perceived as an alternative and not as a complementary, as it really is, approach to the organismal one in the study of evolution. The divide between the organismal level and the molecular level became deep when it was proposed that different causal processes predominate at each level: selection at the organismal level, drift and selection at the molecular level. Thus, one of the problems seemed to be a proper understanding of the role of drift. Dietrich gives an account of drift as a process, based on work by him and by Roberta Millstein, different from the account of drift described by Pigliucci in this book. Understanding drift as a causal process interacting with selection according to Dietrich was a key point in understanding the divide. Drift was strongly linked to molecular evolution and was perceived as opposed to selection, although both processes take place at the molecular level. A useful suggestion made in the 1980s was to assume neutrality as the norm at the molecular level, because most DNA sequences are not selected, and when this is not the case to turn to selection for explanations. Currently, molecular evolution is the outcome of the complex interplay of drift and selection acting upon all kinds of sites: from strictly neutral to strongly selected.

In the following chapter, *Educational Lessons from Evolutionary Properties of the Sexual Genome,* John Avise turns the focus to gene-centered evolution in sexually reproducing organisms that explains several observations made at their genomes. Avise has the following main aims: to recapitulate conceptual paradigms that have guided our effort to understand biological complexity, to relate these to the ongoing advancements in genomics and to point to the emerging gene-centric view of the sexual genome that differs from viewpoints of the past. Avise provides a detailed account of how the idea that organisms were specially designed was quite widespread before Darwin's time, and how theodicy was invoked to explain the imperfections and malformations observed. He also briefly discusses intelligent design and explains in detail why evolution by natural selection can account for a variety of observations without the need to assume any supernatural cause. Then Avise focuses on the role of selection on genes. Although this is not the only level at which natural selection is considered to operate, it nevertheless is an important one. Sexual reproduction has the consequence that selection can operate at the level of genes because genes in recombining genomes sometimes can increase their odds of survival and proliferation by acting against the "interests" of the genome and the host organism. The realization that natural selection operating at the level of the gene can oppose natural selection operating at the level of the organism was a major conceptual breakthrough that has helped clarify many otherwise enigmatic molecular features of sexual genomes, including the evolution of mobile elements found in them. Mobile elements are responsible for several kinds of disease. Avise describes several other cases of genetic conflict within the sexual genome. He concludes by describing the role of metaphors in genetic, genomic and evolutionary discourse.

These two chapters focus on evolution at the molecular level. Molecular studies provide important information about organismal evolution but also reveal complexities which are not obvious at the organismal level. Thus, one should pay attention to evolution at both the organismal and the molecular level. The latter can be either selective or neutral, i.e. non-selective, evolution. Changes in the frequencies of DNA sequences can be caused by selection processes at the organismal level (e.g. a phenotype is selected and so do the DNA sequences implicated in its development). However, changes in the frequencies of DNA sequences may also be the outcome of drift. There is no need to decide which of the two kinds of processes is more important. What the chapters by Dietrich and Avise show is that evolutionary processes may be more complicated than what we tend to think as well as that one should pay attention to the role of both selection and drift in evolution.

6 Evolution and Development

But does evolution depend solely on changes in DNA sequences? No. In recent years, non-genetic inheritance has been considered as an important factor in evolution, one that is almost entirely neglected in secondary education. This is the topic of the next chapter by Tobias Uller, titled *Non-genetic Inheritance and Evolution*. Uller first provides a brief description of how heredity became an object of scientific study during the nineteenth century, and how this was initially done in the context of development. Then, with the rise of Mendelian genetics emphasis was put on genes and heredity was gradually separated from development. With the modern synthesis, evolution was mathematically formalized and was henceforth described as change in gene frequencies; development became entirely ignored. Mayr's ultimate/proximate distinction and the focus of adaptation studies on genes further enhanced the separation between development and evolution. Eventually, developmental biologists came to consider non-genetic resources transmitted from parents to offspring as an alternative system of inheritance, distinct from DNA. Actually, three such systems of non-genetic inheritance have been proposed: epigenetic (e.g. DNA methylation), behavioral (e.g. parent-offspring interaction), and symbolic (e.g. written language). Uller explains that in many cases we cannot separate genetic from environmental influences, nor can we argue that the former are more important than the latter for development or evolution. Thus, by recognizing non-genetic mechanisms of inheritance an explicit consideration of developmental processes in evolutionary explanations is possible. Then the author describes two consequences of taking development and non-genetic inheritance into account in the study of evolution. The first is that by re-introducing development into evolutionary theory we can ask questions about the origin of adaptations, since an understanding of how pre-existing mechanisms of development can give rise to novel phenotypes will be available. The second is that it becomes clearer how non-genetic inheritance can influence the course of evolution by (i) affecting individual fitness; (ii) modifying the relationship between what is selected and what is inherited; and (iii) modifying selection on future generations. Uller concludes by suggesting that the growing molecular and behavioral evidence for non-genetic inheritance should stimulate a more in-depth treatment of the concept of heredity in evolution education.

Similarities between the developmental processes of different species are evidence of evolutionary relatedness. Generally speaking, organisms exhibit both similarities and differences. Differences are usually due to the accumulation of novel characteristics during the evolutionary process, whereas similarities are often due to common descent, i.e. they are derived from a common ancestor who also possessed them. Such similarities are usually described as homologies. However, the concept of homology is much more inclusive than that, and it is the topic of the chapter *Homology* by Alessandro Minelli and Giuseppe Fusco. The authors start by describing in detail the major steps in the "complex semantic evolution" of this concept, namely: (a) the non-historical concept of homology (shared characters simply being variants of the same archetype), (b) the historical concept of homology (shared characters stemming phylogenetically from a common ancestor), and (c) the proximal-cause concept of homology (characters sharing the same generative gene network module). Minelli and Fusco suggest that the concept of homology as a simple relationship between two structures is inadequate and should be replaced by a context-dependent concept. They thus introduce the factorial concept of homology, according to which homology is not an all-or-nothing relation but can be a matter of degree (e.g. structurally non-homologous characters can be developmentally homologous having independently co-opted the same developmental module present in their most recent common ancestor). Thus they note that homology statements need to include a specification about what aspects of variation constitute a different state of the same thing, or different things. One cannot simply claim for example that bats' and birds' wings are or are not homologous because they are historically homologous as tetrapod forelimbs, but they are not homologous as tetrapod wings. Then the authors explain that neither morphological, nor developmental, not even molecular data alone can reveal homologies. For example, although one can compare the development in two different organisms, the body parts may be formed at different times or at different rates and so comparison is not easy. The authors then explain specific concepts of homology such as serial homology (repetitive structures of the same individual e.g. vertebrae), positional homology (different, non-homologous structures localized in homologous positions in individuals of two species) and special homology (the same homologous structure is localized in non-homologous positions in individuals of two species). Finally, Minelli and Fusco explain how homology is used in scientific practice for inferring phylogenetic relationships, for understanding the origin of evolutionary novelties or for classifying organisms.

In their chapter, Minelli and Fusco repeatedly refer to the relation between the study of development and the study of evolution. This is the object of a tremendous research field which is called evolutionary developmental biology, usually dubbed evo-devo, which is the topic of the next chapter by Alan Love, titled *Teaching Evolutionary Developmental Biology: Concepts, Problems, and Controversy*. The author puts emphasis on the fact that the public image of science often ignores that it is answering questions and solving problems instead of confirming or rejecting hypotheses which guides scientific inquiry. Thus, he turns to evolutionary developmental biology, a very active field of research, to show how this is done. Evo-devo focuses on how development evolves as well as on how it affects the evolution of organismal traits. Love describes the conceptual foundations of evo-devo and the meaning of key terms such as constraints, modularity and evolvability. He then turns to what he considers a genuine controversy in evo-devo, which is about how knowledge is organized, by raising questions about the relation between traditional evolutionary biology focusing on adaptation and population genetics, and evo-devo focusing on variation and development. Love discusses the origin of evolutionary novelties as an example of a genuine controversy. He suggests that explaining their origin, as all problems in biology, constitutes a problem agenda, a list of many different but interrelated questions that require a long-term program of research and contributions from several disciplines. He also describes the dimensions of the structure of problem agendas as well as the criteria of explanatory adequacy they come with, which provide a template for generating an interdisciplinary explanatory framework. These provide a clear picture of which intellectual contributions are necessary for adequate explanations as different novelties at different levels of organization may require different explanatory components in different combinations. Based on these, Love concludes that we need to teach more than one image of science in order to adequately represent its diverse aspects.

Development is a rather neglected topic in secondary biology education. Genes and DNA are important but their importance has perhaps been exaggerated. Development and its complexities should also be taken into account and the three chapters described above point to different reasons for this. Non-genetic inheritance is important as an organism may exhibit a phenotype different than the anticipated one based on its DNA sequences, for instance because some genes are silenced due to epigenetic phenomena. In addition, organisms share similar characters despite differences in their DNA sequences or they may be dissimilar but share similar underlying genetic networks. Finally, it may be the case that it is the evolution of regulatory sequences, and consequently changes in the developmental process, and not the evolution of coding sequences, and thus changes in proteins, which drives the evolution of form. In all these cases, the take home message is that reading DNA sequences (or identifying and sequencing genes) may not be that informative in our attempt to understand evolution. More information is required about when, how and why these DNA sequences are expressed and what the effect of this expression is on the development of the respective organism.

7 Integrating Levels: Taking Ecology and Microbiology Seriously into Account

To achieve a comprehensive understanding of life, one needs to study it at all levels and understand the dynamic interactions between them. Two topics which are crucial for this purpose and yet are not emphasized in biology education are ecology and microbiology, the topics of the next two chapters. In the first one, titled *Philosophical Issues in Ecology*, James Justus notes that although ecology is important for biology in general and for evolutionary theory in particular, less attention has been devoted to philosophically analyzing and teaching ecological concepts. First, he describes that Darwin's theory was shaped by an ecological perspective, as he had emphasized that organisms do not just face a static environment but occupy "places" (niches) which are the product of intraspecific interactions, and are shaped by relationships with other species occupying different niches. It is these factors that explain why some species succeed and others fail in the struggle for existence, and in this sense natural selection can be understood in terms of niche dynamics. Darwin's concept of balance of nature was not static either but was based on the struggle between organisms, while species were balanced at their population levels under the influence of natural selection. Then Justus considers the debate about the character of biological communities, as well as metaphysical issues about their reality. It seems that it is difficult to decide whether communities actually exist, although paleoecological evidence indicates that communities could exist as internally regulated systems which are more than just a sum of individuals. The controversy about what the laws of nature are and whether such laws exist in ecology is the topic that comes next. After describing the prerequisites of identifying something as a law of nature, Justus turns to particular generalizations (called allometries) and discusses whether they deserve to be called laws or not. Finally, Justus tries to clarify a central, but also problematic concept, that of ecological stability. He argues that the concepts of resistance (how much a system changes after being perturbed), resilience (how rapidly it returns to a reference state after being perturbed), and tolerance (whether it will return to that reference state or dynamic after most perturbations) jointly provide an adequate definition of ecological stability. The author concludes that ecological concepts are important for biology education and philosophy of biology and thus more attention should be paid to them.

Another topic that is conceptually very important for understanding evolution in particular and biology in general, the importance of which is not recognized in education, is microbiology. This is the topic of the chapter titled *Small Things, Big Consequences: Microbiological Perspectives on Biology* by Michael J. Duncan, Pierrick Bourrat, Jennifer DeBerardinis and Maureen O' Malley. The authors first briefly describe the organisms studied by microbiology, and note that molecular biology emerged from research on microorganisms, which were later used as tools in biotechnology. Microbes are found everywhere in the living world, around, on or in multicellular organisms. The authors argue that there is a deep reliance and interdependence of all life on microbes, and so philosophy of biology and biology education should pay more attention to them. To show the importance of microbiology, the authors use a microbiological perspective to analyze some core biological concepts: life, biological individuality and the levels of selection. Distinguishing living from non-living entities is difficult in the case of viruses. The authors also discuss biological individuality and how microbiology has changed the way biological individuals (usually multicellular organisms) are viewed. Research in microbiology has shown that biological individuality is more complicated than previously thought, as in many cases it is difficult to individualize a multicellular organism from its microbial symbionts. The more interdependent these are, the more they will function and evolve as a single entity. The authors argue that advances in microbiology have changed the way biological individuality is perceived. They have also changed our perception of the primacy of the organismal level for evolution by natural selection as a large body of empirical work on microbial evolution has shown that group selection has been very important. Eventually, microbiology shows that a commitment to the level of the multicellular organism is too narrow. The authors conclude that any conclusions should be tested against and take into account what we know about microbes.

While there is an ongoing discussion about the teaching of evolution, particularly but not exclusively in the USA, I think that attention should be put to conceptual rather than religious issues. One such suggestion would be to teach and explain ecology and its concepts for some time and in quite some detail before starting to teach evolution. Perhaps students could then realize the complexities of ecosystems and the variety of interactions among organisms or between them and their environment. Evolution is often presented as a linear process in textbooks, usually showing changes in individuals without attention to population and ecosystem dynamics. It is difficult to see how students will understand evolutionary processes if they have not understood the dynamics of ecosystems. Another conceptual issue in understanding evolution is that students do not easily realize that life is much more diverse than what we can observe. Textbooks tend to present prokaryotes as much simpler, and implicitly less diverse, than eukaryotes. However, this is far from true. Not only are prokaryotes very diverse but also they have influenced evolution through what has been described as horizontal gene transfer. Thus, evolution is more complicated than what we think, not proceeding in a linear way as it is usually depicted. Prokaryotes have been evolving on Earth for more than 3.5 billion years and their evolution has affected the evolution of all life forms. Ecology and microbiology are thus important for understanding evolution and biology because they show the immense diversity and the complex dynamics of life.

8 Conceptual Obstacles to Understanding Evolution: Essentialism and Teleology

The next two chapters analyze two important topics which are directly related to understanding evolution: essentialism and teleology. Although the analyses in these chapters are philosophical, the authors have taken into account the respective discussion in conceptual development research and science education. In the first chapter, titled *Essentialism in Biology,* John Wilkins explains what essentialism is and whether biology was or is essentialistic, and if so in what way. Essentialism is the view that things, especially kinds of things, have essences, or sets of properties, that all members of the kind must have. First, Wilkins discusses the popular narrative that pre-Darwinian biology was essentialist, a viewpoint which would have made evolution conceptually impossible since the "essence" of one species could not change into that of another. Darwin supposedly changed this view by showing that evolution was possible and that variation, which was previously considered as inessential "noise", was indeed crucial for the evolutionary process. Wilkins explains that, historically, this is not what happened. First, he explains that the view that there was a scientific Aristotelian essentialism is a mistake due to a casual reading of various twentieth century philosophers. He also explains that the essentialism attributed to Linnaeus is equally misconceived. It seems that the notion that pre-Darwinian biology was essentialistic is a twentieth century invention. Then Wilkins turns to the meaning of the word "essentialism". First, he notes that this word has been given several different meanings, even by the same authors, and identifies six varieties of essentialism: Psychological, Human, Logical, Metaphysical, Scientific, and Biological. Second, he explains that there are three general forms of essentialism available for each type: Constitutive essentialism (a class of objects are what they are because of invariant properties), Diagnostic essentialism (a class of objects is recognizable because all members share some salient properties) and Definitional essentialism (kinds have several necessary and jointly sufficient defining properties). Wilkins explains that biologists before and after Darwin (including Darwin himself) were essentialists in the constitutive sense. He also discusses the notion of natural kinds in detail, and suggests that biological kinds are best thought of as exemplary types. Wilkins concludes with some useful educational considerations drawn from his analysis.

In the next chapter, titled *Biological Teleology: The Need for History,* James Lennox and I provide an analysis of teleology and its relation to biology. First, we give a historical account of teleology starting from Plato and Aristotle, through Ray and Boyle up to Paley and Cuvier. Whereas Plato believed that the natural world is the creation of a divine, rational being, Aristotle defended a natural teleology, free from the Platonic assumption. Long before Darwin's time there was a debate over teleological explanation in natural science that was primarily between those who defended a theistic, creationist teleology and those who opposed the use of any sort of teleology in natural science. What is most interesting is that at the same time the effective scientific use of Aristotelian teleological explanation was bearing fruit in the disciplines of anatomy, physiology and medicine, through the work of William Harvey and others. Darwin was aware of the arguments of William Paley and Georges Cuvier who implemented different teleological approaches, an implicitly Platonic and an explicitly Aristotelian one, respectively. Eventually, Darwin's own explanations are shown to be teleological but in a different sense. This historical analysis prepares the ground for what we consider an important distinction between two types of teleological explanations: (a) teleological explanations based on design, which suggest that a feature exists for some purpose because it was intentionally designed to fulfill it, and (b) teleological explanations based on natural selection, which explain a feature's presence in a population by suggesting that it is selected for its beneficial consequences for the organisms that have it. We then review conclusions from conceptual development research on children's intuitive teleological explanations, and we propose questions for further research. We suggest that animism and creationism can be identified as different types of teleology and we conclude with specific suggestions for education.

Both of these chapters raise important issues which should be taken into account in evolution education and the relevant research. First, the psychological essentialism identified in children and adults is different from the essentialistic views held by past naturalists. Wilkins himself notes in his conclusion that one should be careful not to identify individual conceptual development with the historical development of concepts, especially when the latter are not as linear and straightforward as we may think them to be. The second important point is that there can be different ways by which one view can be described as essentialist and this is not necessarily bad. Thus, it is not accurate to identify the term "essentialist" with students' preconceptions, nor to claim that essentialism in general is an obstacle to understanding evolution. Interestingly enough both of these considerations hold for teleology too. This is important in order to show that teleology per se is not illegitimate in biology; rather it is inherent in all explanations based on natural selection. In this sense, it is not wrong to claim that e.g. humans have hearts in order to pump blood as long as it is clear that this explanation is based on the consequences of natural selection and not on those of some kind of rational design. Thus, it is only design teleology which is problematic and attention should be paid to its underlying assumptions. Now whether students' teleological intuitions are similar to those of Paley or Cuvier is irrelevant. What is important is which causal processes students invoke to explain the origin of features and not whether their explanations are teleological or not.

9 "Proximate" Phenomena: Functions, Mechanisms, Information and the Systemic Approach in Biology

In studying biology, it is important for students to understand the various processes at the cellular and the organismal levels. This is the topic of the next three chapters which focus on the concepts of function, mechanism and information, while the fourth one describes the contemporary systemic approach in the study of life. These concepts are often mentioned in textbooks, but their meaning is not always accurately described, whereas the systemic approach is not always emphasized. The first two chapters offer detailed accounts of functional and mechanistic explanation, respectively; the third chapter explains what information in biology is, putting emphasis on the notion of information as a kind of relation and not a property; the final chapter describes what systems biology is and why it is important to be taught in biology. I describe the respective phenomena as "proximate", using quotation marks to indicate that this is just an instrumental use of the term. As already explained above, the interplay between development and evolution is such that we cannot really retain Ernst Mayr's ultimate/proximate distinction.

The chapter by Arno Wouters, titled *Biology's Functional Perspective: Roles, Advantages and Organization*, focuses on biological functions. Wouters describes what the functional perspective of biology is about, why it is important as well as why it does not necessarily assume design, as is commonly thought. Using a concrete example, that of how emperor penguins recognize each other via their two-voice system, Wouters explains that the functional perspective is like a problem solving approach. The problem is how to stay alive and functions can be seen as solutions to this problem. Consequently, the existence of the features which perform these functions is explained on the basis of the contribution they make to their bearers' staying alive. However, Wouters notes, this is not all. Whether an organism will manage to stay alive or not depends not only on its individual features but also on their arrangement as a whole and on the coordination of their activities. He also notes that functional explanations are independent from assumptions about the origin of the respective features. In this sense, the concept of function is distinct from the concept of adaptation, and functional explanations are distinct from explanations based on natural selection. Wouters also explains why it is a misunderstanding to think that the functional perspective rests on an analogy between function and design, as well as that functions do not explain the presence or structure of their bearer.

Many philosophers have argued that it is not appropriate to think of organisms as machines (see for example the chapter by Brigandt in this volume). However, in his chapter *Understanding Biological Mechanisms: Using Illustrations from Circadian Rhythm Research,* William Bechtel argues that there is nothing wrong with that. He suggests that in their attempt to explain a phenomenon, in many contexts what researchers are actually looking for is the mechanism responsible for it, and employ particular strategies for this purpose. Mechanistic research has some distinctive features, relying on diagrammatic representations and particular strategies for discovering mechanisms, such as identifying the parts that constitute the mechanism. Mechanistic explanations are those developed in order to explain a phenomenon by identifying the working parts of the respective mechanism, i.e. the parts that perform the various operations that go into producing the phenomenon. This approach of decomposing a mechanism into its component parts has been characterized as reductionist. However, studying the phenomenon at a lower level may not provide an adequate explanation since components may operate differently on their own than when they belong to a whole. Moreover, in order to explain a phenomenon additional knowledge of the organization of higher levels, of which the whole mechanism is just a part, is required. The key tasks in developing mechanistic explanations are: (a) delineating the phenomenon, (b) identifying and decomposing the responsible mechanism, and (c) recomposing and situating the mechanism. After having provided a detailed characterization of mechanisms, Bechtel uses circadian rhythms (endogenously controlled oscillations of approximately 24 h in many physiological processes and behaviors) as an example. He describes the various strategies that have been employed in understanding the mechanism underlying this phenomenon: identifying, decomposing, recomposing and situating a mechanism. Bechtel concludes that encouraging an understanding of mechanisms in biology education may contribute to a better understanding of biological phenomena.

But how are functions and mechanisms controlled or coordinated? An intuitive answer would be that the information for this is "coded" in DNA. But is it so? The idea that DNA contains some kind of information within its structure is often the take-home message for students in biology courses. However, it is not as simple as that, as Alfredo Marcos and Robert Arp explain in their chapter *Information in the Biological Sciences*. First, the authors provide a historical introduction on the concept and nature of information. Then, they turn to "bioinformation" and provide several examples of information at work in the biological sciences: DNA and protein synthesis, microorganisms and their environment, neural communication, visual perception, interactions among the components of organisms' internal environments or between themselves and their external environments. After discussing opposing views about the nature of bioinformation, such as whether information talk in biology is metaphorical or not, Marcos and Arp note that information is a distinctive characteristic of organisms, related to the concept of function. They also argue that bioinformation is best understood as a relationship between entities. DNA is "informational" only in relation to a given cellular context, and it is misguiding to locate information within a particular molecule without any reference to its context. In particular, bioinformation should be conceived of as a triadic relationship, involving three entities: a message, a receiver, and a system of reference which the message informs the receiver about. In this sense, a segment of DNA molecule could be an example of a message, its system of reference could be an mRNA molecule to be synthesized, and the receiver could be the appropriate part of the cell. The authors also explain the relation between the concept of information and other concepts such as correlation, form, entropy, order, organization, complexity, and knowledge. They conclude with a short presentation of bioinformatics and a description of teaching as an informational process.

It should be clear that the "proximate" phenomena addressed in these three chapters – functions, mechanisms and information in biology – cannot be studied at the molecular or cellular or organismal level alone. In contrast, what is required and what is actually currently implemented is a systemic approach. This is the topic of the chapter titled *Systems Biology and Education* by Pierre-Alain Braillard. Although it is difficult to give a definition of systems biology, it is important to note that it makes use of mathematical and computational models to describe the complexity of biological systems. Braillard argues that systems biology challenges the traditional view of biology as it affects not only its explanatory practices but also its philosophical foundations. Starting from reductionism, the author explains that molecular biology is characterized by explanatory reductionism which mainly has to do with discovering molecular mechanisms involved in the production of biological phenomena. This reductionism made genes seen as the fundamental causes of phenotypes, orienting research towards the identification of "genes for". However, it has become clear that even if it is found that some genes are implicated in some process, in order to explain that process we need to understand how all cellular components (genes, proteins, RNA, metabolites, etc.) and mechanisms are interconnected and affect each other (this is also the topic of the next three chapters). Thus, reductionist approaches have limited outcomes because they cannot fully explain the influence of the context on molecular mechanisms and cannot fully represent the complexity of regulatory mechanisms through which biological functions emerge. Systems biology can provide a better understanding due to its focus on multi-level modeling of networks and mechanisms. But to achieve this, scientists must be biologically knowledgeable as well as able to use methods and approaches from other fields such as physics and computer science. As this is difficult for one person, building interdisciplinary groups with experts from various fields within which the transfer of knowledge will be possible is required. However, this is neither simple nor necessarily efficient, and Braillard describes concrete examples to illustrate this. He also explains why systems biology has implications for some fundamental issues related to research aims (e.g. whether biologists should try to uncover general principles) and methods (e.g. whether simple diagrams can be used to represent phenomena). The author concludes that systems biology offers the opportunity to replace the dominant oversimplified, reductionist description of biological phenomena with a richer and finer wide picture.

Many important pedagogical points are made in these chapters. One has to do with the analogy that people make between organisms and machines. Although a functional approach is useful in teaching about the structures and properties of organisms, it should be made clear to students that there is no underlying assumption of intentional design behind that. We might thus claim that the function of the heart is to pump blood, and also make clear that this does not automatically imply intention or design. I strongly believe that in teaching we should be explicit about the origin of features and functions, not because this is a philosophical requirement but because of students' tendency to explain structure and function in terms of plans and purposes. Otherwise, we should refrain from using machine talk. Bechtel is right that machine talk has been fruitful in science so far; but scientists have a clear understanding of the underlying assumptions whereas students do not. Bechtel makes another very important point in his chapter that scientists often have to intervene in nature to elicit the patterns which they aim to explain. The example he gives is that of microscopic observation where the specimens are dehydrated and chemically modified by stains and fixatives. Bechtel notes that what one observes through a microscope is not the actual tissue but the product of these manipulations, which are often quite significant. This is a very important point because students do not often realize that scientists intervene in natural phenomena (e.g. with experiments) and so how reliable the data obtained is might be questioned. This is also relevant to how sequence data is obtained through DNA analyses. In describing various examples of information in biology, Marcos and Arp start with, perhaps, the most commonly cited example in biology: genetic information as supposedly contained in DNA. In explaining why DNA is an informational molecule, they point to the important distinction between genome sequencing (the methods and technologies used to determine the specific order of nitrogenous bases in DNA or RNA) and genome annotation (the methods and technologies used to identify the locations of genes and to determine what they do). Thus, it is one thing to read a message and another to understand what it is about. What is more important is that this message is not identified with the DNA sequence. DNA alone does not include information; it is informational only in a cellular context and in this sense, genetic information is not exclusively genetic as Marcos and Arp note. Consequently, as Braillard argues biology education must move beyond genes to also include in an explicit manner, the contexts in which they operate. This is important for various reasons, which are the focus of the three chapters that follow.

10 Genetics: Beyond Mendel and Genetic Determinism

Anyone who has ever attended a high school genetics class has probably heard of Gregor Mendel and his "laws", perhaps the most widely known ones in biology. However, in their chapter titled *Putting Mendel in His Place: How Curriculum Reform in Genetics and Counterfactual History of Science Can Work Together,* Annie Jamieson and Greg Radick explain that many Mendelian concepts, such as the concept of dominance, are misconceived and as a result the "Mendelian" genetics taught in schools and universities are inconsistent with our contemporary knowledge. Mendelian genetics, as nowadays taught, refer to fixed hereditary properties which are determined by genes. Thus, students learn that there are "genes for" properties e.g. black color and white color. When an heterozygote exhibits e.g. black color, then the trait and the respective allele are called dominant. Jamieson and Radick explain that these descriptions are wrong because they overlook the complexities of development and because this model is not sufficient to describe inheritance in all kinds of traits. "Mendelian" inheritance is mistakenly taught as the norm, because it is not. The authors then suggest that one should turn to the work of W.F.R. Weldon, who expressed concerns about the dogmatic nature of Mendelism and adopted a much more contextualized interpretation of dominance. Weldon's interpretation might be adopted in basic genetics instruction and might help avoid the problems raised by current approaches. Weldon studied hybrid pea varieties himself, and concluded that peas actually ranged from greenish yellow to yellowish green, as well as from smooth to wrinkled. So, what Weldon did was to draw attention to the natural variability which was ignored in Mendel's model, and eventually to suggest that scientists should take this into account. Weldon's critique was the starting point of the "Mendelian-biometrician debate". William Bateson published the first textbook on Mendelian genetics as a response, which eventually served as a template for genetics textbooks until today, with Mendel's experiments being the first step, and all the rest being an extension of Mendel's work. In contrast, Weldon put emphasis on statistics and interactions during embryo development. Jamieson and Radick suggest an alternative approach along Weldonian lines: teach that gene-environment interactions are pervasive and primary, and then teach about Mendel and his peas. In doing so, the focus would be on development rather than inheritance, since the latter is meaningless if the complexities of development are not taken into account. Mendelian genetics could still be taught, but as a special case, not as the norm.

As already mentioned, one consequence of teaching Mendelian genetics in high school genetics courses seems to be the widespread view that there exist "genes for" traits. Thus, people tend to think that particular alleles, such as those which controlled the green or the yellow color of seed in edible peas, must also control all characteristics. For a long time, this is how genes have been conceptualized and they have usually been identified with particular segments on chromosomes. Recent developments have shown that this is inadequate to capture the complexity of inheritance and development, and so other gene concepts have been proposed. Eventually, it seems that the concept of a "gene for", but perhaps even the concept of "gene" are no longer valuable to describe the plethora of phenomena that scientists study nowadays. After describing the various gene concepts used by scientists since the term was first coined in 1909, in our chapter *Against "Genes For": Could an Inclusive Concept of Genetic Material Effectively Replace Gene Concepts?* Richard Burian and I point to inconsistencies in the use of gene definitions, and provide a classification of gene concepts. Then we propose that the concept of gene should be replaced by the concept of genetic material, at least in educational contexts and in the public discourse about genetics. We consider this concept as more inclusive as it is compatible with the actual complexities of phenomena such as epistasis, pleiotropy, alternative splicing etc. Nucleic acids are the genetic material and this concept can capture both coding and regulatory sequences, as well as all kinds of interactions between nucleic acids and other molecules in cellular contexts. We conclude that this is a way both to provide an accurate description of phenomena and to challenge the widespread notion of "genes for" that enhances notions of strong genetic determinism which seem to be very intuitive.

The distinction between genetic and environmental influences has long been described as the distinction between "nature" and "nurture". However, as David Moore explains in his chapter titled *Current Thinking about Nature and Nurture,* this is a mistaken distinction. Moore first explains that although the idea of eugenics was largely rejected after World War II, its underlying conceptual framework, based on Galton's distinction between the influence of nature and nurture, was largely retained. The author explains that it makes no sense to try to answer questions such as how much nature or nurture influence the development of a characteristic. Instead, the question that should be driving research is how genetic factors and aspects of an organism's environment interact to produce the organism's traits during development. Moore notes that both genetic and non-genetic components are essential for biological processes to take place. Then he turns to the concept of heritability and explains that it is mistakenly perceived to be a measure of the relative importance of nature and nurture to trait development. Heritability estimates provide no information about influences on trait development, but only inform how one can account for variation in a population. One excellent example that Moore gives of why it is not possible to separate nature from nurture is the development of queens and workers in bees. Although these are genetically identical individuals, they are very different in terms e.g. of size and structure (queens are larger and possess ovaries). Their nutrition has long been characterized as the cause of their differences, but although it might alone account for the difference in size, it cannot do so for the difference in possessing ovaries. Hence, it seems that both workers and queens have the genetic potential to develop ovaries but only the latter do so because their DNA interacts with factors from the environment to produce them. Environmental factors can have temporal but also long-term effects on genetic activity, by changing the genes themselves or by changing what they are doing. Moore also discusses the concept of developmental plasticity and provides several examples of it. He concludes by suggesting that instead of teaching Mendelian genetics with Punnett squares which can be perceived to support genetic determinism, teachers could adopt a pedagogical approach that encourages the study of the *emergence* of phenotypes during development.

As the chapters by Uller, Minelli and Fusco, and Love suggested that development should be taken into account in the study and the teaching of evolution, the chapters just presented suggest that development should also be taken seriously into account in the study and teaching of genetics. These chapters also serve as a call to action for revising the teaching of genetics. Mendelian genetics certainly is a useful heuristic tool, and secondary students do not need to learn all details. However, it is one thing to teach science in the school context in a simplified form and another to enhance intuitions about strong genetic determinism and misconceived notions about what genes are and what they can do. Genes do not control anything on their own but operate within cellular environments which affect their expression. If two people used the same recipe (genes) to cook some food, the outcome (phenotype) could be very different although they might have both followed the same recipe. The expression of the information in the recipe (genes) depends on the cook (developmental system) that will implement it. Consequently, development should be taught alongside genetics, as developmental processes may produce outcomes different that those expected by just reading the DNA sequences.

11 Biology and Ethics

Last, but not least, comes ethics. I should note at this point that biology teachers and educators are not supposed to, and perhaps should not, teach ethics. However, I think that they should be prepared to discuss the ethical implications of the biological phenomena they are teaching. I deliberately put the chapters of ethics at the end of the book for two reasons. The first is that any discussion about ethics requires knowledge of the respective scientific content. Thus, discussion of science concepts should precede any discussion of ethics so that students are well informed. The second reason is that this is a way to show that many biological topics raise ethical issues. There often are important ethical implications in many kinds of biological research, especially when they involve manipulation of any kind, and this should be clear to students. This does not mean that we should be afraid of biological research; rather, we should always consider its ethical implications before initiating it.

Lisa Gannett argues, in her chapter titled *Genomics and Society: Why "Discovery" Matters*, that the commercialized social context within which research in genomics is carried out has raised concerns about whether the objectivity of scientists is compromised. Are scientists aiming at answering questions and producing a better understanding of genetics and genomics? Or are they simply trying to fulfill the requirements of industries and institutions which fund their research? Gannett explains that philosophers of science have in the past argued that the objectivity of science is not affected by social values. To support this, several distinctions were drawn: between theory and practice, between the context of discovery and the context of justification, and between facts and values. Gannett reviews the various criticisms of these distinctions, and she concludes that they are misconceived as theory is embedded in practice, discovery matters for justification, and facts and values are entangled. She then turns to the concept of biogeographical ancestry which has been introduced as a substitute for "race" in population genomics. This serves as a case study to illustrate how the idea that science is value-free is not confirmed in this case as research in genomics is conducted in a commercialized social context. Gannett concludes with particular implications for biology education.

In the next chapter, *Philosophical Issues in Human Pluripotent Stem Cell Research,* Andrew Siegel focuses on human pluripotent stem cell research. So far, the cells that have been used in research are human embryonic stem cells, the collection of which requires the destruction of human embryos. This has caused fierce opposition to this kind of research and overall the topic has been quite controversial. Research of this kind has implications for metaphysics, ethics, and political philosophy, raising many questions such as when a human life begins, if human embryos have a moral status, whether there is a moral distinction between creating embryos for research and creating them for reproductive ends, the ethics of creating human/non-human chimeras, and the challenge of developing public policy in a society characterized by opposing views on these topics. Siegel suggests that biology education should go beyond a presentation of biological properties of stem cells to address the philosophical questions that surround the respective research. Students should not consider this research in isolation from the social context in which it occurs. In contrast, they should be made aware of the disagreements about when a human life begins or about the moral status of human embryos, as well as of the ethical issues around the creation of human/non-human chimeras and the production of human gametes from human pluripotent stem cells. It is suggested that researchers are also citizens, and so they should take part in policy making in a way that all different views are respected.

Similar issues exist more generally for all kinds of biomedical research and this is the topic of the next chapter, *Ethics in Biomedical Research and Practice*, by Anya Plutynski. The author distinguishes between ethical questions which are "intrinsic" to biomedical research, e.g., questions about what kind of research is ethically acceptable, and questions "extrinsic" to research, e.g. about how biomedical research is funded. First, she briefly explains how different ethical theories perceive what is moral or immoral, using Kantian and utilitarian ethics as examples. Plutynski explains that research ethics is the branch of biomedical ethics that concerns the responsible conduct of research, including the ethical treatment of human and non-human subjects, the conflicts of interest raised and how they might be avoided, the acknowledgment of all contributors, and the social responsibility of scientists. One example of intrinsic issues concerns the ethics of research on human and non-human subjects. After WWII, a set of norms for research was codified because it seemed necessary to protect vulnerable populations from abuse. However, until recently vulnerable populations such as prisoners and soldiers participated in biomedical research that posed serious harms and had very little benefit, both for them and for science. Plutynski also discusses ethical issues that emerge from clinical trials and relevant clinical research on drug efficiency. An example of "extrinsic" issues has to do with the fact that the vast majority of biomedical research focuses on diseases that affect the wealthy. Biomedical research is not only conducted in the government sector, but also in the private sector. This raises questions about potential conflicts of interest. In addition, researchers should be careful in selecting and designing their research project in order to refrain from wasting research funds, especially those coming from public sources. Finally, researchers should not be influenced by commercial gain. Plutynski concludes that addressing these ethical issues is important to biology education, because biomedical research must be discussed within its social and ethical context.

The final chapter, *Environmental Ethics*, by Roberta Millstein explains that questions about what is of value in the natural environment or how we ought to behave towards it, raised by conservation biology, environmental science, and ecology, are very important and need to be addressed in biology education. She suggests that environmental ethics has much to teach about these questions. Millstein identifies three primary areas in which environmental ethics can contribute to biology education. The first has to do with the question about what our moral community includes: humans only, all life, or whole ecosystems? Answers to this question should be explicit and students should think about arguments for and against them. The second area has to do with the application of these answers to actual environmental issues and problems. Students need to be aware of how the different answers concerning what the moral community includes can imply conflicting suggestions for how we should act. Thus, it is important to think about how such conflicts can be resolved. The third area in which environmental ethics can contribute focuses on central concepts such as biodiversity, sustainability, species, and ecosystems. Exploring and evaluating the various conceptions associated with these concepts could make students more reflective and thoughtful citizens. Millstein suggests that for these reasons biology educators should incorporate these topics into their teaching.

Acknowledgements I am indebted to all authors for the high quality chapters they contributed to this book. Working with them has been an intellectually stimulating experience. I am also indebted to all reviewers of these chapters, many of whom were also contributing authors. These include Denis Alexander, André Ariew, Robert Arp, Wallace Arthur, Stefan Artman, John Avise, Francisco Ayala, Bill Bechtel, Ingo Brigandt, John Hedley Brooke, Lindley Darden, David Depew, Laura Franklin-Hall, Patrick Forber, Lisa Gannett, Eva Jablonka, James Justus, Maureen O'Malley, Roberta Millstein, Robert Nola, Massimo Pigliucci, Erik Peterson, Angela Potochnik, Anya Plutynski, Richard Richards, Elliott Sober, Andrew Siegel and Tobias Uller. I am also very grateful to Michael Ruse who wrote a foreword for this book. Finally, I want to thank Dick Burian and Jim Lennox who invited me to co-author their chapters with them. It has been a great privilege and experience to have worked with them.

I am of course grateful to the publishing editor Ties Nijssen as well as to the editors of this series Charles Wolfe, Philippe Huneman and Thomas Reydon for making the publication of this book possible. Thomas Reydon has worked with me on this project from its inception and, being a philosopher of science with a good sense of what science education is about, he has been enormously helpful all the way. I am also grateful to Megan Faurot who helped finalizing the glossary. The glossary terms were written by the authors of the relevant chapters and were edited by me for consistency. Editing this book owes a lot to my previous experience as a guest editor for thematic issues for the journal *Science & Education* and to Michael Matthews, editor-in-chief of that journal. Finally, discussions with colleagues and friends over the years, about philosophy of science and how it relates to science education, have been extremely useful. I thank Alexandros Apostolou, Norm Lederman, Michael Matthews, Giorgos Malamis, Bill McComas, and Thomas Reydon for making me think hard about these topics.

Last but not least, I owe a lot to my wife, Katerina, and our children, Mirka and Giorgos, for their love and support. Without these, working on any book project would have been impossible. As I am finishing editing this book, I am also finishing writing my first book which I will dedicate to them. So, I thought that if I may dedicate this edited book to someone, it should be someone else to whom I also owe a lot. Thus, I dedicate this book to my father, Giorgos, who encouraged and supported my intellectual life ever since I was a child.

What Is Life?

Carol E. Cleland and Michael Zerella

1 Introduction

Science education helps students understand scientific facts, principles, and theories, and also helps them appreciate science as an ongoing process of critical investigation grounded in observation. Philosophical exploration of the question "what is life?" is particularly useful regarding the latter goal, especially since students readily can participate in the process. In so doing, they can come to appreciate how scientists often pursue their research in the face of deep uncertainty over the nature of what they are studying. Along the way, they also address the former goal of science education because philosophical exploration of the nature of life draws upon many of the facts, principles and theories from a wide range of biological sciences, including some important new fields like astrobiology (Sullivan and Baross 2007) and artificial life (Adami et al. 1998), as well as new research into origins of life (Orgel 1998; Shapiro 2007).

An answer to "what is life?" is important because it can have practical implications for ongoing biological research. Investigation of life itself, which includes research into the origins and extent of life in the universe and the possibilities of creating life in a laboratory,[1] must start with at least a tentative understanding of

[1] A good overview of recent advancements and challenges in discovering the origins of life can be found in Orgel (1998), Gasteland et al. (2006), and Shapiro (2007). Boden (2003) has an excellent discussion of how an answer to "What is life?" affects the search for life on other planets. Attempts to create artificial life are reviewed in Adami et al. (1998) and Bedau (2003).

C.E. Cleland (✉)
Department of Philosophy, Center for Astrobiology,
University of Colorado, Boulder, CO, USA
e-mail: carol.cleland@colorado.edu

M. Zerella
Department of Philosophy, University of Colorado, Boulder, CO, USA
e-mail: mjzerella@gmail.com

K. Kampourakis (ed.), *The Philosophy of Biology: A Companion for Educators*, History, Philosophy and Theory of the Life Sciences 1, DOI 10.1007/978-94-007-6537-5_2,

what life is. For without some understanding of their subject matter, scientists will have a difficult time identifying their goals and planning their research strategies. Given the importance of an answer to this question, some scientists and philosophers have gone so far as to propose specific definitions of life; see Popa (2004, Appendix B) for a list of the most popular. In this chapter, however, we argue that reliance on definitions to guide scientific research is a mistake, and we suggest more appropriate, alternative strategies. We begin in Sect. 2 by exploring logical and philosophical difficulties associated with a definitional approach to understanding the nature of life. As we explain, relying on a definition can do more harm than good because definitions of life fully determine what counts as a living thing before scientists have gathered enough information to warrant such a generalization. Section 3 builds on Sect. 2 by examining scientific limitations of our current understanding of life, which is founded upon a single, possibly unrepresentative example of life, namely, familiar Earth life. Finally, in Sect. 4 we explore a variety of options for expanding our knowledge of living systems, ideally through discovery of novel or extraterrestrial forms of life. Our exploration of the limitations of a definitional approach to understanding the nature of life illustrates an important characteristic of science: Science is not dogmatic but rather is an ongoing process of exploration that is always open to revision.

2 Concepts and Definitions: From Philosophy to Science

To understand why definitions do not provide good scientific answers to questions about natural phenomena such as life, we first must clarify some terminology. Definitions are primarily concerned with language and concepts, that is, with words and the "meanings" (qua concepts[2]) associated with them. Since it is crucial that we maintain the distinction between a word as a string of symbols and the meaning of that word, we will enclose the word itself in single quotes. For example, a dictionary may define 'line' as linear path and 'cause' as that which produces an effect. Dictionary definitions run into problems, however, because they usually are explicitly or implicitly circular, defining words in terms of synonyms ('linear' in the case of 'line') or close cognates ('effect' in the case of 'cause'). As a consequence, they can't enhance the comprehension of someone who doesn't already understand the term being defined.

The best definitions are not circular, and hence are genuinely informative. A widely used illustration in beginning college logic classes is defining 'bachelor' as unmarried, human male. This definition is not circular. The concept of being a bachelor is not included in the concepts of being unmarried or human or male; one can grasp each of the latter concepts without understanding the word ('bachelor') being defined. Further, the definition (supposedly) fully determines membership in the

[2]The idea that meanings are just concepts in the head is no longer widely accepted by philosophers but, for our purposes here, we can ignore this complication; it doesn't affect the point that we are making here about definitions.

class of things to which 'bachelor' applies; nothing lacking these characteristics can qualify as a bachelor and anything having all of them is *ipso facto* a bachelor. Definitions of this sort are clearly superior to dictionary definitions. Indeed, they represent an ideal to which one should aspire in formulating truly informative, conceptually rigorous definitions. We will henceforth designate such definitions as 'ideal' to distinguish them from other forms of definition.[3]

A different form of definition that is very popular among scientists is the operational definition. Operational definitions supply a means of testing whether an item falls under the term being defined without saying what it is to be a thing of the kind concerned. A good illustration is operationally defining 'acid' as a liquid that turns litmus paper red, which is known to be a reliable indicator that a given liquid is acidic. But this is the case only because scientists already know the chemical characteristics of acids and why acids turn litmus paper red. Therefore, operational definitions do not provide good answers to questions like 'What is an acid?' or 'What is life?' because they presuppose that one already knows the intrinsic nature of the items (e.g., acids or living things) concerned. This is especially true when we do not understand the intrinsic nature of the thing being defined, as is the case with life.

Philosophers are mostly interested in ideal definitions, and they construct such definitions using a process of conceptual analysis. Most concepts are built from other concepts, and philosophical analysis reveals these constituent concepts. For example, when analyzing the concept of a hammer, one sees that it includes other concepts, such as that of nail, swinging, pounding, heavy, etc. All of these sub-concepts, along with several others, are relevant to appropriate understanding of hammers. Taken together, they provide a full description of the items (actual hammers) falling under the concept hammer associated with the word 'hammer'. The philosophical challenge, then, is to find a combination of concepts that fully determine (describe) the extension of 'hammer,' where the extension is the class of things falling under the concept. In other words, philosophers try to find a set of descriptions that are *necessary* and *sufficient* for something being a hammer. Necessary conditions are those that an item *must* have in order to fall under the concept in question. For example, a necessary condition for being a hammer is being solid, since no liquid or gaseous things can be hammers. Sufficient conditions are conditions that ensure that the item really does fall under the concept in question. Determining a sufficient condition for application of a concept often is a bit trickier than finding necessary conditions, but here is an initial attempt for hammers. Perhaps in order to be a hammer, it is sufficient that an object be solid, durable, have at least one broad surface and one graspable surface, and be light enough to wield yet heavy enough to drive nails. If this is the correct definition of hammer, then we have determined the individually necessary and jointly sufficient conditions for being a hammer. Every hammer has (at least) these features, and everything that has these features is a hammer.

[3]See Cleland and Chyba (2002, 2007) for more on the different forms of definitions and why none are not up to the task of explicating the nature of life.

For the most part, conceptual analysis is an *a priori* process. It usually (but not always) can be done merely by thinking carefully about the concepts we already possess, as in the hammer and bachelor examples. Then, if we find necessary and sufficient conditions for the application of the concept, our analysis fully determines proper use of that concept. Scientists, on the other hand, typically (but not always) rely more heavily on *a posteriori* knowledge, which is knowledge we only can possess after exploring nature. For example, scientific understanding of the nature of water (viz, that it is H_2O) required a great deal of empirical investigation; it was not discovered merely by thinking hard about the pretheoretical (before molecular theory) human concept of water.[4]

The distinction just sketched between acquiring an understanding of a category of things by means of conceptual analyses and acquiring it by means of scientific investigation is best fleshed out in terms of a famous philosophical distinction between *natural kinds* and *artificial kinds*. Scientists usually study natural kinds, where a natural kind is a category of things (objects or processes) that exists on its own, independent of any human conceptions of it. For example, stars are thought to constitute a natural kind because even if humans had never existed, there still would be stars, and their properties would be unaffected by the absence of humans. Hammers, by contrast, do not constitute a natural kind because they were designed and constructed by humans for human purposes. Without us, and our occasional need to pound nails, there would be no such thing as hammers. When studying artificial kinds like hammers, we need only consult our *a priori* concepts because we are analyzing something whose characteristics depend exclusively on human needs and purposes. The characteristics of natural kinds, by contrast, are not established by humans, and so scientists rely on *a posteriori* information. And since scientists are always discovering new features of the natural kinds they study, our *a posteriori* understanding is always subject to revision. Indeed, a scientist may even discover that what she took to be a natural kind is not one after all, or that what she thought were distinct natural kinds are in fact a single natural kind. A good illustration of the former is the gemstone jade, which was once thought to comprise a natural kind and later discovered to consist of minerals having two different molecular compositions. On the basis of their molecular structures, geologists differentiate between "jadeite" and "nephrite", but jewelers continue to call both minerals "jade" because they share the ornamental properties of interest to them. In short, scientists *discovered* that jeweler's jade is actually two distinct natural kinds (for a discussion of natural kinds, see Wilkins, this volume).

Empirical data alone does not tell us what the world is like. While careful observations in the lab or field can show that our preconceptions are problematic, they do

---

[4]We are not overlooking the fact that, in the hammer example, a person must become familiar with hammers and their uses in order to possess the relevant concepts. The point is that a typical English-speaking adult could analyze the concept of hammer without doing any extra empirical (a posteriori) research into the material things that qualify as hammers. By contrast, scientists typically must conduct empirical investigations in order to achieve an understanding of the natural categories (e.g., water) that interest them.

not necessarily show how our understanding of a class of natural phenomena should be changed. As a consequence, scientists frequently have what are in essence philosophical debates over the proper interpretation of their data and what it tells them about the phenomena they are studying. In efforts to settle such debates scientists sometimes propose definitions. To understand the delicate interplay between empirical and conceptual considerations in scientific research and why definitional approaches are so problematic, students need to understand how an *a priori* investigation of concepts is conducted.

Philosophy often is portrayed as abstract and distant, but students readily can participate in philosophical analysis, as long as they have some experience with the concept being analyzed. Consider again the bachelor example mentioned earlier. Clearly this is an artificial kind and is dependent on human concerns. Thus, as long as students are old enough to understand the general social structure of marriage, they should be able to engage in *a priori* investigation of the key features of all bachelors. For example, students often propose on their own, after some small debate, that bachelors are unmarried males. This is a good first approximation since being unmarried and male appear to be necessary conditions for being a bachelor, but more careful consideration of the concept reveals that some refinement is needed in order to provide the full set of conditions that are individually necessary and jointly sufficient for being a bachelor. The most common way of refining a conceptual analysis is to consider what individuals are in the extension of the concept. As mentioned above, the extension of a concept includes all and only things that are appropriately described by the concept, so our analysis of 'bachelor' must cover all bachelors, but no non-bachelors. Accordingly, if students suggest that bachelors are unmarried males, they can be prompted to consider whether, for example, they should include all males of any species. If not, then the extension of their original analysis is too broad (included far too many individuals) and must be narrowed to include only humans. Earlier, we said that being an unmarried, adult, human male fully determines what it is to be a bachelor. This is not quite true, however, because even after restricting the account to unmarried, human males, students can be challenged again by asking them whether newborn infant human males are bachelors. If not, then an age restriction must also be included as a necessary condition in the analysis. But even when restricted to adult, unmarried, human males, the analysis may again be tested by other challenging cases, such as widowers, divorcées, unmarried men in long term relationships, men who have taken a vow against marriage (such as various religious leaders), and others. Each new challenging case prompts further debate over additional necessary conditions to exclude unwanted cases. Of course, we also don't want to make our analysis of 'bachelor' so narrow that it inappropriately excludes individuals from the extension that are supposed to be there. For example, sometimes students suggest that a necessary condition of bachelors is that they are messy. While this may be a feature of the bachelor stereotype, it is not really shared by all actual bachelors, so inclusion of 'messiness' in the analysis inappropriately excludes tidy bachelors. In short, the purpose of conceptual analysis is to come up with a description that is, in Goldilocks' oft quoted words, "just right"—not too narrow and not too wide, and thus fits all and only bachelors.

Engaging in this process can show students that critical investigation often reveals that our first impressions sometimes are not able to capture even relatively simple concepts like bachelors.

When we move from philosophical investigation of artificial kinds to scientific investigations of natural kinds, the limitations of our first impressions become even more apparent. Thus the practice of science (like the practice of philosophy) requires a willingness to update and refine one's understanding, even when this upsets long-held beliefs. Adherence to a rigid definition of one's subject matter can undermine that willingness and, as a result, thwart scientific advancement. A fruitful scientific example is the recent relegation of Pluto to "dwarf planet" status after debate over what it is to be a planet in the context of surprising astronomical discoveries. This is somewhat of a bridge example between science and philosophy because the debate was not over what Pluto *is*, in and of itself. Pluto's size, shape, orbital characteristics, and the like were not in dispute. Instead, the debate was over whether Pluto qualifies as a planet, and so this was a debate over the boundaries of a natural kind and involved a kind of conceptual analysis. As with the bachelor concept, students can participate in this process by trying to identify necessary and sufficient conditions for being a planet. In doing so, they must take into account the recent discovery that Pluto is just one of many bodies, including some larger than Pluto, occupying a belt of debris known as the Kuiper belt in the outer regions of our solar system. Consistency demands that we put Pluto in the same category as similar objects in its orbital neighborhood. Students must therefore decide which is more important: counting Pluto along with many other objects in the Kuiper belt as planets, or excluding Pluto and everything else in the Kuiper belt, as planets; another possibility of course would be to conclude, in the face of the discovery of so many large objects beyond Pluto, that 'planet' doesn't designate a natural kind. In the end the scientific community agreed upon necessary and sufficient conditions for being a planet that excludes Pluto and its many neighbors. To be a planet, an object must mainly orbit a star[5] and it must be large enough to gravitationally pull itself into a sphere and clear its orbital path of debris (Tyson 2009). Pluto meets the first criterion but not the other two. Thus, scientists felt compelled to change our long-standing belief that Pluto is the ninth planet of our solar system. In addition to prompting a change in all of our astronomy textbooks, Pluto's demotion has a potentially embarrassing effect on our attempts to contact intelligent extraterrestrial beings. The Pioneer spacecrafts, launched in the early 1970s and just now leaving our solar system, contain pictorial plaques that identify our solar system as one that has nine planets. Any aliens encountering those plaques will either fail to find our planet or will be a bit puzzled by our oddly inaccurate depiction. In this context, we hasten to add that even our current, more empirically adequate, account of planets is open to revision. As scientists continue to discover and study extra-solar planets,

---

[5]Most objects in the universe are in orbit around more than one thing at the same time. For example, Earth's main orbit is around the sun, but Earth, along with the rest of the solar system, also orbits around the center of the Milky Way, which itself is involved in large-scale movements around the other galaxies in our cosmic neighborhood.

some of those planets may have properties that are significantly different from those of solar planets. If so, the scientific community may need to revisit and refine their account yet again. The point is even the best scientific theories are tentative and open to revision in light of new empirical discoveries and theoretical advances.

This brings us back to the question "What is life?" Students can join scientists and philosophers in debating this question, as they did when considering what it is to be a planet. What features of life seem necessary for all life? What features of familiar life could something lack, yet still be a life form? Some researchers are working to create artificial life in the form of robots (e.g. Lipson and Pollack 2000) or computer models (e.g. Lange 1996; Pennock 2007b), prompting us to consider whether metal machines or informational structures (patterns of electronic activity) in a computer could really count as alive, as opposed to merely *simulating* life (Sober 1991; Keller 2002; Bedau 2003; Pennock 2007a)? As with the debate over Pluto, answers to these questions require careful consideration of the *a posteriori* knowledge that scientists have already gathered, and the following sections can be used to help guide student discussions through that process. As the following section reveals, our current state of biological knowledge is woefully inadequate for generalizing about *all* life. In Sect. 4 we provide suggestions for future research that could push our understanding of life as a unitary natural phenomenon forward. Our goal is to help students to discover and appreciate some of the challenges faced by scientists on the forefront of biological research and to help them better understand how science progresses in the face of inadequate bodies of empirical information.

3 Limitations of Our Current Understanding of Life

In order to conduct research into a particular subject, scientists must have some idea what it is they are studying. Thus, while investigating life's origins or searching for life elsewhere in the solar system, or trying to create life in the laboratory, it may be tempting to start with a "definition" of life, if only a tentative one. However, if one uses such a definition to guide attempts to "create" life in the laboratory or search for life on other worlds, one is likely to produce or find only what one is looking for, and if our current scientific account of life is unreliable, then even a tentative definition is likely to seriously mislead us.

To better appreciate the difficulty facing definitional approaches to understanding life, consider an analogous, although somewhat contrived, example from the history of science (Cleland and Chyba 2002, 2007). Suppose that a seventeenth-century scientist attempted to answer the question 'what is water?' by "defining" water. This is before Antoine Lavoisier laid the foundations of molecular theory in the late eighteenth century (Levere 2001), and so our hypothetical scientist knows nothing of molecules. His knowledge of water is limited to sensible properties such as appearance, taste, smell, or its interactions with other substances. Unfortunately many of the substances called "water" by his seventeenth-century contemporaries differ with respect to one or more of these properties. For example, when compared to pure water, muddy water is not transparent, salty water is not tasteless, and brackish water is not odorless.

It is difficult for us, steeped in twenty-first-century chemistry, to fully appreciate the dilemma that our seventeenth-century scientist faces in selecting one or more of the sensible properties associated with water as essential to it. The alchemists, who were medieval chemists, were impressed by water's powers as a solvent, and hence chose solvency to define water. As a consequence they classified nitric acid and mixtures of hydrochloric acid, which are even better solvents than water (and share many of its other sensible properties, e.g., being transparent and a liquid), as "water." But as we now know the alchemists were wrong. Nitric acid is not water, and like muddy water and salty water, mixtures of hydrochloric acid are not pure water. Of course, we now know that water is distinguished from all other chemical substances by its unique chemical composition of H_2O. Being H_2O is what salty water, muddy water, brackish water, distilled water, and even acidic solutions have in common, despite their apparent differences. It is what distinguishes nitric acid, whose molecular composition is HNO_3, from water despite their superficial similarities. Could a seventeenth-century scientist have discovered this? The answer clearly is "no". His scientific understanding of water is based upon its superficial sensible properties, and no amount of reflection on or investigation of these properties with the primitive equipment then available could reveal that water consists of two atoms of hydrogen chemically bonded to an atom of oxygen. To achieve this understanding of water required a new theoretical framework for thinking about chemical substances: molecular theory. Therefore, if our seventeenth century chemist had established a definition of water to guide his research, his incomplete understanding would have seriously restricted his ability to find out what water really is.

In some ways the situation is worse for contemporary scientists seeking an understanding of the nature of life than for our seventeenth-century scientist seeking an understanding of the nature of water. For there are compelling scientific reasons for thinking that life as we know it on Earth today represents a single, potentially unrepresentative, example of life. Molecular biologists have discovered that all known life on Earth descends from a last universal common ancestor, most likely a community of primitive proto-cells (Woese 2004). Much evolution and diversification has happened since then, but all life still uses the same basic molecular building blocks, the same basic metabolic pathways, and the same genetic code. This means that we are dealing with a single example of life. Without additional examples of life, one cannot discriminate features that are universal to life, wherever it might be found, from features derived from mere physical and chemical contingencies on the early Earth, or representing mere quirks of the lucky little bugs that happened to become the ancestors of us all.

How different could life be from familiar Earth life? Biochemists have established that some of the basic molecular building blocks of familiar life could have been at least modestly different without compromising their biological functionality (for more detail, see Benner 1994; Benner et al. 2004). For example, life on Earth could have utilized a different collection of amino acids to synthesize its proteins; familiar life employs the same approximately 20 amino acids, out of over

100 possibilities. Similarly, life could have synthesized nucleic acids (DNA and RNA) out of a different suite of nucleotide bases or used different sugars to construct the sugar-phosphate backbones of nucleic acids.

When we look at the genetic code, we see that all known life on Earth utilizes a triplet coding scheme to store hereditary information on nucleic acids and, moreover, the same triplet of bases (known as a codon) always codes for the same amino acid; there is some redundancy with most amino acids being represented by more than one codon but the same codon never represents more than one amino acid. There is little chemical reason to suppose that life couldn't have mapped codons to different amino acids or utilized a different number of nucleotide bases (two or four, for instance, instead of three) to encode amino acids, particularly if it had employed a different number of amino acids to build its proteins. Furthermore, the process of synthesizing proteins from the information encoded on DNA is carried out in ribosomes, which are minuscule but highly complex molecular machines composed of both RNA and protein (for information in biology see Marcos and Arp, this volume). It is unlikely that the earliest forms of life on Earth utilized something this sophisticated to build their structural and enzymatic material, and even more unlikely that ribosomes represent the only possible chemical mechanisms for performing this crucial biological function. Therefore, it seems that these remarkable molecular similarities among familiar Earth organisms most likely reflect contingent characteristics of their last universal common ancestor, and are not necessary characteristics of all life. To claim that they are necessary for all life or, even worse, to define life as having only these characteristics, would prematurely restrict the extension of the class of living things and run the risk of excluding forms of life differing in critical ways from our own. Recalling the philosophical example from Sect. 2, this would be like observing one bachelor who happens to be messy and concluding that messiness is a necessary condition for being a bachelor.

Even if we are convinced that the specific biomolecules composing familiar life are not necessary for all life, we might still believe that all life must have certain *functional* characteristics in common. For example, perhaps all life must have the capacity to self-organize, to maintain that self-organization for an extended period of time against internal and external perturbations, and to reproduce and transmit heritable changes to progeny. Unfortunately, this does not really avoid the problem because for all we know these pervasive functional characteristics of contemporary Earth life represent unreliable symptoms of more fundamental but as yet unknown properties of life. This sort of thing has happened before in science. Some natural phenomena that once were identified as distinct are now recognized to be the same in virtue of being caused by the same underlying phenomenon: electricity and magnetism were discovered to be manifestations of the same underlying force; protons and neutrons were discovered to both be made of quarks. Similarly, as discussed earlier, different chemical substances (distinct molecular compounds) that share some of the same sensible properties, like being liquid, transparent, and a good solvent (this last being a functional property) were once mistakenly all classified as water. Thus, although we possess an enormous amount of empirical information about life, this body of information is inadequate for generalizing beyond familiar Earth life.

To this point, we have argued that it is a mistake to define life in general based on the single example of familiar life on Earth. Our argument can be extended to show that a narrower definition restricted only to familiar life (by which we mean life that is descended from the same common ancestor as all known life) would face similar problems. One problem comes from borderline cases of familiar life that share some important features of prototypical familiar life but fail to have others. For example, viruses (and, to a lesser extent, prions) are reproduced and can evolve, but they are not enclosed in cells and do not metabolize on their own, leaving their status as familiar living systems ambiguous when compared with prototypical familiar life. What if, as some microbiologists have suggested (e.g. Bandea 2009), viruses are parasites that have evolved from unicellular Bacteria or Archea. Should we rule them out as part of familiar life merely because they do not conform to a rigid definition of familiar life based upon what could turn out to be a mistaken prototype? Surely not.

The problem is exacerbated when one reflects that definitions of familiar life based upon our current experiences with Earth life also face the risk of ruling out other, as yet undiscovered varieties of life that clearly share a common ancestor with us. It seems possible that one or more lineages of familiar life could evolve novel features that are significantly different from other, more common lineages, and if our definition of familiar life does not allow for these novel features, then the newly discovered lineage would be *ipso facto* excluded from the category of familiar life. For example, someone who defines familiar life in terms of the composition of its biomolecules, which some biochemists have suggested, would not be able to admit the infamous "arsenic loving" bacteria (Wolfe-Simon et al. 2011) as living if they actually did substitute arsenic for phosphorous in their major biomolecules (nucleic acids and proteins). It is currently thought that this did not occur. Not everyone agrees, however (Pennisi 2010). Suppose that biochemists discover that these bacteria, which are a strain of a well-known bacterial family (Halomonadaceae), have done this, or that some other microbe sharing a common ancestor with familiar life has evolved such biomolecules. Should we exclude them as familiar life just because they don't fit a popular biochemical definition? Surely this would be a mistake! The point is that even when restricted to known life any proposed definition has a very real chance of incorrectly excluding unknown varieties of familiar life merely because they have evolved surprising new features. Nevertheless, we do know quite a lot about familiar Earth life, and there may be good reasons for developing a list of features common to all life known so far, as long as the list is not treated like a definition. In Sect. 4, we describe such features as "tentative criteria" and discuss their usefulness in the search for new forms of life.

In summary, from a scientific perspective, investigating the nature of life by means of a definition of life is fundamentally misguided. Our experience with life is limited to a single example that we have good scientific reasons for believing could have been at least modestly different. Moreover we have no idea how different life could be from life as we know it so far. Until we encounter forms of life descended from a separate origin we will be in no position to speculate about the possibilities for life considered generally.

4 Searching for Alternative Forms of Life

The conclusions of the previous section appear to leave us in a quandary: How can one search for alternative forms of life given that our experience with life is limited to a single, possibly unrepresentative, example? The problem is exacerbated when one considers that any life we encounter is likely to be microbial (Ward and Brownlee 2000), and therefore much harder to detect than large, complex organisms like trees or mammals (for philosophical issues related to microbial life see Duncan et al., this volume). Put succinctly, in order to formulate a truly general theory of living systems we need unfamiliar forms of life, and yet in the absence of such a theory it is unlikely that we will recognize unfamiliar forms of life as living things, even if we happen to encounter them. It may be tempting at this point to think that artificial life will provide an additional form of life that helps us expand our understanding of life in general. Unfortunately, this is not the case because the status of hard, soft, or wet A-life is founded upon the supposition that certain features of familiar biological life are true of all life, and this is just what is at issue. What we really need are examples of life that are not of our own making, and that brings us back to the difficulty of searching for new forms of life without being completely sure what we're looking for.

Cleland has argued elsewhere (Cleland and Chyba 2007; Cleland 2006) that the solution to this conundrum is not to search for life directly, but instead to search for anomalies. In the context of the search for alternative life, anomalies are physical systems resembling familiar Earth life in provocative ways and yet also differing from it in important and unanticipated ways. But how does one search for anomalies? The best way forward is to utilize selected features of familiar Earth life as tentative criteria for life. Insofar as they are construed as tentative, such criteria are understood to be incomplete and, most importantly, revisable. They are not viewed as defining or delimiting life. Indeed, as discussed earlier, it is possible that someday we will discover that all the characteristics currently thought to be essential to life are little more than potentially unreliable symptoms of more fundamental but as yet unknown properties. The purpose of tentative criteria for life is not to settle the issue of whether a weird physical system discovered on Mars, for instance, is alive, but rather to focus scientific attention on suspicious physical systems—to identify the best candidates for further scientific investigation.

Tentative criteria for life should not be confused with definitions. They do not fully fix the class of living things but rather represent empirically accessible signs of—provisional tests for—life. Instead of selecting one ostensibly universal characteristic of familiar life as more essential than the others (e.g., metabolism, as in metabolic definitions of life, or the capacity to evolve by natural selection, as in Darwinian definitions) they may be employed jointly and in different combinations in a search for novel forms of life. For it is important that criteria used to search for truly novel forms of life include a diversity of disparate features of familiar life since we do not know which characteristics of familiar life are fundamental to all life, wherever it might be found, as opposed to being the result of mere physical and chemical contingencies on the early Earth at the time of the origin of life. Furthermore, tentative criteria need not be universal to familiar Earth life. In fact, we may even want to de-emphasize any features universal to familiar life because features that are common only to life found in certain special kinds of environments could prove more useful for searching for life in analogous extraterrestrial environments. Recognizing a truly novel form of life won't be easy because it will resemble life as we know it in seemingly important ways but also differ from it in seemingly important ways, making it difficult to judge whether it is a living thing or a novel, nonliving physical system. In other words, it will appear as anomalous vis-à-vis our current understanding of life. Investigation of such anomalies is just what we need in order to expand our concept of life beyond familiar Earth life, paving the way to an understanding of the nature of life. Somewhat paradoxically, then, the purpose of tentative criteria for life is not to identify clear-cut cases of life but rather to identify physico-chemical systems that are anomalous for purposes of further scientific investigation.

Geologists, chemists, and other non-biologists can aid in the search for anomalies by helping biologists identify features of Earth life that are non-existent or very uncommon among nonliving geochemical systems on Earth or elsewhere. Such features make good criteria for searching for unfamiliar forms of life because they stand out against a background of nonliving processes. Thus, for instance, the tiny (4–100 nm), chemically pure, prismatic magnetite crystals found in a famous Martian meteorite (ALH84001), recovered in 1984 in Antarctica, were initially cited as providing compelling evidence of fossilized Martian microbes (see Jakosky et al. 2007, pp. 374–378, for a more detailed discussion). This claim was grounded in their striking similarities to magnetite crystals produced by a particular strain of magnetotactic bacteria (MV1) on Earth, and at the time no one knew of an abiotic mechanism for producing tiny magnetic crystals of such chemical purity and uniform geometrical shape under natural conditions; indeed, there is still controversy about whether such a mechanism exists, and even if it does, it is not clear that it can explain the crystals in ALH84001. Significantly, the estimated age of the crystals in ALH84001 coincides with a period (around 3.9 billion years ago) in which Mars is thought to have been wet and geologically active, with a powerful magnetic field. If such crystals cannot be (or are rarely) produced abiotically, then even though they are not universal to life on Earth, they could provide useful criteria for exploring certain extraterrestrial environments for suspicious (anomalous) physical systems.

Now let us turn to a contrasting example that shows the limitations of a definitional or similarly rigid approach compared to a search for anomalies: the Viking missions to Mars, which are the only dedicated search for extraterrestrial life that has thus far been conducted. Of particular interest to us is the "labeled release" (LR) experiment, one of three metabolic experiments performed robotically on Mars by the Viking lander. The LR experiment yielded results that initially seemed positive for life but quickly turned baffling (for a review of all three experiments, see Klein 1978). Martian soil introduced into the test chamber was injected with a radioactively labeled nutrient solution and started releasing radioactively labeled $^{14}CO_2$—just what one would expect from Earth microbes. When the Martian soil sample was subsequently heated to 160 °C for three hours, more than enough to kill Earth microbes, the reaction stopped, strongly suggesting that the initial response had been biological. But when another Martian soil sample was given a second helping of nutrients the anticipated burst of new activity from hungry Martian microbes not only failed to occur, but even more mysteriously $^{14}CO_2$ left over from the initial reaction began disappearing. Scientists were flummoxed. They were facing a genuine anomaly. While closely resembling what one would expect from Earth microbes, the results obtained by the LR experiment also deviated in baffling ways.

One might think that such an intriguing yet inconclusive result would immediately prompt further investigation. However, because the Viking experiments were explicitly designed around an officially sanctioned (chemical metabolic) definition, the results of the LR experiment were interpreted as negative and the failure of the Viking gas chromatograph mass spectrometer (GCMS) to detect any organic molecules was treated as conclusive. Non-biological explanations for the mysterious Viking results have since been developed and the current consensus is that they were produced by unusual states of iron. To this day, however, there is no direct empirical evidence that the Martian surface is strongly oxidizing and, most importantly, NASA has not made testing this hypothesis in future missions a high priority. This illustrates the difference between searching for familiar life and searching for anomalies. Because they did not conform to the favored definition of life, the results of the LR experiments were attributed to a non-biological oxidant, despite the fact that known life is also an oxidant. If the Viking experiments had been designed as a search for anomalies, reaction to the LR experiment would have been quite different. It would have been interpreted as just what it was: an ambiguous result worthy of further empirical investigation.

So far, the examples in this section have involved space exploration. However, the search for alternative forms of life need not necessarily reach beyond Earth. Although it may strike one as highly implausible, it turns out that there are good reasons for entertaining the possibility of alternative forms of life right here on Earth, what Cleland and Copley (2005) call a "shadow biosphere". We're not talking about large, easily identifiable, alternative forms of life. If those existed on Earth, it seems unlikely that they would have escaped our notice. Rather, we are considering the possibility that microbial descendants of an alternative origin of life on Earth may still be with us today. The earliest forms of life were almost certainly microbial and there are good reasons for thinking that the evolution of large multicellular forms of life is the exception rather than the rule (Ward and Brownlee 2000).

While we don't know the specific chemical and physical processes that gave rise to the earliest forms of Earth life, we do know that substantial variations in the molecular building blocks of life were available on the early Earth. Today we are aware of many possible nonbiological sources of sugars, amino acids, and other organic molecules that could serve as the building blocks of proteins and nucleic acids. These sources include electrical discharges through various mixtures of simple gases (Miller 1953, 1955), hydrothermal processes in oceanic volcanic vents (Holm and Andersson 1998; Martin and Russell 2003), geochemical processes involving mineral surfaces (Wächtershäuser 1988; Cairns-Smith et al. 1992; Cody

2004), and even incoming meteorites (Anderson and Haack 2005). These sources provide a variety of small organic molecules (e.g., amino acids) that are not but could have been used to synthesize alternative biomolecules (proteins). If conditions conducive to the emergence of life were present at multiple locations on the early Earth and these locations contained, as they surely did, variations in the basic molecular building blocks of life, then one would expect the earliest protocells to reflect these differences in their molecular compositions. In short, the hypothesis that Earth hosted multiple origins of life and that some of these origins produced molecular variations on familiar life is consistent with our current chemical and biological understanding of life.

Most microbiologists concede the above points but contend that (1) any alternative form of microbial life would be eliminated by our form of life in the ruthless Darwinian competition for vital resources and (2) if such microbes existed, we would have discovered them by now or at least encountered telling signs of them using the sophisticated tools available to contemporary microbiologists. As several researchers (Cleland 2007; Cleland and Copley 2005; Davies and Lineweaver 2005) have discussed, however, these objections to the possibility of a shadow biosphere do not hold up under close scrutiny. Evolution in the microbial world typically does not result in elimination of rare species by dominant species. Microbial communities are more complex than that, with many different varieties of rare microbes existing alongside more common ones, often engaging in cooperation rather than competition. Therefore, even if alternative microbial life is quite rare, it still may be able to survive in complex microbial communities. Nevertheless, if alternative microbes exist, even in small numbers, we might think that microbiologists would have detected them by now. This is not necessarily so. As students quickly find out after observing a sample of pond water or some other environmental sample under a microscope, different varieties of microbes can be very difficult to distinguish and identify. Staining makes certain cellular structures more visible but these structures aren't all that useful for distinguishing among different species. The discovery that prokaryotes actually encompass two quite different types of microbe, Bacteria and Archaea, despite closely resembling each other in gross morphological structure (lacking membrane enclosed organelles, especially a nucleus) underscores just how problematic it is to classify microbes on the basis of structural considerations (Pace 1997). Indeed, the discovery of the Archaea resulted in major revisions to the top of the hierarchy of biology classification with the original five kingdoms of life being supplanted by three domains of life (Bacteria, Archaea, and Eukaryota). The point is as long as alternate life has roughly the same shape and basic internal structure as known microbes, microscopy will not reveal them to be different from familiar life. This means that we may already have observed alternate microbial life without realizing it.

Of course, microscopy is just one of the tools in the microbiologist's toolbox. Individual varieties of microbes often can be separated and identified by cultivating a mixed sample under a variety of different growth conditions. Different species grow under different conditions, so selective cultivation can be a reliable method for identifying the microbial strains present in an unknown sample. However, only a tiny fraction of all microbes can be cultivated successfully, so this technique is very limited when we are trying to identify new microbes (Hugenholtz and Pace 1996). This problem is made worse when we consider that the specific growth requirements of unknown organisms are, of course, unknown, and they may require unexpected nutrients or energy sources. As a consequence, our chances of cultivating a pure sample of alternate life are very low.

When cultivation fails to distinguish and identify all the species present in a sample, microbiologists sometimes turn to genetic analysis. Isolation and amplification of gene variants is a great tool for sifting through complex samples. In fact, it was genetic analysis that revealed the great number and diversity of microbes that were not showing up in growth assays (Hugenholtz and Pace 1996; Pace 1997). However, isolation and amplification of unknown gene variants requires that said variants are compatible with known DNA replication enzymes. It is unlikely that alternate life forms would use the same genetic replication machinery as known life, so current methods of genetic analysis would be of little use in isolating or identifying alternate life (Cleland and Copley 2005). In summary, it appears that none of the tools currently used by microbiologists for identifying organisms in environmental samples could detect an alternative form of microbial life that differed even modestly from familiar life at the molecular level. Thus, the fact that we haven't found any using such techniques does not count much against their existence. As with the search for extraterrestrial life, scientists searching for alternative life on Earth should look for anomalies that cannot be explained by known biological or geochemical processes. This makes the search seem more challenging and uncertain, but it is far better than imposing a deceptive, restrictive definition to guide our investigations.

5 Conclusion

Although biologists know a lot about life on Earth, they cannot yet provide a scientifically compelling answer to "what is life?" The reasons for this illustrate some important features of science that students sometimes overlook. First of all, in Sect. 2, we explained that the answer to "what is life?" should not take the form of a definition. Science must allow for revision of our understanding of natural phenomena in light of new empirical discoveries, and the rigidity of definitions tends to discourage this. If scientists interested in the origins of life, astrobiology, or A-life are guided by a definition of life, they are likely to overlook forms of life that do not conform to it. In Sect. 3 we explained that even if scientists avoid a strict definition of life, their efforts to explain life are severely hampered by the fact that our experience with life is currently limited to a single example. In Sect. 4 we argued that instead of starting with a definition of life, the search for new forms of life is best served by using tentative criteria to identify anomalies, which are phenomena that are not readily explained by known non-biological processes and yet not clearly biological either. An anomaly could turn out to be an indicator of an unanticipated form of life that does not match any of our "definitions" of life. Such phenomena are likely to be overlooked in a definition-based search for life. Perhaps someday scientists will find enough independent examples of life to begin settling on a scientifically compelling, general theory of life. This has happened for other domains of natural phenomena such as chemical substance; as an illustration, molecular theory distinguishes water from other chemical substances by its unique molecular composition, H_2O.

Of course, we must always allow for future revisions to even our most well established theories, which means that all scientific theories should be viewed as tentative. Tentativeness and revisability in light of new empirical information are hallmarks of science as it is actually practiced and critical to its future development. Unfortunately science educators, understandably eager to impart to students well-accepted scientific facts and laws, often neglect this dynamic aspect of scientific practice. Discussion of the challenges associated with answering the question "what is life?" can be a valuable tool for introducing biology students to this important and exciting aspect of science while imparting to them a wealth of scientific information about life as we know it on Earth today.

References

Adami, C., B. Richard, K. Hiroaki, and C. Taylor (eds.). 1998. *Artificial life VI*. Cambridge: MIT Press.

Anderson, A.C., and H. Haack. 2005. Carbonaceous chonderites: Tracers of the prebiotic chemical evolution of the solar system. *International Journal of Astrobiology* 4: 12–17.

Bandea, C. 2009. The origin and evolution of viruses as molecular organisms. *Nature Precedings*. http://hdl.handle.net/10101/npre.2009.3886.1

Bedau, M. 2003. Artificial life: Organization, adaptation and complexity from the bottom up. *Trends in Cognitive Sciences* 7(11): 505–512.

Bedau, M.A., and C.E. Cleland (eds.). 2010. *The nature of life: Classical and contemporary perspectives from philosophy and science*. Cambridge: Cambridge University Press.

Benner, S.A. 1994. Expanding the genetic lexicon: Incorporating non-standard amino acids into proteins by ribosome-based synthesis. *Trends in Biotechnology* 12: 158–163.

Benner, S.A., A. Ricardo, and M.A. Carrigan. 2004. Is there a common chemical model for life in the universe? *Current Opinion in Chemical Biology* 8: 672–689 (Reprinted in Bedau and Cleland, 2010, pp 164–185).

Boden, M. 2003. Alien life: How would we know? *International Journal of Astrobiology* 2(2): 121–129 (Reprinted in Bedau and Cleland, 2010, pp 249–259).

Cairns-Smith, A.G., A.J. Hall, and M.J. Russell. 1992. Mineral theories of the origin of life and an iron sulfide example. *Origins of Life and Evolution of the Biosphere* 22: 161–180.

Cleland, C.E. 2006. Understanding the nature of life: A matter of definition or theory? In *Life as we know it*, ed. J. Seckbach, 589–600. Dordrecht: Springer.

Cleland, C.E. 2007. Epistemological issues in the study of microbial life: Alternative terran biospheres? *Studies in History and Philosophy of Biological and Biomedical Sciences* 38: 847–861.

Cleland, C.E., and C.F. Chyba. 2002. Defining 'life'. *Origins of Life and Evolution of the Biosphere* 32: 387–393.

Cleland, C.E., and C.F. Chyba. 2007. Does 'life' have a definition? In *Planets and life*, ed. W.T. Sullivan and J.A. Baross, 119–131. Cambridge: Cambridge University Press (Reprinted in Bedau and Cleland, 2010, pp 326–339).

Cleland, C.E., and S.D. Copley. 2005. The possibility of alternative microbial life on earth. *International Journal of Astrobiology* 2: 165–173 (Reprinted in Bedau and Cleland, 2010, pp 198–209).
Cody, G. 2004. Transition metal sulfides and the origin of metabolism. *Annual Review of Earth and Planetary Sciences* 32: 569–599.
Davies, P.C.W., and C.H. Lineweaver. 2005. Finding a second sample of life on earth. *Astrobiology* 5: 154–163.
Gasteland, R.F., T.R. Cech, and J.F. Atkins. 2006. *The RNA world*, 3rd ed. New York: Cold Spring Harbor Press.
Holm, N.G., and E.M. Andersson. 1998. Hydrothermal systems. In *The molecular origins of life: Assembling pieces of the puzzle*, ed. A. Brack, 86–99. Cambridge: Cambridge University Press.
Hugenholtz, P., and N. Pace. 1996. Identifying microbial diversity in the natural environment: A molecular phylogenetic approach. *Trends in Biotechnology* 14(6): 190–197.
Jakosky, B.M., F. Westall, and A. Brack. 2007. Mars. In *Planets and life*, ed. W.T. Sullivan and J.A. Baross, 357–387. Cambridge: Cambridge University Press.
Keller, E.F. 2002. *Making sense of life: Explaining biological development with models, metaphors, and machines*. Cambridge: Harvard University Press.
Klein, H.P. 1978. The Viking biological experiments on Mars. *Icarus* 34: 666–674.
Lange, M. 1996. Life, "artificial life," and scientific explanation. *Philosophy of Science* 63(2): 225–244 (Reprinted in Bedau and Cleland, 2010, pp 236–248).
Levere, T. 2001. *Transforming matter: A history of chemistry from alchemy to the buckyball*. Baltimore: Johns Hopkins University Press.
Lipson, H., and J.P. Pollack. 2000. Automatic design and manufacture of robotic life forms. *Nature* 406: 974–978 (Reprinted in Bedau and Cleland, 2010, pp 260–267).
Martin, W., and M. Russell. 2003. On the origin of cells: A hypothesis for the evolutionary transition from abiotic geochemistry to chemoautotrophic prokaryotes, and from prokaryotes to nucleated cells. *Philosophical Transactions of the Royal Society of London* B358: 59–85.
Miller, S.L. 1953. A production of amino acids under possible primitive earth conditions. *Science* 117: 528–529.
Miller, S.L. 1955. Production of some organic compounds under possible primitive earth conditions. *Journal of the American Chemical Society* 17: 2351–2361.
Orgel, L. 1998. The origin of life: A review of facts and speculation. *Trends in Biochemical Sciences* 23: 491–495.
Pace, N.R. 1997. A molecular view of microbial diversity and the biosphere. *Science* 274: 734–740.
Pennisi, E. 2010. What poison? Bacterium uses arsenic to build DNA and other molecules. *Science* 330: 1302.
Pennock, R. 2007a. Models, simulations, instantiations, and evidence: The case of digital evolution. *Journal of Experimental and Theoretical Artificial Intelligence* 19(1): 29–42.
Pennock, R. 2007b. Learning evolution and the nature of science using evolutionary computing and artificial life. *McGill Journal of Education* 42: 211–224.
Popa, R. 2004. *Between necessity and probability: Searching for the definition and origin of life*. Berlin: Springer.
Shapiro, R. 2007. A simpler origin for life. *Scientific American* 2007: 47–53.
Sober, E. 1991. Learning from functionalism: Prospects for strong artificial life. In *Artificial life II*, ed. G. Langton, C. Taylor, J. Farmer, and S. Rasmussen, 749–765. Oxford: Westview Press (Reprinted in Bedau and Cleland, 2010, pp. 225–235).
Sullivan, W.T., and J.A. Baross (eds.). 2007. *Planets and life*. Cambridge: Cambridge University Press.
Tyson, N.D. 2009. *The Pluto files: The rise and fall of America's favorite planet*. New York: W.W. Norton and Co.
Wächtershäuser, G. 1988. Before enzymes and templates: Theory of surface metabolism. *Microbiological Reviews* 52: 452–484.

Ward, P.D., and D. Brownlee. 2000. *Rare earth*. New York: Copernicus.
Woese, C.R. 2004. The archaeal concept and the world it lives in: A retrospective. *Photosynthesis Research* 80: 371–372.
Wolfe-Simon, F., J.S. Blum, T.R. Kulp, G.W. Gordon, S.E. Hoeft, J. Pett-Ridge, J.F. Stolz, S.M. Webb, P.K. Weber, P.C.W. Davies, A.D. Anbar, and R.S. Oremland. 2011. Responses to comments on "A bacterium that can grow using arsenic instead of phosphorus". *Science* 332(6034): 1149.

Biological Explanation

Angela Potochnik

1 Introduction

One of the central aims of science is explanation: scientists seek to uncover *why* things happen the way they do. In biology, explanations have been sought for why offspring generally have the same traits as their parents; for why one area has a greater variety of species than another; for why the patterns on land snails' shells show the type of variation they do; for why shark populations increased in the Adriatic Sea during World War I. Biologists have also sought to understand the process by which plant cells convert sunlight into nutrients; the particular genetic influences on human smoking behavior; and why male seahorses, not females, gestate seahorse embryos. All of these—and many, many more besides—are attempts to explain biological phenomena, phenomena ranging from generalized to highly specific and from subcellular to encompassing vast swaths of the Earth.

Accordingly, a primary project in philosophy of science is providing an account of the nature of explanation, of what it takes to explain something. For over 100 years, philosophers of science have been generating competing accounts of explanation. These accounts provide criteria that are supposed to be essential to explanation, such that any successful explanation will meet those criteria. Accounts are motivated with reference to examples of successful scientific explanations. In the early to mid-twentieth century, much of philosophy of science largely focused on physics. Since then, philosophical treatments of explanation have been both complicated and enriched by attention to explanatory strategies in biology.

In this chapter, I survey biology's influence on philosophical accounts of scientific explanation. This highlights important features of explanatory practice in biology (Sect. 2). I then discuss how the explanatory strategies utilized in biology are integral

A. Potochnik (✉)
Department of Philosophy, University of Cincinnati, Cincinnati, OH, USA
e-mail: angela.potochnik@uc.edu

K. Kampourakis (ed.), *The Philosophy of Biology: A Companion for Educators*, History, Philosophy and Theory of the Life Sciences 1, DOI 10.1007/978-94-007-6537-5_3,

to making sense of other features of scientific practice, such as the continued neglect of some central causal factors (Sect. 3). Finally, I make explicit how these issues bear on biology education (Sect. 4).

2 Biology and Philosophical Accounts of Explanation

A traditional and historically influential view in philosophy of science is that scientific explanations are produced by deriving the phenomenon to be explained from laws of nature. This deductive-nomological (D-N) account suggests that explanations follow a simple pattern: a phenomenon is explained by a set of true sentences from which the phenomenon's description can be derived, and which contains at least one law of nature essential to the derivation (Hempel and Oppenheim 1948; Hempel 1965).[1] For example, Mendel's law of independent assortment and the fact that two genes are located on different chromosomes explain why the different alleles for those two genes are paired with each other in approximately the same number of gametes: according to Mendel's law, each pairing is equally likely (for problems related to the concepts of "dominance" and "gene" see Jamieson and Radick as well as Burian and Kampourakis respectively, this volume).

One feature of the D-N account of explanation that this example violates is that this strategy can only explain phenomena when scientific laws *guarantee* their occurrence. The phenomenon must follow deductively, as a matter of logic, from the law and conditions cited. A companion to the D-N account of explanation was thus developed to apply to statistical cases. This inductive-statistical (I-S) account holds that phenomena can also be explained using an applicable statistical law, so long as the law confers high probability on the phenomenon. Technically, my simple example of explaining using the law of independent assortment is an I-S explanation. Broadly, the idea behind the D-N and I-S approaches to explanation is that a phenomenon is explained by specifying how what we know about the world—our scientific laws—bears on the particular circumstances at hand, which renders the phenomenon expectable. Laws of nature and the circumstances guarantee or render highly probable the phenomenon to be explained.

The D-N and I-S approaches to explanation have largely fallen out of favor among philosophers in recent decades. One prominent criticism is that there seems to be an asymmetry in the explanatory value of derivations that satisfy the D-N conditions of explanation. Salmon (1989) employs the following example as an illustration. By deriving the length of a shadow from the height of a flagpole and the position of the sun, one explains the length of the shadow. But one can equally well derive the height of the flagpole from the length of the shadow and sun's

[1]For the sake of simplicity, I use the word "phenomenon" throughout this chapter to stand in for various conceptions of the target of explanation: events or laws, propositions, explananda, etc. Such distinctions are not central to the aim of this chapter.

position, and it seems this does nothing to explain the height of the flagpole. This and other criticisms are taken to show that derivation is not in itself sufficient for explanation.

Beyond the general difficulties with the D-N and I-S accounts, it seems that many biological explanations do not conform to this view of explanation. For one thing, some phenomena that are acknowledged to be improbable are nonetheless thought to be explained. For example, some genetic mutations are explained by oxidative damage, even though such mutations are rare and oxidants are frequently present. Additionally, there are many biological explanations in which laws, whether deterministic or statistical, seem to play little or no role (Hull 1992). Why does sickle-cell disease result in anemia? The explanation will undoubtedly cite features of the abnormally rigid, sickled red blood cells found in those with sickle-cell disease. It would be at best strained to construe any element of the resulting explanation as a scientific law. Finally, there is plenty of uncertainty regarding even what should qualify as a biological law, and thus whether biology has many, or any, laws to offer (Ruse 1970; Brandon 1997; see Lange this volume). Whether Mendel's "law" of independent assortment, used in the example of D-N explanation above, would qualify as a scientific law is itself dubious.

Setting aside the difficulties with the requirement that any explanation cite a scientific law, as well as the requirement that any explanation confer a high probability on the explained phenomenon, the D-N and I-S approaches do align with some intuitions about what explanations should accomplish. This point was made by Friedman (1974) and Kitcher (1981, 1989). Friedman and Kitcher both argue that an explanation of a phenomenon "unifies" that phenomenon with other scientific beliefs in virtue of providing a pattern of argument from which all can be derived. According to this unification account, an explanation's value stems from its generality, simplicity, and cohesion, as these features together generate the power to unify disparate phenomena. Explanations that cite Mendel's law of independent assortment fare better on this account than the D-N account. Positing the independent assortment of genes (on different chromosomes) is a simple, cohesive explanation that is general enough to explain a variety of phenomena, ranging from a pea plant inheriting a parent's wrinkled peas but not the yellowness of its peas, to there being a 50 % chance that a woman who carries the sex-linked recessive gene for Duchenne muscular dystrophy has a son with the disease, regardless of what other traits he does or does not inherit (not on the X chromosome).

In contrast to the troubles encountered by the D-N and I-S accounts, explanatory practice in biology offers support for a different philosophical view of explanation, namely the causal account. On this view, a phenomenon is explained by the causal factors that brought it about (Scriven 1962; Salmon 1989, 1998; Woodward 2003). This is a natural interpretation of, for example, evolutionary explanations that feature natural selection. The redshank sandpiper (*Tringatotanus*), a bird that feeds on worms in mudflats, exhibits a preference for eating large worms over small worms. This preference is explained by the fact that natural selection favors foraging habits that maximize energy intake; if large worms and small worms are both readily available, then a redshank sandpiper's energy intake is maximized when large worms are chosen, since they yield more ingested biomass (Goss-Custard 1977). Notice, however, that although natural selection is an important cause of the sandpiper's evolved preference, selection does not *guarantee* that the preference will evolve. It is not the sole determiner, but one influence among many (Potochnik 2010a).

Biology has also been used to motivate mechanistic accounts of explanation (Glennan 1996; Machamer et al. 2000; Bechtel and Abrahamsen 2005; Bechtel 2006; see Bechtel this volume). Mechanisms are "entities and activities organized such that they are productive of regular changes from start or set-up to finish or termination conditions" (Machamer et al. 2000, p. 3). Explaining by citing a mechanism thus provides both causal and organizational information. A familiar mechanistic explanation in biology can be given for the organic compounds created via photosynthesis. This style of explanation would cite the initial presence of carbon dioxide and sunlight, then detail the successive reactions among the chemical compounds that eventuate in organic compounds and, as a byproduct, oxygen. Significant debate surrounds the question of how broadly this conception of explanation should be employed, for instance, whether natural selection should be considered a mechanism (Skipper and Millstein 2005; Barros 2008).

Further disagreements regard the proper scope and purpose of biology explanations. Some argue that many or all biology explanations will soon be replaced by explanations that feature molecular biology; this is a form of explanatory reductionism. In large part, this argument and its rebuttal have focused on whether explanations that feature molecular genetics will entirely replace classical genetics (Waters 1990). One of the main arguments employed in defense of the explanatory value of classical genetics is that the explanations it provides are *general* in the right way to be maximally explanatory (Kitcher 1984; Sterelny 1996). Sober (1999) suggests a middle ground, according to which some explanations benefit from generality—they explain by lumping together all similar phenomena—whereas other explanations are designed to be highly specific—they explain by showing what exactly brought about the specific phenomenon, in this particular case.

This distinction between generally applicable explanations and those that track the exact process that brought about a particular instance of a phenomenon evokes another distinction that has been made in the philosophical literature on explanation. Some philosophers distinguish how-possibly explanations from how-actually explanations (Dray 1957; Brandon 1990). As the terminology suggests, a how-actually explanation tracks the actual causal process that brought about a phenomenon, whereas a how-possibly explanation outlines a process that *could have* (but may not in fact have) brought about a phenomenon. How-possibly explanation is one way to conceive of the role of explanations that involve claims not fully supported by evidence (Forber 2010).

To summarize, it seems that some patterns of explanation in biology corroborate a causal understanding of explanation, while other patterns of explanation suggest that mechanisms, where they exist, are explanatory. Also, though the traditional philosophical idea that all explanations cite laws of nature is undermined by biology, some biology explanations nonetheless corroborate the idea that citing general law-like patterns is indeed explanatory. This is further complicated, however, to the extent that biology explanations vary in their portrayal of a pattern shared by many phenomena versus the specific details of a single phenomenon, and relatedly, how closely an explanation is supposed to mirror actual reality.

This variety suggests that it is not a simple matter to find a single principle underlying all explanations that fall within the purview of biology (let alone all explanations in all of science). This introduces the question of how to reconcile the different points that have been made about biological explanation, if indeed they should be reconciled. There are at least two types of responses one could have to this question. One response is to simply acknowledge that a broad range of explanatory styles is present in biology, and then to focus on accurately characterizing that range of styles and the relationships among them. This would be a pluralist approach to scientific explanation, for it would not attempt to reconcile divergent points about explanation in biology. The end result would be a catalogue of different approaches to explanation, with the hope that the approaches described together capture all of explanatory practice (Brigandt 2013).

The habit in philosophy is to consider this sort of pluralism a position of last resort. Simply declaring that there are several approaches without rhyme or reason governing the selection among them should be avoided until all avenues of discovering common principles have been exhausted. The alternative is to try to accommodate the variety of explanatory practices found in biology, features currently captured by different accounts of scientific explanation. This may create the groundwork for a unitary account of biological explanation, in spite of the seeming diversity.

Indeed, various attempts to reconcile different insights into explanation have been made. The unification account is presented by Kitcher (1981, 1989) as a successor view to the D-N account, the basis of which is supposed to be in Hempel's own observations. Strevens (2004) articulates an account of explanation that assimilates the insights of a causal approach to explanation and a unification approach. In Strevens' view, an explanation cites causal information at a sufficiently general, yet cohesive, level of description. There is an array of views regarding the relationship between mechanistic explanation and causal explanation; Skipper and Millstein (2005) view them as competing options, whereas Craver (2007) suggests the mechanistic approach as a way to make sense of the explanatory role of causal relationships.

I will conclude this section with some of my own ideas regarding how to create a unitary account of biological explanation. In my view, a promising start is to base a unitary account of biological explanation on the idea that causal information is explanatory. A causal understanding of explanation, in one version or another, seems to have gained dominance in philosophy of science, especially in philosophy of biology. Yet research in biology amply demonstrates that most biological phenomena result from complex causal processes, with many factors combining and interacting at each step in the process. This renders impractical a simple causal approach to explanation, whereby to explain you simply cite all the causes. It also creates an opportunity to fill out a broadly causal approach to explanation in a way that accommodates other intuitions about biological explanation.

I suggest adopting an insight advocated by the unification account, Strevens' (2004, 2009) causal account, and many other philosophical accounts of explanation. This is the idea that *generality* benefits an explanation. Though proper laws of nature may be few and far between in biology, depicting causal patterns—that is, how certain types of causes tend to bring about certain types of effects, given other conditions—is a more modest way to generate explanations that showcase lawlike behavior. This motivates explanations that ignore some details in order to depict broad causal patterns (Potochnik 2011).

One example of this feature of explanation is the difference between explanations for short-term and long-term evolutionary change. An evolutionary modeling approach termed optimality or optimization modeling accounts for the prevalence of a trait in a population by showing how that trait led to selective advantage (in the environment at hand). Several biologists have shown that this modeling approach can be expected to succeed only with *long-term* evolutionary change, that is, over a large number of generations (Hammerstein 1996; Eshel et al. 1998), whereas a population genetic approach is required for generational evolutionary change. One might thus anticipate that, in virtue of the different causal patterns involved in short-term and long-term evolution, different explanations are warranted. I explore this difference between optimality explanations and population genetic explanations in Potochnik (2010a).

A similar contrast can be drawn between *microevolutionary* and *macroevolutionary* explanations. Microevolution is the evolutionary change within a population, whereas macroevolution is the evolution of species (or even larger clades). Sterelny (1996) argues that this is another instance where different types of phenomena warrant different types of explanations, explanations that vary as to their degree of generality. In his view, whether macroevolution requires a distinct type of explanation comes down to whether it is due to distinct causal influences acting on whole species or clades. This version of explanatory pluralism once again suggests attending to the sort of causal patterns embodied by a phenomenon.

Yet a complication is introduced by the point I made just above, that many biological phenomena result from exceedingly complex causal processes. Consider, for example, the causal processes involved in bringing about the long necks of giraffes. In no particular order, these include, at least, features of ancestral giraffes' environment, including the presence of nutritious leaf matter high up in tall trees; various genetic influences on giraffe morphology; developmental processes, including additional regulator genes, involved in giraffe neck-development; certain genetic mutations arising; competition for resources such that giraffes with a greater reach enjoyed increased rates of survival; and changes in developmental processes resulting in longer necks. All of this causal complexity means that different explanations may focus on different causal patterns. For instance, there may be one explanation of giraffe neck development and a different explanation of selection for lengthened giraffe necks. What causal pattern is explanatory, and thus what parts of the causal process should be mentioned, depends on what one might generally call the *context of explanation*. This is determined by the goals of the research program for which the explanation is generated. Recall from above the debate over reductionism, including whether biological explanations will ultimately all feature molecular biology. The current view is antireductionist, for it suggests that multiple, different explanations will continue to be valuable, insofar as each captures a different causal pattern (Potochnik 2010b).

To summarize, my attempt to integrate different insights into explanation results in the view that biological explanations (1) give causal information, (2) in a way that depicts a broad causal pattern that is (3) explanatory given the particular research goals at hand. This approach accommodates much of the diversity of views about biology explanations surveyed in this section, but it unites them into a single view. It also disputes or neglects some claims, such as the idea that some explanations benefit by maximizing their specificity (Sober 1999), or the idea that explanations generally depict mechanisms. Finally, I must emphasize that my suggested account of explanation is of course one view among many, and the debate surrounding different philosophical views of scientific explanation, and explanation in biology, will not end anytime soon.

3 Explanation and Scientific Practice

In the previous section, we surveyed the range of styles of explanation found in biology and considered a few approaches to making sense of that diversity. Let us now set aside questions surrounding how biological explanations are formulated and focus instead on how the aims of explanation influence other features of scientific practice in biology. This will demonstrate how an accurate understanding of explanatory practice in biology contributes to an understanding of other characteristics of the field. In this section I will focus primarily on evolutionary biology, but I will also indicate points of contact and resonances with other areas of biology.

In contemporary evolutionary biology, genes are important. From the discovery of DNA, to the Human Genome Project, and most recently the Thousand Genome project, genetics—and especially molecular genetics—has received much attention both in biology and society at large. And genes are, of course, absolutely central to the evolutionary process. Though epigenetic inheritance is well documented (see Uller this volume), genetic inheritance remains central to most evolutionary processes (see Avise this volume).

In spite of all of this, many well-regarded models of evolutionary change ignore genes entirely. A prime example of this is evolutionary game theory. This modeling approach is applicable to the long-term evolution of traits with frequency-dependent fitness, i.e., when the fitness of a phenotypic trait depends upon the traits of others in the population.[2] Different phenotypes are represented as different strategies to

[2] A trait's fitness is a measure of the trait's relative contribution to organisms' ability to survive and reproduce. However, the concept of fitness is vexed; see Rosenberg and Bouchard (2010) for an overview of the difficulties.

playing a game, and their fitness is represented as the "payoffs" of those strategies. Evolutionary game theory is used to calculate the equilibrium point for distribution of phenotypes that would result if natural selection acted unimpeded on the population; there may be one such equilibrium, multiple, or none. For example, the vampire bat's behavioral trait of sharing hunting spoils with other vampire bats is conceived of as a strategy, as is the behavior of not sharing. The first trait has a higher payoff—a greater fitness value—when other bats share food in return. Thus one observes reciprocal altruism in the form of food-sharing among vampire bats (Wilkinson 1984).

Most evolutionary game theory models entirely ignore genetic inheritance. Some explicitly incorporate population genetics, featuring one- or at most two-locus inheritance, but this is uncommon, and even then the genetic dynamics are simplified to the point of unreality. This situation is puzzling: genes are acknowledged by all to be crucial causal influences on evolution, and yet they are ignored in many approaches to modeling evolution, with evolutionary game theory as a prime example.

This neglect of important influences is a feature of modeling approaches throughout biology. Population genetics and quantitative genetics both ignore the niceties of complex genetic influences on phenotypic traits, as well as ignoring the environmental sources of fitness upon which game theory focuses. Cutting-edge genetic research sets aside a host of non-genetic factors. For example, Amos et al. (2010) focus on the genetic influences on human smoking behavior, mentioning that of course there are many other causal influences on an individual's decision to smoke or abstain from cigarettes. Models of development tend to ignore entirely evolutionary influences on traits. In recognition of this, Mayr (1961) distinguished between proximate (developmental) and ultimate (evolutionary) causes.[3]

This practice of continued neglect of central causal factors would be mysterious without attending to explanatory aims. Recall that in the previous section, I encouraged thinking of explanations as portraying causal patterns, and I pointed out that complex causal processes necessitate a choice of which causal pattern an explanation should feature. This offers a way to make sense of modeling approaches in biology continuing to neglect many causal influences, some of which are actually crucial to bringing about the phenomenon to be explained. A primary use of models in biology is to provide explanations. Causal factors, including some central ones, are neglected when those factors are not part of an explanation's focal causal pattern.

Neglecting causal factors makes a model more general in the following sense. A causal factor would be represented in a model by including an additional variable

[3]Mayr's distinction between proximate and ultimate causes can be construed as a distinction between explanations of *why* members of a population have some trait (evolutionary/ultimate causes) and explanations of *how* members of the population came to have that trait (proximate/developmental causes). This proximate/ultimate distinction has received a good amount of attention in philosophy of biology. Ariew (2003) reinterprets the distinction as distinguishing between dynamical versus statistical explanations. Laland et al. (2011) argue that Mayr's distinction fails because the types of causes distinguished are interrelated.

or parameter. By omitting that variable or parameter, the model simply says less about the world; it remains mute about that factor, including even whether it *is* a factor. Put another way, the model abstracts away from any causal factor it neglects; it does not represent anything with regard to the factor—not its presence, its value, nor its absence. For example, a population genetic model that does not employ a parameter for effective population size (N_e) is mute on whether and to what degree drift is a significant causal factor. This results in a model that is more abstract than if the neglected causal factor had been represented, and also more general, for the model applies to systems that vary with respect to the neglected causal factor. Continuing the example above, a population genetic model that does not represent drift is more abstract because of that omission. It is also more general, for it applies to genetic change in populations where the significance of drift varies. (Notice, however, that the fidelity of a model that omits N_e will be lower than that of a model that employs N_e whenever drift is a significant influence.)

Models of a phenomenon that represent just one applicable causal pattern and neglect other causal influences are sometimes simplified to the point of unreality. That is, sometimes a dummy variable or parameter is included in a model that no one expects to accurately represent the world. This is the strategic use of idealizations to ignore causal influences. For instance, population genetic models often assume that a population of organisms is infinite in size. This assumption allows the influence of genetic drift to be ignored. Similarly, evolutionary game theory models often simply assume that offspring resemble parents—that like begets like—thereby ignoring the complexities of systems of genetic inheritance. This is, then, an additional feature of biology that explanatory practice helps to make sense of. The aims of explanation account not only for the continued use of simplified, partial models, but even models that are *unrealistic* in many respects.

A variety of philosophers and biologists have appealed to the aims of explanations in order to account for the continued use of abstractions and idealizations in models. Levins (1966) introduced the idea that there are competing aims for models—accuracy, precision, and generality—and that some precision and accuracy may be traded off for a compensatory gain in generality. Weisberg (2006) argues that such a tradeoff is justified by the aim of using models to give explanations. Godfrey-Smith (2006) dubs the resulting way of doing science "model-based science". Finally, Wimsatt (1987) discusses the role of idealizations in particular.

Notice that, although generating explanations motivates the continuing importance of abstract and idealized models, this does not guarantee that all of biology functions this way. Though some precision and accuracy may be sacrificed to the end of building a general model, there may be situations where other tradeoffs are warranted. For instance, some explanations integrate more causal factors than others; an example is models that integrate both game theory and population genetics. Additionally, some models are used for purposes other than generating explanations of real-world phenomena; this is true of many models in theoretical population genetics. Finally, mathematical modeling may not be central to all fields of biology, for example, to physiology.

Setting aside the features of models, another aspect of biological practice that explanatory practice helps account for is many biologists' total commitment to one or another methodological approach, and the intense disagreements that result. Proponents or critics of particular approaches are prone to making sweeping, ideologically loaded claims. Evolutionary game theory is a prime example here as well. The use of game theory in biology has been described as a "leap of faith" (Grafen 1984) and a "worldview" (Brown 2001) by its proponents, and criticized for the same reason by its detractors. Roughgarden (2009) criticizes sexual selection theory on the grounds that it is wrong about what is "basic to biological nature". Many similar sweeping claims can be found in other areas of biology.

That differences in approach are frequently construed as a matter of fundamentally opposed ideologies suggests that different research programs are incompatible insofar as they are committed to different views of biological reality. But in my view, such debates are better understood as arising not from different theoretical commitments at all, but commitments to different explanatory projects. As we have discussed here, models employ abstractions and idealizations in order to focus on targeted features of a phenomenon, at the expense of ignoring or misrepresenting other features. Different modeling approaches thus can seem to be incompatible, for they employ different parameters/variables and opposed assumptions. However, the exact opposite is true. The limitations of such models make the use of multiple approaches essential. Thus, despite the ideologically laden rhetoric biologists often employ, the question to ask about apparently competing modeling approaches is not which grounds a more successful worldview, but which method better serves one's present research aims (Potochnik 2013). And research aims are in large part determined by explanatory goals, that is, by what phenomena and causal patterns that influence them are of primary interest (Brigandt 2013). To return to one of the examples above, evolutionary game theorists focus on the role of natural selection in evolution and set aside non-selective influences, either by ignoring them entirely or by accommodating their influence in model parameters. This need not be the result of a *worldview*—at this point, biologists agree that non-selective influences can crucially shape the evolutionary process. The use of evolutionary game theory is instead best defended on the basis of the aim of explaining selection's influence on evolutionary phenomena.

Though I have argued that ideological positions are not often warranted in biology, I also suspect that the tendency of biologists to adopt such ideological positions indicates something important about biological phenomena. Let us ask: what enables simple differences of explanatory focus to be interpreted as wholly different worldviews? That there are such entrenched proponents and opponents to different approaches indicates that a variety of approaches have some purchase on the evolutionary process. In my view, this reflects the complex causal processes at work in biological phenomena, and the endless variety in how causal factors combine and interact. This further corroborates the suggestion made in Sect. 2 that a philosophical account of biological explanation must accommodate variety in explanations that arise from focusing on different causal patterns.

Put most broadly, explanatory aims account for the continued diversity of approaches in biology, as well as biologists' tendencies to adopt one or a few approaches as their guiding principle/worldview/etc. Explanatory aims also account for why grappling with exceedingly complex causal processes often does not motivate increasingly complex models. Explanations focus on just one among many causal patterns that govern a phenomenon, and this is accomplished by models that abstract and idealize away from other causal factors in order to represent the focal causal pattern. Sometimes the resulting model is simplified to the point of unreality, yet it can still do its job of representing a causal pattern important to the occurrence of the phenomenon to be explained.

4 Conclusion: Teaching About Biological Explanation

So far in this chapter, we have considered what philosophical accounts of scientific explanation can tell us about biology explanation, and how explanatory practice in biology has influenced—and should influence—general accounts of explanation. We have also explored some features of the field of biology that only make sense in light of the aim of generating explanations and particular explanatory strategies. By means of all of this, I hope to have demonstrated that approaches to biological explanation and how they influence scientific practice are important to biology education. In this section I will develop five suggestions for particular ways in which biology education should attend to issues related to scientific explanation. Along the way I will suggest a few advantages that stand to be gained from implementing these suggestions.

4.1 Suggestion 1: Do Not Overly Emphasize Laws When Thinking About Biology Explanations

It is to be expected that discussions in biology will include reference to "laws." Calling something a law is a way to express the idea that certain phenomena proceed according to a more-or-less lawlike pattern. For instance, Mendel's Laws capture some regularities pertaining to genetic transmission. Such references to laws may, for the most part, reasonably set aside the question raised in Sect. 2 regarding whether and to what degree there are laws of biology. In discussions focused on biological phenomena and not intended to describe the field of biology, the term "law" can simply be used loosely. Hence we continue to refer to Mendel's Laws, even though there are clear exceptions to these laws—exceptions that generate their own distinct lawlike patterns.

What should be avoided is taking too seriously references to biological laws. From the fact that there are references to laws within biology, it should not be inferred that the field of biology progresses via the discovery of new laws (see Lange this volume for an examination of what would be required for there to be biological laws). Similarly, it should not be inferred that finding a law is needed in order to explain a phenomenon. It has been thoroughly demonstrated in this chapter that many explanatory projects in biology do not rely on laws. This means that, in biology education, accepted explanations should not be portrayed as citing laws, especially when such a portrayal is somewhat forced. Encountering a range of biological explanations that resist simplification to laws will help prepare students for the vast range of work in biology to which laws are minimally relevant or not relevant at all.

4.2 Suggestion 2: Explicitly Motivate Forms of Explanation That Are Common in Biology

Following on the heels of the first suggestion, the idea here is that biology education is enriched by explicitly attending to features of biological explanations that may seem strange to outsiders, but are in fact quite common explanatory strategies in biology. This involves more than discussing particular explanations, and resisting the temptation to construe them as based on laws. The suggestion additionally involves inviting students to think—critically but openly—about *how* various explanations succeed. I will provide three brief illustrations here, though there are almost certainly additional forms of explanation across biology that deserve such focus.

Recall from above that traditional optimality explanations account for the prevalence of a trait in a population by showing how that trait led to selective advantage (in the environment at hand). Optimality explanations may be understood as a type of functional explanation: the presence of a trait is explained according to the role it plays for an organism (for functions in biology see Wouters, this volume). In evolutionary contexts, this style of explanation is made possible by the assumption that natural selection promotes traits that increase fitness. Thus the fitness-conferring role of a trait is a *causal influence* on the trait's propagation. This is a helpful lens through which to view optimality explanations, for it at once showcases what is fitting about this style of explanation, and also its limitations, or what may be problematic.

Evolutionary game theory models provide an explanatory strategy closely related to that of optimality models. However, the emphasis is shifted from the selective advantage of a trait to points of stability in the shifting proportions of a range of trait values. One prominent approach to evolutionary game theory is fruitfully considered as a type of equilibrium explanation.[4] Equilibrium explanations are, in my view, a type of broadly causal explanations, for they capture some features of causal patterns (cf. Kuorikoski 2007). Yet equilibrium explanations differ from traditional causal explanations, for they entirely omit any information about the causal process that led to the equilibrium point.

---

[4] This approach analyzes games for points of stability, e.g., evolutionary stable strategies (Maynard Smith and Price 1973). A different approach to evolutionary game theory instead specifies population dynamics, e.g., replicator dynamics, and thus results in dynamical models.

Another, very different type of explanation is mechanistic explanation, viz., explaining a phenomenon as the result of a structured series of causal steps (for mechanistic explanations see Bechtel this volume). An example is the molecular explanation of photosynthesis, which traces the series of chemical transformations among macromolecules by which carbon dioxide and sunlight are converted into sugars and other organic compounds. In some regards, this form of explanation is the complete opposite of functional and equilibrium forms of explanation. Whereas equilibrium and functional explanations cite endpoints and neglect processes, mechanistic explanations instead detail the exact steps by which a phenomenon proceeds.

There are, of course, many unresolved questions about the relationship among these forms of explanation and the relative value of each. Some considerations were introduced in Sect. 2, including one possible way to assimilate all forms of biological explanation. Regardless of the theoretical questions about their relationship, though, each form of explanation deserves explicit attention in biology education. Implementing this suggestion will facilitate a broad education in the range of explanatory projects in biology. It will also help spur students to explicitly consider what form of explanation is generated—or attempted—in different and novel research programs. This is facilitated by introducing forms of explanation as tentative, susceptible to reinterpretation or the revision of methodology (see the next suggestion for more on this idea).

4.3 Suggestion 3: Resist the Temptation to Simplify the Diversity of Approaches in Biology and Their Apparent Incompatibility

This chapter has only surveyed a small part of the astounding variety of explanatory projects in biology. This variety of explanatory projects is not surprising, given the vast array of types of phenomena under investigation in different subfields of biology. Some considerations from Sect. 2, and subsequently suggestion 2 above, suggest that there may even be different *strategies* of explanations in biology, viz., explanations with wholly different aims and attributes.

One might be tempted to simplify this picture in the classroom. Introducing a large variety of explanatory projects can undermine generalizations that can be especially useful as heuristics for students. It also takes up additional class time that could be used in other valuable ways. Yet ignoring—or not focusing upon—the immense variety of explanatory projects and explanatory strategies in biology trains students to expect the field of biology to proceed in lockstep, and may result in later suspicion regarding unfamiliar projects or opposed methodology. Exposure to variety should have the opposite effect. This instead facilitates a more nuanced appreciation for the vast range of causal influences and interactions within the purview of biology, and the diverse routes to understanding found throughout the field. For example, recall from above Mayr's distinction between proximate (developmental) and ultimate (evolutionary) explanations. Kampourakis and Zogza (2009) employed that distinction to classify students' explanations, but it could also help students clarify for themselves the elements of biological explanations. In this case, emphasizing the distinction between *why* members of a population have some trait and *how* they came to have the trait leads students to the recognition that evolutionary explanations and developmental explanations contain distinct elements, that they play distinct roles—that is, that one does not preclude the other, and that there are patterns in what sort of causal information is provided by each. Notice that this educational role can be played by Mayr's distinction in spite of criticisms of that distinction, viz., regardless of whether the types of causes in question are in fact often interrelated or the precise construal of the type of explanatory difference (Laland et al. 2011).

Teaching the diversity of explanatory projects and strategies and how those have changed over time should also engender in students an expectation that accepted explanatory strategies change alongside accepted knowledge in the field. For instance, optimal foraging theory was initially met with suspicion, for it was thought that this required too much psychological sophistication of cognitively simple animals. Since, it has been clarified that optimal foraging explanations are evolutionary explanations, with no assumptions made about the means by which organisms' foraging behavior develops. Such shifts in accepted explanatory strategies are a central example of how methodological norms, and not just stores of knowledge, progress in biology. As such, it helps prepare students to think explicitly and critically about methodology, and to see explanatory practice as a central component of the field of biology.

4.4 Suggestion 4: Explicitly Consider the Role of Models—Partial, Unrealistic Representation

Another move that facilitates the explicit and critical analysis of methodology is a teaching focus on the role of models in biology and how that role varies among subfields and research programs. I suggested above that there is no reason to assume that laws are central to biology. In contrast to the circumstances regarding laws, it is clear that constructing models—whether mathematical, physical, or computer—is an important component of many projects in biology: evolutionary game theory and population genetic models, predator-prey models, species abundance models, model organisms, agent-based models; and the list could go on much longer.

Explicitly addressing the role of models in biology should involve, at least, considering the purposes to which models are put, and how those purposes and the features of models vary among different subfields and research programs. There may or may not be much found to be common among mathematical models, physical models, and computer simulations. Another prime emphasis should be the mechanics of abstraction and idealization, and the purposes to which these are put. As we saw in Sect. 3, the continued importance of simplified models of complex phenomena is due, at least in part, to aims of biological explanations. An investigation of the prominent methodological role of models will thus both necessitate and further investigations of the role of explanation in biology.

4.5 *Suggestion 5: Emphasize Methodological Differences Over Seemingly Ideological Differences; Teach That a Plurality of Approaches Is Here to Stay*

This suggestion takes off from the considerations introduced toward the end of Sect. 3. There I argued that a range of issues on which biologists have taken ideological positions—declaring that a research program is the basis of a successful "worldview," or should be taken on faith (or avoided for that reason), etc.—are more profitably considered to be methodological differences. Commitments to different explanatory projects can lead to the endorsement of different background assumptions, abstractions, and idealizations, and hence differing views about the well-foundedness of various modeling approaches. For instance, advocates of evolutionary developmental biology (or "evo-devo", see Love this volume), the subfield of biology devoted to the evolution of developmental processes, view the field as a corrective to traditional evolutionary biology. Yet some statements of evo-devo's role go too far in the opposite direction. According to Müller (2007), the "explanatory weight" belongs to development, not evolution, for evolving developmental systems are "*the* causal basis for phenotypic form" (emphasis added). Evo-devo draws attention to one set of causal influences, and how they interact with selection. This likely is an important, even crucial, part of the evolutionary story, but it does not undermine the importance of evolution. Emphasis in biology education on how methodological differences arise in the field of biology would help the next generation of biologists avoid such arguments over the primacy of one or another approach, refocusing attention on careful development and critique of methodology, etc.

The suggestion to emphasize methodological differences instead of ideological positions is an outgrowth of the first four suggestions made here. Those suggestions began with the idea that a monolithic picture of law-based explanation should be avoided (Suggestion 1), substituting in its place a critical analysis of the range of common forms of explanation in biology (Suggestion 2). That analysis should resist the temptation to simplify the diversity of approaches to explanations or to minimize differences or seeming incompatibilities (Suggestion 3). Careful attention to the features of biological explanations benefits from and reinforces consideration of the role of models in biology and their relationship to explanations (Suggestion 4). All of this arms the student of biology to interpret debates among biologists with an eye to the diversity of projects, and the diversity of methods they motivate (Suggestion 5). A consequence of implementing these components in biology education is the lesson that a plurality of methods in biology is here to stay.

Acknowledgments Thanks to Francis Cartieri for assistance on this project. I also appreciate getting helpful feedback on earlier drafts from Francis Cartieri, Kostas Kampourakis, and an anonymous referee for this volume.

References

Amos, C.I., M.R. Spitz, and P. Cinciripini. 2010. Chipping away at the genetics of smoking behavior. *Nature Genetics* 42(5): 366–368.

Ariew, A. 2003. Ernst Mayr's 'Ultimate/Proximate' distinction reconsidered and reconstructed. *Biology and Philosophy* 18: 553–565.

Barros, D.B. 2008. Natural selection as a mechanism. *Philosophy of Science* 75: 306–322.

Bechtel, W. 2006. *Discovering cell mechanisms: The creation of modern cell biology*. Cambridge: Cambridge University Press.

Bechtel, W., and A. Abrahamsen. 2005. Explanation: A mechanistic alternative. *Studies in History and Philosophy of Biological and Biomedical Sciences* 36: 421–441.

Brandon, R. 1990. *Organism and environment*. Princeton: Princeton University Press.

Brandon, R. 1997. Does biology have laws? The experimental evidence. *Philosophy of Science* 64(4): 444–457.

Brigandt, I. 2013. Explanation in biology: Reduction, pluralism, and explanatory aims. *Science & Education* 22(1).

Brown, J.S. 2001. Fit of form and function, diversity of life, and procession of life as an evolutionary game. In *Adaptationism and optimality*, ed. S.H. Orzack and E. Sober, 114–160. Cambridge: Cambridge University Press.

Craver, C.F. 2007. *Explaining the brain: Mechanisms and the mosaic unity of neuroscience*. Oxford: Oxford University Press.

Dray, W. 1957. *Laws and explanation in history*. London: Oxford University Press.

Eshel, I., M. Feldman, and A. Bergman. 1998. Long-term evolution, short-term evolution, and population genetic theory. *Journal of Theoretical Biology* 191: 391–396.

Forber, P. 2010. Confirmation and explaining how possible. *Studies in History and Philosophy of Biological and Biomedical Sciences* 41(1): 32–40.

Friedman, M. 1974. Explanation and scientific understanding. *Journal of Philosophy* 71(1): 5–19.

Glennan, S. 1996. Mechanisms and the nature of causation. *Erkenntnis* 44(1): 49–71.

Godfrey-Smith, P. 2006. The strategy of model-based science. *Biology and Philosophy* 21: 725–740.

Goss-Custard, J.D. 1977. Predator responses and prey mortality in redshank, Tringatotanus (L.), and a preferred prey, Corophiumvolutator (Pallas). *Journal of Animal Ecology* 46: 21–36.

Grafen, A. 1984. Natural selection, kin selection, and group selection. In *Behavioural ecology: An evolutionary approach*, 2nd ed, ed. J.R. Krebs and N.B. Davies. Oxford: Blackwell Scientific Publications.

Hammerstein, P. 1996. Streetcar theory and long-term evolution. *Science* 273(5278): 1032.

Hempel, C. 1965. *Aspects of scientific explanation*. New York: Free Press.

Hempel, C., and P. Oppenheim. 1948. Studies in the logic of explanation. *Philosophy of Science* 15: 135–175.

Hull, D. 1992. The particular circumstance model of scientific explanation. In *History and evolution*, ed. M. Nitecki and D. Nitecki, 69–80. Albany: SUNY Press.

Kampourakis, K., and V. Zogza. 2009. Preliminary evolutionary explanations: A basic framework for conceptual change and explanatory coherence in evolution. *Science & Education* 18: 1313–1340.

Kitcher, P. 1981. Explanatory unification. *Philosophy of Science* 48(4): 507–531.

Kitcher, P. 1984. 1953 and all that: A tale of two sciences. *Philosophical Review* 93: 335–373.

Kitcher, P. 1989. Explanatory unification and the causal structure of the world. In *Minnesota studies in the philosophy of science*, Scientific explanation, vol. 13, ed. P. Kitcher and W.C. Salmon, 410–505. Minneapolis: University of Minnesota Press.
Kuorikoski, J. 2007. Explaining with equilibria. In *Rethinking explanation*, ed. J. Persson and P. Ylikoski, 149–162. Dordrecht: Springer.
Laland, K., K. Sterelny, J. Odling-Smee, W. Hoppitt, and T. Uller. 2011. Cause and effect in biology revisited: Is Mayr's proximate-ultimate dichotomy still useful? *Science* 334: 1512–1516.
Levins, R. 1966. The strategy of model building in population biology. *American Scientist* 54(4): 421–431.
Machamer, P., L. Darden, and C.F. Craver. 2000. Thinking about mechanisms. *Philosophy of Science* 67(1): 1–25.
Maynard Smith, J., and G. Price. 1973. The logic of animal conflict. *Nature* 246: 15–18.
Mayr, E. 1961. Cause and effect in biology. *Science* 134: 1501–1506.
Müller, G. 2007. Evo-devo: Extending the evolutionary synthesis. *Nature Reviews Genetics* 8: 943–949.
Potochnik, A. 2010a. Explanatory independence and epistemic interdependence: A case study of the optimality approach. *The British Journal for the Philosophy of Science* 61(1): 213–233.
Potochnik, A. 2010b. Levels of explanation reconceived. *Philosophy of Science* 77(1): 59–72.
Potochnik, A. 2011. Explanation and understanding: An alternative to Strevens' depth. *European Journal for Philosophy of Science* 1(1): 29–38.
Potochnik, A. 2013. Defusing Ideological Defenses in Biology. *BioScience* 63(2): 118–123.
Rosenberg, A., and F. Bouchard. Fitness. In *The Stanford encyclopedia of philosophy* (Fall 2010), ed. E. Zalta. http://plato.stanford.edu/archives/fall2010/entries/fitness/.
Roughgarden, J. 2009. *The genial gene: Deconstructing Darwinian selfishness*. Berkeley: University of California Press.
Ruse, M. 1970. Are there laws in biology? *Australasian Journal of Philosophy* 48(2): 234–246.
Salmon, W.C. 1989. *Four decades of scientific explanation*. Minneapolis: University of Minnesota Press.
Salmon, W.C. 1998. *Causality and explanation*. Oxford: Oxford University Press.
Scriven, M. 1962. Explanations, predictions, and laws. In *Minnesota studies in the philosophy of science*, Scientific explanation, space, and time, vol. 3, ed. H. Feigl and G. Maxwell, 170–230. Minneapolis: University of Minnesota Press.
Skipper, R.A., and R.L. Millstein. 2005. Thinking about evolutionary mechanisms: Natural selection. *Studies in History and Philosophy of Biological and Biomedical Science* 36(2): 327–347.
Sober, E. 1999. The multiple realizability argument against reduction. *Philosophy of Science* 66: 542–564.
Sterelny, K. 1996. Explanatory pluralism in evolutionary biology. *Biology and Philosophy* 11: 193–214.
Strevens, M. 2004. The causal and unification accounts of explanation unified—Causally. *Nous* 38: 154–179.
Strevens, M. 2009. *Depth: An account of scientific explanation*. Cambridge: Harvard University Press.
Waters, K. 1990. Why the antireductionist consensus won't survive the case of classical Mendelian genetics. *Philosophy of Science Association Proceedings* 1: 125–139.
Weisberg, M. 2006. Forty years of 'The Strategy': Levins on model building and idealization. *Biology and Philosophy* 21: 623–645.
Wilkinson, G.S. 1984. Reciprocal food sharing in the vampire bat. *Nature* 308: 181–184.
Wimsatt, W.C. 1987. False models as means to truer theories. In *Neutral models in biology*, ed. N. Nitecki and A. Hoffman, 23–55. Oxford: Oxford University Press.
Woodward, J. 2003. *Making things happen: A theory of causal explanation*. Oxford: Oxford University Press.

What Would Natural Laws in the Life Sciences Be?

Marc Lange

1 Introduction

On March 20, 2012, the United States Supreme Court ruled unanimously that patent protection does not extend to a medical test that relies on the fact that if a certain chemical reaches a certain level in the blood of a patient who has received a certain dose of a given drug, then that patient would have certain chances of suffering various possible outcomes if she were treated with that drug. Writing for the Court in *Mayo Collaborative Services v. Prometheus Laboratories* (No. 10–115), Justice Stephen Breyer said: "Einstein could not patent his celebrated law that $E=mc^2$; nor could Newton have patented the law of gravity." (Breyer 2012, pp. 1–2) By the same token, Breyer wrote, Prometheus Laboratories cannot patent a natural law in the life sciences – in this case, "relationships between concentrations of certain metabolites in the blood and the likelihood that a dosage of a thiopurine drug will prove ineffective or cause harm."

This relationship is typical of the results presented in many journal articles in the life sciences. Articles commonly present generalizations concerning the biological properties characteristic of particular species: their characteristic morphology, anatomy, physiology, chemical constitution, environmental tolerances, behavior, development, and so forth. Of course, such generalizations are not merely the focus of *current* biological research; it has obviously been known for a very long time that chickens lay eggs with hard shells, for example.

Research in the life sciences also leads to generalizations concerning *groups* of related species. Tadpoles turn into frogs and caterpillars turn into butterflies, for instance. There are also ecological generalizations about the numbers and distribution of biological species themselves, such as the well-known fact that species density decreases as latitude on the Earth increases, as well as the puzzling fact that

M. Lange (✉)
Philosophy Department, University of North Carolina, Chapel Hill, NC, USA
e-mail: mlange@email.unc.edu

K. Kampourakis (ed.), *The Philosophy of Biology: A Companion for Educators*, History, Philosophy and Theory of the Life Sciences 1, DOI 10.1007/978-94-007-6537-5_4,

regions with greater biodiversity tend to have fewer species with small geographic ranges. (See Brown and Lomolino 1998, pp. 450, 468.)

Do these various biological generalizations constitute laws of nature, as Justice Breyer contends? Biologists seldom refer to them explicitly as laws, though in some cases, they do. That the mammalian heart's stroke volume increases in a certain way with the volume of blood filling the left ventricle has long been known as the Frank-Starling law or (as Starling poetically called it) "the law of the heart" (Starling 1918). Steenstrup (1845, p. 113) identified explicitly as laws certain generalizations about the alternation of generations in jellyfish. In any case, there are plenty of laws *in physics* that are never actually called "laws", such as Maxwell's equations and Archimedes' Principle. In characterizing certain scientific results as laws, philosophers are describing not what the results are commonly called, but the roles that they are called upon to play in science. Laws of nature are standardly thought to differ from "accidents" (i.e., contingent truths that are not laws) in connection with scientific explanations, subjunctive conditionals, necessity, natural kinds, inductive confirmation, and so forth. To ask whether the biological generalizations that I have mentioned function in biology as laws is to ask whether the life sciences call upon these generalizations to play the roles that are supposed to set laws apart from accidents.

There have been other philosophical motivations for thinking about whether the concept of natural law applies to results in the life sciences. Some philosophers have thought that there must be biological laws in order for there to be distinctively biological scientific explanations (for explanations in biology see Potochnik this volume). This view presupposes that all scientific explanations must use natural laws. For instance, all scientific explanations that conform to Hempel and Oppenheim's D-N model (Hempel 1966, p. 51) must appeal to natural laws. I do not accept this presupposition (see also Potochnik this volume). An explanation powered by the principle of natural selection may work without using any laws at all; laws are contingent whereas the principle of natural selection is not. The principle of natural selection – roughly, that an individual having greater fitness is more likely to leave more descendants than a less fit individual – is a logical necessity rather than a contingent truth; for it to be false would be a contradiction. That laws differ from accidents in their explanatory roles does not entail that all scientific explanations must use laws.

By the same token, some philosophers have been prompted to ask whether there are biological laws because they have thought that in the absence of any biological laws, biology would lose its credentials as a science; biology would be merely descriptive. As Ernest Rutherford is supposed to have said, "All science is either physics or stamp collecting." I do not think that biology's scientific bona fides depend upon the existence of distinctively biological laws. My motivation for thinking about what a biological law would be is to see whether the concept of natural law helps us to understand the roles played by various notable biological generalizations.

My aim is not to argue that there are in fact biological laws. Whether there are any is a scientific, not a philosophical question. It is an *open* scientific question in the case of laws of island biogeography, for example. My aim is to understand what island biogeographers are arguing about when they argue about whether there are laws of island biogeography. They do not appear to be arguing about whether to bestow a certain honorific title upon certain discoveries, like whether to give someone a knighthood. Rather, they appear to be arguing about certain facts, just as scientists are supposed to do. What facts are in dispute here – what would make it the case that there are laws of island biogeography? What would laws of the life sciences be?

Many of the same questions arise regarding generalizations in sciences *outside* of the life sciences, such as hydrogeology and stellar astronomy. Many of my comments will therefore concern simply the concept of a law in a "special" science (that is, a scientific field with a scope that is more limited than fundamental physics).

Let us begin, then, by thinking briefly about some of the features of scientific reasoning that philosophers have traditionally believed to be captured by the notion of a law of nature. We will then be in a better position to investigate whether the same work needs doing in the life sciences and whether laws are there to do it.

2 Laws of Nature: The Standard Picture

By philosophical tradition, there are three kinds of facts. First, there are the logical, conceptual, mathematical, and metaphysical necessities (the "broadly logical necessities"): facts that absolutely could not have been otherwise. These do not include the laws of nature. The natural laws' characteristic variety of necessity, natural necessity, is weaker than logical necessity in that the laws of nature (unlike, for instance, the mathematical truths) could have been different. For example, gravity might have declined more rapidly with distance, without violating any broadly logical necessities. The facts that are not broadly logical necessities (the "contingent" facts) divide into two classes: the "natural necessities" (which follow from the laws alone) and the "accidents" (which do not). What distinguishes laws from accidents?

To begin with, an accident just happens to obtain. In Reichenbach's (1947, p. 368) example, a gold cube larger than a cubic meter could have formed, but the proper conditions never happened to arise. In contrast, it is no accident that a large cube of uranium-235 never formed, since the laws governing nuclear chain-reactions prohibit it. In short, things *must* conform to the laws; the laws have a kind of *necessity* (weaker than broadly logical necessity) whereas accidents are just coincidences.

That is to say, had Bill Gates wanted to build a large gold cube, then (I dare say) there would have been one. But even if Bill Gates had wanted to build a large cube of uranium-235, all U-235 cubes would still have been less than a cubic meter. In other words, the laws govern not only what actually happens, but also what would have happened under various hypothetical circumstances. Whatever would have happened had I gone to the store for a quart of milk this morning (for instance) would still have conformed to the natural laws. In other words, the laws underwrite "counterfactual conditionals" (Goodman 1947): if-then statements about what would have occurred under certain hypothetical circumstances. For instance, "If I had gone to the store this morning to buy a quart of milk, then I would have bought a quart of milk and gravity would still have operated" is a counterfactual conditional – one that, I believe, is true. In contrast to a law, an accident would not still have held under certain circumstances that are logically consistent with all of the laws taken together.

Counterfactuals are notoriously context-sensitive. In Quine's (1960, p. 222) example, in some contexts it is correct to say that had Caesar been in command in the Korean War, he would have used the atomic bomb, whereas in other contexts, it is correct to say that he would have used catapults. What is preserved under a counterfactual antecedent (that is, under the circumstance posited by the "If…" part of a counterfactual conditional), and what is allowed to vary, depends somewhat upon our interests in entertaining the antecedent. But in any conversational context, the laws would still have held under any natural possibility – that is, under any circumstance that is logically consistent with the laws. In other words:

> *Nomic Preservation (NP)*: g is naturally necessary if and only if in any context, g would still have held if p had obtained, for every p that is logically consistent with every natural necessity.

For instance, that all bodies travel no faster than the speed of light is a natural necessity if and only if all bodies would still have done so under every circumstance p that is logically consistent with every natural necessity – for instance, even if a body had been accelerated by the Stanford Linear Accelerator operating at full power. This principle will play an important role later.

Let us continue with the simple traditional picture of the scientific roles that laws play. Because of their necessity, laws have an explanatory power that accidents lack (Hempel 1966, p. 56). In Hempel's example, a certain powder burns with yellow flames, not another color, because the powder is a sodium salt and it is a law that all sodium salts, when ignited, burn yellow. The powder *had* to burn yellow, considering that it was a sodium salt – and the "had-to-ness" arises from the kind of necessity that is distinctive of laws. In contrast, we cannot explain why my wife and I have two children by citing the fact that all of the families on our block have two children – since this fact is an accident. Were a childless family to try to move onto our block, they would not encounter an irresistible opposing force or acquire more children. Why do all samples of sodium salts that are actually ignited burn yellow? Because a law compels them to (and this law, in turn, is explained by other laws). In contrast, the fact that all of the families on our block have two children is not explained by the fact that this regularity is *not* a law.

We believe that it would be mere coincidence if all of the coins in my pocket today turned out to be silver-colored. So our discovery that the first coin I check from my pocket is silver-colored fails to justify raising our confidence that the next one will also be silver-colored. A candidate law is confirmed differently, as Goodman (1983, p. 20) emphasized: that one sample of a given chemical melts at a given temperature confirms, of every unexamined sample, that its melting point is also that temperature.

That the same claims play all of these special roles in science would suggest that scientific reasoning draws an important distinction here, which philosophers characterize as the difference between laws and accidents. However, it is notoriously difficult to capture the laws' supposed "special roles" precisely. For example, it would be an accident if all of the figs now hanging on my tree are tasteless. Yet it seems like this hypothesis can be confirmed inductively: the fact that each of the figs that I have sampled so far is tasteless confirms that the other figs are, too. Despite being accidental, this generalization seems to support counterfactuals: presumably, had there been another fig on the tree, it would also have been tasteless. Of course, had the weather or the plant hormone levels earlier in the season been different, then perhaps the figs would now have tasted good. So there are some counterfactual antecedents under which the fig generalization is *not* invariant. But even a law fails to be invariant under certain counterfactual antecedents. For example, the electron would not have been a stable particle had there been a lighter negatively charged lepton (because there would then have been a particle available for it to decay into without violating the conservation of energy, electric charge, and lepton number). Why does this limitation on the range of invariance of the electron generalization not undercut its status as a law? Because the existence of a negatively charged lepton lighter than the electron contradicts the laws specifying the complete inventory of the kinds of particles there are. *NP* requires only that the laws be preserved under certain counterfactual antecedents: those that are logically consistent with the laws.

But this means that *NP* is *using* the notion of lawhood to specify what sets laws apart from accidents as far as their invariance under counterfactual antecedents is concerned. Our original motivation for using the concept of natural law to understand scientific reasoning was that we noticed various scientific roles that only *certain* scientific results could play, and then we used the fact that those results are taken to be laws to account for their power to play these roles. But now it seems that no such roles can be specified without already using the distinction between laws and accidents in the first place. We do not seem to need that distinction after all in order to account for any difference that can be recognized independent of that distinction.

It might seem, then, that we were mistaken in drawing a sharp distinction between laws and accidents (van Fraassen 1989; Giere 1995). Instead, we should simply recognize that different facts have different degrees of resilience under counterfactual antecedents; those facts with broader ranges of invariance are presumably better able to explain, more prone to inductive confirmation, and so forth (Hitchcock and Woodward 2003).

However, as I see it, this view cannot be quite right. As an example, consider a generalization that is the same as Coulomb's law of electrostatics except regarding one arbitrary combination of charges and distance that happens never to be instantiated; concerning that combination, the given generalization predicts an electrostatic force that departs wildly from the prediction made by Coulomb's law.[1] For instance, the generalization might agree with Coulomb's law except for saying that if a charge

[1] For argument's sake, consider Coulomb's law as exceptionless.

of exactly 1.234 statcoulombs is located one centimeter from a charge of exactly 5.678 statcoulombs, then they experience 100 dynes of mutual repulsion – whereas Coulomb's law says that they feel about 7 dynes of repulsion. This deviant generalization has a range of invariance that is nearly as broad as the range possessed by Coulomb's law (being smaller by just one combination of the independent variables). Yet its role in science is not remotely like the role played by Coulomb's law. An actual case conforms to this deviant generalization not because all cases have got to, but because coincidentally its demands in all actual cases agree with those of Coulomb's law and all cases have got to conform to Coulomb's law.

We have thus been led right back to a sharp distinction between a law and a non-law that exhibits some range of invariance. Later, I will show how we do not need to use the distinction between laws and accidents in order to recognize the special kind of resilience under counterfactual antecedents that sets laws apart. But first let us look at the special problems posed by the concept of life-science laws. Let us return to the biological generalizations with which I began. They raise two questions: what would it take for such a generalization to be *true*? And, beyond that, what would make it a *law*? I will take up these two questions in the next two sections.

3 The Problem of Exceptions

The discovery of generalizations concerning particular biological species is the goal of a great deal of biological research. In connection with counterfactuals, explanation, and confirmation, the biological generalizations at which this research aims function much as laws of nature do on the traditional picture. Jane's trachea has cartilaginous rings because Jane is a human being and the human trachea has cartilaginous rings in order to keep it from collapsing between breaths. Explanations of this sort are quite ordinary. At the zoo, a child might point to a bird and ask, "Why did he do that?" An adult might properly reply, "It is a pelican and that is how pelicans eat."

Of course, a generalization of this sort must accord with population thinking in biology. It must leave room for the kind of variation on which natural selection operates. A generalization regarding the members of a given species does *not* presuppose that a biological species involves a normal type where any departure from that type requires special explanation (see also Wilkins this volume). Indeed, biologists use generalizations about the biological properties characteristic of particular species even when they are describing natural *variation* among the members of the same species:

> Within a single species [...] individuals sometimes have the diagnostic characteristics of related species or even genera. The form and number of teeth in mammals are important for classification; yet in a single sample of the deer mouse *Peromyscus maniculatus*, Hooper (1957) found variant tooth patterns typical of 17 other species of *Peromyscus*. Among fossils of the extinct rabbit *Nekrolagus*, Hibbard (1963) found one with the premolar pattern characteristic of modern genera of rabbits; and the *Nekrolagus* pattern is occasionally found in living species (Futuyma 1979, p. 161).

Even while Futuyma is emphasizing the variation among members of the same species, he refers to the dentition "characteristic" or "typical" of a given species.

But what does it mean to say that the *Nekrolagus* tooth pattern is such and such? It obviously does not mean that all members of this species have that tooth pattern. Nor is it a simple statistical generalization; it assigns no particular probability to that tooth pattern. However, the fact that it is not immediately clear what makes this generalization hold should not lead us to conclude that this generalization has no content. After all, there are well-known Darwinian grounds for expecting groups defined by common descent to largely share a host of biological properties (this is relevant to the concept of homology, see Minelli and Fusco this volume). Various mechanisms keep members of the same species clustered together in the space of biological possibility. These mechanisms are expressed by various species concepts: for instance, Mayr's biological species concept emphasizes gene exchange within a species and barriers to reproduction between species. These clustering mechanisms feature in Boyd's (2007) conception of a biological species as a homeostatic property cluster kind. We should not give up on biological generalizations such as "The robins' egg is greenish-blue"; instead, we should try to figure out just what it takes for one of these generalizations to be true.

On my view, "The robins' egg is greenish-blue" functions in certain biological fields as a rule of default reasoning. It tells us what we should expect about a given robin's egg in the absence of specific information suggesting that it is not greenish-blue. The generalization is made true by the fact that by using the corresponding rule of default reasoning, we are led often enough to expectations that are close enough to the truth for certain tacitly understood purposes. The clustering mechanisms that I have just mentioned are responsible for the reliability of these generalizations. A policy of default reasoning that is expressed by one of these generalizations may be sufficiently reliable for the purposes of neurology or physiology or embryology – though perhaps not for the purposes of population genetics or evolutionary biology. Whether the inference rule is sufficiently reliable for the generalization to be true depends upon how readily available "information to the contrary" is and how tolerant of error we can afford to be, considering the relevant purposes. An inference rule's "reliability" is a pragmatic notion, reflecting the purposes for which the rule would be put to use and how much departure from the truth how often those purposes can tolerate.

This explains how it can be true that "The lion is tawny" and also be true that "The lion with a certain gene is white". When we have no reason to believe that Leo possesses a certain rare gene, then we ought to expect Leo to be tawny, but when we have some reason to believe that Leo possesses the gene, then we ought to withhold judgment, and when we have sufficient reason to believe that Leo possesses the gene, then we ought to expect Leo to be white (in the absence of any further relevant information). The default here concerns *our* expectations, not what *a lion* is like in the absence of disturbances to its allegedly natural state.

Whatever the detailed semantics of remarks like "The lion is tawny", the key point is that such generalizations can tolerate variation. Therefore, the existence of natural variation within a species does not make these generalizations false and so does not disqualify them from constituting life-science laws.

Some biological generalizations do not specify a biological property's distribution, but instead have a causal significance. Such a generalization often comes with a ceteris paribus proviso – that is (roughly speaking), with a proviso "all other things being equal". It limits the generalization to cases involving the absence of certain factors. Scientists would typically be hard pressed to enumerate these factors exhaustively, but those who understand the generalization understand which factors are encompassed by "ceteris paribus". They include the major disturbing influences, but they do not encompass all of the kinds of petty disturbing factors there are. Even when all of the major disturbing factors are absent, the expectations licensed by the generalization may fail to obtain. But if large departures from these expectations are infrequent enough, the generalization will be reliable; it will be accurate enough for the relevant purposes and thus not be prevented from qualifying as a law.

Mill expressed this point well – not in connection with a biological science, but in connection with Tidology:

> It may happen that the greater causes, those on which the principal part of the phenomena depends, are within the reach of observation and measurement [...] But inasmuch as other, perhaps many other causes, separately insignificant in their effects, co-operate or conflict in many or in all cases with those greater causes, the effect, accordingly, presents more or less of aberration from what would be produced by the greater causes alone. [...] It is thus, for example, with the theory of the tides. No one doubts that Tidology [...] is really a science. As much of the phenomena as depends on the attraction of the sun and moon [...] may be foretold with certainty; and the far greater part of the phenomena depends on these causes. But circumstances of a local or casual nature, such as the configuration of the bottom of the ocean, the degree of confinement from shores, the direction of the wind, &c., influence in many or in all places the height and time of the tide [...] General laws may be laid down respecting the tides; predictions may be founded on those laws, and the result will in the main [...] correspond to the predictions. And this is, or ought to be meant by those who speak of sciences which are not *exact* sciences. (Mill 1961, pp. 552–553, section 6.3.1)

A "reliable" generalization in one of the inexact sciences must reflect all of the "greater causes." But it can afford to neglect a host of petty influences. For example, classical physics might suffice for the purposes of human physiology or marketing or traffic science; relativistic corrections are negligible.[2]

Let us look at a biological example. Take island biogeography (IB), which deals with the abundance, distribution, and evolution of species living on separated patches of habitat. It has been suggested that all other things being equal, the equilibrium number S of species of a given taxonomic group living on an "island" increases with the island's area A in accordance with a power law: $S=cA^z$. The (positive-valued) constants c and z are specific to the taxonomic group and island group – for instance, Indonesian land birds. According to the theory that MacArthur and Wilson developed to explain the area law, a larger island tends to have larger available habitats for its species, so it can support larger populations of them, making chance extinctions less likely. Larger islands also present larger targets for

[2]Biological controversies often concern the "relative significance" of various factors (Beatty 1995), and these disputes may be understood as concerning which are the "greater causes" that must figure in biological laws (Sober 1997, p. S461).

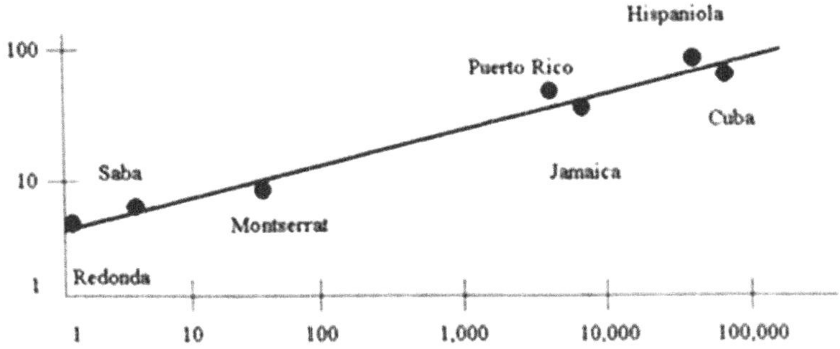

Fig. 1 (Note the log scales) The area (in square kilometers) of various islands in the West Indies is depicted on the x-axis. The number of amphibian and reptilian species on each island is depicted on the y-axis (After MacArthur 1972, p. 104)

stray creatures. Therefore, larger islands have higher immigration rates and lower extinction rates, and so tend to equilibrate at higher biodiversity (Fig. 1).

Nevertheless, a smaller island nearer to the "mainland" may have greater biodiversity than a larger island farther away. This factor is covered by the *ceteris paribus* qualifier to the "area law." Likewise, a smaller island with greater habitat heterogeneity may support greater biodiversity than a larger, more homogeneous island. This factor is also covered by "ceteris paribus." Ecologists may never have enumerated all of these factors, but they tacitly recognize them.[3]

To discover the "area law," ecologists did not need to identify *every* factor that may cause deviations from the equation; the "greater causes" suffice for the area law to yield predictions good enough for various sorts of applications, from the design of nature reserves to serving as a common starting-point for building more accurate ecological models fine-tuned to particular cases. Of course, by the purposes of island biogeography, I do not mean the de facto interests of some particular group of scientists, but partly a normative matter: certain problems ought to be addressed by the practitioners in a given field in view of what else is at issue in that field.

The "area law" may actually not be reliable. Perhaps only a thoroughly case-by-case approach makes even approximately accurate predictions regarding island biodiversity. This is a part of what is in dispute among ecologists who disagree about

[3]Likewise, Hutchinson (1957, p. 417) took the principle of competitive exclusion to have a ceteris paribus qualification requiring the absence of various disturbing factors (such as predation that keeps the competing prey species at such low levels that competition between them is ineffective). These factors are widely understood but need not ever be exhaustively enumerated. Accordingly, Hutchinson said that the principle "is true except in cases where there are good reasons not to expect it to be true." These kinds of "good reasons" are not ad hoc hedges. Rather, these kinds were often anticipated theoretically before any actual cases of them had been discovered. Weber (1999, p. 86) points out other, related generalizations in community ecology, such as "Predation increases the number of coexisting competitors at the next lower trophic level."

whether the area law is an IB law (Kingsland 1985). Of course, my aim is not to settle this scientific dispute, but merely to understand what is in dispute. In any case, island biogeography's status as a scientific field does not depend on its having its own laws.

4 The Problem of Accidentalness

We now arrive at what may be the most influential argument against the existence of life-science laws, namely, that even if some biological generalizations of the sort I have mentioned are true, they are accidental and not naturally necessary.

Certain important biological generalizations (such as the Hardy-Weinberg "law" and the principle of natural selection) are logically or mathematically necessary and so are not laws on the traditional picture. Some biologically relevant facts are really matters of chemistry and so are genuine laws, but not distinctively biological ones. Neither of these categories includes the biological generalizations that we have been discussing. They hold only because certain mutations never occurred or certain selection pressures never operated. Had evolutionary history been replayed from the same initial conditions, the outcome might well have been radically different, since different mutations might have been introduced, random drift in small populations might have led in another direction, and different selection pressures might have been imposed. Gould (1989) famously makes this point in terms of getting a different result from "replaying life's tape". In that event, the robin's egg might not have been greenish-blue. This generalization therefore lacks the proper invariance under counterfactual perturbations to qualify as a law. It lacks natural necessity; it merely expresses "current evolutionary fashions" (Waters 1998, p. 16); it is a coincidence of natural history (Beatty 1981, 1995; Rosenberg 2001a, b, c).

Let us examine this argument more closely. It presupposes *NP*: that a generalization is a law only if its reliability is invariant under every counterfactual antecedent that is logically consistent with the laws. For example, *NP* entails that "All gold cubes are smaller than a cubic meter" does not express a natural necessity because it would not still have held had Bill Gates wanted to build a large gold cube. Likewise, "The robin's egg is greenish-blue" would have failed under certain mutations or environmental conditions. Therefore, by *NP*, it is not a law.

However, this conclusion follows from *NP* only if it is logically consistent with the laws for those mutations or environmental conditions to occur. Undoubtedly, it is consistent with the laws of physics. But to argue that therefore it is logically consistent with all of the laws is to take as a premise that the laws of physics are all of the laws – which amounts to presupposing that there are no life-science laws. But that was supposed to be the argument's conclusion. It is circular to use NP to argue that there are no life-science laws on the grounds that the life-science generalizations are not invariant under all of the counterfactual antecedents that are logically consistent with the laws of physics and hence with all of the laws.

Let us think of *NP* as a general schema. For the laws of physics, it becomes:

> *NP-physics*: *g* is one of the laws *of physics* (or a logical consequence of those laws) if and only if in any context, *g* would still have been reliable if *p* had obtained, for every *p* that is logically consistent with the reliability of the laws *of physics*.

How should *NP* be applied to life-science laws? To require that a life-science law have the same range of invariance as a law of physics is to allow physics to dictate what sort of necessity a life-science law would have to have. It is no wonder that if we employ that standard, we find that there are no life-science laws. On the other hand, we might use the following generalization of NP-physics:

> *NP-field*: *g* is one of the laws *of a given field* (or a logical consequence of those laws) if and only if in any context, *g* would still have been reliable if *p* had obtained, for every *p* that is logically consistent with the reliability of the laws *of that field*.

This principle permits a generalization to qualify as *necessary* for the purposes of some branch of the life sciences even if it is not invariant under the same range of counterfactual antecedents as the laws of physics are invariant under. *NP-field* does not force biology to borrow its sense of natural necessity from physics and so does not stack the deck against biology's autonomy.

NP-field arises naturally from the ideas that we saw earlier about what laws are. To see this, we must at last grapple with the question that we deferred earlier: what kind of invariance must a fact exhibit in order for it to possess natural necessity, allowing it to function as laws do in connection with explanations, counterfactuals, induction, and so forth?

According to *NP*, the laws would all still have held under any counterfactual antecedent that is logically consistent with the laws. No accident is always preserved under all of these antecedents. But as we saw earlier, *NP* cannot save our intuition that the laws possess a special relation to counterfactuals. That is because the range of counterfactual antecedents in *NP* has been tailor-made to suit the laws.

What if, instead of considering the counterfactual antecedents that are logically consistent with the *laws*, we allow every set of truths to pick out its own tailor-made range of counterfactual antecedents – namely, those antecedents that are logically consistent with that particular set? Would the set's members still have held under all of those antecedents? Let us think about it. Take a logically closed set of truths that includes the accident that all gold cubes are smaller than a cubic meter but omits the accident that all of the figs on my tree are tasteless. Take the counterfactual antecedent positing that one or the other of these two accidents had failed to hold. Each of these accidents is logically consistent with this antecedent, so each *could* remain true under this antecedent. But it is not the case that each of them *would* remain true; they cannot *both* remain true. So what would have happened under this antecedent? Would the accident that is *in* the set – the gold cubes accident – still have held? It *might* then have held, but it is not the case that it *would* then have held. Neither of these two accidents takes so much priority over the other that in every conversational context, it would still have held under this counterfactual antecedent pitting it against the other accident.

The same sort of argument could be made regarding any logically closed set of truths that includes *some* accidents but not *all* of them. It is not the case that the set is invariant under every counterfactual antecedent with which it is logically consistent. In contrast, *NP* says that the natural necessities *are* invariant under every counterfactual antecedent with which they are all logically consistent.

Here, then, is my preliminary suggestion for the laws' special relation to counterfactuals. Take a set of truths that is logically closed and is neither the empty set nor the set of all truths. Let us say that such a set is *stable* exactly when every member of the set would still have held under any counterfactual antecedent that is logically consistent with every member of the set. My preliminary suggestion is that a truth is naturally necessary exactly when it belongs to a stable set.[4]

By the definition of stability, a stable set is *maximally* invariant: collectively, its members are invariant under as *broad* a range of counterfactual antecedents as they could collectively be. No set containing an accident can make that boast (except for the set of all truths, for which the boast is trivial since there is *no* counterfactual antecedent that is logically consistent with *all* truths). Intuitively, necessity consists of maximal invariance. Accordingly, a set's stability gives its members a kind of necessity. In particular, the stability of the set spanned by the laws gives its members "natural necessity."

The notion of "stability", then, allows us to draw a sharp distinction between laws and accidents. It gives us a way out of the circle that results from specifying the natural necessities as the truths that would still have held under those counterfactual antecedents that are logically consistent with ... the natural necessities.

For the sake of simplicity, I have been speaking in terms of laws of nature *simpliciter.* But I intend all of this to be implicitly relative to some or another scientific field. Let us make this more explicit and see what a law of a particular field would be. What would it take for a set to be stable for a given field's purposes – that is, what would it take for a set to be maximally invariant as far as the scientific field is concerned? In this way, we will arrive at *NP-field.*

To begin with, the set's members must all actually be *reliable* – in other words, close enough to the truth for that field's purposes. The set's stability for the field's purposes also requires that the set's reliability be invariant under a certain range of counterfactual antecedents. What range is that? Certain claims and counterfactual antecedents lie outside of the field's concerns, and in certain conversational contexts, the field's concerns are irrelevant. So let us take a set that is spanned by claims that *are* of interest to the field and *are* reliable for its purposes. Let us take the contexts where the field's interests *are* what is relevant. Such a set is "stable" for the field's purposes exactly when in every such context the set would still have been reliable under every counterfactual antecedent that is of interest to the field and consistent with the set's reliability. A set that is stable for island biogeography (IB) would then possess as much invariance as it could possess, as far as island biogeography is concerned. The set's members would then be naturally necessary in IB.

To get a grip on this idea, let us apply it to the so-called "area law". Suppose for the sake of argument that the area law is reliable for IB purposes. How broad must

[4]For more detail, see Lange (2009, pp. 32–44).

its range of invariance be for it to qualify as an IB law? How contingent can it afford to be? There are counterfactual antecedents under which the laws of physics *would* still have held, but under which the "area law" would *not* still have held. For example, had Earth always lacked a magnetic field then cosmic rays would have bombarded all latitudes, which might well have prevented life from arising, in which case the number of species on an island would have been zero irrespective of the island's area. Here is another counterfactual antecedent: had evolutionary history proceeded differently so that many species were capable of covering long distances over unfamiliar terrain nearly as safely as short distances over familiar territory. Under this supposition, the "area law" might not still have held, since an island's size as a target for stray creatures might then have made little difference to its immigration rate.

The area law's failure to be invariant under these two counterfactual antecedents does not prevent the area law from belonging to a set that is stable for IB purposes, even though each antecedent is consistent with the laws of physics. The antecedent concerning Earth's magnetic field falls outside of IB's interests. It twiddles with a parameter that IB does not take as a variable. IB is concerned with what Granada's biodiversity would have been like, had its area been larger. But IB is *not* concerned with how species would have been distributed had Earth's basic physical constitution been different – for instance, had Earth failed to have had the Moon knocked out of it by an early cataclysm.[5] Biogeographers need not be geophysicists.

Let us now turn to the other counterfactual antecedent I mentioned, which posits many species capable of covering long distances over unfamiliar terrain nearly as safely as they cover short distances over familiar territory. This antecedent is logically inconsistent with other generalizations that would have to join the "area law" to form an IB-stable set. For example, the "distance law" says that other things being equal, islands farther from the mainland equilibrate at lower biodiversity. Underlying both the area and distance laws are various constraints – for instance, that creatures travel along continuous paths, and that the difficulty of crossing a gap in the creature's habitat increases smoothly with the gap's size (other things being equal). These "continuity principles" (MacArthur 1972, pp. 59–60) would have to join the area and distance laws in an IB-stable set.

The area law might not still have held, had these continuity principles been violated. Yet the area law's range of invariance under counterfactual antecedents may nevertheless enable it to qualify as an IB law because *other* IB laws express these continuity principles, so violations of them are naturally impossible in IB, and so an IB law does not have to remain invariant under counterfactual antecedents that posit these violations. This is like the example we saw earlier involving the electron's stability.[6] How can the electron's stability be a law of physics when the electron would not still have been stable, had there been a lighter negatively charged lepton?

---

[5]Earth's rotation rate would then have been greater, its tides would have been less, and the CO_2 level in its atmosphere would have been greater.

[6]Of course, here I mean "stability" in an entirely different sense from the sense in which the laws of IB would form a set that is "stable" for IB purposes.

Because the kinds of particles there are is also fixed by the laws of physics, so a counterfactual antecedent that posits a lighter negatively charged lepton is logically inconsistent with the laws of physics. The electron's stability can have the range of invariance demanded of a law of physics without being invariant under counterfactual antecedents positing alien kinds of particles.

A set that is IB stable can omit some laws of physics. The *gross* features of the physical laws that are captured by continuity principles like those I have mentioned (along with the other IB laws and the field's interests) may not need to be supplemented by the *fundamental* laws of physics in order to limit the range of counterfactual antecedents under which an IB law must be invariant. For example, the area law would still have held even if some birds had been equipped with modest anti-gravity organs, assisting in takeoffs. The factors affecting species dispersal would then have been no different; for example, smaller islands would still have presented smaller targets to off-course birds and so accumulated fewer strays, other things being equal. Likewise, the area law would still have held even if material bodies had consisted of some continuous rigid substance rather than molecules. The range of invariance associated with the IB laws' stability may in some respects extend *beyond* the corresponding range for the laws of physics; the IB laws do not reflect every *detail* of the fundamental physical laws.

This is an important point. If there are IB laws, then their necessity corresponds to the range of invariance associated with their stability. But that range is not wholly contained within the range of invariance associated with the stability of the laws of physics since it includes some counterfactual antecedents violating the physical laws. Thus, the stability of the laws of physics cannot be responsible for the IB laws' stability. The IB laws do not inherit their necessity from the physical laws. The kind of necessity characteristic of IB laws is not even possessed by the physical laws (since the physical laws are not invariant under all of the counterfactual antecedents within the range of invariance associated with the IB laws's stability). The approximate *truth* of IB laws might well follow from the physical laws and certain initial conditions that are accidents of physics. But the *lawhood* of IB laws – their stability for IB's purposes – *cannot* follow from physical laws and initial conditions. The IB laws's stability depends on their remaining reliable under certain counterfactual antecedents *that violate* physical laws. The physical laws cannot be responsible for the area law's remaining reliable under those counterfactual antecedents.

Therefore, if there turn out to be IB laws, then IB has an important kind of *autonomy*. IB's explanations are irreducible to the explanations of the same phenomena at a more microphysical level.

For example, there are then two different explanations of the number of land-bird species currently inhabiting Grenada. One explanation proceeds on the macro level, using IB laws and Grenada's area, distance from the mainland, and so forth. The other explanation proceeds on the micro level, by explaining the fates of various individual creatures that might have migrated to Grenada and left descendants. This micro account explains why Grenada is inhabited by certain particular species rather than a different combination. This fact cannot be explained by the macro account. Conversely, the micro account fails to reveal that Grenada's biodiversity would have been roughly the same even if the mainland species of birds had been different – indeed, even if some of those species had been made of continuous rigid substance. The IB laws would then still have applied.

As far as IB is concerned, the fact that there are no birds equipped with modest antigravity organs or made of continuous rigid substance is merely an accident of the actual world. The macro outcome is insensitive to this accident. The IB explanation of Grenada's biodiversity supplies this information.[7]

5 Evolutionary Accidents as Laws of Certain Fields of Biology

Let us return to generalizations such as "The robins' egg is greenish-blue" and consider whether such generalizations belong to a set that is stable for the purposes of some branch of biology. If they belong to such a set, then these generalizations may exhibit a range of stability that extends in some respects beyond the range associated with the laws of physics. In that case, the explanations supplied by these branches of biology are irreducible to explanations in terms of natural selection operating on organic chemistry.

Let us look at a biological counterfactual in context. A physician might say that the shooting victim would not have survived even if he had been brought to the hospital sooner, since the bullet punctured his aorta and the human aorta carries all of the body's oxygenated blood from the heart to the systemic circulation. (This fact about the human aorta is invariant under the counterfactual antecedent positing that the victim was brought to the hospital sooner.) Counterfactuals may also arise in connection with functional explanations (see also Wouters this volume); for instance, the human trachea has cartilaginous rings in order to make it rigid and so to keep it from collapsing between breaths. This explanation depends on the fact (expressed by a counterfactual conditional) that there would have been no

[7]What would have happened had either some birds possessed modest anti-gravity organs or the area law been violated? The correct answer is highly context-sensitive (compare: What would have happened had Caesar been in command in the Korean War?). In a context concerned with the sorts of things of interest to fundamental physics, the correct answer is that the law of gravity would still have held, and so the area law would have been violated (perhaps because no living things would have evolved). This result does not undermine the IB laws' stability for IB purposes, since this context does not matter to IB. Likewise, in a context concerned with the abundance, distribution, and evolution of species living on separated patches of habitat, the correct answer is that the area law would still have held and the law of gravity would not. This result does not undermine the fundamental physical laws' stability for the purposes of fundamental physics, since this context is not of interest to fundamental physics. This result does mean, however, that the laws of fundamental physics fail to be stable simpliciter; there is a context where a fundamental physical law would not still have held under a counterfactual supposition that is logically consistent with the fundamental physical laws. These laws are stable for the purposes of fundamental physics just as the area law is stable for IB purposes. It is not the case that the laws of fundamental physics are the real laws, whereas the "area law" is a law merely for IB purposes.

such rings if they did not make the trachea rigid. The rings' presence is not explained, for instance, by their making the trachea's outer surface white, since had cartilage been bright blue instead of white, the human trachea would still have had cartilaginous rings.

Let us think about that last counterfactual. In a context concerned with evolutionary history, it is *incorrect* to say that had cartilage been bright blue instead of white, then the human trachea would still have had cartilaginous rings. For if cartilage had been bright blue, different selection pressures might have acted upon various creatures with cartilaginous parts that are visible to predators. Evolutionary history might then have taken a different path, and so it might not have been the case that the human trachea has cartilaginous rings. Likewise, in a context concerned with molecular structure and the laws of physics, the counterfactual positing that cartilage is bright blue instead of white demands changes of some sort either in the chemical structure of cartilage or in the laws governing light's interaction with molecules. All bets are off as to what the human trachea would have been like then. Nevertheless, in certain contexts, it is correct to say that had cartilage been bright blue instead of white, the human trachea would still have had cartilaginous rings. In these contexts, it is irrelevant how cartilage could have managed to be blue.

We can now reexamine the argument that a generalization such as "The human trachea has cartilaginous rings" expresses an accident of evolution, not a law, because it would not still have held if a certain mutation had occurred or a certain selection pressure been present in the past. This is not the sort of counterfactual antecedent with which certain branches of biology are concerned. Therefore, a generalization's failure to be invariant under such an antecedent does not prevent it from belonging to a set that is stable for certain biological purposes.

Let us pay close attention to the contrast classes here. To explain why the buckeye butterfly has eyespots rather than tasting foul to birds, it is not enough to point out that the eyespot discourages predation. We would also have to identify why this particular defense mechanism evolved rather than a different one. Likewise, laws of island biogeography fail to explain why a certain particular species inhabits Grenada. Nevertheless, IB laws explain why Grenada is inhabited by a certain number of species rather than far fewer. In the same way, the fact that the buckeye butterfly's eyespot discourages predation explains why the butterfly has the eyespot rather than having no eyespot but otherwise being more or less as it actually is.[8]

The fact that the buckeye butterfly has eyespots is not a law of evolutionary biology. Nevertheless, it can possess necessity for the purposes of certain branches

[8]Had the buckeye butterfly tasted foul to birds, then it might not have sported eyespots. Here we have a counterfactual antecedent under which "The buckeye butterfly has eyespots" is not invariant. However, this result does not undermine the stability (for the purposes of certain branches of biology) of a set of reliable species-specific generalizations. That is because the counterfactual antecedent "Had the buckeye butterfly tasted foul to birds" is itself logically inconsistent with some member of that set (namely, that the buckeye butterfly does not taste foul to birds). (This is like the earlier example involving the law that the electron is stable.)

of biology by virtue of belonging to a set that is stable for those purposes.[9] Take the explanation that the vulture has no feathers on its head and neck because the vulture feeds by sticking its head and neck deep inside the bodies of carrion, so any feathers on its head and neck would become matted and dirty. This explanation is independent of the details of the laws of physics. Putnam uses a similar example to defend the irreducibility of macro explanations. He asks why a cubical peg, a bit less than an inch on a side, cannot fit into a round hole 1 in. in diameter. Putnam writes:

> The explanation is that the board is rigid, the peg is rigid, and as a matter of geometric fact, the round hole is smaller than the peg [...] That is a correct explanation whether the peg consists of molecules, or continuous rigid substance, or whatever. (Putnam 1975, p. 296)

Now a peg (or vulture) made of continuous rigid substance would violate laws of physics. But the same functional explanation would apply to it. That distinctive range of invariance reflects the irreducibility of this kind of explanation to anything that could be supplied, even in principle, by the laws of physics.

6 Conclusion

In teaching the life sciences, educators generally succeed in making the point that one of the great discoveries of modern biology is that living things are wholly physical. They are made of the same fundamental stuff as non-living things. There is no vital spark, life energy, or special kind of substance that is present exclusively in living things. The same laws of physics apply to living things as to non-living things. We can explain how living things manage to carry out various life activities by examining their biochemistry and applying the same laws that govern chemical reactions in the test tube. Living things cannot violate the conservation of energy or the second law of thermodynamics.

When teachers take this approach, they may inadvertently suggest to their students that any scientific explanations supplied by ecological generalizations or by generalizations concerning the biological properties characteristic of particular species are mere placeholders for more fundamental biochemical accounts that really do the explaining. In the foregoing sections, we have seen one way for this view to be mistaken. A biological generalization can possess a distinctive variety of necessity – can be a biological law. Associated with this distinctive necessity is a range of invariance under counterfactual antecedents that is broader in some respects than the range of invariance exhibited by the fundamental physical laws. A biological law can therefore supply explanations that cannot be replaced, even in principle, by the most fundamental physical explanations. This conclusion runs

---

[9]Brandon (1997, p. S456) and Schaffner (1993, pp. 121–122; 1995, p. 100) appear to be after roughly the same idea in referring to "historical accidentality [...] 'frozen into' a kind of quasi-nomic universality" and thus able, certain contexts, to support counterfactuals in the manner of law.

contrary to many forms of reductionism. A biological law can thus supply a distinctive kind of scientific understanding.

Higher levels of organization (organismal, ecological, sociological) can bring with them new explanations that are irreducible to the explanations available at the most fundamental level. That the life sciences can supply a distinctive kind of understanding, even though living things are made of nothing but physical stuff, should be an important lesson of education in the life sciences. It reveals why there is such a thing as biology in the first place. It is not merely impractical to study organisms, populations, communities, and ecosystems at the molecular level. Even if we could study them entirely at that level, we should not do so. That is because we would thereby miss out on distinctive ways of understanding biological phenomena.

References

Beatty, J. 1981. What is wrong with the received view of evolutionary theory? In *PSA 1980*, vol. 2, ed. P.D. Asquith and R.N. Giere, 397–426. East Lansing: Philosophy of Science Association.

Beatty, J. 1995. The evolutionary contingency thesis. In *Theories and rationality in the biological sciences*, ed. G. Wolters and J. Lennox, 45–81. Pittsburgh: University of Pittsburgh Press.

Boyd, R. 2007. Homeostasis, species, and natural taxa. In *Philosophy of science: An anthology*, ed. Marc Lange, 406–426. Malden: Blackwell.

Brandon, R. 1997. Does biology have laws? The experimental evidence. *Philosophy of Science* 64(Proc.): S444–S457.

Breyer, S.J. 2012. Opinion of the Court: Mayo Collaborative Services v. Prometheus Laboratories, Inc. http://www.supremecourt.gov/opinions/11pdf/10-1150.pdf.

Brown, J.H., and M.V. Lomolino. 1998. *Biogeography*, 2nd ed. Sunderland: Sinauer.

Futuyma, D. 1979. *Evolutionary biology*. Sunderland: Sinauer.

Giere, R. 1995. The skeptical perspective: Science without laws of nature. In *Laws of nature*, ed. F. Weinert, 120–138. Berlin: deGruyter.

Goodman, N. 1947. The problem of counterfactual conditionals. *Journal of Philosophy* 44: 113–128.

Goodman, N. 1983. *Fact, fiction and forecast*, 4th ed. Cambridge: Harvard University Press.

Gould, S.J. 1989. *Wonderful life*. New York: Norton.

Hempel, C.G. 1966. *Philosophy of natural science*. Englewood Cliffs: Prentice-Hall.

Hibbard, C.W. 1963. The origin of the P3 pattern of *Sylvilagus*, *Caprolagus*, and *Lepus*. *Journal of Mammalogy* 44: 1–15.

Hitchcock, C., and J. Woodward. 2003. Explanatory generalizations, part II: Plumbing explanatory depth. *Nous* 37: 181–199.

Hooper, E.T. 1957. Dental patterns in mice of the genus Peromyscus. *University of Michigan Museum of Zoology Miscellaneous Publications No. 99.*

Hutchinson, G.E. 1957. Concluding remarks. *Cold Spring Harbor Symposia on Quantitative Biology* 22: 415–427.

Kingsland, S. 1985. *Modeling nature*. Chicago: University of Chicago Press.

Lange, M. 2009. *Laws and lawmakers*. New York: Oxford University Press.

MacArthur, R. 1972. *Geographic ecology*. Princeton: Princeton University Press.

Mill, J.S. 1961. *A system of logic*. London: Longmans Green.

Putnam, H. 1975. Philosophy and our mental life. In *Mind, language, and reality: Philosophical papers*, vol. 2, ed. H. Putnam, 291–303. Cambridge: Cambridge University Press.

Quine, W.V.O. 1960. *Word and object*. New York: Wiley.
Reichenbach, H. 1947. *Elements of symbolic logic*. New York: Macmillan.
Rosenberg, A. 2001a. Reductionism in a historical science. *Philosophy of Science* 68(2): 135–163.
Rosenberg, A. 2001b. How is biological explanation possible? *The British Journal for the Philosophy of Science* 52(4): 735–760.
Rosenberg, A. 2001c. On multiple realization and the special sciences. *Journal of Philosophy* 98(7): 365–373.
Schaffner, K. 1993. *Discovery and explanation in biology and medicine*. Chicago: University of Chicago Press.
Schaffner, K. 1995. Comments on Beatty. In *Theories and rationality in the biological sciences*, ed. G. Wolters and J. Lennox, 99–106. Pittsburgh: University of Pittsburgh Press.
Sober, E. 1997. Two outbreaks of lawlessness in recent philosophy of biology. *Philosophy of Science* 64(Proc.): S458–S467.
Starling, E.H. 1918. *Linacre lecture on law of the heart*. London: Longmans, Green, and Co.
Steenstrup, J.J.Sm. 1845. *On the alternation of generations*. London: Ray Society.
van Fraassen, B.C. 1989. *Laws and symmetry*. Oxford: Clarendon.
Waters, C.K. 1998. Causal regularities in the biological world of contingent distributions. *Biology and Philosophy* 13(1): 5–36.
Weber, M. 1999. The aim and structure of ecological theory. *Philosophy of Science* 66: 71–93.

The Nature of Evolutionary Biology: At the Borderlands Between Historical and Experimental Science

Massimo Pigliucci

> Even today a good many distinguished minds seem unable to accept or even to understand that from a source of noise natural selection could quite unaided have drawn all the music of the biosphere.
>
> Jacques Monod

1 On the Scientific Status of Evolutionary Theory

The scientific status of evolutionary theory seems to be more or less perennially under question. I am not referring here (just) to the silliness of young Earth creationism (Pigliucci 2002; Boudry and Braeckman 2010), or even of the barely more intellectually sophisticated so-called Intelligent Design theory (Recker 2010; Brigandt this volume), but rather to discussions among scientists and philosophers of science concerning the epistemic status of evolutionary theory (Sober 2010). As we shall see in what follows, this debate has a long history, dating all the way back to Darwin, and it is in great part rooted in the fundamental dichotomy between what French biologist and Nobel laureate Jacques Monod (1971) called chance and necessity—i.e., the inevitable and inextricable interplay of deterministic and stochastic mechanisms operating during the course of evolution. In turn, this discussion reaches as far as our very concept of what counts as a science and why, with the perennial struggle between primarily or exclusively experimental sciences (e.g., physics) on one hand and chiefly historical sciences (e.g., paleontology) on the other. While the two issues (deterministic vs. stochastic phenomena and experimental vs. historical sciences) are

M. Pigliucci (✉)
Philosophy Program, The Graduate Center, City University of New York, New York, NY, USA
e-mail: massimo@platofootnote.org

K. Kampourakis (ed.), *The Philosophy of Biology: A Companion for Educators*, History, Philosophy and Theory of the Life Sciences 1, DOI 10.1007/978-94-007-6537-5_5,

logically distinct, as we shall see they overlap in practice, and for philosophically interesting reasons.

Hull (1973) pointed out that evolutionary theory got under the fire of philosophers of science from the onset, with the publication of Darwin's (1859) masterpiece. In particular, Darwin's famous "long argument" laying out the foundations of the new field of evolutionary biology, was seen as ill-fitting with both of the major philosophical views of how science works that were being debated in late nineteenth century Victorian England. This is known as the great induction debate, and featured a who's who of early philosophy of science, with John Stuart Mill and William Whewell (the man who coined the term "scientist" in 1834), both attempting to improve on the model of scientific reasoning articulated by their predecessor, Francis Bacon. Famously, Bacon (1620/2000) had published a provocatively titled book, *The New Organon*, in which he set out to criticize what he took as Aristotle's view (expressed in the latter's *Organon*) that science proceeds by way of deductive reasoning. Bacon thought that deduction wouldn't do, because the premises of any deductive argument have to be arrived at by way of empirical evidence, which means that at bottom science relies on a different type of reasoning, induction.

The problem is that there are several types of induction, and the one Bacon mostly focused on, enumerative induction, suffers from severe problems of its own. Essentially, enumerative induction is the everyday practice of generalizing from a small number of instances to a broader context, a practice that Hume (1748/1952) quickly realized is without independent logical foundations and open to spectacular failures. Hence, the debate between Mill and Whewell on how to improve on Bacon's proposal. Mill thought that enumerative induction could be strengthened by invoking a law of universal causation, according to which similar effects are produced by similar causes, an example of parsimony applied to scientific explanations. Mill also added eliminative induction, sometimes referred to as "strong inference" (Platt 1964), where alternative hypotheses are eliminated in succession, based on stringent tests or crucial experiments.

Both Bacon and Mill gave preeminence to data over hypotheses, while Whewell's approach emphasized the idea that a scientific investigation has to start with some hypothesis, even if approximate, to serve as a heuristic guide to data collection and further hypothesizing. Whewell thought that progress was made by consilience, a situation where the data consistently point to one particular hypothesis being true over its rivals. This way of reasoning is also known with the somewhat strange name of abduction, or inference to the best explanation.

Darwin got in trouble with both Mill and Whewell because they both regarded Darwin's work as an example of deductive reasoning, not of induction (of any kind). According to Mill and Whewell's understanding, what Darwin had done was to arrive at a hasty generalization based on a small number of observations, proceeding then (deductively) to derive consequences from them, and finally collecting data for decades to back up his hasty conclusions. This superficially appeared to be almost a textbook example of what Bacon said one should not do in science. Evolutionary biology, particularly when compared to the already successful Galilean and Newtonian physics, had already gotten off on the wrong foot. Needless to say, Darwin was taken aback by all this, particularly by Whewell's criticism. Indeed, in apparent frustration, Darwin wrote to a friend: "How odd it is that anyone should not see that all observation must be for or against some view if it is to be of service!" (Darwin, F. 1903, p. 240), a sentiment that is actually perfectly consonant with Whewell's idea of inference to the best explanation (and, indeed, is also compatible with Mill's eliminative induction, to a point). In fact, Michael Ruse (1975, 2000) has persuasively argued that Darwin consciously tried to develop his theory in accordance with the best philosophy of science of his time, particularly following the views of Whewell and John Herschel.

The reason these early skirmishes are important is because they set the stage for much discussion over the following century and a half about the goals and methods of evolutionary biology. Moreover, as we shall see below, the distinction between eliminative induction and inference to the best explanation parallels, roughly, the distinction between the methods of the experimental and historical sciences as determined by their different epistemologies.

2 The Fisher-Wright Debates and the Importance of Stochastic Events in Evolution

The next big battle for the soul of evolutionary biology was the famous, decade-long, debate between two of the founding fathers of population genetics, Ronald Fisher and Sewall Wright. The history of this debate, and the role it played in population genetics, is both complex and fascinating, and it has been examined in detail elsewhere (Provine 2001). Fisher was convinced of the absolute preeminence of natural selection in shaping organismal evolution, so much so that he consciously modeled his famous fundamental theorem (Frank and Slatkin 1992; Okasha 2008) after the second principle of thermodynamics, one of the most successful laws formulated within the dominant experimental science, physics.

Wright, on the other hand, was attracted by the complexity and messiness of biology, and his emphasis was always on non-linear, non-additive genetic effects (epistasis, pleiotropy) as well as, most famously, on the role of genetic drift in evolution. Wright built drift into his innovative theory of "shifting balance", meant to account for how deterministic selection and stochastic drift combine to allow populations to explore new "peaks" in what Wright defined as the "adaptive landscape" (Pigliucci 2008). The status of shifting balance in evolutionary theory is itself under perennial discussion (Wade 1992; Coyne et al. 1997), though there seems to be an emerging sense that the mechanism, while theoretically feasible, is unlikely to play a major role in the evolution of actual biological lineages. Still, the concept of drift has become incorporated into standard population genetic theory, and has even been the focus of detailed, and still unsettled, philosophical analyses (e.g., Pigliucci and Kaplan 2006; Millstein et al. 2009).

Indeed, in a recent attempt at a major shift in perspective, Lynch (2007) has forcefully pushed the idea that a great deal of genome evolution at the structural level (e.g., how gene networks are formed and change over time) is likely the result of drift, not of selective mechanisms, as usually assumed. Whether and to what extent Lynch is correct remains to be seen, but the "problem" of distinguishing drift from selection in now standard in biological research (e.g., Chapuis et al. 2008; Hofer et al. 2008).

It is important to understand that "drift," although often portrayed as a "force" affecting the equilibrium state of natural populations (together with the other four canonical "forces" that can cause displacement from Hardy-Weinberg equilibrium, selection, mutation, migration and recombination: Hartl and Clark 1997, pp. 48–52; see also Depew this volume), actually is a beast of a very different kind from natural selection. Perhaps the best way to make the point clear is to use the standard analogy between genetic drift and the kind of statistical drift one commonly observes when examining series of coin tosses of equal length. Assuming that the coins are fair (i.e., there is no weight biasing the outcome toward either tail or head), the expected outcome of a series is, of course, 50–50 %. This, however, is true only asymptotically, as the series reaches infinite length—a direct analogy with the oft-made assumption of "infinite" (i.e., not affected by drift) population size in population genetic models. The shorter the series, the more likely it is that the actual outcome is going to deviate (in either direction, with random frequency) from the expected one. The likelihood and intensity of these deviations become increasingly high as the coin toss series becomes shorter and shorter. It is in a very similar sense that biologists speak of drift getting "stronger" in smaller and smaller populations.

But notice a rather odd thing about this account of drift: if one focuses on individual coin tosses, no quantity is changing at all. For each individual toss the chances of landing tail or head is always the same (*ceteris paribus* as far as environmental conditions are concerned, including the strength and mode of the toss). There is no mystery here, of course: drift is a phenomenon that manifests itself only when a *population* of objects is concerned; it is undefined at the individual level. But this is different from all the other "forces" that can push a population off Hardy-Weinberg equilibrium: mutations, recombination, migration and selection happen at the *individual* level, though of course their *outcomes* may be conveniently measured statistically for entire populations. There is, therefore, something strange going on when biologists want to know the relative strength of selection vs. drift in a population, considering that they are comparing mechanisms acting at two different levels of the biological hierarchy.

Jonathan Kaplan and I (Pigliucci and Kaplan 2006) have suggested a way to conceptualize what is happening here, an approach that might be helpful as part of a general understanding of the roles of chance and necessity in evolution. Instead of thinking of drift as a force antagonistic to selection, we can conceptualize it as a measure of the "error" surrounding the expected evolutionary change caused by selection. So, if the target of selection is a particular phenotypic value of a particular trait (in a particular environment, of course), then there is a probability distribution that tells us how likely the population actually is to land on that phenotypic target.

The smaller the population (the higher the drift), the broader the error will be, with increasing chance that the population will end up anywhere in an expanding circle centered on the actual target. This is a different, and we think more satisfying, account of how selection and drift interact, because it gets away from the "force" metaphor (another reason for doing which is that drift does not have the properties of a vector, so it cannot reasonably be represented as a force; see also Depew this volume about whether natural selection should be described as a force). It also makes explicit that we can answer the drift vs. selection question only in terms of *outcomes* at the population level, the question being meaningless at the individual level because individuals experience selection, but not drift.

3 Gould and the Project for a Nomothetic Evolutionary Biology

The latter part of the twentieth century saw the opening of a different front in the ongoing discussion about the relative role of chance events in evolutionary biology, as well as about the status of the discipline as historical and yet scientifically fully mature. The main charge was led by Stephen Gould and his associates, with a series of papers that unleashed decades of debates and new research—a voluminous output that makes for what philosopher Imre Lakatos called a successful (as opposed to a degenerate) research program in the light of his treatment of the nature of science (Lakatos 1978).

The opening salvo by Gould and company was the famous 1972 paper on punctuated equilibria (Eldredge and Gould 1972), where the standard Darwinian view of gradual evolution was challenged and, by implication, the role of natural selection in shaping long-term evolution somewhat curtailed. Gould then developed that theme in his highly influential book, *Ontogeny and Phylogeny* (Gould 1977), where the idea is put forth that developmental constraints play a major role in both antagonizing and sometimes even facilitating natural selection. Because constraints are themselves the result of past history, and hence also of stochastic events, this helped shape an alternative and broader view of evolutionary phenomena, one where organismal biology (including of course paleontology) would reclaim a place at the high table, so to speak (Prothero 2009) in virtue of the renewed focus on historicity and chance events. Indeed, Gould even partially rethought the role of natural selection itself by introducing a new term in the evolutionary jargon, exaptation, to indicate situations where currently advantageous traits result from selection co-opting previously existing structures, which themselves evolved either for other functions or as the result of constraints (Gould and Vrba 1982; see also Forber this volume). In a sense, exaptation is what one gets when chance and necessity work sequentially.

Gould's most conscious attempt at articulating a new view for paleontology in particular, and for historical biological science in general, came with his paper on nomothesis (Gould 1980). In philosophy of science nomothetic means law-like (see Lange this volume), which is somewhat puzzling because Gould's emphasis throughout his career was on the importance of stochastic, not deterministic causes in evolution. But Gould was also aware that one needs to combine stochasticity and determinism if one wishes to obtain a reasonably complete view of biological phenomena. His attempt at nomothesis took the shape of a novel approach to the study of why certain forms and not others appear throughout the history of any given lineage. The basic observation is obvious enough: not all conceivable forms of, say, shelled animals are in fact found in the fossil record. Why not? There is more than one conceivable answer to the question. Perhaps some forms did evolve, briefly, but went extinct. Or, perhaps, some shell shapes may simply be intrinsically disadvantageous, regardless of the specific environment, and therefore are constantly selected against. Finally, forms would be selected in favor if they could be generated, but perhaps the genetic-developmental systems, characteristic of certain lineages, simply are incapable of producing the right kind of variation.

The idea began with paleontologist David Raup (Raup and Gould 1974), who formulated a theoretical space of shell forms (whimsically called by Gould "the cube of life") and then filled it in with actual living or fossil shells to see which parts of the space have historically been occupied and how frequently. What is important to notice here is that Raup and Gould did not construct the space of forms *a posteriori*, i.e. from actual observations of shells. It was derived a priori using a simple equation that describes the growth curve of a shell. The equation has three basic parameters, which constitute the three axes of the cube of life, and it is the variation of these parameters that generates all hypothetical shell forms. Only afterward the researcher plots actually observed shell types on the same space, and it is the comparison and differences between the hypothetical space and the real organisms that is of interest. Of particular relevance, of course, are the areas of the cube of life occupied by theoretically possible, but never realized, shell shapes. Once these are identified, the scientist can bring in the result of functional analyses (for instance about buoyancy, or strength of the shell in response to predators, etc.) to account for at least some of the mismatches between hypothetical and actual spaces.

A particularly interesting example of the application of this approach is recounted in detail by McGhee (2006). The work was again started by Raup (1967), who explored a mathematical-geometrical space of ammonoid forms defined by two variables: W, the rate of expansion of the whorl of the shell; and D, the distance between the aperture of the shell and the coiling axis. As McGhee shows in his detailed discussion of this example, Raup arrived at two simple equations that can be used to generate pretty much any shell morphology that could potentially count as "ammonoid-like," including shells that—as far as we know—have never actually evolved in any ammonoid lineage. Raup then moved from theory to empirical data by plotting the frequency distribution of 405 actual ammonoid species in W/D space and immediately discovered two interesting things: first, the distribution had an obvious peak around $0.3<D<0.4$ and $W\approx 2$. Consider that this kind of peak is not a direct measure of fitness or adaptation, it is simply a reflection of the actual occurrence of certain forms rather than others. Second, the entire distribution of ammonoid forms was bounded by the $W=1/D$ hyperbola, meaning that few if any species crossed that boundary on the morphospace. This was interesting, since the 1/D line represents the limit in morphospace where whorls still overlap with one another. This means that for some reason very few ammonites ever evolved shells in which the whorls did not touch or overlap.

Raup's initial findings were intriguing, but they were missing a sustained functional analysis that would account for the actual distribution of forms in W/D space. Why one peak, and why located around those particular coordinates? Here is where things become interesting and the morphospace approach delivers much more than just heuristic value. John Chamberlain, a student of Raup, carried out experimental work to estimate the drag coefficient of the different types of ammonoid shells. His first results (Chamberlain 1981) clarified why most actual species of ammonoids are found below the W = 1/D hyperbola: as one would expect, shells with overlapping whorls have a significantly lower drag coefficient, resulting of course in more efficiently swimming animals.

However, Chamberlain also found something else rather intriguing: the experimental data suggested that there should be *two* regions of the W/D morphospace corresponding to shells with maximum swimming efficiency, while Raup's original frequency morphospace detected only one peak. It seemed that for some reason natural selection evolved one peak, but not the other. Four decades had to pass from Raup's initial paper for the mystery of the second peak to be cleared up: the addition of 597 new species of ammonoids to the original database showed that indeed the second peak had also been occupied! This is a rather spectacular case of confirmed prediction in evolutionary biology, not exactly a common occurrence, particularly in paleontology, and a superb vindication of Raup's and Gould's idea that historical disciplines where stochastic events play a major role can still be nomothetic in an interesting and sometimes even decidedly surprising way.

4 The Modern Study of Chance vs. Necessity

Part of the problem with the antithesis between chance and necessity—to put it as Monod did—or stochasticity and determinism, is that the best we can usually do is to study their relative importance only retroactively, attempting to determine after the fact the respective contributions of selection, drift and constraints on already realized evolutionary trajectories. During their long sparring careers, Richard Dawkins and Stephen Gould have often argued about what would happen if one could "rewind the tape of life," to use Gould's metaphor (Beatty 2006; Sterelny 2007). The outcome of such a thought experiment was very different according to the two antagonists: Dawkins suggested that we, as *Homo sapiens* might not end up evolving again, but *surely* some sort of hyper-intelligent, big-brained biped would. Gould, on the contrary, suggested that for all we know not even vertebrates might have re-evolved, depending on minor vagaries of the Burgess Shale fauna, back in the Middle Cambrian (540 million years ago). Or perhaps planet Earth may never have gotten past a biosphere made entirely of bacteria and blue algae. Of course, the problem is that there actually is no way to rewind the tape of life, so that any such discussion amounts to speculation largely reflecting the intellectual prejudices of the speculators.

However, recent research on long-term laboratory evolution in microorganisms has offered us the possibility to replay short segments of the tape of life, over and over again while manipulating the conditions to see what happens. A full review of that literature is beyond the scope of this chapter, but I will briefly discuss a couple of examples, to provide a flavor of the possibilities opened up by this approach—arguably a powerful window into the relative importance of randomness and determinism, at least as far as simple and rapidly evolving organisms are concerned.

The first example comes from work done on *Chlamydomonas reinhardtii*, a single celled green alga, by Graham Bell's group at McGill University (Collins et al. 2006). The group began with populations of *Chlamydomonas* that had previously been selected for survival at high levels of CO_2, imposing on them new selective pressures to adapt them either to current or to Pleistocene CO_2 levels (i.e., lower and much lower than the starting point levels, respectively). Bell and colleagues consciously patterned their experiment after Gould's "tape of life" metaphor, even using the phrase in the title of the paper. Their results showed that *Chlamydomonas* had no trouble re-adapting to current CO_2 levels, or indeed to Pleistocene levels. In that sense, the tape of life could indeed be rewinded with similar results. However, the process also produced a range of phenotypes in the replicate populations, an outcome that Collins et al. interpreted as indicative of the importance of stochastic events in shaping the details of the evolutionary trajectories. As they put it: "There was no effect of selective history on adaptation [...] phenotypes were evolutionarily reversible. [However] Adaptation produced a range of phenotypes, suggesting that chance rather than selective history contributes to the divergence of replicate populations in this system" (Collins et al. 2006, p. 1392).

The approach taken by Woods et al. (2006), within the context of work done by Richard Lenski's lab at Michigan State University, was different yet aiming at the same problem of replicability of evolutionary outcomes. Lenski and colleagues began with 12 identical lines of the bacterium *Escherichia coli*, and let them evolve under the same conditions for about 20,000 generations. The idea was to track the molecular evolution of four pre-identified gene sequences to see if and how they would diverge in the replicated lines. Woods et al. found that two of the genes had substitutions in all populations, while the other two genes had substitutions in several but not all populations. Statistical tests were congruent with the conclusion that the parallel changes in the candidate genes were driven by natural selection, but the researchers found few cases in which the exact same mutations were substituted—thereby documenting a significant role also for stochastic events.

Both experiments were of course likely to underestimate the relevance of stochasticity, because they were carried out by design under uniform environmental conditions. Yet, even under those conditions chance clearly affected the evolutionary outcome. It also has to be noted that one of Gould's major sources of interference with natural selection, developmental constraints, could not play any role in either experiment, for the simple reason that neither *Chlamydomonas* nor *E. coli* have any development to speak of. Regardless, Woods et al. found differences among the replicate populations that were not found in a conceptually similar experiment in viruses, presumably because the *genetic* constraints on viral sequences are stronger than those acting on the much larger genome of *E. coli*. This is consistent with the general idea that larger genomes (and eventually actual developmental systems) allow more room for stochastic events, as well as for a more complex interplay of chance and necessity. At the same time, though, it is hard to imagine how similar experiments could be done on multicellular eukaryotes, the sort of organisms that are really at the center of Dawkins-Gould type debates. Nonetheless, these and more recent (e.g., Saxer et al. 2010) experiments clearly provide us with an empirical, if limited, handle on the question.

5 The Philosophical Context: Cleland's Analysis

Much of the preceding discussion—like any discussion on the nature of science—reflects a certain philosophy of what it means to be doing experimental or historical science. It is time to explicitly deal with that philosophy. An excellent framework for it is provided by a paper published by Carol Cleland (2002; see also Cleland 2011 for her recent elaboration and rebuttal to critics) on the epistemic differences between historical and experimental sciences. Cleland's pivotal idea is that the two types of science are separated by what she calls an asymmetry of overdetermination. Building on previous work by David Lewis, she explains that "the basic idea is that localized present events overdetermine their causes and underdetermine their effects." She elucidates the concept by considering the example of a crime being investigated. Once committed, a crime leaves a number of historical traces, no matter how careful the perpetrator was in erasing as many of them as possible. All it takes for a criminal investigator to figure out what happened is a relatively small number of traces that clearly enough point toward a particular sequence of events. The investigator would then be using a type of induction known as inference to the best explanation to pinpoint the culprit (the same one we have seen argued for by Whewell early on). Conversely, the simple act of *not* committing the crime obviously instantly erases the possibility of any historical trace to be left around. Few currently available clues *overdetermine* a past event, while so many futures are possible given a particular current state of things that the latter *underdetermines* the range of futures.

Cleland cashes in this asymmetry of overdetermination by arguing that—contrary to popular wisdom (and to the opinion of many practicing scientists)—there is nothing inherently epistemically superior about experimental over historical science. This is because of two consequences of overdetermination. On the one hand, while experimental scientists have the ability to strictly control the conditions of their experiments, it is that very strictness that limits the scope of applicability of their results: as soon as one widens the settings of a given experiment, different factors begin to interact with each other in complex ways, quickly leading to a large number of possible future outcomes; in other words, predictability is purchased at the expense of generality. On the other hand, while historical traces constantly decay through time, and may disappear forever, the historical scientist often needs only a small amount of them to arrive at a sufficiently accurate reconstruction of what happens—just like the criminal investigator in the hypothetical example of the impossibility of a perfect crime.

To make things more concrete, Cleland's account makes sense of some surprising limitations of experimental science, as well as some spectacular successes of historical science. In the first case, it is notable, for instance, that non-equilibrium thermodynamics (for example in its applications to atmospheric physics and climate science) quickly reaches a limit in terms of predictive ability, where complex mathematical models are incapable of generating more than very approximate statistical predictions about the future behavior of complex systems, predictions often accompanied by rapidly expanding margins of error. In the second case, however, we have for example the success of paleontologists in determining that an extraterrestrial body of massive proportions hit the Earth 65 million years ago, contributing to the extinction of countless numbers of species, chief among them the dinosaurs. The impact was suspected once geologists discovered a worldwide thin layer of iridium in rocks datable to the K/T (Cretaceous/Tertiary) boundary. This led to a search for a crater, the remnants of which were eventually identified off the Yucatan peninsula via satellite imagery. From there, geologists could calculate the size and direction of the impact, and therefore make fresh predictions concerning additional historical traces, for instance those left by the tsunamis that must have hit the western coast of Mexico as a result of the asteroid crush. Sure enough, those traces were found, leading to even more confidence in the conclusion that "the crime" had indeed taken place in the way it had been hypothesized.

Cleland's framework is particularly helpful for a re-evaluation of the chance vs. necessity issue in evolutionary biology. By its very nature, evolutionary studies are *both* experimental (consider again the examples from the Bell and Lenski labs discussed above), as well as historical (Gould's and Raup's reconstruction of morphospaces affecting the evolution of shelled animals). While biologists may not be able to conduct meaningful experiments on *the same* systems for which they have abundant historical information (*E. coli* are certainly different from ammonoids), the fact that they do have access to vast repositories of historical information *and* that they can conduct controlled experiments, albeit within limits, puts evolutionary biology in a position of epistemic advantage over purely historical and possibly even purely experimental sciences, as long as its dual historical/experimental nature is properly understood and adequately exploited by its practitioners.

6 Conclusion: Chance and Necessity Within the Extended Synthesis

Jacques Monod's (1971) analysis of evolutionary theory and the philosophy of biology—with his dualism of chance and necessity—appeared at a strange juncture in the history of biology and of philosophy of science. In the late 1960s and early

1970s evolutionary biologists were satisfied with the Modern Synthesis of the 1930s and 1940s (Huxley 1942/2010), and both the explosion of evo-devo (evolution of development: Love 2009; Love this volume) and of evolutionary genomics (Lynch 2007) were still to come. At the same time, philosophy of science was largely dominated by the philosophy of physics, with philosophy of biology emerging as a mature field only in the 1980s and 1990s.

The situation has, of course, changed dramatically since. Not only is philosophy of biology arguably the dominant sub-field in philosophy of science, or at the very least a major player, but evolutionary biology finds itself again facing a moment of re-evaluation of the basic structure of its theory. There are no serious suggestions that the original Darwinian core, or its expanded outlook within the Modern Synthesis, are in danger of being undermined or rejected, *pace* the creationists. However, an increasing number of biologists and philosophers are convinced that a new, Extended Synthesis is taking place in evolutionary biology (Pigliucci and Müller 2010).

It remains to be seen what sort of Extended Synthesis will emerge over the next decade or so, but one of the elements that seems pretty much certain to be a part of it is in fact a serious and more nuanced consideration of the balance between deterministic and stochastic events in evolution. This is arguably in good part a result of Gould's legacy (particularly of his still somewhat under-appreciated last book: Gould 2002), but also of the onset of evo-devo, which puts constraints at the forefront of its agenda (Müller 2007), and of increasing evidence that large aspects of genomic evolution are more likely the result of stochastic events than of fine-tuning orchestrated by natural selection (Lynch 2007).

Gould (2002) in particular argued that we can follow the evolution of evolutionary theory by considering how our ideas about natural selection—one of the two cardinal concepts that make up the original Darwinian core (the other one being common descent)—change in light of three criteria: agency, efficacy and scope. *Agency* refers to where natural selection acts: genes, individuals, groups, species, etc. In this respect, the moves from core Darwinism to the Modern Synthesis and to an Extended Synthesis have seen the agency of selection increase, as more levels of biological organization are recognized as being possible targets of selection (Okasha 2006). *Efficacy* refers to the relative power of natural selection when compared to other evolutionary mechanisms. Here the story is different, since the Modern Synthesis introduced several additional evolutionary mechanisms, reflected in the core mathematical treatment of population genetics (Hartl and Clark 1997). Particularly due to the work of Sewall Wright and his followers, it is clear that drift is here to stay as a major counter to the efficacy of selection. Gould's emphasis on developmental constraints falls in the same category, the two efforts resulting in an augmented role of chance in evolution. Finally, *scope* is the degree to which microevolutionary phenomena can be extrapolated to macroevolutionary ones. Notoriously, we saw no change here when biologists moved from Darwinism to the Modern Synthesis, particularly after the so-called "hardening" of the Synthesis that involved a much reduced role of paleontology. But the Extended Synthesis is bound to take seriously the work of Gould and his followers (Jablonski 2005), resulting in a partial decoupling of micro- and macro-evolution, a decoupling realized by the intervention of stochastic processes (which make it impossible to simply extrapolate macroevolutionary patters from microevolutionary phenomena), and hence by a surprisingly anti-reductionist role of stochasticity (after all, asteroids wiping out entire ecosystems are hardly the stuff of molecular biology).

These are exciting times for evolutionary biology, both in terms of empirical discoveries (evo-devo, comparative genomics) and conceptual advances (e.g., discussions of evolvability, emergent complexity, and the like: Brigandt 2007). And while it is true, as Monod said, that "even today a good many distinguished minds seem unable to accept or even to understand that from a source of noise natural selection could quite unaided have drawn all the music of the biosphere," that understanding is getting richer and deeper, and it still hinges on taking seriously the dichotomy and interaction between randomness and determinism.

What, then, are the implications of all of this for the teaching of evolutionary biology? The standard approach, at both pre-college and introductory college levels, is to teach students largely about micro-evolution, probably because of the inherent prestige of population genetics—the most mathematical of the sub-branches of evolutionary biology. Students are therefore exposed to the centrality of natural selection as an evolutionary mechanism, as well as to the focal idea that evolution can be defined in terms of changes in gene frequencies.

What does not get much attention is the broader picture of evolution offered by a consideration of macro-evolutionary patterns and processes, and particularly an explicit emphasis on stochasticity and on the consequences of rare but high impact events that take place during the history of life. Indeed, Catley (2006) has identified this as a major deficiency in the teaching of evolution, and has proposed a radical shift toward a much more balanced exposure of students to both micro- and macro-evolutionary concepts. The sort of historical and philosophical discussions about chance vs. necessity that I have briefly outlined in this chapter provide precisely the sort of additional intellectual exposure that Catley is advocating.

Moreover, the standard approach to teaching biology is notoriously missing in conceptual and philosophical themes of the type developed, for instance, through Cleland's analysis of the asymmetry of overdetermination. This is unfortunate not just from the standpoint of general educational principles (e.g., that science education—especially for students who are not likely to pursue science at a more advanced level—should be about the "big picture" and not a host of technical details). It is a deficiency highlighted by empirical evidence that a better appreciation of the nature of science itself is correlated with and facilitates the development of better understanding of evolutionary biology (Nadelson and Sinatra 2010). It may very well be, as Dobzhansky famously put it, that nothing (in the teaching of) biology makes sense except in the light of evolution. But it makes increasingly less sense to teach evolution within a narrow perspective and without proper historical and philosophical contexts.

References

Bacon, F. 1620/2000. *The new organon*. Cambridge: Cambridge University Press.
Beatty, J. 2006. Replaying life's tape. *Journal of Philosophy* 103: 336–362.
Boudry, M., and J. Braeckman. 2010. Immunizing strategies and epistemic defense mechanisms. *Philosophia* 39(1): 145–161.
Brigandt, I. 2007. Typology now: Homology and developmental constraints explain evolvability. *Biology and Philosophy* 22: 709–725.
Catley, K.M. 2006. Darwin's missing link—A novel paradigm for evolution education. *Science & Education* 90: 767–783.
Chamberlain, J.A. 1981. Hydromechanical design of fossil cephalopods. *Systematics Association Special Volume* 18: 289–336.
Chapuis, E., G. Martin, and J. Goudet. 2008. Effects of selection and drift on G matrix evolution in a heterogeneous environment: A multivariate Qst–Fst test with the freshwater snail *Galba truncatula*. *Genetics* 180: 2151–2161.
Cleland, C.E. 2002. Methodological and epistemic differences between historical science and experimental science. *Philosophy of Science* 69: 474–496.
Cleland, C.E. 2011. Prediction and explanation in historical natural science. *The British Journal for the Philosophy of Science* 62: 551–582.
Collins, S., D. Sultemeyer, and G. Bell. 2006. Rewinding the tape: Selection of algae adapted to high CO_2 and current and Pleistocene levels of CO_2. *Evolution* 60: 1392–1401.
Coyne, J.A., N.H. Barton, and M. Turelli. 1997. A critique of Sewall Wright's shifting balance theory of evolution. *Evolution* 51: 643–671.
Darwin, C. 1859. *On the origin of species by means of natural selection: Or, the preservation of favoured races in the struggle for life*. London: John Murray.
Darwin, F. 1903. *More letters of Charles Darwin. A record of his work in a series of hitherto unpublished letters*. London: John Murray.
Eldredge, N., and S.J. Gould. 1972. Punctuated equilibria: An alternative to phyletic gradualism. In *Models in paleobiology*, ed. T.J.M. Schopf, 82–115. San Francisco: Freeman.
Frank, S.A., and M. Slatkin. 1992. Fisher's fundamental theorem of natural selection. *Trends in Ecology & Evolution* 7: 92–95.
Gould, S.J. 1977. *Ontogeny and phylogeny*. Cambridge: Harvard University Press.
Gould, S.J. 1980. The promise of paleobiology as a nomothetic, evolutionary discipline. *Paleobiology* 6: 96–118.
Gould, S.J. 2002. *The structure of evolutionary theory*. Cambridge: Harvard Press.
Gould, S.J., and E.S. Vrba. 1982. Exaptation—A missing term in the science of form. *Paleobiology* 8: 4–15.
Hartl, D.L., and A.G. Clark. 1997. *Principles of population genetics*. Sunderland: Sinauer.
Hofer, T., N. Ray, D. Wegmann, and L. Excoffier. 2008. Large allele frequency differences between human continental groups are more likely to have occurred by drift during range expansions than by selection. *Annals of Human Genetics* 73: 95–108.
Hull, D. 1973. *Darwin and his critics: The reception of Darwin's theory of evolution by the scientific community*. Cambridge: Harvard University Press.
Hume, D. 1748/1952. *An enquiry concerning human understanding*. Charleston, SC: Forgotten Books.
Huxley, J. 1942/2010. *Evolution: The modern synthesis: The definitive edition*. Cambridge: MIT Press.
Jablonski, D. 2005. Evolutionary innovations in the fossil record: The interaction of ecology, development and macroevolution. *The Journal of Experimental Zoology* 304B: 504–519.
Lakatos, I. 1978. *The methodology of scientific research programmes*. Cambridge: Cambridge University Press.
Love, A. 2009. Marine invertebrates, model organisms, and the modern synthesis: Epistemic values, evo-devo, and exclusion. *Theory in Bioscience* 128: 19–42.

Lynch, M. 2007. *The origins of genome architecture*. Sunderland: Sinauer.
McGhee, G.R. 2006. *The geometry of evolution: Adaptive landscapes and theoretical morphospaces*. Cambridge: Cambridge University Press.
Millstein, R.L., R.A. Skipper, Jr., and M.A. Dietrich. 2009. (Mis)interpreting mathematical models: Drift as a physical process. *Philosophy & Theory in Biology*. http://hdl.handle.net/2027/spo.6959004.0001.002.
Monod, J. 1971. *Chance and necessity: An assay on the natural philosophy of modern biology*. New York: Knopf.
Müller, G. 2007. Evo-devo: Extending the evolutionary synthesis. *Nature Reviews Genetics* 8: 943–949.
Nadelson, L.S., and G.M. Sinatra. 2010. Shifting acceptance of evolution: Promising evidence of the influence of the understanding evolution website. *The Researcher* 23: 13–29.
Okasha, S. 2006. *Evolution and the levels of selection*. Oxford/New York: Oxford University Press.
Okasha, S. 2008. Fisher's fundamental theorem of natural selection—A philosophical analysis. *The British Journal for the Philosophy of Science* 59: 319–351.
Pigliucci, M. 2002. *Denying evolution. Creationism, scientism, and the nature of science*. Sunderland: Sinauer.
Pigliucci, M. 2008. Sewall Wright's adaptive landscapes: 1932 vs. 1988. *Biology and Philosophy* 23: 591–603.
Pigliucci, M., and J. Kaplan. 2006. Natural selection and fitness: after the "force" metaphor. In *Making sense of evolution. The conceptual foundations of evolutionary biology*, 13–35. Chicago: University of Chicago Press.
Pigliucci, M., and G. Müller. 2010. *Evolution, the extended synthesis*. Cambridge: MIT Press.
Platt, J.R. 1964. Strong inference. *Science* 146: 347–353.
Prothero, D. 2009. Stephen Jay Gould: Did he bring paleontology to the "High Table"? *Philosophy & Theory in Biology*. http://hdl.handle.net/2027/spo.6959004.0001.001.
Provine, W.B. 2001. *The origins of theoretical population genetics*. Chicago: University of Chicago Press.
Raup, D.M. 1967. Geometric analysis of shell coiling: Coiling in ammonoids. *Journal of Paleontology* 41: 43–65.
Raup, D.M., and S.J. Gould. 1974. Stochastic simulation and evolution of morphology—Towards a nomothetic paleontology. *Systematic Zoology* 23: 305–322.
Recker, D. 2010. How to confuse organisms with mousetraps: Machine metaphors and intelligent design. *Zygon* 45: 647–664.
Ruse, M. 1975. Darwin's debt to philosophy: An examination of the influence of the philosophical ideas of John F.W. Herschel and William Whewell on the development of Charles Darwin's theory of evolution. *Studies in History and Philosophy of Science Part A* 6: 159–181.
Ruse, M. 2000. Darwin and the philosophers: Epistemological factors in the development and reception of the origin of species. In *Biology and epistemology*, ed. R. Creath and J. Maienschein, 3–26. Cambridge: Cambridge University Press.
Saxer, G., M. Doebeli, and M. Travisano. 2010. The repeatability of adaptive radiation during long-term experimental evolution of *Escherichia coli* in a multiple nutrient environment. *PLoS One* 5: e14184.
Sober, E. 2010. Natural selection, causality, and laws: What Fodor and Piattelli-Palmarini got wrong. *Philosophy of Science* 77: 594–607.
Sterelny, K. 2007. *Dawkins vs. Gould: Survival of the fittest*. Cambridge: Icon.
Wade, M.J. 1992. Sewall Wright: Gene interaction and the shifting balance theory. In *Oxford surveys in evolutionary biology*, vol. 8, ed. D.J. Futuyma and J. Antonovics, 35–62. Oxford: Oxford University Press.
Woods, R., D. Schneider, C.L. Winkworth, M.A. Riley, and R.E. Lenski. 2006. Tests of parallel molecular evolution in a long-term experiment with *Escherichia coli*. *Proceedings of the National Academy of Sciences of the United States of America* 103: 9107–9112.

Evolutionary Theory and the Epistemology of Science

Kevin McCain and Brad Weslake

1 Introduction

Evolutionary theory is a paradigmatic example of a well-supported scientific theory. In this chapter we consider a number of objections to evolutionary theory, and show how responding to these objections reveals aspects of the way in which scientific theories are supported by evidence. Teaching these objections can therefore serve two pedagogical aims: students can learn the right way to respond to some popular arguments against evolutionary theory, and they can learn some basic features of the structure of scientific theories and evidence.

We begin, in Sect. 2, with some general remarks about epistemology (the theory of knowledge) in order to help frame our discussion. After these brief remarks, in Sect. 3 we turn to the objections to evolutionary theory. In Sect. 4 we describe the epistemological lessons of these objections, and in Sect. 5 we conclude.

The authors contributed equally to this chapter

K. McCain (✉)
Department of Philosophy, University of Alabama, 900 13th Street South, Birmingham, AL 35294-1260, USA
e-mail: mccain@uab.edu

B. Weslake
Department of Philosophy, University of Rochester, Rochester, NY, USA
e-mail: bradley.weslake@rochester.edu

K. Kampourakis (ed.), *The Philosophy of Biology: A Companion for Educators*, History, Philosophy and Theory of the Life Sciences 1, DOI 10.1007/978-94-007-6537-5_6,

2 Epistemological Background

2.1 *The Traditional Account of Knowledge*

It is widely accepted by philosophers that in order to know some proposition, it must be true and you must believe it. However, to have knowledge one must satisfy more than these two conditions. To see this, consider the following case:

> Sally makes a wild guess that there is now an even number of stars in the Milky Way galaxy and forms the belief that there is an even number of stars in the Milky Way galaxy on the basis of her guess.

Assume for the sake of illustration that Sally's guess happens to be true. Does Sally know that there is an even number of stars in the Milky Way galaxy? Surely not. After all, Sally has no reason to believe as she does. She is simply guessing, and she is aware that she has no good reasons for her belief. Not only does Sally fail to know that there is an even number of stars in the Milky Way galaxy, she is not even reasonable in believing that there is. Given her lack of evidence, the rational thing for Sally to do is to refrain from believing that there is an even number of stars and refrain from believing that there is not an even number of stars—in other words, she should suspend judgment concerning the number of stars. Since Sally has a true belief in this case, but fails to have knowledge, something more must be needed for knowledge in addition to true belief.

The fact that knowledge is not simply true belief has been recognized by philosophers since Plato. But what else is needed? In the *Meno*, Plato claims that what must be added to true belief is something that tethers one's belief to the truth. Many philosophers understand this "tether" to be the idea that one must have evidence that supports one's belief in order to have knowledge.[1] Thus we arrive at what is sometimes called the traditional account of knowledge: one has knowledge of some proposition when the proposition is true and one believes the proposition on the basis of sufficiently strong evidence.[2]

2.2 *Evidence and Knowledge*

Of course, the traditional account of knowledge gives rise to an important question: how much evidence is required for knowledge?

---

[1]Some philosophers, *e.g.* Goldman (1999), would object to the idea that all knowledge requires evidence. However, most philosophers, Goldman included, accept that in order to have knowledge one's belief must be justified in some sense. Whatever that sense is, it must be capable of capturing everything we say below about the relationship between evidence and knowledge.

[2]See Feldman (2003) for further elaboration of the traditional account of knowledge as well as complications, such as the Gettier problem, that we are setting aside.

One answer to this question is that one must have infallible evidence for a proposition in order to know that the proposition is true. Infallible evidence for a proposition is evidence that guarantees that the proposition is true. In other words, given infallible evidence, the proposition couldn't possibly be false. Although it is tempting to think that knowledge requires infallible evidence, this cannot be correct. At least, it cannot be correct if we have knowledge of much of anything. For the requirement of infallible evidence for knowledge leads to a thoroughgoing skepticism. We all accept that our perceptual experiences can fail to be accurate. That is to say, sometimes we are subject to illusions or hallucinations. For example, in the famous Müller-Lyer illusion, lines of the same length appear to be different lengths. Illusions such as this demonstrate that our perceptual experiences do not provide us with infallible evidence. Moreover, it is very plausible that, for any given perceptual experience, it *might* be misleading. So, for example, if you must have infallible evidence in order to have knowledge, you do not know that you are reading this book. After all, it is possible that you have the same evidence that you do (perceptual experiences) and yet are not reading this book. Thus, if infallible evidence is required for knowledge, we lack perceptually grounded knowledge of the world around us.[3] This extreme skepticism is highly implausible. Thus, infallible evidence cannot be required for knowledge. Instead, some less demanding, fallible standard of evidence must be required for knowledge.

This is progress, but we are still left with a question: How much fallible evidence is required for knowledge? This is a difficult question that is still debated by epistemologists. Fortunately, for our purposes it is not necessary to give a precise answer. It is enough that we recognize that some standard of fallible evidence, rather than infallible evidence, is required for knowledge. That said, we note that a plausible rough characterization of the strength of fallible evidence required for knowledge is the "criminal standard" of evidence.[4] According to this standard, in order to know a proposition is true one must have evidence that makes the truth of the proposition beyond a reasonable doubt. So, according to the criminal standard, knowledge requires evidence that is much weaker than infallible evidence, yet stronger than merely a good reason to believe. Although we grant that 'beyond a reasonable doubt' is somewhat vague, we will adopt this standard as a working guide for the degree of evidence required for knowledge.

At this point there is one final question about evidence and knowledge that it is worth briefly considering: What is it to be good evidence? There are many different kinds of evidence, so it would be a monumental task to try and say exactly what it is in virtue of which evidence is good evidence. In light of this, we will here only offer some brief remarks about good evidence. There are a variety of sources that provide good evidence. They include: logical and mathematical proofs, the sensory and introspective experiences of normally functioning humans, the testimony of

[3]For more on the nature of perceptual knowledge and the sort of evidence it requires see Alston (1993), Chisholm (1957), and Huemer (2001).

[4]This standard for the evidence required for knowledge is suggested by Conee and Feldman (2004).

trustworthy sources, and controlled experiments. As is all too familiar, one's overall evidence can become better or worse with respect to a particular proposition. An obvious example is learning that one's evidence is defective. For example, when one has evidence from testimony and subsequently learns that the testifier is biased, this information can weaken or cancel the evidence from testimony. Similarly, learning that the results of an experiment are yielded by biased sampling can weaken or cancel the evidence provided by the experiment. On the other hand, ensuring that samples are fair and that testifiers are not biased can improve the strength of evidence that those sources provide. Finally, an important feature of evidential support is that the evidence for a proposition is much stronger when, other things being equal, the individual pieces of evidence come from independent sources. For example, when two consumer reports claim that a certain car is best, the evidence you gain is stronger when unrelated agencies publish the reports than it is when the same agency publishes both reports. These remarks are far from exhaustive, but they will suffice for our purposes.[5]

3 Objections to Evolutionary Theory

In this section we develop a characterization of the nature of science by responding to a set of objections that have been raised against evolutionary theory. We emphasize that these objections are not taken seriously by any scientists or philosophers of science. Moreover, they have recently been superseded in the public imagination by the arguments of intelligent design advocates (see Brigandt, this volume). We have chosen these objections not because they are plausible but because they are instructive. As we will argue in the remainder of this section, each misses something important about the nature of science.

Before we begin, we should clarify what we mean by evolutionary theory. We will say more about the structure of evolutionary theory overall below, but for now we can understand the theory to be composed of the following two hypotheses:

[5]It is worth noting that although we have been discussing what is required for knowledge, similar considerations apply to rational acceptability. Any time one lacks sufficient evidence for a proposition it is not rational for her to believe the proposition. Some may doubt this claim because they distinguish between things that someone should have *beliefs about* and things that someone should *believe in*. That is to say, they think that some propositions (the kind we have beliefs *about*) require evidence in support of them before they are rationally acceptable and some propositions (the kind we have beliefs *in*) do not require evidence for acceptability. Perhaps there is such a distinction between kinds of propositions and the evidential requirements for rationally accepting them (we doubt it). However, even if there are propositions that one should *believe in* without the need for supporting evidence, scientific theories are not among them. Scientific theories are accepted or rejected on the basis of evidence. If the evidence does not support a given scientific theory, then the theory should be rejected. So, when we discuss the rational acceptability of evolutionary theory we are assuming that rational acceptability requires having good evidence in support of the theory.

1. *Common Descent.* All living organisms descended from a single common ancestor.
2. *Natural Selection.* Natural selection has been an important cause of the history of life.

The second hypothesis is vague, but is adequate for our purposes.

3.1 Evolution Is a Mere Theory

One criticism that is often leveled against evolutionary theory is that it is merely a theory, not a scientific fact. But what does this claim mean, and why should we believe it?

One defense of this claim is that evolutionary theory has not been "proved".[6] The thought seems to be that a scientific fact is the sort of thing that has been proved true, whereas a theory is something that has not been proved true.[7] Of course, defending the claim that evolutionary theory is a mere theory on the grounds that it has not been proved true raises an important question of its own—what is required for something to be proved? Clearly, it cannot be the case that in order for something to be accepted as a scientific fact we must possess a proof of the sort that can be found in logic or mathematics. This sort of proof involves showing that a conclusion follows deductively (it cannot possibly be false if the premises are true) from axioms that are self-evidently true. Proofs of this sort, if they exist at all, can only exist for truths of reason, which can be known without making observations of the universe around us. As we will show below, science does involve some claims that we can know in this way. But this cannot be all that science involves, because if it were, we would only be in a position to endorse very few scientific facts. Much of science makes claims about the universe around us as it happens to be. These claims are neither self-evidently true nor do they follow deductively from anything self-evidently true; instead they must be supported by observational evidence. So, it cannot be that proof of the logical or mathematical sort is required for something to be more than a mere theory.

Let us consider another way to make the distinction between a mere theory and a scientific fact. Perhaps the idea is that scientific facts are things that we know to be true, while mere theories are not known to be true. Now we distinguished above between three necessary conditions for knowledge (the proposition must be true, one must believe the proposition, and one's believing the proposition must be based on sufficient evidence). A proponent of this objection therefore must argue that at least one of the three conditions for knowledge is not met in the case of evolutionary theory. However, only the third condition is a plausible candidate for attack.

[6]See Watson (1976).

[7]Critics of evolution have sometimes been aided here by the careless statements of biologists, who have claimed that evolution has been demonstrated, or is indubitable (Kitcher 1982, p. 31).

It is simply false or irrelevant to claim that evolutionary theory is not believed to be true. On one hand, it is straightforwardly false that supporters of evolutionary theory do not believe it. On the other hand, while it is true that some who object that evolutionary theory is only a theory do not believe it, this fact is irrelevant. For evolutionary theory to be known requires merely that *someone* believe it, while satisfying the other conditions for knowledge. It does not matter that critics of evolutionary theory fail to believe it.

Arguing that evolutionary theory is not true would be an effective way for objectors to show that it is merely a theory, on the interpretation of "merely a theory" under discussion. Unfortunately, there is no shortcut to the truth. The only way to argue that evolutionary theory is not true is to show that the evidence does not support it.

So the real question concerning whether evolutionary theory is only a theory is about evidence. Objectors can attempt to argue that evolutionary theory is only a theory because we do not have sufficient evidence to know that it is true.

We argued above that we should be fallibilists about the evidence required for knowledge. As we noted, this leaves us with difficult questions concerning exactly what strength of evidence is required for knowledge. Rather than trying to first settle this issue before determining whether evolutionary theory is only a theory, it is more practical to simply look at the evidence there is in support of evolutionary theory and see how that evidence compares to the evidence we have in support of other things that we know.

The first point to be made here is simply that there is an enormous amount of evidence in support of evolutionary theory.[8] Moreover, the evidence comes from a very diverse array of independent sources. Regarding *Common Ancestry*, there is the evidence from the geographic distribution of species, the fossil record, the molecular record, embryology, comparative anatomy, and so on.[9] Regarding *Natural Selection*, there is the evidence from experimental tests of evolution under natural selection in natural and laboratory contexts, together with work on mathematical models and computer simulations.[10]

To describe all this evidence and explain how it bears on the hypotheses that compose evolutionary theory is obviously far beyond the scope of this chapter.[11] But it is helpful to consider two contrasts that may be made. One contrast is with knowledge claims made in ordinary life, and another contrast is with knowledge claims made elsewhere in science.

---

[8] Sometimes those who claim that evolution is only a theory seem to mean that evolution is a mere unsupported speculation, so we note in passing that the existence of this evidence is sufficient to show that the objection in this form is false. We thank Elliott Sober for reminding us of this interpretation.

[9] Good popular surveys of this evidence can be found in Coyne (2009) and Dawkins (2009). Carroll (2006) gives a popular survey of the evidence from the molecular record.

[10] For recent surveys of selection experiments, see Garland and Rose (2009) and Kawecki et al. (2012). For a survey of work involving computer simulations, see Adami (2006). There are myriad introductions to mathematical models of evolution, but for a concise survey of population genetics see Gillespie (2004).

[11] For an excellent treatment of the structure of the evidence, see Sober (2008).

Compare the evidence in support of evolutionary theory to the evidence that we have for propositions we normally take ourselves to know. For example, we typically think that we can know things via the testimony of others either directly or through reading things that others have written. Consider the following sort of situation:

> You want to see a particular movie at the local cinema. You check the showtimes in a local newspaper and it says that the movie starts at 7pm.

Most people accept that in this situation you may come to know that the movie starts at 7 pm by reading the newspaper. How does the evidence that you have in this situation stack up to the evidence in support of evolutionary theory? Intuitively, the evidence in support of evolutionary theory is much better (both in terms of quantity and strength) than the evidence you have for the showtime for the movie. After all, you only consulted a single newspaper and you do not have evidence about the typical showtimes of the local cinema. Further, it is not uncommon for local newspapers to occasionally have printing errors, nor is it uncommon for cinemas to change their showtimes. Despite the chance of error, it is plausible that you can know the showtime in this sort of situation. So it is also plausible that the strength of evidence in support of evolutionary theory is sufficient for knowledge.

Now compare the evidence for evolutionary theory with the evidence that warrants belief in other scientific theories. We can be said to know many scientific theories, but we will take as an example the discovery of electrons.[12] This example provides a nice analogy with evolutionary theory. Like evolutionary theory, electron theory can be considered the conjunction of a number of independent hypotheses, *e.g.* that electrons are negatively charged, that they are much less massive than atoms, that they are of a single type, that they are constituents of all atoms. And like evolutionary theory, the type of evidence that elevated electron theory to the status of knowledge consisted in strong independent evidence for these core hypotheses. In the case of electron theory, the independent evidence included the deflection expected for negative charge in both electric and magnetic fields, the independent determination in these two fields of the ratio of mass to charge, and the robustness of these results across a wide range of experimental procedures and different gases. After J. J. Thomson presented these results in 1897, it was no longer reasonable to doubt the existence of electrons.

Now just as with evolutionary theory, there have since been radical changes in other aspects of the electron theory together with an impressive stability of these core hypotheses. The electron, like evolution, is now very far beyond a reasonable doubt. Our point is that if we compare the degree of evidence for electron theory at the time of Thomson with the degree of evidence for evolutionary theory now, it is extremely implausible that electron theory met, but evolutionary theory fails to meet, the evidential standards for knowledge.

---

[12] Our presentation of the discovery of electrons follows the lucid discussion of Norton (2000) and Bain and Norton (2001), who also consider the example in light of philosophical disputes concerning the very possibility of scientific knowledge.

These considerations show that it is plausible that if we have the sort of knowledge that we typically take ourselves to have, both in everyday life and in science, we can have knowledge of evolutionary theory too. So one cannot plausibly claim that evolutionary theory is only a theory on the grounds that the evidence in support of evolutionary theory is insufficient. The evidence in support of evolutionary theory is comparatively stronger than the evidence we have in support of things that it is uncontroversial that we know. Thus, this objection to evolutionary theory rests on a misunderstanding of the degree of evidence required for something to be known.

There is a final point to be made about the objection that evolution is merely a theory. This phrase is often used by objectors to evolutionary theory as if there were some simple, easily determined property of evolutionary theory that makes it fit to be described as a mere theory. But even if it were true—contrary to the argument we have presented—that evolutionary theory failed to live up to the evidential standards for knowledge, there is no shortcut to this result. To argue that evolutionary theory fails to meet the evidential standards for knowledge would require detailed criticism of the entire range of evidence for the theory. This detailed criticism has yet to materialize.

3.2 *Evolution Is not Falsifiable*

A second criticism that has been leveled at evolutionary theory is that it is not falsifiable.[13] This criticism traces back to an influential conception of the distinction between scientific and non-scientific theories due to Karl Popper (1963). According to this conception, a theory is scientific if and only if it is falsifiable. A theory is falsifiable, in turn, if and only if an observational consequence can be derived from the theory. The idea is that if such a consequence can be derived, then an observation designed to determine whether the consequence obtains provides a genuine test of the theory, and if the test fails then the theory must be false.

This criticism of evolutionary theory is therefore supposed to establish that evolutionary theory is not a scientific theory at all. The argument is as follows:

1. A theory is scientific if and only if it is falsifiable.
2. Evolutionary theory is not falsifiable.
3. Therefore, evolutionary theory is not scientific.

Given the definition of falsifiability introduced above, the critic of evolutionary theory claims that since it is not possible to derive an observation with the potential to establish that evolutionary theory is false, evolutionary theory is not falsifiable. Thus, the objection goes, evolutionary theory is not scientific at all.

There are two straightforward responses to this criticism. A first response is to reject the first premise of the argument. According to this response, falsifiability does not distinguish between scientific and non-scientific theories. A second response is

[13] See Gish (1979) and Morris (1974).

to reject the second premise of the argument. According to this response, evolutionary theory is falsifiable after all. The first response is the correct one to make, as can be seen by considering the second response.

It is sometimes said by biologists (often under the influence of Popper) that evolutionary theory is falsifiable.[14] Legend has it that the biologist J.B.S Haldane nominated the discovery of a rabbit fossil in Precambrian strata as something that evolutionary theory predicts will not occur. So let us consider whether such a discovery would falsify evolutionary theory. Notice that such a prediction cannot be deduced from *Common Descent* and *Natural Selection* alone. These propositions have no direct bearing whatsoever on where rabbits will be found (or, for that matter, on whether rabbits even exist). Rather, it is only in conjunction with other, more specific hypotheses that evolutionary theory makes predictions about where rabbits will be found. Philosophers of science refer to those other hypotheses as *auxiliary hypotheses*. The auxiliary hypotheses in this case include those that specify the evolutionary relationships between species, and those that specify the time-periods over which evolution has occurred. Since it is only in conjunction with auxiliary hypotheses that evolutionary theory makes predictions, the failure of such a prediction would not entail that evolutionary theory is false. Rather, at best it would entail that one of the cluster of hypotheses from which the prediction was derived is false. This point does not depend on anything special about fossil rabbits—*all* predictions of evolutionary theory are made by clusters of hypotheses, not by *Common Descent* and *Natural Selection* alone. In sum, evolutionary theory is not falsifiable, since it does not in isolation entail any predictions whatsoever.

It is important to be clear about this point. We have suggested that evolutionary theory is not falsifiable, in Popper's sense. This does not entail that evolutionary theory cannot be tested, but rather that falsifiability is an inadequate theory of testability (we discuss the idea that scientific theories should be testable in the following section). It also does not entail that the choice of which hypothesis to abandon in the face of a failed prediction is arbitrary.[15] Our point is simply that no scientific theories are falsifiable in Popper's sense, and that this is a problem for Popper, not for our theories.

This leads us back to the first response. Does the unfalsifiability of evolutionary theory entail that it is not scientific? It is widely accepted by philosophers of science that it does not. For the feature of evolutionary theory in virtue of which it is not falsifiable is not unique, but rather is a feature of scientific theories in general. Hypotheses do not confront the world alone, but in groups.[16] This point was clearly made in response to Popper by both Lakatos (1970) and Putnam (1974), who pointed out that Newton's theory of universal gravitation is unfalsifiable for exactly this reason. Newton's theory of universal gravitation alone implies nothing at all

[14] For a survey of Popper's influence on biologists, see Hull (1999).

[15] The problem of how to distribute belief in a cluster of hypotheses that are incompatible with an observation is known as the Quine-Duhem problem.

[16] For the classic presentation of this conception of the relationship between theories and predictions, see Duhem ([1914] 1954).

about the motions of bodies, for it says nothing about where bodies are located, and about what other forces are acting. So it is not falsifiable. Moreover, as Lakatos and Putnam point out, when predictions made by Newton's theory in conjunction with accepted auxiliary hypotheses of these sorts failed, scientists frequently responded by rejecting the auxiliary hypotheses rather than the theory—and they were reasonable to do so. Likewise, it is possible that if a fossil rabbit were discovered in the Precambrian, it would be some of the auxiliary hypotheses to evolutionary theory that it would be reasonable to reject. This impugns the scientific status of neither Newton's theory nor Darwin's. It is false that a theory is scientific if and only if it is falsifiable.[17]

3.3 *Evolution Makes no Predictions*

We have argued that falsifiability is a poor criterion for judging whether a theory is scientific. However, our discussion also suggests a much better criterion for scientific theories, one that captures what might lead one to erroneously latch onto falsifiability. This criterion is that scientific theories must be testable. In order to be testable a theory must (in conjunction with appropriate auxiliary hypotheses, a qualification we will hereafter leave implicit) make predictions that can either be borne out or not. So, for a theory to be scientific, it must make predictions. This is another point on which evolutionary theory has been criticized. It has been claimed that evolutionary theory is not scientific because it does not make predictions.[18]

Although this criticism does at least challenge evolutionary theory from the standpoint of an accurate conception of scientific theories, it is hopelessly mistaken. Not only is evolutionary theory testable in virtue of the predictions it makes, it is one of the best tested theories in the history of science. While even a cursory survey is far beyond the scope of this chapter, we will mention two nice examples for the sake of illustration. First, after inspecting an orchid species from Madagascar with a foot-long nectary, Darwin predicted the existence of a moth with an extraordinarily long tongue, on the basis that orchids and moths had evolved together: "in Madagascar there must be moths with probosces capable of extension to a length of between 10 and 11 in.!" (Darwin 1862, p. 198). Even though Darwin remarked, in a later edition of the same work, that he had been ridiculed for this prediction by some entomologists, his prediction was finally confirmed in 1903.[19] Of course, this illustrates a prediction the failure of which would hardly have been damaging to the core hypotheses of evolutionary theory. Second, there is a prediction that also

[17] Popper was aware of these objections, and revised his account of falsifiability in response. We will not discuss Popper's response, though we note that interpretations of falsifiability that increase the plausibility of the first premise of the argument under discussion also decrease the plausibility of the second premise. For other objections to Popper see Sober (2008, pp. 129–131).

[18] See Gish (1979) and Morris (1974).

[19] A brief presentation of this story can be found in Kritsky (1991).

provides a lovely illustration of the power of independent lines of evidence. When the molecular structure of the gene was discovered and it was appreciated that phylogenies (descriptions of the genealogical relationships between species) could be constructed on their basis, it was an open question whether the older phylogenies constructed on the basis of morphological evidence would agree with the new phylogenies constructed from molecular evidence. Here are Zuckerkandl and Pauling (1965, p. 102), writing before the evidence was in:

> It will be determined to what extent the phylogenetic tree, as derived from molecular data in complete independence from the results of organismal biology, coincides with the phylogenetic tree constructed on the basis of organismal biology. If the two phylogenetic trees are mostly in agreement with respect to the topology of branching, the best available single proof of the reality of macro-evolution would be furnished. Indeed, only the theory of evolution, combined with the realization that events at any supramolecular level are consistent with molecular events, could reasonably account for such a congruence between lines of evidence obtained independently [...]

Zuckerkandl and Pauling put things too strongly, for as we have suggested, the success of this prediction would certainly not *prove* that the theory is true. Nevertheless, failure of the prediction would have been as dramatic as the discovery of a Precambrian rabbit, in the sense that many well-confirmed hypotheses would have had to come under scrutiny (Godfrey-Smith 2003, pp. 73–74). As it happens, the phylogenies constructed from molecular data are congruent with the older phylogenies to a remarkable degree.[20,21]

Sometimes critics of evolutionary theory have something more specific in mind when they claim that evolutionary theory makes no predictions. This is the so-called "tautology problem", according to which evolutionary theory is not testable because it is equivalent to the thesis that the fittest organisms survive (in the sense that they leave more offspring).[22] It is then claimed that the fitness of an organism is defined in terms of survival (in the sense that it is identified with number of offspring). Putting the thesis and definition together, the theory therefore is said to amount to the claim that the organisms that leave more offspring leave more offspring. This claim is tautologous, that is, it is true in virtue of logical form alone (the logical form here is something like: if A is b, then A is b). Tautologous claims cannot be tested, since they are necessarily true. Hence, the tautology problem.

The first thing to notice about this objection is that evolutionary theory cannot be reduced to the proposition that the fittest organisms survive. As we have presented the theory, it involves the hypotheses of *Common Descent* and *Natural Selection.*

---

[20] For a survey see Patterson et al. (1993), who write of the morphologists: "none of their well-supported phylogenies is overthrown by molecular data". Of course, there is a huge amount of dispute concerning the precise details of the relationships between species. But the basic structure of the tree of life is supported by a vast array of independent lines of evidence (Cracraft and Donoghue 2004). Indeed, intense disagreement on the branches coupled with broad agreement on the trunk provides a nice metaphor for the state of evolutionary theory itself.

[21] There are interesting epistemological issues concerning just what propositions are tested by this evidence, and why. For discussion of these issues see Sober (2008, Chapter 4).

[22] See Morris (1974) and Wilder-Smith (1981).

Neither of these hypotheses is tautologous. If there is a problem here, it must concern *Natural Selection*. According to *Natural Selection*, natural selection has been an important cause of the history of life. If natural selection reduces to the proposition that those who leave more offspring leave more offspring, would *Natural Selection* be untestable? It would not. For natural selection to cause a change in a population, there must be fitness differences in that population. If fitness is identified with number of offspring, then this is equivalent to there being differences in number of offspring in the population. Whether there are such differences is a testable proposition, as is whether such differences have been an important cause of the history of life. So even if the claim that the fittest survive were untestable, it would not follow that the proposition we have called *Natural Selection* is untestable. That a theory contains a tautology does not entail that the theory as a whole is a tautology (Sober 2000a, p. 71).

As it happens, it is also a mistake to think that fitness is always defined in terms of actual number of offspring. In many contexts, fitness is instead defined in terms of *expected* number of offspring. A simple thought experiment exhibits the attractions of this alternative definition (Scriven 1959, p. 478). Suppose that two identical twins with identical behavioural dispositions happen to be sitting next to one another, and a tree falls on one of them. The other goes on to reproduce. Should we assign different fitnesses to them? If fitness is identified with actual number of offspring we must, but if fitness is identified with expected number of offspring we can assign them the same fitnesses, and chalk up the difference in offspring to chance. It is considerations of this sort that have led philosophers to propose the so-called *propensity interpretation* of fitness, according to which fitness is identified with the propensity to leave a certain number of offspring (Mills and Beatty 1979; Sober 2000b).

There is a final point to be made about the role of fitness in evolutionary theory. Much of the work of modern evolutionary theory involves the construction of mathematical models of evolutionary processes, which involve assigning fitnesses of various different types to various different entities (Orr 2009). These models can be used to discover various surprising results, for example that if the fitnesses of two types vary with time, the type that dominates over the long term can depend not only on mean fitness, but on variance in fitness. Results such as these are analogous to tautologies in being necessarily true, since they are produced by mathematical proof. So evolutionary theory does contain some truths of reason, which cannot be tested. What *can* be tested is whether these mathematical models apply to actual biological populations. That evolutionary theory makes use of mathematical truths in this way is obviously a virtue rather than a defect of the theory.

3.4 Evolution Has Been Falsified

The final charge against evolutionary theory we will consider is that it has been falsified. Amusingly, some critics of evolutionary theory have argued both that is unfalsifiable and that it has been falsified.[23] It cannot be that both are true: if a theory has been falsi*fied* then it was falsi*fiable*. We argued above that evolutionary theory is not falsifiable, and these arguments also suffice to show that evolutionary theory has not—indeed, that it cannot have—been falsified. Perhaps what is intended is the weaker thesis that evolutionary theory has not been well tested. Again, we have argued that this is a mistake. Evolutionary theory has been very well tested, partly in virtue of making a host of successful predictions such as those we described above. There is, however, a different argument that might lie behind the idea that evolution has been falsified. The argument is that evolutionary theory has made some predictions that have failed, and therefore that it should not be believed. It is this argument we will consider in this section.

We made the point earlier that evolutionary theory only makes predictions in conjunction with auxiliary hypotheses, and that in this respect the theory is typical. We also noted that in the case of Newton's theory of universal gravitation, when predictions made by the theory in conjunction with appropriate auxiliary hypotheses failed, frequently the reasonable response was for scientists to abandon the auxiliary hypotheses rather than the theory. A classic example is provided by the discovery of Neptune. Newton's theory in conjunction with accepted auxiliary hypotheses about the planets failed to accurately predict the orbit of Uranus. Rather than rejecting Newton's theory, scientists revised the auxiliary hypotheses to include the postulation of an additional planet, Neptune, which observations later confirmed. This example is already sufficient to show that the fact that a theory has made failed predictions does not entail that the theory should not be believed.

It is time to reveal an idealisation that we have been making about the nature of evolutionary theory. We have been treating the theory as consisting simply in the conjunction of the hypotheses of *Common Descent* and *Natural Selection*, and as surrounded by a host of auxiliary hypotheses that enable the theory to be tested. This is a picture that is associated with Lakatos (1970), who called these the "hard core" and "protective belt" of what he called scientific research programmes. But this is an oversimplification. In fact, evolutionary theory at any point in time consists in an enormous number of hypotheses. These hypotheses are located on a continuum from the more fundamental to the less fundamental, and are supported to different degrees. Different biologists pick out different subsets of these hypotheses as specifying evolutionary theory, depending on their purposes. For example, Ayala (1985, p. 59) distinguishes between three subsets of hypotheses. *Common Descent* and *Natural Selection* are members of the first subset. The second subset contains hypotheses about the specific relationships between species. The third subset contains hypotheses about the processes responsible for evolutionary changes. According to Ayala, the first subset contains the most fundamental claims of evolutionary theory, while those in the third subset are more peripheral. On the other

[23]In particular see Gish (1979), Morris (1974) and Wysong (1976) for creationist attacks on evolutionary theory that claim both that evolutionary theory is not falsifiable and that it has been falsified. See Kitcher (1982) and Sober (2000a) for discussion of the fact that creationists have leveled these inconsistent objections.

hand, in an influential textbook on evolutionary theory, Futuyma (2009, pp. 9–11) identifies 16 "fundamental principles of evolution" that include principles concerning the genetic basis of evolutionary change and the processes responsible for speciation (the origin of new species). There is no reason to privilege any particular subset of these hypotheses as providing the real essence of evolutionary theory (Hull 1988). Ayala and Futuyma, in highlighting different subsets of the hypotheses accepted by most biologists, are not in disagreement.

This is important to recognize, since there is a strategy frequently used by critics of evolutionary theory to make it seem as if biologists themselves accept that evolutionary theory has been falsified. The strategy is to find some hypothesis that has been rejected, to claim that this hypothesis formed an essential part of evolutionary theory, and then conclude that evolutionary theory has been rejected. An interesting example of this strategy, since it strikes at *Common Descent* itself, is provided by the response of some critics to the discovery of so-called lateral gene transfer in prokaryotes. Some biologists have taken the discovery of lateral gene transfer to cast doubt on the existence of a single common ancestor of all life on Earth, understood as a single cell containing ancestors of all genes present today.[24] Instead, it has been proposed that the common ancestor of life was a community of cells (Woese 1998). Critics of evolutionary theory have in turn seized this proposal as a refutation of a core hypothesis of evolutionary theory, *Common Ancestry* (see Doolittle 2009). According to these critics, biologists themselves have admitted that evolutionary theory is false.

So they have, *if* we take the exact statement of *Common Ancestry* to entail a single common cellular ancestor, and *if* this particular formulation of *Common Ancestry* is essential to the nature of evolutionary theory. But there is no reason to treat evolutionary theory as having an essence of this sort. If it turns out that the evidence supports the proposal that the common ancestor of life was a community of cells, we can replace *Common Ancestry* with that very similar hypothesis while preserving the bulk of the other hypotheses that compose evolutionary theory. To describe this possibility as involving the falsification of evolutionary theory is absurd, for the same reason it would be absurd to describe the quantum theory of the electron as involving the falsification of the existence of electrons, on grounds that it involved the reformulation of various classical hypotheses.

More generally, once we appreciate the vast range of hypotheses that compose evolutionary theory at any point in time, we are in a position to appreciate how it can be that biologists reject evolutionary hypotheses of various sorts all the time, while the core of the theory exhibits the remarkable stability we described above. The refinement and replacement of hypotheses on good evidential grounds is a mark of healthy science, not a sign of a failed theory.

---

[24] See Doolittle (2000) for a popular survey, and for philosophical reflections a special issue of *Biology & Philosophy* on "The Tree of Life" edited by M. A. O'Malley (Vol. 25, No. 4, September 2010).

4 The Evidence for Evolution

We have explained why various objections to evolutionary theory are unfounded and have explored some of the epistemic principles at work in scientific inquiry brought to light by consideration of these objections. In this final section we will briefly describe a principle of scientific inference that can be used to characterize the way in which evolutionary theory is supported by the evidence. This principle is inference to the best explanation.

In simplest terms inference to the best explanation involves the idea that explaining phenomena, just like predicting phenomena, provides evidence in support of a hypothesis. More precisely, the idea is that given a set of candidate hypotheses for some phenomena, the hypothesis that best explains the phenomena is the likeliest to be true. Of course, like all evidence, this evidence can be weakened or cancelled by other evidence.

There are difficult questions to ask about this principle, concerning what makes for good explanations, how it fits with general theories of confirmation, and why there should be a connection between good explanation and truth. Regardless of the answers to these questions, it is indisputable that inference to the best explanation is in fact deployed across a wide range of scientific disciplines (Glymour 1984, p. 173). It is also at the heart of some of the most famous episodes in the history of science, for example Antoine Lavoisier's argument against the phlogiston theory in favor of the oxygen theory of combustion, Christiaan Huygens' argument in support of the wave theory of light, and, of course, Charles Darwin's argument for evolution.[25] In addition to being widely used in the sciences, inference to the best explanation is employed in everyday life: by jurors hearing a trial, by a doctor forming a diagnosis on the basis of a patient's symptoms, by someone's determining what is wrong with her computer. Indeed, inference to the best explanation is so pervasive in our reasoning that some have argued that it is a basic belief forming method for humans.[26]

As stated above, the principle merely says that the best explanation is the likeliest to be true. This does not entail that the best explanation should always be believed, for the likeliest may still be very unlikely. Warranted belief in the best explanation therefore requires in addition that the best explanation be good enough to warrant belief. That is to say, in order for a hypothesis to be legitimately inferred as true because it is the best explanation, it needs to meet certain minimal standards.[27] Of course, it is a difficult task to make these minimal standards explicit. As with the degree of evidence required for knowledge, it is more practical to work with examples.

This brief characterization of inference to the best explanation is sufficient for us to see how it warrants accepting evolutionary theory on the basis of the evidence. For not only does evolutionary theory provide the best explanation for the evidence, it has no serious rivals. As Theodosius Dobzhansky famously said "nothing in

[25] See Thagard (1978) for a description of these arguments.

[26] See Enoch and Schechter (2008).

[27] For more on this see Lipton (2004).

biology makes sense except in the light of evolution."[28] That is to say, there is no rival theory that can provide anywhere near as good an explanation of the evidence as evolutionary theory.[29] The fact that evolutionary theory has no serious rivals makes it easy for evolutionary theory to be the best available explanation—it is the best by default. Of course as we noted above, being the best available explanation is not sufficient for being accepted as true. The best available explanation must also be good enough. Is evolutionary theory a good enough explanation?

It is not an exaggeration to say that evolutionary theory offers one of the best explanations in the history of science. The sheer breadth of phenomena explained by the theory is astounding. As Ernst Mayr (1970, p. 1) writes, evolutionary theory "is quite rightly called the greatest unifying theory in biology. The diversity of organisms, similarities between kinds of organisms, patterns of distribution and behavior, adaptation and interaction, all this was a bewildering chaos of facts until given meaning by the evolutionary theory." Similarly, Philip Kitcher (1982, p. 50) writes that "the questions that evolutionary theory has addressed are so numerous that any sample is bound to omit important types." The explanatory and predictive power of evolutionary theory is simply incredible. Evolutionary theory provides an extremely good explanation of an enormous range of evidence, and there are no serious rival explanations. By inference to the best explanation, it is a paradigmatic case of a scientific theory that it is rational to accept.

5 Conclusions

We have explored a variety of objections to evolutionary theory. In every case we have shown that the objection is either clearly false or involves a misunderstanding of the nature of evidential support and scientific knowledge. In the process, we illuminated some of the epistemic principles that are at the heart of scientific inference, and showed how they are employed to establish the rational acceptability of evolution. We conclude by summarizing our central points—all of which are important for biology education because they help students to understand why evolutionary theory is a legitimate scientific theory and to understand some general features of good scientific theories:

- Knowledge requires true belief appropriately supported by evidence.
- Knowledge does not require infallible evidence. A working standard for knowledge is that one's evidence in support of a proposition (or theory) must make it beyond a reasonable doubt.
- The evidence for a theory is much stronger when, other things being equal, the individual pieces of evidence come from independent sources.

[28] Dobzhansky (1973).

[29] So-called intelligent design theories do not constitute genuine rivals (Sober 2008, Chapter 2; Brigandt, this volume).

- Rational acceptability of a scientific theory cannot *require* the sort of proof that one can give in logic or mathematics, as this would render most scientific theories rationally unacceptable.
- Rational acceptability of any particular scientific theory should not require a level of evidence that cannot be met by other scientific theories, nor should it be so stringent that it would lead to a thoroughgoing skepticism.
- Scientific theories are tested in conjunction with auxiliary hypotheses, so a failed prediction does not entail that a theory is false. Hence, falsifiability is a poor criterion for scientific theories.
- In order to be scientific a theory should, in conjunction with auxiliary hypotheses, make testable predictions.
- The fact that a theory contains some tautologies or mathematical truths does not render the theory untestable.
- In order to be rationally acceptable at least some of a scientific theory's predictions must have been borne out.
- The fact that a theory explains relevant phenomena better than its rivals provides evidence in support of the theory.
- The fact that a theory is the best available explanation is not enough to support accepting the theory. The theory must also be a good explanation in its own right.

Acknowledgements We would like to thank Matt Frise, Kostas Kampourakis, Lynnette Lounsbury, H. Allen Orr, and Elliott Sober for their helpful comments.

References

Adami, C. 2006. Digital genetics: Unravelling the genetic basis of evolution. *Nature Reviews Genetics* 7(2): 109–118.

Alston, W.P. 1993. *The reliability of sense perception*. Ithaca: Cornell University Press.

Ayala, F. 1985. The theory of evolution: Recent successes and challenges. In *Evolution and creation*, ed. Mc.Mullin. Ernan, 59–90. Notre Dame: University of Notre Dame Press.

Bain, J., and J.D. Norton. 2001. What should philosophers of science learn from the history of the electron? In *Histories of the electron: The birth of microphysics*, Dibner Institute Studies in the History of Science and Technology, ed. Jed Z. Buchwald and Andrew Warwick, 451–466. Cambridge, MA: MIT Press.

Carroll, S.B. 2006. *The making of the fittest: DNA and the ultimate forensic record of evolution*. New York: W. W. Norton.

Chisholm, R. 1957. *Perceiving*. Ithaca: Cornell University Press.

Conee, E., and R. Feldman. 2004. Making sense of skepticism. In *Evidentialism*, 277–306. New York: Oxford University Press.

Coyne, J.A. 2009. *Why evolution is true*. New York: Viking.

Cracraft, J., and M.J. Donoghue (eds.). 2004. *Assembling the tree of life*. Oxford: Oxford University Press.

Darwin, C. 1862. *On the various contrivances by which British and foreign orchids are fertilised by insects*. London: John Murray.

Dawkins, R. 2009. *The greatest show on earth: The evidence for evolution*. New York: Free Press.

Dobzhansky, T. 1973. Nothing in biology makes sense except in the light of evolution. *The American Biology Teacher* 35: 125–129.

Doolittle, W.F. 2000. Uprooting the tree of life. *Scientific American* 282(6): 90–95.
Doolittle, W.F. 2009. The practice of classification and the theory of evolution, and what the demise of Charles Darwin's tree of life hypothesis means for both of them. *Philosophical Transactions of the Royal Society B: Biological Sciences* 364(1527): 2221–2228.
Duhem, P. 1954/1914. *The aim and structure of physical theory*. Princeton: Princeton University Press.
Enoch, D., and J. Schechter. 2008. How are basic belief-forming methods justified? *Philosophy and Phenomenological Research* 76(3): 547–579.
Feldman, R. 2003. *Epistemology*. Upper Saddle River: Prentice-Hall.
Futuyma, D.J. 2009. *Evolution*, 2nd ed. Sunderland: Sinauer Associates.
Garland Jr., T., and M.B. Rose. 2009. *Experimental evolution: Concepts, methods, and applications of selection experiments*. Berkeley: University of California Press.
Gillespie, J.H. 2004. *Population genetics: A concise guide*, 2nd ed. Baltimore: Johns Hopkins University Press.
Gish, D.T. 1979. *Evolution? The fossils say no!* San Diego: Creation-Life Publishers.
Glymour, C. 1984. Explanation and realism. In *Scientific realism*, ed. Jarrett Leplin, 173–192. Berkeley: University of California Press.
Godfrey-Smith, P. 2003. *Theory and reality: An introduction to the philosophy of science*, 2nd ed. Chicago: University of Chicago Press.
Goldman, A. 1999. Internalism exposed. *Journal of Philosophy* 96(6): 271–293.
Huemer, M. 2001. *Skepticism and the veil of perception*. Lanham: Rowman & Littlefield Publishers, Inc.
Hull, D.L. 1988. *Science as a process: An evolutionary account of the social and conceptual development of science*. Chicago: University of Chicago Press.
Hull, D.L. 1999. The use and abuse of Sir Karl Popper. *Biology and Philosophy* 14(4): 481–504.
Kawecki, T.J., et al. 2012. Experimental evolution. *Trends in Ecology and Evolution* 27(10): 547–560.
Kitcher, P. 1982. *Abusing science: The case against creationism*. Cambridge, MA: MIT Press.
Kritsky, G. 1991. Darwin's Madagascan Hawk Moth prediction. *American Entomologist* 37(4): 206–210.
Lakatos, I. 1970. Falsification and the methodology of scientific research programmes. In *Criticism and the growth of knowledge, proceedings of the international colloquium in the philosophy of science*, vol. 4, ed. Imre Lakatos and Alan Musgrave, 91–196. Cambridge: Cambridge University Press.
Lipton, P. 2004. *Inference to the best explanation*, 2nd ed. New York: Routledge.
Mayr, E. 1970. *Populations, species, and evolution*. Cambridge, MA: Harvard University Press.
Mills, S.K., and J.H. Beatty. 1979. The propensity interpretation of fitness. *Philosophy of Science* 46(2): 263–286.
Morris, H.M. 1974. *Scientific creationism*. San Diego: Creation-Life Publishers.
Norton, J.D. 2000. How we know about electrons. In *After Popper, Kuhn and Feyerabend*, Recent issues in theories of scientific method, Australasian studies in history and philosophy of science, vol. 15, ed. Robert Nola and Howard Sankey, 67–97. Dordrecht: Kluwer.
Orr, H.A. 2009. Fitness and its role in evolutionary genetics. *Nature Reviews Genetics* 10(8): 531–539.
Patterson, C., D.M. Williams, and C.J. Humphries. 1993. Congruence between molecular and morphological phylogenies. *Annual Review of Ecology and Systematics* 24(1): 153–188.
Popper, K. 1963. *Conjectures and refutations*. London: Routledge and Kegan Paul.
Putnam, H. 1974. The 'corroboration' of theories. In *The philosophy of Karl Popper*, Library of living philosophers, vol. 14, ed. Schilpp Paul Arthur, 221–240. La Salle: Open Court.
Scriven, M. 1959. Explanation and prediction in evolutionary theory: Satisfactory explanation of the past is possible even when prediction of the future is impossible. *Science* 130(3374): 477–482.
Sober, E. 2000a. *Philosophy of biology*, 2nd ed. Boulder: Westview Press.

Sober, E. 2000b. The two faces of fitness. In *Thinking about evolution: Historical, philosophical, and political perspectives*, ed. Singh Rama Shankar et al., 309–321. Cambridge: Cambridge University Press.

Sober, E. 2008. *Evidence and evolution: The logic behind the science*. Cambridge: Cambridge University Press.

Thagard, P. 1978. The best explanation: Criteria for theory choice. *Journal of Philosophy* 75(2): 76–92.

Watson, D.C.C. 1976. *The great Brain Robbery*. Chicago: Moody Press.

Wilder-Smith, A.E. 1981. *The natural sciences know nothing of evolution*. San Diego: Creation-Life Publishers.

Woese, C. 1998. The universal ancestor. *Proceedings of the National Academy of Sciences* 95(12): 6854–6859.

Wysong, R.L. 1976. *The creation-evolution controversy*. Midland: Inquiry Press.

Zuckerkandl, E., and L. Pauling. 1965. Evolutionary divergence and convergence in proteins. In *Evolving genes and proteins*, ed. Vernon Bryson and Henry J. Vogel, 97–166. New York: Academic.

Conceptual Change and the Rhetoric of Evolutionary Theory: 'Force Talk' as a Case Study and Challenge for Science Pedagogy

David Depew

1 Conceptual Schemes and Darwin's Interacting Metaphors

Especially in its first edition, Darwin's *Origin of Species* maintained a subtle and delicate balance between the *functional* and *goal-directed* quality we see in living things, an element of *chance* in how they originate, and the *deterministic* role of environmental agents in shaping them. Rhetorically, the text achieves this balance because Darwin's analogy between natural selection and the breeder's art of selectively retaining offspring with desired characteristics is complemented and constrained by his treatment of Malthusian population pressure as a force as ineluctable as gravity. Darwin included in the first edition of the *Origin* a slightly amended version of an early note in which he portrayed natural selection as "a force like a hundred thousand wedges trying [to] force every kind of adapted structure into the gaps in the economy of nature, or rather forming gaps by thrusting out weaker ones" (Barrett 1987, p. 135; Darwin 1966 [1859], p. 67). The sentence disappeared from subsequent editions on grounds of delicacy, but the thought remained (Peckham 1959, p. 150).

Darwin's stress on 'force talk' had the effect of shifting the accent in evolutionary theorizing from inner tendencies to externally impinging pressures. The drive to reproduce is endogenous. So is the tendency of offspring to vary. But because external, environmental forces rather than internal drives determine which variants will have a future there is an element of sheer contingency about whether on any occasion heritable variants will be available to enhance adaptedness. Widespread extinction testifies to the fact that sometimes they are not. On Darwin's theory, then, both chance and force, which the ancients regarded as the very opposite of

D. Depew (✉)
Project on the Rhetoric of Inquiry, University of Iowa, Iowa, IA, USA
e-mail: david-depew@uiowa.edu

K. Kampourakis (ed.), *The Philosophy of Biology: A Companion for Educators*, History, Philosophy and Theory of the Life Sciences 1, DOI 10.1007/978-94-007-6537-5_7,

teleology, are required to ascribe objective functions and goal-directedness (and not just their appearance) to organisms. Intentions are not required. Indeed, they are ruled out.

The interaction between Darwin's invocation of the breeder's art and his image of a world filled to the brim with impinging and constantly re-equilibrating natural forces created an interpretive framework in which these forces, preeminently Malthusian population pressure, recast artificial selection as subordinate to and ultimately explainable by natural selection and in which adaptations, without losing their functional goal-directedness, arise from purely natural processes (Darwin 1966, p. 75). In arguing in this image-, simile-, analogy- and metaphor-filled way Darwin seems to have anticipated, in practice if not in theory, the twentieth-century insight that metaphors can have explanatory effects if a *comparandum,* what is being compared, and a *comparans,* what it is compared to, fuse into a single new meaning (Black 1962; Hesse 1963; Ricoeur 1978; Lakoff and Johnson 1980). By treating images of force and purpose in this interactive way Darwin intensified the explanatory effect of his metaphors. By adding still more images, such as nature's tangled bank and the branching tree of life, he compounded the effect further (see Beer 2000; Gould 1989; Depew 2009 on Darwin's metaphors).

To confirm just how delicate Darwin's explanatory balance is all one has to do is look at his book's immediate reception. To a man, his friends and enemies misapprehended it. They misapprehended it, I think, because they would not allow Darwin's metaphors to interact.

Almost before the ink had dried on the page clerics and physicists who were usually at odds with each other about the relation between religion and science were ganging up to dismiss Darwin's analogy between natural selection and laws of nature such as gravity. Natural laws, argued the physicist and philosopher of science William Herschel, range over homogeneous quantities, not over a heterogeneous collection of chance occurrences. Herschel's quip that natural selection is a "law of higgledy-piggledy" [thoroughly jumbled] hurt Darwin greatly. He had tried hard to make the *Origin* conform to Herschel's methodological norms, according to which hypotheses are confirmable only if we are able independently to identify as real the causes they postulate (*verae causae*) (Hodge 1977; but see also Ruse 1975, 2000). Herschel's contemptuous remark, Darwin reported to Charles Lyell, was "a great blow and discouragement" (Darwin to Lyell, 10 December 1859, DCD entry 2575).[1]

For their part, Darwin's clerical-naturalist enemies willfully exaggerated the role he assigned to chance in order to protect intelligent design from the naturalistic gloss that the force metaphor imparted to it. William Bowen, for example, took natural selection to mean that given enough time chance would come up all by itself with something as meaningful as Newton's *Principia.* The point is that this is highly unlikely and so, then, is natural selection. Darwin complained to Asa

[1]Unless otherwise noted, references to letters to and from Darwin are to the Darwin Correspondence Database [DCD] http://www.darwinproject.ac.uk

Gray, his American correspondent and Bowen's colleague at Harvard, about this misreading:

> It is monstrous [...] that [Bowen] should argue against the possibility of accumulative variation and actually leave out entirely selection[...] The chance that an [...] improved pouter-pigeon should be produced by accumulative variation without man's selection is almost infinity to nothing; and so with natural species without selection (Darwin to Gray, 26 November, 1860; DCD entry 2998).

To this day, however, creationists are still recycling Bowen's bad reading and, in attempting to defend laws that range only over homogeneous quantities, physicists are still giving them cover by making chance, not natural selection, the author of adaptations[2] (for the concept of adaptation see Forber, this volume).

So much for Darwin's enemies. What about his friends? In 1866 Alfred Russel Wallace, Herbert Spencer, and Thomas Henry Huxley teamed up to urge Darwin to replace 'natural selection' with 'survival of the fittest' in future editions of *Origin.* They felt that the purely *ex post facto* process suggested by this phrase would undercut the creationist objection that natural selection is so haunted by intentional design that it merely substitutes one theology for another (Wallace to Darwin, 2 July 1866, DCD entry 5140). In the 3rd and subsequent editions, Darwin complied, but only half-heartedly. Perhaps he feared that eliminating the analogy between natural and artificial selection would shift the accent to chance and force in a way that was bound to undercut the genuine purposiveness of adaptations. If so this would have been a loss to Darwin but not to his defenders. Wallace never fancied the analogy with the breeders' art anyway. Perhaps this was because his version of natural selection ranges over groups, lineages, and races, not over the traits of individuals whose good (in contrast to the good of breeders and those whose purposes they serve) is in Darwin's view the end that natural selection supports (Darwin 1966, pp. 83–84). Huxley, for his part, cared even less for biological end-directedness. He was a determinist. He objected as much to the value-laden implications of 'survival of the fittest' as to those of 'natural selection.' The very fact that Huxley, Darwin's bulldog, advised Darwin to weaken or even abandon his gradualism shows that he never really understood this process very well at all, at least as Darwin construed it (Huxley to Darwin, 23 November 1859, DCD entry 2544; Depew 2011b).

Most ominously for the future of Darwinism, the phrase 'survival of the fittest' allowed Spencer to assimilate natural selection to his own already published theory of evolution. In fact, Spencer was the author of the phrase. He used it in his 1864 *Principles of Biology* to cast Malthusian population pressure in the role of forcing embryos to equilibrate with their environments ('adapt') or die. Over time, Darwin gave more and more ground on the direct effect of environments. He confided to

---

[2]One can see the over-reliance on chance on display in Steve Fuller's inaccurate description of natural selection. "Chance mutations," he writes, "are the driving force of evolution." So natural selection consists of "compounded historical accidents" (Fuller 2007, pp. 31, 48). It is not surprising that Fuller testified on the side of intelligent design in Kitzmiller vs. Dover (Pa.) Area School Board (2005). If natural selection is anything as chancy as Bowen, Fuller, or the physicist Steven Weinberg imagine it to be, intelligent design can seem a sensible alternative (Weinberg 1992).

Lyell in 1867 that in the *Origin* he had "underrated [...] the effects of the direct action of external conditions in producing varieties" (Darwin to Lyell, Oct 4, 1867, DCD, entry 5640). But Darwin, unlike Spencer, never confused the force of environments in inducing variant responses in individuals with the process of adaptation by natural selection. For this reason, he never confused selection *against* organisms that fail to "adapt to" stern environmental exigencies with gradual selection over trans-generational time *for* adaptive traits. His reason is clearly expressed in the large manuscript from which he 'abstracted' the text of the *Origin*:

> Seeing how absolutely necessary whiteness is in the snow-covered arctic regions [...] we might attribute the absence of color [in a white bird] to a long course of selection. But it may be that whiteness is the direct effect of cold and that the struggle for life has only so far come into play that colored animals in arctic regions live under a great disadvantage (Stauffer 1975, p. 377).[3]

The point is that natural selection properly conceived evolves genuinely functional goal-directed adaptive traits only by working over much time and many generations on small variants in traits that first arise independently of the utility they subsequently acquire as they move toward fixation. It was for just this reason that Darwin repeatedly proclaimed in the *Origin* that his theory would be falsified if its gradualist axiom were to be proven unsound (Darwin 1966, pp. 189, 194, 471). By contrast, the effect of environments in "preserving" from predation white birds that are in Spencer's sense "fit" to "survive" in arctic regions is so strong that it quickly eliminates the variation necessary to make the whiteness of arctic birds a cumulative effect of relative degrees of whiteness in many birds over many generations. It is on just this point that we can appreciate Darwin's exquisite balance between chance, force, and purposiveness. We may concede that in this example whiteness has the *effect* of protecting arctic birds. But unlike cases of genuine adaptation by natural selection we would not be entitled to conclude that arctic birds are white *because* whiteness incrementally has this protective effect. There is no evolved connection between cause and effect as there is in Darwin's theory of adaptive natural selection. So the trait is not irreducibly there for the sake of some adaptive good, goal, or function (Lennox 1994; Lewens 2004, p. 29; Depew 2008; see also Lennox and Kampourakis, this volume).

In Spencer's version, or rather distortion, of natural selection as survival of the fittest Darwin's theory actually disappears. But ominously Spencer's interpretation co-optively defined 'Darwinism' in the last decades of the nineteenth century. Nor was his eliminative view of selection challenged when early Mendelians

[3]The passage also shows that neither did Darwin confuse selection for traits with selection of traits. The whiteness of arctic birds is (possibly) a "direct effect" of cold not of the protection it affords against prey. If we choose to call this selection at all, it is selection of whiteness, not for it. The philosopher Elliott Sober illustrates the distinction by imagining a sorting mechanism that allows balls of several sizes to fall through holes of various sizes. Balls of each size are painted a different color. There is selection for size, but only selection of color (Sober 1984, pp. 99–100).

discredited his assumption that somatic modifications are inherited. On the contrary, by assigning 'the creative factor' in the evolutionary process to spontaneous mutations in the germ line many self-proclaimed Darwinians retained Spencer's notion of natural selection as a pruning force.

I say this was ominous because it was at just this time that the image of Darwinism that still prevails in our public culture was fixed. The issue is not whether any practicing biologists actually lived up to the dog-eat-dog image of 'Social Darwinism' that orators such as William Jennings Bryan circulated and of which the Progressive historian Richard Hofstadter made so much in his book *Social Darwinism in American Life* (Hofstadter 1944). The issue is whether Darwinian biologists of the day, unlike Darwin himself, understood natural selection as either favoring or eliminating organisms whose traits are or are not adapted *from the outset*. Vernon Kellogg's judicious report about the state of the question between Darwinians and Mendelians in his 1907 *Darwinism Today* suggests that they did indeed understand it as merely a pruning tool. "Natural selection," he wrote, "is the saving of one or ten by the actual killing of a 1,000 or 10,000 because in the struggle for existence the variations of the one or ten are of sufficient advantage to have life-or-death-determining value" (Kellogg 1907, p. 35). It is just this picture that undergirds the persistent notion in our culture that selection is an eliminative force that discriminates among whole organisms rather than the slightly variant traits they bear and that adaptations are nothing but retained accidents.

Such was the unhappy result of leaving the metaphor of force unattended by the analogy to artificial selection or breeding, as Darwin's false friends in fact left it.[4]

2 Natural Selection as the Creative Factor in Evolution: The Significance of the Modern Evolutionary Synthesis

We have discovered something important. Natural selection isn't a simple, self-evident idea that either does or does not explain the relevant empirical data. The history of the Darwinian research tradition abundantly shows that natural selection

[4]The elimination of Darwin's natural teleology is encoded in the stress Darwinians placed on natural selection as a mechanism. Spencer writes that "survival of the fittest" is Darwin's 'natural selection' "express[ed] in mechanical terms" (Spencer 1864, I, p. 445). Wallace liked to compare natural selection to a steam engine fitted out with a Watt automatic governor. In contrasting Spencer with Darwin I take myself to be in agreement with, and reliant on, Gayon (1995). I do not mean to contravene Gayon's claim that for Darwin natural selection is not metaphorical, but real (Gayon 1995, p. 269). I merely put his point otherwise. Darwin's interactionist and explanatory sense of metaphor was itself a casualty of his caving in to allies who viewed metaphor as mere decoration. In acknowledging 'survival of the fittest' in the 3th and subsequent editions of the Origin Darwin says that natural selection falls on the metaphorical side of a distinction between literal and metaphorical senses. (For his variant ways of making this concession, see Peckham 1959, p. 165.) But this distinction severely disrupts the "one long argument" of the text.

explains data through the mediating role of conceptual frameworks and that these differ considerably (Depew and Weber 1995). There are, as a result, many conceptions of natural selection. We have also discovered that conceptual frameworks announce themselves in images, similes, metaphors, and analogies that interpret the process of differential retention of heritable variation and bring it to bear on cases (Depew 2012). Darwin's own framework relied on several interacting metaphors that together made natural selection the creative author of adaptations and so a rival of intelligent design. By contrast, the notion that natural selection is the executioner of unfit organisms rather than a process that gradually shapes variant traits into adapted organisms passed itself off as Darwinism for a long time, in the process allowing it to serve as backing for unrestricted capitalism, racist imperialism, and eugenics.

In this section, I argue that Darwinism would have gone the way of other nineteenth century ideologies if the creative role Darwin assigned to natural selection had not been revived in the middle decades of the twentieth century by the population genetical theory of natural selection and the Modern Evolutionary Synthesis that was built on it. This resurrection was made possible by what is called the probability revolution in science (Gigerenzer 1990). The probability revolution, which was just beginning to find its sea legs in Darwin's day, and affected him to some degree, refers to the recoding of all of scientific knowledge, from sub-atomic to social, in statistical terms. This was not a matter merely of using statistics and probability to find the objective value of variables, as in the so-called "error law." It was a matter of defining the very *objects* of a science as statistical arrays over which calculable probabilities range. Not all or each member of a given population may have a certain property, but the ensemble does. This shift is famously visible in the transition from neo-classical to statistical and subsequently to quantum mechanics. In the social sciences the effect of the probability revolution can be seen in the transition from the "average man" of nineteenth century thought to the recognition that societies are inherently, and for that reason happily, diverse: full of variation, the degree of which can be measured by statistical-probabilistic analysis.

Population genetical Darwinism shares the statistical and probabilistic spirit with other twentieth century sciences. It arose in the 1930s as a way of solving problems that had been accumulating for Darwinians for a long time—how, for example, could natural selection ever get anywhere if small variations are constantly being 'swamped' by existing genes?—and in particular of mediating a long and increasingly tiresome quarrel between (Mendelian) mutationists and (Darwinian) selectionists (Provine 1971). Its seminal figure was the statistician Ronald Aylmer Fisher. For Fisher fitness is a measure not of differential death rates, as in earlier forms of Darwinism, but of genetic contribution to the next generation measured by the comparative birth rates of closely related populations that possess slightly variant properties. Fitness for Fisher is so indirectly connected to the myriad underlying organic causes of these differences that we should leave the latter out of it.[5] Consciously using statistical mechanics and thermodynamics as a model—natural selection plays probabilistically over the different fitness of various genotypes in freely interbreeding populations in the same way entropy plays over energy differences in arrays of atoms and molecules, even if it moves in the opposite direction (Fisher 1930, pp. 36–37)—Fisher demonstrated on purely probabilistic grounds that genotypes distributed in populations do not spontaneously regress to an average fitness, as Darwin's half-cousin Francis Galton, the most influential post-Darwinian, had assumed. They remain distributed in populations until affected, as they in fact constantly are, by selection and other processes. Accordingly, chance genetic mutations with very small effects on phenotypes can be propelled through large interbreeding populations if they enhance, however slightly, the average reproductive output of populations that have them compared to closely related, usually adjacent, populations that do not. Indeed, Fisher claimed that natural selection, which he defined as the differential retention of additive heritable genetic variation, adapts populations to environments at a rate directly proportioned to the amount of additive genetic variation that these populations contain (Fisher 1930, p. 36).

In saying this, Fisher did not mean to reduce biology to physics. The very fact that he has selection moving in an ordering direction against the strong current of a disordering entropic tendency shows that he was proposing an analogy based on similarities between dynamical models in two very different sciences. Yet, far from standing in the way of a more mature evolutionary theory Fisher's analogical shift, and related shifts by evolutionary biologists who followed his lead, but who challenged this or that aspect of his interpretive framework, managed to make Darwinism more scientifically respectable than it had ever been before.

In the course of doing so, Fisher and his successors rescued Darwin's idea that natural selection, not mutation, is the "creative factor" in evolution. Because Fisher's analogy between natural selection and statistical physics turned Malthusian population pressure into a special case, and a limiting one at that, it opened up a conceptual space in which differential death could be replaced with differential reproduction as the measure of comparative fitness (Gayon 1995). In the population genetical theory of natural selection the threat of violent death does not hang nearly as imminently over living things as it does in earlier forms of Darwinism, Darwin's only a little

[5] Some philosophers of biology are not as insouciant as Fisher in believing that the many underlying context-dependent causes of fitness can be omitted from any definition of the concept of fitness that purports to understand its role in a theory of adaptive natural selection (Sober 1984; Brandon 1990, p. 13). To be sure, these philosophers are understandably as jumpy as Fisher about identifying fitness solely with the aggressive traits with which the popular mind still associates it. But they take fitness to be identical with relative adaptedness and so think it depends by definition on an array of underlying differences as its components. These do not have to be stated unless we are interested in analyzing one or another of them (for a recent discussion of fitness, see Ariew and Lewontin 2004).

less than Spencer's. All populations have to do is out-reproduce closely related neighbors. George Gaylord Simpson, one of the seminal figures of the Modern Synthesis (which from the 1940s through 1960s applied population-genetical principles to problems about sub-species, speciation, and the evolution of higher taxa) put this point the following way:

> The modern concept of natural selection is [...] different from Darwin's. Darwin emphasized the survival of favored (or the early death of unfavored) individuals. The survivors were, for the most part, the 'fittest,' in the sense of qualification for success in competition, the 'struggle for existence.' That is still the usual nonscientific understanding of the process, but to specialists in the study of evolution 'natural selection' now means the average production of more offspring by such organisms in a population as are distinguished by any particular heredity factors. 'Fitness' is now defined solely as relative success in reproduction (Simpson, Introduction to F. Darwin 1959, p. xi.)

This framing, although it fails to note the importantly different uses to which Spencer and Darwin put the idea of Malthusian population pressure, sees clearly how population thinking explains better than Darwin did how an initially decoupled, and in this sense chancy, relation between variation and functional adaptedness becomes ever tighter as reproduction-enhancing heritable genetic variations amplify their representation in populations over trans-generational time by a process that is better compared to positive than to negative feedback (Brandon 1990, p. 188, n.22, speaks of selection as "feeding forward").

Ernst Mayr, the most ardent, long-lived, and rhetorically persistent of the original advocates of the Modern Synthesis, argued that this "two step" theory, as it is sometimes called—spontaneous or random genetic variation followed by selection on interbreeding organisms seen as genetically and phenotypically integrated dynamical wholes—revived Darwin's view of adaptive natural selection as gradual, adaptive, and creative (Mayr 1988, p. 99). What Mayr actually produced and disseminated to the public, however, is something of a hybrid between Darwin's Darwinism and population-genetical Darwinism. Mayr appealed to what he called "population thinking" to eliminate the tropes of force and design that interacted in Darwin's account but tended after him to separate into religious defenses of design and secularist defenses of natural forces that bear down on organisms, killing off some of them. At the same time, Mayr adverted to Darwin's conception of himself as a robustly common sense naturalist (not coincidentally, Mayr thought of himself the same way) to damp down any implication that population genetics, especially what Mayr called "bean bag genetics," is the formal object of evolutionary science rather than its ancillary tool (Mayr 1980, p. 18). As a result, Fisher's analogy with statistical mechanics and thermodynamics is nowhere to be seen or heard. In this way, Mayr represented Darwin and Darwinism as liberated from their shady past; their ideological sins, from Darwin's complicity with *laissez-faire* capitalism to the support of earlier Darwinians for imperialism, racism, and eugenics, would disappear into history's forgotten dustbin with the elimination of natural selection as an eliminative executioner of the unfit. To be sure, Mayr's hybrid tended to retroject onto a far-sighted founder ideas that could only have come into their own after much time, debate, and difficulty. He implies, for instance, that Darwin, in his ability to honestly recognize what he didn't know, judiciously left open a blank space for the mechanism of inheritance that in the fullness of time would be filled by molecular genetics (Mayr 1978, 1991, p. 7). In doing so, Mayr underestimated the effect of Darwin's own views about inheritance on the core of his evolutionary theory, which, as his toleration of Lamarckism suggests, was not committed to the neo-Darwinian principle that inherited factors are entirely screened off from environmental influences (Hodge 1985). At the same time, Mayr could not resist scolding Darwin for failing to stick with his original inclination to imagine speciation as typically occurring at the stressed edge of a species' range, where Mayr himself thought it took place, and for opting instead for so-called sympatric speciation, which occurs squarely in the fit and fecund middle of a range (Mayr 1988, pp. 206–207). We can learn little from Mayr about why Darwin himself believed that his theory became coherent enough to be published only when he made that shift.

The overall effect of these and other retrospective adjustments has been to smooth out the history of the Darwinian tradition and to imply that much of its scientifically and ideologically choppy past is not really part of the story at all. But Mayr is not alone in his tendency toward teleological storytelling from an anachronistic and generally triumphalist starting point. As a general rule, the more enthusiastic Darwinians are about this or that new version of the theory of natural selection the more they hope to confer authority on what they are promoting by attributing it *in nuce* to Darwin. The bigger the idea is, in fact, the more anachronistic is the tale.

A safer and more accurate way of defending the claim that adaptive natural selection is a creative process is to acknowledge that conceptual frameworks vary, that population-genetical Darwinism frames natural selection quite differently from the way Darwin framed it, and that the conceptual history of Darwinism is indeed choppy and likely to remain so (Depew and Weber 1995). In this way we will be able to see in a more accurate way than Mayr that population-genetical models and metaphors, even if they are abstract and mathematical, made the case for natural selection's creativity in a way that was far less internally prone than Darwin's to lapse into dog-eat-dog views of natural selection as a pruning force. The achievement of population-genetical Darwinism will be even more praiseworthy when we realize that it was different from Darwin's Darwinism, better than that of Darwin's immediate heirs, and might be succeeded by something more explanatorily fecund still.

Among its superiorities is that the statistical-probabilistic Darwinism of the Modern Synthesis gave clarity and weight to Darwin's conviction that where there is no variation there can be no natural selection and where there is no natural selection there can be no adaptation. Statistical framing makes this point more technically and restrictively than Darwin did. If all the variation were removed from, rather than remaining latently hiding within, a line of mice destined for laboratory experimentation we should not conclude that the heritability of their uniform traits is 100 %. On the contrary, it is zero. Heritability applies to the causes of differences (see also Moore, this volume). If there are no differences there can be no heritability. Suppose we start breeding these genetically uniform mice with each other. They begin to show differences. These differences, which are the formally constituted objects of population genetics, must be ascribed entirely to the effect of the environment. Selection and adaptation are thus made relative to the inter-defined variables of width and depth of variation, degree of heredity, and similarity of environment. The difficulty of translating into ordinary discourse the technical terms in which evolutionary scientists since Fisher have cast the theory of natural selection is immense. In popular discourse the universality with which a trait is distributed in a species—I don't say 'population' because popular discourse does not really possess that concept—is a presumptive sign of heredity. In population genetics it is not. Mayr's effort to produce a hybrid suitable for public consumption is understandable. Still, loss of connection to the public sphere is more than compensated by enhanced authority in the technical-scientific sphere, even if the public fails to grasp how different the new Darwinism is from the old Darwinism that they have long been predisposed to hate.

This gain in the scientific prowess of the theory of natural selection can be seen most readily in the way in which the new framing eliminates or reduces anomalies that had been dragged along by earlier, technically weaker concepts in Darwinian thinking. Population-genetical Darwinism does not, for example, have to rely, as Darwin did, on artificial and sexual selection to lend genuine goal-directedness and functionality to adaptations. On the contrary, the new framing rescues sexual selection from the obloquy to which it had long been subjected—Do peahens have taste? Are peacocks Victorian gentlemen? (Kellogg 1907)—by treating it as a kind of adaptive natural selection. It can do so because it does not have to conceive of natural selection as a death-dealing power, as Spencer and Darwin's Malthusianism led them to do, and so does not have to think of sexual selection, since it is clearly about fostering life, as something quite different from natural selection. One might even say that the emphasis on reproductive success in Darwin's conception of sexual selection is transferred in the new view to natural selection itself.

Perhaps more importantly, the new conceptual framing does not rely on an impinging force like Malthusian population pressure at all. All that is required for differential reproduction is that a sub-population has a slight reproductive advantage over another. In contrast to earlier theories, in which fitness was far too closely tied to violence and competition in zero-sum Malthusian conditions, even cooperative traits can be components of fitness if they enhance the probability of reproductive success. In fact, much of evolutionary science in the past 40 years has been focused on using the principles of the Modern Synthesis to explain, not explain away, the fact of cooperation in animals and humans (Sober and Wilson 1998). In a purely Malthusian world, cooperation is difficult and at best strategic, and agency or the ability to do things is scarce. By contrast, when the pioneers of the Modern Evolutionary Synthesis used metaphors that figuratively had populations climbing "adaptive hills" and "crossing adaptive valleys" until they hit upon "a new adaptive peak"—a combination of genotypes that is more reproductively successful than in closely related sub-populations—they were stressing the agency of populations, and implicitly of individual organisms, in order to distance themselves from images in which under the stern imperatives of Malthusian population pressure organisms are forced on pain of death to behave in unpleasant ways in Spencer-like environments that by forcing them to "root, hog, or die" monopolize all the causality (Dobzhansky 1937, p. 190, following Wright 1932).

There is also the matter of naturalism versus intelligent design. The statistical-probabilistic framework of the Modern Evolutionary Synthesis is less dependent than earlier Darwinisms on comparing the apparent design of natural selection with the design of artifacts in order to lend purposive functions to purely natural processes. I don't think there is much of an analogy between organisms and machines in the first place, at least any machines we actually have, including computers (see Lewens 2004; Pigliucci and Boudry 2011; but see Bechtel, this volume). But even if "design without a designer" were a better trope than it is, the great mid-twentieth century population geneticist Theodosius Dobzhansky saw a deeper and more subtle disanalogy. His point is about the *explananda* of evolutionary theory, or what it explains, more than about its *explanantia,* or what does the explaining. The probabilistic reframing of evolutionary theory brings into view phenomena about living systems that are invisible, dimly lit, or sometimes even inconceivable when natural selection is seen as offering alternative explanations for identically the same phenomena that natural theologians like William Paley thought could be explained only by intelligent design and its attendant analogy between organisms and artifacts (see Ayala and Avise in this volume; see also Brigandt, this volume for the modern version of the intelligent design movement). It was not until the remarkable stability of sex ratios, for example, was referred to the statistical notion of frequency dependence that this phenomenon became visible enough to be explained, as Fisher explained it, by fairly simple mathematics (Fisher 1930). It was not until statistical science made accessible the possibility of genetic drift—the possibility that in small, relatively isolated populations genes are more likely to go to fixation by chance—that Sewall Wright could see how drift might come to the aid of natural selection in the process of speciation (Wright 1932; Dobzhansky 1937). It was not until the phenomenon of hybrid vigor was reframed in terms of how genotypes are distributed in populations that one could see why, and how, natural selection tends to preserve genetic variation by selectively favoring heterozygotes (Dobzhansky 1937; Depew 2011a).

When at the end of his life Dobzhansky famously wrote, "Nothing in biology makes sense except in the light of evolution," he had in mind phenomena such as sex ratios, genetic drift, dominance and recessiveness, chromosomal diploidy, and heterozygote superiority (Dobzhansky 1973). He was saying that statistical-probabilistic Darwinism does more than give better explanations for phenomena that are descriptively neutral between theories that appeal to intelligent design and those that don't. Statistical-probabilistic Darwinism brings into view phenomena that are not accessible to the intelligent design paradigm *at all* or, if they are, are badly misdescribed by it. Since it takes evolution even to see such phenomena, explanations of them are not about whether they evolve, but how. Implicit in Dobzhansky's suspicion that evolutionary biologists who like to see natural selection as design without a designer (as Richard Dawkins, for example, does in *The Blind Watchmaker* in his attempt to persuade the public of Darwinism's truth by appealing to design and creationist tropes already familiar to them [Dawkins 1986]) run a risk of taking too limited a view of the very things they are commissioned to explain. The design paradigm, even when naturalized, confines Darwinism to too narrow a range of evolved and evolving phenomena (Dobzhansky 1970, p. 4).

In sum, mid twentieth-century Darwinism did not rely, as Darwin did, on the interacting metaphors of force or design. Nonetheless, it managed to give even greater voice than Darwin to the "paramount power of natural selection" in evolving genuinely functional, goal-directed, well-adapted populations and organisms. Thus it is no accident that the classical texts of the Modern Synthesis speak of natural selection, mutation, genetic drift, and gene flow as factors, processes, and agents, *not as forces or mechanisms* (Dobzhansky1937, pp. 186, 191; Mayr 1942, pp. 292–293).[6] The statistical-probabilistic reframing of evolutionary theory allows us to see how combinations of these factors or agents can propel variant genotypes through populations in ways that invite us to say that successful genotypes not only have good effects, but are successful *because* they have good, adaptive effects and not merely because their numbers have increased. Accordingly, they invite us to entertain what Darwin sometimes entertained: a modest and wholly natural form of biological teleology.[7] This approach makes natural selection as creative as Darwin took it to be without flirting with the design analogy or with forces that deal death to organisms in the way fly swatters deal death to flies.

3 When Metaphors Become Excessive: Game Theory and the Return of Design and Force

The most convincing proof that conceptual frameworks, as well as the images, similes, analogies, metaphors, and other tropes that are their signatures, are at work in evolutionary theories, and probably in scientific theories generally, is that they appear empirically empty, arbitrary, metaphysical, and ideologically contaminated as soon as they begin to outrun the phenomena they perspicuously describe, organize, and explain. When they "transcend the bounds of sense," as Kant would have put it, tropes that may once have yielded real insight into some matters of fact begin to appear as nothing more than the interpretive, or even metaphysical, preferences of this or that evolutionary biologist or school. Oddly, when this occurs, advocates of this or that conceptual framing and trope usually continue to defend it. Dogmatism

[6]The makers of the Modern Synthesis called mutation, selection, drift, and gene flow "factors" because they thought of themselves as contributing to the resolution of a long-running "factors of evolution debate" that had been unfolding ever since Spencer wrote a book with that title in 1886 (Pearce 2010).

[7]On selectionist explanations of adaptation as naturally teleological, see Wright (1976) and Brandon (1981, 1990). On Darwin's inclination to say this, see Gray (1963), p. 237, Darwin to Gray, June 5, 1874, DCD, entry 9483; see also Lennox (1993, 1994), Depew (2008), and Lennox and Kampourakis, this volume.

of this sort does nothing to damp down popular resistance to Darwinism, the origins of which lie in earlier episodes of just such dogmatism. Not surprisingly, conflicts of this sort often lead to renewed calls for empirical purity and humility in science. Let the facts speak for themselves, it is said. The trouble is that without a conceptual framework to identify the objects and processes that a theory countenances facts don't usually speak for themselves at all. The temptation to exceed a productive ratio of fact to conceptual scheme is for this reason endemic. Disciplines that enjoy robust practices of criticism and debate, however, tend to correct themselves over time. I include evolutionary biology among them (Depew 2012).

Historians and philosophers of science since Thomas Kuhn have shown how a dialectical cycle arises between criticism and the growth of scientific knowledge. In any healthy professionalized or professionalizing science new theories can be counted on to arise to account for well-attested phenomena that old theories cannot explain or explain well. When they in turn exceed their empirical warrants they are challenged by even newer theories. In just this way the Modern Evolutionary Synthesis, in the course of its relatively short life-span, has produced a "shifting balance theory" (Wright 1932), a "balancing selection theory" (Dobzhansky 1937), a "kin selection theory" (Hamilton 1964; Williams 1966), a "selfish gene theory" that builds on kin selection and supports "adaptationism" (Dawkins 1989/1976), a "punctuated equilibrium theory" that opposes it (Eldredge and Gould 1972), a theory of selection at multiple levels that says everyone is right in a way (Gould 1980; Eldredge 1985), a "niche construction theory" that attempts to take back the agency of organisms from selfish genes (Odling-Smee et al. 2003), and a "developmental systems theory" that hopes to succeed the Synthesis altogether (Oyama et al. 2001). As these competing theories contest each other, the conceptual diversity we have already spotted between Darwin's Darwinism and the Darwinism of the Modern Evolutionary Synthesis manifests itself as much if not more *within the Modern Synthesis.* Population-genetical Darwinians have long since agreed that natural selection is a creative process in the sense specified above, not mere elimination. But the metaphors cherished by different *versions* of the Modern Synthesis, as well as the paradigmatic examples prized by each—the beaks of Darwin's finches, the darker pepper moths found on soot-blackened trees, the benefits of a heterozygote mix of variant red blood cells in protecting against malaria, the cooperative effects of the lopsided haplodiploid genetic system of social insects–show that the Darwinian research tradition is still as alive with contestation as it ever was, and better off for it.

One might insist that these so called theories are only hypotheses or research programs, and so have at best heuristic or suggestive value. In fact, they are somewhat stronger than that. They identify what their advocates take to be the kinds of entities and processes over which evolutionary theory ranges. One may like to think that after criticism what remains are only framework-free empirical facts. When hypotheses about matters of fact are validated they do indeed disappear in just this way. By contrast, however, when theories find strong empirical support they not only remain theories, but become even more theoretical. This is so because theories are held together by conceptual frameworks around which an array of facts are described, ordered, and explained. Remove the theoretical scaffolding and you won't see the facts at all, let alone understand them.

It is the relativity of fact to frame that tempts advocates of this or that theory to exceed its empirical bounds. Enthusiasts of this or that theory often fantasize that they have in hand a *complete* theory, a theory that will parsimoniously explain *all* the phenomena that fall within the scope of a science and will be able to resolve *all* of its outstanding problems. This shows that theories in the sense I am identifying are often written in the future anterior tense, as if they were presenting themselves as having already done what they prospectively propose to do: derive from principles, systematically order, and store for textbook-like retrieval a large amount of empirical knowledge that has been acquired by imposing on phenomena a distinctive interpretive scheme, ontology, or way of "seeing as" that works through its attendant descriptive metaphors. This way of looking at the process of science also explains why partisans of this or that theory are so frequently inclined to expel errant colleagues from the field. Their "final theory," even if they acknowledge that is not yet quite complete, coincides in fantasy with the boundaries of the discipline itself. So they imagine that dissenting colleagues are not in the field at all. The dissenters, as you might well imagine, usually suspect more than a little question begging in such grandiose power plays, and often go public in order to blow the whistle.

Richard Dawkins's selfish gene theory is a good example of a Darwinian theory that turns explicitly on its conceptual framework or ontology and so exhibits the discursive pattern summarized in the previous paragraphs (Dawkins 1989/1976). In the genetic Darwinism of the 1940s through the 1960s organisms were seen not only as necessarily members of interbreeding populations but ontologically as integrated wholes able to respond to constant environmental change because their genomes have been harmoniously balanced by natural selection. Selfish gene theory thinks of organisms as related to genotypes in a less integrated, more atomistic way. It regards them as assemblies of separately evolved phenotypic traits, each of which has been selected for because by interacting well with the environment it increases the likelihood that the segments of DNA that code for it will leave more representatives in the next generation than they otherwise would. That genes are inherently more-making is rooted in the supposed fact that that's just what DNA, whether it codes for traits or not, does: replicate. Genes are tropologically selfish, then, because considered as chunks of DNA that tend to remain intact through meiotic division they replicate themselves whenever they can without regard to the overall good of the organisms that contain them.

Selfish gene theory not only relies on the mechanism of DNA and gene replication discovered by Crick and Watson, but explains phenomena about this mechanism, such as the selective neutrality of many variant alleles that code for structural proteins and the fact that a great deal of DNA doesn't code for structural traits at all, that the older, organism-centered theorists found so anomalous that at first they treated these observations as "non-Darwinian evolution." Simply by seeing organisms and their phenotypes as coming to be for the sake of genes rather than the other way around selfish gene theory reduces or removes these anomalies, tightens up genetic Darwinism's relation to the new molecular genetics, which didn't even exist when the Synthesis was formulated, and as a bonus explains phenomena such as social cooperation that were so at odds with what organisms are supposed to be like according to the Modern Synthesis that seemingly ad hoc hypotheses such as group selection had to be invoked to accommodate them. Selfish gene theory explains cooperation among social insects in a way that had been anticipated in the 1960s by W. D. Hamilton (1964) and G. C. Williams (1966). Organisms can be self-sacrificing if their genes are selfish. After all, they are not highly unified substantial beings but, as Dawkins colorfully puts it, ontologically no more integrated than "clouds in the sky or dust storms in the desert" (Dawkins 1989, p. 34). This being so, nothing stands in the way of genes clubbing together for mutual benefit even if the result is decreased fitness for the organisms that contain them, such as we see in sterile castes of insects. This idea is indeed anomalous for the original version of the Modern Synthesis. Selfish gene theory does not deny that the overall relationship among the genotypes and phenotypes of organisms or insect societies is balanced and harmonious. In this way and others it regards itself as a version of the Modern Synthesis, not its gravedigger. It is just that this balance is a result of summing over independent transactions between DNA and environments, not a phenomenon that has evolved *because* of the flexible agency and developmental plasticity it confers on organisms, as Dobzhansky, Mayr, and Simpson supposed. Understandably, the clash between holistic and atomistic versions of population-genetical Darwinism gave rise in the 1980s to a more "pluralistic" theory that made much of the hierarchical nature of organic structure. Selfish genes can be embedded in that structure to help to explain the tensions and trade-offs that make the genome more a repository of contingent histories than we like to imagine (see also Avise, this volume). The point, however, is that balancing selection, selfish genes, and hierarchical expansion signify three different theories, not modest tinkerings with the same 'synthetic theory' that has been around since the late 1930s.

Selfish gene theory did not arise in a vacuum. It is a more ontologically committed example of a widespread tendency among recent Darwinians to supplement population-genetical calculations with game-theoretical "payoff matrices," which were originally developed during the cold war to manage nuclear standoff (hence the imagery of 'hawks' and 'doves'; see Hamilton 1964; Williams 1966; Trivers 1971; Maynard Smith 1982; Wilson 1975; but also Sober and Wilson 1998). Images of selfishness and design come with this territory. Selfishness comes into it because you cannot use payoff matrices to explain the evolution of cooperative phenomena without assuming that each game player is trying to optimize a set of limited, contested, zero-sum goods and that the best way to go about it is to try to get as much as you can for yourself by spending as few of your resources as possible. The design idea gets into the act as follows. On the game-theoretical framing, we assume until we find otherwise that each evolved trait, behavioral no less than morphological, is as well adapted to its environment as the parts of an efficient machine. Since each trait is envisioned as a maximally functional part of a maximally functional whole, we can envision the strategy of the genes as like that of engineers. We can use this model to explore how particular traits of particular organisms are adapted, thereby displacing the received criteria for predicating the concept of adaptedness, which ask us to trace the evolutionary history of a trait to see whether it arose because it has an adaptive effect, with an engineering conception that is indifferent to history. The trope of natural selection as "design without a designer" tends for these reasons to ignore suspicions of the makers of the Modern Synthesis about the analogy between organisms and artifacts. In their well-known critique of the "adaptationist programme" Stephen Jay Gould and Richard Lewontin rehearse these complaints and add a few more (Gould and Lewontin 1979).

The atomizing, adaptationist, crypto-creationist presumptions of selfish gene theory did not come out of nowhere. Dawkins is heir to the Oxford School of Population Genetics, which in the 1940s and 1950s gave empirical weight to population genetic Darwinism by proving that earlier systematists had been wrong in presuming that the traits that mark off species such as Darwin's finches are non-adaptive (Lack 1947). In turn, population genetics at Oxford was itself heir to an earlier Oxford school of naturalists whose members had taken to arguing at the end of the nineteenth century that Darwin's theory is a very good way of letting natural laws do what Paley had God doing directly, namely, optimally adapting organisms on a trait-by-trait basis (England 2001). In doing so, they displaced Darwin's analogy with the breeder's art with an analogy between organisms and functional machines, a trope that is far less prominent in the *Origin* than we are accustomed to think. Selfish gene theory follows suit. It does so because of its ontology. The parts of machines are indeed independently designed, functionally subordinated, and potentially replaceable. All that seems to have changed at Oxford over the course of a century is that its spokesmen have become as keen on blocking the inference from design to a designer as their Edwardian predecessors were keen on having that inference go through.

Because they are committed to the design analogy selfish genes theorists are understandably more than a little sensitive on the subject of God. Their urgent need to block the inference to a designer arises from seeing organisms as exhibiting apparent design and the analogy with artifacts in the first place. Sure enough, soon a new breed of "intelligent design theorists" who were tolerably acquainted with molecular genetics was fastening upon this metaphorical framing to argue, as had the original Oxford school naturalists, that we are entitled to move from apparent design to an intelligent designer after all (Behe 1996; see Brigandt, this volume). Rather than abandoning the tropology that figures organisms and their parts as machines adapted by natural selection, however, or conceding in good positivist fashion that selfish gene theory pays its way solely by accounting empirically and parsimoniously for phenomena that are recalcitrant to orthodox, organism-centered population genetics, Dawkins and his allies rose to the bait and engaged fiercely with the new creationists. They ratcheted up the ontological commitments of selfish gene theory until it became a manifesto of science's supposed commitment to metaphysical materialism, secular Enlightenment, and, ultimately, atheism (Dawkins 2006). It is just here we can see selfish gene theory completely outrunning the empirical data that it set out to explain and indeed still illuminates. Unmoved when improved versions of group selection more congenial to organism-centered

Darwinism became available,[8] or by the growing possibility that there is little or no junk DNA after all (Pink et al. 2011), or by cascading discoveries showing that, far from being autonomous causal agents, genes are activated, and even individuated, by developmental processes (Bateson and Gluckman 2011; Stern 2011), selfish gene theory has become a virtual confession of the sin with which its neo-creationist enemies tax it. Its metaphysical commitment to materialism precedes, and so undermines, its claim to be empirical science. It fails to realize that conceptual frameworks have no standing beyond the facts they collect, order, and illuminate. It is in just this way that selfish gene theory so nicely illustrates the discursive inflation to which scientific theories are generally prone because of their built-in duality of fact and frame. Its adepts, as Wittgenstein said, have been captured by a picture.[9]

How then, can a self-respecting evolutionary biologist insure that the design-without-a-designer trope that comes with his or her game theoretical tools resists the natural-theological potential that, as the rise of the neo-Paleyan intelligent design movement shows, lurks within it? I suggest that one response to this rhetorical problem has been to re-commission the old idea of natural forces to counterbalance the design trope and to keep natural selection natural. Rather than speaking of natural selection, mutation, genetic drift, and gene flow as 'agents,' 'processes,' or 'factors' that combine to push populations away from their default (Hardy-Weinberg) equilibrium distribution of genotypes, as their intellectual forebears did, it is now fashionable to compare these factors to the way Newtonian forces deflect mass away from its default inertial motion.[10] As a result selection, drift and other factors are now commonly spoken of as 'evolutionary forces.' It is seldom realized that to do so is to part company with the makers of the Modern Synthesis, whose use of the tropological resources of statistical physics inclined them to distance themselves from the classical, force-based Newtonian paradigms to which Darwin, Wallace, and Spencer looked. Just as Darwin attempted to keep the analogy between artificial

[8]In contrast to E. O. Wilson, who has recanted his earlier embrace of genocentric kin-selection theory in favor of the trait-group selectionism of D. S. Wilson (Nowak et al. 2010).

[9]To be fair, it should be noted that Dobzhansky, too, had a weakness for metaphysics, although more on the idealist than the materialist side. From the start, he argued that natural selection is creative because it evolves adaptations for adapting and evolving. It does so by preserving variation that may prove useful in changed and changing environments and by favoring the evolution of phenotypic plasticity (Dobzhansky 1937). But toward the end of his life Dobzhansky went further. He extended his conception of natural selection's creativity to evolution's creativity in a way that was only a whisker away from the religiously inspired, progress-oriented orthogenesis of Pierre Teilhard de Chardin (Dobzhansky 1970, pp. 430–431).

[10]The origins of this analogy have little to do with the reason for its subsequent diffusion. The analogy was initially invoked by analytical philosophers of biology who were trying to decide whether the Hardy-Weinberg equilibrium formula is a law of nature. David Hull, in his pioneering Philosophy of Biological Science, evoked Newton's laws as a model, but followed the founders of the Modern Synthesis in calling natural selection, mutation, drift, and gene flows agents and processes (Hull 1974, p. 58). It was the philosopher Elliott Sober who developed the Newtonian analogy further by talking about mutation, selection, drift, and gene flow as forces (Sober 1984). He should not be held too responsible for the rhetorical uses to which this idea has been put.

and natural selection fully natural by invoking Malthusian forces, so contemporary Darwinians speak of the basic concepts of population genetics as forces to prevent the design analogy from going transcendental.

To be sure, philosophers of biology are now having second thoughts about this trope. How can genetic drift, a purely chance process, be called a force? Moreover, if drift is to be aggregated with natural selection in the way forces are combined in Newtonian physics natural selection cannot be a force either (Matthen and Ariew 2002; Pigliucci and Kaplan 2006; on drift see also Pigliucci, this volume; Dietrich, this volume).[11] These conceptual liabilities suggest that the reason for casting natural selection as a force, or sometimes a mechanism, are rhetorical, not conceptual. They have to do with keeping natural selection natural by damping down the tendency of game theoretical models, with their evocation of optimal strategies, to inflate into the ascription of intentions and rational choice. 'Force talk' gives plausible deniability to Darwinians who, having embraced the metaphor of efficient design, need to block the intelligent design inference, especially after the rise of new creationists. In this respect, contemporary Darwinism begins to look, rhetorically at least, more like Darwin's Darwinism than like the Darwinism of the Modern Synthesis in its heyday.

4 Force and Design as a Problem for Science Pedagogy

In contemporary pedagogy, 'force talk,' whose uses I have been tracing as a way of exploring the role of conceptual schemes and rhetorical strategies in evolutionary biology, is not restricted to portraying natural selection as a force. In an interesting article, Nehm et al. report that a textbook widely used to introduce biology to college students who seek careers in the biological and health sciences makes repeated use of the notion that pressures and forces coming from the environment cause evolutionary change by generating variation (Nehm et al. 2010; see also Nehm and Reilly 2007). This literal-minded interpretation of 'selection pressure,' they report, is even more evident in the lectures of teachers who must explain the book. The idea seems to be rather easily taken up by students, even though it wreaks havoc with the theory of natural selection. Nehm et al., in analyzing written descriptions of what students say they are learning, report that many of them are so focused on causal

[11] Ariew's and Matthen's critique of 'force talk' is aimed at knocking the causal wind out of population genetics in order to assign causal explanatory power (as opposed to the merely descriptive uses of population genetics) to the developmental process of individual organisms as the source of variation and the locus of adaptation. The rise of evolutionary developmentalism ("evo-devo") might well be nurturing a paradigm shift from populations back to organisms (Walsh 2006; Pigliucci and Kaplan 2006; for "evo-devo" see Love, this volume). Still, it is worth noting that the population-genetical paradigm it wishes to displace is not as wedded to the force metaphor as these critics presume and so does not rise and fall with it. As I have shown, the metaphor has a recent origin, not an ancient pedigree.

pressures exerted on organisms that in practice they forget that natural selection works on random genetic variation, thereby reviving the shades of Lamarck and Spencer. Perhaps unwittingly, efforts to keep natural selection natural by portraying it as a force are bringing back nineteenth century ways of thinking about evolution that today's evolutionary scientists would surely disavow.

I do not find as much overt pressure or force talk in this book as Nehm and his colleagues do, at least in the edition that I read (Campbell et al. 2009, 6th ed.). But I do find key concepts defined or more often simply used in a way that invites the sort of response Nehm and his co-authors report. To its credit, this textbook, which must prepare students for anatomy, physiology, and other aspects of "functional biology," consistently frames organisms in an evolutionary way. Nor is it genocentric. In fact, it is so organism-centered that in generalizing Mendel's rules to "gene pools" and giving instructions on how "mechanisms" like natural selection disturb the resulting Hardy-Weinberg equilibrium in calculable ways it conveys only a weak sense of what the populations of population biology actually are (p. 258). It oscillates between implying that we can see what is going on in individual organisms better if we look at them as members of groups and saying (in a way that fails to distinguish units of evolution like species from units of selection like groups, organisms, or genes) that populations rather than organisms are uniquely "the units of evolution" (p. 264). Is that why the text fails to make a categorical enough distinction between adaptation to environments and, say, acclimatization to high altitudes? More troubling is its claim that "the commonly used phrases 'struggle for existence' and 'survival of the fittest' are misleading if we take them to mean direct competitive contests" (p. 269). In going on to say that "more subtle and passive" processes like mimicry also count (as what?) it gives the impression that the phrases 'survival of the fittest' and 'struggle for existence' may be retained if they are more broadly defined (p. 269). This is unfortunate. Not coincidentally perhaps, the text also says that natural selection is "more of an editing process than a creative mechanism" (p. 259). This is clearly to invite regression to forms of Darwinism that preceded the Modern Synthesis. The letter of this book may be consistent with the fruits of population genetical Darwinism. But in making good on its promise to support Dobzhansky's famous claim that "Nothing in biology makes sense except in the light of evolution," the spirit of the evolutionary theory it teaches is actually a Darwinized version of Spencerism that Dobzhansky would have repudiated.[12]

Or, rather, the text fails to guard against that interpretation. Research in science education has shown that students are not empty vessels waiting to be filled. They

[12] Ironically, the introductory biology textbook that some of the same authors have written for non-majors does a better and more coherent job of inviting students to imaginatively project themselves into an evolutionary and ecological perspective. Because it does not have to teach all the concepts that majors and future health professionals will need it is not burdened with smoothing out the inherent heterogeneity of these concepts, which were developed at different stages in the history of biology and often contain in the way they are named and described traces of the conceptual frameworks in which they first arose.

come with "passing theories" about almost everything they encounter in school. The task of education is to help students clarify, amend, and in almost all cases replace the inadequate accounts they bring to the scene of inquiry with better ones. It has been shown that the best way to do this is to have students try to articulate the views they already have and to keep writing about the process by which their initial views are being displaced and replaced by more adequate ones. Otherwise, they will tend to assimilate new facts and language to the same old way of understanding (for relevant topics, see Vosniadou 2008).

This seems to me especially true in education about evolution. Our culture is strewn with the litter and wreckage of old evolutionary theories and their encounters with religion and politics. It requires only the slightest stimulus for a textbook to reignite Spencerian, Lamarckian, or intelligent design ideas in students. It takes even less effort for them to turn these images, once revived, into full-blown Lamarckism re-written in the language of Darwinism or, if they are as bothered about 'the survival of the fittest' as their great-grandparents were, to reject evolution altogether, or, more commonly, to split the difference by thinking of evolution by natural selection as "design on the installment plan" (see Smith 2010). In the course of doing any of these things, students may think they are being spoon fed a theory that in reality they are probably bringing, mostly unconsciously, to the text, the classroom, and the lab. The fact that they typically keep quiet about it should not be taken as a good thing.

One can hardly expect students to understand the theory of natural selection if they are exposed to incoherent versions of it. It is even less seemly to blame them for misunderstanding it. The first imperative in re-energizing evolutionary pedagogy, accordingly, is to ensure that textbook writers and teachers acquire a genuinely Darwinian view of natural selection and adaptation. This is not as easy as it sounds. Natural selection, which exquisitely balances elements of chance, purposiveness, and environmental determinism, has eluded the grasp of very knowledgeable people. Even harder is to grasp the number of questions that remain even when the concept has been correctly analyzed. These open questions give rise to the pluralism about Darwinian theories that we have reviewed in the previous sections of this paper. If there is one recommendation that would put textbook writers, teachers, and students on the right track, however, it would be to advise them to embrace an etiological or natural historical approach to evolved traits and to flee from thinking of adaptedness as optimal engineering design. I agree with Lennox and Kampourakis on this point (Lennox and Kampourakis, this volume; Kampourakis 2013). This approach, unlike the currently fashionable history-free adaptationism, will show students why genuinely adaptive traits are genuinely there for a reason and will do so without suggesting that they were designed this way, either by God or by natural selection (Kampourakis and Zogza 2007, 2008, 2009).

I wonder whether it might also be helpful in steering students away from errors whether textbooks and teachers should also build into their exposition of evolutionary concepts some of the intellectual etiology or cultural history of those very concepts. The idea is to avoid the tendency of textbooks to eliminate the history of the ideas they repackage. One might learn, for example, that

Darwin, who is usually figured in biology textbooks as a mythical hero to whom we might attribute virtually everything that has subsequently been confirmed, did not suddenly intuit either the fact of evolution or his way of explaining it on his youthful voyage on the *Beagle*, but made his discoveries over an extended period of time in the midst of intense doubts and controversies (*pace* the impression left by Campbell et al. 2009, pp. 256–57; see McComas 1997). One might learn, too, that Darwin was not the only Darwinian and that different people are responsible for most of the sometimes ill assorted concepts they are being asked to master, as well as for lending support to them by finding paradigmatic examples that are all too easily over-extended. Dobzhansky should be mentioned, for example, or better yet pictured, when heterozygote superiority and the case of sickle cell anemia are discussed. But why not also admit that neither he nor Mayr would have been a fan of kin selectionism and that one of its early champions, E. O. Wilson, has now dissociated himself from it? Why not display Dawkins, Wilson, Gould, Lewontin and others in debate (Ruse 1999)? Why not acknowledge that, while it is true that over time deposits in the bank account of evolutionary knowledge keep growing larger and paying more handsome dividends, some of what students before their time learned has been withdrawn from the account and that the same thing might happen to some of what they are learning?

This is, I admit, dangerous. But a strategy of teaching evolutionary biology that acknowledges controversy as well as textbook-like accumulation and in doing so acquaints students with the interesting people who have contributed to evolutionary knowledge, even when they also cherished some very wacky notions, might be worth the risk if it helps bring students' own tacit presuppositions to the surface so that by means of dialogue with teachers and fellow students they can reflectively replace them with clearer and more empirically adequate ideas.

To say such things is, of course, to fly in the face of deeply embedded conventions of science textbook writing. These virtually require a rhetoric in which facts are accumulated in a value-free, impersonal way. To depart from these conventions is to risk causing rather than curing whatever skepticism students bring to the study of science. Still, there are good reasons for bending the genre rules for evolutionary biology textbooks, or at least for initiating pedagogical experiments in which the results of conventional and non-conventional teaching models are systematically compared to see what happens. These reasons have to do with the charged emotional field in which students encounter evolutionary biology. The first is that students are prone to misinterpret the idea of natural selection because they are already attuned to false or inadequate commonplaces about it that are constantly recycling in the cultural background. Better to get these commonplaces out into the open. Second, biology textbooks and teachers are likely to make a serious mistake if they try to appeal to these commonplaces by amending them rather than replacing common sense approaches to biology's subject, organisms, with conceptual frameworks that ask students to grasp such abstract objects as statistically defined populations by statistical methods. Such departures from common sense, requiring imaginative, logical, and mathematical leaps on the part of learners, are the very stuff of mature sciences. Students learn them in physics and chemistry. Why not in biology too?

Acknowledgments I wish to thank André Ariew, Deirdre Egan, Kostas Kampourakis, and Jaime Sabel for substantial and collegially offered help in developing the ideas in this chapter. I developed its basic picture of Darwinism's history in my long collaboration with Bruce Weber.

References

Ariew, A., and R.C. Lewontin. 2004. The confusions of fitness. *The British Journal for the Philosophy of Science* 55: 347–363.
Barrett, P. (Ed). 1987. *Charles Darwin's notebooks*, ed. P. Barrett. Chicago: University of Chicago Press.
Bateson, P., and P. Gluckman. 2011. *Plasticity, robustness, and evolution*. Cambridge: Cambridge University Press.
Beer, G. 2000. *Darwin's garden plots*. Cambridge: Cambridge University Press. X. 1st ed. 1985.
Behe, Michael. 1996. *Darwin's black box*. New York: Free Press.
Black, M. 1962. *Models and metaphors*. Ithaca: Cornell University Press.
Brandon, R. 1981. Biological teleology: Questions and explanations. *Studies in History and Philosophy of Science* 12: 91–105.
Brandon, R. 1990. *Adaptation and environment*. Princeton: Princeton University Press.
Campbell, N., J. Teece, M. Taylor, E. Simon, and J. Dickey. 2009. *Biology: Concepts and connections*, 6th ed. San Francisco: Pearson Benjamin Cummings.
Darwin, F. (ed.). 1959. *Life and letters of Charles Darwin*. New York: Basic Books. With introduction by George Gaylord Simpson.
Darwin, C. 1966. *On the origin of species. Facsimile of the 1st edition*, 1859, London: J. Murray, ed. E Mayr. Cambridge, MA: Harvard University Press.
Dawkins, R. 1986. *The blind watchmaker*. New York: Norton.
Dawkins, R. 1989/1976. *The selfish gene*. Oxford/New York: Oxford University Press.
Dawkins, R. 2006. *The god delusion*. Boston: Houghton Mifflin.
Depew, D. 2008. Consequence etiology and biological teleology in Aristotle and Darwin. *Studies in History and Philosophy of Biological and Biomedical Sciences* 38(4): 379–390.
Depew, D. 2009. The rhetoric of Darwin's origin of species. In *The Cambridge companion to the origin of species*, ed. M. Ruse and R. Richards, 237–255. Cambridge: Cambridge University Press.
Depew, D. 2011a. Adaptation as process: The future of Darwinism and the legacy of Theodosius Dobzhansky. *Studies in the History and Philosophy of Biology and the Biomedical Sciences* 42: 89–98.
Depew, D. 2011b. Accident, adaptation, & teleology in Aristotle, Empedocles, and Darwinism. In *Biological evolution: Facts and theories: A critical appraisal after 150 years after the origin of species*, ed. G. Auletta, M. Leclerc, and R. Martinez, 461–478. Rome: Gregorian and Biblical Press.
Depew, D. 2012. The rhetoric of evolutionary theory. *Biological Theory*. doi:10.1007/s13752-012-0054-2.
Depew, D., and B. Weber. 1995. *Darwinism evolving: Systems dynamics and the genealogy of natural selection*. Cambridge, MA: MIT Press.
Dobzhansky, T. 1937. *Genetics and the origin of species*. New York: Columbia University Press.
Dobzhansky, T. 1970. *Genetics of the evolutionary process*. New York: Columbia University Press.
Dobzhansky, T. 1973. Nothing in biology makes sense except in the light of evolution. *The American Biology Teacher* 35: 125–129.
Eldredge, N. 1985. *Unfinished synthesis*. New York: Oxford University Press.
Eldredge, N., and S.J. Gould. 1972. Punctuated equilibria: An alternative to phyletic gradualism. In *Models in paleobiology*, ed. T.J.M. Schopf, 82–115. San Francisco: Freeman.

England, R. 2001. Natural selection, teleology, and the logos: From Darwin to the Oxford Neo-Darwinists, 1859–1909. *Osiris*, 270–287, 2nd Series, vol. 16, Science in theistic contexts: Cognitive dimensions. Chicago: University of Chicago Press.
Fisher, R.A. 1930. *The genetical theory of natural selection*. Oxford: Oxford University Press.
Fuller, S. 2007. *Science v. religion? Intelligent design and the problem of evolution*. Cambridge: Polity Press.
Gayon, J. 1995. Sélection naturelle ou survie des plus aptes? Éléments pour une histoire du concept de fitness dans la théorie évolutionniste. In *Nature, histoire, société: Essais en hommage à Jacques Ro er*, ed. C. Blanckaert, J.L. Fischer, and R. Rey, 263–287. Paris: Klincksieck.
Gigerenzer, G. (ed.). 1990. *The empire of chance*. Cambridge: Cambridge University Press.
Gould, S.J. 1980. Is a new and general theory of evolution emerging? *Paleobiology* 6: 119–130.
Gould, S.J. 1989. The wheel of fortune and the wedge of progress. *Natural History* 19: 14–21.
Gould, S.J., and R.C. Lewontin. 1979. The spandrels of San Marco and the Panglossian paradigm: A critique of the adaptationist programme. *Proceedings of the Royal Society of London Series B* 205(1161): 581–598.
Gray, A. 1963. *Darwiniana*. Cambridge, MA: Harvard University Press. Originally published in 1876.
Hamilton, W.D. 1964. Genetic evolution of social behavior. *Journal of Theoretical Biology* 7(1): 16–52.
Hesse, M. 1963. *Models and analogies in science*. London: Sheed and Ward.
Hodge, M. 1977. The structure and strategy of Darwin's 'long argument'. *British Journal for the History of Science* 10: 237–246.
Hodge, M. 1985. Darwin as lifelong generation theorist. In *The Darwinian heritage*, ed. D. Kohn, 207–243. Cambridge: Cambridge University Press.
Hofstadter, R. 1944. *Social Darwinism in American thought*. Philadelphia: University of Pennsylvania Press.
Hull, D. 1974. *Philosophy of biological science*. Englewood Cliffs: Prentice Hall.
Kampourakis, K. 2013. Teaching about adaptation: Why evolutionary history matters. *Science & Education* 22(2): 173–188.
Kampourakis, K., and V. Zogza. 2007. Students' preconceptions about evolution: How accurate is the characterization as "Lamarckian" when considering the history of evolutionary thought? *Science & Education* 16(3–5): 393–422.
Kampourakis, K., and V. Zogza. 2008. Students' intuitive explanations of the causes of homologies and adaptations. *Science & Education* 17(1): 27–47.
Kampourakis, K., and V. Zogza. 2009. Preliminary evolutionary explanations: A basic framework for conceptual change and explanatory coherence in evolution. *Science & Education* 18(10): 1313–1340.
Kellogg, V. 1907. *Darwinism today*. New York: Holt.
Lack, D. 1947. *Darwin's finches*. Cambridge: Cambridge University Press.
Lakoff, G., and M. Johnson. 1980. *Metaphors we live by*. Chicago: University of Chicago Press.
Lennox, J.G. 1993. Darwin was a teleologist. *Biology and Philosophy* 8: 405–421.
Lennox, J.G. 1994. Teleology by another name: A reply to Ghiselin. *Biology and Philosophy* 9: 493–495.
Lewens, T. 2004. *Organisms and artifacts*. Cambridge, MA: MIT Press.
Matthen, M., and A. Ariew. 2002. Two ways of thinking about fitness and natural selection. *Journal of Philosophy* 99: 55–83.
Maynard Smith, J. 1982. *Evolution and the theory of games*. Cambridge: Cambridge University Press.
Mayr, E. 1942. *Systematics and the origin of species*. New York: Columbia University Press.
Mayr, E. 1978. Evolution. In *Evolution: A scientific American book*, 2–11. San Francisco: Freeman.
Mayr, E. 1980. Prologue: Some thoughts on the history of the evolutionary synthesis. In *The evolutionary synthesis*, ed. E. Mayr and W. Provine, 1–48. Cambridge, MA: Harvard University Press.
Mayr, E. 1988. *Toward a new philosophy of biology*. Cambridge, MA: Harvard University Press.
Mayr, E. 1991. *One long argument: Charles Darwin and the genesis of modern evolutionary thought*. Cambridge, MA: Harvard University Press.

McComas, W.F. 1997. The discovery and nature of evolution by natural selection: Misconceptions and lessons from the history of science. *The American Biology Teacher* 59(8): 492–500.
Nehm, R.H., and L. Reilly. 2007. Biology majors' knowledge and misconceptions of natural selection. *BioScience* 57: 263–272.
Nehm, R., M. Rector, and M. Huy. 2010. 'Force talk' in evolutionary explanation: Metaphors and misconceptions. *Evolution Education and Outreach* 3: 605–613.
Nowak, M., C. Tarnita, and E.O. Wilson. 2010. The evolution of eusociality. *Nature* 466: 1057–1062.
Odling-Smee, J., K. Laland, and M. Feldman. 2003. *Niche construction*. Princeton: Princeton University Press.
Oyama, S., P. Grifftihs, and R. Gray (eds.). 2001. *Cycles of contingency: Developmental systems and evolution*. Cambridge, MA: MIT Press/Bradford Books.
Pearce, T. 2010. From 'circumstances' to 'environment': Herbert Spencer and the origins of the idea of organism–Environment interaction. *Studies in History and Philosophy of Biological and Biomedical Sciences* 41: 241–252.
Peckham, M. (ed.). 1959. *The origin of species: A variorum text*. Philadelphia: University of Pennsylvania Press.
Pigliucci, M., and M. Boudry. 2011. Why machine-information metaphors are bad for science and science education. *Science & Education* 20(5–6): 453–471.
Pigliucci, M., and J. Kaplan. 2006. *Making sense of natural selection*. Chicago: University of Chicago Press.
Pink, R., K. Wicks, D.P. Caley, E.M. Punch, L. Jacobs, and D.R.F. Carter. 2011. Pseudogenes: Pseudo-functional or key regulators in health and disease. *RNA* 17: 792–798. doi:10.1261/rna.2658311.
Provine, W. 1971. *The origins of theoretical population genetics*. Chicago: University of Chicago Press.
Ricoeur, P. 1978. *The rule of metaphor*. London: Routledge.
Ruse, M. 1975. Darwin's debt to philosophy: An examination of the influence of the philosophical ideas of John F.W. Herschel and William Whewell on the development of Charles Darwin's theory of evolution. *Studies in History and Philosophy of Science* 6: 159–181.
Ruse, M. 1999. *Mystery of mysteries: Is evolution a social construction?* Cambridge, MA: Harvard University Press.
Ruse, M. 2000. Darwin and the philosophers: Epistemological factors in the development and reception of the origin of species. In *Biology and epistemology*, ed. R. Creath and J. Maienschein, 3–26. Cambridge: Cambridge University Press.
Smith, M. 2010. Current status of research in teaching and learning evolution. II. Pedagogical issues. *Science & Education* 19: 539–571.
Sober, E. 1984. *The nature of selection*. Cambridge, MA: MIT Press. 2nd ed. 2000.
Sober, E., and D. Wilson. 1998. *Unto others*. Cambridge, MA: Harvard University Press.
Spencer, H. 1864. *Principles of biology*. London: John Chapman.
Stauffer, R. (ed.). 1975. *Charles Darwin's natural selection. Second part of his 'big species book'*. Cambridge: Cambridge University Press.
Stern, D.L. 2011. *Evolution, development, and the predictable genome*. Greenwood Village: Roberts & Company Publishers.
Trivers, R.L. 1971. The evolution of reciprocal altruism. *The Quarterly Review of Biology* 46(1): 35–57.
Vosniadou, S. (ed.). 2008. *International handbook of research on conceptual change*. New York/London: Routledge.
Walsh, D.M. 2006. Organisms as natural purposes: The contemporary evolutionary perspective. *Studies in History and Philosophy of Biological and Biomedical Sciences* 37: 771–791.
Weinberg, S. 1992. *Dreams of a final theory*. New York: Pantheon.
Williams, G.C. 1966. *Adaptation and natural selection*. Princeton: Princeton University Press.
Wilson, E.O. 1975. *Sociobiology: The new synthesis*. Cambridge, MA: Harvard University Press.
Wright, S. 1932. The roles of mutation, inbreeding, crossbreeding, and selection in evolution. *Proceedings of the Sixth International Congress on Genetics* 1: 355–366.
Wright, L. 1976. *Teleological explanations*. Berkeley/Los Angeles: University of California Press.

Debating the Power and Scope of Adaptation

Patrick Forber

1 Adaptation and Adaptationism

Adaptationism refers to an unruly family of views about the power, prevalence, methodological privilege or explanatory importance of natural selection (see also Dietrich, this volume). Yet the contemporary debate over perils and advantages of this collection of views, including the effort to disentangle them, is peculiarly shaped around a single paper: "The spandrels of San Marco and the Panglossian paradigm: A critique of the adaptationist programme" (Gould and Lewontin 1979), known now simply as the *Spandrels* paper. By using architectural and anthropological examples with rhetorical flourish, Gould and Lewontin criticized the inference from current utility to evolutionary origin: "One must not confuse the fact that a structure is used in some way [...] with the primary evolutionary reason for its existence and conformation"(1979, p. 587). The problem that confronts the inference is that structures, objects, traits, and characters can be co-opted and put to uses different than their evolutionary origin.[1] Although analogies between artifacts and organisms can be misleading in a variety of ways (Kampourakis 2013), a simple one can help clarify the difference between origin and current use. I have co-opted my office chair to serve as an end table for stacks of paper and coffee mugs, and indeed this is the primary reason I keep an extra chair in my office, but that does not change the fact that the chair is supposed to serve as a seat for a person, not an end table for a disorganized philosopher. It is testament to the importance of this inference pattern to evolutionary biology, from current utility to origin, that it raises such broad concerns about evidence, method, and explanation.

[1]Indeed, Gould and Vrba (1982) found this phenomenon so prevalent in evolutionary history that they recommended a new term for adaptations co-opted to play a different role: *exaptation*.

P. Forber (✉)
Department of Philosophy, Tufts University, Medford, MA, USA
e-mail: Patrick.Forber@tufts.edu

K. Kampourakis (ed.), *The Philosophy of Biology: A Companion for Educators*, History, Philosophy and Theory of the Life Sciences 1, DOI 10.1007/978-94-007-6537-5_8,

Before investigating the controversy over adaptation*ism*, we need to address the concept of *adaptation*. Let me start with some clarifications. The concept of adaptation has a rich history that reaches back before Darwin (see, e.g., Amundson 1996). The concept also suffers from a process/product ambiguity: it can refer to the process of natural selection adapting organisms to their environments, or to the evolutionary products of natural selection. Although both uses play important roles in evolutionary biology, I will focus primarily on the product sense of adaptation.

What, then, is an adaptation? Prima facie, an adaptation is a trait evolved via natural selection because it plays a particular role in the organism's struggle for existence, and plays this role well. Consider the following adaptive hypothesis: the bioluminescent light organ of the Hawaiian bobtail squid is an adaptation for camouflage against the night sky in their aquatic environment. Predators often lurk below, looking for shadows of passing prey against the sparkling night sky. The light organ provides bioluminescence to eliminate the shadow, allowing the squid to blend in with the stars and moon above. This adaptive hypothesis involves a rich set of empirical commitments about squid lifeways, their evolutionary history, and the nature of their environment. The squid should be nocturnal, face a predation risk from below, regulate their bioluminescence to blend in with the ambient light levels, and so on (see Nyholm and McFall-Ngai 2004). How to test the hypothesis is an important thread in the adaptationism debate (see below). How to determine the extent of the empirical commitments of an adaptive hypothesis is, in part, a matter of definition. One question about adaptation makes clear that the prima facie definition needs to be made more precise: can there be adaptations without selection?

If the definition of adaptation is *historical* then the answer must be no. Sober provides a canonical defense and formulation of the historical definition: "*A* is an adaptation for task *T* in population *P* if and only if *A* became prevalent in *P* because there was selection for *A*, where the selective advantage of *A* was due to the fact that *A* helped perform task *T*" (1984, p. 208). Thus, to say that some trait is an adaptation is simply to say that there is a history of selection for trait *A* (to do *T* in *P*). The historical definition has a number of advantages and has achieved something of a consensus (Lewens 2007; Kampourakis 2013). Despite this, there is an alternative definition that focuses not on the history of natural selection, but on the performance of an organism in its current environment.

Gould and Lewontin opt for such an *ahistorical* definition of adaptation in the *Spandrels* paper. They claim that adaptation is "the good fit of organisms to their environment," and that selection can be "decoupled" from adaptation by other processes, such as developmental plasticity, that can produce the striking lock and key fit between a trait and the environment (Gould and Lewontin 1979, p. 592). Reeve and Sherman (1993) give one of the most thorough defenses of an ahistorical definition. They formu late the definition as follows: "An adaptation is a phenotypic variant that results in the highest fitness among a specified set of variants in a given environment" (Reeve and Sherman 1993, p. 9).[2] The rationale for removing history from the definition of

[2]Evolutionary biologists take fitness to be a measure of reproductive success, usually expected number of offspring. However, there are a variety of fitness measures available and determining the correct interpretation of fitness is notoriously complex (see, e.g., Ariew and Lewontin (2004) or Beatty (1992)).

adaptation is that we usually can determine fitness differences among variation in a population, but we usually do not have access to the evolutionary history. Getting that access to history is but one goal of evolutionary research. Another is explaining why certain traits do better than others in a given environment, irrespective of history. Reeve and Sherman argue that focusing on these "phenotypic existence" questions structures evolutionary research in the right way: first we establish current utility, then we can evaluate whether and how that counts as evidence for past selection processes.

Must adaptations always play their role well? In short, no. Consider Gould's (1980) example of the panda's thumb. The so-called thumb evolved from a wrist bone, forming a notch that pandas use to strip leaves from bamboo shoots. Pandas spend most of their day occupied with this activity. An opposable thumb would do better at this task, but due to the contingencies of history and the constraints of the ancestral morphology this evolutionary innovation was unavailable to panda ancestors. Another set of examples, discussed originally by Darwin, concern the contrivances of orchids. Different orchid species have converged on various petal arrangements, but achieved these arrangements via the evolution of different developmental pathways. Some species evolved an efficient unfolding, whereas others evolved a more tortuous pathway to achieve the same result. Such developmental disparity evolved despite the species sharing a common ancestor.[3] Judgments about how well a trait may play its role depend on contrasts with biological structures that, due to the contingency of history, are not available to particular lineages. Notice that the panda's thumb and the contrivances of orchids still count as adaptations on the historical definition, for there are histories of natural selection for those traits. However, they only count as adaptations on the ahistorical definition when we artificially constrain the set of variants to just those historically available. That is, we make a covert appeal to history to get the classification right.

That adaptations need not play their role well is a primary reason for adopting the historical definition, thus divorcing the concept of adaptation from any notion of design. Yet, as Lloyd (2007, p. 52) argues, some notion of design is often connected to the concept, and this leads her to distinguish between a *product of selection* (historical) sense and an *engineering* (ahistorical) sense of adaptation. The engineering sense of adaptation is perhaps better construed as *adaptedness*. As Brandon (1990) makes clear, the relative adaptedness of a trait must be evaluated within a specific selective environment, and it is differential adaptedness that drives evolution by natural selection. I find the historical definition of adaptation to be the best explication of the concept of adaptation, reserving the concept of adaptedness to capture ahistorical notions of performance, fitness advantage, and fit to the environment. So I will follow consensus and adopt the historical definition: a trait *A* is an adaptation (to do task *T* in population *P*) if and only if there is a history of direct selection for that trait *A* (to do *T* in *P*). Notice that the claim some trait is an adaptation applies to a single population. A trait being an adaptation in one lineage (e.g., the light organ for camouflage in bobtail squid) does not entail that similar traits are adaptations in other lineages (e.g., light organs in deep sea fishes); these are separate empirical claims.

[3] Beatty (2004) and Beatty and Desjardins (2009) provide excellent discussion of this case in support of their analysis of why evolutionary history matters.

While adaptationism, in its many guises, is intimately connected with natural selection, the debate does not engage the nature of the process. Some version of a standard recipe is presupposed: evolution by natural selection occurs when there is variation in some trait, when that variation translates into differences in evolutionary fitness, and when that variation is heritable. It turns out that this standard recipe makes some interesting idealizations, and it does not line up exactly with more mathematical treatments of natural selection (Godfrey-Smith 2009). I suspect there are links between foundational issues about the nature of natural selection and aspects of the adaptationism debate, a suspicion I explore below. Yet much of the debate is independent of the foundational concern, and I will focus on those aspects first.

2 Different Kinds of Adaptationism

There are a number of ways to disentangle the knot of issues tied up in the adaptationism debates and these ways differ in the grain of resolution they impose on the debate. The most important coarse-grained distinction separates *empirical* from *methodological* issues. We might ask an empirical question about the prevalence of adaptation across the biological world, or about the power of the process of natural selection. Or we might ask methodological questions about how we should go about investigating the world. While the answers to the empirical questions may inform our methods for investigating the world, there is some important separation. We need to first adopt methods of investigation in order to answer the empirical questions and so we should expect broader concerns to enter into debate over methodological views.

The distinction between empirical and methodological strands is apparent in the *Spandrels* paper itself. Part of Gould and Lewontin's argument attempts to show that evolutionary factors other than natural selection are often in play. Neutral molecular evolution and constraints imposed by genetic and developmental structure are two important alternatives they cite (see also Dietrich, this volume; Love, this volume). Another part of their argument is independent of these empirical concerns. They criticize the methods deployed by so-called adaptationists to test their hypotheses, arguing that they fail to provide adequate evidence for natural selection.

There are a variety of finer-grained typologies for distinguishing different flavors of adaptationism (see, e.g., Amundson 1988, 1990; Sober 1996; Godfrey-Smith 2001; Lewens 2009; Orzack and Forber 2010). For my purposes, Godfrey-Smith (2001) provides a useful way to untie that knot that distinguishes between three kinds of adaptationism. The three kinds are (quoted from Godfrey-Smith 2001, pp. 336–337)[4]:

[4]While the definitions are directly quoted from Godfrey-Smith (2001), I have changed to order to fit my exposition.

> *Empirical adaptationism*: Natural selection is a powerful and ubiquitous force,[5] and there are few constraints on the biological variation that fuels it. To a large degree, it is possible to predict and explain the outcome of evolutionary processes by attending only to the role played by selection. No other evolutionary factor has this degree of causal importance. [...]
>
> *Methodological adaptationism*: the best way for evolutionary science to analyze biological systems is to invoke adaptation and natural selection; alternatives to natural selection need not be considered in most cases. [...]
>
> *Explanatory adaptationism*: The apparent design of organisms, and the relations of adaptedness between organisms and their environments, are the *big questions*, the amazing facts in biology. Explaining these phenomena is the core intellectual mission of evolutionary theory. Natural selection is the key to solving these problems—selection is the *big answer*. Because it answers the biggest questions, selection has unique explanatory importance among evolutionary factors.

In short, the empirical kind makes a claim about the world, the methodological kind makes a claim about how best to conduct evolutionary inquiry, and the explanatory kind makes a claim about the philosophical priority of certain explanations. While there are natural lines of support between positive and negative positions on the three kinds of adaptationism, they are logically independent (Godfrey-Smith 2001). So, for example, even if adaptations are rare we might still reasonably think that looking for them first is a good method, or that explaining these rare adaptations is the intellectual mission of evolutionary biology.

In the sections that follow, I will discuss issues raised by each kind of adaptationism with the goal of revealing the importance and diversity of lessons one can learn from scientific controversy.

3 Making Claims About the World

The thesis of empirical adaptationism makes a claim about the world, and so to resolve controversy on this front we simply need to check the world. Unlike the other kinds, we can *test* empirical adaptationism. Of course, how to test the thesis is a difficult issue and involves making some methodological commitments. So there are covert connections between empirical and methodological theses that prevent controversy from disappearing completely, connections that will surface over the next two sections. Despite complications, there is a clear and useful proposal on how to go about checking the world to see if empirical adaptationism is true.

Orzack and Sober (1994a, b, 1996) develop an ensemble test of (empirical) adaptationism. The ensemble test works by identifying what it means for natural selection to provide a sufficient explanation for a single trait, then evaluating whether this is true for most biological traits (see Potochnick, this volume for explanation in biology). To get clear on the test we need to analyze what it means to count as a sufficient explanation, and how an evaluation of the entire wide and wild biological world is supposed to go.

[5]See Depew this volume on this topic.

Consider three claims of increasing strength about some trait *T* of an individual member of some population (quoted from Orzack and Sober 1994, p. 362):

> (U) Natural selection played some role in the evolution of *T*. (U stands for ubiquitous since we believe this proposition applies to most traits.)
>
> (I) Natural selection was an important cause of the evolution of *T*. (I stands for important.)
>
> (O) Natural selection is a sufficient explanation of the evolution of *T*, and *T* is locally optimal. (O stands for optimal)

Notice that these claims do not deny the operation of other evolutionary processes. Drift, mutation, constraints and other factors are always at play to some degree in any natural population (see Pigliucci, this volume; Dietrich, this volume for the concept of drift). Instead, these claims focus on the relative importance of natural selection versus other non-selective factors for the evolution of trait *T*, and identify important benchmarks on this spectrum.

Determining relative importance involves assessing the sufficiency of an evolutionary explanation for trait *T* that invokes only natural selection. Obviously, a sufficient explanation is not a complete explanation. Complete explanations are rarely useful—they include every detail, no matter how minute—and so we often seek explanations that invoke one or a few factors that made the difference. Thus, Orzack and Sober devote a majority of their analysis to providing an account of when natural selection *alone* made the difference; that is, to distinguishing (O) from (I) for a given trait *T*. Briefly, they argue that the appropriate test for (O) contrasts the predictions of an uncensored evolutionary model with a censored model. An uncensored model includes *all* evolutionary factors, from natural selection to mutation pressures, drift, constraints, etc. A censored model, on the other hand, only includes natural selection. Optimality models, an important tool for representing fitness differences among traits in a given environment, are examples of censored models.[6] If the predictions of the censored model fit the observations (in a rigorous statistical way), then (O) is taken to be true.

Brandon and Rausher (1996) raise two interesting objections to Orzack and Sober's analysis that track contours in the adaptationism debate. The first objection concerns the inclusion of the claim that *T* is locally optimal in the formulation of (O). The use of optimality models is generally associated with adaptationist views; the very idea that evolution optimizes traits to the prevailing environmental conditions seems to presume the truth of empirical adaptationism. While optimality models do presume that natural selection is operating, sophisticated application of the models cannot be dismissed in this way. Such models must incorporate underlying constraints, and they can provide quantitative predictions that can be tested rigorously. As Orzack and Sober (1994b, 1996) argue, mere qualitative fit between model and data is not rigorous enough. So, for example, in the squid bioluminescence case, observing that squid increase bioluminescence in the presence of predators provides a qualitative test of the adaptive hypothesis. Observing that squid finely tune

[6]For canonical defenses of optimality modeling see Parker and Maynard Smith (1990) and Seger and Stubblefield (1996).

bioluminescence to ambient light levels, and that this degree of fine-tuning correlates with decreased predation risk provides a better, quantitative test. Jones and Nishiguchi (2004) conducted just such a test and their results support the adaptive hypothesis. Furthermore, while optimality modeling may be a natural tool for an adaptationist, many uses of the modeling technique in biology need not make strong adaptationist commitments (Potochnik 2009).

A second, related objection concerns the contrast between censored and uncensored models as a testing protocol for determining whether natural selection provides a sufficient evolutionary explanation for some trait *T*. As Brandon and Rausher put the problem, "…failing to reject the selection (O-type) model says nothing about whether there are other, I-type or U-type models that might also fail to be rejected when their predictions are compared with observations" (1996, p. 190). This raises deep issues about the nature of scientific testing that I will discuss in more detail in the next section, but let me mention two points here. First, the effectiveness of a testing protocol is limited by the alternative hypotheses considered. Second, testing protocols should balance complexity of the model and fit to data.[7] This is because more complex models face a higher risk of *over-fitting* the data and compromising predictive accuracy. Recall that the uncensored model includes all the evolutionary factors and the censored model includes only selection, excluding many other factors. Thus, the comparison between uncensored and censored models involves a difference in complexity that can complicate the test. As Godfrey-Smith (2001, p. 344) suggests, perhaps a better testing protocol would contrast models of comparable complexity.

In short, determining whether natural selection provides a "sufficient explanation" for some trait is truly difficult. The difficulties arise in the implementation of the testing protocol for individual traits. Bracketing these for a moment let me focus on how the analyses of individual traits function in the overall test of adaptationism. Orzack and Sober sharpen the (empirical) thesis as a generalization of (O): "Natural selection is a sufficient explanation for most non-molecular traits, and these traits are locally optimal" (1994, p. 364). This is an empirical claim about the relative frequency of (sufficient) natural selection explanations for traits, and so requires an ensemble of individual cases to test.

An ensemble test is no easy undertaking; we need to conduct a large number of evolutionary studies and resolve the explanation for each individual (non-molecular) trait studied. As Orzack and Sober (1994, p. 378) recognize, this test must overcome an ambiguity that faces any ensemble test: how to count the instances. A standard statistical approach to the problem would recommend a random sample of independent data points. But this recommendation is hard to follow. We do not have a good grasp on the sample space of *all* biological traits. Indeed, how to individuate biological traits is one strand of the adaptationist controversy. Part of the *Spandrels* critique attacks the strategy of "atomizing" the organism into optimized traits (Gould and Lewontin 1979, p. 585). Further, in biology the data points are not independent due to common ancestry, a problem familiar in phylogenetics (Felsenstein 1985).

[7] Sober (2008) provides a precise analysis of testing in evolutionary biology that takes these model selection issues seriously.

To illustrate, recall the bobtail squid with its bioluminescent light organ. We can adopt a coarse-grained description of the trait: bioluminescence is an adaptation for camouflage against the night sky. Yet this description can be made more fine-grained. The light organ is a complex structure made of many mechanisms that foster the colonization and control of the symbiotic bioluminescent bacteria. The squid must acquire the right bacteria at the right time to develop a functioning light organ. In some sense these fine-grained features of the light organ are independent, for selection may have had varying roles in the evolution of each component; (O) may be true for some structures, whereas (I) or (U) may be true for others. But in another sense these fine-grained traits share an evolutionary history, for they are all features of individuals belonging to a single lineage. There is also a hierarchical part-whole relationship between the coarse and fine-grained trait descriptions. Wilkins and Godfrey-Smith (2009) argue that differentiating between various "grains of resolution" helps mitigate some of the disagreements over empirical adaptationism, for the thesis may be true at more coarse-grained levels of description and false at more fine-grained levels.

The contrast between morphological and molecular levels may provide just such an example. Notice that the Orzack-Sober ensemble test focuses on non-molecular traits. While the adaptationist debate has tended to focus on the morphological traits of organisms, accelerating research in molecular evolution has raised some of the same issues (see Dietrich, this volume). But at the molecular level neutral evolution is a viable alternative, so we must consider another claim about the role of selection: (N) Natural selection played no role in the evolution of *T* (N denotes neutral). Even supposing (empirical) adaptationism is true at the morphological level, the case for adaptationism at the molecular level is an open question (see Orzack and Forber 2010).

While the strategy for evaluating empirical adaptationism is clear—we simply check the world—how to implement that strategy is a genuine challenge. The implementation raises issues that penetrate deep into evolutionary theory, concerning the standards for identifying traits, for evaluating the role of selection in the evolution of those traits, and for assessing the relative frequency of selection actually shaping aspects of the biological world. And the truth of empirical adaptationism is very much an open question. Despite the rhetorical element in the *Spandrels* paper and replies (see, e.g., Queller 1995), discussion about how to evaluate the empirical thesis produced scientific progress of a sort. As Rose and Lauder (1996) put it, the need to formulate a "post-spandrels adaptationism" led to both conceptual and practical innovations for investigating the truth of (empirical) adaptationism. Some of those innovations are methodological, and to those questions I now turn.

4 Getting the Method Right

The thesis of methodological adaptationism makes a claim about how best to investigate the world. Of course, assessing the methodological thesis depends, in part, on the way the world is. If empirical adaptationism were true, if natural selection provided a sufficient evolutionary explanation for most traits, then looking for adaptations first would be an effective method. But the assessment also depends, in part, on broad concerns about the nature of scientific methodology, and on more precise concerns about testing practices in evolutionary biology.

One broad concern about methodology involves whether there should be a single consensus method for evolutionary investigation. Mayr (1983) replies to the *Spandrels* paper in this way, arguing that biologists should look first and foremost for adaptive explanations and only resort to non-adaptive explanations when all else fails. Looking for non-adaptive explanations first would, on Mayr's view, be bad evolutionary science and impede epistemic progress.

This version of the thesis is problematic. As Kitcher (1993) argues, dividing our cognitive labor among competing research programs actually encourages progress towards our epistemic goals. Given that certainty is a luxury we seldom have in science, allowing diversity among methods, programs, and agendas of members of the scientific community hedges our epistemic bets. Such diversity increases the power of the community to solve outstanding problems (see also Potochnick, this volume). Forcing the community to adhere to a single method destroys this advantage. Thus, strong methodological adaptationism, interpreted as a thesis about how the community of evolutionary biologists should conduct their investigations, is probably false.

The need for diversity of methods is compatible with a weaker form of (methodological) adaptationism. Mayr's methodological recommendation could be correct for *some* evolutionary biologists. In fact, the point that Kitcher makes about the social structure of science provides support for the claim that some individual evolutionary biologists *should* accept methodological adaptationism, so long as other evolutionary biologists are following alternative programs that focus on (say) developmental and genetic constraints. The recently emerging field of evolutionary-developmental biology is arguably such an alternative program (see Love, this volume). Of course, the existence of a diversity of methodologies raises the possibility of conflicts between them. Ideally we will have the capabilities to integrate the products of different methodologies, but how exactly we should resolve such conflicts when they occur is an open question.

A more narrow methodological concern involves what testing protocols within evolutionary biology should look like. The discussion of the Orzack-Sober test of adaptationism raised connected issues about what the testing protocol for selection as a sufficient evolutionary explanation for some trait T should be, and whether optimality models provide the right tool for such a test. This issue can be connected to a more general moral about the nature of testing: *evidence is contrastive*. For a testing protocol to provide evidence *for* some hypothesis, it must provide evidence *against* rival hypotheses. To clarify the consequences of this point, let me approach it from a more philosophical angle.

Any scientific discipline must confront the problem of *underdetermination of theory by evidence*. This is a problem for theory choice. Evidential support is the primary criterion for making these choices. Yet how should we proceed when all the available evidence we have *fails* to discriminate between competing theories?

We might appeal to pragmatic concerns or other virtues of the theories, such as simplicity. While that is a plausible response to the problem, let me, for the sake of the argument, hold to the empiricist commitment that evidence and evidence alone should guide theory choices. How, then, should an empiricist proceed?

In principle, the solution is obvious: gather more data. In practice, it is precisely those factors that generate the problem that make executing the solution so difficult. We lack perfect epistemic access to the world. In evolutionary biology the problem is particularly acute since one primary aim of the discipline is to reconstruct the deep past. A significant part of the science endeavors to create better and more precise access to evolutionary history (Forber and Griffith 2011; see also Pigliucci, this volume). This search for evidence is guided by the formulation of competing evolutionary accounts with the goal of uncovering data that will discriminate between the rival hypotheses. If we do not have the right rival hypotheses on the table then the search may not provide the right evidential support.

One way to interpret the *Spandrels* critique is that it accuses adaptationists of ignoring rival evolutionary hypotheses that invoke drift or constraint (Forber 2009). This is subtly different than the methodological point in response to Mayr. Getting the set of rival hypotheses right is a prerequisite for testing adaptive hypotheses properly. Without contrasting an adaptive hypothesis against a non-adaptive hypothesis, there is a risk that the apparent evidential support for the adaptive hypothesis may be misleading.

To illustrate the problem, consider the testing protocol proposed by Sober (2008) to contrast directional selection (plus some drift) with pure drift (no selection) for a quantitative phenotypic trait. Sober works with the length of polar bear fur. The protocol assumes an optimality model that specifies the optimum mean fur length for the polar bear environment. Suppose we send an expedition into the far north and they return with observations that fit the predictions of the optimality model.[8] Sober (2008, p. 200) argues that this fit confirms the selection hypothesis over the pure drift hypothesis because such observations are much more likely given selection for fur length. Notice the contrastive element to the test.

If we augment the set of rivals, the observations may no longer univocally support the selection hypothesis. Let us suppose further that there is a plausible constraint hypothesis: fur length, due to features of polar bear development, is deeply entrenched and there is little or no variation among the population. Then the precise fit between the trait and the environment may be due instead to the migration of polar bears north after the last ice age. The precise fit that favors selection over drift will not discriminate between selection and constraint. Both hypotheses predict, with high probability, the observed match between phenotype and environment. While this toy example of a constraint is perhaps a bit far-fetched, developmental constraints certainly affect the evolutionary trajectories of complex organisms. The squid's light organ is clearly an adaptation, but explanations of some fine-grained features of the organ may be better explained by developmental constraint. We need

[8] Assume the fit is statistically rigorous. The protocol also assumes that sufficient evolutionary time has elapsed for selection to operate (Sober 2008, pp. 199–200).

to enrich the testing protocol to find evidence that will support selection over drift *and* developmental constraint.[9]

Disagreements about methodology are particularly sharp when evolutionary inquiry turns to human behavior. The research programs of sociobiology and evolutionary psychology offer adaptive evolutionary explanations for features of contemporary human behaviors and cognitive abilities that have generated enormous controversy (Kitcher 1985; Laland and Brown 2002; Buller 2005; Downes 2010). Part of the controversy concerns normal scientific issues about the nature of paleobiological evidence and the structure of our minds. But another part concerns the methodologies behind these programs, with many critics identifying a connection to (methodological) adaptationism that introduces biases into efforts to "evolutionize" the mind.[10] More sophisticated approaches to the evolution of cognition emphasize the subtle interaction between selection and non-adaptive processes (e.g., Sterelny 2003, 2012), and these show much more promise.

In sum, controversy over (methodological) adaptationism concerns grand issues about how to structure scientific inquiry, and more pedestrian issues about how to test adaptive hypotheses. While the grand issues may not intersect with daily scientific practice, the issues about testing certainly do. The emphasis on pervasive problems of evidence for evolutionary biology, and how testing protocols should be revised to handle these problems, is a beneficial feature of the debate, making adaptationism worthy of careful study.

5 A Philosophy of Nature

Explanatory adaptationism is the most contentious thesis, for it makes claims about the primacy of certain questions for making sense of the world. It is not about the prevalence of adaptations, nor about methodology for evolutionary inquiry. It is instead about adopting a particular conceptual lens through which we understand and explain the biological world. Dennett (1995) is one of the most enthusiastic proponents of the thesis, arguing that Darwin's theory of evolution by natural selection extends beyond revolutionizing biology, generating far-reaching implications for our overall worldview.

A problem for evaluating explanatory adaptationism is that it appears to be a sort of aesthetic claim. The striking interest we take in the apparent design of organisms seems to be a fact about us, not a fact about the biological world or the nature of evolutionary inquiry (Amundson 1988, 1990; Godfrey-Smith 2001). An illuminating

---

[9]Sober's protocol is based on the law of likelihood and so can easily be extended to incorporate constraint hypotheses, so long as such hypotheses specify an appropriate likelihood function. Pigliucci and Kaplan (2000) have an alternative protocol for contrasting selection and constraint that compares the probabilities of transition between possible forms.

[10]Lloyd (2005) makes an interesting argument of this kind about how methodological biases obscured evolutionary research on human female orgasms.

way to contrast the explanatory thesis with its empirical and methodological relatives invokes a triad due to later work by Godfrey-Smith. When reflecting on the nature of the scientific enterprise we can distinguish between science, philosophy of science, and philosophy of nature. The contrast between science and philosophy of science should be relatively clear. We can ask questions about the world (science) or we can ask questions about the nature of the scientific enterprise (philosophy of science). There is interplay between the products of science and philosophy of science, for the latter seeks to develop normative strategies for evaluating the quality of science. This cannot be done without both consulting and revising the practice. Philosophy of nature, however, stands apart. As Godfrey-Smith describes it:

> This is the project of taking science as developed by scientists, and working out what its real message is, especially for larger questions about our place in nature. So we aim to use scientific work to inform our view of the world, but we do not determine this view using science in its 'raw' form (2009, p.3).

With some simplification, we can treat empirical adaptationism as a thesis within the scope of science, and methodological adaptationism as a thesis within the scope of philosophy of science. Explanatory adaptationism, then, is primarily a thesis about philosophy of nature.

One may legitimately question the relevance of philosophy of nature, and whether it has, or should have, any interaction with scientific practice. One common view is that while philosophy of nature may be important for packaging and exporting the products of science to the broader public, it does not have anything of value to offer the practicing scientist. That is, it is part of science education, not science. This common view is right to emphasize the connection to science education, but science ignores philosophy of nature at its peril. Philosophy of nature provides perspective on the science, a perspective that students come to internalize as they become the next generation of scientists. Such perspective can interact with practice by making certain features of the world more or less perspicuous, by emphasizing certain causal factors and downplaying others.

Consider an example of philosophy of nature interacting with science. This example involves a foundational concept: *the replicator*. This concept rose to prominence with *The Selfish Gene* (Dawkins 1976), a popular presentation and extension of concepts found in Williams (1966). Roughly, replicators are entities of which copies are made and that can influence the rate or probability of being copied. Genes are the canonical replicators. Talk of replicators pervades biology and evolutionary biology, and there is a legacy of controversy over whether the concept unduly privileges genes, gene action, and genetic inheritance. I do not want to rehearse that here. Instead, let me point to an interesting feature of the replicator concept: the integral use of agential language—strategies, plans, interests, agendas—at the level of the gene to describe evolutionary phenomena. Such language treats genes as metaphorical agents. This metaphor can have enormous effects on the science, effects that can be problematic because genes are not agents in any literal sense. Godfrey-Smith (2009, pp. 10–14) calls the unregulated use of this language *Darwinian paranoia*, arguing that the peculiarly powerful agential language has obscured core Darwinian ideas. The focus on replicators in popular presentations of evolutionary science has also misrepresented the science, for much of evolution proceeds without using any of this agential language. Of course, many find such language indispensable to evolutionary theorizing. Haig (1997) generalizes the concept of the selfish gene to that of the social or strategic gene. Dennett (2011) responds directly to Godfrey-Smith, embracing agential language for genes as the primary way we make sense of complex evolutionary ideas. This debate about the foundational consequences of agential language is particularly relevant to biological education, for most of us learn the basic evolutionary concepts in agential terms.

Another way philosophy of nature may interact with science involves the export and trafficking of concepts between disciplines within the broad field of biology. A prominent evolutionary biologist once famously remarked, "nothing in biology makes sense except in the light of evolution" (Dobzhansky 1973). The general idea is that evolutionary considerations frame questions in more proximal biological sciences, such as ecology and developmental biology, in an essential way. Griffiths (2009) puts a sophisticated twist on explanatory adaptationism by treating it as a thesis of this kind. More precisely, he argues that much of ecology and developmental biology needs to consider forward-looking evolutionary consequences when doing research because these considerations are necessary to ground the concept of biological function. Thus, evolution by natural selection provides something different than the big answer to the question about apparent design in nature; it provides the backdrop that structures whole biological disciplines.

Perhaps explanatory adaptationism is an aesthetic claim, or perhaps it is something more. Regardless, resolving debate over this thesis requires going beyond the usual scope of scientific inquiry. Our stance on explanatory adaptationism depends on views about how we tend to think, and how those tendencies interact with scientific practice. Such a stance also has consequences for biological education: should we teach evolution emphasizing the importance and ubiquity of natural selection? Or, should we draw attention to the roles of contingency, constraint, and historical accident? How we teach the science contributes to our perspective on philosophy of nature, and that perspective can (and indeed, often does) influence how research is conducted and disseminated in the future.

6 Science in Action

The adaptationism controversy is such a fascinating and instructive episode of science because it raises a truly diverse set of questions. There are empirical questions about the prevalence and power of natural selection. These questions are open. Science continues to investigate evolutionary lineages, providing better resources to answer these questions. There are methodological questions about how to test evolutionary hypotheses, and how to structure the community of evolutionary biologists. These questions are unresolved. Far from a source or sign of problems, continuing discussion and debate on methodological issues fashions better testing protocols, advancing the scientific discipline. There are questions about the importance and status of foundational concepts in evolutionary science. These questions are dangerously philosophical. Yet they are relevant to the practice of biology, for they concern overall perspectives on the science that scaffold the education of future biologists. And while the rhetoric of the debate sometimes distracts from the core issues, it illustrates an indelible feature of science: it is an enterprise undertaken by humans, with human motivations, and bearing the influences of their culture and society (Shapin 2010). The debate over adaptationism is far from an unproductive distraction from real science—it is science in action.

References

Amundson, R. 1988. Logical adaptationism. *The Behavioral and Brain Sciences* 11: 505–506.

Amundson, R. 1990. Doctor Dennett and Doctor Pangloss: Perfection and selection in psychology and biology. *The Behavioral and Brain Sciences* 13: 577–584.

Amundson, R. 1996. Historical development of the concept of adaptation. In *Adaptation*, ed. M.R. Rose and G.V. Lauder, 11–53. San Diego: Academic.

Ariew, A., and R.C. Lewontin. 2004. The confusions of fitness. *The British Journal for the Philosophy of Science* 55: 347–363.

Beatty, J. 1992. Fitness: Theoretical contexts. In *Keywords in evolutionary biology*, ed. E.F. Keller and E.A. Lloyd, 115–119. Cambridge: Harvard University Press.

Beatty, J. 2004. Chance variation: Darwin on orchids. *Philosophy of Science* 73: 629–641.

Beatty, J., and E.C. Desjardins. 2009. Natural selection and history. *Biology and Philosophy* 24: 231–246.

Brandon, R. 1990. *Adaptation and environment*. Princeton: Princeton University Press.

Brandon, R., and M.D. Rausher. 1996. Testing adapationism: A comment on Orzack and Sober. *American Naturalist* 148: 189–201.

Buller, D. 2005. *The adapting mind: Evolutionary psychology and the persistent quest for human nature*. Cambridge: MIT Press.

Dawkins, R. 1976. *The selfish gene*. Oxford: Oxford University Press.

Dennett, D. 1995. *Darwin's dangerous idea*. New York: Simon and Schuster.

Dennett, D. 2011. Homunculi rule: Reflections on Darwinian populations and natural selection by Peter Godfrey-Smith. *Biology and Philosophy* 26: 475–488.

Dobzhansky, T. 1973. Nothing in biology makes sense except in the light of evolution. *The American Biology Teacher* 35: 125–129.

Downes, S. M. 2010. Evolutionary psychology. In *The Stanford encyclopedia of philosophy*, E.N. Zalta, ed. URL: http://plato.stanford.edu/archives/fall2010/entries/evolutionary-psychology/.

Felsenstein, F. 1985. Phylogenies and the comparative method. *The American Naturalist* 125: 1–15.

Forber, P. 2009. *Spandrels* and a pervasive problem of evidence. *Biology and Philosophy* 24: 247–266.

Forber, P., and E. Griffith. 2011. Historical reconstruction: Gaining access to the deep past. *Philosophy and Theory in Biology* 3. URL:http://www.philosophyandtheoryinbiology.org.

Godfrey-Smith, P. 2001. Three kinds of adaptationism. In *Adaptationism and optimality*, ed. S.H. Orzack and E. Sober, 335–357. Cambridge: Cambridge University Press.

Godfrey-Smith, P. 2009. *Darwinian populations and natural selection*. New York: Oxford University Press.

Gould, S.J. 1980. *The Panda's thumb: More reflections in natural history*. New York: Norton.

Gould, S.J., and R.C. Lewontin. 1979. The spandrels of San Marco and the Panglossian paradigm: A critique of the adaptationist programme. *Proceedings of the Royal Society London, Series B, Biological Sciences* 205: 581–598.
Gould, S.J., and E.S. Vrba. 1982. Exaptation—a missing term in the science of form. *Paleobiology* 8: 4–15.
Griffiths, P.E. 2009. In what sense does 'nothing makes sense except in the light of evolution'? *Acta Biotheoretica* 57: 11–32.
Haig, D. 1997. The social gene. In *Behavioral ecology: An evolutionary approach*, ed. J.R. Krebs and N.B. Davies, 284–304. London: Blackwell.
Jones, B.W., and M.K. Nishiguchi. 2004. Counterillumination in the Hawaiian bobtail squid, *Euphrymnascolopes* Berry (Mollusca: Cephalopoda). *Marine Biology* 144: 1151–1155.
Kampourakis, K. 2013. Teaching about adaptation: Why evolutionary history matters. *Science & Education* 22(2): 173–188.
Kitcher, P. 1985. *Vaulting ambition*. Cambridge: MIT Press.
Kitcher, P. 1993. *The advancement of science*. Oxford: Oxford University Press.
Laland, K.N., and G.R. Brown. 2002. *Sense and nonsense: Evolutionary perspectives on human behaviour*. Oxford: Oxford University Press.
Lewens, T. 2007. Adaptation. In *The Cambridge companion to the philosophy of biology*, ed. D.L. Hull and M. Ruse, 1–21. Cambridge: Cambridge University Press.
Lewens, T. 2009. Seven types of adaptationism. *Biology and Philosophy* 24: 161–182.
Lloyd, E.A. 2005. *The case of the female orgasm: Bias in the science of evolution*. Cambridge: Harvard University Press.
Lloyd, E.A. 2007. Units and levels of selection. In *The Cambridge companion to the philosophy of biology*, ed. D.L. Hull and M. Ruse, 44–65. Cambridge: Cambridge University Press.
Mayr, E. 1983. How to carry out the adaptationist program? *The American Naturalist* 121: 324–334.
Nyholm, S.V., and M.J. McFall-Ngai. 2004. The winnowing: Establishing the *squid-vibrio* symbiosis. *Nature Reviews Microbiology* 2: 632–642.
Orzack, S. H., and P. Forber. 2010. Adaptationism. In *The Stanford encyclopedia of philosophy*, E.N. Zalta, ed. URL: http://plato.stanford.edu/archives/fall2010/entries/adaptationism/.
Orzack, S.H., and E. Sober. 1994a. Optimality models and the test of adaptationism. *American Naturalist* 143: 361–380.
Orzack, S.H., and E. Sober. 1994b. How (not) to test an optimality model. *Trends in Ecology and Evolution* 9: 265–267.
Orzack, S.H., and E. Sober. 1996. How to formulate and test adaptationism. *The American Naturalist* 148: 202–210.
Parker, G.A., and J. Maynard Smith. 1990. Optimality theory in evolutionary biology. *Nature* 348: 27–33.
Pigliucci, M., and J. Kaplan. 2000. The fall and rise of Dr. Pangloss: Adaptationism and the spandrels paper 20 years later. *Trends in Ecology and Evolution* 15: 66–70.
Potochnik, A. 2009. Optimality modeling in a suboptimal world. *Biology and Philosophy* 24: 183–197.
Queller, D. 1995. The spaniels of St. Marx and the Panglossian paradox: A critique of a rhetorical programme. *The Quarterly Review of Biology* 70: 485–489.
Reeve, H.K., and P.W. Sherman. 1993. Adaptation and the goals of evolutionary research. *The Quarterly Review of Biology* 68: 1–32.
Rose, M.R., and G.V. Lauder. 1996. *Adaptation*. San Diego: Academic.
Seger, J., and J. Stubblefield. 1996. Optimization and adaptation. In *Adaptation*, ed. M.R. Rose and G.V. Lauder, 93–124. San Diego: Academic.
Shapin, S. 2010. *Never pure: Historical studies of science as if it was produced by people with bodies, situated in time, space, culture, and society, and struggling for credibility and authority*. Baltimore: The John Hopkins University Press.
Sober, E. 1984. *The nature of selection*. Chicago: The University of Chicago Press.

Sober, E. 1996. Evolution and optimality: Feathers, bowling balls, and the thesis of adaptationism. *Philosophic Exchange* 26: 41–55.
Sober, E. 2008. *Evidence and evolution: The logic behind the science*. Cambridge: Cambridge University Press.
Sterelny, K. 2003. *Thought in a hostile world*. Oxford: Blackwell Publishing.
Sterelny, K. 2012. *The evolved apprentice: How evolution made humans unique*. Cambridge: MIT Press.
Wilkins, J., and P. Godfrey-Smith. 2009. Adaptationism and the adaptive landscape. *Biology and Philosophy* 24: 199–214.
Williams, G.C. 1966. *Adaptation and natural selection: A critique of some current evolutionary thought*. Berkeley: University of California Press.

Biology and Religion: The Case for Evolution

Francisco J. Ayala

Darwinism appeared, and under the guise of a foe, did the work of a friend. It has conferred upon philosophy and religion an inestimable benefit, by shewing us that we must choose between two alternatives. Either God is everywhere present in nature, or He is nowhere.

Aubrey Moore (1891)

What I find even more surprising, and less understandable, is the way in which the 'disguised friend' of Darwinism, and more generally of evolutionary ideas, has been admitted (if at all) only grudgingly [...] into the parlors of Christian theology. I believe it is vital for this churlishness to be rectified [...] if the Christian religion (indeed any religion) is to be believable and have intellectual integrity.

Arthur Peacocke (1998)

1 Introduction

The theory of biological evolution is the central organizing principle of modern biology (Dobzhansky 1973). Evolution provides a scientific explanation for why there are so many different kinds of organisms on Earth and gives an account of their similarities and differences (morphological, physiological, and genetic). It accounts for the appearance of humans on Earth and reveals our species' biological connections with other living things. It provides an understanding of the constantly evolving bacteria and viruses and other pathogenic organisms, and enables the development of effective new ways to protect ourselves against the diseases they cause. Knowledge

F.J. Ayala (✉)
Department of Ecology and Evolutionary Biology, University of California,
Irvine, CA 92697, USA
e-mail: fjayala@uci.edu

K. Kampourakis (ed.), *The Philosophy of Biology: A Companion for Educators*, History, Philosophy and Theory of the Life Sciences 1, DOI 10.1007/978-94-007-6537-5_9,

of evolution has made possible improvements in agriculture and medicine, and has been applied in many fields outside biology; for example, software engineering, where genetic algorithms seek to mimic selective processes, and chemistry, where the principles of natural selection are used for developing new molecules with specific functions.

Science has demonstrated again and again, beyond reasonable doubt, that living organisms evolve and diversify over time, and that their features have come about by natural selection, a process that accounts for their design. Two meanings of the noun *design* are given in the Random House *Thesaurus*, College Edition (1984): "sketch, drawing, outline, plan blueprint, diagram"; and "pattern, motif, form, arrangement". It is evident that organisms exhibit pattern, motif, form, and arrangement. This is the meaning in which I will use "design" in this chapter, except when referring to religious authors claiming intentionality.

There are many people of faith in the United States and elsewhere who think that science, particularly the theory of evolution, is contrary to the teachings of the Bible and to religious beliefs, such as creation by God. Well before the formulation of the theory of evolution, religious authors over the centuries used the "argument-from-design" to demonstrate rationally, without reference to faith or divine revelation, the existence of God, as the author of the design of organisms. These authors use the term design in its alternative meaning of sketch, drawing, outline, plan, blueprint, and diagram. A plan or blueprint calls for intentionality. Theologians and other religious authors, over the centuries, argued that organisms are intentionally designed. Darwin discovered natural selection, a process that can account for the design of organisms without postulating intentionality.

The religious argument from design asserts, first, that organisms evince to have been designed; second, that only God could account for the design. The argument from design was advanced, in a variety of forms, in Classical Greece and early Christianity. Its most extensive formulation is due to William Paley in his *Natural Theology* (1802). The eye—as well as all sorts of organs, organisms, and their interactions—manifests to be the outcome of design and not of chance, thus it shows to have been created by God. In the 1990s, the design argument was revived in the United States by several authors. The flagellum used by bacteria for swimming and the immune system of mammals, as well as some improbability calculations, were advanced as evidence of "intelligent design," on the grounds that chance processes could not account for the phenomena to be explained (see Brigandt, this volume).

In *The Origin of Species*, Darwin (1859) advanced a scientific explanation of the design of organisms. The adaptations of organisms are outcomes not of chance, but of a process that, over time, causes the gradual accumulation of features beneficial to organisms, whenever these features increase the organisms' chances of surviving and reproducing (see Forber, this volume). There is design in the living world: eyes are designed for seeing, wings for flying, and kidneys for regulating the composition of the blood. The design of organisms comes about not by intelligent design, but by a natural process, which is creative through the interaction of chance and necessity.

Organisms are pervaded by imperfections, dysfunctions, cruelties, and even sadism. The theory of evolution accounts for these mishaps by natural selection, as the outcomes of natural processes, so that they need not be attributed to God's explicit design. The theory of evolution perceived by some people of faith as contrary to religion, may thus be acknowledged as their "disguised friend." The theory of evolution accounts for the design of organisms, but also for the dysfunctions, oddities, cruelty, and sadism that pervade the world of life, so that these deficiencies need not be attributed to specific agency by the Creator, which might implicitly amount to blasphemy.

2 The Argument from Design

The argument from design to demonstrate the existence of God has been repeatedly used throughout the history of Christianity, but in different versions. The argument form design is a two-tined argument. The first prong of the argument asserts that there is design in the universe. The second prong affirms that only God, an omnipotent and omniscient Creator, could account for the perfection and universality of the design.

The first prong comes, importantly, in at least two forms. One version refers to the order and harmony of the universe as a whole; as for example, in St. Augustine (1998): "The world itself, by the perfect order of its changes and motions and by the great beauty of all things visible" (pp. 452–453); or in St. Thomas Aquinas (1905): "It is impossible for contrary and discordant things to fall into one harmonious order except under some guidance, assigning to each and all parts a tendency to a fixed end. But in the world we see things of different natures falling into harmonious order" (p. 12). The second version of the first prong refers to the living world, the intricate organized complexity of organisms, as formulated, among others, by William Paley (1802) and the modern proponents of intelligent design (ID).

The second prong of the argument from design has been formulated in, at least, three important versions. One formulation of the Designer appears in Classical Greece, including Plato (1997), who postulates the existence of a Demiurge, a creator of the universe's order, who is a universal and impersonal ordering principle, rather than the personalized Judeo-Christian God. Plato's Demiurge is an orderer of the world who accounts for the world's rationality, but not necessarily for its creation. A second version of the Designer is the familiar one of the Judeo-Christian God, as formulated by Paley and other Christian philosophers and theologians (e.g., Aquinas 1905, 1964; Swinburne 1994), who is a "person," the creator and steward of the universe, who creates a world from nothing and is omniscient, omnipotent, omnibenevolent, and is provident for humans (see also Lennox and Kampourakis, this volume). Proponents of ID have in recent years formulated a third version of the second prong of the argument: an unidentified Designer who may account for the order and complexity of the universe, or who may simply intervene from time to time in the universe so as to design organisms and their parts, because the complexity of organisms, it is claimed, cannot be accounted for by natural processes. According to ID proponents, this intelligent designer could be, but need not be, God. The intelligent designer could be an alien from outer space or some other creature, such as a "time-traveling cell biologist," with amazing powers to account for the universe's design (Behe 1996; Dembski 2002; Johnson 1993, 2002; Meyer 2009).

An elaborate formulation of the argument from design was *The Wisdom of God Manifested in the Works of Creation* (1691) by the English clergyman and naturalist John Ray (1627–1705). Ray regarded as incontrovertible evidence of God's wisdom that all components of the universe—the stars and the planets, as well as all organisms—are so wisely contrived from the beginning and perfect in their operation. The "most convincing argument of the Existence of a Deity," writes Ray (1691), "is the admirable Art and Wisdom that discovers itself in the Make of the Constitution, the Order and Disposition, the Ends and uses of all the parts and members of this stately fabric of Heaven and Earth" (p. 33).

The design argument was advanced, in greater or lesser detail, by a number of authors in the seventeenth and eighteenth centuries (see, e.g., Arp 1999; Klocker 1968). John Ray's contemporary Henry More (1614–1687) saw evidence of God's design in the succession of day and night and of the seasons: "I say that the Phenomena of Day and Night, Winter and Summer, Spring-time and Harvest [...] are signs and tokens unto us that there is a God [...] things are so framed that they naturally imply a Principle of Wisdom and Counsel in the Author of them. And if there be such an Author of external Nature, there is a God" (More 1662, p. 38). Robert Hooke (1635–1703), a physicist and eventual Secretary of the Royal Society, formulated the watchmaker analogy: God had furnished each plant and animal "with all kinds of contrivances necessary for its own existence and propagation [...] as a Clock-maker might make a Set of Chimes to be a part of a Clock" (Hooke 1665, p. 124). The clock analogy, among other analogies such as temples, palaces, and ships, was also used by Thomas Burnet (1635–1703) in his *Sacred Theory of the Earth* (1691), and it would become common among British natural theologians of the time.

On the Continent, the Dutch philosopher and theologian Bernard Nieuwentijdt (1654–1718) developed, at length, the argument from design in his three-volume treatise, *The Religious Philosopher*, where, in the Preface, he introduces the watchmaker analogy (Nieuwentijdt 1718/2007). Voltaire (1694–1778), like other philosophers of the Enlightenment, accepted the argument from design. Voltaire asserted that in the same way as the existence of a watch proves the existence of a watchmaker, the design and purpose evident in nature prove that the universe was created by a Supreme Intelligence (Voltaire 1967, pp. 262–270).

The most elaborate formulation of the argument from design is William Paley's *Natural Theology* (1802). Paley was an influential writer of works on Christian philosophy, ethics, and theology, such as *The Principles of Moral and Political Philosophy*, published in 1785, and *A View of the Evidences of Christianity*, published in 1794. With *Natural Theology*, Paley sought to update Ray's *Wisdom of God Manifested in the Works of the Creation* (1691). But Paley could now carry the argument much further than Ray, by taking advantage of one century of additional biological knowledge. Paley's (1802) keystone claim is that there "cannot be design without a designer; contrivance, without a contriver; order, without choice; [...] means suitable to an end, and executing their office in accomplishing that end, without the end ever having been contemplated" (pp. 15–16). *Natural Theology* is a sustained argument for the existence of God based on the obvious design of humans and their organs, as well as the design of all sorts of organisms, considered by themselves and in their relations to one another and to their environment. Paley's first analogical example in *Natural Theology* is the human eye. Early in Chapter 3, Paley points out that the eye and the telescope "are made upon the same principles; both being adjusted to the laws by which the transmission and refraction of rays of light are regulated" (p. 20). Specifically, there is a precise resemblance between the lenses of a telescope and "the humors of the eye" in their figure, their position, and the ability of converging the rays of light at a precise distance from the lens—on the retina, in the case of the eye.

Paley makes two remarkable observations, which enhance the complex and precise design of the eye. The first observation is that rays of light should be refracted by a more convex surface when transmitted through water than when passing out of air into the eye. Accordingly, "the eye of a fish, in that part of it called the crystalline lens, is much rounder than the eye of terrestrial animals. What plainer manifestation of design can there be than this difference? What could a mathematical instrument maker have done more to show his knowledge of [t]his principle [...]?" (p. 20). The second remarkable observation made by Paley in support of his argument is dioptric distortion: "Pencils of light, in passing through glass lenses, are separated into different colors, thereby tinging the object, especially the edges of it, as if it were viewed through a prism. To correct this inconvenience [...] a sagacious optician ... [observed] that in the eye the evil was cured by combining lenses composed of different substances, that is, of substances which possessed different refracting powers." (pp. 22–23) The telescope maker, accordingly, corrected the dioptric distortion "by imitating, in glasses made from different materials, the effects of the different humors through which the rays of light pass before they reach the bottom of the eye. Could this be in the eye without purpose, which suggested to the optician the only effectual means of attaining that purpose?" (p. 23).

Natural Theology has chapters dedicated to the human frame, which displays a precise mechanical arrangement of bones, cartilage, and joints; to the circulation of the blood and the disposition of blood vessels; to the comparative anatomy of humans and animals; to the digestive tract, kidneys, urethra, and bladder; to the wings of birds and the fins of fish; and much more. For 352 pages, *Natural Theology* conveys Paley's expertise: extensive and accurate biological knowledge, as detailed and precise as was available in the year 1802. After detailing the precise organization and exquisite functionality of each biological entity, relationship, or process, Paley draws again and again the same conclusion: only an omniscient and omnipotent Deity could account for these marvels of mechanical perfection, purpose, and functionality, and for the enormous diversity of inventions that they entail.

In 1829, nearly three decades after the publication of *Natural Theology*, Francis Henry Egerton (1756–1829), the eighth Earl of Bridgewater, bequeathed the sum of 8,000 pounds sterling with instructions to the Royal Society that it commission eight treatises that would promote natural theology by setting forth "The Power, Wisdom and Goodness of God as manifested in the Creation." Eight treatises were published in the 1830s, several of which artfully incorporate the best science of the time and had considerable influence on the public and among scientists. One additional treatise, never completed, was authored by the notable mathematician and pioneer in the field of calculating machines, Charles Babbage (1791–1871). In *The Ninth Bridgewater Treatise: A Fragment*, published in 1838, he seeks to show how mathematics may be used to bolster religious belief. One of the Bridgewater treatises, *The Hand, Its Mechanisms and Vital Endowments as Evincing Design* (1833), was written by Sir Charles Bell, a distinguished anatomist and surgeon, famous for his neurological discoveries, who became professor of surgery in 1836 at the University of Edinburgh. William Buckland, Professor of Geology at Oxford University, notes in *Geology and Mineralogy* (1836) the world distribution of coal and mineral ores, and proceeds to point out that they had been deposited in a remote part, yet obviously with the forethought of serving the larger human populations that would come about much later. Another geologist, Hugh Miller in *The Testimony of the Rocks* (1857), formulated what I call the *argument from beauty*, which allows that it is not only the perfection of design, but also the beauty of natural structures found in rock formations and in mountains and rivers that manifests the intervention of the Creator.

In the 1990s and beyond, several authors in the United States have revived the argument from design, notably biochemist Michael Behe (1996), theorist William Dembski (1995, 2002), law professor Phillip Johnson (1993, 2002), and Stephen C. Meyer (2009), among others. These authors typically avoid explicit reference to God, so that the "theory" of intelligent design could be taught in the public schools, as an alternative to the theory of evolution, without incurring conflict with the U.S. Constitution, which forbids the endorsement of any religious beliefs in public institutions. These modern proponents, at times, claim that the Intelligent Designer need not be God, but could be a space alien or some other intelligent superpower unknown to us. The folly of this pretense is apparent to anyone who takes the time to consider the issue seriously. It is nothing but a vulgar charade. (For an extended discussion of the modern version of intelligent design, see Brigandt, this volume).

3 Darwin

Charles Darwin (1809–1882) occupies an exalted place in the history of Western thought, deservedly receiving credit for the theory of evolution. In *The Origin of Species*, published in 1859, he laid out the evidence demonstrating the evolution of organisms. However, Darwin accomplished something much more important for intellectual history than demonstrating evolution. Indeed, accumulating evidence for common descent with diversification may very well have been a subsidiary objective of Darwin's masterpiece. Darwin's *Origin of Species* is, first and foremost, a sustained effort to solve the problem of how to account scientifically for the adaptations or design of organisms. Darwin seeks to explain the design of organisms, their complexity, diversity, and marvelous contrivances as the result of natural processes.

In *The Origin of Species*, Darwin brings about the evidence for evolution mostly because evolution is a necessary consequence of his theory of design, and because the evolution of organisms displays the explanatory power of his theory of natural selection to account for the diversity of organisms, as well as their adaptations. The evolution of organisms was commonly accepted by naturalists in the middle decades of the nineteenth century. The intellectual challenge was to explain the origin of distinct species of organisms and how new ones come to be adapted to their environments. This Darwin accomplished with his theory of natural selection. Natural selection occurs because individuals having more useful traits, such as more acute vision or swifter legs, survive better and produce more progeny than individuals with less favorable traits. The beneficial traits, thus, gradually increase in frequency over the generations, while the less beneficial or harmful ones eventually get eliminated. Diversification occurs because different organisms become adapted to different environments.

Darwin's evidence for the evolution of organisms went much further than what was generally known by his scientific contemporaries. Moreover, Darwin shows that the evidence is consistent with his theory of natural selection. For example, he showed that different parts of organisms evolve at different times and rates, rather than the gradual evolution of each individual as a whole, which was the generally accepted view and was postulated, for example, by Lamarck in his *Zoological Philosophy* of 1809, and by Robert Chambers in his *Vestiges of Natural Creation* of 1844, two works that were largely speculative providing little or no empirical evidence in support of their theories. Darwin also affirms that some organisms may not evolve at all when they happen to be adapted to unchanging environments. Indeed, "Some of the most ancient Silurian animals, as the Nautilus, Lingula, etc., do not differ much from living species" (*Origin*, ch. IX, p. 306). The strength of the evidence accumulated by Darwin in *The Origin of Species* greatly contributed to extend the acceptance of evolution by Darwin's contemporaries.

Darwin and other nineteenth-century biologists found compelling evidence for biological evolution in the comparative study of living organisms, in their geographic distribution, and in the fossil remains of extinct organisms. Since Darwin's time, the evidence from these sources has become stronger and more comprehensive, while biological disciplines that have emerged recently—genetics, biochemistry, ecology, animal behavior (ethology), neurobiology, and especially molecular biology—have supplied powerful additional evidence and detailed confirmation. Accordingly, evolutionists are no longer concerned with obtaining evidence to support the fact of evolution, but rather are concerned with finding out additional information of the historical process in cases of particular interest. Moreover and most importantly, evolutionists nowadays are interested in understanding further and further how the process of evolution occurs.

4 The Fossil Record

Nevertheless, important discoveries continue, even in traditional disciplines, such as paleontology. Skeptical contemporaries of Darwin asked about the "missing links," particularly between apes and humans, but also between major groups of organisms, such as between fish and terrestrial tetrapods or between reptiles and birds. Evolutionists can now affirm that these missing links are no longer missing. Indeed, the known fossil record has made great strides over the last century and a half. Many fossils intermediate between diverse organisms have been discovered over the years. Two examples are *Archaeopteryx*, an animal intermediate between reptiles and birds, and *Tiktaalik*, intermediate between fishes and tetrapods.

The first *Archaeopteryx* was discovered in Bavaria in 1861, 2 years after the publication of Darwin's *The Origin*, a discovery that was noted by Darwin in the last two editions of *The Origin*. Other *Archaeopteryx* specimens have been discovered in the past 100 years. The most recent, the tenth specimen so far recovered, was described in December 2005. *Archaeopteryx* lived during the Late Jurassic period, about 150 million years ago, and exhibited a mixture of both avian and reptilian traits. All known specimens are small, about the size of a crow, and share many anatomical characteristics with some of the smaller bipedal dinosaurs. Its skeleton is reptile-like, but *Archaeopteryx* had feathers, clearly shown in the fossils, with a skull and a beak like those of a bird. *Archaeopteryx* is now considered an early bird. The recently described *Haplocheirus sollers*, 15 million years older than *Archaeopteryx*, is more nearly intermediate between dinosaurs and birds (Stone 2010; Choiniere et al. 2010).

Paleontologists have known for more than a century that tetrapods (amphibians, reptiles, birds, and mammals) evolved from a particular group of fishes called lobe-finned. Until recently, *Panderichthys* was the known fossil fish closest to the tetrapods. *Panderichthys* was somewhat crocodile shaped and had a pectoral fin skeleton and shoulder girdle intermediate in shape between those of typical lobe-finned fishes and those of tetrapods, which allowed it to "walk" in shallow waters, but probably not on land. In most features, however, *Panderichthys* was more like a fish than like an amphibious tetrapod. *Panderichthys* is known from Latvia, where it lived some 385 million years ago (the mid-Devonian period).

Until very recently, the earliest tetrapod fossils that are more nearly fishlike were also from the Devonian, about 376 million years old. They have been found in Scotland and Latvia. *Ichthyostega* and *Acanthostega* from Greenland, which lived more recently, about 365 million years ago, are unambiguous walking tetrapods, with limbs that bear digits, although they retain from their fish ancestors such characteristics as true fish tails with fin rays. Thus, the time gap between the most tetrapodlike fish and the most fishlike tetrapods was nearly 10 million years, between 385 and 376 million years ago.

Recently several specimens have been discovered of a fossil that has been named *Tiktaalik*, which goes a long way toward breaching this gap; it is the most nearly intermediate between fishes and tetrapods yet known. Several specimens have been found in Late Devonian river sediments, dated about 380 million years ago, on Ellesmere Island in Nunavut, Artic Canada. *Tiktaalik* displays an array of features that are just about as precisely intermediate between fish and tetrapods as one could imagine and exactly fits the time gap as well (see Daeschler et al. 2006; Shubin et al. 2006).

The missing link between apes and humans is not, either, missing any longer. The fossils that belong to the human lineage after its separation from the ape lineages are called hominids (or hominins). Not one, but hundreds of fossil remains from hundreds of individual hominids have been discovered since Darwin's time and continue to be discovered at an accelerated rate. The oldest known fossil hominids are 6–7 million years old, come from Africa, and are known as *Sahelanthropus, Orrorin*, and *Ardipithecus*. These ancestors were predominantly bipedal when on the ground and had very small brains. Some *Ardipithecus* lived about 4.4 million years ago, also in Africa. Numerous fossil remains from diverse African origins are known of *Australopithecus*, a hominid that appeared between 3 and 4 million years ago. *Australopithecus* had an upright human stance but a cranial capacity of less than 500 cc, comparable to that of a gorilla or chimpanzee. The skull of *Australopithecus* displayed a mixture of ape and human characteristics. Other early hominids partly contemporaneous with *Australopithecus* include *Kenyanthropus* and *Paranthropus*; both had comparatively small brains. *Paranthropus* represents a side branch of the hominid lineage that became extinct.

Along with increased cranial capacity, other human characteristics have been found in *Homo habilis*, which lived between about 2 and 1.5 million years ago in Africa and had a cranial capacity of more than 600 cc, and in *Homo erectus*, which evolved in Africa sometime before 1.8 million years ago and had a cranial capacity of 800–1,100 cc. Shortly after its emergence in Africa, *H. erectus* spread to Europe and Asia, even as far as the Indonesian archipelago and northern China. *Homo erectus* fossils from Java have been dated at 1.81 and 1.66 million years ago, and from Georgia between 1.6 and 1.8 million years ago.

The transition from *H. erectus* to *H. sapiens* may have started around 400,000 years ago. Some fossils of that time appear to be "archaic" forms of *H. sapiens*. The species *Homo neanderthalensis* appeared in Europe more than 200,000 years ago and persisted until 30,000 years ago. The Neandertals have been thought to be ancestral to anatomically modern humans, but comparisons of DNA from Neandertal fossils with living humans indicate that *H. neanderthalensis* may have been a separate species that became extinct.

5 Molecular Evolution

Molecular biology, a discipline that emerged in the second half of the twentieth century, nearly 100 years after the publication of *The Origin of Species*, has provided the strongest evidence yet of the evolution of organisms. Molecular biology proves the fact of evolution in two ways: first, by showing the unity of life in the nature of DNA and the workings of organisms at the level of enzymes and other protein molecules; second, and most important in practice for evolutionists, by making it possible to reconstruct evolutionary relationships that were previously unknown, and to confirm, refine, and time all evolutionary relationships from the last universal common ancestor up to all living organisms. The precision with which these events can be reconstructed is one reason why the evidence from molecular biology is so useful to evolutionists and so compelling.

The molecular components of organisms are remarkably uniform—in the kinds of molecules that are present, as well as in the ways in which these molecules are assembled and used. In all microorganisms, plants, animals, and humans, the instructions that guide the development and functioning of organisms are encased in the same hereditary material, DNA, which provides the instructions for the synthesis of proteins. The thousands of enormously diverse proteins that exist in organisms are synthesized from different linear combinations, in sequences of variable length, of 20 amino acids, the same 20 in all proteins and in all organisms. Yet several hundred other amino acids exist, such as those that are found in a variety of plants, and a virtually infinite number of them could be synthesized. Moreover, the genetic code, by which the information contained in the DNA of the cell nucleus is passed on to proteins, is virtually the same in all organisms. Similar metabolic pathways—sequences of biochemical reactions—are used by the most diverse organisms to produce energy and to make up the cell components.

The unity of life reveals the genetic continuity and common ancestry of all organisms. There is no other rational way to account for their molecular uniformity, given that numerous alternative structures and fundamental processes are in principle equally likely.

DNA and proteins have been called "informational macromolecules" because they are long linear molecules made up of sequences of smaller units—nucleotides in the case of DNA, amino acids in the case of proteins—that embody evolutionary information in their particular sequence, similarly as particular sequences of letters and words convey semantic information (see Marcos and Arp, this volume). Comparing the sequence of the components in two macromolecules establishes how many units are different. Because evolution usually occurs by changing one unit at a time, the sequence differences between two organisms are an indication of their recency of common ancestry. Thus, the inferences from paleontology, comparative anatomy, and other disciplines that study evolutionary history can be tested in molecular studies of DNA and proteins by examining the sequences of nucleotides and amino acids. The authority of this kind of test is overwhelming: each of the thousands of genes and thousands of proteins contained in an organism provides an independent test of that organism's evolutionary history (see also Dietrich, this volume).

Molecular evolutionary studies have three notable advantages over comparative anatomy and the other classical disciplines: precision, universality and multiplicity. First, *precision* because molecular information is readily quantifiable. The number of units that are different is easily established when the sequence of units is known for a given macromolecule in different organisms. It is simply a matter of aligning the units (nucleotides or amino acids) between two or more species and counting the differences. The second advantage is *universality*: comparisons can be made between very different sorts of organisms. There is very little that comparative anatomy can say when, for example, organisms as diverse as yeasts, pine trees, and human beings are compared, but there are numerous DNA and protein sequences that can be compared in all three. The third advantage is *multiplicity*. Each organism possesses thousands of genes and proteins, every one of which reflects the same evolutionary history. If the investigation of one particular gene or protein does not satisfactorily resolve the evolutionary relationship of a set of species, additional genes and proteins can be investigated until the matter has been settled.

6 The Problem of Evil

Christian scholars for centuries struggled with the problem of evil in the world. The Scottish philosopher David Hume (1711–1776) set the problem succinctly with brutal directness: "Is he [God] willing to prevent evil, but not able? Then he is impotent. Is he able, but not willing? Then, he is malevolent. Is he both able and willing? Whence then is evil?" (Hume 1935). If the reasoning is valid, it would follow that God is not all-powerful or all-good. Christian theology accepts that evil exists, but denies the validity of the argument.

Traditional theology distinguishes three kinds of evil: (1) moral evil or sin, the evil originated by human beings; (2) pain and suffering as experienced by human beings; (3) physical evil, such as floods, tornados, earthquakes, and the imperfections of all creatures. Theology has a ready answer for the first two kinds of evil. Sin is a consequence of free will; the flip side of sin is virtue, also a consequence of free will. Christian theologians have expounded that if humans are to enter into a genuinely personal relationship with their maker, they must first experience some degree of freedom and autonomy. A "virtuous" life can only come about as an outcome of free will, as many Christians see it. Christian theology also provides a good accounting of human pain and suffering. To the extent that pain and suffering are caused by war, injustice, and other forms of human wrongdoing, they are also a consequence of free will; people choose to inflict harm on one another. On the flip side are good deeds by which people choose to alleviate human suffering.

What about earthquakes, storms, floods, droughts, and other physical catastrophes? Enter modern science into the theologian's reasoning. Physical events are built into the structure of the world itself. Since the seventeenth century, humans have known that the processes by which galaxies and stars come into existence, the planets are formed, the continents move, the weather and the change of seasons, and floods and earthquakes occur are natural processes, not events specifically designed by God for punishing or rewarding humans. The extreme violence of supernova explosions and the chaotic frenzy at galactic centers are outcomes of the laws of physics, not the design of a fearsome deity.

Before Darwin, theologians still encountered a seemingly insurmountable difficulty. If God is the designer of life, whence the lion's cruelty, the snake's poison, and the parasites that secure their existence only by destroying their hosts? Evolution came to the rescue. John Haught (1998), a contemporary Roman Catholic theologian, has written of "Darwin's gift to theology." The Protestant theologian Arthur Peacocke has referred to Darwin as the "disguised friend," by quoting the earlier theologian Aubrey Moore, who in 1891 wrote that "Darwinism appeared, and, under the guise of a foe, did the work of a friend" (Peacocke 1998). Haught and

Peacocke are acknowledging the irony that the theory of evolution, which at first had seemed to remove the need for God in the world, now has convincingly removed the need to explain the world's imperfections as failed outcomes of God's design.

Indeed, a major burden was removed from the shoulders of believers when convincing evidence was advanced that the design of organisms need not be attributed to the immediate agency of the Creator, but rather is an outcome of natural processes. If we claim that organisms and their parts have been specifically designed by God, we have to account for the incompetent design of the human jaw, the narrowness of the birth canal, and our poorly designed backbone, less than fittingly suited for walking upright. Modern proponents of ID would do well to acknowledge Darwin's revolution and accept natural selection as the process that accounts for the design of organisms, as well as for the dysfunctions, oddities, cruelties, and sadism that pervade the world of life.

One difficulty with attributing the design of organisms to the Creator is that imperfections and defects pervade the living world. Consider the human eye. The visual nerve fibers in the eye converge to form the optic nerve, which crosses the retina (in order to reach the brain) and thus creates a blind spot, a minor imperfection, but an imperfection of design, nevertheless; squids and octopuses do not have this defect. Did the Designer have greater love for squids than for humans and, thus, exhibit greater care in designing their eyes than ours? It is not only that organisms and their parts are less than perfect, but also that deficiencies and dysfunctions are pervasive, evidencing incompetent rather than intelligent design. Consider the human jaw. We have too many teeth for the jaw's size, so that wisdom teeth need to be removed and orthodontists can make a decent living straightening the others. Would we want to blame God for this blunder? A human engineer would have done better.

Evolution gives a good account of this imperfection. Brain size increased over time in our ancestors; the remodeling of the skull to fit the larger brain entailed a reduction of the jaw, so that the head of the newborn would not be too large to pass through the mother's birth canal. The birth canal of women is, nevertheless, much too narrow for easy passage of the infant's head, so that thousands upon thousands of babies (several hundred thousand per year by some estimates) and many mothers die during delivery. Surely we don't want to blame God for this dysfunctional design or for the children's deaths. Science makes it understandable, a consequence of the evolutionary enlargement of our brain. Females of other primates do not experience this difficulty. Theologians in the past struggled with the issue of dysfunction because they thought it had to be attributed to God's design. The theory of evolution, much to the relief of theologians, provides an explanation that convincingly attributes defects, deformities, and dysfunctions to natural causes.

More disturbing yet has to be the following consideration. About 20 % of all recognized human pregnancies end in spontaneous miscarriage during the first 2 months of pregnancy. This misfortune amounts at present to more than 20 million spontaneous abortions worldwide every year (more than 100 million children are born in the world each year; 20 % of them amount to more than 20 million). Do we want to blame God for the deficiencies in the pregnancy process? Many people of faith would rather attribute this monumental mishap to the clumsy ways of the evolutionary process than to the incompetence or deviousness of an intelligent designer.

Examples of deficiencies and dysfunctions in all sorts of organisms can be listed endlessly, reflecting the opportunistic, tinkerer-like character of natural selection, which achieves imperfect, rather than intelligent, design. The world of organisms also abounds in characteristics that might be called "oddities," as well as those that have been characterized as "cruelties," an apposite qualifier if the cruel behaviors were designed outcomes of a being holding onto human or higher standards of morality. However, the cruelties of biological nature are only metaphoric cruelties when applied to the outcomes of natural selection.

Examples of "cruelty" involve not only the familiar predators tearing apart their prey (say, a small monkey held alive by a chimpanzee biting large flesh morsels from the screaming monkey), or parasites destroying the functional organs of their hosts, but also, and very abundantly, between organisms of the same species, even between mates. A well-known example is the female praying mantis that devours the male after coitus is completed. Less familiar is that, if she gets the opportunity, the female praying mantis will eat the head of the male *before* mating, which thrashes the headless male mantis into spasms of "sexual frenzy" that allow the female to connect his genitalia with hers. In some midges (tiny flies), the female captures the male as if he were any other prey and with the tip of her proboscis she injects into him her spittle, which starts digesting the male's innards which are then sucked by the female; partly protected from digestion are the relatively intact male organs that break off inside the female and fertilize her. Male cannibalism by their female mates is known in dozens of species, particularly spiders and scorpions. The world of life abounds in "cruel" behaviors: numerous predators eat their prey alive; parasites destroy their living hosts from within; and, as noted, females of many species of spiders and insects devour their mates (Ayala 2007; California Academy of Sciences 2007; Judson 2002; Rönn et al. 2007).

Should God be held accountable? The argument has been advanced by some critics, that the process of evolution by natural selection does not discharge God's responsibility for the dysfunctions, cruelties, and sadism of the living world, because for people of faith God is the Creator of the universe and thus would be accountable for its consequences, direct or indirect, immediate or mediated. If God is omnipotent, the argument would say, He could have created a world where such things as cruelty, parasitism, and human miscarriages would not occur. One possible answer is to claim that God's deeds are inscrutable and humans are not entitled to seek understanding of God's purposes, much less to bring His actions into account. This answer may seem to many unsatisfactory, because it simply evades the question instead of answering it. Theologians who see theology as "faith seeking understanding" are unlikely to be satisfied with such elision.

Other answers are, however, possible. One explanation that may be acceptable to some religious believers, but perhaps not all, would go along the following lines of reasoning. Consider, first, human beings, who perpetrate all sorts of misdeeds and sins, even perjury, adultery, and murder. People of faith believe that each human being is a creation of God, but this does not entail that God is responsible for human crimes and misdemeanors. Sin is a consequence of free will; the flip side of sin is virtue, as pointed out above. The critics might say that this account does not excuse God, because God could have created humans without free will (whatever these "humans" may have been called and been like). But one could reasonably argue that "humans" without free will would be a very different kind of creature, a being much less interesting and creative than humans are. Robots are not a good replacement for humans; robots do not perform virtuous deeds (Ayala 2008).

Before modern physical science came about, God (in some religious views) caused rain, drought, volcanic eruptions, etc. to reward or punish people. This view entails that God would have caused the tsunami that killed 200,000 Indonesians a few years ago. That would seem incompatible with a benevolent God. However, we now know that tsunamis and other "natural" catastrophes come about by natural processes. Natural processes don't entail moral values. Some critics might say, "that does not excuse God, because God created the world as it is. God could have created a different world, without catastrophes." Yes, according to some belief systems, God could have created a different world. But that would not be a creative universe, where galaxies form, stars and planetary systems come about, and continents drift causing earthquakes. The world that we have is creative and more exciting than a static world.

Turn now to badly designed human jaws, parasites that kill millions of people, and a poorly designed human reproductive system that accounts for millions of miscarriages every year in the world. If these dreadful happenings come about by direct design by God, God would seem responsible for the consequences. If engineers design cars that explode when you turn on the ignition key, they are accountable. But if the dreadful happenings come about by natural processes (evolution), there are no moral implications, because natural processes don't entail moral values.

Nevertheless, some would say the world was created by God, so God is ultimately responsible; God could have created a world without parasites or dysfunctionalities. But a world of life with evolution is much more exciting; it is a creative world where new species arise, complex ecosystems come about, and humans have evolved. The Anglican theologian Keith Ward states the case even in stronger terms, arguing that the creation of a world without suffering and moral evil is not an option even for God: "Could [God] not actualize a world wherein suffering is not a possibility? He could not, if any world complex and diverse enough to include rational and moral agents must necessarily include the possibility of suffering [...] A world with the sorts of success and happiness in it that we occasionally experience is a world that necessarily contains the possibility of failure and misery" (Ward 2007).

7 Natural Selection and Design

An engineer has a preconception of what a design is supposed to achieve, and will select suitable materials and arrange them in a preconceived manner so that the design fulfills the intended function. On the contrary, natural selection does not operate according to some preordained plan. It is a purely natural process resulting from the interacting properties of physicochemical and biological entities. Natural selection is simply a consequence of the differential survival and reproduction of living beings. It has some appearance of purposefulness because it is conditioned by the environment: which organisms survive and reproduce more effectively depends on which variations they happen to possess that are useful or beneficial to them, in the place and at the time where they live.

Natural selection does not have foresight; it does not anticipate the environments of the future. Drastic environmental changes may introduce obstacles that are insuperable to organisms that were previously thriving. In fact, species extinction is a common outcome of the evolutionary process. The species existing today represent the balance between the origin of new species and their eventual extinction. The available inventory of living species describes nearly two million species, although at least ten million are estimated to exist. But we know that more than 99 % of all species that have ever lived on Earth have become extinct (Aitken 1998).

Creationists and proponents of intelligent design point out the incredible improbability of chance events, such as mutation, in order to account for the adaptations of organisms (e.g., Meyer 2009). These arguments are irrelevant because evolution is not governed by chance processes (see also Brigandt this volume). Rather, there is a natural process (namely, natural selection) that is not random, but oriented and able to generate order or "create." The traits that organisms acquire in their evolutionary histories are not fortuitous but, rather, determined by their functional utility to the organisms, designed, as it were, to serve their life needs (see Lennox and Kampourakis, this volume). Natural selection preserves what is useful and eliminates what is harmful. Without hereditary mutations, evolution could not happen because there would be no variations that could be differentially conveyed from one to another generation. But without natural selection, the mutation process would yield disorganization and extinction because most mutations are disadvantageous. Mutation and selection have jointly driven the marvelous process that, starting from microscopic organisms, has yielded orchids, birds, and humans.

The theory of evolution conveys chance and necessity jointly enmeshed in the stuff of life; randomness and determinism interlocked in a natural process that has spurted the most complex, diverse, and beautiful entities that we know of in the universe: the organisms that populate the Earth, including humans who think and love, endowed with free will and creative powers, and able to analyze the process of evolution itself that brought them into existence.

8 Coda

Scientists and religious authors have written eloquently about their awe and wonder at the history of the universe and of life on this planet, explaining that they see no conflict between the evidence for evolution and their belief in God. Moreover, authorities of diverse religious denominations have issued statements affirming the compatibility between the tenets of their faith and the acceptance of biological evolution.

Science and religion concern different aspects of the human experience. Scientific explanations are based on evidence drawn from examining the natural world and rely exclusively on natural processes to account for natural phenomena. Scientific explanations are subject to empirical tests by means of observation and experimentation and are subject to the possibility of modification and rejection. Religious faith, in contrast, does not depend on empirical tests and is not subject to the possibility of rejection based on empirical evidence. The significance and purpose of the world and human life, as well as issues concerning moral and religious values, are of great importance to many people, perhaps a majority of humans, but these are matters that transcend science.

It may be fitting to conclude by referring to the two quotations at the beginning of this essay, by Protestant theologians, Aubrey Moore and Arthur Peacocke, and adding a statement from Pope John Paul II in his address of October 22, 1996 to the Pontifical Academy of Sciences:

> New scientific knowledge has led us to realize that the theory of evolution is no longer a mere hypothesis. It is indeed remarkable that this theory has been progressively accepted by researchers, following a series of discoveries in various fields of knowledge. The convergence, neither sought nor fabricated, of the results of work that was conducted independently is in itself a significant argument in favor of this theory.

References

Aitken, G.M. 1998. Extinction. *Biology and Philosophy* 13: 393–411.

Aquinas, T. 1905. Of God and his creatures. In *Summa contra gentiles*, ed. J. Rickaby, 241–368. London: Burns & Oates.

Aquinas, T. 1964. Existence and nature of God. In *Summa theologiae*, ed. Blackfriars, 10–249. New York: McGraw-Hill.

Arp, R. 1999. The *quinque via* of Thomas Hobbes. *History of Philosophy Quarterly* 16: 367–394.

Augustine. 1998. *The city of God*, ed. R. Dyson. Cambridge: Cambridge University Press.

Ayala, F.J. 2007. *Darwin's gift to science and religion*. Washington, DC: Joseph Henry Press.

Ayala, F.J. 2008. Darwin's gift to science and religion: Commentaries and responses. *Theology and Science* 6: 179–196.

Babbage, C. 1838. *The ninth Bridgewater treatise. A fragment*. London: John Murray.

Behe, M. 1996. *Darwin's black box: The biochemical challenge to evolution*. New York: The Free Press.

Bell, C. 1833. *The hand, its mechanisms and vital endowments as evincing design*. Whitefish: Kessinger Publishing, LLC.

Buckland, W. 1836. *Geology and mineralogy. Considered with reference to natural theology*. London: William Pickering.

Burnet, T. 1691. *Sacred theory of the earth*. London: R. Norton.

California Academy of Sciences. 2007. Suicide mating. *California Academy of Sciences* 5: 16–17.

Chambers, R. 1844. *Vestiges of the natural history of creation*. London: John Churchill.

Choiniere, J.N., X. Xu, J.M. Clark, C.A. Forster, Y. Guo, and F. Han. 2010. A basal Alvarezsauroid Theropod from the early Late Jurassic of Xinjiang, China. *Science* 327: 571–574.

Daeschler, E.B., N.H. Shubin, and F.A. Jenkins Jr. 2006. A Devonian tetrapod-like fish and the evolution of the tetrapod body plan. *Nature* 440: 757–763.

Darwin, C. 1859. *On the origin of species by means of natural selection*. London: John Murray.

Dembski, W. 1995. *The design inference: Eliminating chance through small probabilities*. Cambridge: Cambridge University Press.
Dembski, W. 2002. *No free lunch: Why specified complexity cannot be purchased without intelligence*. Lanham: Rowman & Littlefield.
Dobzhansky, Th. 1973. Nothing in biology makes sense except in the light of evolution. *The American Biology Teacher* 35: 125–129.
Haught, J.F. 1998. Darwin's gift to theology. In *Evolutionary and molecular biology: Scientific perspectives on divine action*, ed. R.J. Russell, W.R. Stoeger, and F.J. Ayala, 393–418. Vatican City State/Berkeley: Vatican Observatory Press/Center for Theology and the Natural Sciences.
Hooke, R. 1665. *Micrographia, or, some physiological descriptions of minute bodies made my magnifying glasses with observations and inquiries thereupon*. London: Martyn & Allestry.
Hume, D. 1935. *Dialogues concerning natural religion*, ed. N. K. Smith, 244. Oxford: Oxford University Press.
Johnson, P. 1993. *Darwin on trial*. Downers Grove: InterVarsity Press.
Johnson, P. 2002. *The wedge of truth: Splitting the foundations of naturalism*. Downers Grove: InterVarsity Press.
Judson, O. 2002. *Dr. Tatiana's sex advice to all creation*. New York: Henry Holt.
Klocker, H. 1968. *God and the empiricists*. New York: The Bruce Publishing Company.
Lamarck, J.-B. 1809. *Philosophie Zoologique*. Paris: Duminil-Lesueur. English Trans. Elliot, H. 1914. The Zoological Philosophy. London: Macmillan.
Meyer, S.C. 2009. *Signature in the cell: DNA and the evidence for intelligent design*. New York: Harper-Collins.
Miller, H. 1857. *The testimony of the rocks. Or, geology in its bearings on the two theologies, natural and revealed*. Boston: Gould and Lincoln.
Moore, A. 1891. *Lux mundi*, 12th ed. London: John Murray.
More, H. 1662. An antidote against atheism. In *A collection of several philosophical writings of Dr. Henry More*, ed. F.I. MacKinnon, 1925. London: Flesher and Morden.
Nieuwentijdt, B. 1718/2007. In *The religious philosopher, or the right use of contemplating the works of the creator in the most amazing structure of the heavens with all their furniture*. Trans. J. Chamberlayne. Whitefish: Kessinger Publishing, LLC.
Paley, W. 1802. *Natural theology, or evidences of the existence and attributes of the deity collected from the appearances of nature*. New York: American Tract Society.
Peacocke, A.R. 1998. Biological evolution: A positive appraisal. In *Evolutionary and molecular biology: Scientific perspectives on divine action*, ed. R.J. Russell, W.R. Stoeger, and F.J. Ayala, 357. Vatican City State/Berkeley: Vatican Observatory Press/Center for Theology and the Natural Sciences.
Plato. 1997. Timaeus. In *Plato: Complete works*, ed. J. Cooper. Indianapolis: Hackett Publishing Company.
Ray, J. 1691. *The wisdom of God manifested in the works of creation*. London: Wernerian Club.
Rönn, J., M. Katvala, and G. Arnqvist. 2007. Coevolution between harmful male genitalia and female resistance in seed beetles. *Proceedings of the National Academy of Sciences USA* 104: 10921–10925.
Shubin, N.H., E.B. Daeschler, and F.A. Jenkins Jr. 2006. The pectoral fin of *Tiktaalik roseae* and the origin of the tetrapod limb. *Nature* 440: 764–771.
Stone, R. 2010. Bird-dinosaur link firmed up, and in brilliant Technicolor. *Science* 327: 508.
Swinburne, R. 1994. *The Christian God*. Oxford: Oxford University Press.
Voltaire. 1967. Voltaire, François-Marie Arouet de. In *Encyclopedia of philosophy*, vol. 8, 262–270. London: Macmillan.
Ward, K. 2007. *Divine action. Examining God's role in an open and emergent universe*, 44–45. West Conshohocken: Templeton Foundation Press.

The Implications of Evolutionary Biology for Religious Belief

Denis R. Alexander

1 Introduction

Religious beliefs are extremely varied, ranging from the historical monotheistic belief systems of the three Abrahamic faiths to monistic systems of thought such as Taoism and Confucianism, if indeed these should be classified as religions. According to the editors of the World Christian Encyclopedia there are 19 major world religions which are subdivided into a total of 270 large religious groups, and thousands of smaller ones, including 34,000 distinct Christian groups (Barrett et al. 2001). No wonder that anthropologists have a hard time coming up with precise definitions of 'religion' that everyone agrees on.

The implications of contemporary biology for religious beliefs therefore vary widely depending upon the religion in question. This chapter will focus on the Abrahamic faiths in general, and the Judaeo-Christian tradition in particular, partly due to length constraints, but mainly because biology as a recognizable discipline emerged in Europe over the past few centuries, and it was therefore Christianity with which it mainly related. Even talking about a 'relationship' in this historical context is ambiguous because what we now call biology was termed natural history in earlier centuries and it was but one component of the wider programme of natural philosophy, of which natural theology was an integral component. Those engaged in such scientific pursuits were called natural philosophers and they saw their Christian faith as a motivation for exploring the wisdom and power of God in creation, theology providing a worldview within which their science was integrated.

John Ray (1701), for example, 'father of British natural history', who first introduced the idea of a 'species' to scientific literature, used to give some of his lectures in the chapel of Trinity College, Cambridge, whilst there as a fellow, because he saw

D.R. Alexander (✉)
The Faraday Institute for Science and Religion, St. Edmund's College,
University of Cambridge, Cambridge, UK
e-mail: dra24@hermes.cam.ac.uk

K. Kampourakis (ed.), *The Philosophy of Biology: A Companion for Educators*, History, Philosophy and Theory of the Life Sciences 1, DOI 10.1007/978-94-007-6537-5_10,

the practice of science as an act of worship. Ray's highly influential book *The Wisdom of God Manifested in the Works of Creation* (Ray 1701) went through five large editions within 20 years of its publication. In it he tells us that:

> the treasures of nature are inexhaustible [...] Some reproach methinks it is to learned men that there should be so many animals still in the world whose outward shapes is not yet taken notice of or described[...]if man ought to reflect upon his Creator the glory of all his works, then ought he to take notice of them all....

And 'taking notice of them all' was precisely what John Ray did, publishing, for example, three great volumes containing 2,610 folios on the classification of 18,600 plants in the years 1686–1704.

Small wonder that when Richard Bentley [1662–1742], Master of Trinity College, gave the first Boyle lecture on natural theology in 1692, 'A Confutation of Atheism', he used arguments drawn 'from the structure of animate bodies and the origin and frame of the world' (Alexander 2001). Natural theology continued to be a first cousin of the scientific enterprise, at least in Britain, until well into the nineteenth century, with the works of Archdeacon William Paley, such as *The Evidences of Christianity* (Paley 1795), required reading for generations of Cambridge undergraduates. Darwin's *On the Origin of Species* (Darwin 1859) has been dubbed 'the last great work of Victorian natural theology' (Durant 1985) and in his autobiography (Darwin 1958) Darwin reports that

> ....I am convinced that I could have written out the whole of the Evidences with perfect correctness, but not of course in the clear language of Paley. The logic of this book, and, as I may add, of his Natural Theology, gave me as much delight as did Euclid. The careful study of these works [...] was the only part of the academical course which, as I then felt, and as I still believe, was of the least use to me in the education of my mind.

Yet it was precisely Darwin's theory of natural selection which provided a mechanism to explain the complexity and diversity of the living world, features which Paley had tended to ascribe more directly to the guiding and designing hand of providence.

Darwin was a deist during the period when he was writing the *Origin* but later self-described himself as an agnostic once his friend Thomas Henry Huxley had invented the word in 1869. But Darwin was never an atheist and always maintained that belief in God was compatible with commitment to evolutionary theory. His eventual drift from faith was caused, it is thought, by the sad experience of losing three of his family whilst still in infancy or childhood, together with doubts about several tenets of Christian faith (Brooke 2010). Meanwhile by the 1870s evolution was adopted by the leading Christian thinkers in the mainstream denominations on both sides of the Atlantic, by which time earlier critical voices had largely faded from view, at least those voices that were able to gain themselves a hearing in the public domain. For example, Darwin's own mentor at Cambridge, the clerical Professor of Geology Adam Sedgwick (1785–1873), published a critical review of the *Origin* and wrote to Darwin that "I have read your book with more pain than pleasure".[1]

[1]Adam Sedgwick to Darwin, Nov. 24th, 1859. http://www.darwinproject.ac.uk/entry-2548. Accessed Sep. 16th, 2012.

Sedgwick felt that Darwin had departed from the sound inductive approach of true science and was concerned that human moral dignity would be undermined by evolution. But Sedgwick was of the old school and the Anglican voices that proved to be more influential in the long run are well illustrated by Frederick Temple (1885), the future archbishop of Canterbury, who in his Bampton lectures (Temple 1885) argued that the activity of God was to be discerned throughout the laws governing the natural world, laws responsible for the evolutionary process, not in the gaps in current scientific knowledge.

The parting of the ways between science and faith, certainly as far as the English-speaking world is concerned, came about not as a result of evolution, as is sometimes thought, but due to the increasing professionalization of science that characterized the closing decades of the nineteenth century (Turner 1978). Until this time science had been largely carried out by interested amateurs, gentlemen of leisure (Darwin himself being a typical example) with the financial patronage or investments that allowed them the freedom to pursue their interests. Many clerics were in this category and the clerical natural philosophers were prominent in their leadership of British scientific institutions, such as the British Association for the Advancement of Science, during the first half of the nineteenth century. But by 1900 dramatic changes had occurred in the social structures of science. Led by Thomas Henry Huxley, a group of 'scientists', as they now called themselves, campaigned for the professionalization of the scientific community, eager to wrest from the established Church the financial resources and intellectual prestige that it then enjoyed. There were now far more salaried positions in the sciences, which themselves were becoming more specialized, and there was a proliferation of scientific societies with their associated journals. Enthusiasm for natural theology had waned and clerics were now far less represented in scientific leadership and in education.

It is important to understand this historical background in order to appreciate the framework within which much of the discussion about the relationship between religion and contemporary biology is carried out (Alexander 2001). Underlying the discussion is the assumption that a triumphalist science has steadily but inevitably replaced the religious types of explanation that characterized earlier eras. Such a metanarrative is given some credence by the story of how Paleyan design was replaced by scientific mechanism – natural selection – thereby rendering the religious interpretation redundant. In reality, however, the interaction of biology with religious belief has been historically, and continues to be in contemporary discourse, highly complex, and does not readily lend itself to metanarratives (Brooke 1991).

1.1 Models for Relating Science and Religion

Models can act as useful heuristic tools for exploring the complex relationships between science and religion, providing that it is not imagined that any one model will encompass the whole story. When people refer to the impact of biology on religious thought, they often frame their discussion within the understanding of a particular model, tacitly or overtly, and so it is good to be aware of the range of models on offer. Four main models will be briefly summarized here, though there are many more, and the four models listed below are often given different names.

1.1.1 The Conflict Model

The Conflict model proposes, as the name suggests, that science and religion are in fundamental opposition, and that this has always been the case. The idea is clearly expressed by Worrall when he writes that "Science and religion are in irreconcilable conflict [...] There is no way in which you can be both properly scientifically minded and a true religious believer" (Worrall 2004). It is worth noting both the descriptive and normative elements in such an assertion. Richard Dawkins is a strident supporter of the conflict model, stating that "I pay religions the compliment of regarding them as scientific theories and [...] I see God as a competing explanation for facts about the universe and life" (Dawkins and Ward 1995). In this view, therefore, science and religion are competing for the same kind of territory. In his book *The Blind Watchmaker* (1986) Dawkins recounts how evolutionary design replaced the religiously inspired Paleyan notion of design, and it is this 'replacement theory' which appears to undergird his position.

The conflict model is well sustained by the fact that in the USA, the country that currently leads the world in science and technology, about 40 % of the population hold to creationist beliefs (Miller et al. 2006). Over the past two decades an anti-Darwinian movement known as Intelligent Design (ID) has achieved popularity in the USA, claiming that certain biological entities are too complex to have come about by 'chance', therefore pointing to 'design' as a purported alternative (see Brigandt, this volume). Both creationism and ID have led to high-profile court cases over what should be taught in US secondary schools. In the more secularised European context, where in any case educational curricula are established nationally rather than by local school boards, as in the USA, creationist/ID movements have attracted relatively less attention. Nevertheless, the huge influence of the US media plus coverage in science journals has ensured that such local conflicts achieve wide international coverage. Furthermore, a UK poll carried out in 2009 revealed that 10 % of the population apparently adheres to creationist views (Spencer and Alexander 2009).

Sociologically the conflict model is maintained on one hand by those who espouse a triumphalist science in which ultra-Darwinian narratives are deployed to show how evolution now 'explains' all that religion previously purported to explain, and on the other hand by fundamentalist religious believers who look to religious texts (be it the Bible or the Qur'an) to espouse a so-called 'creation science' which, in practice, involves the rejection of much that the scientific community takes for granted. For the media, 'conflict sells', be it books or TV programmes, and there is therefore a bias to give these contributions to the 'conflict model' more exposure. Arguably both the extreme poles in the discussion need each other for their very existence and it has often been suggested that the robust attacks on religion by the so-called 'new atheists' have done much to stimulate the popularity of creationism and ID (Kitcher 2011).

One feature of the 'conflict model' is far less prominent than it used to be and that is the idea of the historical 'warfare' between science and religion, a thesis prominent in books written in the late nineteenth and early twentieth centuries. The revisionist historians of the last few decades have put that idea to rest, pointing out the many and varied ways in which religious beliefs have stimulated and shaped the assumptions and methods of what we now call modern science. As Stephen Shapin comments in his book *The Scientific Revolution* (1996):

> In the late Victorian period it was common to write about 'the warfare between science and religion' and to presume that these two bodies of culture must always have been in conflict. However, it has been a very long time since these attitudes have been held by historians of science (p. 195).

Equally many of the myths about the history of science that litter the historical literature, many of them used to support the conflict model in earlier eras, have now been revealed for what they are (Numbers 2009), making it less likely that attempts will be made to support the conflict model by appeals to history. For example, only those rather ignorant of the extensive academic literature on the subject would be tempted to cite the 'Galileo affair' as representing a clash between science and religion (McMullin 2005).

1.1.2 The NOMA Model

The 'NOMA' Model (Non-Overlapping Magisteria) was popularized by the late Stephen Jay Gould in his book *Rocks of Ages* (Gould 2002). Gould maintained that science and religion operate within separate compartments addressing quite different kinds of questions, and therefore there can be no conflict between them virtually by definition. In addition, Gould held that science deals with matters of fact whereas religion addresses questions of ethics, value and purpose. Gould was not the first to hold such a view, but we will use his convenient 'NOMA' label here.

Science and religion do indeed ask rather different kinds of question about the world. Science is interested in finding mechanistic explanations, those that elucidate how things become as they are, or operate as they do. Science seeks broad generalisations that describe the properties of matter in a way that allows accurate predictions. Science values mathematics highly and looks for mathematical expressions of data whenever feasible. Experimental testing and reproducibility are critical in the scientific method. Religion, by contrast, is interested in asking ultimate questions; in Leibniz's famous aphorism: 'Why is there something rather than nothing?' Religion wishes to know why science is possible in the first place. In the words of Stephen Hawking: 'What breathes fire into the equations?' Why does the universe go to all the bother of existing? Does life have any ultimate meaning or purpose? Does God exist? How ought we to act in the world? Gould was right – science and religion do indeed ask different kinds of questions.

There are, however, some significant problems with the NOMA model. Gould himself fatally undermined his own model by writing entertaining essays on key figures in the history of science whose thinking was greatly influenced by their religious beliefs.[2] The constant traffic of ideas between science and religion over the centuries, interactions that continue to the present day, does not support the idea that these human activities lie in completely separate realms (Brooke 1991; Brooke and Cantor 1998). One only has to think of iconic figures in the history of science, such as Isaac Newton, who wrote an estimated 2.5 million words on theology, the single largest subject category in his manuscript corpus. Newton was convinced that scientific laws were only possible because of the ever-present actions of a law-giving God, writing in Query 31 of his *Opticks* that 'if there be an universal life and all space be the sensorium of a thinking being who by immediate presence perceives all things in it [...] the laws of motion arising from life or will may be of universal extent' (Westfall 1971) arguing that 'there exists an infinite and omnipresent spirit in which matter is moved according to mathematical laws' (Brooke 1988).

Although it is true, as the NOMA model suggests, that science and religion ask distinct types of question about reality, nevertheless it is the same reality that is being addressed in both cases. Science owes its success to the restricted nature of its questions. Nevertheless, even that limited repertoire uncovers facts that, to many scientists, have religious significance. In addition, the partitioning of facts to the world of science and values to the world of religion is inconsistent with religious claims. Most religious believers would maintain that they hold to facts about the world, such as the fact that the human psyche is sinful or that rape is wrong. These are not scientific facts but, arguably, facts nonetheless. The neat facts/values partition does not therefore work in practice.

1.1.3 Fusion Models

Fusion models represent the polar opposite of the NOMA model in that they tend to blur the distinction between scientific and religious types of knowledge altogether, or attempt to utilise science in order to construct religious systems of thought, or vice versa. The plural 'models' is necessary because the various strategies for achieving fusion are very diverse. There are many examples of this approach from the world of biology, some of them tending towards mysticism.

For example, the French Jesuit priest and palaeontologist Teilhard de Chardin was a Lamarckian evolutionist who proposed a grand religious evolutionary theory, inspired by biological evolution, in which the whole living world evolves towards an 'omega point' which, Teilhard de Chardin suggested, is the ultimate end-point of the evolutionary process in which all things find their ultimate fulfillment in Christ (Teilhard de Chardin et al. 1959). Here the ideas and language of Lamarckian

---

[2]E.g. Gould, S. J. on the Revd. Thomas Burnet, author of the seventeenth century work *The Sacred Theory of the Earth* in *Ever Since Darwin*, Penguin Books (1980), ch. 17 pp. 141–146.

evolutionary theory have been used to construct a great theological metanarrative involving a fusion of scientific and religious ideas.

A more contemporary and very different example is given by Stuart Kauffman in his book *Reinventing the Sacred* (Kauffman 2008) in which the author argues that qualities of divinity that we hold sacred, such as creativity, meaning and purpose, act in a secular society as the dynamic equivalent of transcendent feelings. Emergent properties of the biosphere are such as to arouse awe and wonder which, in turn, constitute the key elements of a religion free of traditional concepts of God.

Fusion models tend to be promoted by individual thinkers who influence others who in turn may adhere to various religious traditions, but the ideas in themselves rarely become institutionalized. One reason is that if religious ideas are linked closely to particular scientific theories, these will inevitably change or even disappear with the passage of time, and the religious inferences then die with them. Those who marry their metaphysics to contemporary scientific theories may soon find themselves widows.

Fusion models can also lead to a rather confusing mix of scientific and religious language in which explanatory levels become conflated. Intelligent Design is an example of a fusion model idea in that it attempts to introduce philosophical and religious concepts of 'design' into scientific discourse as if the 'inference to design' represented an actual scientific explanation for something (which it does not). Labeling a complex biological entity, such as a bacterial flagellum, as 'designed', does not generate experiments that could in principle falsify such a notion, so the idea does not belong to science (see Brigandt, this volume).

1.1.4 The Complementarity Model

The Complementarity Model takes a different approach, maintaining that science and religion are addressing the same reality from different perspectives, providing explanations that are not in any kind of rivalry to each other, but rather are complementary. The language of complementarity was originally introduced by the physicist Niels Bohr to describe the relationship between the particle and wave descriptions of matter; it was necessary to hold on to both understandings simultaneously to do justice to the data (Bohr 1950). Since Bohr's time, the idea of complementarity has been greatly extended within the science-religion field to encompass any entity that requires explanations at multiple levels in order to explain its complexity adequately. The classic example is provided by the multiple descriptions required to understand the human individual at the various levels of analysis provided by disciplines such as biochemistry, cell biology, physiology, psychology, anthropology and ecology. None of these scientific descriptions is rival to any others – all are required for our understanding of the complexity of human beings in the context of their environment. A similar complementary relationship exists between brain and mind. Scientific descriptions of neuronal events that occur during brain activity are complementary to the 'I' language of personal agency that reflects the thoughts of the conscious human mind. Ignoring one level at the expense of the other impoverishes our understanding of human personhood.

Within the language of complementarity, religion provides a further set of explanations, beyond the ability of science to adjudicate, that relate to the factual questions of ultimate purpose, value and meaning. There is nothing in such religious explanatory levels that need be in rivalry with the scientific explanatory levels: the descriptions are complementary. Just as it is possible, in principle, to use brain imaging to describe the neuronal activity in the brain of a scientist as she assesses data from her laboratory, pondering the significance of those data for the current theory under investigation, so equally is it possible to carry out the same experiment on someone (it could be the same person) in a different context as they assess evidence for a religious belief. But in neither case could the scientific data generated by brain imaging be used to justify (or not) the ensuing conclusions, which have to be based on the rational assessments made by the person involved. Those personal assessments, and the brain activity described by the scientist that occurs during that process, provide complementary insights into what is arguably a single reality. But both accounts are essential to do justice to the phenomenon.

The complementarity model is distinct from the NOMA model in its insistence that the narratives provided by science and religion are valid within their own particular frameworks of reference. There is no need, in principle, to privilege one narrative over another, nor to insist that one type of narrative speaks only of 'facts', whereas another speaks only of 'values'. Each narrative needs to be assessed within its own framework.

Two main criticisms have been levelled against the complementarity model. The first is that it can too readily slide into a form of the NOMA model by default, thereby escaping the hard task of bringing apparently irreconcilable data together into a unified theory. This is a valid criticism. It was addressed by Donald MacKay who cautioned against using the concept to relate two phenomena which are in fact contradictory. MacKay suggested that complementary explanations are justified 'Only when we find both are necessary to do justice to experience' (MacKay 1988). The principle must never therefore be invoked as a cloak for intellectual laziness or to retard the pursuit of better ways of explaining the relationships between different types of experience.

The second criticism is that the model can give the impression that science is the realm of objective truth and facts, whereas religion is the realm of subjective convictions and values. Yet there is no reason in principle why complementary moral and religious descriptions cannot be seen as factual as scientific descriptions. For example, as already suggested, we may accept as moral facts such claims as that rape and cannibalism are wrong. If we accept such statements as moral facts, then it does not seem irrational to argue that such moral or religious dimensions in our complementary descriptions of reality can be as factual as the various scientific levels of description.

1.2 Ontological and Methodological Reductionism

Many disciplines within the biological sciences are reductionist in their approach, not least genetics and molecular biology, sciences which are essential for an understanding of evolutionary processes. To appreciate the implications of contemporary evolutionary biology for religious belief, it is important to distinguish between two distinct brands of reductionism.

Methodological reductionism is an essential research strategy for the biological sciences. To find out how something works, we take it to bits. The challenge in the investigation of living systems is then to discern how the myriad components operate together to generate an operational complex whole. For example, systematic research programmes exist to delete genes using 'knockout technology' from the genomes of inter-breeding mouse colonies with the aim of identifying the functions of the approximately 21,000 genes that comprise the mammalian genome. This is an immensely powerful and fruitful research strategy, made more powerful by the ability to delete gene expression in specific tissues in a time-dependent way. For example, deletion of a gene called *Caml* just after birth revealed its important role in hearing in mice (Bryda et al. 2012) and selective deletion of a gene encoding a key signaling pathway in the head and neck of mice gave valuable insights into the development of a carcinoma (Bian et al. 2009). But all researchers are aware that the outcome of such experiments is often strikingly unpredictable. Making firm predictions about the effects of perturbing single components of complex systems is notoriously difficult. Yet this in itself does not render the approach of methodological reductionism invalid, only highlights the challenge of interpretation. Those engaged in such research are also well aware that their ultimate goal is to elucidate the way in which the living system under scrutiny functions as a whole.

Ontological reductionism, by contrast, is a philosophy parasitic upon methodological reductionism that claims that the components of systems provide the 'real and only valid' explanation. For example, the late Francis Crick proclaims in his book *The Astonishing Hypothesis* (Crick 1994) that: "The Astonishing Hypothesis is that 'You', your joys and your sorrows, your memories and your ambitions, your sense of personal identity and free will, are in fact no more than the behaviour of a vast assembly of nerve cells and their associated molecules". In fact it is quite common to read in more popular presentations of the latest findings in the neurosciences that the neuronal events, perhaps revealed by brain imaging, provide the 'real story' of what is happening, with the tacit if not explicit implication that the 'I' story provided by the human agent is epiphenomenal. The assumption of ontological reductionism is often flagged up by phrases such as "no more than", as in the Crick quotation above, or "nothing but". In fact so prevalent is the latter phrase in popular discourse about the brain that the neuroscientist Donald MacKay once dubbed such displays of ontological reductionism as "nothing buttery".

Although it is rare to find ontological reductionism promoted by academic philosophers of biology, it provides a popular framework within which the latest scientific results may be presented, sometimes dubbed 'scientism', the idea that scientific explanations provide the only valid and reliable forms of truth-telling, and that the reductionist account is the only one that should be taken seriously (Van Woudenberg 2012). It is often the case that biological accounts are presented as if they provide the 'real' story of what is going on and that once the evolutionary story has been told, there are no other narratives to recount at other complementary levels of discourse, such as "why does this amazing, complex, evolutionary narrative, which has brought minds into being which can ponder their own evolutionary history, exist anyway?"

There are several research areas of contemporary biology that have implications for religious belief. Of these perhaps evolution is the most dominant and is therefore chosen as the exemplar in what follows. However, in focusing on this example, it is good to remember that the points made frequently apply to interactions of religion with other biological disciplines, such as the neurosciences.

2 Darwinian Evolution and Religion

The implications of evolution for religious belief (see also Ayala, this volume) have been the main focus of discussions between biology and religion ever since 1859, although the use of evolution in support of a wide range of disparate ideologies over this period has done much to complicate those discussions (Alexander and Numbers 2010). What often happens in the history of science is that a new scientific theory gains public attention due to its explanatory power, but then its prestige becomes used by various interest groups in support of causes that have little or nothing to do with the biology itself. A process of 'social transformation' occurs in which 'Scientific Theory X' becomes transformed in the public consciousness into 'Social Meaning Y'. Evolution, for example, has been used in support of capitalism, socialism, communism, racism, militarism, eugenics, feminism, atheism, theism, and other political and social ideas, many of them mutually exclusive (Alexander and Numbers 2010). This helps to explain the rise of creationism in the USA during the course of the twentieth century. Concerns about the supposed evolutionary doctrine of 'might is right'[3] espoused by the Kaiser's officers during the First World War together with the possible import of such ideas into the US to threaten liberal democratic political causes all played their part in promoting the rise of American creationism (Numbers 2006).

The thicket of misunderstandings that surround the topic of evolution renders education in this field challenging. Misunderstandings often arise from the ideological weight with which the word 'evolution' is loaded in public discourse. Some of the issues with particular relevance to religious belief are considered below.

2.1 *Hermeneutics*

Many are puzzled by the odd phenomenon of a country, the USA, that leads the world in science and technology, which has around 40 % or more of its population rejecting the theory of evolution. Many cultural, political and religious reasons

[3]Referring to the idea that just as the strong are better able to survive and reproduce in the process of evolution, so this provides some kind of justification for humans to behave in a similar way.

underlie this strange phenomenon, many of which have been helpfully discussed by Ronald Numbers in his book *The Creationists* (Numbers 2006). Since one of those reasons is Biblical hermeneutics, the methods whereby texts are interpreted, a brief summary of how changing views on hermeneutics have impacted on the growth of US creationism may be useful.

In traditional Jewish and Christian hermeneutics, the early chapters of Genesis were interpreted figuratively to a large extent. The early Church Father, Origen [c.185–c.254], wrote with respect to the interpretation of Genesis 1 and 2:

> What man of intelligence, I ask, will consider that the first and second and the third day, in which there are said to be both morning and evening, existed without sun and moon and stars, while the first day was even without a heaven? And who could be found so silly as to believe that God, after the manner of a farmer 'planted trees in a paradise eastward in Eden' [...] I do not think anyone will doubt that these are figurative expressions which indicate certain mysteries through a semblance of history (Origen 1936).

In his commentary entitled "The Literal Interpretation of Genesis" [410], Augustine also adopted a figurative interpretation of the days of Genesis, seeing God's creative activity as having two different aspects:

> Some works belonged to the invisible days in which he created all things simultaneously, and others belong to the days in which he daily fashions whatever evolves in the course of time from what I call the primordial wrappers (6.6.9).

The 'invisible days' in Augustine's understanding were the days as described in Genesis 1, which he understood not chronologically but as a kind of inventory of all God's acts of creation which were performed simultaneously. This single act of creation then brought forth, in due course, all the rest of the diversity of the created order. All the potentiality of the created order was encompassed within those original 'primordial wrappers', like seeds with the potential to grow into something quite different.

Augustine's figurative understanding of the 'days' of Genesis in some ways echoed those of the philosopher Philo, an Alexandrian Jew who lived from around 15–10 BC to AD 45–50. In his book *On the Account of the World's Creation Given by Moses and Allegorical Interpretation of Genesis II, III*, Philo writes of the days of Genesis 1 in highly symbolic terms:

> He says that in six six days the world was created, not that its Maker required a length of time for His work, for we must think of God as doing all things simultaneously, remembering that "all" includes with the commands which He issues the thought behind them. Six days are mentioned because for the things coming into existence there was a need of order [...] For it was requisite that the world, being most perfect of all things that have come into existence, should be constituted in accordance with a perfect number, namely six. (Philo et al. 1929)

In the medieval period the Christian tradition of reading Biblical texts as if there were various layers of meaning was widespread. The text was seen like an onion in which layers of meaning could be peeled off layer by layer. The 'commonsense' reading of the text was for the spiritually unenlightened, whereas the deeper understandings were reserved for the more learned. This in turn led to an exotic flowering of allegorical interpretations whereby the natural world, for example, became imbued with theological meaning (Harrison 1998). The pelican came to symbolize

Christ's atonement because of the widespread belief that parental pelican blood could bring to life its dead offspring. The phoenix came to represent the resurrection. Hugh of St Victor (c.1078–1141) declared that "the whole sensible world is like a book written by the finger of God" in which each of the creatures is a figure "not invented by human decision, but instituted by the divine will to manifest the invisible things of God's wisdom".[4] Symbolic and allegorical interpretations were imposed upon the reading of both nature and the Bible.

The Reformation saw a distinct shift in Biblical hermeneutics towards more straightforward and less allegorical readings of the text, associated with the translation of the Bible into vernacular languages and its widespread printing and distribution to become widely accessible, for the first time, to the masses. This increased the tendency to read texts such as Genesis 1 in more literalistic terms, particularly in the hands of Martin Luther. Commenting on Augustine's more allegorical approach, Luther wrote:

> I ask you, dear reader, what need is there of those obscure and most foolish allegories when this light is so very clear [...] Do they not smother the true meaning and replace it with an idea which is not merely useless but disastrous? [...] For we have the Holy Spirit as our Guide. Through Moses, He does not give us foolish allegories, but He teaches us about most important events.[5]

Calvin gave much greater scope to the idea of 'accommodation' in his writings, the idea that God speaks in the Bible in the kind of everyday language that is accessible (or 'accommodating') to any reader and is therefore not about science. As Calvin wrote, Moses "adapted his writing to common usage." The Bible was "a book for laymen" and "he who would learn astronomy and other recondite arts, let him go elsewhere." Calvin also wrote that:

> The Holy Spirit had no intention to teach astronomy; and, in proposing instruction meant to be common to the simplest and most uneducated persons, he made use by Moses and the other prophets of popular language [...] the Holy Spirit would rather speak childishly than unintelligibly to the humble and unlearned.[6]

So for Calvin it was not the role of Biblical texts to teach science, and such a stance was supported by the "two books" idea that was widespread in the early modern period, knowledge of God being obtained from the Scriptures as the Book of God's Word, whereas Nature represents the Book of God's Works. Robert Boyle, for example, wrote in his 1674 tract *The Excellence of Theology Compared with Natural Theology* (Boyle 1674) that:

> As the two great books, of Nature and of Scripture, have the same author; so the study of the latter does not at all hinder an inquisitive man's delight in the study of the former.

[4]Cited by P. Harrison in Berry, R.J. (ed), *Lion Handbook on Science and Christianity,* Oxford: Lion, page 57.

[5]J.P. Pelikan and H. Lehmann, ed., 'Luther's Works' American Edition, Volume 1, 'Lectures on Genesis, Chapters 1–5' St. Louis, Concordia Publishing House; Philadelphia, Fortress Press, 1955, p.1.

[6]Calvin. Commentary on Psalm 136:7; Calvin's Commentaries, Vol. 12: Psalms, Part V, tr. by John King, [1847–50]. Edinburgh: Calvin Translation Society.

The Christian natural philosophers who dominated the growth of science in the early modern period were typically accommodationists in their interpretation of Scripture and this was a key stance that helped avoid the inappropriate interpretation of Biblical texts as if they were scientific texts.

The widespread knowledge of the Bible in the English-speaking world, including education in basic hermeneutics, may well have played a role in the rapid acceptance of Darwinism following 1859 by mainstream denominations. No church leader in Darwin's time would have thought that the Earth was other than very old, even though it was not yet known how old. Darwin did not face young Earth creationism in the reception of his theory nor, amongst most church leaders at least, a wooden literalistic handling of the Genesis text. The conservative Anglican view of the time is well expressed by the Revd. Richard Main writing in 1862: 'Some school-books still teach to the ignorant that the Earth is 6,000 years old [...] No well-educated person of the present day shares that delusion'.[7] Having said that, Darwin himself expressed surprise (in 1861) that the date of 4004 BC estimated by Archbishop Ussher (of Armagh in the Church of Ireland, published in 1650) for the creation of the world was not actually found in the Bible.[8] This date was still to be found printed in the margin of Genesis in some older printings of the Bible even in Darwin's time, although the fact that the Earth was ancient was already well-established by the early nineteenth century (Rudwick 2004).

The stance of harmonization as well as accommodationism both played important roles in the Christian reception of evolution. The harmonisers tried to impose a contemporary scientific understanding on to Genesis 1, not very successfully it has to be said, but at least to their own satisfaction, whereas the accommodationists tended to view the early chapters of Genesis in more figurative and theological terms. For example, the widely read nineteenth century Scofield Reference Bible (1909) popularized the 'gap theory', the harmonising idea that a long period of geological time could be inserted between the first two verses of Genesis chapter 1, an idea that goes back to the Scottish theologian Thomas Chalmers [1780–1847] and before. This enabled Christians to accept the geological evidence for a very old Earth without contradicting Genesis 1. Accommodationists were typified more by the writers of *Lux Mundi* (1889), a collection of essays by Anglo-Catholic theologians based in Oxford, who adopted a more figurative and allegorical understanding of the early chapters of Genesis.

The Biblical hermeneutics that eventually helped to undergird today's young Earth creationism in the USA came not through reactions to Darwinism but as a result of reactions against the perceived influence of liberal German protestant theology in US theological seminaries in the late nineteenth and early twentieth centuries. This in turn led to the writing of a series of 90 essays published in 12 volumes as *The Fundamentals* published in the period 1910–1915 which were

[7]Cited in Berry, R.J. (ed), *Lion Handbook on Science and Christianity,* Oxford: Lion, page 137.

[8]Charles Darwin in a letter to Leonard Horner dated 20th March 1861 [Darwin Correspondence Project, http://www.darwinproject.ac.uk/entry-3094].

intended to act as a bulwark against the inroads of liberal theology (Numbers 2006). *The Fundamentals* gave rise to our contemporary term 'fundamentalist', although ironically many of the essayists were not actually fundamentalist as the term is currently used today, being well-known for their acceptance of evolutionary theory. For example, R.A. Torrey, who edited the last two volumes in the series, once referred to Darwin as 'the greatest scientific thinker of the nineteenth century' and with reference to the interpretation of the creation story in Genesis, Torrey wrote that 'Anyone who is at all familiar with the Bible and the Bible usage of words knows that the use of the word "day" is not limited to periods of 24 h. It is frequently used of a period of time of an entirely undefined length'. In the same vein another frequent contributor to *The Fundamentals*, James Orr, reminded his readers that the 'Bible was never given us in order to anticipate or forestall the discoveries of modern twentieth century science' (Livingstone 1987).

It was not until the work of the early twentieth century self-styled 'geologist' Seventh Day Adventist George McReady Price that the text of Genesis 1 began to be interpreted as a scientific text, leading to his book published in 1923 entitled *New Geology* which attempted to critique the fossil data based on the geological column and replace it with a 'Mosaic geology' derived from the Bible. Outside of his Adventist community Price's book had little influence. As the historian George Marsden points out, before 1960, what is today known as 'creation science' had only meager support even among fundamentalist communities in the United States (Marsden 1992). But this situation changed dramatically in the early 1960s with the publication of *The Genesis Flood* (Whitcomb and Morris 1961) written by a lecturer in civil engineering from the Rice Institute called Henry Morris together with a young theologian called John Whitcomb. To their great surprise, and the surprise of the public at large, Price's once marginalised beliefs now began to enjoy a wide readership and notoriety. Within a quarter of a century *The Genesis Flood* went through 29 printings and sold over 200,000 copies. The interest stimulated by this book led to the formation of the Creation Research Society (CRS) in 1963 which required its members to sign a statement of belief accepting the inerrancy of the Bible, the special creation of "all basic types of living things", and a worldwide deluge. Unlike the 1920s creationist movement, which was largely devoid of any support from scientists, the CRS made a particular point of recruiting scientists onto its committee and into its membership. CRS founders referred to themselves as "scientific creationists" (Numbers 2006).

To a very large degree the present prevalence of creationism in the USA therefore depends upon a particular hermeneutics that has only become embedded in certain Christian communities relatively recently. The hermeneutics espoused by the creationists is strikingly modernist, as illustrated by this comment from one of the authors of *The Genesis Flood*, the late Henry Morris, who claims that 'the Scriptures, in fact, do not need to be "interpreted" at all, for God is well able to say exactly what he means'. This, for Morris, entailed that the days of Genesis 1 are taken as literal days of 24 h, with the chapter in general being interpreted as if it were a scientific text (Morris 1984). The inappropriate attribution of scientific value and reliability to an ancient text is a typically modernist attitude characterized by the assumption that scientific knowledge displays a privileged position. It is important to realize from an educational perspective that for the 40 % or more of Americans who reject evolution, their Biblical hermeneutics entails that if they accept evolution then, in their minds, they would need to give up important elements of their faith. For example, creationists believe that Adam and Eve lived around 10,000 years ago and were the genetic progenitors of all humans living today. Adam was literally created out of dust by a miracle.[9] If this were not so, then, creationists believe, this would undermine the special role of humankind in God's creation and subvert the Christian understanding of the Fall. This explains why Christians of this persuasion, and indeed many Muslims, are so resistant to the overwhelming evidence in support of evolution. Unless the hermeneutical problem is addressed, then little educational progress will be made in this community on the topic of evolution.

2.2 *What Creation Means*

Evolution was accepted by Christian denominations on both sides of the Atlantic remarkably soon after the publication of *The Origin of Species* in 1859. The historian James Moore writes that "with but few exceptions the leading Christian thinkers in Great Britain and America came to terms quite readily with Darwinism and evolution" (Moore 1979), and the American historian George Marsden reports that "… with the exception of Harvard's Louis Agassiz, virtually every American Protestant zoologist and botanist accepted some form of evolution by the early 1870s" (Marsden 1984). It was in fact Agassiz, not known for his Christian commitment, who opposed evolution.

Why was evolution not a significant problem for these late nineteenth century scientists and church leaders? One reason is that their theological matrix was shaped by the traditional Christian understanding of creation, which starts with the claim that everything that exists apart from God has only come into existence because God has brought it into existence. God is the ground of all existence, and in this view 'existence' refers to anything that exists, material or immaterial – the laws of nature, quantum vacuums, mathematical principles, the elements of the periodic table, or whatever it might be. If it exists and is not God then it must by definition be part of the created order, in this view.

The words 'create' or 'creation' in a theological context are frequently misunderstood. Part of the problem comes from the way in which the word 'create' is used in the English language. When human beings make things they work with already existing material to produce something new. The human act of creating is not the complete cause of what is produced; but God's creative act is the complete cause of what is produced. As the theologian William Carroll writes:

> God's causality is so different from the causality of creatures that there is no competition between the two, that is, we do not need to limit, as it were, God's causality to make room for

[9]Genesis 2:7.

> the causality of creatures. God causes creatures to be causes [...] Creation is not essentially some distant event; rather, it is the on-going complete causing of the existence of all that is. At this very moment, were God not causing all that is to exist, there would be nothing at all. Creation concerns first of all the origin of the universe, not its temporal beginning (Carroll 2012).

These comments echo the thirteenth century theologian Thomas Aquinas who wrote: "Over and above the mode of becoming by which something comes to be through change or motion, there must be a mode of becoming or origin of things without any mutation or motion, through the influx of being".[10] So in this traditional Christian view, the term 'creation' refers primarily to ontology, the existence of things and the meanings of their existence.

Another aspect of the theological understanding of creation that has had, and continues to have, a positive impact on the widespread acceptance of evolution in Christian communities (though more so outside of the USA), is the emphasis on God's 'immanence' in the created order. This refers to God's continued upholding and sustaining of the created order. The term 'creation' refers not primarily to the temporal origin of the universe (though certainly it includes that), but rather to God's immanence in its on-going existence, giving rise to the nomic regularity that renders the properties of matter coherent, thereby making science possible. The work of a scientist, therefore, within this world-view, entails the investigation of God's created order, continually sustained by God. The term *creatio continua* has traditionally been used to refer to this emphasis on God's on-going creative activity.

That this traditional Christian understanding of creation played an important role in the initial reception of Darwinism is apparent from the very first written extant evidence that we have, to be found in the extensive collection of Darwin's Correspondence located at the Cambridge University Library. Darwin exchanged letters with nearly 2,000 correspondents in the course of his life, of whom around 200 were clergyman, some personal friends, many of whom provided Darwin with biological data for his publications. Darwin sent out several pre-publication copies of the *Origin of Species*. One of them was to his friend the Reverend Charles Kingsley, who wrote a thank-you letter back to Darwin, dated 18 Nov 1859, 6 days before the book's official publication date. Kingsley was at the time vicar of Eversley, but became the Regius Professor of Modern History at Cambridge the following year. In thanking Darwin for his kind gift, Kingsley wrote that "All I have seen of it awes me", going on to remark that he didn't believe in the fixity of species anyway, and then making a comment that Darwin liked so much that he quoted from it in the Second Edition of the Origin: "I have gradually learnt to see", writes Kingsley, "that it is just as noble a conception of Deity, to believe that he created primal forms capable of self development [...] as to believe that He required a fresh act of intervention to supply the lacunas[11] which he himself had made". It is

[10] Aquinas, T. On Separated Substances, c.9.

[11] 'Lacunas' means 'gaps'.

intriguing that the very first extant response to Darwin's theory was an extremely positive one and came from an Anglican cleric.

A further example comes from Aubrey Moore, Anglican cleric, Fellow of St. John's College, Oxford, and Curator of the Botanical Gardens. In his contribution to a volume of essays authored by Anglo-catholic theologians entitled *Lux Mundi* [1889], Moore claimed that there was a special affinity between Darwinism and Christian theology, remarking that "Darwinism appeared, and, under the guise of a foe, did the work of a friend". The reason for this attraction, claimed Moore, was based on the intimate involvement of God in His creation, for

> There are not, and cannot be, any Divine interpositions in nature, for God cannot interfere with Himself. His creative activity is present everywhere. There is no division of labour between God and nature, or God and law. [...] For the Christian theologian the facts of nature are the acts of God (Moore 1979, p. 261).

It was therefore the understanding of God as immanent in the created order that helped to facilitate the widespread acceptance of evolution in the Church in the late nineteenth century and by the same token also subverts the 'god-of-the-gaps' understanding of God as creator. This refers to the attempt to locate the special activity of God in creation to gaps in our present scientific knowledge. Such gaps, of which there are many, are often cited as arguments for God: "science has not/could not explain that, so God must have done it". Such arguments are found frequently in the Intelligent Design literature where, it is claimed, the origin of life or the complexity of the cell could not have come about through gradual physical processes, leading to a 'designer-of-the-gaps' form of the argument. Inevitably as science extends its explanatory scope with the passage of time, this notion of God or the 'designer' shrinks accordingly.

It should already be clear that the 'god-of-the-gaps' is very different from the traditional understanding of God as creator. And what is of interest in the present context is the way in which Darwin's clerical contemporaries such as Kingsley and Moore specifically welcomed Darwin's new theory because it rendered unnecessary such "lacunas" or gaps in which God then had to "intervene". By placing renewed emphasis on the immanence of God in creation, Moore shows how the very notion of continual stages of "intervention" is incoherent because "God cannot interfere with Himself". If "the facts of nature are the acts of God" then it is clear that Christians need have no hidden investments in scientific ignorance. This theological emphasis, allied with a renewed emphasis on the teaching of Augustine about creation, a renewal that took place from the late nineteenth century onwards, has ensured that the great majority of Christians, at least outside of the USA, have continued to accept that Darwinian evolution is the best explanation for the origins of all biological diversity on the planet, including humankind.

Such acceptance should not, however, imply that evolution provides no philosophical or theological challenges at all to theology, and a few of these are considered below by way of example.

2.3 The Role of Chance

One objection that is frequently raised against evolution by creationists is that it represents a 'chance' process that is incompatible with the idea of a creator God who has plans and purposes for the world in general and for humanity in particular. But this is based on a misunderstanding of the mechanisms involved in the evolutionary process. Richard Dawkins wrote in the Preface to his book *The Blind Watchmaker* (Dawkins 1986) that his purpose in writing the book was 'to destroy this eagerly believed myth that Darwinism is a theory of "chance"'. The reason for this is that the contemporary understanding of evolution is that it is a two-step process. In the first step variation is introduced into the genome by a wide range of random mechanisms, random in the sense that they occur without the benefit or otherwise of the organism in question; in the second step the slightly different genotypes generate slightly different phenotypes such that the reproductive fitness of each individual organism is tested out in the workshop of life, the process known as natural selection. So although the first step does indeed involve 'chance' in the sense that genetic variation occurs without the organism's well-being in mind, on the other hand natural selection is a very stringent sieve and in the evolutionary process as a whole it is necessity that has the upper hand.

As it happens, the introduction of variation into genomes is not a random process as there are some regions within the DNA in which change is much more likely to occur than in other regions. For example, studies on bacteria, in which mutation rates can be measured more easily due to their rapid rates of division, have revealed mutational 'hot-spots' at which mutational change may be up to a thousand-fold higher than average age (Alexander 2011, p. 68). But even if mutational change in the DNA was truly random, this would not change the conclusion that the process taken as a whole is highly constrained.

Taking the overall trajectory of evolutionary history it is clearly the case that 'chance events' such as mass extinctions due to climate change or volcanic activity have played important roles. Furthermore, the fate of animal or plant populations is often determined by local stochastic events, such as a disease pandemic or drought. But whereas it is clear that such events are critical in describing the precise contours and timing of evolutionary history, on the other hand such events are bound to happen sooner or later given the vast time-spans involved, and it is the planetary constraints that define the many different ecological niches on Earth that eventually determine the products of the evolutionary process.

The extent of the constraints involved in the evolutionary process have recently been highlighted in a number of fascinating studies. For example, 'convergence' refers to the way in which the same biological adaptations emerge repeatedly in independent evolutionary lineages. At the phenotypic level these can be very striking (Conway Morris 2003). The hedgehog tenrecs of Madagascar were long thought to be close relatives of 'true' hedgehogs, because their respective morphologies are so similar, but it is now realized that they belong to two quite separate evolutionary lineages and have 'converged' independently upon the same adaptive solutions, complete with spikes. The convergence of mimicry of insects and spiders to an ant morphology has evolved at least 70 times independently. The technique of retaining the egg in the mother prior to a live birth is thought to have evolved separately about 100 times amongst lizards and snakes alone. Compound and camera eyes taken together have evolved many times during the course of evolution. Convergence does not entail that a different set of genes evolves separately each time to generate such adaptations, although new genetic variants can be involved, but rather that there is a selection pressure to use the genomic resources already available. For example, the gene that encodes the protein Pax-6 provides a dramatic example of a gene that has been involved in photosensitivity and eye building for more than 500 million years (Alexander 2011, p. 128). Pax-6 is a transcription factor, a protein that regulates the switching on and off of many other genes, and is expressed in different kinds of eyes right across the animal kingdom, in vertebrates, arthropods, annelids, and mollusks. The ubiquitous presence of the gene which encodes Pax-6 in so many different genomes means that it is ready to be used for 'eye construction duties' as required.

There are many striking examples of evolutionary convergence at the molecular level, for example in the photosynthetic pathway known as the C4 pathway. Photosynthesis, the process whereby plants derive energy from the sun, 'fixing' carbon dioxide and giving off oxygen in the process, normally generates a molecule containing 3 carbon (C) atoms as its first product, so the process is known as C3 photosynthesis. About 95 % of the biomass of all photosynthesizing life on the planet uses this C3 pathway, which is most efficient when abundant carbon dioxide is available, temperatures are moderate and water is plentiful (Gowik and Westhoff 2012).

C4 photosynthesis, used by crops such as maize and sugarcane, generates a molecule containing 4 atoms of carbon as its first product, and comes into its own under rather different conditions when carbon dioxide levels fall, and the climate is hot and dry. The leaves of plants are organised differently when they photosynthesise this way in order to concentrate the carbon dioxide at a particular spot. The C4 pathway evolved in grasses about 30 million years ago, probably in response to a declining level of carbon dioxide in the atmosphere. What is most striking in the present context is that C4 photosynthesis has evolved independently more than 40 times in at least 18 different plant families, one of the most striking examples of molecular convergence yet described. This has evolved through the parallel evolution of several different genes encoding the enzymes that are required for the C4 pathway. One of these genes (known as PEPC) evolved C4 functionality at least 8 times in different grasses independently, involving similar or identical changes in 21 different codons in their genomes (Besnard et al. 2009).

Similar striking examples of PEPC gene convergence have been found in flowering plants as well. The sedge family contains more than 1,500 species that use C4 photosynthesis, which appears to have evolved 5 times independently within this family (Besnard et al. 2009). In this case the PEPC gene has acquired identical or similar sets of mutations in at least 16 different codons that have been under strong natural selection. Identical changes in the same codons were also found in various grasses and eudicots, a quite distinct group of flowering plants that includes familiar friends like the buttercup and dandelion. In other words, in many cases precisely the same mutation occurred to convert one amino acid to another at precisely the same spot in the PEPC amino acid sequence.

It therefore seems that evolution is like a search engine exploring 'design space' and that the same solutions to the same adaptive challenges crop up again and again independently during evolutionary history.[12] The point here is not that this fact per se has theological implications, but instead it tends to undermine the claim that evolutionary history illustrates 'pure chance' and could have happened quite otherwise, thereby being incompatible with the idea of a creator God who has intentions and purposes for the trajectory of biological diversity. On the contrary, the highly constrained characteristics of actual evolutionary history, coupled with the striking overall increase in complexity over its 3.8 billion year time-span, appear to be facts rather compatible with the idea of a sense of directionality, taking the process as a whole. A narrative in which intelligent minds that can comprehend the wonders of the universe eventually emerge from the unpromising beginnings of a molten planet being bombarded by meteorites, does seem to require some kind of metascientific explanation. It is not for nothing that Simon Conway Morris gave his book on convergence the provocative title *Life's Solution – Inevitable Humans in a Lonely Universe* (Conway Morris 2003). The finding of life on other planets, if and when it occurs, will have a significant impact on such considerations. Given that the elements essential for life as we know it, such as carbon, phosphorus, nitrogen and oxygen, are formed in the furnaces of second-generation stars, from which they are then scattered round the universe when those stars then develop into exploding supernovae, it seems very likely that life elsewhere in the universe will be carbon-based life since silicon has some disadvantages in playing such a role. Boringly for sci-fi enthusiasts it seems that life on other planets might appear somewhat similar to our own given that the biochemistry of the universe is very likely to be similar. What will be truly fascinating is to find out whether life is DNA/RNA/protein based, like our own, and whether the genetic code is similar or even identical. Equally fascinating will be to determine the evolutionary history of life on planet X, which may display its own convergent history. To know that, we may have to wait a few more years, unfortunately.

2.4 Evolution and Morality

Biology has been claimed to impinge upon human morality in a wide variety of ways, in turn implying significance for religious belief. The fact that several of these suggested ways have mutually exclusive implications for belief indicates that the question may not be that straightforward. Four examples have been chosen here to represent this luxuriant foliage of ideas.

[12] The term 'design space' here should not be confused with 'Intelligent Design'. 'Design space' simply points to the fact that there is a finite number of ways of achieving evolutionary fitness within a given environment and the evolutionary search engine will keep finding similar fitness-generating adaptations again and again.

First, it has been suggested that biology is the last place one should look if one wishes to construct a civilized human morality. This proposal is found, for example, in the writings of Darwin's bulldog, Thomas Henry Huxley, who wrote: "Let us understand, once for all, that the ethical progress of society depends not on imitating the cosmic process, still less in running away from it, but in combating it" (Huxley 1894). The biologist George Williams comments that "An unremitting effort is required to expand the circle of sympathy for others. This effort is in opposition to much of human nature" (Williams 1988). Richard Dawkins is blunt: "Let us try to teach generosity and altruism because we are born selfish" (Dawkins 1989). In this view, it is human freedom which allows us to rise above the dictates of our genes, and we need to act, at least much of the time, in opposition to our innate biological tendencies.

Second, in stark contrast, it has been proposed by commentators such as the philosopher Michael Ruse, drawing from Hume, that

> The naturalistic approach, locating morality in the dispositions produced by the epigenetic rules, makes our sense of obligation a direct function of human nature. We feel that we ought to help others and to cooperate with them, because of the way that we are. That is the complete answer to the origins and status of morality. There is no need to invoke [...] some Platonic world of values. Morality has neither meaning nor justification, outside the human context. Morality is subjective. [...] In a sense, therefore, morality is a collective illusion foisted upon us by our genes. Note, however, that the illusion lies not in the morality itself, but in its sense of objectivity (Ruse 1986).

In this view evolution has produced human individuals characterized by a universal morality which feels objective, even though in reality it is not. The naturalistic fallacy, the claim that "ought" can be derived from "is", is "side-stepped" by Ruse (his terminology) by subsuming the feeling of "ought" into psychology.

There are a number of problems with this position which have been discussed in detail elsewhere (Alexander 2001). For example, in practice the moral belief systems practiced by different societies are startlingly varied. Whereas it is the case that some minimalist level of cooperation is essential to maintain any kind of animal social group, human or otherwise, in existence, beyond that bare minimum lies a vast array of diversity. If there were, for example, some innate disposition not to be cruel to innocent people, "foisted on us by our genes", then it seems odd that this disposition appears to exert such a weak effect. In pre-colonial New Guinea, male homicide rates have been estimated to be in the range 20–35 % (Wrangham 2004), or around 300-fold higher than the worldwide average current homicide rate. Traditionally amongst the Yanomamo of the Amazon basin 50 % of the men engage in killing and the homicide rate amongst males is around 30 %. The Waorani tribe of Ecuador may have the highest rate of homicide of any society known to anthropology. In one extensive study it was found that 42 % of all deaths were accounted for by one Waorani killing another Waorani (Larrick et al. 1979), homicides which accounted for 54 % of male and 39 % of female deaths at all ages. This has been associated with lower reproductive fitness amongst the male warriors in contrast with their less warlike tribal brethren (Beckerman et al. 2009). Of course in each of these cases there are particular social, economic and geographical reasons why the homicide rates are so high, but that is just the point. Innate gene-dependent morality of the kind that Ruse proposes is not observed in practice and if the conflicting data are dismissed as 'exceptions', then all that tends to happen is that the theory begins to die the death of a thousand qualifications. What so often happens is that the standards of polite (or not so polite) western academic society are taken as the 'norm', and then biological arguments are invoked to explain why the norm exists. Ruse wants to derive from Darwinian theory a universally shared moral understanding with common "standards and values cherished by decent people of all nations" (Ruse 1986, p. 272). But evolution must apply to the whole human population and it is the sheer diversity of actual moral practice in the population as a whole which tends to subvert overly bold claims about the role of evolution in the generation of an innate, universally held, and genetically undergirded human morality.

A third approach to the question as to how evolution impacts on morality is exemplified by the suggestion of Martin Nowak of Harvard University that cooperation plays a central role in evolutionary history (Nowak 2006; Traulsen and Nowak 2006; Nowak et al. 2010). In his book *Supercooperators* (Nowak and Highfield 2012) Nowak shows how *H. sapiens* is the species par excellence for cooperation, not as some universal inviolable human behaviour that always applies, but as an important element in the repertoire of human behaviours that has rendered our own species so able to flourish in such a wide range of habitats. Such a claim is more modest than Ruse's proposal that evolution generates a universal morality and it is both supported by a wide range of evidence and consistent with some elegant mathematics. At the same time it should not be thought that cooperation can be identified with human altruism which, in its highest form, goes well beyond the scope of biological cooperation. In humans, motivation, together with intellectual pondering, are both important elements in moral choice, elements that contribute to the distinctiveness of altruism when compared with the kind of cooperation commonly found in animal populations.

A fourth type of discussion on the relation of biology to morality in the context of religion has been generated by the increasing ability to describe the biological processes that occur during, for example, moral decision-making. Brain-scanning of people as they meditate, pray or have religious experiences have been incorporated into a field known as 'neurotheology', a field characterized by an implicit ontological reductionism. The results have often been over-hyped in the media; earlier reports of a 'god-spot' in the brain have been discredited. What is becoming apparent is what one might expect: the aspect of religious practice that is under investigation is reflected in increased brain activity (inferred by increased blood flow) in areas of the brain that display comparable patterns of activity whilst the subject is engaged in similar thinking of a non-religious nature. For example, scans of subjects engaged in prayer show increased activity of brain areas involved in social cognition, the authors of one study concluding "that praying to God is an intersubjective experience comparable to 'normal' interpersonal interaction" (Schjoedt et al. 2009). Moral deliberation likewise involves brain areas that are similarly activated in other forms of non-moral cogitation (Schaich Borg et al. 2011).

What these four areas of research and discussion indicate is that the evolved human brain has sophisticated capacities for moral decision-making, but it is noteworthy that in each instance the value judgments about what is right and wrong are made based on principles or assumptions that do not derive from the biological investigation itself, be it of evolutionary mechanisms or of brain functioning. Those who argue that morality should rise above the dictates of our biology, as in the first example described above, are ipso facto making moral judgments about what kind of morality they have in mind. They have a strong point; there are many aspects of the evolutionary process that we should probably not wish to emulate. Those who wish to identify morality with our deepest human intuitions, drawing upon evolution to justify those intuitions, as in the second example, still have to invoke value-judgments about which intuitions are deemed to represent acceptable human moral behaviour and which are not. Consideration of the human population as a whole suggests that not all human intuitions are to be equally applauded. The third example, insofar as it is used as a justification for human morality, likewise involves identifying cooperation as a positive human attribute and making a value judgment to assign a particular moral status to that attribute. This entails rejecting other adaptive animal behaviors such as infanticide, cannibalism and rape. The fourth rather different kind of example highlights the fact that it is possible to give a description at the neuroscientific level of human agents as they perform their daily tasks. Using the complementarity model, such a description involves the "it" language of the neuroscientist investigator, complementary to the "I" language of the conscious personal agent, as discussed above. Both narratives are required to do justice to the human activity in question, and there is no reason why the "it" account should be privileged over the "I" account.

A Christian perspective on the evolution of morality is likely to focus on the emergence of moral decision-making capacities with a concomitant highlighting of personal human responsibility. In that respect, at least, it allies itself with the first of the four examples illustrated above. Christian theology does not entail that God's moral character, nor God's expectations for human morality, are revealed through the processes of creation, so Christians will bring value judgments to the table drawn from other resources, which in practice is what all people do whatever their particular world-view might be. In some cases those value judgments (drawn, for example, from the Ten Commandments and from Jesus' teaching in the Sermon on the Mount) will find themselves allied with important features of the evolutionary process, such as cooperation and care for one's family. Other manifestations of the evolutionary process, however, common through the animal kingdom, such as murder, rape, incest and infanticide, will be rejected. In fact, the radical lifestyle to which Jesus calls his disciples is characterized to a large extent by its promotion of positive human qualities that stand in sharp contrast with such a list (Matthew chapters 5–7).

3 Conclusions

Religious belief has played an important positive role in the emergence of modern science in general and in the development of the biological sciences in particular. There is therefore some irony in the reflection that biology is now often seen as representing a challenge to faith. It is easy to see how such a misunderstanding could arise. With nearly half the population of the country currently leading the world in science and technology nevertheless rejecting biology's central integrating theory, it is hardly surprising that the impression has been given that Christians in particular are hostile to biology. However, as this chapter has hopefully made clear, this feature of Christian life and culture is a very recent one, with particular geographical and political contours, and supported by a Biblical hermeneutic alien to traditional norms. Arguably its popularity is maintained by those who seek to invest evolution with a secular rhetoric. If science education can avoid such rhetoric, then creationism may begin to seem less appealing.

The challenge to education is to provide a nuanced view of the discussion between biology and religion which avoids stereotypes and unnecessary polarization. History can help in this regard. There is no reason why the robust theological matrix which nurtured the emergence of modern science should not be of equal value today in restoring positive attitudes towards the biology that finds its roots in the Christian natural philosophers of the early modern period. For example, mention can be made of those, such as John Ray and Carl von Linnaeus, who established the groundwork for the animal and plant classification system used by biologists to the present day, both of whom saw their science as reflecting the power and wisdom of God in the created order. More recently it is interesting to note that two of the key founders of the neo-Darwinian synthesis, Theodosius Dobzhansky and Ronald Fisher, were committed Eastern Orthodox and Anglican Christians, respectively. In the broader scientific enterprise, many of the methods and tools that are now taken for granted as part of modern science have deep theological roots. The scientific enterprise is open to those of any faith or none. But it is important in our education that the indebtedness of science to faith be made clear, not least so that those with faith commitments can realize that by doing science they are following in a long tradition in which people of faith have made vital contributions. Indeed, science is their natural home.

References

Alexander, D.R. 2001. *Rebuilding the matrix: Science and faith in the 21st century*. Oxford: Lion, pp 141.

Alexander, D.R. 2011. *The language of genetics – An introduction*. Philadelphia: Templeton Foundation Press.

Alexander, D.R., and R.L. Numbers. 2010. *Biology and ideology from Descartes to Dawkins*. Chicago: University of Chicago Press.

Barrett, D.B., G.T. Kurian, and T. Johnson. 2001. *World Christian encyclopedia: A comparative survey of churches and religions in the modern world*, 2nd ed. Oxford: Oxford University Press.

Beckerman, S., P.I. Erickson, J. Yost, J. Regalado, L. Jaramillo, C. Sparks, M. Iromenga, and K. Long. 2009. Life histories, blood revenge, and reproductive success among the Waorani of Ecuador. *Proceedings of the National Academy of Sciences of the United States of America* 106(20): 8134–8139.

Besnard, G., A.M. Muasya, F. Russier, E.H. Roalson, N. Salamin, and P.A. Christin. 2009. Phylogenomics of C(4) photosynthesis in sedges (Cyperaceae): Multiple appearances and genetic convergence. *Molecular Biology and Evolution* 26(8): 1909–1919.

Bian, Y., A. Terse, J. Du, B. Hall, A. Molinolo, P. Zhang, W. Chen, K.C. Flanders, J.S. Gutkind, L.M. Wakefield, and A.B. Kulkarni. 2009. Progressive tumor formation in mice with conditional deletion of TGF-beta signaling in head and neck epithelia is associated with activation of the PI3K/Akt pathway. *Cancer Research* 69(14): 5918–5926.

Bohr, N. 1950. On the notions of causality and complementarity. *Science* 111(2873): 51–54.

Boyle, R. 1674. *The excellency of theology, compar'd with natural philosophy, as both are objects of men's study*. London: T. N. for Henry Herringman.

Brooke, J.H. 1988. The God of Isaac Newton. In *Let Newton Be!* ed. J. Fauvel, R. Flood, M. Shortland, and R. Wilson, 168–183, 172. Oxford: Oxford University Press.

Brooke, J.H. 1991. *Science and religion: Some historical perspectives*. Cambridge: Cambridge University Press.

Brooke, J.H. 2010. Darwin and religion: Correcting the caricatures. *Science & Education* 19: 391–405.

Brooke, J.H., and G.N. Cantor. 1998. *Reconstructing nature: The engagement of science and religion*. Edinburgh: T&T Clark.

Bryda, E.C., N.T. Johnson, K.K. Ohlemiller, C.L. Besch-Williford, E. Moore, and R.J. Bram. 2012. Conditional deletion of calcium-modulating cyclophilin ligand causes deafness in mice. *Mammalian Genome* 23(3–4): 270–276.

Carroll, W.R. 2012. Aquinas and contemporary cosmology: Creation and beginnings. *Science and Christian Belief* 24(1): 5–18.

Conway Morris, S. 2003. *Life's solution: Inevitable humans in a lonely universe*. Cambridge: Cambridge University Press.

Crick, F. 1994. *The astonishing hypothesis: The scientific search for the soul*. New York/Toronto: Scribner.

Darwin, C. 1958. *The autobiography of Charles Darwin, and selected letters*. New York: Dover Publications.

Darwin, C. 1859. *On the origin of species by means of a natural selection, or the preservation of favored races in the struggle for life*. London: Murray.

Dawkins, R. 1986. *The blind watchmaker*, 1991. London: Penguin.

Dawkins, R. 1989. *The selfish gene*. New ed. Oxford: Oxford University Press.

Dawkins, R., and L. Ward. 1995. *River out of Eden: A Darwinian view of life*. London: Weidenfeld & Nicolson.

Durant, J. 1985. *Darwinism and divinity: Essays on evolution and religious belief*. Oxford: Blackwell.

Gould, S.J. 2002. *Rock of ages*. New York: Ballantine Books.

Gowik, U., and P. Westhoff. 2012. The path from C3 to C4 photosynthesis. *Plant Physiology* 155(1): 56–63.

Harrison, P. 1998. *The bible, Protestantism, and the rise of natural science*. Cambridge: Cambridge University Press.

Huxley, T.H. 1894. *Evolution and ethics*. New York: D. Appleton & Co.

Kauffman, S.A. 2008. *Reinventing the sacred: A new view of science, reason and religion*. New York: Basic Books.

Kitcher, P. 2011. Militant modern atheism. *Journal of Applied Philosophy* 28(1): 1–13.

Larrick, J., J. Yost, J. Kaplan, G. King, and J. Mayhall. 1979. Patterns of health and disease among the Waorani Indians of Eastern Ecuador. *Medical Anthropology* 3: 147–189.

Livingstone, D.N. 1987. *Darwin's forgotten defenders: The encounter between evangelical theology and evolutionary thought*. Edinburgh: Scottish Academic Press.

MacKay, D.M. 1988. *The open mind*. Leicester: IVP.

Marsden, G.M. 1984. Understanding fundamentalist views of science. In *Science and creationism*, ed. A. Montagu. Oxford: Oxford University Press.

Marsden, G.M. 1992. Literal interpretations. *Nature* 360: 637–638.

McMullin, E. 2005. *The church and Galileo, Studies in science and the humanities from the Reilly Center for Science, Technology, and Values*. Notre Dame: University of Notre Dame Press.

Miller, J.D., E.C. Scott, and S. Okamoto. 2006. Science communication. Public acceptance of evolution. *Science* 313(5788): 765–766.

Moore, J.R. 1979. *The post-Darwinian controversies: A study of the protestant struggle to come to terms with Darwin in Great Britain and America, 1870–1900*. Cambridge: Cambridge University Press.
Morris, H.M. 1984. *The biblical basis for modern science*. Grand Rapids: Baker Book House.
Nowak, M.A. 2006. Five rules for the evolution of cooperation. *Science* 314(5805): 1560–1563.
Nowak, M.A., and Roger Highfield. 2012. *SuperCooperators: Altruism, evolution, and why we need each other to succeed*, 1 Free Press hardcoverth ed. New York: Free Press.
Nowak, M.A., C.E. Tarnita, and E.O. Wilson. 2010. The evolution of eusociality. *Nature* 466(7310): 1057–1062.
Numbers, R.L. 2006. *The creationists: From scientific creationism to intelligent design*. Expanded ed. Cambridge, MA/London: Harvard University Press.
Numbers, R.L. 2009. *Galileo goes to jail: And other myths about science and religion*. Cambridge, MA: Harvard University Press.
Origen. 1936. *On first principles, Book IV*, Ch. 3. Trans. G. Butterworth. London: SPCK.
Paley, W. 1795. *A view of the evidences of Christianity in three parts*. Philadelphia: Thomas Dobson, at the Stone-house, no.41.
Philo of Alexandria. 1929. *Philo. With an English translation by F. H. Colson and G. H. Whitaker*: 10 vol. London/New York: William Heinemann/G. P. Putnam's Sons.
Ray, J. 1701. *The Wisdom of God manifested in the Works on the Creation ... The third edition, very much enlarg'd throughout. [With a portrait.]*. London: Sam Smith & Benj. Walford.
Rudwick, M.J.S. 2004. *The new science of geology: Studies in the earth sciences in the age of revolution*. Aldershot: Ashgate Variorum.
Ruse, M. 1986. *Taking Darwin seriously: A naturalistic approach to philosophy*. Oxford: Blackwell.
Schaich Borg, J., W. Sinnott-Armstrong, V.D. Calhoun, and K.A. Kiehl. 2011. Neural basis of moral verdict and moral deliberation. *Social Neuroscience* 6(4): 398–413.
Schjoedt, U., H. Stodkilde-Jorgensen, A.W. Geertz, and A. Roepstorff. 2009. Highly religious participants recruit areas of social cognition in personal prayer. *Social Cognitive and Affective Neuroscience* 4(2): 199–207.
Spencer, N., and D.R. Alexander. 2009. *Rescuing Darwin: God and evolution in Britain today*. London: Theos.
Teilhard de Chardin, P., B. Wall, and J. Huxley. 1959. *The phenomenon of man*. London: Collins.
Temple, F. 1885. *The relations between religion and science*. London: Macmillan.
Traulsen, A., and M.A. Nowak. 2006. Evolution of cooperation by multilevel selection. *Proceedings of the National Academy of Sciences of the United States of America* 103(29): 10952–10955.
Turner, F.M. 1978. The Victorian conflict between science and religion: A professional dimension. *Isis* 69(248): 356–376.
Van Woudenberg, R. 2012. Limits of science and the Christian faith. *Science and Christian Belief* 24(2): 129–148.
Westfall, R. 1971. *Force in Newton's Physics*, 397. London: Macdonald.
Whitcomb, J.C., and H.M. Morris. 1961. *The genesis flood. The biblical record and its scientific implications*. Philadelphia: Presbyterian & Reformed Publishing Co.
Williams, G.C. 1988. Huxley's Evolution and ethics in sociobiological perspective. *Zygon* 23: 384–385.
Worrall, J. 2004. Science discredits religion. In *Contemporary debates in philosophy of religion*, ed. M.L. Peterson and R.J. Van Arragon. Oxford: Blackwell.
Wrangham, R. 2004. Killer species. *Daedalus* 133: 25–35.

Intelligent Design and the Nature of Science: Philosophical and Pedagogical Points

Ingo Brigandt

1 Introduction

In the United States, creationists and evangelical Christians have threatened high school instruction in evolutionary biology for decades, even in public schools (where religious views may not be taught due to the constitutional separation of state and church). Similar worrisome trends have more recently started in other Western countries, exacerbated by the promotion of the label 'intelligent design theory' (Numbers 2009). While this alleged theory has hardly any intellectual content and does not pose a scientific threat to evolutionary theory, intelligent design ideas and more generally alleged arguments against evolution are known to many students. For this reason it is important for teachers to develop their classroom instruction in evolutionary theory with the knowledge that some students may be hesitant to accept evolution due to religious reasons or because they are exposed to erroneous claims about evolutionary theory. At the very least, the teaching of evolutionary theory has to bring forward considerations that can serve as implicit responses to common objections to evolution.[1] It may also be fruitful to directly address intelligent design and why its 'arguments' fail, presented not as a rejection of intelligent design (or even religion) but as a critical thinking lesson for students. More generally, beyond teaching particular evolutionary facts it is worthwhile to make students reflect on, and teach them about,

[1] While traditional classroom instruction thoroughly covers different aspects of microevolution, using non-human animals as examples, it is essential to present more examples about macroevolutionary transformations, including the evolution of humans. This stems from the fact that young children can more easily conceive of microevolutionary changes than of macroevolutionary changes (Samarapungavan 2011) and a person can use multiple epistemologies, leaving room for the possibility that while using a scientific epistemology for microevolution, students may use a non-scientific epistemology when thinking about human origins (Evans et al. 2011).

I. Brigandt (✉)
Department of Philosophy, University of Alberta, 2-40 Assiniboia Hall,
Edmonton, AB T6G2E7, Canada
e-mail: brigandt@ualberta.ca

K. Kampourakis (ed.), *The Philosophy of Biology: A Companion for Educators*, History, Philosophy and Theory of the Life Sciences 1, DOI 10.1007/978-94-007-6537-5_11,

the *nature of science*. Understanding what the aim of scientific explanation is, how empirical methods function, and how science makes progress, gives students a much better appreciation of what science is and how it works—which in itself should be a goal of science education. It also has the side-effect of making plain to students what virtues evolutionary biology has over intelligent design.

This essay discusses intelligent design (ID) from the perspective of the philosophy of science, drawing several implications for science education.[2] I proceed from concrete biological issues to more general issues about the nature of science. Section 2 engages Michael Behe's irreducible complexity argument against evolution, highlighting why the ID portrayal of organisms as designed machines is not only at odds with contemporary biology but prevents an understanding of how organisms can evolve. A long-standing objection to evolution is that the formation of complex structures by means of processes involving chance is too improbable to be credible. This small probability argument has recently been developed and promoted by ID theorist William Dembski, and in Sect. 3 I show why it is easy to explain to students why such arguments are fallacious, connecting it to issues about the nature of scientific explanation. Science's commitment to explanations only in terms of natural causes—called 'methodological naturalism'—has been criticized by ID proponents on the grounds that it is presumed by scientists without valid justification and that it entails atheism. Section 4 lays out why neither is the case, and this discussion of why scientists have good reasons to use empirical methods has implications for the nature of science and how to demarcate science from pseudoscience. I broaden the scope yet again in Sect. 5 by highlighting the need for philosophers to construe scientific approaches as practices based on institutional factors and values, and to assess them in terms of the socially embedded activities of their practitioners. By implication, instructors should not just present science as a set of facts and theories, but convey that science is a practice, as this puts students in a position to see much clearer why evolutionary biology differs from intelligent design. The last section summarizes my overall discussion, emphasizing the various pedagogical points made about biology education. This is a long essay, but the four main sections can be read independently of each other.

2 Irreducible Complexity and Organisms as Machines

2.1 Behe's Irreducible Complexity Argument Against Evolution

A prominent intelligent design argument against evolution is based on the notion of irreducible complexity, explicitly introduced by ID proponent and biochemist Michael Behe in *Darwin's Black Box: The Biochemical Challenge to Evolution* (1996). He states his central idea as follows:

[2] ID proponents have only leveled arguments against evolutionary theory, and there is no intelligent design theory that makes predictions and explains phenomena. For this reason, 'ID proponent' has to henceforth refer to someone endorsing the 'intelligent design' label, and more concretely someone who is part of the intelligent design movement (Sect. 5).

> By *irreducibly complex* I mean a single system composed of several well-matched, interacting parts that contribute to the basic function, wherein the removal of any one of the parts causes the system to effectively cease functioning. An irreducibly complex system cannot be produced directly [...] by slight, successive modifications of a precursor system, because any precursor to an irreducibly complex system that is missing a part is by definition nonfunctional. (Behe 1996, p. 39)

Behe often illustrates this idea with a simple example—the mousetrap. A mousetrap has the following parts: a base plate, a spring, a hammer (doing the killing), a bar that holds the hammer in place before the trap is activated, and the catch that holds the bait and releases the holding bar and hammer upon being touched. Given the way these parts are arranged, the mousetrap can be used to catch mice; but if any single part is missing, it is not functional any longer. Applied to the biological realm, the argument is that an evolutionary origin of an organismal system (without the influence of an intelligent designer) would require ancestral precursor systems that have been favored by natural selection, yet any precursor to an irreducibly complex system missing a part is non-functional.

This idea against the natural origin of complex organisms is not completely new, as it was already part of William Paley's (1802) watchmaker argument, which asserted that one may infer the presence of a designer from a watch found on a heath, given that the parts of the watch are arranged in a purposeful fashion and that it would not function if the parts were randomly assembled (see also Ayala, this volume; Avise, this volume; Lennox and Kampourakis, this volume).[3] However, the novelty of Behe's account is that he points to *molecular* systems within organisms. Systems that Behe claims to be irreducibly complex include the vertebrate immune system (suggesting a design influence during vertebrate evolution), the blood clotting cascade, and the cell's vesicular transport. To be sure, the icon of intelligent design has been the bacterial flagellum, the tail-like protrusion that by its motion propels the bacterial cell so as to permit motility. The central aspect for Behe is the flagellum's anchor point inside the cell wall, which consists of a few dozen proteins that are arranged in such a way that some of them rotate as in a motor, creating the flagellum's motion.

Behe's irreducible complexity argument has convincingly been criticized by many biologists and philosophers (Sarkar 2007; Shanks 2004; several of the contributions in Young and Edis 2004). I discuss this matter not because another argument against Behe is needed, but because seeing why he fails reveals how evolution works and how it is to be taught. Several have pointed out that even if upon removing a system's part it cannot fulfill its *current* function, it may well be able to perform a different, possibly simpler function—a function that may have been important for the ancestor, so that the system with fewer parts is a candidate for an ancestral precursor system. To illustrate this in the case of the bacterial flagellum as found in *Escherichia coli*, consider another bacterium, *Yersinia pestis*, which is the cause of the bubonic plague. Not dissimilar to a flagellum, *Y. pestis* also has a thin long

---

[3]One difference is that while Paley argued against a natural origin of organisms by mere chance, Behe argues (and has to argue) against an evolutionary origin by natural selection.

structure protruding from the cell wall; however, it does not move as a flagellum would. The reason is that the structure's anchor point in the cell wall consists of only a subset of the protein types present in the flagellum's base in *E. coli*, so that it cannot generate rotary motion. Still, though it has fewer components than a flagellum motor, the structure in *Y. pestis* does fulfill a function important for this microorganism. Being located in the cell wall it permits the transport of virulence factors from inside the cell into the long hollow structure attached to the cell, which functions as a syringe, injecting toxins into mammalian cells to suppress their immune response. Behe's irreducible complexity argument ignores that the primary functions of biological structures can change over the course of evolution, and a function essential for one species may not be relevant for another.

Y. pestis is an extant species, so that the structure in its cell wall is of course not the historical precursor of any other species. But a similar structure could have been the actual precursor of the flagellum motor in *E. coli*. More generally, comparing related structures in several extant species provides important clues to their evolution. Shared structures in extant species are often homologies, suggesting how ancestral conditions may have looked (Minelli and Fusco, this volume). While some ID proponents have claimed that among the 42 protein components of the flagellum, about 2/3 are unique to this system and not found in other systems, actually homologies to other proteins have been identified for all but 1/3 of the components. Moreover, since half of the components are missing in one or the other extant species, a functional flagellum is possible even with missing components. There are only two proteins (i.e., 5 % of components) that are indispensable and with no known homologies to other proteins (Pallen and Matzke 2006; for the immune system see Bottaro et al. 2006). Needless to say, this picture of the evolution of the bacterial flagellum is incomplete. But future comparative studies will add to the account, and most importantly, Behe and other ID proponents have not offered any explanation of how the flagellum evolved. Behe assumes that evolutionary descent with modification—albeit with the additional influence of an intelligent designer—has occurred, but he does not lay out at what time such interventions happened and what protein changes they yielded. Indeed, if his irreducible complexity argument was sound, given that there is not just 'the' bacterial flagellum, but that the protein composition of flagella differs across bacterial taxa, Behe would be forced to claim that many intelligent interventions have occurred during bacterial evolution. Yet he simply proclaims 'design', without attempting or intending to offer an explanation of the structural similarities and dissimilarities observed in extant species.

Of the points made so far, two are relevant to biology education. First that it is valuable to highlight to students the conceptual issue that the particular functions which enable an organism to survive and reproduce, and are favored by natural selection, are context-dependent and vary across species and evolutionary time. Second, rather than making inferences based on the study of one species, sound evolutionary biology uses the comparative method and the best evolutionary explanation is the one that yields an account of the features of many extant species.

2.2 Why Organisms Should Not Be Portrayed as Machines

Apart from the fact that the removal of a system's parts may lead to a system that can perform a different function, there is another problem with Behe's irreducible complexity argument. For in the above quote he tacitly presupposes that any ancestral precursor system has *fewer* parts than the descendant. But it may well have more parts, and exhibit *redundancy*, i.e., some of its parts can be eliminated or some activities can be deactivated without any loss in function (so that systems with redundancy are not irreducibly complex). Despite Behe's claim that an irreducibly complex system cannot evolve, such a system can be obtained if one starts out with a system exhibiting redundancy, and then removes all redundant parts and activities. One common evolutionary way to generate novel structures and functions on the molecular level is gene duplication. Upon duplication of a gene, there are two identical copies G and G′. They still have the same function A (e.g., coding for a certain protein or activating the expression of certain genes), so that the system exhibits redundancy. For this reason, it often happens that one of the copies is destroyed by mutation. If mutations do not destroy, but increasingly modify one of the copies, say G′, the gene may eventually acquire a new function B, which could have some beneficial role for the organism (while G still has function A). Then a new gene G′ with a new function B has evolved. Should both functions A and B eventually become essential for the survival of later descendants, the evolutionary outcome is an irreducibly complex system.

Behe and other ID proponents are fond of likening cells to artifacts and its components to machines, by terming cellular structures as 'highways', 'factories', and 'assembly lines'. DNA is conceived as a blueprint, where gene expression is like the reading of a computer punched card (Pigliucci and Boudry 2011). Behe uses the mousetrap to illustrate his notion of irreducible complexity. Needless to say, all these machine metaphors are used to create the impression that biological systems are designed, similar to artifacts. Apart from this being rhetorical rather than logical support for intelligent design, Behe's irreducible complexity argument—that organisms are machines that break down if one of their parts is removed—is empirically false. For the molecular systems he points to are not irreducibly complex, and organismal systems often exhibit redundancy (Shanks and Joplin 1999). In the case of *robustness* in gene regulatory networks and developmental processes, a gene may well be involved in an important function, yet a deactivation of this gene (e.g., in a knockout study) hardly leads to any phenotypic difference, as the organism compensates for this situation by activating other genes (Brigandt in press-a; Edelman and Gally 2001; Mitchell 2009; Wagner 2005). Organisms can flexibly react to potentially harmful disturbances, even genetic modifications. This has important evolutionary consequences.

Evolvability is a biological system's ability to evolve (see also Love, this volume). More specifically, evolutionary developmental biologists use this term to refer to an organism's capacity to generate viable, heritable variation

(Hendrikse et al. 2007; Kirschner and Gerhart 1998; Wagner 2005).[4] Morphological change can take place only when there is heritable phenotypic variation, on which natural selection acts. Genetic mutations occur in a random fashion, but due to an organism's mode of development, this random genetic variation translates to a structured *phenotypic* variation, where the heritable phenotypic variation generated tends to be more viable and functional than if it was generated in a random fashion. An account of evolvability aims at explaining how this is possible, as this is vital for understanding how sufficiently rapid morphological change is possible. A mere appeal to long periods of time being available is unconvincing as an explanation of how complex structures could have evolved if not supplemented with an explanation of why sufficiently large amounts of phenotypic variation tend to be functional.

Upon modification of an artifact like Paley's watch, either no significant change results or the artifact breaks down. If organisms were artifacts as Behe contends, they would not be able to evolve. Marc Kirschner and John Gerhart address this issue in *The Plausibility of Life: Resolving Darwin's Dilemma* (2005), which lays out their account of evolvability (which they dub a theory of facilitated phenotypic variation) in a manner accessible to a general audience. They point to different features enabling evolvability, such as weak regulatory linkage, compartmentation, and exploratory behavior. A cellular or developmental process exhibits exploratory behavior if it is able to generate many, if not an unlimited number, of outcome states, any of which can be physiologically stabilized if it is adaptive to the organism. One example is how microtubules generate the shape of eukaryotic cells, by each of the many microtubules growing and shrinking (exploring), until the length of some of them is stabilized by a signal from outside the cell. In this fashion, many cell shapes can be produced in an individual organism, with remodeling of a cell being possible. Another instructive example is the development of the limb of land-living vertebrates. Apart from several skeletal elements and muscles, the limb needs blood vessels and nerves. The positions of the latter are not represented in some organismal blueprint; instead, their anatomy emerges by means of exploratory developmental processes, with new nerves (and blood vessels) growing from the body core toward the developing limb, guided by chemical signals and their current surrounding milieu, with those nerves that do not find a target degenerating by cell death (Kirschner and Gerhart 2005, Ch.5).

One advantage of this mode of development is that it creates the regular functional outcome even if the developmental process is temporarily disturbed. It also facilitates evolutionary modification. The size and placement of limbs differs significantly in different vertebrates. If the placement of a limb changes in evolutionary time, it is not necessary to respecify the new positions of the developed bones, muscles, blood vessels, and nerves—all of which have to be at the right place for the limb to function—on an alleged organismal blueprint. Instead, these structures

[4]For a historical discussion of the concept of evolvability and its relation to the concept of developmental constraint see Brigandt (in press-b), and for a connection to the phenomenon of homology see Brigandt (2007).

adjust to the new situation accordingly by means of exploratory developmental processes. This shows that it is possible that a *simple* genetic change (e.g., changing the position where the limb starts to develop) can lead to a coordinated, complex phenotypic modification, involving *many simultaneous* phenotypic changes. In general, Kirschner and Gerhart (2005) point to mechanisms that permit physiological adaptation and developmental robustness, where a functional developmental outcome is created even in the face of an environmental change or a developmental disturbance. Such developmental aspects of organisms have evolutionary implications. For they not only ensure that a functional phenotype is produced upon an environmental change, but they also make it likely that a *functional* phenotype results from a *genetic* change, so that evolutionary modification is enabled (see also Wagner 2005).

In the eighteenth and early nineteenth century, debates about reductionism in physiology and embryology were typically phrased in terms of 'mechanism' versus 'organicism' (Brigandt and Love 2008). Mechanists assumed that developmental and physiological processes could potentially be explained by a framework relying primarily on the physical contact of bodily particles, broadly in line with Newtonian mechanics. Mechanists were favorable toward viewing organisms as complicated machines governed by the laws of physics and chemistry. Organicists, in contrast, were unconvinced that a mechanical framework sufficed for the explanation of life processes. As evidence, they pointed to development and regeneration. The freshwater hydra, for example, can regenerate into several full organisms even if cut into pieces. In sea urchins, splitting the blastomere or taking some of its cells away can in some cases still lead to a normally developed embryo. This was seen as a clear disanalogy between organisms and machines.

Within a twentieth century framework, organisms can be conceived as machine-like if one uses the human artifact metaphors of genetic 'information' and organisms developing from a genetic 'blueprint'. Among other things, this image has been promoted in the widely influential popular science book *The Selfish Gene*, with Richard Dawkins asserting that the "argument of this book is that we, and all other animals, are machines created by our genes" (Dawkins 1989, p. 2). Dawkins conflates the legitimate *evolutionary* idea that genes have a past involving natural selection that makes them evolutionary adaptations for certain functions with the problematic *developmental* idea that every organism is a "machine built by [...] genes" (p. 44)—suggesting genetic determinism (on this issue see Moore, this volume; Jamieson and Radick, this volume; Burian and Kampourakis, this volume).[5] The notions of genetic information, blueprints and programs have been rightly criticized on the grounds that they are empty metaphors that do not provide a mechanistic explanation of development while creating the illusion of explanatory

[5]Dawkins's (1989) presentation also construes organisms as largely passive machines (controlled by genes), while portraying genes as active agents that have desires (selfish aims) and carry out actions (building organisms). However, while it may make the material more attractive, the anthropomorphizing of nature in classrooms can have negative effects on students' epistemological development (Evans et al. 2011).

understanding (Robert 2004). The information metaphor erroneously suggests that the function of molecular genes is context-independent ('if the information for making a phenotype is in the gene, the gene will produce it in any context'). To the extent that there is biological information underlying development, it does not reside in genes alone. The activation of genes and the production of their products in different cells emerges from the interaction of molecular genes, various non-DNA molecules inside the cell, and the neighboring cells, so that rather than development being controlled by an organism's DNA as a *central* agent (every cell has a separate set of DNA anyway), the generation and modification of biological information in development is a temporally dynamic and spatially *distributed* process (Brigandt in press-a; Stotz 2006; Wagner 2005; see also Marcos and Arp, this volume).

Talk about molecular machines can be repeatedly found in contemporary molecular and cellular biology (Alberts 1998). While this may get at some features of cellular systems, such metaphors at the same time obscure many features that reveal cellular and organismal systems to be unlike machines (Kirschner et al. 2000). In the context of explanations of development, already eighteenth century organicists pointed to regeneration and robust development as being at variance with an organism-as-machine picture. But it has more recently become clear that this is essential for an understanding of *evolution*. My above discussion of evolvability explained why physiological adaptability and robustness in development are the reasons why organisms can generate heritable phenotypic variation that tends to be functional, so as to permit evolutionary change by natural selection. Thereby viewing organisms as flexible developmental systems rather than machines is the key to understanding morphological evolvability, so that machine metaphors are not just biologically inadequate, but also harmful for science education (Brigandt 2013; Pigliucci and Boudry 2011; but see Bechtel, this volume).

In his irreducible complexity arguments, Behe focuses on molecular or biochemical pathways—a reductionist vision ignoring the larger context. Even if it is the case that the removal of some parts leads to a breakdown of this specific pathway, due to redundancy or robustness, the larger system may compensate for it so as to avoid detrimental effects to the organism. The irony is that whereas ID proponents often charge biologists with endorsing a materialist and reductionist view of living creatures, in fact their metaphor-based representations of organisms as designed machines (that would break down if modified by random mutation) are guilty of an empirically false reductionism. While neo-Darwinists, like Dawkins, who focus on population genetics have sympathies for viewing organisms as designed machines (a commonality with ID proponents even if they assume that natural selection was the designer), many evolutionary biologists who attempt to understand organismal evolvability and the evolutionary origin of morphological novelty have moved away from a machine vision of organisms. They see evolutionary developmental biology (evo-devo) as allowing for an interdisciplinary approach that offers integrative explanations appealing not just to the molecular level but to the interaction of entities on several levels of organization (on the non-reductionist epistemology of evo-devo see Brigandt 2010, 2013, in press-a; Love 2008, 2013, this volume).

The main lesson for biology education to be derived from this section's critique of Behe's irreducible complexity claims is that teachers should, wherever possible, avoid describing organismal features using machine and information metaphors, as they prime the false inference that organisms were designed by an intelligent agent, and prevent a proper understanding of how organismal development works and why flexibility and robustness in development make morphological evolution possible.

3 Small Probability Arguments and the Nature of Explanation

A very common idea brought forward against evolution and in favor of intelligent design is that organisms are so complex and consist of so many individual traits that their origination by an unguided process involving chance (such as naturalistic evolution) is *extremely improbable*, so improbable that intelligent design must have occurred. Such small probability arguments against the possibility of evolution have been raised by creationists for decades, but they have also more recently been employed by intelligent design proponents (Berlinski 2008; Gauger and Axe 2011; Sewell 2000, 2001). In his more recent book *The Edge of Evolution* (2007), Michael Behe points, among other examples, to the structural fit among different interacting proteins, arguing that several mutations in different proteins must have occurred to generate such a function-enabling fit, but the probability of this happening decreases exponentially with the number of mutations required.[6] One of the most prominent intelligent design proponents, the mathematically trained theologian William Dembski, has developed the most sophisticated version of this probabilistic argument against naturalistic evolution. In *The Design Inference: Eliminating Chance through Small Probabilities* (1998a), Dembski develops his 'explanatory filter', that first seeks to eliminate the possibility that an event has occurred as a matter of natural regularity, and then to rule out that it was due to chance, so as to conclude that the event came about by design. Dembski presents a universal probability bound of 1 in 10^{150}, where an event more improbable than this can be assumed to not have arisen by chance. He obtains this number by multiplying the number of particles in the known universe, the maximal rate of change in physical states, and the age of the universe, multiplying again with one billion. In later work, Dembski (2002b) invokes mathematical information theory and introduces the notion of complex specified information, where in line with his earlier account, 'complex' refers to an extremely improbable event. Dembski's account is more complicated than this,[7] but the details of his mathematical account do not concern us here and have been

[6]Section 2.2 pointed out that exploratory behavior and other aspects of developmental processes permit several coordinated and instantaneous phenotypic changes to result from a simple genetic change.

[7]For instance, Dembski does not infer design simply from an event being extremely improbable, but from it being improbable and specified (exhibiting a pattern), although he has not offered a consistent account of specificity.

rigorously criticized by others (Elsberry and Shallit 2011; Felsentein 2007; Fitelson et al. 1999; Häggström 2007; Olofsson 2008; Sarkar 2007), with some critics pointing to Dembski's extensive use of irrelevant mathematical formalism, which may impress his intended audience while concealing the actual incoherence of his account (Perakh 2004; Sarkar 2011).

Luckily, small probability arguments for design can be shown to be problematic without much mathematical sophistication, as they all are based on a basic fallacy. In contrast to Behe's notion of irreducible complexity, small probability arguments are less tied to concrete biology, but I discuss them here as the small probability fallacy is so common that it must be addressed by science and mathematics teachers. Beyond direct attacks against evolution, similar arguments occur in the context of the idea of a fine-tuned universe and the strong anthropic principle, i.e., the argument that since conscious life can occur only when the basic physical constants are within a very narrow range, the universe and its constants must have been designed. Small probability arguments are so common and even educated people are prone to fallacious reasoning involving probabilities, that this is something to pay attention to when teaching students about probability.

3.1 Why Small Probability Arguments Are Fallacious

The basic argument from small probabilities can be reconstructed as follows:

(1) The evolution of complex biological features (be it anatomical structures, be it genetic information) solely by means of Darwinian processes is extremely improbable.
(2) Therefore, Darwinian evolutionary theory is probably false (given that there are complex biological features).
(3) Therefore, intelligent design is probably true.

There are several obvious issues with this argument. First, premise (1) can be challenged. Often a small probability is just asserted, but not calculated. If a probability is derived, the calculation may misrepresent the process of evolution by assuming that it is a purely random process. This is the case with the common argument that the naturalistic evolution of organisms is as absurd as a Boeing 747 being assembled by a tornado going through a junkyard. Such probability calculations ignore that mutations occur, not just in a single genome, but in thousands of organisms within a species at the same time, and that most importantly, natural selection retains the best variants, so that evolution does not have to randomly start in every generation from scratch.[8] However, while many probability assertions can be shown

[8]The discussion on evolvability in Sect. 2.2 mentioned further relevant aspects of the evolutionary process.

to be faulty, some version of (1) is the case, as a specific outcome of the evolutionary process is unlikely. Second, statement (2) does not entail statement (3). Even if the current version of evolutionary theory is false, another theory based on purely natural processes may be true, so that the probable truth of intelligent design does not follow. Still, if (2) was the case, i.e., if current evolutionary theory was probably false, this alone would be very damaging for evolutionary biology. Statement (2) is a claim that no evolutionist is willing to accept.

For this reason, my discussion focuses on the fact that *(1) does not entail (2)*. The small probability argument starts out with the legitimate statement that the evolution of complex biological structures, given only Darwinian processes, is very unlikely. In mathematical terms:

(1) $P(\text{complex structures}|\text{Darwinian evolution}) \approx 0$

However, what intelligent design proponents want to conclude, and must argue, is that the truth of evolutionary theory is very unlikely given that we have evidence about the presence of complex biological structures. That is:

(2) $P(\text{Darwinian evolution}|\text{complex structures}) \approx 0$

Yet the conditional probabilities $P(O|H)$ and $P(H|O)$ are very different probabilities. Moreover, they can have completely different values. According to Bayes's formula, $P(H|O) = P(O|H) \cdot P(H)/P(O)$. Thus, even if, as asserted by premise (1), $P(O|H)$ is extremely small and close to 0, $P(H|O)$ can be close to 1, depending on $P(H)$ and $P(O)$. As a result, (1) does not entail (2), and the small probability argument is fallacious based on the confusion of two conditional probabilities.[9]

This fallacy has been further analyzed by Elliott Sober (2008), who explains why it appears to be such a compelling line of reasoning, as it is a probabilistic analogue of falsification (Brigandt 2011). Strict falsification is a valid deductive inference, based on the logical principle of modus tollens. If hypothesis H deductively predicts that observable event O will *not* happen but it is observed that O is the case, then hypothesis H is shown to be false. That is, from $H \rightarrow$ not-O and O one may infer that not-H. The small probability argument is a probabilistic analogue of this, starting not with premise $H \rightarrow$ not-O (if hypothesis H is true, then O is false), but with the weaker claim that $P(O|H) \approx 0$ (assuming hypothesis H to be true, O is very unlikely). Combined with observation O, the intended conclusion is not that hypothesis H is false, but H is *probably* false. However, while deductive falsification is a valid inference, Sober is at pains to argue that there is *no probabilistic analogue of falsification*. Not even an inductive or probabilistic inference is possible. From the fact that

[9] A similar conflation of two distinct conditional probabilities can occur not only in small probability arguments against evolution, but also in more direct 'arguments' for intelligent design. Inferring that an irreducibly complex or machine-like object is likely to have been designed on the grounds that (human) designers frequently produce irreducibly complex and machine-like objects is a fallacy. For while the premise is that $P(\text{machine-like object}|\text{designed})$ is high, the conclusion states that $P(\text{designed}|\text{machine-like object})$ is high.

an observation O is extremely unlikely according to hypothesis H (though O turns out to be the case), *nothing* can be said about the probability or improbability of hypothesis H.

Here is the reason why any small probability argument inferring (2) from (1) is fallacious. This can fortunately be made plain to students without mentioning the above philosophical analysis that the argument is a probabilistic analogue of falsification. Very small probabilities mean little, as such events can be easily generated. Assume that a given coin is fair, and that our hypothesis H is that the coin is fair, so that it asserts that the probability of heads and tails is each ½, i.e., $P(h) = ½$ and $P(t) = ½$. Consider five tosses of this coin and a particular outcome (a certain sequence of heads and tails): $P(h,t,t,h,t) = ½ \cdot ½ \cdot ½ \cdot ½ \cdot ½$, which is equal to 1 in 2^5. For 70 tosses the probability of a particular outcome P(t,h,t,...) is 1 in 2^{70}, and for 500 tosses the probability P(t,t,h,...) is 1 in 2^{500}, which is smaller than 1 in 10^{150} and thus smaller than Dembski's universal probability bound.[10] Inferring the falsity of the hypothesis 'coin is fair' because of this extremely small probability would be fallacious; we cannot even infer that the hypothesis is *probably* false, as by assumption it is true. In fact, this hypothesis assigns a high probability to some events (one coin toss = ½) and an extremely low probability to other events (500 tosses of the coin)—but we cannot infer that the hypothesis is at the same time probably true and probably false. Both a *true* hypothesis (coin is fair) and a *false* hypothesis (coin is biased with $P(h) = ¾$) can assign a very small probability to *one and the same* event (500 tosses of the coin), which makes plain that nothing can be inferred about the probable truth or probable falsity of the hypothesis asserting the small probability. The problem with Dembski's universal probability bound is not that the number he provides is still too large, but that there cannot be any such bound!

Small probabilities have a strange psychological effect on us and can even mislead educated persons into fallacious inferences.[11] For this reason, this issue ought to be clarified when teaching probability theory to high school students. Arbitrarily small probabilities result if one considers the conjunction of different events, and the particular outcome of a sequence of many evolutionary events (such as all mutations in a lineage leading from a remote ancestor to an extant descendant) is no exception. Since complex events (involving many individual events) with small probabilities happen all the time in nature, a small probability suggests neither that the hypothesis postulating this probability is probably false, nor that some intelligent intervention must have taken place.

[10] If one does not want to toss a coin 500 times, using two decks of cards likewise yields an outcome whose probability is smaller than the universal probability bound.

[11] In addition to persons being poor at reasoning with probability and detecting patterns where there are none, Elsberry and Shallit (2011) point to cognitive science studies according to which humans have agency attribution systems, which may be biased toward overdetection of agency.

3.2 Comparative Testing and the Nature of Explanation

Likelihoods of the form P(observations|hypothesis) as occurring in premise (1) can matter, but only if *several* rival hypotheses are compared. If $P(O|H_1)>P(O|H_2)$, observations O favor hypothesis H_1 over hypothesis H_2. Thus, even if $P(O|H_1)$ is an extremely small probability, it may still be higher than the probability assigned by an alternative hypothesis, and possibly higher than the various likelihoods $P(O|H_i)$ assigned by all other relevant hypotheses. It is well-known that in science, alternative hypotheses often happen to be in competition, but the point here is that a scientific hypothesis often cannot be tested in isolation but *must* be tested *relative to* other hypotheses (Sober 1999, 2007). The fact that P(complex structures|Darwinian evolution) is extremely small does not tell us anything about Darwinian evolutionary theory. It does not make evolutionary theory implausible—as creationists and ID proponents falsely claim—nor does it make ID theory plausible. What intelligent design proponents would have to show is that P(observations|intelligent design)> P(observations|Darwinian evolution).

Now the question is how to assess P(observation|intelligent design) for some given observation. At this point intelligent design proponents face a dilemma. To portray ID as a non-religious theory and to avoid having to confess that God is the assumed designer, ID theory is often described as the hypothesis that at *some* point in the remote past *some* intelligent agent influenced the history of life in *some* way. But this version of ID does not predict any observation, and does not even assign a probability to observations. Intelligent design proponents routinely claim that ID does make testable predictions, for instance the presence of complex specified information in living systems, the occurrence of irreducible complexity, the increase of biological complexity across time, and that DNA, even that considered to be junk DNA, is functional (Meyer 2009; Wells 2011). However, while all these observational claims are *consistent* with intelligent design theory, they do not follow from intelligent design theory as construed here, whereas an actual prediction has to *follow* from the theory predicting it. If its proponents construe ID in a vacuous fashion like the one above, no concrete prediction can be made from it, in fact, not even a probability P(observation|intelligent design) can be assigned. Thus, by trying to portray ID as a scientific (in the sense of non-religious) theory, its proponents have rendered intelligent design untestable!

Predictions are made and probabilities can be assigned only if ID is made more concrete by a specification of the intentions and abilities of the designer, but this is not an option for those who want to create the illusion that ID is not a religious approach.[12] Intelligent design proponents routinely claim that their design inference

[12] If ID is made more concrete so that predictions result, there is still the question of whether it fits the known evidence to a higher degree than evolutionary theory. Young Earth creationism of course makes concrete, testable claims (e.g., about the age of the Earth and the occurrence of a worldwide flood), which have been shown to be false.

is analogous to how human agency is inferred in forensic science, archaeology, and physical anthropology, suggesting that since the latter are scientific inferences, so is the ID inference. However, in forensics it is possible to distinguish between a non-human cause of a fire and arson because for each cause its mode of operations and its specific effects are well known, so that the plausibility of each possible scenario (hypothesis), given the evidence, can be assessed. The same applies for paleoarchaeologists determining whether the marks on stones are due to non-human natural causes or due to the agency of ancestral humans—such an inference gets off the ground only because scientists know what marks are left by natural processes such as erosion, and why and how humans modify certain stones. In sum, its proponents portray ID as a modest approach, which merely attempts to infer the existence of some kind of design from natural phenomena. However, no such inference to design can be made without (a) providing a specification of the nature of this design and the operation of the designer (Sarkar 2011; Sober 2008), and (b) showing that intelligent design fits the evidence better than other relevant theories, in particular evolutionary theory (Elsberry and Shallit 2011; Sober 2008).

So far I have phrased the point that science can work only by comparing different hypotheses (two contemporary rivals, or an earlier and later version of a theory) in terms of prediction: one has to determine whether H_1 predicts observation O to a higher degree than H_2 does, i.e., $P(O|H_1) > P(O|H_2)$. But the same point can also be cast in terms of explanation: the question is always whether one approach offers a better explanation of a phenomenon than another approach. In this context, ID proponents have been criticized for putting nothing forward but illicit 'God of the gaps' arguments, i.e., pointing to phenomena for which science does not have a satisfactory explanation and using this as evidence for a supernatural influence (Scott 2004). Creationists and ID proponents are fond of making 'arguments' against evolutionary theory and pointing to aspects of extant species for which no detailed evolutionary explanation is available. But this is irrelevant as long as no intelligent design explanation of this phenomenon is put forward. ID proponents do not bother to offer explanations; in Sect. 2.1 we have seen that Behe and others do not attempt to offer an explanation of how the biological features they allege to be irreducibly complex have originated in time. They could not offer an explanation, as the vacuous hypothesis that somehow some kind of intelligent influence was involved does not explain at all. Similar to the above mentioned erroneous claim that ID makes predictions, creationists may feel that something having been designed offers an explanation (or a better explanation) of complex biological features. But this is an illusion, as an explanation has to lay out why an entity exists at a certain time *rather than* failing to exist, and why it has the properties it has *rather than* having different properties. The mere appeal to that entity being designed does not shed any light on this.

This is a lesson about the *nature of scientific explanation* that can and should be conveyed to students. While ID proponents suggest that making inferences from evidence is the essence of science, the central aim of science is to put forward explanations. An explanation of a phenomenon has to shed light on why it is the way it is rather than otherwise. Explanations are typically incomplete, where for some phenomena no explanation is currently available. But science strives to make explanations more complete and revise and improve upon past explanations. The adequacy of a proposed explanation for a phenomenon must always be assessed in terms of how it *compares* to other attempts to explain the phenomenon, including past explanations. Science in general, and evolutionary biology in particular, develops explanations in ever increasing detail, whereas ID proponents do not undertake anything like this.

This section implies that science education needs to explain to students why an event being extremely improbable, given the mechanisms postulated by a scientific theory, does not undermine this theory in any way. While ID proponents phrase their approach in terms of making inferences from observations, the real issue in biology is explaining observable phenomena, where rival explanations of a phenomenon are to be compared. Evolutionary explanations are often incomplete, but improve over time, whereas intelligent design does not have a positive explanatory agenda. While ID proponents pick on a few observations and claim that one can infer design from it, evolutionary theory offers explanations of a vast array of phenomena.

4 Methodological Naturalism and the Nature of Science

An important characteristic of science is its commitment to methodological naturalism, which is broadly speaking the scientific approach. *Methodological naturalism* asserts that science ought to make claims about natural (in the sense of material) phenomena only, as its claims have to be backed up by empirically accessible evidence. Science explains by appeal to natural causes, as opposed to invoking supernatural causes. This is a commitment pertaining to the methods of science, but also embodies a limitation of the scope of scientific claims, and thus the basic aims of science. Methodological naturalism does *not* claim that no supernatural phenomena exist, it merely claims that science cannot study the supernatural. *Metaphysical naturalism*, in contrast, claims that only natural, i.e., material, phenomena exist. The latter is a not a tenet about the methods or aims of science, but a metaphysical tenet referring to what does and does not exist (Sarkar 2007; Shanks 2004). (In popular evolution vs. intelligent design debates metaphysical naturalism is often called 'philosophical naturalism', which is a bad term as in philosophy many different varieties of naturalism are distinguished.) The reason this distinction is so important is that while metaphysical naturalism entails atheism, methodological naturalism does not have any religious implications—though intelligent design proponents have tried to muddy the waters by claiming that methodological naturalism slides into inherently atheist metaphysical naturalism (Forrest 2011). The fact that many scientists, including evolutionary biologists, are religious believers shows that science and its commitment to methodological naturalism do not amount to atheism.

Methodological naturalism provides a clear way to distinguish between theistic evolutionism and intelligent design. Theistic evolutionists believe that the cosmos and the laws of nature were created by God, but that subsequently all material processes have unfolded due to natural laws without any divine intervention, so that material, worldly phenomena (including the history of life) are to be explained using the standard resources of science—i.e., a commitment to methodological naturalism (Lamoureux 2008). Intelligent design proponents, in contrast, assume that there had to be some direct influence by a supernatural agent *during* the history of the world, and definitely during the history of organismal life. William Dembski states that "theistic evolution is no different from atheistic evolution, treating only undirected natural processes in the origin and development of life" (Dembski 1998b, p. 20). Even though ID proponents attempt to portray intelligent design as a good scientific approach that uses empirical evidence, through its insistence that the history of life is partially to be explained by the influence of a supernatural intelligence, ID rejects methodological naturalism, and thus is actually opposed to the scientific approach. In fact, ID proponents have heavily criticized theistic evolutionists (Johnson and Lamoureux 1999), with Dembski asserting that "theistic evolution remains intelligent design's most implacable foe" (Dembski 2002a).

Since many high school students tend to view evolution and religion as being in conflict, it is important to convey to them that science does not take a stance on religious matters (no matter whether the label 'methodological naturalism' is used or whether this is more simply phrased as a lesson about the methods, nature, and limits of science). Students can fruitfully be taught how there is a common ground in science which permits scientifically minded persons to either be religious or atheist, whereas only ID proponents and creationists view science and religion in conflict (for the relation between evolutionary biology and religion see Ayala, this volume; Alexander, this volume). Once it is clarified that science and religion not only use different epistemologies, but also concern different domains regarding the human condition, students will have a conceptual framework through which they can reconcile their religious beliefs with the evolutionary biology taught to them (Sinatra and Nadelson 2011).

4.1 Why Methodological Naturalism Is Not an a Priori Commitment

In addition to claiming that it slides into metaphysical naturalism, creationists have directly objected to methodological naturalism on the alleged grounds that scientists simply presuppose it without justification. Science has to presuppose its methods before being able to use them to conduct research. But if methodological naturalism is just presupposed, then the supernatural is by sheer assumption excluded from the realm of science. The creationist objection continues that just as naturalistic scientists can avail themselves of a 'philosophy' (epistemology) as a starting point, so creation scientists may use their epistemological point of departure—the fact that the bible is the inerrant word of God—and proceed from there in their interpretation and formation of beliefs about the biological world. This criticism of methodological naturalism has been put in a more sophisticated fashion by some intelligent design proponents, who have claimed that methodological naturalism is an *a priori* philosophical commitment (Beckwith 2003a, b; Johnson 1991). Whereas a posteriori knowledge is knowledge obtained based on experience and the empirical investigation of the world, a priori knowledge is obtained without the involvement of any experience or investigation of the observable world. Logical and mathematical principles have typically been considered by philosophers to be knowable a priori, and metaphysical and theological principles have been other traditional candidates for a priori knowledge (if they are knowledge at all). If scientists endorsed methodological naturalism a priori, they would use it to test hypotheses, but they could not *empirically* test the methodological naturalism presupposed or empirically support it. In what follows, I discuss why scientists' endorsement of methodological naturalism is not a priori, as apart from showing the claims by ID proponents to be wrong, it prepares subsequent lessons for how to demarcate science and non-science.

That scientists do not endorse methodological naturalism *a priori* is shown by the fact that their understanding of what methodological naturalism involves and what counts as a 'natural' phenomenon has *changed* substantially over the course of history. Several centuries ago it was assumed that natural philosophy (as science was called back then) could, in its study of the natural world, appeal to the divine realm. The astronomer Johannes Kepler, a proponent of the new heliocentric system, wondered why the solar system had six planets, rather than more or less (only six planets where known at this point). His *Mysterium Cosmographicum* (1596) proposed a mathematical explanation based on the fact that there are only five perfect solids, motivated by the conviction that the heavenly bodies were arranged by an elegant plan of God the mathematician. Even if not combined with theological considerations, nowadays such a purely mathematical or metaphysical explanation is deemed unscientific, and the number of bodies in the solar system or in the universe is not even deemed to be a central astronomical *question*.

Views about what scientific *methods* are reliable, what can be observed, and what can be empirically tested have changed over the history of science. When the telescope was developed, its use for the purpose of astronomy was initially challenged based on the idea that naked eye observation was the way to obtain valid knowledge. Due to the simple lenses available, early telescopes made the observer see some stars double or even suggested heavenly bodies where there were none. It took astronomers decades to learn which visual observations with a telescope actually represented features of the cosmos (the same holds for the introduction of the microscope; Hacking 1983), but the method of telescopic observation eventually became universally accepted. Likewise, while the scientific consensus once held that the study of a person's facial features permits inferences about her intellectual abilities, personality features, and criminal tendencies, this method has been soundly rejected, resulting in phrenology—once a reputable scientific approach—being nowadays considered pseudoscience. Likewise, views about what counts as a 'natural' phenomenon, and what is *physically possible* and impossible have changed in history based on new views about what the laws of nature are and what kinds of entities exist. Is it possible to penetrate a massive body like a ghost? In the past this may have seemed absurd, but nowadays it is clear that this is possible as there are elementary particles that penetrate massive bodies all the time. Spontaneous generation was the idea of a simple living organism (e.g., a little worm) emerging from inanimate matter in a short period of time. This biological view was generally held for centuries, until 200 years ago, and it was assumed that the spontaneous generation of new simple organisms was an everyday occurrence. Why spontaneous generation is virtually impossible has become clear with the advent of the cell theory of organisms in the nineteenth century. Such a change in views about what natural phenomena there are and what is physically possible is important in the context of methodological naturalism as it entails a change of which phenomena and causes a scientific explanation may or may not appeal to.

Methodological naturalism includes a number of concrete commitments about what counts as a scientific question, what methods are valid, what natural causes there are, and what qualifies as a scientific explanation. Methodological naturalism is endorsed by science, but it is not an a priori 'philosophical' commitment, as past construals of methodological naturalism came to be rejected and replaced by revised construals of what methods are empirically reliable, what causes exist, and what explanations are valid. Indeed, science will continue to revise and improve its understanding of methodological naturalism. These various past revisions were done for good empirical reasons, as hinted at in the above examples.[13] As a result, current scientists endorse methodological naturalism because of its historical track record; it is accepted *a posteriori* based on past experience. This shows that scientists are justified in endorsing methodological naturalism, including its current construal that excludes appeal to supernatural features. Far from being an arbitrary 'philosophical' commitment, the current construal and use of methodological naturalism is *justified* by the historical reasons for changing past construals of methodological naturalism. These considerations about methodological naturalism also have implications for how to possibly demarcate science from pseudoscience.

4.2 Demarcation and the Nature of Science

One strategy for pointing to the inadequacy of creationism and intelligent design is to argue that it is not a scientific approach, based on a *demarcation* account that distinguishes science and pseudoscience (or sound science and junk science).

[13] In the case of the use of the telescope for astronomical observations, even though this method was controversial upon its introduction, some reasons for its increased acceptance were that repeated observations gave consistent orbits and that telescope-based predictions on the future positions of planetary bodies were borne out. Moreover, it was demonstrated that the telescope reliably represented distant objects on Earth, whose properties could be verified by naked eye observation. Using lamps, Galileo showed that, unlike telescopic observations, naked eye observations overestimated the size of distant bright lights against a dark background, so that he was in a position to explain the inconsistency of the apparent size of the planets and stars viewed by naked eye vs. telescope. Thus, a previously accepted scientific method (naked eye observation) can be used to show the reliability and even superiority of a new method (telescopic observation), yielding an empirical justification for a change in method.

The 1982 ruling in *McLean v. Arkansas Board of Education*, which found the teaching of creationism in public schools of the US state of Arkansas to be unconstitutional, was, among other things, based on such a demarcation account. However, it has turned out to be difficult to put forward valid demarcation criteria. An idea that enjoys wide popularity among scientists is that the essential feature of a scientific theory is that it is falsifiable. Below in this section and in Sect. 5, I make plain why falsifiability should not be the primary demarcation criterion as it is insensitive to the empirical context and focuses on theories rather than scientific practices. However, there are also other initial problems with it. Any hypothesis that has been falsified—including Nazi race theory or a pseudoscientific claim—is a falsifi*able* hypothesis, which shows that it is moot to use falsifiability as a demarcation criterion, given that one does not want to give credence to long discarded hypotheses by still calling them 'scientific'. Furthermore, statistical and probabilistic theories are widespread in science, occurring even in physics, and, due to the involvement of population genetics, evolutionary theory is a statistical theory. Yet any probabilistic theory is unfalsifiable (and would thus count as 'unscientific'), because, while assigning probabilities to various events, such a theory does not predict that certain events must happen, events which could then be shown to be failed predictions.[14]

The 1982 *McLean v. Arkansas* judgment relied on the testimony of philosopher of science Michael Ruse, who in addition to using falsifiability, employed other considerations to determine whether a theory is scientific, such as being testable against the empirical world and explaining by reference to laws of nature (Ruse's testimony, his defenses of his demarcation account, and criticisms of it by others are reprinted in Ruse 1988). Ruse's demarcation account has been found wanting by some philosophers, most prominently Larry Laudan, who apart from criticizing particular demarcation criteria by counterexamples from the history of science, concluded that in general "the problem of demarcation between science and non-science is a pseudo-problem (at least as far as philosophy is concerned)" (Laudan 1983, p. 124). While the idea of demarcating science from pseudoscience suggests the existence of a *one-dimensional* scale from 'scientific' to 'unscientific', my view is

[14]Another drawback is that the notion of falsifiability stems from Karl Popper's (1959) falsificationism. This general doctrine of confirmation assumes that while it is possible to conclusively *dis*confirm (falsify) a theory, there is no incremental confirmation of theories by evidence. Falsificationism maintains that it is not rationally possible to inductively increase one's degree of belief in a theory as evidence accumulates. In fact, one cannot have any degree of confidence in its truth—one may only believe that a particular theory has shown to be false. However, even if a theory is incompatible with some observations, scientists may very well continue using the theory if it is supported by other lines of evidence and if there is no better alternative available, as opposed to rejecting it as 'falsified'. More importantly, the only way to rationally justify one's actions is with reference to factual beliefs for which one has some support, so there has to be some degree of positive belief in theoretical claims. Since scientists do have rational support for their (limited) endorsement of a theory (e.g., mechanics) and since scientists and policy makers use this theory for further action (e.g., building space rockets), philosophers of science have generally rejected falsificationism (Godfrey-Smith 2003).

that approaches claiming to be science have to be judged and compared based on a variety of important considerations that are better kept separate than merged into the single feature of 'being scientific' (Hoyningen-Huene 2008). In this sense, offering an account of demarcation, or providing a definition of science, is not a central *aim* of philosophy of science. Still, there are important questions about the nature of science and the credentials of particular approaches claiming to be sound science in the philosophical vicinity.[15] And even if many considerations can be used to assess theoretical approaches, intelligent design falls short of all of them, unlike evolutionary theory (Thagard 2011).

Sahotra Sarkar (2011) favors showing intelligent design to be intellectually problematic independently of an account of demarcation, as demarcation criteria that are context-independent and assume scientific approaches to be historically static are bound to fail. The latter is well taken, but it leaves the option of using 'demarcation' considerations that are *context-dependent and historically dynamic*. Robert Pennock (2011) emphasizes that a commitment to methodological naturalism was used as one demarcation criterion in the 2005 *Kitzmiller v. Dover Area School District* trial (which ruled intelligent design to not be science), as ID proponents admit that they reject methodological naturalism. Combine this with my above point that scientists' understanding of what methodological naturalism involves has changed over the course of history. For the purposes of Pennock's trial testimony, it sufficed to show that intelligent design proponents want to explain biological phenomena by appeal to *divine* influences, but my notion of methodological naturalism laid out in the previous subsection is richer by including a variety of considerations about what phenomena currently count as natural, what kinds of explanations are currently permitted, and what methods are currently deemed to be valid. For instance, I have pointed out that while appeal to spontaneous generation was once scientifically legitimate, it is not any longer. Thus, a contemporary approach that explains by invoking spontaneous generation is scientifically flawed, even if it does not appeal to divine causes. Rather than offering a universal philosophical account of science, any account of what science involves, what is scientifically legitimate, and what makes an approach scientifically dubious—at a certain point in history—has to be based on a variety of considerations that are taken from the concrete scientific traditions in the relevant historical period.

In addition to changing across *historical* time, an account of what science consists in has to be context-dependent in a second fashion—it may differ across different *domains* of science. Ruse's testimony in the 1982 court case included the idea that science explains by reference to laws of nature. This may hold true for physics,

[15]Here is an analogy: While biological attempts to define 'life' exist, such definitions are fraught with difficulties (see Cleland and Zerella, this volume). This is not a problem, as biology offers many insights into the features of living organisms independently of a definition of life. The biology of viruses offers empirical understanding even if it is not settled whether or not a virus is a living entity. Biologists pursue various aims and address concrete questions about particular organism groups, but defining life is not an aim of biology. In the same vein, philosophy can address various normative issues about scientific approaches without aiming at a definition of 'science'.

but quantitative generalities that could be called laws of nature can hardly be found in molecular, cellular, and developmental biology (see also Lange, this volume). In these domains, explanations in terms of molecular mechanisms are used, so that how scientific explanations look and whether they involve laws can differ from domain to domain (Bechtel, this volume; Bechtel and Abrahamsen 2005; Brigandt 2013; Potochnik, this volume). In fundamental physics, relativity theory and quantum mechanics have not been reconciled yet, and applications in physics use many models that may make mutually incompatible idealizations. In contrast, a database of telephone numbers may be free of any inconsistencies and its data may have a degree of precision clearly exceeding measurements in experimental physics. Of course, this does not entail that this database is more scientific than physics, as it concerns a different domain of knowledge. Thus, considerations about the features characterizing 'science' have to be relative to historical periods and scientific domains or disciplines.

Finally, there is another way in which history matters in judging the credentials of a scientific approach (apart from criteria of legitimate science changing across time), namely, the past track record of the approach (Hoyningen-Huene 2008). While some scientific approaches may have started in a promising fashion (e.g., phrenology), they later failed to generate new insights, degenerated, and became abandoned. Such a consideration of the advance of a scientific approach gains traction when two rival approaches are compared. A striking difference between evolutionary biology and creationism/intelligent design is that only the former has steadily improved its explanations and closed gaps in our knowledge. In fact, creationists and ID proponents are not interested in advancing our knowledge of the origin and change of organisms, as they primarily aim to put forward 'arguments' against evolution without developing rival explanations (as pointed out in Sects. 2.1 and 3.2).

In summary, there is no universal and unchanging philosophical account of the nature of science. Rather than treating science as a monolithic whole, there are many different scientific fields with differing standards of evidence and explanation. Any adjudication of an approach claiming to be sound science (e.g., intelligent design or alternative medicine) has to be based on a variety of factors that are specific to the particular empirical domain. Falsifiability is a poor demarcation criterion because, apart from erroneously suggesting that a single consideration (promising a yes-or-no answer) will suffice, falsifiability, as a universal criterion, cannot capture considerations that differ across fields or history. Furthermore, falsifiability focuses on a particular theory and thus an approach at a single point in time, rather than evaluating the past development and future promise of the approach.

The pedagogical lesson of this section is that one should make plain to students that science addresses a restricted domain and does not speak on religious matters, but that it has a solid method to gain knowledge about the empirical realm. Teachers have to avoid conveying to students a monolithic picture of science that is exclusively modeled on current science or even an area of current science. Ideally, the diversity of scientific disciplines and their different scientific characteristics are to be addressed (Brigandt 2013; Love 2013; Pigliucci, this volume). It is important to point out to students that methodological and explanatory standards and the criteria for a scientific approach have changed through time. They have not done so in an arbitrary fashion, as scientific standards have improved and past developments offer a *justification* for the current standards and conception of science. As a result, even though past scientists appealed to religious considerations, there are good reasons why this is not legitimate for any contemporary approach, including intelligent design.

5 Practices and Values: Epistemic and Social

5.1 Construing Science as Epistemic and Social Practices

Growing out of logical positivism, several decades ago philosophy of science tended to construe science in terms of various factual claims, in particular observation statements and theoretical claims. Confirmation was construed as a logical relation between observation statements and theories. While this justification of theories by evidence was seen as an objective procedure, it was assumed that scientific discovery need not always be a rational procedure, so that discovery was not a matter for philosophy, but rather for psychology and history of science. This situation has changed, leading to a broader picture of what science involves and what philosophers of science must take into account. Apart from various factual claims about the natural world, the aims pursued by scientists (e.g., which phenomena are currently deemed in need of explanation) are an important part of science, as the recognition of an explanatory problem prompts various scientific efforts devoted toward it, and a change in what are deemed to be the relevant scientific problems accounts for the historical dynamics of a field and different trends among disciplines (Brigandt 2012, 2013). Nowadays, scientific discovery and the employment and refinement of various experimental methods are of concern to philosophers. More generally, the study and assessment of *scientific practice*—in fact, various scientific practices—is a central topic for philosophy, in particular naturalistic philosophy of science.[16] Modern science generates and validates knowledge in a collaborative fashion, involving various institutional factors, such as funded projects and the peer-review system, making it necessary for philosophers to take into account the social dimensions of science (Downes 1993; Solomon 2001; see also Gannett, this volume). Against the traditional rational–social dichotomy that views social factors as subverting objective science, Helen Longino (2002) argues that

[16]Footnote 15 argued that just like developing a definition of life is not a genuine aim of biology, so philosophy of science need not aim at a definition of science. The biology–philosophy analogy can be extended further. Similar to biology making progress by understanding various life *processes*, philosophers should not ask what science is, but analyze how science *works* and judge the credentials of intellectual traditions based on their epistemic and social practices.

certain social factors are constitutive of scientific rationality, such as mechanisms of establishing intellectual authority, publically recognized standards, diversity of perspectives, and venues for criticism.

Apart from the fact that it offers a more faithful portrayal of science, conveying science as an investigative practice has benefits for science education (Brigandt 2013; Love 2013). Especially in secondary education, the traditional focus is on presenting scientific facts and theories, and, given the rapid progress of science, it turns out to be impossible to teach even those recent scientific ideas that are well-supported. Rather than exclusively teaching the *content* of science, it is fruitful to give students an idea of the *practice* of science, which would provide an understanding of how scientific knowledge is generated, validated, and subsequently revised—a lesson that has validity even if some of the content taught in classrooms is already outdated. By conceiving of scientific activity directed at scientific questions and problems, it becomes plain to students that scientific explanations are initially incomplete but improved over time, and that assessing an explanation involves comparing it with other explanations targeting the same question—entailing that intelligent design's arguments against incomplete evolutionary explanations would be cogent only if ID offered better explanations. Given the collaborative nature of modern science, there can be substantial support for a scientific theory such as anthropogenic climate change even if no individual scientist—lest a science teacher in the classroom—can present it. Students should also be aware of the collaborative practice and what counts as consensus in science, and who qualifies as having expertise in a certain topic.

Studying and evaluating a scientific approach in terms of its epistemic and social practice is particularly important in the case of intelligent design (Brigandt 2011, Sect.4). In the previous section I have argued that rather than using a universal demarcation criterion, the credentials of an approach claiming to be science have to be assessed based on various concrete considerations that are specific to a scientific domain in the respective historical period. I have also pointed to the need to pay attention to the approach's past track record and progress. A consideration of a theory (as a specific set of tenets) cannot achieve this—by implication the same holds for any alleged demarcation criterion that applies to theories only, such as falsifiability. To assess the *future* promise of a current theoretical approach (and how it compares to rival approaches), it is necessary to analyze the epistemic and social practice of its current practitioners. This reveals most strongly why intelligent design does not measure up with evolutionary biology.[17] Richard Duschl and Richard Grandy (2011) argue that in secondary school classrooms, science should not exclusively be construed as the inferring of predictions and testing of

[17] Arguing that the question is not whether theories are scientific, but whether epistemic practices are so, Chinn and Buckland (2011) compare the practices of evolutionary biologists, young Earth creationists, intelligent design proponents, and the nineteenth century scientist-creationists of Darwin's era. The latter's practice turns out to be more scientific than the practice of contemporary intelligent design proponents.

hypotheses (the hypothetico-deductive model of science), but in terms of their dialogic practice model of scientific method, which gives students a much better appreciation of why intelligent design proponents do not participate in crucial epistemic and social practices characteristic of science. I add that relying too much on the hypothetico-deductive model as a vision of science also plays into the intelligent design strategy of (falsely) claiming that ID makes predictions and is simply about making inferences from the evidence. The latter contributes to the misperception among students that intelligent design is scientific; and the hypothetico-deductive model obscures the necessity for ID to offer an alternative explanation of organismal diversity.

What follows are some aspects of the practice of ID proponents highlighting how they differ from most evolutionary biologists. Rather than developing explanations of biological phenomena, ID proponents promote alleged arguments against evolution, most of which were already developed by traditional creationists. Many academic ID proponents are lawyers or engineers, but only a few of them are biologists. A handful of the latter (such as Michael Behe) conducts bona fide scientific research and publishes in the peer-reviewed literature. However, none of these papers concern intelligent design. In fact, ID proponents do not even have enough scientific-looking material to keep alive the online journals founded and run by themselves.[18] Even though the ID arguments against evolution have been repeatedly debunked, ID advocates keep promoting them to their non-academic audience—so that failing to accommodate criticism is a feature of their practice. Not only is there currently no content-laden ID theory, but these practices of ID proponents show that no such theory is forthcoming.

There are reasons for this. The actual growth of scientific knowledge is not required given the primary aim of ID proponents, an aim which is to embed Christianity into all aspects of society.[19] This includes the ID proponents in academia, most of whom reject the idea of common ancestry for religious reasons. European scholars often underestimate the financial and political power of creationists and the religious right in the US, as well as the magnitude this organized threat poses to science education. The label 'intelligent design' was developed by American creationists in an attempt to have creationism taught in public schools, where religious views may not be promoted due to the separation of state and church

[18]See http://www.arn.org/odesign/odesign.htm (1996–2000) and http://www.iscid.org/pcid.php (2002–2005). The latter was abandoned just after *Kitzmiller v. Dover* ruled the teaching of intelligent design to be unconstitutional. In late 2009, largely the same group of editors set up a new journal http://bio-complexity.org, though in the last 3 years only 11 articles or reviews (all but one research article co-authored by the editors) have appeared.

[19]From an internal Discovery Institute memo leaked to the public: "Governing Goals: To defeat scientific materialism and its destructive moral, cultural and political legacies. To replace materialistic explanations with the theistic understanding that nature and human beings are created by God. [...] Twenty Year Goals: [...] To see design theory permeate our religious, cultural, moral and political life." (http://www.antievolution.org/features/wedge.pdf)

mandated by the US constitution (Forrest and Gross 2004). In December 2005, the teaching of intelligent design in public schools was ruled to be unconstitutional in *Kitzmiller v. Dover Area School District.* As a result, ID proponents have backed off from calls to teach ID (some now falsely claiming that they never advocated its teaching), and instead come to lobby for 'teaching the controversy', 'critical analysis of evolution', and 'academic freedom' in public schools—which does not make a difference given that there is no ID theory apart from alleged arguments against evolution—continuing the creationist strategy of undermining the teaching of evolution (Forrest 2010).

ID proponents are very active, but their activities are not so much devoted to scientific research, but to political and legal campaigns in the public arena. Financially supported by a conservative think tank called the Discovery Institute, ID supporters attempt to influence local school boards and state school boards (who are in charge of public school curricula), state politicians (to pass legislation that enables false critiques of evolution being taught), and the media. Given this, what matters is how ID is perceived by the general public. Creating the public impression of ID being a scientific approach and there being a scientific controversy about evolution—even if it is easy for academics to see that this is not the case—suffices to further the social goals of the ID movement. The inflation of credentials is one strategy. The Discovery Institute's 'Scientific Dissent from Darwinism' list features more than 700 persons claimed to be scientists, yet many of them obtained an academic degree in the past but are not active in research, and none of them are working in evolutionary biology, so that the list is populated with non-experts claimed to be otherwise.[20] Speaking to a scholarly audience or in the legal arena, ID proponents claim their approach is not tied to any religious assumptions; yet when speaking to their supporters they more frankly hail ID as part of a culture war about religion. ID advocates publicly misrepresent legitimate criticism of their views and actions, crying censorship, to the point of falsely claiming to have been removed from academic positions for criticizing evolutionary theory.[21] While the scientific publications of ID proponents do not support intelligent design, they are still advertised to intelligent design's non-academic followers as providing scientific support for ID.

In summary, any epistemological analysis of a scientific approach (or an approach claiming to be science) has to encompass considerations about its epistemic practice, including how the persons developing the approach interact with others individuals, and how their practices are institutionally and socially embedded. The fact that intelligent design, unlike evolutionary biology, will not improve in the future cannot be seen in terms of the current theories of each approach, but only in terms of the practices of the two communities and their epistemic and social aims.

[20] http://www.dissentfromdarwin.org

[21] Compare the statements made in the 'documentary' *Expelled: No Intelligence Allowed* with http://www.expelledexposed.com

5.2 *How Epistemic and Social Values Matter*

The thrust of this section's argument so far has been the necessity of construing and evaluating a theoretical approach in terms of its epistemic and social practice. But the discussion has also hinted at the relevance of *epistemic values*, such as intellectual honesty and the uptake of criticism. One of the particularly striking features of the intelligent design movement is the disingenuousness and underhanded tactics of several of its advocates. Epistemic values are not only relevant for studies in philosophy of science (Brigandt 2011, Sect.4), but they likewise ought to be addressed in science education. The many instances of research misconduct show that it does not suffice to train students in the use of scientific methods (narrowly construed), but that they have to be taught what ethical scientific conduct involves. More to the point of this chapter, making students aware of the relevance of epistemic values and standards of conduct in science may contribute to students approaching intelligent design (should they come across it) in these terms as well, as opposed to merely considering arguments for an alleged ID theory.

In addition to epistemic values, social values (including ethical and political values) may matter for discussions by contemporary philosophers of science, since research is to be assessed not only in terms of whether it is methodologically sound, but also in terms of whether this research is ethical and what consequences it will have for society (Douglas 2009; Fehr and Plaisance 2010; Kitcher 2001; Kourany 2010; Tuana 2010; Gannett, this volume).[22] Social values are likewise relevant to a portrayal of what science involves in the context of science education, to the extent that such values matter to practicing scientists. The reason that scientists publicly advocate the scientific consensus on anthropogenic climate change is not just due to this scientific view being well-supported by evidence, but primarily because scientists employ environmental and social values that entail that unchecked global warming has dire consequences. This has parallels to other cases, including intelligent design. A few decades ago the tobacco industry succeeded in preventing public recognition that smoking causes cancer, and averted state regulations against smoking for some while. Though the scientific opinion tended to go against the tobacco industry, it sufficed for the latter's purposes to merely spread doubts on this issue in the public mind by some industry-employed scientists. Nowadays many companies use this strategy to conceal the actual scientific evidence on the potential harmfulness of their products, and creating public doubt about the scientific consensus on anthropogenic climate change has been used by corporations opposed to regulations countering global warming (Michaels 2008; Oreskes and Conway 2010). Needless to say, even though the consensus among evolutionary biologists is solid, a primary

[22] I have argued in Sect. 4.2 that whereas a framing of philosophy of science in terms of a demarcation of science from pseudoscience erroneously suggests that approaches can be evaluated on a one-dimensional scale from scientific to unscientific, theoretical approaches differ in many respects. Considerations about ethical, social, and political consequences are yet other philosophical considerations that can be brought to bear in the assessment of epistemic traditions.

strategy of the ID movement is to create the public illusion that there is a debate about evolution and that evolutionary theory is poorly supported. Scientists react to such activities *outside* of science precisely because of their social-political consequences. This yields a dual task for science education: to teach sound science (against efforts to undermine science instruction) and to teach that scientists pay attention to the social implications of their research.

In Sect. 5.1 I have argued that an *epistemic* evaluation of intelligent design and its future (non-)promise benefits from knowledge of the ID movement's social aims—though this does not require judging these social aims. Of course, one can also engage in a *social* and political evaluation of the ID movement. The ID movement's ultimate aim is to re-Christianize largely secular Western societies and to more generally impose a socially conservative agenda upon them. In the US, this includes denial of equality and basic rights, particularly opposition to legal abortions and equal rights for homosexuals, as witnessed by the calls of several academic ID proponents that universities be permitted to hire only heterosexual professors. Barbara Forrest (2011) points to ideological affinities and institutional relations between several ID proponents (including William Dembski) and Christian Reconstructionism, which pursues a repressive social order by shaping public policy and legislation in terms of biblical principles (and hence is also called Theocratic Dominionism). At the very least, the ID movement's immediate social aim—a reprehensible aim at that—is to undermine the teaching of evolution in high schools. But this is part of a broader assault on school curricula, which includes the attempt to not include climate change (in line with the strong free-market and deregulation ideology of US conservatism) or embryonic stem cell research (in line with conservative objections to such research that are tied to its anti-abortion agenda). It would also involve the inclusion of a revisionist American history that downplays slavery and its impact on contemporary racism, and attempting to portray the US as a Christian nation by presenting its foundation and its constitution as a religious achievement, rather than a secular enterprise that set up a political and judicial system independently of religious commitments.

In the case of intelligent design and the ID movement, it may well be possible to give an assessment in terms of epistemic values without recourse to social values (the epistemic assessment demands knowledge of ID's social agenda, but does not require judgment of the agenda). However, there are cases in which epistemic and social considerations are entwined, so that it is impossible for a scientist or a philosopher of science to judge the epistemic credentials of some research *without* using ethical, social, or political values. Consider a hypothesis, the tentative endorsement of which by scientists will have policy implications, as will failure to endorse the hypothesis. Evidence never fully supports a hypothesis, so that, given uncertainty, scientists have to consider the social consequences of tentatively endorsing an actually false hypothesis, and the consequences of withholding endorsement of an actually true hypothesis, as has been prominently argued by Heather Douglas (2009). When there is evidence based on animal studies that a new pesticide which is to be approved by regulatory bodies is highly carcinogenic for humans, not asserting that the pesticide is dangerous (on the grounds that the evidence—as always—falls short of certainty) will have very harmful consequences to humans, should the substance be carcinogenic. In contrast, tentatively recommending that the pesticide be considered carcinogenic (so that this new pesticide is not approved) will have much more benign societal consequences, given that there are other pesticides available. In this scenario, it is sensible for scientists to recommend that the pesticide be deemed carcinogenic, and such scientific advice is partially influenced by social values—in fact, an epistemic decision of whether to endorse the hypothesis cannot responsibly be made independently of considerations about social consequences.

A stronger joint epistemic-social agenda for science and philosophy of science stems from Janet Kourany's (2010) call for a *socially responsible science*, motivated from the perspective of feminist philosophy of science. Studies that focus on male primates and men (no matter their veracity) can lead to such a biased and misleading picture of primate and human social organization and the role of females, that it has harmful effects on the condition of contemporary women. This shows that (selectively) obtaining well-confirmed items of knowledge is neither enough to achieve an adequate scientific account, nor socially responsible science (Anderson 1995). Women have traditionally been excluded from and still are underrepresented in many drug trials, because, among other things, their menstrual cycle has been seen as a confounding factor. But precisely due to their differing endocrinology, the results of the drug's efficacy, dosage, and side-effects do not carry over to women. As a result, doctors either have to withhold available drugs from women or prescribe them without knowing the involved risks or how to dose them properly (Goering 1994; Kim et al. 2010). Restricting scientific claims about such drugs to men—only to declare an 'epistemically' valid account—would still be research that excludes a relevant social group from the benefits of research. The better alternative is to acknowledge that *epistemic* considerations about how to design drug trials have to answer to *social* considerations. As an instance of exemplary socially responsible research, Kourany (2010) points to the studies by Carolyn West (2002) on domestic violence against African-American women, which aims at studying (and improving) the particular condition of these women but without promoting the stereotype that black men are inherently more violent. These social aims influence what counts as a proper epistemic approach in terms of choice of concepts, selection of study subject, data collection and analysis, and dissemination of results.

Let me conclude with a remark on education. I have indicated that, beyond the evidence for anthropogenic climate change, climate scientists publicly advocate the knowledge of their field because they employ social values according to which the consequences of global warming are socially harmful. The present point is that scientists may well have reasons (independently of their scientific expertise) for holding the social values prompting the public promotion of climate science. A primary motivation for religious belief is that it is deemed to give meaning and moral guidance, and the resistance to accepting the teaching of evolution among some high school students is often due to their perception that evolutionary biology, and more generally naturalistic science, promotes an amoral or moral relativist worldview. Of course, moral nihilism (or relativism) vs. religiously based moral dogma is a false dichotomy, and with Kantian ethics and utilitarianism, moral philosophy has created rigorous justifications of ethical principles that do not require recourse to the divine. Many high school teachers (especially in the US) will shy away from addressing ethics in the classroom, as it is easily deemed to be an encroachment on the personal views of students—but so can be the teaching of evolution. While it is not an issue for science classrooms, in those school systems where there are classes on ethics, it is useful to convey that ethics is not so much about particular moral principles, but about how to justify them. Just as students must learn about ambiguous evidence for scientific claims and that such claims can be revised based on increased evidential support, they must understand that there can be rational disagreement about ethical and social values, and that there are means to adjudicate such issues. We have good arguments for why past moral assumptions were wrong and why current ethical standards are superior.

Given that in many students' minds evolution is tied to a materialistic, amoral worldview, it is desirable that students be taught that not only are there adequate and inadequate ways to confirm factual and scientific claims, but there are also better and worse ways to justify ethical tenets.

6 Summary: Lessons for Science Education

Michael Behe's irreducible complexity argument against evolution fails because he ignores that precursor systems could have performed a different function. Structures in different extant species exhibit homologies and comparative studies show how structures have actually evolved, highlighting the need to teach students the comparative method as well as many examples of macroevolutionary transformations (beyond microevolutionary theories). Behe also focuses on isolated molecular systems, ignoring their context and, more generally, the redundancy and robustness of organismal systems. Systems need not break down or lead to detrimental effects for the organism when they are modified or some of their parts are removed. The notion of a genetic blueprint is a metaphor that fails to actually explain the process of development and obscures its flexibility. Robust development and physiological adaptability—which are the opposite of irreducible complexity and organisms being like Paley's watch—have come to be seen as the key to understanding morphological evolvability, so that biology instruction should avoid portraying organisms and organismal systems using information and machine metaphors.

Small probability arguments against evolution are common, and have recently been developed and promoted by William Dembski. Since they are fallacious and made even beyond the context of evolution, science and mathematics instructors should address them. The probability of the occurrence of complex structures assuming that naturalistic evolution occurred (which is indeed very low) must not be conflated with the probability of naturalistic evolution given the occurrence of complex structures: $P(\text{observation}|\text{hypothesis}) \neq P(\text{hypothesis}|\text{observation})$. In fact, from $P(\text{observation}|\text{hypothesis})$ being extremely small, nothing can be inferred about the probability of the hypothesis—it could be high or low. It is easy to see why, as both a true and a false hypothesis entail small probabilities for some observations if the conjunction of many random events is considered. Likelihoods of the form P(observation|hypothesis) can be relevant if several hypotheses are compared. So the real question is whether P(observations|Darwinian evolution) < P(observations|intelligent design), requiring intelligent design to put forward its positive account (the latter probability is undefined if intelligent design merely claims there has been some intelligent influence on some entities at some point in history) The lesson for science education is the need to highlight that hypothesis testing in science involves comparing several rival hypotheses, in which the one offering the best explanation is chosen. Both the irreducible complexity and the small probability arguments are merely arguments against evolution, but do not create any explanation of biological phenomena.

ID proponents have falsely claimed that methodological naturalism (science's study of natural phenomena and explanation by natural causes only) amounts to metaphysical naturalism, and thus atheism. They have also contended that methodological naturalism is an a priori philosophical commitment, and thus without scientific justification. However, that it is not endorsed *a priori* is shown by the fact that the construal of methodological naturalism (and thus what science actually endorses) has changed in the history of science. Over the past centuries there has been significant modification in views about what observational, experimental, and inferential methods are reliable, what can be empirically ascertained, what natural phenomena there are, what laws of nature obtain, what is empirically possible and impossible, and thus what qualifies as a scientific explanation or theory. The implication for science education is that there is not an unchanging, overarching scientific method or nature of science, and that the reasons for why the methods of science have been improved provides contemporary scientists with an *empirical* justification for why the current version of methodological naturalism is endorsed and used in scientific practice. My discussion has also shown that any assessment of the credentials of an approach (which often has problematically been phrased in terms of deciding whether it is science or pseudoscience) must be based on a variety of concrete considerations that may differ across scientific domains and fields.

Beyond the past philosophical focus of construing science in terms of theories, nowadays philosophers of science study the process of discovery and various concrete research practices, so as to assess intellectual approaches also in terms of their socially constituted epistemic practices. This is likewise relevant to biology education, as the difference between intelligent design and evolutionary biology is most marked in the practices of their respective proponents. Given their primarily social aims (undermining the teaching of evolution in high schools so as to make secular societies religious), ID proponents do not develop explanations of biological phenomena and properly react to scholarly criticism, but instead focus their efforts on the public and the political arena, promoting debunked arguments against evolution so as to create the public impression that there is scientific disagreement about evolutionary theory. Current philosophy of science assesses research, not only in terms of whether it is methodologically rigorous, but also in terms of what societal consequences it has and whether it is socially responsible. The intelligent design movement is not to be excluded from scrutiny in terms of social values.

This chapter's primary recommendation for science education is to not only teach the content of scientific theories, but also convey the aims and practice of science. Making students reflect on the nature of science in general attends them to what science actually involves. This pedagogy also has benefits in concrete scientific domains, as it endows students with a better appreciation of how evolutionary biology works, why it does not aim to compete with religious beliefs, what the merits of evolutionary explanations are, and why intelligent design falls short.

Acknowledgments I am indebted to Kostas Kampourakis, Alessandro Minelli, and an anonymous referee for comments on earlier versions of this essay. I thank Emma Kennedy for her thorough copyediting of the manuscript. This work was funded by the Social Sciences and Humanities Research Council of Canada (Standard Research Grant 410-2008-0400).

References

Alberts, Bruce. 1998. The cell as a collection of protein machines: Preparing the next generation of molecular biologists. *Cell* 92: 291–294.

Anderson, Elizabeth. 1995. Knowledge, human interests, and objectivity in feminist epistemology. *Philosophical Topics* 23: 27–58.

Bechtel, William, and Adele Abrahamsen. 2005. Explanation: A mechanist alternative. *Studies in History and Philosophy of Biological and Biomedical Sciences* 36: 421–441.

Beckwith, Francis J. 2003a. *Law, Darwinism, and public education: The establishment clause and the challenge of intelligent design*. Lanham: Rowman & Littlefield.

Beckwith, Francis J. 2003b. Science and religion twenty years after *McLean v. Arkansas*: Evolution, public education, and the new challenge of intelligent design. *Harvard Journal of Law and Public Policy* 26: 455–499.

Behe, Michael J. 1996. *Darwin's black box: The biochemical challenge to evolution*. New York: Free Press.

Behe, Michael J. 2007. *The edge of evolution: The search for the limits of Darwinism*. New York: Free Press.

Berlinski, David. 2008. *The Devil's delusion: Atheism and its scientific pretensions*. New York: Crown Forum.

Bottaro, Andrea, Matt A. Inlay, and Nicholas J. Matzke. 2006. Immunology in the spotlight at the Dover 'Intelligent Design' trial. *Nature Immunology* 7: 433–435.

Brigandt, Ingo. 2007. Typology now: Homology and developmental constraints explain evolvability. *Biology and Philosophy* 22: 709–725.

Brigandt, Ingo. 2010. Beyond reduction and pluralism: Toward an epistemology of explanatory integration in biology. *Erkenntnis* 73: 295–311.

Brigandt, Ingo. 2011. Critical notice of *Evidence and evolution: The logic behind the science* by Elliott Sober, Cambridge University Press, 2008. *Canadian Journal of Philosophy* 41:159–186.

Brigandt, Ingo. 2012. The dynamics of scientific concepts: The relevance of epistemic aims and values. In *Scientific concepts and investigative practice*, ed. Uljana Feest and Friedrich Steinle. Berlin: de Gruyter.

Brigandt, Ingo. 2013. Explanation in biology: Reduction, pluralism, and explanatory aims. *Science & Education* 22: 69–91.

Brigandt, Ingo. in press-a. Evolutionary developmental biology and the limits of philosophical accounts of mechanistic explanation. In *Explanation in biology: An enquiry into the diversity of explanatory patterns in the life sciences*, eds. Pierre-Alain Braillard and Christophe Malaterre. Berlin: Springer.

Brigandt, Ingo. in press-b. From developmental constraint to evolvability: How concepts figure in explanation and disciplinary identity. In *Conceptual change in biology: Scientific and philosophical perspectives on evolution and development*, ed. Alan C. Love. Berlin: Springer.

Brigandt, Ingo, and Alan C. Love. 2008. Reductionism in biology. In *The Stanford encyclopedia of philosophy*, ed. Edward N. Zalta. http://plato.stanford.edu/entries/reduction-biology.

Chinn, Clark A., and Luke A. Buckland. 2011. Differences in epistemic practices among scientists, young earth creationists, intelligent design creationists, and the scientist-creationists of Darwin's era. In *Epistemology and science education: Understanding the evolution vs. intelligent design controversy*, ed. Roger S. Taylor and Michel Ferrari, 38–76. New York: Routledge.

Dawkins, Richard. 1989. *The selfish gene*, 2nd ed. Oxford: Oxford University Press.

Dembski, William A. 1998a. *The design inference: Eliminating chance through small probabilities*. Cambridge: Cambridge University Press.

Dembski, William A. 1998b. Introduction: Mere creation. In *Mere creation: Science, faith & intelligent design*, ed. William A. Dembski, 13–30. Downers Grove: InterVarsity Press.

Dembski, William A. 2002a. Naturalism's argument from invincible ignorance: A response to Howard Van Till. http://www.designinference.com/documents/2002.09.Van_Till_Response.htm. Accessed 1 Apr 2011.

Dembski, William A. 2002b. *No free lunch: Why specified complexity cannot be purchased without intelligence*. Lanham: Rowman & Littlefield.

Douglas, Heather. 2009. *Science, policy, and the value-free ideal*. Pittsburgh: University of Pittsburgh Press.

Downes, Stephen M. 1993. Socializing naturalized philosophy of science. *Philosophy of Science* 60: 452–468.

Duschl, Richard A., and Richard E. Grandy. 2011. Demarcation in science education: Toward an enhanced view of scientific method. In *Epistemology and science education: Understanding the evolution vs. intelligent design controversy*, ed. Roger S. Taylor and Michel Ferrari, 3–19. New York: Routledge.

Edelman, Gerald M., and Joseph A. Gally. 2001. Degeneracy and complexity in biological systems. *Proceedings of the National Academy of Sciences USA* 98: 13763–13768.

Elsberry, Wesley, and Jeffrey Shallit. 2011. Information theory, evolutionary computation, and Dembski's 'complex specified information'. *Synthese* 178: 237–270.

Evans, Margaret E., Cristine H. Legare, and Karl S. Rosengren. 2011. Engaging multiple epistemologies: Implications for science education. In *Epistemology and science education: Understanding the evolution vs. intelligent design controversy*, ed. Roger S. Taylor and Michel Ferrari, 111–139. New York: Routledge.

Fehr, Carla, and Kathryn Plaisance. 2010. Socially relevant philosophy of science: An introduction. *Synthese* 177: 301–316.

Felsentein, Joe. 2007. Has natural selection been refuted? The arguments of William Dembski. *Reports of the National Center for Science Education* 27(3–4): 20–26.

Fitelson, Branden, Christopher Stephens, and Elliott Sober. 1999. Review: How not to detect design. *Philosophy of Science* 66: 472–488.

Forrest, Barbara. 2010. It's Déjà vu all over again: The intelligent design movement's recycling of creationist strategies. *Evolution: Education and Outreach* 3: 170–182.

Forrest, Barbara. 2011. The non-epistemology of intelligent design: Its implications for public policy. *Synthese* 178: 331–379.

Forrest, Barbara, and Paul R. Gross. 2004. *Creationism's Trojan horse: The wedge of intelligent design*. Oxford: Oxford University Press.

Gauger, Ann K., and Douglas D. Axe. 2011. The evolutionary accessibility of new enzyme functions: A case study from the biotin pathway. *BIO-Complexity* 2011: 1–17.

Godfrey-Smith, Peter. 2003. *Theory and reality: An introduction to the philosophy of science*. Chicago: University of Chicago Press.
Goering, Sara. 1994. Women and underserved populations: Access to clinical trials. In *"It Just Ain't Fair": The ethics of health care for African Americans*, ed. Annette Dula and Sara Goering, 182–192. Westport: Praeger.
Hacking, Ian. 1983. *Representing and intervening: Introductory topics in the philosophy of natural science*. Cambridge: Cambridge University Press.
Häggström, Olle. 2007. Intelligent design and the NFL theorems. *Biology and Philosophy* 22: 217–230.
Hendrikse, Jesse L., Trish E. Parsons, and Benedikt Hallgrímsson. 2007. Evolvability as the proper focus of evolutionary developmental biology. *Evolution and Development* 9: 393–401.
Hoyningen-Huene, Paul. 2008. Systematicity: The nature of science. *Philosophia* 36: 167–180.
Johnson, Phillip E. 1991. *Darwin on trial*. Downers Grove: InterVarsity Press.
Johnson, Phillip E., and Denis O. Lamoureux (eds.). 1999. *Darwinism defeated? The Johnson-Lamoureux debate on biological origins*. Vancouver: Regent College Publishing.
Kim, Alison M., Candace M. Tingen, and Teresa K. Woodruff. 2010. Sex bias in trials and treatment must end. *Nature* 465: 688–689.
Kirschner, Marc W., and John C. Gerhart. 1998. Evolvability. *Proceedings of the National Academy of Sciences USA* 95: 8420–8427.
Kirschner, Marc W., and John C. Gerhart. 2005. *The plausibility of life: Resolving Darwin's dilemma*. New Haven: Yale University Press.
Kirschner, Marc W., John C. Gerhart, and Tim Mitchison. 2000. Molecular 'vitalism'. *Cell* 100: 79–88.
Kitcher, Philip. 2001. *Science, truth, and democracy*. Oxford: Oxford University Press.
Kourany, Janet A. 2010. *Philosophy of science after feminism*. Oxford: Oxford University Press.
Lamoureux, Denis O. 2008. *Evolutionary creation: A Christian approach to evolution*. Eugene: Wipf & Stock Publishers.
Laudan, Larry. 1983. The demise of the demarcation problem. In *Physics, philosophy, and psychoanalysis: Essays in honor of Adolf Grünbaum*, ed. Robert S. Cohen and Larry Laudan, 111–127. Dordrecht: Reidel.
Longino, Helen E. 2002. *The fate of knowledge*. Princeton: Princeton University Press.
Love, Alan C. 2008. Explaining evolutionary innovation and novelty: Criteria of adequacy and multidisciplinary prerequisites. *Philosophy of Science* 75: 874–886.
Love, Alan C. 2013. Interdisciplinary lessons for the teaching of biology from the practice of evo-devo. *Science & Education* 22: 255–278.
Meyer, Stephen C. 2009. *Signature in the cell: DNA and the evidence for intelligent design*. New York: HarperOne.
Michaels, David. 2008. *Doubt is their product: How industry's assault on science threatens your health*. Oxford: Oxford University Press.
Mitchell, Sandra D. 2009. *Unsimple truths: Science, complexity, and policy*. Chicago: University of Chicago Press.
Numbers, Ronald L. 2009. That creationism is a uniquely American phenomenon. In *Galileo goes to jail and other myths about science and religion*, ed. Ronald L. Numbers, 215–223. Cambridge, MA: Harvard University Press.
Olofsson, Peter. 2008. Intelligent design and mathematical statistics: A troubled alliance. *Biology and Philosophy* 23: 545–553.
Oreskes, Naomi, and Erik M. Conway. 2010. *Merchants of doubt: How a handful of scientists obscured the truth on issues from tobacco smoke to global warming*. New York: Bloomsbury Press.
Paley, William. 1802. *Natural theology, or evidences of the existence and attributes of the deity, collected from the appearances of nature*. London: R. Faulder.
Pallen, Mark J., and Nicholas J. Matzke. 2006. From *The Origin of Species* to the origin of bacterial flagella. *Nature Reviews Microbiology* 4: 784–790.

Pennock, Robert T. 2011. Can't philosophers tell the difference between science and religion? Demarcation revisited. *Synthese* 178: 177–206.

Perakh, Mark. 2004. *Unintelligent design*. Amherst: Prometheus Books.

Pigliucci, Massimo, and Maarten Boudry. 2011. Why machine-information metaphors are bad for science and science education. *Science & Education* 20: 453–471.

Popper, Karl. 1959. *The logic of scientific discovery*. New York: Basic Books.

Robert, Jason. 2004. *Embryology, epigenesis, and evolution: Taking development seriously*. Cambridge: Cambridge University Press.

Ruse, Michael (ed.). 1988. *But is it science? The philosophical question in the creation/evolution controversy*. Buffalo: Prometheus Books.

Samarapungavan, Ala. 2011. Ontological assumptions about species and their influence on students' understanding of evolutionary biology. In *Epistemology and science education: Understanding the evolution vs. intelligent design controversy*, ed. Roger S. Taylor and Michel Ferrari, 77–94. New York: Routledge.

Sarkar, Sahotra. 2007. *Doubting Darwin? Creationist designs on evolution*. Oxford: Blackwell.

Sarkar, Sahotra. 2011. The science question in intelligent design. *Synthese* 178: 291–305.

Scott, Eugenie C. 2004. *Evolution vs. creationism: An introduction*. Berkley: University of California Press.

Sewell, Granville. 2000. A mathematician's view of evolution. *The Mathematical Intelligencer* 22: 5–7.

Sewell, Granville. 2001. Can anything happen in an open system? *The Mathematical Intelligencer* 23: 8–10.

Shanks, Niall. 2004. *God, the devil, and Darwin: A critique of intelligent design theory*. Oxford: Oxford University Press.

Shanks, Niall, and Karl H. Joplin. 1999. Redundant complexity: A critical analysis of intelligent design in biochemistry. *Philosophy of Science* 66: 268–282.

Sinatra, Gale M., and Louis Nadelson. 2011. Science and religion: Ontologically different epistemologies? In *Epistemology and science education: Understanding the evolution vs. intelligent design controversy*, ed. Roger S. Taylor and Michel Ferrari, 173–193. New York: Routledge.

Sober, Elliott. 1999. Testability. *Proceedings and Addresses of the American Philosophical Association* 73: 47–76.

Sober, Elliott. 2007. What is wrong with intelligent design? *The Quarterly Review of Biology* 82: 3–8.

Sober, Elliott. 2008. *Evidence and evolution: The logic behind the science*. Cambridge: Cambridge University Press.

Solomon, Miriam. 2001. *Social empiricism*. Cambridge, MA: MIT Press.

Stotz, Karola. 2006. Molecular epigenesis: Distributed specificity as a break in the central dogma. *History and Philosophy of the Life Sciences* 28: 527–544.

Thagard, Paul. 2011. Evolution, creation, and the philosophy of science. In *Epistemology and science education: Understanding the evolution vs. intelligent design controversy*, ed. Roger S. Taylor and Michel Ferrari, 20–37. New York: Routledge.

Tuana, Nancy. 2010. Leading with ethics, aiming for policy: New opportunities for philosophy of science. *Synthese* 177: 471–492.

Wagner, Andreas. 2005. *Robustness and evolvability in living systems*. Princeton: Princeton University Press.

Wells, Johnathan. 2011. *The myth of junk DNA*. Seattle: Discovery Institute Press.

West, Carolyn. 2002. Black battered women: New directions for research and black feminist theory. In *Charting a new course for feminist psychology*, ed. Lynn Collins, Michelle Dunlap, and Joan Chrisler, 216–237. Westport: Praeger.

Young, Matt, and Taner Edis (eds.). 2004. *Why intelligent design fails: A scientific critique of the new creationism*. New Brunswick: Rutgers University Press.

Molecular Evolution

Michael R. Dietrich

1 Introduction

Within a decade of James Watson and Francis Crick's announcement of the double helical structure of DNA, biologists, physicists, and chemists began to ask fundamental questions about how biological molecules had evolved. As proteins and later RNA and DNA sequences became the objects of evolutionary inquiry, the study of molecular evolution coalesced as a new discipline within biology bringing with it significantly new forms of data, inference, modeling, and concepts (Suarez and Barahona 1996; Dietrich 2008).

For many evolutionary biologists, molecular biology represented a mixed blessing. New techniques, such as electrophoresis, were greeted with enthusiasm as they promised access to elusive evidence of genetic variation in natural populations (Beatty 1987a; Lewontin 1991). At the same time, molecular biology was perceived in opposition to evolutionary biology and the "molecular wars" that ensued pitted molecular and organismal evolutionary biologists in a struggle for authority and resources (Wilson 1994; Hagen 1999; Smocovotis 1996; Dietrich 1998). Molecular evolution attempted to span this divide by building a bridge between Neo-Darwinian evolutionary biology, on the one hand, and biochemistry and molecular biology, on the other. However, controversy quickly enveloped molecular evolution itself as some of its early adherents championed the prevalence of neutral mutations that were not subject to natural selection alone, but instead allowed random drift to play a significant role in evolution.

In retrospect, biologist Alexey Kondrashov describes the impact of the introduction of neutrality in dramatic terms:

> Once upon a time, the world seemed simple when viewed through the eyes of evolutionary biologists. All genomes were tightly controlled by various forms of natural selection. [...]

M.R. Dietrich (✉)
Department of Biological Sciences, Dartmouth College, Hanover, NH, USA
e-mail: michael.dietrich@dartmouth.edu

K. Kampourakis (ed.), *The Philosophy of Biology: A Companion for Educators*, History, Philosophy and Theory of the Life Sciences 1, DOI 10.1007/978-94-007-6537-5_12,

> This idyllic world began to crumble in 1968, when Kimura made his modest proposal that most allele substitutions and polymorphisms do not substantially affect an organism's fitness and are governed, not by positive or balancing selection, but by random drift (Kondrashov 2005)

The supposed tyranny of panselectionism and the valiant challenge of neutrality were themes that Motoo Kimura included in his own account of the neutral theory (Kimura 1983).[1] To be fair, the neutral theory of molecular evolution was championed by Jack King, Thomas Jukes, and Motoo Kimura beginning in 1968 (King and Jukes 1969; Kimura 1968). Like some biochemists who were developing the first comparisons of molecular similarity and difference, King, Jukes, and Kimura became convinced that not all changes in proteins or DNA were subject to natural selection (Zuckerkandl and Pauling 1965; Dietrich 1994). In the 1950s, neutral mutations had been acknowledged as a possibility by leading evolutionary biologists, such as Theodosius Dobzhansky, but the presence of large numbers of neutral mutants was not taken seriously and so genetic drift, which would describe their evolutionary fate, was not given much weight as an alternative to natural selection (Dobzhansky 1955). In the late 1960s, King, Jukes, and Kimura marshaled emerging molecular evidence to argue that most observed differences in protein sequences and DNA sequences were the result of neutral mutations subject only to random drift. To older evolutionary biologists who had worked hard to establish neo-Darwinian evolutionary biology with its emphasis on natural selection, the ideas of neutrality and genetic drift were particularly vexing. King and Jukes fanned the flames of dissent by calling their advocacy of neutral mutations and random drift at the molecular level, Non-Darwinian Evolution. The resulting neutralist-selectionist controversy in molecular evolution lasted for at least 20 years pitting King, Jukes, Kimura, and Tomoko Ohta against selectionists such as Theodosius Dobzhanksy, Richard Lewontin, and John Gillespie, to name only a few (Dietrich 1994).

Putting aside the contentious history of molecular evolution, for philosophers its rise allows us to pose fundamental questions about domains, levels, and causal processes in science. The introduction of molecular data greatly expanded the domain of phenomena to which evolutionary biology could be applied. However, that expansion was accompanied by a significant diversification in the kinds of causes used within evolutionary biology that divided its domain between molecular and organismal phenomena. Where natural selection was assumed to predominate at the organism level, genetic drift and selection are claimed to both be important causal factors in molecular evolution. The recognition of drift and neutral molecular evolution weakened the panselectionist perspective that had been typical of most of evolutionary biology and in practice fostered the widespread use of neutral null models leading to a profound methodological reversal in evolutionary biology.

---

[1] The meanings of panselectionism will be explicated later in this essay. I believe that Kimura understood it to refer to the claim that natural selection is the most important factor in evolution. Panselectionism is broader than adaptationism because it encompasses all forms of selection, not just those that produce adaptations.

2 Dividing the Domain of Evolutionary Biology

The rise of molecular evolution does not represent a Kuhnian revolution with a shift to a new scientific paradigm. Molecular evolution is a hybrid discipline borrowing problems, techniques, concepts, and theories from existing fields and combining and adapting them to fit new forms of molecular data. Scientific change in the history of molecular evolution was piecemeal and multi-dimensional in the sense that there was not a single shift in worldview, but a diverse set of drivers that motivated change and created a myriad of distinct novelties. At a very basic level, however, we can say unequivocally that the introduction of molecular data altered the domain of evolutionary biology.

Every field of science has a domain of phenomena that it attempts to explain (Shapere 1977; see Potochnick, this volume for explanation in biology). One of the great virtues of Darwin's theory of evolution by natural selection is that it explains a very broad range of biological phenomena and is thus heralded as having incredible unifying power (Kitcher 1981). At first glance, the flood of molecular phenomena from the 1960s onward should have only increased the unifying power of evolutionary biology. After all, it now potentially explained a much broader set of phenomena than ever before. However, as the domain of evolutionary phenomena grew, important differences between molecular and organismal evolution began to emerge, as we will see below in the case of the molecular clock.

In today's post-genomic era, the most common form of data for molecular evolution are DNA sequences. The billions of available base pairs stand in sharp contrast to the early days of molecular evolution when information about DNA content, similarity, or difference had to be indirectly inferred from immunological comparisons, hybridization studies, or protein sequences. Even with these relatively inexact stand-ins for DNA data, molecular evolutionists were able to estimate genetic distances and rates of evolution. These measures allowed them to do unprecedented work in molecular systematics and to postulate the existence of molecular clocks (Morgan 1998).

The term "molecular clock" refers to the approximate rate constancy observed for a type of molecule. Emile Zuckerkandl discovered the phenomena in 1962 when he compared differences among mammalian hemoglobins (Zukerkandl and Pauling 1962, 1965; Zuckerkandl 1963). Zuckerkandl presented his work at the 1962 Wenner-Gren Foundation meeting on molecular anthropology, where luminaries of evolutionary biology, such as Theodosius Dobzhansky, Ernst Mayr, and George G. Simpson, took the opportunity to try to bring him in line with Neo-Darwinism. If evolution is governed predominantly by natural selection and natural selection is modulated by changing environments, then one would expect that the rate of evolution would vary just as the relevant environment varies. Observations of clocklike rates of evolution, thus, presented an anomaly for the Neo-Darwinian perspective since the variable rates at the molecular level seemed to contradict the variable rates of change at the morphological level.

Despite the best efforts of Dobzhanksy, Mayr, and Simpson, they could not contain this anomaly. The result is that, in the late 1960s and 1970s, the perceived division between molecular and organismal levels entered evolutionary biology and began to divide the domain of evolution. What drove the divide was not just the contrast between rate constancy at the molecular level and rate variability at the organismal level that was at issue in the debates over the molecular clock. The molecular – organismal divide was strengthened by the proposal that different causal processes predominated at each level – at the organismal level, selection was the dominant force, while at the molecular level, drift occurred along with selection. In 1971, biochemist Allan Wilson described the molecular – organismal divide as follows:

> Molecular evolution proceeds in a rather regular fashion with respect to time. By contrast, organismal evolution is classically considered to be an irregular process, some species (e.g., placental mammals) changing rapidly, while others (e.g., frogs) change slowly. The paradox may be resolved by postulating the random fixation of mutations producing amino-acid substitutions that have no effect, favorable or unfavorable, on protein function and hence no effect on anatomy, physiology or behavior. (Wilson 1971)

The presence of neutral mutations at the molecular level and not the organismal level provided the key difference needed to explain the difference in rates of evolution.

Where molecular biologists have often been described as making reductionist arguments with regard to organismal phenomena, the division of the domain of evolutionary biology was not understood as implying that the organismal level could be reduced to the molecular level. In fact, quite the opposite was understood by organismal evolutionists, such as Dobzhanksy, Mayr, and Simpson, and molecular evolutionists, such as Kimura, Zuckerkandl, and Wilson. Evolutionary biologists saw the molecular and organismal levels as compatible and complementary. This does not mean the molecular clock and neutral theory were not controversial. It means that expanding the domain of phenomena in evolutionary biology defied easy unification (Smocovitis 1996; Mitchell and Dietrich 2006). A crucial contributor to this divide are the different forms of causal processes that were postulated as operating at different levels by molecular evolutionists.

3 The Causes of Molecular Evolution

The subdivision of evolutionary phenomena into the molecular and the organismal levels supported the causal diversification of evolutionary explanations. Explanations at the organismal level were understood as predominantly selectionist and explanations at the molecular level were understood as predominantly neutralist. This does not mean that selection did not play an important role at the molecular level. Neutralist explanations acknowledged significant roles from both selection and drift. The neutral theory was very controversial, however. In part this controversy was fueled by the difficulty of empirically distinguishing between drift and selection as causes of evolution.

From the outset, advocates of the neutral theory claimed that one of its virtues was that it made testable quantitative predictions (Crow 1969). Frustratingly, those early tests were not definitive. Distinguishing drift and selection was not as easy as it first seemed (Beatty 1987b).

Philosophically, untangling drift and selection requires that we distinguish causal processes from their outcomes and realize that drift and selection are causally active at both the molecular and organismal levels (Dietrich and Millstein 2008). In 2002 and 2005, Roberta Millstein articulated a philosophical distinction between process and outcome as a means of explicating significant differences between biological advocacy of drift and selection. "Causal process" here refers to series of causally connected physical states occurring over time, while "outcome" refers to the effect of that process. Biologists often appeal to both process and outcome without necessarily marking that distinction. Consider Kimura's definition of drift:

> By random genetic drift I mean random fluctuation of gene frequencies in a population caused by random sampling of gametes in reproduction. In any sexually reproducing species, the total number of individuals is not only finite, but also can be regarded as a random sample chosen from a much larger collection of male and female gametes (or 'gene pool') produced by the parental generation. The amount of fluctuation in gene frequencies (that is, proportion of various alleles) is expected to be larger, the smaller the population. (Kimura 1983, p. 37)

For Kimura, the "process" of drift is the random sampling of gametes and the "outcome" is the random fluctuation of gene frequencies. The problem is that drift is not the only process that can produce the outcome of random fluctuation of gene frequencies. Selection in a changing environment, for instance, could produce this effect. Unless there is an outcome that is uniquely produced by drift, outcomes are not a reliable means to detect or define drift (Millstein and Dietrich 2008). As a result, Millstein advocates defining drift and selection as causal processes.

Both drift and selection can be understood as sampling processes. Kimura understood the process of drift as random gamete sampling, for instance. John Beatty distinguished drift from selection by arguing that the "random" sampling in a drift process should be thought of as "indiscriminate sampling," while the sampling process inherent in selection is a discriminate sampling process (Beatty 1984). Put another way, in selection, the heritable physical differences between entities are causally *relevant* to differences in reproductive success (Millstein 2002, 2005). In drift, heritable physical differences are causally irrelevant to the sampling inherent in the reproductive process.

The distinction between process and outcome is especially important for understanding a distinction the Motoo Kimura and his collaborator, Tomoko Ohta, drew between strictly neutral, effectively neutral, and nearly neutral mutations. By definition, strictly neutral mutations are those where there is no selection operating at all, so drift is the only relevant process. From 1968 on, however, Kimura and Ohta acknowledged that there were probably significant numbers of effectively neutral, also known later as nearly neutral mutations. These mutations behaved very similarly to strictly neutral mutations in that they showed very similar outcomes to strictly neutral mutations that were subject only to drift. In fact, these nearly neutral mutations were subject to very weak selection and to drift. This has led some biologist to claim that "nearly neutral" really means "a little bit selective" (Dover 1997, p. 91).

From an outcome point of view, nearly neutral mutants have drift-like outcomes. From a process point of view, nearly neutral mutants are subject to processes of both drift and selection, but unlike most organismal cases nearly neutral molecular mutants are so weakly selected that the process of drift has a much stronger effect (Dietrich and Millstein 2008).

In the 1990s, Tomoko Ohta began to champion the Nearly Neutral Theory of molecular evolution, in part because she thought that nearly neutral outcomes could be distinguished from strictly neutral outcomes (Ohta 1992, 2002; Ohta and Gillespie 1996). Francisco Ayala and his coworkers had measured genetic variability in *Drosophila* and found a large number of relatively rare alleles (Ayala et al., 1974). Ohta realized that the strictly neutral theory would not predict such a high number of these rare alleles. However, if most of the presumed neutral alleles were allowed to be very slightly deleterious alleles, then more rare alleles would be expected. These rare alleles would be slightly deleterious or nearly neutral, so they would still be subject to random drift as a strictly neutral mutant would, but they would also be subject to very weak selection. Ayala's discovery of rare alleles thus became an outcome associated with near neutrality.

Drift has become strongly linked to molecular evolution because of the neutral theory. The prevalence of drift as a phenomenon or an outcome does not mean that drift alone is the causal process operating to produce those outcomes. While there may be cases of drift processes acting alone or selection processes acting alone, most cases of molecular evolution are best explained in terms of combinations of processes of drift and selection, such as is the case with nearly neutral mutants in natural populations (Dietrich and Millstein 2008).

4 The End of Panselectionism

The neutral theory is commonly juxtaposed to panselectionism – the idea that natural selection is the most important factor in biological evolution and as a result formulating hypotheses of selection and adaptation was the best method for an evolutionary biologist (see Kimura 1983; Provine 1988; Dietrich 2006; for adaptation see also Forber, this volume). Indeed, when Stephen Jay Gould and Richard Lewontin attacked panselectionism and adaptationism in their paper, "The Spandrels of San Marco and the Panglossian Paradigm" (1979), Ernst Mayr responded that drift could not be reliably detected, so starting with drift hypotheses was never going to be a successful method, especially relative to the search for selection in nature (Mayr 1983). Mayr's confidence was grounded in a history of what he saw as problematic attempts to establish drift as a cause of evolution for morphological traits (see Millstein 2008; Gould 1983). Even at the molecular level, definitive tests of drift had not been established by 1983. That all started to change in 1984, as Martin Kreitman began to apply evolutionary analysis to natural variation in DNA sequences (Kreitman 1983). As DNA sequence data became readily available, statistical tests for selection using DNA data became accepted tools in molecular evolution. Because these tests often used neutral hypotheses, by the 1990s, both proponents and critics of the neutral theory recognized that neutrality, not selection, was a useful starting hypothesis when analyzing DNA sequences in evolutionary biology (Kreitman 2000; Beatty 1987b; Crow 1987).

This methodological shift toward neutrality represents a significant reversal from the selectionist approach. Treating the neutral theory as a null hypothesis spelled the end of methodological panselectionism (Mitchell and Dietrich 2006). According to biologist Roger Selander: "All our work begins with tests of the null hypothesis that variation in allele frequencies generated by random drift is the primary cause of molecular evolutionary change" (Selander 1985, p. 87). Neutrality becomes a starting place because, according to Selander, it is preferable "to begin with the simplest model" and then determine "a baseline for further analysis and interpretation" (Selander 1985, p. 88). Beginning with neutrality as a null hypothesis is not a commitment to neutrality and drift as necessary features of a good evolutionary explanation. Borrowing a distinction regarding adaptationism from Peter Godfrey Smith, we can distinguish empirical, methodological, and explanatory senses of panselectionism. The neutral null model argues against methodological panselectionism, e.g., "The best way for scientists to approach biological systems is to look for evidence of natural selection in some form." The use of neutral null models requires no prior commitment vis-à-vis empirical panselectionism ("Natural selection is a powerful and ubiquitous force, and there are few constraints, except general and obvious ones, on the biological variation that fuels it.") or explanatory panselectionism (explaining the effects of selection is "the core intellectual mission of evolutionary theory") (Godfrey Smith 2001, pp. 335–357).

That said, the rise of statistical tests of selection do have consequences for both empirical and explanatory panselectionism. At the height of the neutralist-selectionist controversy in the 1970s, Christopher Wills offered a defense of "naïve panselectionism" where he claimed that "virtually any change in amino acid composition of any protein molecule produces a molecule of slightly different properties and therefore of slightly different selective value from the original" (Wills 1973, p. 23). No biologist today would defend such an extreme claim. In fact, Martin Kreitman has argued that "Kimura's theory of neutrally evolving mutations is the backbone for evolutionary analysis of DNA sequence variation and change" because a "substantial fraction" of the genome is best modeled as selectively neutral, because selective neutrality is a "useful null hypothesis," and because "statistical analysis of (potentially) neutral variation in a gene (or other region of the genome) can be informative about selection acting at linked sites" (Kreitman 2000, pp. 541–542; Hudson et al. 1987). Kreitman's approach embraces the idea that neutral and selected variation is to be expected in a natural population, and that the function of statistical tests of selection is to locate and characterize selected sites in nucleotide sequences. Such a view argues against the universality of selection at the molecular level and rejects empirical panselectionism. Moreover, the adoption of these tests argues against explanatory adaptationism as well. The existence of significant proportion of neutral sites makes it untenable to argue that the primary explanatory goal of molecular evolution can be that of explaining selection.

Kondrashov was right when he argued that the idyll of selection has crumbled, at least, if he meant the idyll of panselectionism. At the molecular level, neutral variation is accepted and is the effect of random drift. Although neutral null hypotheses are now common starting places in molecular analysis, panneutralism has not replaced panselectionism. Instead, molecular evolution is understood in terms of the complex interplay of drift and selection acting upon sites ranging from strictly neutral to strongly selected.

5 Conclusion

The availability of genomic data and the success of statistical tests of selection that use genomic data have contributed to increased engagement with molecular evolution and tests of selection in particular. This increased activity has produced a shift in the place of molecular evolution within the evolutionary biology curriculum. Consider the subsequent chapters of Douglas Futuyma's textbooks on evolution. In the editions spanning 1986–2009, molecular evolution has always had its own chapter toward the end of the book. But, in later editions, material on molecular evolution has been increasingly incorporated into other chapters. In Futuyma's 1986 *Evolutionary Biology*, for instance, the chapter on drift has three pages on the neutral theory. In the 1998 edition of the same textbook, the neutral theory section had expanded to seven pages with additional pages on coalescents. By the 2005 textbook, Kimura has been added to the introductory chapter, molecular clocks and molecular systematics to the classification chapter, and statistical test of selection to the chapter on evolutionary genetics. By the 2009 edition, even the chapter on adaptation has examples of the molecular evolution of crystallin proteins while the molecular evolution chapter has steadily increased in size (Futuyma 1986, 1998, 2005, 2009). Even if an instructor chooses to skip the chapter on molecular evolution, evolution at the molecular level has become unavoidable in more general chapters on selection, classification, and of course drift.

Molecular evolution offers an opportunity to explore the basic structure of evolutionary biology. Comparison between molecular and organismal evolution shed light on how the domain of evolutionary biology is defined, how it has grown, and how it has been subdivided. The different roles of drift and selection at the organismal and molecular levels raise important questions about the nature of causal processes and their reliable detection. Lastly, the success of statistical tests using genomic data reveal an important methodological and empirical reversal in evolutionary biology that marks the end of panselectionism and the beginning of a more pluralistic approach to evolution in terms of both neutrality and selection. Articulating how processes of drift and selection operate at the molecular level and how molecular and organism phenomena can be integrated are challenges now facing evolutionary biology as a result.

Acknowledgement I am grateful for the very useful commentary provided by this collection's editor and reviewers, as well as, the detailed input and collaboration of Roberta Millstein.

References

Ayala, F., M. Tracey, L. Barr, J. McDonald, and S. Perez-Salas. 1974. Genetic variation in natural populations of five *Drosophila* species and the hypothesis of the selective neutrality of protein polymorphisms. *Genetics* 7: 226–235.

Beatty, J. 1984. Chance and natural selection. *Philosophy of Science* 51: 183–211.

Beatty, J. 1987a. Weighing the risks: Stalemate in the classical/balance controversy. *Journal of the History of Biology* 20(1987): 289–319.

Beatty, J. 1987b. Natural selection and the null hypothesis. In *The latest on the best*, ed. John Dupre. Cambridge: MIT Press.

Crow, J.F. 1969. Molecular genetics and population genetics. *Proceedings of the Twelfth International Congress of Genetics* 3: 105–113.

Crow, J.F. 1987. Neutral models in molecular evolution. In *Neutral models in biology*, ed. Matthew H. Nitecki and Hoffman Antoni. New York: Oxford University Press.

Dietrich, M.R. 1994. The origins of the neutral theory of molecular evolution. *Journal of the History of Biology* 27: 21–59.

Dietrich, M.R. 1998. Paradox and persuasion: Negotiating the place of molecular evolution within evolutionary biology. *Journal of the History of Biology* 31: 85–111.

Dietrich, M.R. 2006. From Mendel to molecules: A brief history of evolutionary genetics. In *Evolutionary genetics: Concepts and case studies*, ed. Charles W. Fox and Jason B. Wolf, 3–13. New York: Oxford University Press.

Dietrich, M.R. 2008. Molecular evolution. In *A companion to philosophy of biology*, ed. Sohotra Sarkar and Anya Plutynski, 157–168. Cambridge: Blackwell.

Dietrich, M.R., and R.A. Millstein. 2008. The role of causal processes in the neutral and nearly neutral theories. *Philosophy of Science* 75: 548–559.

Dobzhansky, T. 1955. A review of some fundamental concepts and problems in population genetics. *Cold Spring Harbor Symposia on Quantitative Biology* 20: 1–15.

Dover, G. 1997. There's more to life than selection and neutrality. *Bioessays* 19: 91–93.

Futuyma, D.J. 1986. *Evolutionary biology*, 2nd ed. Sunderland: Sinauer Publishing.

Futuyma, D.J. 1998. *Evolutionary biology*, 3rd ed. Sunderland: Sinauer Publishing.

Futuyma, D.J. 2005. *Evolution*, 1st ed. Sunderland: Sinauer Publishing.

Futuyma, D.J. 2009. *Evolution*, 2nd ed. Sunderland: Sinauer Publishing.

Gould, S.J. 1983. The hardening of the modern synthesis. In *Dimensions of Darwinism*, ed. M. Grene. Cambridge: Cambridge University Press.

Gould, S.J., and R.C. Lewontin. 1979. The Spandrels of San Marco and the Panglossian Paradigm: A critique of the adaptationist programme. *Proceedings of the Royal Society of London, Series B* 205: 581–598.

Hagen, J. 1999. Naturalists, molecular biologists, and the challenges of molecular evolution. *Journal of the History of Biology* 32: 321–341.

Hudson, R., Kreitman, M., and M. Aquadé. 1987. A test of neutral molecular evolution based on nucleotide data. *Genetics* 116: 153–159.

Kimura, M. 1968. Evolutionary rate at the molecular level. *Nature* 217: 624–626.

Kimura, Motoo. 1983. *The neutral theory of molecular evolution*. Cambridge: Cambridge University Press.

King, J.L., and T.H. Jukes. 1969. Non-Darwinian evolution. *Science* 164: 788–798.

Kitcher, P. 1981. Explanatory unification. *Philosophy of Science* 48: 507–531.

Kondrashov, A. 2005. Evolutionary biology – Fruitfly genome is not junk. *Nature* 437: 1106.

Kreitman, M. 1983. Nucleotide polymorphism at the alcohol dehydrogenase locus of *Drosophila melanogaster*. *Nature* 304: 412–417.
Kreitman, M. 2000. Methods to detect selection in populations with application to the human. *Annual Review of Genomics and Human Genetics* 1: 539–559.
Lewontin, R.C. 1991. Twenty-five years ago in *Genetics*: Electrophoresis in the development of evolutionary genetics: Milestone or millstone? *Genetics* 128: 657–662.
Mayr, E. 1983. How to carry out the adaptationist program. *American Naturalist* 121: 324–333.
Millstein, R.A. 2002. Are random drift and natural selection conceptually distinct? *Biology and Philosophy* 17(1): 33–53.
Millstein, R.A. 2005. Selection vs. drift: A response to Brandon's reply. *Biology and Philosophy* 20(1): 171–175.
Millstein, R.A. 2008. Distinguishing drift and selection empirically: 'The Great Snail Debate' of the 1950s. *Journal of the History of Biology* 41: 339–367.
Mitchell, S., and M.R. Dietrich. 2006. Integration without unification: An argument for pluralism in the biological sciences. *The American Naturalist* 168: S73–S79.
Morgan, Greg. 1998. Emile Zuckerkandl, Linus Pauling, and the molecular evolutionary clock, 1959–1965. *Journal of the History of Biology* 31: 155–178.
Ohta, T. 1992. The nearly neutral theory of molecular evolution. *Annual Review of Ecology and Systematics* 23: 263–286.
Ohta, T. 2002. Near-neutrality in evolution in genes and gene regulation. *PNAS* 99: 16134–16137.
Ohta, T., and J. Gillespie. 1996. Development of neutral and nearly neutral theories. *Theoretical Population Biology* 49: 128–142.
Provine, W.B. 1988. Progress in evolution and the meaning of life. In *Evolutionary progress*, ed. M. Nitecki, 49–74. Chicago: University of Chicago Press.
Selander, R. 1985. Protein polymorphism and the genetic structure of natural populations of bacteria. In *Population genetics and molecular evolution*, ed. T. Ohta and K. Aoki, 85–106. Tokyo: Japan Scientific Societies.
Shapere, D. 1977. Scientific theories and their domains. In *The structure of scientific theories*, ed. F. Suppe, 518–565. Urbana: University of Illinois Press.
Smith, P.G. 2001. Three kinds of adaptationism. In *Adaptationism and optimality*, ed. S. Hecht and E. Sober, 335–357. Cambridge: Cambridge University Press.
Smocovitis, V.B. 1996. *Unifying biology: The evolutionary synthesis and evolutionary biology*. Princeton: Princeton University Press.
Suarez, E., and A. Barahona. 1996. The experimental roots of the neutral theory of molecular evolution. *History and Philosophy of the Life Sciences* 18: 55–81.
Wills, C. 1973. In defense of naïve Pan-selectionism. *The American Naturalist* 107: 23–34.
Wilson, A. 1971. Albumin evolution in frogs: A test of the evolutionary clock hypothesis. *Proceedings of the National Academy of Sciences* 68: 3127–3129.
Wilson, E.O. 1994. *Naturalist*. Washington, DC: Island Press.
Zuckerkandl, E. 1963. Perspectives in molecular anthropology. In *Classification and human evolution*, ed. S. Washburn. Chicago: Aldine Publishing.
Zuckerkandl, E., and L. Pauling. 1962. Molecular disease, evolution, and genetic heterogeneity. In *Horizons in biochemistry*, ed. M. Kasha and B. Pullman, 189–225. New York: Academic.
Zuckerkandl, E., and L. Pauling. 1965. Evolutionary divergence and convergence in proteins. In *Evolving genes and proteins*, ed. V. Bryson and H. Vogel, 97–166. New York: Academic.

Educational Lessons from Evolutionary Properties of the Sexual Genome

John C. Avise

1 Introduction

With respect to biology, the histories of philosophy, religion, and science in essence have been the histories of mankind's attempts to understand why organic systems work as well as they do. Paradoxically, such understanding might seem at face value to have spiraled downward ever since Darwin and Mendel in the sense that we have many more unanswered questions today than we did in bygone times when almost everyone "knew" that supernatural agents orchestrated the organic world. My favorite metaphor about this paradoxical relationship between science and faith involves a balloon. Think of leading-edge inquiry in any scientific discipline as occurring at the outer surface of a balloon that encompasses contemporary knowledge. As the balloon expands via objective discovery processes, so too does its interface with the unknown, thereby exposing ever-broader horizons that beg further investigation. Thanks to technological and interpretative breakthroughs, the empirical balloon of molecular genetics is now bloating at a pace that is almost unprecedented in the history of any scientific discipline, so the cliché that science can be humbling (as well as enabling) is especially true today in the field of genomics. Sometimes it seems that each passing day brings genomic discoveries that challenge what we thought we knew (Table 1).

My goals in this chapter are to: (a) recapitulate conceptual paradigms that have given compass to mankind's effort to understand biological complexity; (b) consider how these traditional paradigms translate into the ongoing scientific revolution in genomics; and (c) emphasize ramifications of an emerging gene-centric view of the sexual genome that departs rather dramatically from all biological viewpoints that had gone before. Readers wishing to pursue these topics in greater depth should consult two of the author's books (Avise 1998, 2010a).

J.C. Avise (✉)
Department of Ecology and Evolutionary Biology, University of California,
Irvine, CA 92697, USA
e-mail: javise@uci.edu

K. Kampourakis (ed.), *The Philosophy of Biology: A Companion for Educators*, History, Philosophy and Theory of the Life Sciences 1, DOI 10.1007/978-94-007-6537-5_13,

Table 1 A few examples of 'paradigms lost' through startling discoveries about the genome[a]

(a)	The Central Dogma (that information flows only from DNA → RNA → protein) proved to be violated routinely with the discovery of reverse transcriptases that catalyze RNA → DNA.
(b)	Conventional wisdom that only proteins can be organic catalysts was proved wrong when RNA catalysts (ribozymes) were discovered.
(c)	The notion that most of the genome codes polypeptides was dismantled with the discovery of introns.
(d)	A standard metaphor of genes being arranged like tight beads along each chromosomal string was abandoned after geneticists came to appreciate the abundance of repetitive DNA.
(e)	The notion that all DNA in sexual species is routinely subject to recombination had to be revised with the elucidation of transmission genetics in cytoplasmic genomes such as mitochondrial DNA.
(f)	The notion that an allele was structurally autonomous had to be revised after the discovery of the phenomena of gene conversion and concerted evolution.
(g)	The traditional view that genes collaborate for the collective good of the individual had to be abandoned with the insight that DNA sequences in effect often behave selfishly.
(h)	Traditional concepts of gene homology had to be modified to accommodate the reality of different types of shared ancestry such as paralogy versus orthology.
(i)	The conventional use of particular DNA sequences to reconstruct a species phylogeny had to be revised when an appreciation was gained of a fundamental distinction in sexual species between gene trees and organismal phylogenies.
(j)	Conventional thought about the stationarity of DNA sequences was thrown out following the discovery of ubiquitous mobile elements.
(k)	Heritable changes in gene expression not attributable to alterations of DNA sequence per se have opened biologists' eyes to an array of previously underappreciated "epigenetic" phenomena.
(l)	Many examples of horizontal gene transfer across species have challenged the conventional wisdom that phylogenies invariably can be depicted as non-reticulate branched trees.

[a]These and countless lesser revelations in the evolutionary-genomic sciences might be viewed as inspirational or disturbing depending in part on each person's educational outlook. Science is not well suited for those who demand eternal truths

2 Religion and Science Before Genetics

2.1 *Natural Theology: The Argument from Design*

In the thirteenth century, the Dominican scholar Saint Thomas Aquinas invoked the "argument from design" (design implies a designer) as one of his Five Ways to prove the existence of a Creator. The design argument has a long and distinguished intellectual pedigree in science and philosophy. For example, in the fourth century BCE the Greek mathematician and philosopher Plato mused at length about how a conscious power must be responsible for organic complexity, and across the ensuing two millennia countless scientists and theologians expressly sought to glorify God's handiwork through their studies of nature. Late in the seventeenth century, the Christian naturalist John Ray encapsulated this approach in *The Wisdom of God* (1691), which in effect became a prequel to minister William Paley's *Natural*

Theology (1802) that further elaborated what many laypeople and intellectuals long had deemed to be self-evident: that nature offers powerful testimony to God's omnipotence (see also Lennox and Kampourakis, this volume). In his youth, Darwin apparently read both Ray and Paley (Barrett et al. 1987; Birkhead 2010) and became "charmed and convinced of the long line of argumentation" (see Darwin 1887) for intelligent design in their two eloquent treatises. Thus when Darwin boarded the H.M.S. Beagle in 1831, he too was a natural theologian at heart.

Just 2 years earlier (in 1829), the last will and testament of Reverend Francis Henry (Earl of Bridgewater) had directed the president of the Royal Society of London to identify and fund authors willing to write books "on the power, wisdom, and goodness of God as manifested in Creation." Between 1833 and 1840, Henry's directive eventuated in eight works (some in multiple volumes) that became known as the Bridgewater Treatises: *The Moral and Intellectual Constitution of Man* (Chalmers); *Chemistry, Meteorology, and the Function of Digestion* (Prout); *The History, Habits, and Instincts of Animals* (Kirby); *The Hand* (Bell); *Geology and Mineralogy* (Buckland); *The Physical Condition of Man* (Kidd); *Astronomy and General Physics* (Whewell); and *Animal and Vegetable Physiology* (Roget). These books now are remembered not for their scientific merit but rather because they were the last grand endeavors from natural theologians prior to the Darwinian revolution.

2.2 Theodicy and the Counterargument to Sentient Design

A conceptual dilemma for natural theologians always has been how to vindicate a God who would craft a biological world that is rife with imperfection as well as beauty. Therein lies the theodicic challenge, which refers to the difficulty of rationalizing the actions of a well-meaning omnipotent Deity who created grossly defective biological objects. Theodicy (from the Greek roots *theós* for God and *diki* for justice) also has a lengthy pedigree. The word was coined in 1710 by the German philosopher and mathematician Gottfried Leibniz in *Theodicic Essays on the Benevolence of God, the Free Will of Man, and the Origin of Evil*, but Leibniz certainly was neither the first nor last to wrestle with "the problem of evil". Throughout the ages, people have pondered why a caring all-powerful God countenances grotesque biological flaws and permits so much organismal (including human) suffering. In *Dialogues Concerning Natural Religion* (1779), the Scottish philosopher-historian David Hume managed to précis both the argument from design (natural theology) and its antipode (theodicy) in a pithy exchange between two fictional characters, Cleanthes and Philo. First, Cleanthes proclaimed, "the Author of Nature is somewhat similar to the mind of man though possessed of much larger faculties proportioned to the grandeur of the work he has executed [...] By this argument alone, do we prove at once the existence of a Deity"; to which Philo simply responded, "What surprise must we entertain when we find him a stupid mechanic." (part VI, passages 166 and 167).

Hume could not have anticipated that within a century, the age-old dilemma registered in the theology-theodicy dialogue would be rendered largely moot by the Darwinian revolution.

2.3 Scientific Challenges to Natural Theology

In 1514, Poland-born Nicolas Copernicus began to question prevailing anthropocentric interpretations of the universe. Using mathematical arguments based on detailed observations of planetary movements, Copernicus concluded that Earth and several other celestial bodies revolve around the sun and thereby constitute a heliocentric system (perhaps one of many in the cosmos). No longer could the Earth and its inhabitants be viewed quite so comfortably as the focus of all Creation. The importance of the Copernican revolution lay not in the proof of heliocentrism but rather in the introduction of a powerful but then-radical epistemology not shackled to sacred texts or religious revelations. For nearly the first time, the sciences (in this case physics and astronomy) had begun to wrest from theologians some substantive measure of intellectual authority regarding nature's mechanistic operations. More than three centuries later, Charles Darwin would extend the scientific ethos into biological arenas by showing that natural forces amenable to objective analysis had shaped organismal phenotypes (morphologies, physiologies, and behaviors) that traditionally had been ascribed to supernatural agencies. The Darwinian revolution went on to transform the life sciences in analogous fashion to how the Copernican revolution had transformed the physical sciences (see also Ayala, this volume).

An Augustinian friar and contemporary of Darwin was the second most important figure in the history of biology. Gregor Mendel (1865) discovered the particulate nature of hereditary factors [but see the historian Olby (1985) for a different interpretation] that later (in 1909) would be named genes. Mendel's scientific breakthroughs in the mid-1800s went unappreciated during his lifetime but they provided the other half of the puzzle that during the first half of the twentieth century enabled a union of Darwinian and Mendelian principles (Provine 1971) into a so-called "modern synthesis" (now a bit worn and tattered and about the edges) that still provides the foundation for much of the biological sciences (see also Depew, this volume).

Not everyone was persuaded by the science, of course, and many people even today view Darwinism as anathema. In some countries including the United States, religious disapproval or even wholesale rejection of evolutionary thought is a huge stumbling block against incorporating evolutionary science into biology curricula (Scott 2004). This situation is sadly ironic in at least two educational regards: (a) evolution is a core science without which "nothing in biology [otherwise] makes sense" (Dobzhansky 1973); and (b) Darwin's discovery of natural selection could be deemed a tremendous blessing in disguise for religion because it removes the need to explain the world's imperfections as failed outcomes of God's design (Ayala 2007).

Thus, when fundamentalists preach that evolution and religion are incompatible, they ignore the more uplifting possibility that the evolutionary sciences could partner with religion in mankind's broader struggle to understand the ultimate nature of nature. Indeed, in some respects evolution might even be theology's salvation. Rather than blaspheme God for shoddy engineering, theologians after Darwin and Mendel could put the proximate blame for biological flaws on insentient natural selection and hereditary mechanisms. No longer need priests, ministers, and clerics agonize why a Creator God is the world's leading abortionist and mass murderer, nor question God's motives for debilitating innocents with horrific disabilities, nor anguish about the motives of an interventionist Deity who permits so much evil and suffering in His biological flocks. Evolution by natural causes can emancipate religion from many such theodicic dilemmas. Darwin simply discovered a scientifically decipherable process of nature (natural selection) that seems to eliminate the need to invoke direct supernatural intervention for apparent organismal design. Whereas many fundamentalists reject Darwinian notions as heretical, many scientists and scholars in the evolutionary-genetic era have welcomed the opportunity to explore uncharted waters that lie in the traditional gulf between science and religion and that thereby lap the shores of both (see also Ayala this volume; Alexander this volume).

There are many additional reasons why science and religion need not be archenemies. For example, even if all biological outcomes proved to be fully consistent with natural laws and intelligible processes amenable to scientific scrutiny, a non-excluded theological interpretation is that nature's ground-rules somehow were set into motion by a God. In 1973, the evolutionary geneticist Theodosius Dobzhansky issued a clear deistic statement: "I am a creationist *and* an evolutionist. Evolution is God's, or Nature's method of creation." This is the sort of deity that Albert Einstein tried to comprehend in his explorations of energy and matter, and it is clearly the kind of God that Darwin (1859) had in mind when he mentioned the Creator in the closing paragraph of the second edition of *The Origin of Species*. When scientists explore the nature of life including the human condition, they are not necessarily atheistic but they do strive to avoid the subjective and metaphysical explanations of theism by focusing instead on hypotheses that can be analyzed dispassionately and tested critically.

3 Science and Religion in the Genetics Era

3.1 Natural Theology Revisited

On the theological front, recent decades have been witness to the birth and growth of the Intelligent Design (ID), the latest reincarnation of religious creationism and natural theology (Numbers 2006). Proponents of ID insist that complex biological outcomes such as bacterial cells and humans beings offer incontrovertible evidence for purposeful design and direct craftsmanship by a supernatural force, and they often go further to claim that such arguments are based in science rather than faith or revelation. In the United States, ID advocates have been plaintiffs in several high-profile courtroom cases (see NAS 2008) that to date have ruled that ID is a religious movement without scientific merit and as such should not be mandated equal time in science classrooms of public schools. Few biologists in the modern genetic era openly subscribe to natural theology ala Ray or Paley, but one notable exception is biochemist Michael Behe (1996) who in *Darwin's Black Box* issued a challenge to evolution by arguing that complicated biotic traits such as the vertebrate eye, the bacterial flagellum, or the sexual genome are "irreducibly complex" and could only have been constructed *ex nihilo* for their current functions by an intelligent engineer who in effect must be an interventionist Creator God (see Brigandt this volume).

3.2 Theodicy Revisited

On the scientific front, genetic findings in recent decades have extended the age-old theodicy dilemma to previously unexplored inner workings of the cell. Biologists now know that despite its many intricate features, the genome of humans (like those of other sexual species) also is riddled with structural and operational deficiencies ranging from the subtle to the egregious (Table 2).

These molecular defects register not only as deleterious mutational departures from some hypothetical genomic ideal but also as universal architectural flaws. Thus, whereas the theodicic challenge traditionally arose in the context of observable phenotypes such as serious disease conditions and overt medical disabilities that plague humanity, in times past a theological rejoinder always could be entertained that God's handiwork someday might be found in the finer details of human biological operations. However, now that human genomes have been dissected in astounding detail (Lander et al. 2001; Venter et al. 2001), that possibility is no longer scientifically tenable. Furthermore, rampant imperfection inside the human genome is hardly unexpected if indeed non-sentient evolutionary processes generally are in charge of biological outcomes (Table 3). Thus contrary to Behe (1996), molecular findings offer only a Lilliputian challenge to evolution when compared to their Gargantuan challenge to intelligent design.

Again, however, the more important point is that evolutionary-genetic science interpreted properly could in principle help open-minded theologians escape the conundrums of ID and thereby return religion to its rightful realm—not as the interpreter of the biological minutiae in our physical existence (as Behe seems to wish), but rather as a philosophical counselor on much grander matters such as ethics, morality, the soul, spirituality, sacredness, and other such issues of "ultimate concern" to nearly everyone (Dobzhansky 1967). For this reason too, it seems logical that evolution should be welcomed not only into houses of worship but also into academic curricula of the humanities as well as the biological sciences.

Table 2 Some of the many lines of evidence (often overlapping) for suboptimal design in sexual genomes, including those of humans

(a)	Mutations with oft-disastrous consequences for a person's health arise de novo in protein-coding genes as well as in essentially all other classes of loci.
(b)	Many older mutations with deleterious effects continue to segregate in populations long after their evolutionary origin.
(c)	Much of the molecular complexity of the human genome is gratuitous rather than functionally effective.
(d)	Much of the molecular complexity of the human genome is astonishingly wasteful of cellular resources.
(e)	Breakdowns in gene regulation routinely underlie genetic disabilities ranging from inborn errors of metabolism to various cancers acquired during an individual's lifetime.
(f)	Malfunctions in gene-based energy metabolism often underlie physiological deterioration.
(g)	Many DNA sequences ranging from short to lengthy routinely proliferate at organismal expense or at best at organismal indifference.
(h)	Many genes are deceased in the sense that they no longer perform an active function for the cell.
(i)	Even when not overtly deleterious, many alternative alleles scattered throughout the genome are selectively neutral or nearly so.
(j)	Genomes themselves in effect recognize that they are flawed, as gauged by the fact that they have evolved sophisticated yet far-from-infallible repair apparatuses.
(k)	Many genomic features go well beyond merely poor design and into the realm of downright ludicrous design by almost any engineering standard.
(l)	Many suggestions for improvement of genomic design can be imagined readily even by mere mortals with just a modicum of intelligence.
(m)	Genetic disabilities strike even the most innocent and helpless among us, and indeed are especially likely to target embryos, fetuses, and the elderly.
(n)	Everyone is afflicted with at least some genetic ailments at one stage or another of life.
(o)	Senescence and death themselves are inevitable and have evolutionary-genetic etiologies.

See Avise (2010a) for details and elaboration

3.3 Natural Selection: The New Holy Grail

Given that natural selection in the modern era in some ways became a surrogate for God's hand as a proximate sculptor of biological design, perhaps it is not too surprising that scientists ever since Darwin have embarked on missions to understand the operation of natural selection with a zeal almost reminiscent of the fervor with which natural theologians had pursued their earlier quest to understand the mind of God. For example, much of the modern synthesis involved developing (Fisher 1930; Wright 1931; Haldane 1932) and then translating (Dobzhansky 1951) mathematical models that describe how natural selection operates in conjunction with other evolutionary processes to effect genetic changes in populations through time (see Depew this volume). Population-genetic theories remain essential for interpreting molecular data, including those that later would emerge from detailed analyses of

Table 3 Some reasons why evolution often yields sub-optimal biological outcomes

(a)	Natural selection is a non-sentient natural process, as uncaring and dispassionate as gravity.
(b)	Natural selection is not all-powerful, but instead is just one in a nexus of evolutionary forces, others of which can override the adaptation-promoting power of natural selection in particular instances and thereby yield products that fall far short of designer perfection.
(c)	Random mutations continually arise, most of which are either deleterious or fitness-neutral.
(d)	Harmful mutations (especially those that are only slightly deleterious individually) often fly below the radar screen of purifying natural selection, especially in small populations.
(e)	Genetic drift can alter the genetic composition of populations in ways that are uncorrelated with adaptive benefits.
(f)	Sexual selection on particular traits often operates in direct opposition to natural selection.
(g)	Genetic correlations and conflicts are common such that deleterious alleles linked to host-beneficial alleles at other loci can hitchhike with the favorable alleles and thereby at least temporarily escape eradication by purifying selection.
(h)	Pleiotropy and fitness tradeoffs are common, meaning that a genotype often has multiple phenotypic consequences some of which benefit and others may harm the organism.
(i)	Natural selection acts not only at the organismal level but also at the level of genes, so selfish DNA sequences can persist and proliferate in a sexual genome without enhancing the wellbeing of the host population.
(j)	Phylogenetic constraints are ubiquitous and natural selection at any point in time can only work with the genetic diversity presented by lineages that have survived from the past.

After Avise (2010b)

sexual genomes (Lynch 2007). Indeed, ever since molecular technologies were introduced to population biology in the mid-1960s (see Avise 2006), the fields of molecular ecology and evolution have been preoccupied with elucidating the multifarious roles of selection in shaping "the genetic basis of evolutionary change" (Lewontin 1974).

Interestingly, however, an audacious theory that arose near the dawn of the molecular revolution focused on natural selection's scientific antithesis. Architects of the neutrality theory (Kimura 1968; King and Jukes 1969; Kimura and Ohta 1971) proposed that most sequence changes at the level of DNA are attributable not to natural selection but rather to the random fixation of fitness-neutral (selectively equivalent) genetic variants that arise via mutational processes (see Dietrich this volume). Across the ensuing decades, the neutrality-selection controversy has resurfaced time and again as biologists contemplated each new type of molecular data provided by the latest laboratory method. Research always proceeded on two fronts: testing mathematical or statistical predictions of neutrality theory against observed magnitudes or patterns of molecular variation; and addressing functional properties of particular genotypic traits by critical observations or experiments.

Several points should be made clear. Neutralists did not deny the conclusive empirical evidence for high genetic variation at the molecular level but rather they questioned the selective relevance of such variability. They did not suggest that most genes are dispensable (of course they are not) but rather that different alleles at a locus often are functionally equivalent. Neutralists never denied that highly deleterious mutations tend to be eliminated or kept at low frequency by purifying selection but rather they focused on genetic variants that escape selective elimination. Finally, they did not challenge adaptive Darwinian evolution for organismal morphologies and behaviors but rather they explicitly confined their attention to variation at the molecular level (Kimura 1983).

Neutrality theory seemed radical when it was introduced in the late 1960s, but within two decades and continuing today it had become molecular evolution's gigantic null hypothesis: the most straightforward way to interpret molecular variability and the basic theoretical construct whose predictions must be falsified before invoking balancing or other forms of selection to explain a particular molecular polymorphism or evolutionary outcome. This is not to say that all molecular evolution is neutral; the truth undoubtedly resides somewhere between the extreme poles of pan-selectionism and pan-neutrality.

3.4 Levels of Natural Selection

Another recurring scientific debate during the twentieth century addressed the question: At what level or domain does natural selection operate with greatest efficacy? Although the notion of group selection occasionally gained some traction (Wynne-Edwards 1962), most evolutionary biologists remain convinced by the Darwinian argument that natural selection normally acts not for the benefit of an extended group or species but rather via fitness differences among individuals (Williams 1992). In 1964, William Hamilton introduced a more palatable adjustment to individual selection (or perhaps to group selection) when he advanced the concept of kin selection based on inclusive fitness. Whereas genetic fitness traditionally had been defined as an individual's personal reproductive success, inclusive fitness incorporated the novel insight that copies of an individual's genes may be transmitted indirectly through genetic kin as well as directly through genetic parenthood. Kin selection is now accepted as a plausible route for the evolution of at least some otherwise enigmatic traits (such as extreme sociality in ants and other hymenopteran insects). Hamilton's spotlight on the fate of replicate copies of genes also offers a useful segue here into another biological domain for selection: the differential proliferation of particular DNA sequences inside sexual genomes.

4 Neo-Darwinian Selection at the Level of Genes

In any sexual species, unlinked genes have quasi-independent evolutionary trajectories because the Mendelian processes of segregation and independent assortment (due to meiosis and syngamy) tend to shuffle alleles during each round of organismal reproduction, thereby partially randomizing genetic associations across loci and affording little opportunity for any allele to establish stable long-term relationships with particular compatriots. Yet the routine dissolution of potential cliques of genes would seem to run counter to the desirability of evolving coadapted genic alliances that might be of benefit to the host organism. Ergo the longstanding evolutionary question (Turner 1967; Maynard Smith 1977): Why does the sexual genome not congeal? This enigma is tantamount to the monumental puzzle of why sex (as opposed to asexuality or parthenogenesis) is so prevalent in much of the biological world. Standard evolutionary answers appeal to various fitness advantages that arise via the genetic variety and adaptive flexibility that genetic recombination promotes (Bell 1982; Maynard Smith 1978).

4.1 Consequences for Genomic Architecture and Operations

Sexual reproduction also has major consequences for the level at which natural selection operates because recombination in effect decouples the fates of different DNA sequences within the sexual genome. As a result, genes in recombining genomes sometimes can increase their odds of survival and proliferation by acting in disharmony with the broader collective of genes (and hence against the interests of the genome and the host organism). Richard Dawkins (1976) elaborated and popularized the image of the "selfish gene," a concept that is now well ensconced in evolutionary thought. The realization that natural selection operating at the level of the gene can oppose natural selection operating at the level of the organism was a major conceptual breakthrough that has helped to clarify many otherwise enigmatic molecular features of sexual genomes. Indeed, if all forms of life forever had been strictly asexual, then genomes undoubtedly would be structured very differently than they are today because there would have been no evolutionary conflict of interest between loci, no conflict between what is best for the gene and what is best for the organism, and no opportunity for the evolution of selfish genetic elements.

Sexual genomes, however, are riddled with evidence for selfish DNA. Perhaps the most compelling testimony comes from the ubiquity and abundance of mobile elements (also known as jumping genes) that have proliferated to great numbers in the genomes of nearly all plant and animal species. These DNA sequences evolved the capacity to produce and distribute copies of themselves across multiple chromosomal locations within a cell lineage, thereby enhancing their prospects for transmission across the generations and eventually accumulating to vast numbers in most sexual genomes. In the human genome, for example, active or deceased mobile elements outnumber functional protein-coding genes by approximately 100 to 1 and altogether constitute at least 45 % of our DNA.

Researchers distinguish several categories of jumping genes. Some are cut-and-paste elements that move by excision and insertion of DNA whereas others locomote via a copy-and-paste mechanism of reverse transcription from RNA intermediates. Within the latter category are several subclasses (including *LINE*s, *SINE*s, and *LTR*s) each of which in turn is composed of subfamilies of elements. For example, in humans *L1* is the largest subclass of *LINE*s, with each intact *L1* element being approximately 6,000 base pairs long and with much smaller pieces of more than 500,000 such elements collectively comprising about 17 % of the human genome; and *Alu* sequences constitute a major subcategory the *SINE*s, with each *Alu* sequence being about 300 base pairs long and with more than one million copies of *Alu* cluttering the human genome.

Many mobile elements house genes that encode proteins for reverse transcription and integration into a new chromosomal site whereas others hijack the necessary enzymes from other mobile elements or from the host genome. Mobile element replication can be a sloppy molecular process, so many jumping genes have lost bits and pieces that compromise their competency to code for the proteins that once enabled their own intra-genomic movements. Degenerate mobile elements are called non-autonomous elements, in contradistinction to active mobile elements that proliferate within the host genome much like quasi-autonomous and self-serving intracellular parasites (Hickey 1982).

In addition to being parasitic on cellular resources of the host organism, any newly arisen mobile element may cause serious harm by several mechanisms (e.g. Cordaux et al. 2006): by landing in an exon in which case it can ruin the protein encoded by a functional gene; by jumping into an intron-exon boundary in which case it can alter how RNA is processed; by inserting into a regulatory region in which case it also can disrupt proper gene expression; or by inserting almost anywhere in which case it can cause genetic instabilities including deletions of useful parts of the host genome. Apart from such examples of "insertional" mutagenesis, old as well as new mobile elements often promote genomic disruptions via non-allelic homologous recombination that also can disrupt cellular operations and result in serious metabolic disorders (Lupski 1998).

In short, mobile elements routinely invite and trigger genetic disasters for their hosts. For example, some expressions of heart disease, colon cancer, breast cancer, hemophilia, diabetes, and numerous other life-threatening conditions in humans are known to result in various instances from the activities of mobile elements (Prak and Kazazian 2000; Hedges and Deininger 2007). Although most data on mobile elements in humans have been acquired only recently, the list of serious metabolic disorders associated with these proliferate DNA sequences already is long and rapidly growing (see Avise 2010a). Still, any such list provides only a minimal estimate of the collective toll of jumping genes on human health because most of the serious medical difficulties undoubtedly arise so early in embryonic life as to cause miscarriages of undocumented etiology. Indeed, most mobile elements are especially active in cells of the germline, so many of their deleterious effects probably register in gametic deaths and lowered fertility.

From an evolutionary perspective, the ubiquity of transposable elements relates to their spreading nature, some of which reflects multiple historical invasions of germlines by infectious retroviruses but much of which also reflects each element's subsequent selfish proliferation within the genome. In any sexual species, a DNA sequence that gains a capacity to disperse copies of itself across chromosomal sites in germline cells almost inevitably enhances its prospects for passage to succeeding generations, even when the element harms its host. If the host instead were asexual, a mobile element would gain no transmission advantage by dispersing copies of itself across a clonal genome. Thus, mobile elements also can be regarded as sexually transmitted genomic diseases.

As with any host/parasite association, evolutionary games of give-and-take are played across time. Namely, host genomes come under selection to evolve mechanisms that silence or suppress any harmful mobile element activities, and selfish mobile elements are under selection to avoid such strictures. Also, some degree of self-policing by mobile elements might be expected because it is not in the element's selfish interest to harm its host. The net long-term result of such co-evolutionary contests is likely to be a truce or balance wherein mobile elements and their host genomes manage to live together with varying mixes of amicability and hostility.

Sometimes host organisms can even profit from mobile elements in at least two ways. First, mobile elements are powerful mutagenic agents (as mentioned earlier), and mutations are the ultimate source of genetic variation that is necessary for continued evolution. However, it is doubtful that jumping genes evolved expressly for their mutagenic behavior because natural selection lacks foresight (and also because many if not most random mutations are deleterious). Second, host genomes occasionally convert parasitic DNA into host-beneficial functions. Many mobile elements carry DNA sequences that have the capacity to regulate gene expression, so host genomes sometimes manage to capitalize upon ("exapt" or capture) the regulatory potential latent in mobile elements for the cell's own purposes. Such arrangements also can benefit the mobile element directly, which after its functional conversion then experiences selection for evolutionary maintenance as an integral part of the host genome.

The notion that the function of a trait often shifts during evolution was well appreciated by Darwin, but it was not until much later that the terms "exaptation", "cooptation", (Gould and Vrba 1982) and "co-opted adaptation" (Buss et al. 1998) were introduced to encapsulate the sentiment that a character shaped by natural selection for one adaptive function might later assume an altogether different role. Thus, even if it proves to be true that many mobile elements in the human genome now play useful roles in cellular operations, this does not necessarily mean that they evolved to perform these functional tasks from the outset. Instead, they might be exaptations that originated as selfish genetic elements but later were secondarily captured into host-beneficial services (for the concept of adaptation see Forber this volume; for the concept of function see Wouters this volume).

Although mobile elements clearly have been important evolutionary drivers of the sexual genome, they are merely the most conspicuous among many structural and operational properties of sexual genomes that motivate modern evolutionary thought about "genes in conflict" (Burt and Trivers 2006). Table 4 outlines several other recently discovered categories of selfish DNA.

4.2 Evolving Genomic Metaphors

Metaphors can be powerful images that capture complex ideas and educate audiences to new (and old) ways of thinking (Keller 2002), and perhaps nowhere has

Table 4 Several additional lines of evidence (apart from mobile elements described in the text) for genetic conflict within the sexual genome

(a)	Autosomal killers: Known in various animals, fungi, and plants, these are nuclear genetic elements that kill or disable gametes that do not carry copies of themselves, thereby eliminating potential competitors for successful fertilization events.
(b)	Selfish sex chromosomes: These are sex-linked genetic elements that distort the segregation process in ways that bias in favor of their own transmission to the next organismal generation.
(c)	Genomic imprinting: This is a common phenomenon in mammals and plants wherein a gene involved in an embryo's acquisition of maternal resources is expressed in progeny when inherited from one parent but not from the other. Such imprinting appears to be the evolutionary outcome of genetic conflict between the effects of natural selection on genes expressed in fetuses opposed by the effects of natural selection on genes expressed in mothers (Haig 1993).
(d)	Selfish mitochondrial DNA: Because mtDNA normally is inherited maternally rather than biparentally, it often plays by different evolutionary ground rules than those for autosomal genes and accordingly has evolved several properties and tactics (such as promoting male sterility) that bring it into conflict with the consensus interests of nuclear genes.
(e)	Biased gene conversion: These are recently discovered molecular processes (often related to DNA repair) by which particular DNA sequences convert alternative sequences or alleles to their own structure.
(f)	Female drive: This is a specific form of selfish genetic behavior in which a gene or chromosome engaged in meiosis in females disproportionately inserts itself (at the expense of competitors) into a fertilizable egg or ovule as opposed to the soon-to-be discarded polar bodies.
(g)	B chromosomes: These are additional chromosomes (distinguished from the normal or "A" chromosomal set) that are not a necessary part of the genome but nonetheless are maintained in populations because they possess "self-accumulation" mechanisms.
(h)	Genomic exclusion: This is a category of drive-like genetic situations in which individuals discard (rather than transmit to offspring) some or all of the genes inherited from one parent, thereby giving a huge selective advantage to genes from the alternative parent and once again fostering conflicts within the genome.

See Burt and Trivers (2006) for elaboration and for more examples

that been more evident than in the field of evolutionary genetics (Condit 1999; Brandt 2005). The sciences of heredity and evolution are steeped in evocative metaphors (Table 5), so much so that it often becomes difficult to distinguish simile from facsimile, or to disentangle what is real from our perception of what reality might be. Indeed, the most effective metaphors in science in effect become immortalized when they get incorporated so fully into conventional wisdom and language as to be accepted as the actual truth rather than as utilitarian caricatures of whatever reality might be. Furthermore, metaphors themselves can and do evolve under the force of new evidence.

For example, one traditional set of metaphors pictured each genome as carefully coded text in a book of life that was scripted (either by natural selection or perhaps by God) expressly for organismal wellbeing. But recent molecular discoveries about the sexual genome have challenged this standard textual imagery in many ways. Contrary to earlier notions, DNA sequences are not always like intelligible words

Table 5 Examples of the many metaphors that are used routinely in genetics and evolutionary biology[a]

Evolutionary and population-genetic metaphors
Tree of life
DNA as hereditary sap in the tree of life
Phylogenetic branches
Evolutionary pathways
Fabric of life
Web of life
Life's evolutionary ladder
Streams of heredity
Gene pool
Gene flow
Genetic drift
Genetic draft or hitchhiking
Gene or genomic metaphors
Textual metaphors (e.g., genetic book of nature; encyclopedia of life; genetic code; genetic sentences; the genetic alphabet and its letters; the language of DNA; genetic text or scripture).
Genetic engineering, biotechnology
Germplasm, tissue banks, gene banks
Biological atlases (or maps, cookbooks, recipes, instructions, operating manuals)
Informational or occupational metaphors (e.g., DNA blueprints; genetic programs; gene circuits; gene batteries; housekeeping genes; jumping genes; selfish DNA; junk DNA; parasitic DNA; developmental switches; ontogenetic pathways).

For any such metaphor, a useful classroom exercise for students and teachers alike is to address the following types of questions: (a) how and when did the metaphor originate?; (b) how well does it capture our current understanding of biological reality?; (c) to what extent is it helpful or misleading?; (d) might it promote or inhibit further scientific inquiry? and (e) can you think of a different or perhaps better metaphor for the topic in question?

[a]Nelkin (2001) categorized contemporary genetic metaphors in popular discourse into four groups centered on notions of: essentialism (genes as essences of personal identity); religiosity (genes as sacred entities); fatalism (genes as determinants of destiny) and commerciality (genes as commodities)

tightly woven into coherent chromosomal sentences and paragraphs, all working smoothly and collaboratively to translate genetic code in life's instruction manual into meaningful cellular operations. Instead, protein-coding exons and other functional DNA sequences now seem like small islands suspended in rivers of intronic and extra-genic gibberish that flow through each species' hereditary channels. The precise volume of genomic flotsam (junk DNA, selfish DNA, parasitic DNA, or neutral DNA) mixed with host-beneficial DNA in various evolutionary watersheds remains to be determined by monitoring the ebb and flow of genomic operations in each species. Whole-genome sequencing (which has recently become almost routine) is merely a first step toward such characterizations. Far more difficult is the laborious follow-up challenge of genomic annotation—assigning functional roles or otherwise attaching useful biological information to each gene or other segment of genomic sequence (Stein 2001). Genomic annotation (characterizing the genome functionally rather than just structurally) in humans and other model and non-model species is an immense enterprise that will keep geneticists fully occupied for the foreseeable future (see Marcos and Arp this volume for information in biology; see Burian and Kampourakis this volume for a proposal to replace gene concepts).

To help motivate such laborious genomic analyses as well as to tie them to evolutionary thinking, lively new metaphors might be useful too (Avise 2001). One such adaptable genomic metaphor would liken each sexual genome to a community of genes whose behaviors mirror those of humans entangled in a network of social arrangements. These behaviors would include cheating and arbitration (conflict resolution) as well as various expressions of cooperation for the common organismal good. Another such metaphor might envision each sexual genome as a miniature intracellular ecosystem extended through time, complete with different niches for genes that act much like parasites, symbionts, commensals, agents of disease, or that assume other roles traditionally reserved for organisms and species in natural macroscopic biological communities.

5 Conclusions

The field of evolutionary genetics lies at a unique crossroads between science, religion, philosophy, and education. Not only have organisms and their genomes evolved through time but so too have human perceptions about the etiology of biological design. Whereas pre-Darwinian religious philosophers typically focused on natural theology and its theodicic challenges, later scientific perspectives based on Darwinian and Mendelian principles in effect removed the requirement for divine intervention to justify biological outcomes including the complex molecular architectures of sexual genomes. In recent years, a social movement known as intelligent design (ID) has reintroduced natural theology in a form that needlessly seems to pit science against religion over matters of ultimate concern (such as the origins of sophisticated biological traits). Although ID (like traditional creationism) is strictly a religious movement, it has had negative impacts on science education in the United States (and elsewhere) through its overt hostility to evolutionary principles that otherwise provide a unifying conceptual foundation for all of the biological (including medical) sciences. Furthermore, within the scientific arena itself, several genetic controversies continue to swirl regarding precisely how selective processes shape genomic structure and genomic operations. The history of the science-religion interface as well as the ongoing trajectory of evolving scientific notions about the sexual genome both provide rich educational material for curricula in the humanities and in the biological sciences.

Acknowledgments The author's research and writing are supported by funds from the University of California at Irvine.

References

Avise, J.C. 1998. *The genetic Gods: Evolution and belief in human affairs*. Cambridge, MA: Harvard University Press.

Avise, J.C. 2001. Evolving genomic metaphors: A new look at the language of DNA. *Science* 294: 86–87.

Avise, J.C. 2006. *Molecular markers, natural history, and evolution*, 2nd ed. Sunderland: Sinauer.

Avise, J.C. 2010a. *Inside the human genome: A case for non-intelligent design*. Oxford: Oxford University Press.

Avise, J.C. 2010b. Footprints of non-sentient design inside the human genome. *Proceedings of National Academy of Sciences of the United States of America* 107: 8969–8976.

Ayala, F.J. 2007. *Darwin's gift to science and religion*. Washington, DC: Joseph Henry Press.

Barrett, P.H., P.J. Gautrey, S. Herbert, D. Kohn, and S.A. Smith. 1987. *Charles Darwin's notebooks, 1836–1844*. Cambridge: Cambridge University Press.

Behe, M. 1996. *Darwin's black box: The biochemical challenge to evolution*. New York: Free Press.

Bell, G. 1982. *The masterpiece of nature: The evolution and genetics of sexuality*. Berkeley: University of California Press.

Birkhead, T.R. 2010. How stupid not to have thought of that: Post-copulatory sexual selection. *Journal of Zoology* 281: 78–93.

Brandt, C. 2005. Genetic code, text, and scripture: Metaphors and narration in German molecular biology. *Science in Context* 18: 629–648.

Burt, A., and R. Trivers. 2006. *Genes in conflict: The biology of selfish genetic elements*. Cambridge, MA: Harvard University Press.

Buss, D.M., M.G. Haselton, T.K. Shackelford, L. Bleske, and J.C. Wakefield. 1998. Adaptations, exaptations, and spandrels. *The American Psychologist* 53: 533–548.

Condit, C.M. 1999. *The meaning of the gene*. Madison: University of Wisconsin Press.

Cordaux, R., R.J. Hedges, S.W. Herke, and M.A. Batzer. 2006. Estimating the retrotransposition rate of human *Alu* elements. *Gene* 373: 134–137.

Darwin, C.D. 1859. *On the origin of species by means of natural selection*. London: Murray.

Darwin, F (Ed.). 1887. *Charles Darwin's autobiography*. New York: Basic Books 1959 Edition.

Dawkins, R. 1976. *The selfish gene*. Oxford: Oxford University Press.

Dobzhansky, T. 1951. *Genetics and the origin of species*, 3rd ed. New York: Columbia University Press.

Dobzhansky, T. 1967. *The biology of ultimate concern*. New York: New American Library.

Dobzhansky, T. 1973. Nothing in biology makes sense except in the light of evolution. *American Biology Teacher* 35: 125–129.

Fisher, R.A. 1930. *The genetical theory of natural selection*. Oxford: Clarendon Press.

Gould, S.J., and E.S. Vrba. 1982. Exaptation—a missing term in the science of form. *Paleobiology* 8: 4–15.

Haig, D. 1993. Genetic conflicts in human pregnancy. *The Quarterly Review of Biology* 68: 495–532.

Haldane, J.B.S. 1932. *The causes of evolution*. London: Longmans & Green.

Hamilton, W.D. 1964. The genetical evolution of social behavior. *Journal of Theoretical Biology* 7: 1–52.

Hedges, D.J., and P.L. Deininger. 2007. Inviting instability; transposable elements, double-strand breaks, and the maintenance of genomic integrity. *Mutation Research* 616: 46–59.

Hickey, D.A. 1982. Selfish DNA: A sexually-transmitted nuclear parasite. *Genetics* 101: 519–531.

Keller, E.F. 2002. *Making sense of life: Explaining biological development with models, metaphors, and machines*. Cambridge, MA: Harvard University Press.

Kimura, M. 1968. Evolutionary rate at the molecular level. *Nature* 217: 624–626.

Kimura, M. 1983. *The neutral theory of molecular evolution*. Cambridge: Cambridge University Press.

Kimura, M., and T. Ohta. 1971. *Theoretical aspects of population genetics*. Princeton: Princeton University Press.

King, J.L., and T.H. Jukes. 1969. Non-Darwinian evolution: Random fixation of selectively neutral mutations. *Science* 164: 788–798.

Lander, E.S., et al. 2001. Initial sequencing and analysis of the human genome. *Nature* 409: 860–921.

Lewontin, R.C. 1974. *The genetic basis of evolutionary change*. New York: Columbia University Press.

Lupski, J.R. 1998. Genomic disorders: Structural features of the genome can lead to DNA rearrangements and human disease traits. *Trends in Genetics* 14: 417–422.

Lynch, M. 2007. *The origins of genome architecture*. Sunderland: Sinauer.

Maynard Smith, J. 1977. Why does the genome not congeal? *Nature* 268: 693–696.

Maynard Smith, J. 1978. *The evolution of sex*. Cambridge: Cambridge University Press.

Mendel, G.J. 1865. Versuche ueber Pflanzenhybriden, Verhandlungen des Naturforschenden Vereins (Bruenn) 4:3–47.

NAS, National Academy of Sciences. 2008. *Science, evolution, and creationism*. Washington, DC: National Academies Press.

Nelkin, D. 2001. Molecular metaphors: The gene in popular discourse. *Nature Reviews Genetics* 2: 555–559.

Numbers, R. 2006. *The creationists: From scientific creationism to intelligent design*. Cambridge, MA: Harvard University Press.

Olby, R. 1985. *The origins of Mendelism*. Chicago: University of Chicago Press.

Paley, W. 1802. *Natural theology; or evidence of the existence and attributes of the deity, collected from the appearances of nature*. London: Fauldner.

Prak, E.T.L., and H.H. Kazazian. 2000. Mobile elements and the human genome. *Nature Review Genetics* 1: 134–144.

Provine, W.B. 1971. *The origins of theoretical population genetics*. Chicago: University of Chicago Press.

Ray, J. 1691. *The wisdom of god manifested in the works of creation*. London: Smith.

Scott, E.C. 2004. *Evolution versus creationism*. London: Greenwood Press.

Stein, L. 2001. Genome annotation: From sequence to biology. *Nature Review Genetics* 2: 493–503.

Turner, J.R. 1967. Why does the genome not congeal? *Evolution* 21: 645–656.

Venter, J.G., et al. 2001. The sequence of the human genome. *Science* 291: 1304–1351.

Williams, G.C. 1992. *Natural selection: Domains, levels, and challenges*. New York: Oxford University Press.

Wright, S. 1931. Evolution in Mendelian populations. *Genetics* 16: 97–159.

Wynne-Edwards, V.C. 1962. *Animal dispersion in relation to social behavior*. Edinburgh: Oliver and Boyd.

Non-genetic Inheritance and Evolution

Tobias Uller

> *I feel that the study of the history of a field is the best way of acquiring an understanding of its concepts. Only by going over the hard way by which these concepts were worked out – by learning all the earlier wrong assumptions that had to be refuted one by one, in other words by learning all past mistakes – can one hope to acquire a really thorough and sound understanding. In science one learns not only by one's own mistakes but also by the history of the mistakes of others.*
>
> Ernst Mayr (1982, p. 20)

1 Introduction

Ernst Mayr – one of the twentieth century's leading biologists and a major figure in the development of evolutionary biology – made this statement in his 1982 book *The Growth of Biological Thought* (Mayr 1982). Mayr's words are an important reminder that things that we consider obvious today emerged from scientific discoveries (and sometimes battles) of the past; little if any scientific knowledge should be taken for granted no matter how trivial it may seem to us now.

The quotation also suggests that it is helpful for both students and professional biologists if the concepts and theories that form the basis for our understanding of biology are put into the context of how the field has developed over time, including its cultural and social dimensions (Matthews 1994). This will help to make sense of the collection of facts and organise them into a coherent and robust theory (Lederman 2007). It is also arguably more fun to learn things if we understand why

T. Uller (✉)
Edward Grey Institute, Department of Zoology, University of Oxford, Oxford OX13PS, UK
e-mail: tobias.uller@zoo.ox.ac.uk

K. Kampourakis (ed.), *The Philosophy of Biology: A Companion for Educators*, History, Philosophy and Theory of the Life Sciences 1, DOI 10.1007/978-94-007-6537-5_14,

particular problems have stimulated the curiosity of biologists or how technical advances made new discoveries possible.

Finally, Mayr's statement should also be a reminder that by looking back we have an opportunity to see how our current understanding of a field has been shaped by the cumulative, but sometimes irregular, growth of knowledge. Indeed, one aim of this chapter is to show how the development of particular conceptual frameworks during the history of biology, to which Mayr himself made important contributions, now may hamper further progress. However, my main aim is to show how paying attention to the history and philosophy of biology will help us understand how the growing molecular, physiological, and behavioural evidence for non-genetic inheritance fits into evolutionary theory, and how teaching our students about these phenomena can contribute to their development of a deeper understanding of how evolution works.

2 The Emergence of the Modern Concept of Inheritance in Biology

The concept of inheritance or heredity is fundamental to biology. From a modern perspective, it may therefore come as a surprise to learn that the biological concept of heredity as the passing of traits from parents to offspring is a fairly recent innovation. Although people have observed that characters tend to run in families since ancient times (Stubbe 1972; Smith 2006), the biological notion of heredity as transmission of traits from parents to offspring did not fully emerge until the early nineteenth century (Amundson 2005; Müller-Wille and Rheinberger 2007, 2012). The new concept of heredity was made possible by the integration of several different attempts to conceptualize observed patterns of variation, including consistent differences between sheep breeds, the common occurrence of otherwise rare features in 'noble' families (such as the 'Habsburg jaw'), and the skin colour of babies born to parents of different colouration (Müller-Wille and Rheinberger 2007, 2012). These and other examples showed that hereditary dispositions were common not only among species, but also within species and populations. They also showed that both the mother and the father influence character expression in the offspring and contributed to the increasing conceptualization of biological heredity as involving transmission of particles from parents to offspring. However, this did not mean that heredity could be reduced to the transmission of those particles. Instead, throughout most of the nineteenth century heredity was seen as a process whereby similar forms arose in successive generations because of between-generation recurrence of the causes of growth and differentiation, i.e., development (Amundson 2005, pp. 140–142). This was also Darwin's view (Darwin 1868; Hodge 1985). He envisioned that particles ('gemmules') produced by all organs of both parents migrate to the germ cells and combined in the embryo, which formed the basis for the similarity of parents and offspring. Darwin and the vast majority of other biologists at the time thus allowed for acquired characters to influence the expression of traits in the subsequent generation. Heredity was a part of epigenesis, not a separate phenomenon from development as we teach biology today (see Amundson 2005 pp. 139–158 for a review).

The end of the nineteenth century saw several important contributions that dealt with inheritance (e.g., Darwin 1868; Galton 1876; Weismann 1893). Two developments in particular were important for the separation of development and heredity. The first was August Weismann's separation of the germ line from the rest of the body, the soma (Weismann 1893). Weismann reasoned that because embryonic development results in differentiation, each cell would come to contain only a subset of the initial 'determinants'. Thus, inheritance of a full set of units necessary for embryogenesis required this set to be put aside before development begins. This is the germ line, through which hereditary continuity is ensured. Weismann further postulated that changes in the soma would not affect the constitution of the germ plasm (based on what he saw as an absence of evidence to the contrary), which makes the inheritance of acquired characters impossible. However, Weismann did accept that the environment could affect the germ plasm directly, which is one way by which evolutionarily relevant variation can arise (another would be the combination of determinants through sexual reproduction).

The second important development was the rediscovery of Mendel's work in 1900, which further prompted biologists to consider the transmission of heritable factors as separate from development, the study of which was soon coined *genetics* (Falk 2009, pp. 44–57). Nevertheless, there was substantial resistance to the separation of development and heredity because it seemed to imply preformation – the idea that the adult form is already present in the embryo (see also Radick and Jamieson this volume). For example, Riddle (1909, p. 347) considered that "...[the] Mendelian interpretation and description of heredity is a bar to the progress of studies in inheritance and development; with an eye seeing only *particles*, and a speech only symbolizing them, there is no such thing as the study of a *process* possible". This was not an isolated response to Mendelism; the leading embryologist T. H. Morgan expressed similar concerns with the scientific basis for particulate inheritance and viewed "the problem of heredity as identical to the problem of development" (Morgan 1910, p. 449).

Morgan soon changed his mind, however (Allen 1978, 1983). Only a few years after he had taken full part in the attack on the Mendelian factors he published a book that outlined in some detail the genetic research program based on the Mendelian chromosome theory of heredity (Morgan et al. 1915). The empirical evidence from studies of sex determination and his own work on fruit fly mutants had finally convinced him that not only are hereditary factors located on the chromosomes stably transmitted between parents and offspring through the germ line, but also that the presence or absence of a particular factor can justifiably be considered the cause of the character. Aided by the concepts of genotype and phenotype introduced by Johannsen (1909), Morgan and his collaborators equated heredity with transmission genetics and development with the study of the physiological effects of particular genes. The former field flourished during the first decades from 1910 (e.g., Morgan et al. 1915), whereas the latter had to await further development in molecular biology (Falk 2009, pp. 171–177).

3 The Modern Synthesis and Evolution as Change in Gene Frequencies

The success of transmission genetics is easy to understand. Here was a tool that allowed researchers to focus on the inheritance of characters without having to worry about the messy problem of how characters arise in ontogeny. Heredity could be studied simply by conducting carefully controlled breeding experiments and score phenotypes without paying attention to embryology. Morgan's group also pioneered the use of crosses between individuals differing in more than one trait to develop linkage maps (e.g., Sturtevant 1913), which was an important step towards establishing the physical reality of genes. The usefulness of the approach was vindicated in animal and plant breeding, where breeders now had a quantitative theory on which to base their work (Lerner 1950, pp. 12–32).

But the Mendelian chromosome theory of heredity had another advantage. It allowed formalization of the evolutionary process as changes in gene frequencies (Fisher 1930; historical overview in Provine 1971). By assuming that the combination of alleles at a single or several loci determined the phenotype, evolutionary biologists could show mathematically how natural selection would increase the favourable combinations and reduce the unfavourable ones, giving rise to adaptations. This could most obviously be done for discrete phenotypes but the combination of alleles at many loci could also be used to model evolution of continuously distributed traits, like height. Empirical work, by Muller in particular, also showed how genes could mutate (reviewed in Falk 2009; Carlson 2011), and thus provided evolutionary biologists with an explanation for the origin of variation that was necessary for long-term evolution. This set an end to a long debate about whether evolution largely proceeded via gradual change due to natural selection on small-scale variation versus large-scale mutations, with selection on small genetic changes coming out on top (Provine 1971). Together with an emerging general acceptance that macro-evolution, i.e., changes above the species level, could be explained by micro-evolutionary processes (i.e., natural selection and drift), this resulted in the Modern Synthesis of evolutionary biology, marked by the publication of several synthetic and highly influential books (e.g., Dobzhansky 1937; Huxley 1942; Mayr 1942; Simpson 1944; Stebbins 1950). The central importance of the transmission genetic view of heredity in the development of evolutionary thought and construction of the Modern Synthesis is evident from the move to re-define evolution as changes in gene frequencies rather than from the perspective of organismal characters (Dobzhansky 1937; see also Depew this volume).

To understand the current status of non-genetic inheritance in evolutionary thought, it is also important to outline some further developments in biology since the Modern Synthesis. The first is the discovery of DNA, which gave the gene a molecular basis (Waters 1994). This vindicated the gene concept in evolutionary theory since it provided a mechanistic understanding of the basic concepts of genetics, such as the immutability of the heritable material, its replication (in somatic cells) and reduction (in meiosis), and gene linkage. More recently, developments in molecular genetics have made the gene concept more problematic (e.g., Neumann-Held and Rehmann-Sutter 2006; Griffiths and Stotz 2007; Burian and Kampourakis this volume), but the basic features of DNA have allowed evolutionary theory to retain its population genetic framework, and typically consider a gene to correspond to a section of DNA. Thus, progress in molecular genetics during the twentieth century generally strengthened the central position of genes in heredity, development and evolution.

In addition to the 'materialization' of the gene, some important developments in evolutionary theory contributed to a further separation of development from evolutionary biology. Despite being critical of the 'bean-bag genetics' of mathematical geneticists like Haldane and Fisher, Ernst Mayr strongly defended and promoted the Mendelian genetic basis for adaptive evolution by natural selection (Mayr 1942). In an attempt to put to end what he saw as a continuing misunderstanding of the evolutionary process by embryologists, physiologists and natural historians, he made use of a distinction between proximate and ultimate causes (Mayr 1961). Proximate causes, Mayr said, are those studied by developmental biologists or physiologists and are aimed at understanding how organisms work. Ultimate causes, on the other hand, are those that address why organisms function the way they do – they are historical explanations that are concerned with evolution of particular features. In Mayr's view, most of the criticisms of natural selection miss the mark because the alternative solutions that are provided refer to proximate, not ultimate, causes (Mayr 1961, 1993). But because proximate causes are by definition not evolutionary causes, they cannot be invoked in evolutionary explanations. Mayr's distinction between proximate and ultimate was a consequence of his view of development as execution of a genetic program that itself is a result of past natural selection (Mayr 1961). However, the distinction was embraced by most evolutionary biologists and helped to further cement the separation of development from evolution since it made explicit that evolutionary change did not require anything beyond transmission and population genetics (Laland et al. 2011, 2012).

The gene-centric view was further promoted by important advances in the study of adaptive evolution. Hamilton (1964) showed that focusing on the gene, rather than the phenotype, could make sense of traits that otherwise seemed incompatible with natural selection, such as the sterile castes of social insects. This also led to the suggestion that the gene in fact was the fundamental 'unit of selection' (Williams 1966), a view that makes it very difficult to assign a role for development, let alone non-genetic inheritance, in any evolutionary explanation since all non-genetic parts of individual bodies are considered only as a vehicle for the genotype (Dawkins 1976). Modern evolutionary theory allows for selection to occur on multiple levels (genes, individuals, groups; Okasha 2006), but typically consider individual organisms to be the unit of selection if the aim is to provide explanations for why individual organisms appear to be designed. Nevertheless, the gene-eye's view is a very common approach to thinking about adaptive evolution, with substantial explanatory power (see e.g., the literature on social evolution and sex allocation reviewed in West 2009; Bourke 2011; see also Avise, this volume).

4 The Return of the Organism: There Is More to Inheritance than Genes – But Does it Matter?

Although textbooks in evolutionary biology may give the impression that the gene-centric view of inheritance and evolution was unchallenged after the Modern Synthesis, this is not really true. Despite that the geneticists felt confident that they had correctly located the hereditary factors, i.e. genes, to the nucleus, developmental biologists continued to stress cytoplasmic factors in heredity. As pointed out by, for example Jan Sapp, Eva Jablonka and Marion Lamb, non-genetic inheritance continued to be described in organisms across kingdoms, including bacteria, fungi, plants, and animals (Sapp 1987; Jablonka and Lamb 1995; see also essays in Gissis and Jablonka 2011).

Of course all biologists, including geneticists, were aware that the mother provides the offspring with more than DNA (just take a look at any egg). This was not the reason for the debate. Instead, it was the interpretation of the significance of this observation that differed. Some developmental biologists, impressed with how embryos could often develop according to a species-specific trajectory despite dramatic changes in the genome (e.g., inversions), tended to view all maternally transmitted resources and templates as "inheritance". In particular, many considered that the most fundamental characters of species (those that did not vary and therefore could not be studied by geneticists) were due to cytoplasmic inheritance. For example, here is an excerpt from the first chapter in an edited book on maternal effects in development (Newth and Balls 1979):

> In general my prejudice will suggest that each zygote genome expresses itself within an egg organization embedded within the maternal phenotype physiology, most of which is so causally remote from any genome that it is best considered as a structured ambience inherited by each embryo in addition to its own genome (Cohen 1979, p. 9)

Thus, even long after the Modern Synthesis had been completed, some developmental biologists considered the non-genetic resources transmitted between parents and offspring as being "so causally remote from any genome" that it is best seen as an alternative system of inheritance. This feeling is echoed by a more recent extensive treatment of non-genetic inheritance and its evolutionary implications – *Evolution in Four Dimensions* – by geneticists Eva Jablonka and Marion Lamb (Jablonka and Lamb 2005). Jablonka and Lamb divide mechanisms of heredity, by which they intend to capture ways to transmit information between cells or multi-cellular organisms (Jablonka 2002), into four different categories, or systems: genetic, epigenetic, behavioural, and symbolic (Box 1).

The genetic system is DNA and its associated system for duplication that ensures that genes are faithfully copied to daughter cells during mitosis and meiosis. This is what is recognized by the Modern Synthesis and forms the basis for evolutionary theory. The cellular epigenetic inheritance system consists of molecular mechanisms that contribute to maintaining gene expression in daughter cells, but without changing the DNA itself. There are several different mechanisms, the most familiar probably being the complex set of molecules that are associated with the DNA molecule (Box 1; the reader is referred to Jablonka and Lamb 2005 and Jablonka and Raz 2009 for details on this and other epigenetic inheritance mechanisms). The behavioural inheritance system relies on more extensive overlap between generations whereas symbolic inheritance can allow information to remain latent across one or several generations.

Most evolutionary biologists have not been impressed by the argument from developmental biologists regarding cytoplasmic inheritance or, for that matter, the more recent attempts by Jablonka and others. The counter-argument goes as follows. We know that cytoplasmic factors in the egg are important for early development and we know that the egg is produced by the maternal phenotype. And because the maternal phenotype is the product of both genes and environment, the amount or location of maternal factors in the egg can be influenced by genetic variation or environmental maternal effects. Up to this point the Modern Synthesis and Developmental Biology can agree. But, the evolutionary biologist says, although we could speak of cytoplasmic factors as 'inheritance' and although their variation may influence offspring development, those changes will not be inherited by subsequent generations unless they have a genetic basis. Environmental maternal effects are simply environmental noise, carried from one generation to the next but no further. On the other hand, mutations in genes that contribute to maternal effects will faithfully be transmitted down generations, thereby allowing selection to act on any phenotypic variation arising from *genetic* maternal effects. Thus, non-genetic mechanisms of heredity should be seen as genetic adaptations, not as separate systems of inheritance. The same argument is used to dismiss the relevance of acquired traits for evolution more generally. If the genome is a blueprint for development, it is also a blueprint for all forms of non-genetic inheritance; heredity becomes directly linked to selection via differential transmission of genetic variants, and development is reduced to expression of the inherited program as adaptations (see Badyaev 2011 for summary and critique).

The genetic program or blueprint model of development is therefore central for the dismissal of the evolutionary relevance of non-genetic inheritance. However, the use of transmission genetics as an explanation of development in terms of a genetic program, with genes being direct causes of traits, has not gone unchallenged. Both philosophers and biologists have extensively criticised the use of 'genetic programs' by pointing out that if genes carry information about the development of a particular character so too do all other components of development (e.g., Lehrmann 1970; Oyama 1985; Nijhout 1990; Griffiths and Gray 1994; West-Eberhard 2003; see also Marcos and Arp, this volume). That is, for every instance where we can refer to a difference in phenotype, say, from P to P', as being caused by changes in a gene X, and from that infer that gene X carries information about P', we must also admit that it would be equally valid to say that a difference between phenotype P and P' caused by a change in environment Y means that Y carries information about P'. Thus, genetic and environmental inputs are on a par as causes of development.

This inter-changeability of genetic and environmental causes in development is readily seen in nature. For example, whether a mammalian zygote develops as male or female depends (under species-typical circumstances) on whether the genome contains a gene known as SRY that is located on the Y chromosome. If this gene is present (as it normally will be in XY individuals) the embryo develops testes and if it is absent (as in XX individuals) the embryo develops ovaries (Waters et al. 2007). In this sense sex is genetically 'determined'. But in some reptiles (crocodiles, most turtles and some lizards), there is little or no genetic variation in sex determination. Instead, whether an embryo develops testes or ovaries depends on the temperature it experiences during gonad differentiation (Warner 2011). Thus, sex is environmentally 'determined'. Genetic and environmental inputs are therefore interchangeable as 'determinants' of sex (Uller and Helanterä 2011).

This 'parity thesis' is a powerful argument for why there is something wrong with the genetic program as a shorthand for development. However, that genes and environments are on a par as causes in development does not necessarily mean that they are on a par with respect to evolution. To see why genes may be special in evolution if not in development, we can make use of Jablonka's and Lamb's description of inheritance as ways to transmit information and ask if information is generated and transmitted in fundamentally different ways for genetic and non-genetic inheritance (Shea et al. 2011). The concept of information in this context is not without problems (e.g., Oyama 1985), but one interpretation is that "a source becomes an informational input when an interpreting receiver can react to the form of the source (and variations in this form) in a functional manner" (Jablonka 2002; Jablonka and Lamb 2006). This puts the study of inheritance systems into a broader context of signals and communication (Skyrms 2010; Shea in press; see also Marcos and Arp, this volume).

Where does the information in genetic and non-genetic inheritance come from? Consider a genetic locus with alleles that influence the expression of a particular phenotype. If the alleles are stably transmitted down generations, allele frequencies will start to correlate with the selective contexts that favour one phenotype over another (Leimar et al. 2006). Thus, an allele carries correlational information about a selective regime because of past episodes of natural selection. In contrast, for most other mechanisms of heredity, what is transmitted carries information about selection on offspring because the parental phenotype has responded to an adaptively relevant feature of the environment. For example, to the extent that maternally transmitted hormones affect offspring development in an adaptive way, hormone levels carry correlational information about the offspring environment as a result of parental responses to its own environment, not because of past selection on stably transmitted variants of the hormone (Shea et al. 2011; Shea in press). Furthermore, the genetic inheritance system – DNA and its associated machinery of replication and repair mechanisms – seems to have been shaped by natural selection to have the (meta-)function to ensure stable inheritance of phenotypes down generations (Bergstrom and Rosvall 2011). It seems unlikely that this is the case for, for example, trans-generational transmission of information via hormones. Genetic inheritance therefore has features that typically are not shared by non-genetic inheritance, and which makes the former an inheritance system in a more strict sense (Shea et al. 2011).

Although there are problems with this perspective that remain to be solved (e.g., Godfrey-Smith 2011), an information perspective may thus provide a basis for establishing why DNA (or 'genes') have a privileged position in evolutionary theory even if there is nothing special about it as a cause in development (Shea 2011). Furthermore, it shows that it certainly is possible for non-genetic mechanisms of inheritance to have similar functions as DNA (Shea et al. 2011). For example, studies of plants and unicellular organisms suggest that epigenetic variants ('epi-alleles') can remain stable over many, perhaps hundreds, of generations (Jablonka and Raz 2009). This means that epigenetic inheritance alone could form the basis for adaptive evolution in the same way as genetic inheritance and thus represents an inheritance system in the strict sense that is implied for genes (Uller 2012). This is a very interesting expansion of our understanding of heredity, but may not require a substantial revision of our understanding of evolution even if trans-generational epigenetic inheritance turns out to be common. Indeed, if epigenetic variants are stably inherited, they become very similar to genes in formal evolutionary models (e.g., Helanterä and Uller 2010; Tal et al. 2010). However, there may be interesting differences between genetic and (relatively stable) epigenetic inheritance in, for example, the mutation rate (typically higher for epigenetic inheritance), timing and directionality of mutations (potentially non-random), and the degree of blending (perhaps likely). Importantly, theoretical work shows that it is possible to derive fundamental concepts in evolutionary theory – such as Hamilton's rule – without assuming particulate inheritance at all (Gardner 2011). Thus, as long as the inheritance of phenotypes can be assured in the long run, evolutionary change is possible under a range of different mechanisms of inheritance (see also Mameli 2004; Godfrey-Smith 2009).

5 Re-integrating Heredity and Development in Evolutionary Theory

Although the presence of epigenetic inheritance in principle shows that there are alternative systems of inheritance that can form the basis for adaptive evolution, I suggest that the main conceptual advantage of recognizing non-genetic mechanisms of inheritance is that it stimulates an explicit consideration of developmental processes in evolutionary explanations. If heredity is a component of development, developmental mechanisms must not only figure in ontogenetic (proximate) explanations, but also potentially in evolutionary (ultimate) explanations. Proximate mechanisms become relevant for evolutionary problems (Laland et al. 2011, 2012; see also Love this volume). Simplification is necessary in science, of course, and it will be appropriate to reduce development and heredity to transmission genetics under some circumstances. Indeed, the explanatory power and predictive ability of existing models are testament to the validity of this approach. But it is not easy to assess a priori when and why ignoring development is justified and only a careful

Box 1: Mechanisms of non-genetic inheritance

All development starts with an egg produced by the mother. Thus, everything in the egg – macro- and micro-nutrients, messenger RNAs – except for the DNA constitutes non-genetic inheritance in a broad sense. The DNA itself is wrapped up in a complex of molecules (chromatin) that influence how easily transcription factors can bind to the DNA and hence the expression of genes. The best understood of these epigenetic mechanisms is DNA methylation, in which a methyl group is attached to a cytosine base (in vertebrates typically on CpG dinucleotides). Epigenetic mechanisms like methylation can be stable during mitosis and allow groups of cells to develop and maintain different levels of gene expression even if they share the same genes. This stability is fundamental to enable cell differentiation in multi-cellular organisms. Recent research has shown that epigenetic modifications also can remain stable through meiosis and hence be transmitted across generations (Jablonka and Raz 2009).

In live-bearing organisms like mammals, the mother does not only produce an egg but can modify the transfer of resources and developmental templates to the offspring throughout embryonic development. For example, the maternal diet influences development of olfactory bulbs in the developing embryo, which in turn shapes the offspring's attraction to different smells (Robinson and Méndez-Gallardo 2011). Furthermore, behavioural interactions between parents and offspring after birth or hatching can be crucial for development of species-specific phenotypes in many organisms. Changes in parent-offspring interactions in mammals can lead to behavioural variation that lasts until adulthood and in some cases this variation remains stable down several generations via a combination of behavioural and epigenetic mechanisms (Rosenblatt 2010; Champagne and Curley 2012). These and other examples (Jablonka and Lamb 2005) give non-genetic inheritance a 'Lamarckian' flavor, since phenotypes acquired in one generation could be transmitted to subsequent generations. In humans, cultural inheritance may also involve the use of symbols that enable information to be latent over very many generations before it may reappear and influence, for example, behaviour. The invention of the written word of course substantially enhanced the potential for previous generations to transmit information to future generations and thus forms an important part of cultural inheritance.

Jablonka and Lamb (2005) separated these mechanisms into three non-genetic systems of inheritance; epigenetic, behavioural, and symbolic. The classification is based on the mechanisms by which traits are 'transmitted' between generations and not their evolutionary implications; in fact all mechanisms can have similar evolutionary consequences (Helanterä and Uller 2010; see main text for discussion).

assessment of what a developmental perspective can provide will show when it can safely be omitted. Here, I focus on two consequences of taking development and non-genetic inheritance seriously.

Firstly, re-introducing development into evolutionary theory allows us to ask questions about the origin of adaptations. Because all evolutionary change begins with developmental change, a description of the causes of evolution requires an analysis of how pre-existing mechanisms of development can give rise to novel phenotypes. This observation dates back to early criticism of natural selection as an explanation for why organisms seem to exhibit design (e.g., Mivart 1871). Even if natural selection weeds out the unfit, it cannot explain the 'origin of the fittest'. Instead, this requires studies of how genetic or environmental change modifies existing developmental trajectories in ways that result in the expression of potentially beneficial phenotypes. Hence, both the processes that are responsible for the origin of adaptations within generations and those that facilitate their retention and spread across generations are *explanans* and *explananda* for evolutionary biologists.

Mary Jane West-Eberhard (2003, 2005) has described the origin of adaptations as a two-step process (for earlier treatments, see e.g., Baldwin 1902; Waddington 1942; Schmalhausen 1949). Firstly, organisms encounter novel genetic (via mutation, hybridization etc.) or environmental input during development. Secondly, the immediate effect of the novel input is accommodated by mutual adjustment of different parts of the body. This inherent flexibility of development may allow even drastic changes to produce a reasonably functional phenotype, as is evident from extreme cases of developmental abnormalities, such as the skeletal and muscular changes in animals that are born without front legs and hence are forced to walk upright (West-Eberhard 2003, 2005).

Phenotypic accommodation is not evolution, however. But West-Eberhard and others have pointed out that the degree and form of phenotypic accommodation often vary genetically (and sometimes epigenetically) between individuals and hence are heritable. Selection can therefore modify the regulation and form of genetically variable phenotypic accommodations over generations, a process known as genetic accommodation (West-Eberhard 2003; Moczek 2007). Under this scenario, genes are still the mode of transmission of phenotypes down generations, but the causes of adaptive evolution are partly, or even largely, the processes that produced the adaptive developmental variation in the first place. Furthermore, because phenotypic accommodations may not be random with respect to past function, developmental plasticity may bias evolutionary change in particular directions.

Discussions of the role of developmental plasticity for expression of novel phenotypes have focused on the direct effect of genetic or environmental input on organisms within a single generation (e.g., Baldwin 1902; Gottlieb 1992; West-Eberhard 2003). However, non-genetic inheritance allows maternal and paternal phenotypic accommodations to have carry-over effects on offspring development, and can thereby initiate evolutionary divergence in developmental trajectories (Badyaev 2009; Badyaev and Uller 2009; Uller 2012). For example, recent research on spadefoot toads suggests that changes in maternal body condition under interspecific competition promotes adaptive character displacement in feeding strategies of tadpoles (Pfennig and Martin 2009). Furthermore, non-genetic inheritance will tend to increase the recurrence of novel phenotypes, which makes them more likely to be selected. Increased recurrence also increases the likelihood that there will be genetic variation in individual accommodations and thus enhances the scope for evolution via genetic accommodation (Uller 2012). For example, persistence of a novel food preference in mammals is facilitated by mechanisms that enable offspring to copy their parents' diet, which results in the incorporation of novel food types into development in each generation, thereby promoting population divergence in diet (Avital and Jablonka 2000). The recurrence of novel diet preferences via non-genetic inheritance may thus facilitate genetic divergence between populations. Epigenetic and behavioural mechanisms of inheritance may also allow environmentally induced phenotypic variation to be transmitted more or less stably down lineages even after the initial stimulus has ceased to exist, further increasing their recurrence (Jablonka and Lamb 2005). Thus, non-genetic inheritance forms a natural part of a developmental perspective on evolutionary change by connecting the origin of adaptations with their retention and spread within populations (Badyaev and Uller 2009; Uller 2012).

Secondly, a wide range of theoretical models from different fields have shown that non-genetic inheritance can influence the course of evolution by (i) affecting individual fitness; (ii) modifying the relationship between what is selected and what is inherited; and (iii) modifying selection on future generations. Good examples can be found in the literature on cultural evolution, niche construction, and maternal effects (e.g., Cavalli-Sforza and Feldman 1981; Odling-Smee et al. 2003; Boyd and Richerson 2005; Cheverud and Wolf 2009; see Danchin et al. 2011 for overview).[1] For example, since offspring not only inherit genes from their parents, but also their environment, activities of past generations can modify selection on future generations. This creates the potential for feedback between phenotypes and selection, which can speed up the rate of adaptation or lead to evolutionary inertia, result in the fixation of alleles that would be deleterious in the absence of non-genetic inheritance, and promote phenotypic divergence within and between populations.

The differences in the biological motivation for these models have resulted in them being treated as special cases. However, it has recently been shown that the majority, if not all, models can be formalized in a unifying framework based on the Price Equation[2] (Day and Bonduriansky 2011; see also Helanterä and Uller 2010).

[1]These examples share the common feature that, when parents transmit more than genes to their offspring, what is selected and what is transmitted to future generations can be decoupled, and phenotypes in one generation may change the developmental context (and hence the selective regime) of future generations. When only genes are transmitted, selection directly controls all forms of heritable variation, which precludes investigation of many interesting forms of feedback between organism and environment.

[2]The Price Equation is a mathematical formulation about change in a population from one generation to the next (Price 1970; see Gardner 2008 for a non-technical introduction). It is usually written in a form that separates the change in a particular trait value into two terms. The first describes the change due to selection, whereas the second describes the expected difference in trait value between parents and offspring (e.g., based on the mechanism of inheritance; Helanterä and Uller 2010; Day and Bonduriansky 2011).

The Price Equation, originally named after George Price who developed it in the early 1970s (Price 1970), already forms the foundation for formalization of evolutionary change (Gardner 2008). Thus, exploration of this set of implications of non-genetic inheritance does not require development of novel theoretical tools, but simply a realization that non-genetic inheritance can affect the rate and direction of evolution because it affects the distribution or selection of phenotypes.

6 Non-genetic Inheritance in Evolution: Recapitulation and Implications for Evolution Education

Heredity and its evolutionary implications well exemplify how we can learn from history. First, it is sobering to realise that a concept that seems so entrenched in the thinking of both practicing biologists and the general public is relatively recent; the emergence and domination of heredity-as-transmission-genetics resulted from conceptual and empirical advances in biological research. The rationale for the arguments and the tactics employed by the different parties form a basis for understanding not only the development of genetics as a field (Allen 1983; Falk 2009; Kampourakis 2013a), but also why Lamarckian and non-genetic inheritance was seen as being of minor importance or even impossible (Sapp 1987; Jablonka and Lamb 1995; Bonduriansky 2012). Ignoring non-genetic inheritance was easy once development had been written out of evolutionary theory.

The conflicting views of heredity-as-development and heredity-as-transmission genetics are more than a historical battle, however. The consequences of severing development and heredity also lie at the heart of the modern debate of how non-genetic inheritance can affect evolution. On the one hand, the many different ways in which the parents reconstruct the developmental niche of their offspring show that similarity between parents and offspring cannot be reduced to transmission genetics. The genome is not best seen as a blueprint or a developmental program, but rather as one developmental resource among others. However, this resource has some particular features and an evolutionary history that make it, if not unique, at least of special importance in evolution. The privileged position of the genome in evolution arises from its modular organization that allow almost open-ended modification of the structure of developmental resources and templates, stability during ontogeny and across generations, and its capacity to carry information as a result of past selection (e.g., Sterelny et al. 1996). We have seen that these properties probably are shared, to a greater or lesser degree, also with some kinds of epigenetic and behavioural inheritance. This in itself does not pose a fundamental problem to evolutionary theory. As has been emphasized by several authors (e.g., Jablonka and Lamb 2005; Helanterä and Uller 2010), stably inherited epigenetic variants will essentially behave as genes in evolutionary models (although mutation rates, directionality of mutations, and the 'open-endedness' of variation may differ from standard assumptions in genetics). The same may apply also to some other mechanisms of inheritance (Shea 2009).

However, it would be premature to assume that the challenge from non-genetic inheritance arises only from acknowledging that DNA is not the only inheritance system *sensu strictu*. Although non-genetic inheritance may not change the end-point of evolutionary change, it can affect the manner and the speed at which the population gets there (Odling-Smee et al. 2003; Day and Bonduriansky 2011). Such evolutionary implications of non-genetic inheritance are not trivial, in particular in a human-dominated world where environments, and hence selection, often vary rapidly and substantially in time and space. The increasing interest in the effects of non-genetic inheritance on the rate and direction of evolution can perhaps be traced to a more general recognition that ecological and evolutionary change can occur on the same time scale (e.g., Schoener 2011). The effects of non-genetic inheritance on the response to selection can be captured by expanding existing models of evolution; thus, this aspect of non-genetic inheritance does not necessarily require the substantial revision of evolutionary theory that is sometimes called for, but nevertheless is an important omission that requires attention in more advanced treatments of evolutionary change. Another feature of non-genetic inheritance that has been ignored until recently, but that does not require substantial revision of evolutionary theory, is adaptive trans-generational plasticity (Uller 2008). Important as it may be for how organisms adapt to and cope with environmental change, its evolution can be modelled by expanding existing models of adaptive evolution of within-generation phenotypic plasticity (e.g., Uller and Pen 2011).

But we have seen that non-genetic inheritance perhaps should motivate some more fundamental modifications or additions to current evolutionary theory. In particular, non-genetic inheritance calls for re-assessing the rationale for Morgan's separation of heredity and development. This is because a transmission genetics perspective on evolution makes it difficult to address two vital components of evolutionary change. First, it reduces the developmental origin of adaptations to genetic mutation and assigns all explanatory power of adaptation to natural selection. This gives the impression that proximate mechanisms have nothing to do with evolution and thus detracts the focus from the origin of adaptive variation via developmental bias or plasticity. Second, writing development and non-genetic inheritance out of evolutionary theory makes it more difficult to address the importance of the effect that organisms have on selection. Organisms do evolve as a result of genetic change, but environments and selective contexts evolve as well (Lewontin 1983; Odling-Smee et al. 2003). Moreover, non-genetic inheritance and plasticity increases the range over which the actions of organisms can have evolutionary consequences, by biasing or increasing recurrence of phenotypic variation.

There is no doubt that the Modern Synthesis has been extraordinarily successful in expanding our understanding of evolution. However, a concept of heredity that only assumes the presence of genes may not remain as useful as it once was. Indeed, it could be argued that setting aside development was the right thing to do for evolutionary biologists at the time, but that this now hampers further progress (e.g., West-Eberhard 2003; Badyaev 2011; Laland et al. 2011). We have come a long way with respect to understanding the complexities of development since the Modern

Synthesis and it is perhaps now time to make use of these insights in construction of an expanded evolutionary theory that better captures biological reality (Pigliucci and Müller 2010).

This expansion poses several difficult challenges in evolution education. I will limit myself to three problems regarding non-genetic inheritance that are likely to raise concerns among teachers of evolution in higher education.

It may be argued that the separation of development and heredity and the concept of 'genes for traits' are so ingrained in everyday thinking that it will be difficult to overcome without risking students falling into simplistic and erroneous Lamarckian or teleological notions of evolutionary change, confusing proximate explanations for ultimate ones, or failing to understand the important role of natural selection. This may be a justified concern. However, at least two strong counter-arguments could be made. Firstly, the 'genes for traits' perspective is also problematic for students' understanding of genetics (Mills Shaw et al. 2008; Dougherty 2009; Moore this volume; Burian and Kampourakis this volume). It is therefore desirable that developmental biology plays a greater role also in genetics education. Secondly, studies show that students' understanding of natural selection already is very limited and that their explanations of adaptations often invoke a final goal to which the trait contributes (e.g., Alters and Nelson 2002; Kampourakis and Zogza 2009; Smith 2010; Kampourakis 2013b). It is not necessarily the case that a developmental approach would make the situation worse. On the contrary, because a developmental approach attempts to also explain the origin of adaptive phenotypes, and not just their maintenance or spread in a population, it may help students to fill what seems to be an intuitive gap between population-genetic processes occurring within contemporary populations and evolutionary diversification, without having to resort to teleological reasoning. It would be interesting to know whether incorporating teaching of the developmental basis of adaptive evolution would reduce the extent to which students attribute evolutionary change to (future) need. This could very well deserve specific study.

It could also be argued that models including non-genetic inheritance are too complex to be of use in anything but advanced undergraduate or postgraduate education. Indeed, genetic models are simpler than models that also include non-genetic inheritance. However, teaching of population genetics in undergraduate classes typically focuses on models that omit many complexities of the genetic system, such as epistatic interactions between loci. This does not mean that those complexities are ignored. Indeed, the concept of epistasis is a fundamental component of population genetics even if we may initially chose to avoid explicitly bringing it into our models to keep things simple (the same applies to the teaching of quantitative genetics). Furthermore, acknowledging non-genetic inheritance does not change the fundamentals of evolution, i.e., the basic ingredients of variation, heredity, and natural (or sexual) selection. One potentially useful approach in undergraduate courses would therefore be to discuss what heredity is, what kind of heredity would allow evolution by natural selection, followed by showing how transmission genetics simplifies 'inclusive heredity' (Danchin et al. 2011) into a few principles ('narrow heredity') that help us to formalize the evolutionary process. This approach enables teachers to retain the basic conceptual and mathematical tools of evolutionary theory, while allowing students to understand that 'narrow heredity' is a simplified heuristic whose assumptions can be relaxed as the course proceeds. A historical perspective not only becomes natural to include under this approach, but is also necessary to understand current debates and to be able to make informed judgements of (often exaggerated) claims made in the scientific and popular press (e.g., regarding the implications of epigenetic inheritance). An excellent example of how historical and philosophical issues can be integrated in textbooks is Gilbert and Epel (2009).

Finally, it is only recently that the evolutionary implications of non-genetic inheritance, or development more generally, have been explored in light of modern evolutionary and developmental biology (see Love, this volume). It is therefore challenging for teachers to decide what to include in their courses as some concepts and theoretical and empirical results will have only transient effects on the field or even turn out to be flawed. This problem is not specific for evolutionary biology, of course, and as a result textbooks tend to take a conservative view of their field. Without making any strong recommendations, it seems that the choice will be affected by at least two major considerations (in addition to the teacher's personal biases in interest and background).

The first is what is the aim of the course or the education? Taking development and non-genetic inheritance seriously in evolutionary biology may have particular educational value for students that also study developmental biology and ecology, as it will provide points of contact between subjects that typically are perceived (and taught) as if they were isolated from each other. Non-genetic inheritance may be less important for courses where the primary aim is a very basic understanding of evolution as a fact, or the concept of adaptive evolution by natural selection.

The second aspect to consider is the available empirical evidence. There is substantial evidence that non-genetic inheritance (in the broad sense) is important for development, that it is involved in the generation of phenotypic variation, and that it contributes to cultural evolution (e.g., Jablonka and Lamb 2005; Boyd and Richerson 2005). However, there are still relatively few clear empirical examples that non-genetic inheritance can bias, facilitate, or constrain evolutionary change (this is of course always true before new ideas or approaches gain momentum). Thus, teachers risk either rejecting an important conceptual advance in evolutionary biology on the basis of a perceived lack of empirical evidence, or overly enthusiastically embracing novel ideas that may not stand up to scrutiny. Recent textbooks of evolution that takes a more developmental approach are useful as a guide for some of the topics covered in this chapter (e.g., Gilbert and Epel 2009; Arthur 2010), but choosing an appropriate middle ground puts substantial pressure on teachers to stay up to date with the rapidly expanding specialist literature in the field.

This latter concern notwithstanding, I suggest that the growing molecular and behavioural evidence for non-genetic inheritance should stimulate a more in-depth treatment of the concept and biological reality of heredity in evolution education (see also Jamieson and Radick this volume; Burian and Kampourakis this volume; Moore this volume). This includes how and why heredity came to be reduced to transmission genetics and the heuristic value of this for the formalization of evolutionary change. This will enable teachers to acknowledge the reality of non-genetic inheritance, make explicit the assumptions that go into its omission from evolutionary theory (in the form of the Modern Synthesis), and provide pointers or further treatment of what those omissions mean in terms of loss of explanatory power or missing targets of inquiry. Ultimately, this should help students to see both the strengths and weaknesses of current evolutionary theory, while making informed judgements about the evolutionary implications of the examples of epigenetic, behavioural, and ecological inheritance that they encounter in their developmental and ecology courses.

Acknowledgements I am grateful to Kostas Kampourakis for the invitation to contribute to this volume and Sinead English, Heikki Helanterä, Kostas Kampourakis, Eva Jablonka, and Nick Shea for comments on the chapter. My work on non-genetic inheritance is funded in part by the European Union's Seventh Framework Programme (FP7/2007-2011) under grant agreement n° 259679.

References

Allen, G.E. 1978. *Thomas Hunt Morgan: The man and his science*. Princeton: Princeton University Press.

Allen, G.E. 1983. T.H. Morgan and the influence of mechanistic materialism on the development of the gene concept 1910–1940. *American Zoologist* 23: 829–843.

Alters, B.J., and C.E. Nelson. 2002. Perspective: Teaching evolution in higher education. *Evolution* 56: 1891–1901.

Amundson, R. 2005. *The changing role of the embryo in evolutionary thought. Structure and synthesis*. New York: Cambridge University Press.

Arthur, W. 2010. *Evolution: A developmental approach*. Oxford: Wiley Blackwell.

Avital, E., and E. Jablonka. 2000. *Animal traditions. Behavioural inheritance in evolution*. New York: Cambridge University Press.

Badyaev, A.V. 2009. Evolutionary significance of phenotypic accommodation in novel environments: An empirical test of the Baldwin effect. *Philosophical Transactions of the Royal Society B: Biological Sciences* 364: 1125–1141.

Badyaev, A.V. 2011. Origin of the fittest: Link between emergent variation and evolutionary change as a critical question in evolutionary biology. *Proceedings of the Royal Society B: Biological Sciences* 278: 1921–1929.

Badyaev, A.V., and T. Uller. 2009. Parental effects in ecology and evolution: Mechanisms, processes and implications. *Philosophical Transactions of the Royal Society B: Biological Sciences* 364: 1169–1177.

Baldwin, J.M. 1902. *Development and evolution*. New York: Macmillan & Co. Ltd.

Bergstrom, C.T., and M. Rosvall. 2011. The transmission sense of information. *Biology and Philosophy* 26: 159–176.

Bonduriansky, R. 2012. Rethinking heredity, again. *Trends in Ecology & Evolution* 27: 330–336.

Bourke, A.F.G. 2011. *Principles of social evolution*. New York: Oxford University Press.

Boyd, R., and P.J. Richerson. 2005. *The origins and evolution of cultures*. New York: Oxford University Press.

Carlson, E.O. 2011. *Mutation. The history of an idea from Darwin to genomics*. Cold Spring Harbor: CSH Press.

Cavalli-Sforza, L.L., and M.W. Feldman. 1981. *Cultural transmission and evolution: A quantitative approach*. Princeton: Princeton University Press.

Champagne, F.A., and J.P. Curley. 2012. Genetic and epigenetics of parental care. In *Evolution of parental care*, ed. N. Royle, P.T. Smiseth, and M. Kölliker, 304–324. Oxford: Oxford University Press.
Cheverud, J.M., and J.B. Wolf. 2009. The genetics and evolutionary consequences of maternal effects. In *Maternal effects in mammals*, ed. D. Maestripieri and J.M. Mateo, 11–37. Chicago: Chicago University Press.
Cohen, J. 1979. Maternal constraints in development. In *Maternal effects in development*, ed. D.R. Newth and M. Balls, 1–28. New York: Cambridge University Press.
Danchin, É., A. Charmantier, F.A. Champagne, A. Mesoudi, B. Pujol, and S. Blanchet. 2011. Beyond DNA: Integrating inclusive inheritance into an extended theory of evolution. *Nature Reviews Genetics* 12: 475–486.
Darwin, C. 1868. *The variation of animals and plants under domestication*. London: John Murray.
Dawkins, R. 1976. *The selfish gene*. Oxford: Oxford University Press.
Day, T., and R. Bonduriansky. 2011. A unified approach to the evolutionary consequences of genetic and nongenetic inheritance. *The American Naturalist* 178: E18–E36.
Dobzhansky, T. 1937. *Genetics and the origin of species*. New York: Columbia University Press.
Dougherty, M.J. 2009. Closing the gap: Inverting the genetics curriculum to ensure an informed public. *American Journal of Human Genetics* 85: 6–12.
Falk, R. 2009. *Genetic analysis. A history of genetic thinking*. Cambridge: Cambridge University Press.
Fisher, R.A. 1930. *The genetical theory of natural selection*. Oxford: Clarendon Press.
Galton, F. 1876. A theory of heredity. *Journal of the Anthropological Institute* 5: 329–348.
Gardner, A. 2008. The Price equation. *Current Biology* 18: R198–R202.
Gardner, A. 2011. Kin selection under blending inheritance. *Journal of Theoretical Biology* 284: 125–129.
Gilbert, S.F., and D. Epel. 2009. *Ecological developmental biology. Integrating epigenetics, medicine and evolution*. Sunderland: Sinauer Associates Inc.
Gissis, S.B., and E. Jablonka. 2011. *Transformations of Lamarckism: From subtle fluids to molecular biology*. Cambridge, MA: MIT Press.
Godfrey-Smith, P. 2009. *Darwinian populations and natural selection*. New York: Oxford University Press.
Godfrey-Smith, P. 2011. Senders, receivers, and genetic information: Comments on Bergstrom and Rosvall. *Biology and Philosophy* 26: 171–181.
Gottlieb, G. 1992. *Individual development and evolution. The genesis of novel behavior*. New York: Oxford University Press.
Griffiths, P.E., and R.D. Gray. 1994. Developmental systems and evolutionary explanation. *Journal of Philosophy* 91: 277–304.
Griffiths, Paul E., and Karola Stotz. 2007. Gene. In *Cambridge companion to philosophy of biology*, ed. Michael Ruse and David Hull, 85–102. Cambridge, MA: Cambridge University Press.
Hamilton, W.D. 1964. The genetical evolution of social behavior. *Journal of Theoretical Biology* 7(1): 1–16.
Helanterä, H., and T. Uller. 2010. The Price equation and extended inheritance. *Philosophy and Theory in Biology* 2: 1–17.
Hodge, M.J.S. 1985. Darwin as a lifelong generation theorist. In *The Darwinian heritage*, ed. D. Kohn, 207–243. Princeton: Princeton University Press.
Huxley, J. 1942. *Evolution. The modern synthesis*. London: Allen and Unwin.
Jablonka, E. 2002. Information: Its interpretation, its inheritance, and its sharing. *Philosophy of Science* 69: 578–605.
Jablonka, E., and M. Lamb. 1995. *Epigenetic inheritance and evolution. The Lamarckian dimension*. New York: Oxford University Press.
Jablonka, E., and M. Lamb. 2005. *Evolution in four dimensions. Genetic, epigenetic, behavioral and symbolic variation in the history of life*. Cambridge, MA: MIT Press.
Jablonka, E., and M.J. Lamb. 2006. The evolution of information in the major transitions. *Journal of Theoretical Biology* 239: 236–246.

Jablonka, E., and G. Raz. 2009. Transgenerational epigenetic inheritance: Prevalence, mechanisms, and implications for the study of heredity and evolution. *The Quarterly Review of Biology* 84: 131–176.
Johannsen, W. 1909. *Elemente der exacten Erblichkeitslehre*. Jena: Gustav Fischer.
Kampourakis, K. 2013a. Mendel and the path to genetics: Portraying science as a social process. *Science & Education* 22(2): 293–324.
Kampourakis, K. 2013b. Teaching about adaptation: Why evolutionary history matters. *Science & Education* 22(2): 173–188.
Kampourakis, K., and V. Zogza. 2009. Preliminary evolutionary explanations: A basic framework for conceptual change and explanatory coherence in evolution. *Science & Education* 18: 1313–1340.
Laland, K.N., K. Sterelny, J. Odling-Smee, W. Hoppitt, and T. Uller. 2011. Cause and effect in biology revisited: Is Mayr's proximate-ultimate dichotomy still useful? *Science* 334: 1512–1516.
Laland, K.N., J. Odling-Smee, W. Hoppitt, and T. Uller. 2012. More on how and why: Cause and effect in biology revisited. *Biology & Philosophy*, (in press).
Lederman, N.G. 2007. Nature of science: Past, present, and future. In *Handbook of research on science education*, ed. S.K. Abell and N.G. Lederman, 831–880. Mahwah: Lawrence Erlbaum Associates.
Lehrmann, Daniel. 1970. Semantic and conceptual issues in the nature-nurture problem. In *Development and evolution of behavior*, ed. R. Aronson, E. Tobach, D.S. Lehrman, and J.S. Rosenblatt, 17–52. San Francisco: Freeman.
Leimar, O., P. Hammerstein, and T.J.M. Van Dooren. 2006. A new perspective on developmental plasticity and the principles of adaptive morph determination. *American Naturalist* 167: 367–376.
Lerner, I.M. 1950. *Population genetics and animal improvement*. Cambridge: Cambridge University Press.
Lewontin, R.C. 1983. Gene, organism, and environment. In *Evolution from molecules to men*, ed. D.S. Bendall, 273–285. Cambridge: Cambridge University Press.
Mameli, M. 2004. Nongenetic selection and nongenetic inheritance. *The British Journal for the Philosophy of Science* 55: 35–71.
Matthews, M.R. 1994. *Science teaching. The role of history and philosophy of science*. New York: Routledge.
Mayr, E. 1942. *Systematics and the origin of species*. New York: Columbia University Press.
Mayr, E. 1961. Cause and effect in biology. *Science* 134: 1501–1506.
Mayr, E. 1982. *The growth of biological thought. Diversity, evolution and inheritance*. Cambridge, MA: The Belknap Press of Harvard University Press.
Mayr, E. 1993. Proximate and ultimate causations. *Biology and Philosophy* 8: 93–94.
Mills Shaw, K.R., K. Van Horne, H. Zhang, and J. Boughman. 2008. Essay contest reveals misconceptions of high school students in genetics content. *Genetics* 178: 1157–1168.
Mivart, St G. 1871. *On the genesis of species*. New York: D. Appleton & Co.
Moczek, A.P. 2007. Developmental capacitance, genetic accommodation, and adaptive evolution. *Evolution & Development* 9: 299–305.
Morgan, T.H. 1910. Chromosomes and heredity. *American Naturalist* 44: 449–496.
Morgan, T.H., A.H. Sturtevant, H.J. Muller, and C.B. Bridges. 1915. *The mechanisms of Mendelian heredity*. New York: Henry Holt & Co.
Müller-Wille, S., and H.-J. Rheinberger. 2007. *Heredity produced. At the crossroads of biology, politics, and culture*, 1500–1870. Cambridge, MA: MIT Press.
Müller-Wille, S., and H.-J. Rheinberger. 2012. *A cultural history of heredity*. Chicago: Chicago University Press.
Neumann-Held, E.M., and C. Rehmann-Sutter. 2006. *Genes in development. Re-reading the molecular paradigm*. Durham: Duke University Press.
Newth, D.R., and M. Balls. 1979. *Maternal effects in development*. New York: Cambridge University Press.

Nijhout, H.F. 1990. Metaphors and the role of genes in development. *BioEssays* 12: 441–446.
Odling-Smee, F.J., K.N. Laland, and M.W. Feldman. 2003. *Niche construction. The neglected process in evolution*. Princeton: Princeton University Press.
Okasha, S. 2006. *Evolution and the levels of selection*. New York: Oxford University Press.
Oyama, S. 1985. *The ontogeny of information. Developmental systems and evolution*. 2nd ed. 2000. Durham: Duke University Press.
Pfennig, D.W., and R.A. Martin. 2009. A maternal effect mediates rapid population divergence and character displacement in spadefoot toads. *Evolution* 63: 898–909.
Pigliucci, M., and G.B. Müller. 2010. *Evolution. The expanded synthesis*. Cambridge, MA: MIT Press.
Price, G.R. 1970. Selection and covariance. *Nature* 227: 520–521.
Provine, W.B. 1971. *The origins of population genetics*. Chicago: University of Chicago Press.
Riddle, O. 1909. Our knowledge of melanin color formation and its bearing on the Mendelian description of heredity. *Biological Bulletin* 16: 316–350.
Robinson, S.R., and V. Méndez-Gallardo. 2011. Amniotic fluid as an extended milieu intérieur. In *Developmental science, behaviour, and genetics*, ed. K.E. Hood, C.T. Halpern, G. Greenberg, and R.M. Lerner, 234–284. Malden: Wiley-Blackwell.
Rosenblatt, J.S. 2010. Behavioral development during the mother-young interaction in placental mammals. In *Developmental science, behaviour, and genetics*, ed. K.E. Hood, C.T. Halpern, G. Greenberg, and R.M. Lerner, 205–233. Malden: Wiley-Blackwell.
Sapp, J. 1987. *Beyond the gene. Cytoplasmic inheritance and the struggle for authority in genetics*. New York: Oxford University Press.
Schmalhausen, I.I. 1949. *Factors of evolution. The theory of stabilizing selection*. Philadelphia: Blakeston Co.
Schoener, T.W. 2011. The newest synthesis: Understanding the interplay of evolutionary and ecological dynamics. *Science* 331: 426–429.
Shea, N. 2009. Imitation as an inheritance system. *Philosophical Transactions of the Royal Society B: Biological Sciences* 364: 2429–2443.
Shea, N. 2011. Developmental systems theory formulated as a claim about inherited representations. *Philosophy of Science* 78: 60–82.
Shea, N. (in press). Two modes of transgenerational information transmission. In *Signaling, commitment, and emotion*, ed. B. Calcott et al. Cambridge, MA: MIT Press.
Shea, N., I. Pen, and T. Uller. 2011. Three epigenetic information channels and their different roles in evolution. *Journal of Evolutionary Biology* 24: 1178–1187.
Simpson, G.G. 1944. *Tempo and mode in evolution*. New York: Columbia University Press.
Skyrms, B. 2010. *Signals. Evolution, learning and information*. New York: Oxford University Press.
Smith, J.E.H. 2006. *The problem of animal generation in early modern philosophy*. New York: Cambridge University Press.
Smith, M.U. 2010. Current status of research in teaching and learning evolution II: Pedagogical issues. *Science & Education* 19: 539–571.
Stebbins, G.L. 1950. *Variation and evolution in plants*. New York: Columbia University Press.
Sterelny, K., K.C. Smith, and M. Dickison. 1996. The extended replicator. *Biology and Philosophy* 11: 377–403.
Stubbe, H. 1972. *History of genetics, from prehistoric times to the rediscovery of Mendel's laws*. Cambridge, MA: MIT Press.
Sturtevant, A.H. 1913. The linear arrangement of six sex-linked factors in *Drosophila*, as shown by their mode of association. *The Journal of Experimental Zoology* 14: 43–59.
Tal, O., E. Kisdi, and E. Jablonka. 2010. Epigenetic contribution to covariance between relatives. *Genetics* 184: 1037–1050.
Uller, T. 2008. Developmental plasticity and the evolution of parental effects. *Trends in Ecology & Evolution* 23: 432–438.
Uller, T. 2012. Parental effects in development and evolution. In *Evolution of parental care*, ed. N. Royle, P. Smiseth, and M. Kölliker, 247–266. Oxford: Oxford University Press.

Uller, T., and H. Helanterä. 2011. From the origin of sex-determining factors to the evolution of sex-determining systems. *The Quarterly Review of Biology* 86: 163–180.
Uller, T., and I. Pen. 2011. A theoretical model of the evolution of maternal effects under parent-offspring conflict. *Evolution* 65: 2075–2084.
Waddington, C.H. 1942. Canalization of development and the inheritance of acquired characters. *Nature* 150: 563–565.
Warner, D.A. 2011. Sex determination in reptiles. In *Hormones and reproduction of vertebrates, volume 3: Reptiles*, ed. D.O. Norris and K. Lopez, 1–38. Amsterdam: Academic.
Waters, C.K. 1994. Genes made molecular. *Philosophy of Science* 61(2): 163–185.
Waters, P.D., M.C. Wallis, and J.A.M. Graves. 2007. Mammalian sex – Origin and evolution of the Y chromosome and SRY. *Seminars in Cell & Developmental Biology* 18: 389–400.
Weismann, A. 1893. *The germ-plasm. A theory of heredity*. London: Walter Scott.
West, S.A. 2009. *Sex allocation*. Princeton: Princeton University Press.
West-Eberhard, M.J. 2003. *Developmental plasticity and evolution*. New York: Oxford University Press.
West-Eberhard, M.J. 2005. Phenotypic accommodation: Adaptive innovation due to developmental plasticity. *Journal of Experimental Zoology. Part B, Molecular and Developmental Evolution* 304B: 610–618.
Williams, G.C. 1966. *Adaptation and natural selection*. Princeton: Princeton University Press.

Homology

Alessandro Minelli and Giuseppe Fusco

1 Concept(s) of Homology

1.1 Common Usages of the Term Homology

Homology is one of the terms most widely employed in biology. Together with species, gene and a few others, it is likely to occur in texts devoted to the most diverse biological disciplines, from morphology to systematics to molecular genetics. However, the frequent occurrence of this term in such a diversity of contexts and the very long tradition of its recurrence in biology should not be construed as a proof that the scientific community agrees on a definition of homology. On the contrary, a less than critical attitude towards its variegated use is likely to cause dangerous misunderstandings.

Leaving aside a number of more technical contexts where homology and related terms take specific meanings, some of which will be explained in this chapter, there are three main contexts in which a reader is confronted with the term in a range of items ranging from elementary textbooks in biology to research articles in developmental genetics.

A first usage documents the survival, in educational texts especially, of the once fashionable contrast between *homology* and *analogy*, usually accompanied by traditional examples such as the foreleg of a quadruped and the wing of a bird as examples of homologous body parts, while the same bird wing compared to a butterfly wing exemplifies analogy.

A. Minelli (✉) • G. Fusco
Department of Biology, University of Padova, Padova, Italy
e-mail: alessandro.minelli@unipd.it; giuseppe.fusco@unipd.it

K. Kampourakis (ed.), *The Philosophy of Biology: A Companion for Educators*, History, Philosophy and Theory of the Life Sciences 1, DOI 10.1007/978-94-007-6537-5_15,

More recent or more updated texts usually ignore analogy and present homology with the terminology introduced in biological systematics by the phylogenetic school championed by Hennig (1966), which, for a given set of taxa, distinguishes between ancestral (*plesiomorphic*) shared traits (*symplesiomorphies*) from derived (*apomorphic*) shared traits (*synapomorphies*). This basic set of concepts and terms is completed by *homoplasy*, covering trait similarities due to convergence, parallelism or reversal to an ancestral condition.

Third, homology is sometimes used as equivalent to structural similarity. This is a grossly unfortunate usage, which should be ignored, were it not for its very frequent occurrence, especially as applied to nucleic acid and polypeptide sequences. In this chapter, when not specifically stated, we will not consider this meaning.

The vast modern literature on homology includes, among many others and in addition to the works cited elsewhere in this chapter the studies of Boyden (1943, 1947), A. Remane (1955), Inglis (1966), Jardine (1967, 1969), de Beer (1971), Osche (1973), Voigt (1973), Ghiselin (1976, 2005), Riedl (1980), Sudhaus (1980), Van Valen (1982), Roth (1984, 1988, 1991), Patterson (1988), Rieppel (1988, 1992, 2005), Bock (1989), Michaux (1989), J. Remane (1989), Schmitt (1989, 1995), Striedter and Northcutt (1991), Donoghue (1992), Haszprunar (1992), Panchen (1992), Goodwin (1993), Young (1993), Hall (1994, 1995, 2003), McKitrick (1994), Minelli and Schram (1994), Minelli (1996), Sluys (1996), Wray and Abouheif (1998), Müller and Newman (1999), Butler and Saidel (2000), Laubichler (2000), Mindell and Meyer (2001), Müller (2001, 2003), Brigandt (2002, 2003), Rutishauser and Moline (2005), Griffiths (2006), Brigandt and Griffiths (2007), Kleisner (2007), Szucsich and Wirkner (2007), Sommer (2008), Ereshefsky (2010, 2012), Scholtz (2010) and Ramsey and Peterson (2012).

The sometimes confusing terminology about homology and related concepts is summarized in Appendix 1.

1.2 Historical Overview

The concept of homology is a traditional pillar of comparative anatomy and, more generally, of comparative biology. However, this concept and those related to it have witnessed a deep transformation and diversification since their first introduction in the biological literature. We can distinguish four steps (Fig. 1) in this complex semantic evolution.

1.2.1 Non-historical Concept of Homology

The roots of the concept (cf. Peters 1922; Boyden 1943; Panchen 1994, 1999) are usually traced back to the great French school of comparative anatomy of the early nineteenth century although, in the writings of Etienne Geoffroy Saint-Hilaire and other authors, what later became known as homology was called instead analogy, while the term homology was often used (for example by Serres 1827) to connote what we now call serial homology (see Sect. 4.1).

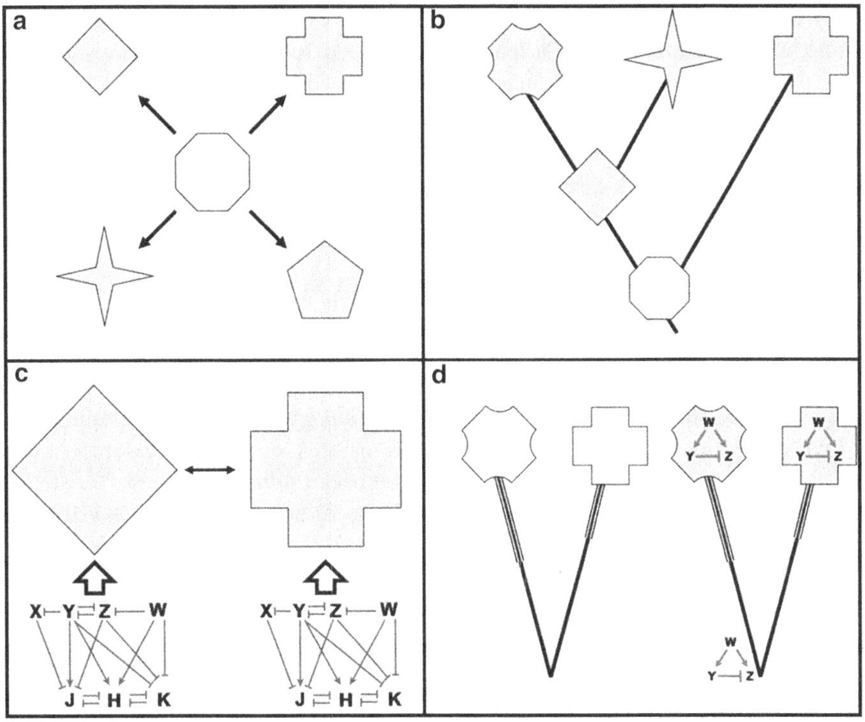

Fig. 1 Four concepts of homology. (**a**) Non-historical concept of homology. *Solid figures* are homologous because they are variants of the same archetype (*empty figure*). (**b**) Historical concept of homology. *Solid figures* are homologous because they stem phylogenetically from the same figure in the most recent common ancestor (*empty figure*) with which they are homologous too. (**c**) Proximal-cause concept of homology. The two *solid figures* are homologous because they share the same generative gene network module. (**d**) Factorial concept of homology. The two *empty figures* are structurally non-homologous (*left*) because they cannot be traced back to a structure present in the most recent common ancestor, having evolved independently from different ancestral structures (*striped lines*), but at the same time they are developmentally homologous (*solid figures, right*) because they have independently co-opted the same developmental module present in their most recent common ancestor

Circumscription of the terms 'homology' and 'analogy' that definitely opens to their modern use is due instead to Owen (1843), who defined as homologue "*the same organ in different animals under every variety of form and function*" (p. 379) and as analogue "*a part or organ in one animal which has the same function as another part or organ in a different animal*" (p. 374).

This is evidently a ***non-historical concept of homology*** (Fig. 1a), i.e. one not committed to evolutionary thinking. "Sameness" in this context is the result of an idealistic, or essentialistic concept of body plan (Ghiselin 2005; for "Essentialism in Biology" see Wilkins, this volume). This is the reason why, when Darwin (1859) used homology to support his theory of descent with modification, he did not beg the question. The specific examples he chose were merely cases of structural similarity not justified by functional necessity.

While the meaning of the term 'homology' has evolved through the subsequent biological literature, and has a long acknowledged status in evolutionary biology (see the next section), the term 'analogy' has remained associated to a pre-Darwinian concept of similarity between traits in organisms with broadly different, if not 'incommensurable' body-plans. Outside discussions in the domain of history and philosophy of biology, its use is thus strongly discouraged. Note that the term 'homoplasy' (see the next section) is not a modern synonym of 'analogy'.

1.2.2 Historical Concept of Homology

With the advent of evolutionary thinking, it seemed quite obvious to reinterpret the "sameness" in Owen's definition as similarity due to common ancestry. This is indeed the core of the so-called ***historical concept of homology*** (Fig. 1b), as exemplified by Mayr's (1969, p. 85) definition: "homologous features (or states of features) in two or more organisms are those that can be traced back to the same feature (or states) in the common ancestor of those organisms" and reformulated by Bock (1974) in the following terms: "Features (or conditions of a feature) in two or more organisms are homologous if they stem phylogenetically from the same feature (or the same condition of the feature) in the immediate common ancestor of these organisms."

Conversely, a relation of similarity between two traits in two or more organisms that do not derive from the same trait in their most recent common ancestor is termed ***homoplasy***.

The historical concept of homology is further articulated into the notions of *apomorphy* and *plesiomorphy* developed by Hennig (1966). An ***apomorphy*** is a trait that is homologous among the members of a taxon and is also in a derived (*apomorphic*) condition with respect to that in a reference common ancestor. A ***plesiomorphy*** is a trait that is homologous among the members of a taxon and is in the same primitive (*plesiomorphic*) condition in which it is found in a reference common ancestor (Fig. 2). For instance, the feather, as an epidermis derivative, is a bird apomorphy within the clade Amniota, while the same character is a Passeriformes plesiomorphy within the clade of birds. This distinction, complemented by the notion that only apomorphies shared by more than one taxon (synapomorphies) are informative for phylogenetic inference, is the basis of the cladistic method of phylogenetic inference and cladistic taxonomy (see Sect. 5.1).

1.2.3 Proximal-Cause Concepts of Homology

Other researchers, however, especially those interested in developmental biology rather than in taxonomy or phylogenetic reconstructions, conceived homology as a relation between traits that share the same developmental causes, or generative mechanisms. An example is the notion that considers homologous those traits that share the same genetic basis or, more generally, the same basis of information, be it genetic or epigenetic (Osche 1973; Van Valen 1982; Roth 1984, 1988; Minelli 1996; Minelli and Peruffo 1991). We introduce here the umbrella term ***proximal-cause concepts of homology*** for this set of concepts (Fig. 1c).

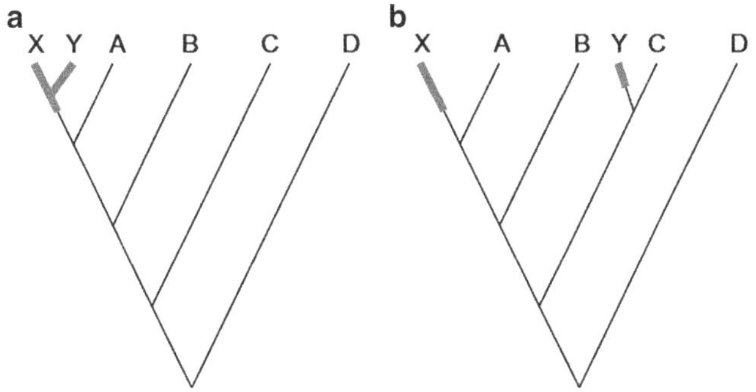

Fig. 2 The same distribution of character states (*gray* or *black*) in six terminal taxa under two different phylogenetic scenarios. (**a**) Homology. The states in taxa X and Y are homologous, because they are derived from the same condition in the most recent common ancestor of the two taxa. Since this is also a derived condition with respect to that at the root of the cladogram, the *gray* state is an apomorphy, while the *black* state in taxa A to D is a plesiomorphy because it represents the primitive condition. (**b**) Homoplasy. The derived states in X and Y are homoplastic, because they are obtained through independent transitions from the primitive state

Instances of proximal-cause homology do not necessarily overlap with cases of historical homology. Developmental genetics studies have shown how often 'obviously' homologous structures are under the control of different genes, or result from different ontogenetic processes. At the very least, a single gene is too weak a unit to provide unquestionable support to homology, and this seems true, as we shall see, even in those cases in which the expression of a single gene appears to be critically important for the entire construction of an eye or a heart. Wray (1999) did not hesitate to say that between homologous genes and homologous structures can exist all sorts of evolutionary dissociations (see also Wray and Abouheif 1998).

A less reductionist perspective than a notion of homology strictly based on gene expression leads to Wagner's *biological concept of homology* (Wagner 1989a, b, 1994, 1996, 1999): "*Structures from two individuals or from the same individual are homologous if they share a set of developmental constraints, caused by locally acting self-regulatory mechanisms of organ differentiation*" (Wagner 1989a: 62). The greater inclusiveness of this definition stems from the fact that shared developmental constraints do not necessarily require shared developmental pathways or shared genes (Wagner 1989a).

Anticipated in this definition is the concept of *module* that Wagner and other authors developed in the following years (Wagner 1996; see also Wagner and Altenberg 1996; Schlosser 2002; Schlosser and Wagner 2004; Love and Raff 2006; Brigandt 2007). A module is intended as a unit, more or less extensive and complex, which boasts its own autonomy in ontogenetic terms, being under the control of a network of genes closely related by mutual epistatic relationships, but with very few connections to the gene networks that control other modules. It is important to note that in the general case these modules do not correspond to body parts with a distinct topographic and/or functional identity, like wings, fingers, and eyes, and do not coincide with the usual characters that the morphologist, or the systematist, typically uses as units of description and comparison (e.g., Minelli and Fusco 1995).

Despite their limits (further discussed in the following section), the proximal-cause concepts of homology are those that allow to compare different structures within the same individual organism (see Sect. 4.1).

1.2.4 Factorial Concept of Homology

It has become clear that units (modules) ascribed to different levels of biological organization (e.g., genes, mechanisms of development, morphological structures) evolve to an extent largely independent from each other, sometimes providing conflicting pictures of homology relationships (Müller and Wagner 1996; Abouheif 1997). Abouheif (1997) remarked that the mixed message that seems to stem from research on homology conducted at different levels must be understood as an indication of the relative nature of homology and the need to contextualize research.

At the basis of the proximal-cause concepts of homology there is the belief that one can explain homologous characters as the outcome of homologous causal mechanisms (Spemann 1915; Atz 1970; Hodos 1976; Roth 1984).

However, as noted by Striedter and Northcutt (1991), any character at any level of organization is subject to several kinds of change during evolution, including change in its causative (generative) mechanisms. The same authors suggests a *hierarchical view of homology*, whereby at any level of biological organization we can recognize homologies between characters, even if at a different level the two characters would cease to be homologous. For example, in terms of morphology, nobody would arguably dispute that the alimentary canal is homologous throughout the vertebrates. However, the ontogenetic precursors of this canal are very different in the different lineages. For example, it derives from the roof of the embryonic primitive gut (the archenteron) in the lampreys and salamanders, from yolk cells in the legless gymnophione amphibians, and from the lower layer of the blastoderm in the amniotes.

Striedter and Northcutt (1991) also recognized that the causal relationships between the various levels of biological organization are complex. The most obvious example of this complexity is *pleiotropy*. Or, as the authors note, the same morphological structure can be involved in a large number of different behaviors (Gans 1974; Liem and Wake 1985) and a same behavior can involve many different morphological structures. Consequently, there is no simple correspondence between characters recognizable at different levels of organization. We would arguably regard the songs produced by the different species of acridid grasshoppers as homologous, but not all of them produce these songs by rubbing the femur of the hindlegs against the forewings. When producing songs, a species known as *Calliptamus italicus* moves indeed its hindlegs with the same rhythmic pace as its relatives, but the femurs of these legs do not come in touch with the wings, and the song is produced instead by rubbing one mandible against the other.

A first conclusion we can draw from this first order of questions (more examples in Ereshefsky 2007 and Brigandt and Griffiths 2007) seems to be that the traditional notion of homology as a simple relationship between two structures, which exist or may simply not exist, depending on the characters and organisms involved in the comparison, is inadequate and must be replaced by a context-dependent notion, requiring case-after-case specification of the scope and level at which the comparison takes place.

In the spirit of the relatively recent field of studies of evolutionary developmental biology (see Love this volume), this leads us to the ***factorial*** (or ***combinatorial***) ***concept of homology*** (Minelli 1998) (Fig. 1d). The central issue here is that homology cannot be an all-or-nothing relation (two structures are either homologous or non-homologous; e.g., Striedter and Northcutt 1991; Bolker and Raff 1996). Because evolutionary change is a continuous process, based on the remolding of pre-existing features along with the underlying genetic networks that regulate and control their development, homology should rather be treated as relative, or partial (Roth 1984; Haszprunar 1992; Shubin and Wake 1996; Meyer 1998; Minelli 1998, 2003; Abouheif 1999; Wake 1999; Pigliucci 2001). The relation of homology becomes a 'matter of degree', although this is generally complex, i.e. multidimensional, or more easily expressed in terms of quality rather than quantity. For example, clearly homologous structures (in a historical sense), can differ radically from the point of view of the developmental paths that carry them out, for example, segments of different groups of arthropods (Abouheif 1997). Conversely, the development of apparently non-homologous structures (again in a historical sense), like the legs and horns of certain beetles, can be under homologous genetic control (Moczek and Rose 2009).

A second conclusion is that a factorial concept of homology goes beyond any hierarchical view of homology. It is true that for complex traits, describable as an assemblage of lower-lever traits, the latter may well be homologous, while this does not necessarily entail homology between the higher level structures. For instance, all the feathers of a bird are homologous, while not all the body parts covered with feathers are homologous. However, beyond this trivial meaning, the use of a hierarchical description of the factors contributing to the expression of a trait is misleading as it is at odds with the results of modern developmental genetics. For instance, for a regulatory gene network involved in the development of a given morphological trait, not only it cannot be described as sub-parts of the trait (hierarchical belonging), but it cannot even be described as a more basic level of its realization (hierarchical causation), because of the biunivocal relationship between gene expression and the material growth and patterning of the trait in question. A factorial description of homology relationships between different features contributing to the expression of a trait, that is not constrained by/committed to a description in terms of hierarchical relationships, seems to be superior.

An inescapable final conclusion is that the idea that characters can 'remain themselves', i.e. homologous throughout an indefinite number of possible alternative states that follow one another in the course of evolution, is probably based on an idealistic interpretation of how organisms evolve (Minelli et al. 2006b).

1.3 Logical Analysis

What would be an ontological characterization of homology? Is it a relation (of similarity/sameness)? A state (of similarity/sameness)? A qualifier (of similarity/sameness)? A quantifier (of similarity/sameness)? The answer depends on the definition adopted, which is not univocal, as it will be shown along this chapter. However, there is a possible core on which most readers would agree.

Homology is a biunivocal relation between two or more entities (like friendship). This can be expressed in different ways, e.g., *A is homologous to B*, or *A is a homologue of B*, or *A and B are homologues*. The relation is symmetric (if *A* is homologous, to some extent, to *B*, then *B* is homologous, to the same extent, to *A*) and reflexive (*A* is homologous to *A*). When homology is intended as an all-or-nothing relation (disregarding its factorial nature), this is also transitive (if *A* is homologous to *B* and *B* is homologous to *C*, then *A* is homologous to *C*) so that, in mathematical terms, it qualifies as an *equivalence relation*.

In some contexts the relation of homology needs to be semantically circumscribed before any analysis. Consider this classic example: bird and bat wings are historically homologous *as* tetrapod forelimbs, but are not homologous *as* tetrapod wings. Thus a proposition on the homology of bird and bat wing can be true or false depending on whether we are comparing them as appendages or as wings. Homology statements need to include a specification about what aspects of variation constitute a different state of the same thing, and what aspects identify different things.

As we will see soon, this is just one instance in which 'sameness', without further specifications, qualifies as a vague concept that can assume different meanings and can have different extensions and implications depending on the context where it is applied and the way is qualified by adjectives. The core of the problem is the meaning for a trait of remaining itself, thus preserving its identity, while at the same time changing across evolutionary or developmental time or body space (Wagner and Stadler 2003; Brigandt 2007; Wagner 2007).

The next three sections provide an analytical review of this fact, while some schematic classifications of different concepts of homology and similarity are presented in Appendixes 2 and 3.

2 Sameness Across Evolutionary Time

The classic comparative context where homology applies is comparisons of organism traits (or states of a trait) across different taxa that are interpreted in an evolutionary context.

2.1 Classic Criteria to Recognize Homology

The traditional views on homology were summarized by Adolf Remane (1952) who listed the following ***criteria for the identification of homology***:

Main criteria:

1. Position – homologous structures occupy equivalent positions

2. Special quality of the structures – two structures agreeing in a number of structural details are homologous and this can be said with much more confidence, the more complex and numerous details are coincident
3. Continuity of form – in spite of their dissimilarity, two structures can be considered equivalent if we know other structures that represent intermediate forms in a continuum between the two structures compared

Accessory criteria:

1. Even in the case of simple structures it is legitimate to suspect the existence of a relationship of homology, if these structures are present in a large number of species
2. The likelihood of a relationship of homology increases with the number of additional homologies found between the two organisms compared
3. The likelihood of a relationship of homology decreases with the number of distantly related species where the same character is present

At the time it was published, Remane's work appeared to many as a definitive crystallization of the results of a long process of study, application and rethinking of notions derived from the comparative method of the great comparative anatomists of the first half of the nineteenth century. However, since the late 1960s different approaches to the concept of homology emerged, which were only in part the result of a reformulation of older concepts. The topic has been discussed from a historical perspective by Breidbach (2003), Williams (2004), Hoßfeld and Olsson (2005), Brigandt (2006) and Ghiselin (2006).

2.2 *Development-Based Criteria to Recognize Homology*

That the study of development can provide an important and often decisive key in our attempts to establish homologies, is a notion we find already in the works of Geoffroy Saint-Hilaire (1807), who identified the centres of ossification as the modules, or structural units, of which the skull of vertebrates consists.

The homology of the germ layers can be arguably traced through the expression patterns of specific markers like the products of the genes *snail* and *twist* (see e.g. Sommer and Tautz 1994), whose expression is an early marker of the mesoderm, or *serpent*, a gene which specifies endoderm versus ectoderm in the gut of *Drosophila* (Reuter 1994). However, homologous structures in different animals are sometimes produced from different germ layers (Hall 1998).

There are several studies in which the study of ontogeny is used to reconstruct the homologies between e.g. the limbs of tetrapod vertebrates (Hinchliffe and Hecht 1984; Rieppel 1993a, b; Burke and Feduccia 1997; Hinchliffe and Vorobyeva 1999), the appendages of arthropods (Grygier 1994; Williams 1999), or the skeletal plates of echinoderms (Hendler 1978, 1988). It should also be noted, however, that since the late nineteenth century cases are known where morphological structures that appear to be homologous differ significantly in terms of ontogeny (see for example

Wilson 1894; Remane 1952; de Beer 1958, 1971; Sander 1983, 1989; Roth 1984; Hanken 1986; Henry and Raff 1990; Wray and Raff 1990; Striedter and Northcutt 1991). We must thus conclude that developmental biology is far from being an infallible guide in our search for homologies (Young and Wagner 2011).

However, processes (Sattler 1994; Gilbert and Bolker 2001) or behaviors (Wenzel 1992; Greene 1999), rather than structures, are reasonable candidates for homology, whereas the occasionally floated idea of *homology of function* (e.g., Love 2007) seems rather to point to what is currently understood as homoplasy by parallelism or convergence.

2.3 Molecules and Homology

Growing dissatisfaction with the often contradictory evidence offered by morphology has prompted many researchers to searching for homologies based on molecules. However, as soon as they began using molecules for comparative purposes, many of the problems that had long troubled the sleep of comparative morphologists were not slow to recur. Fitch (1970) was among the first to emphasize the need to distinguish between homologous and analogous proteins, but also to remark the problems of comparison due to the frequent presence, in the same individual, of two (or more) similar molecular sequences resulting from gene duplication – something comparable, in a sense, to structures in serial homology (see Sect. 4.1).

Whenever a structural component is liable to evolve by duplication, as it is the case for genes and their products, different kinds, or subtypes of homology can be recognized (Fig. 3). ***Orthologous*** genes are those homologues that are present in different organisms and have evolved from a common ancestral gene throughout speciation events. As, by definition, the phylogeny of orthologous genes coincides with the phylogeny of species, orthologous genes are of particular interest for phylogenetic inference. ***Paralogous*** genes instead have evolved from a common ancestral gene by a gene duplication event and are present now in the same organism or in different organisms. If this gene duplication event took place before a speciation event, paralogous genes are retained in the diverging genomes of the daughter species. ***Xenologous*** genes are homologues found in different species because of interspecies (horizontal) transfer of genetic material (Fitch 1970; Holland 1999). In general, phylogenies of paralogous and xenologous genes from different species do not match with species phylogeny.

But when can we say that two macromolecules are homologous? In molecular biology, the term homology has been very frequently used in the mere sense of similarity, and often expressed in percent value to record the amount of identical units (nucleotides, amino acids) in identical position between two strings. Reeck et al. (1987) reacted to this rampant abuse of the term, and strongly stressed the need to speak of homology only when there are enough clues to believe that two molecules are copies, however divergent, of the same molecule present in a common ancestor of the organisms in which the two molecules are found – that is, the notion

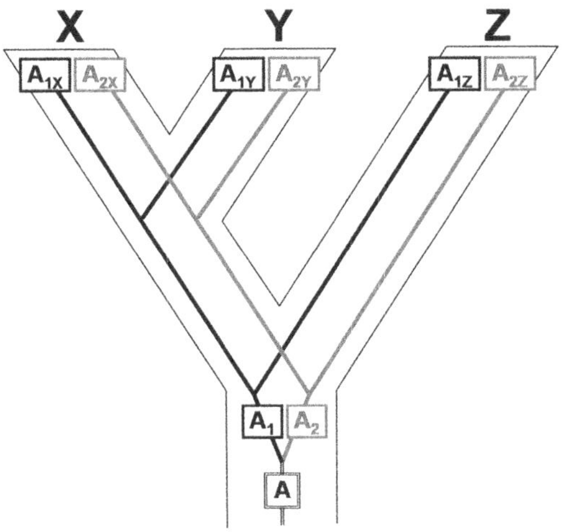

Fig. 3 Phylogeny of six genes (A_{1X}, A_{2X}, A_{1Y}, A_{2Y}, A_{1Z} and A_{2Z}) within the phylogeny of the three species that carry them (X, Y and Z). A duplication event of the ancestral gene A in the common ancestor of the three species produced two paralogous copies (A_1 and A_2). Subsequently, each of them diverged repeatedly within the whole lineage, according to the pattern of speciation events. Thus, in the three terminal taxa each gene can have either a orthology or a paralogy relationship with any of the other five genes. For instance, A_{1X} is orthologous to A_{1Y} and A_{1Z} and is paralogous to A_{2X}, A_{2Y} and A_{2Z}. Note that for a set of orthologous genes (e.g., A_{1X}, A_{1Y} and A_{1Z}) genes phylogeny coincides with the species phylogeny, while for a set of genes from different species that include paralogy relationships (e.g., A_{1X}, A_{2Y} and A_{1Z}) this relation may not hold (A_{1X} is more closely related to A_{1Z} than to A_{2Y}, while species X is more closely related to species Y than to species Z)

of homology must be applied in the historical, phylogenetic sense. A similar malpractice also transpires in morphometrics, despite Bookstein's (1994) remark that there is no legitimation to calling homologous the reference points (landmarks) arbitrarily used to compare biological forms with the techniques of geometric morphometrics.

Beyond the non trivial problem represented by the homology between genes as such, there is a second level of analysis, involving the possible homology of the temporal and spatial patterns of expression of these genes in the context of ontogenetic trajectories of the species involved in the comparison, which does not follow by logical implication from the homology at the level of gene sequence.

The first considerations about the genetic basis of homology, such as those of Boyden (1935) and Kosswig (1961), were steeped in the traditional naive notion according to which between genes and phenotypic traits there would be simple and direct causal relations, apart from the cases – long regarded in genetics as infrequent – of epistasis (the influence of a locus on the phenotypic effect of a separate locus) and pleiotropy (the influence of a locus on separate phenotypic characters). However, with the advancement of our understanding of the temporal and spatial patterns of expression of many genes and their actual mechanisms of action, this notion has been gradually abandoned for models where the focus is on complex interactions within networks of genes and their products.

One wonders what the specific role of these genes is in relation to the complex structures that are in some way under their control. More and more often it turns out that the mechanisms of action of genes that appear to play a key role in the morphogenesis of a complex structure are nothing more than generic control mechanisms responsible, for example, for the orientation of the mitotic spindle or for a process of dichotomous branching.

Dickinson (1995) and Wray (1999) admitted an evolutionary dissociation between homologous genes and homologous structures. Often, genes that appear to be involved in the morphogenesis of more or less equivalent structures in phylogenetically very distant organisms may actually perform very general jobs, so it can be very risky to use their role in these developmental processes as indicating homology of the structures eventually produced. Even more dramatically, Akam (1999) observed that it does not make any difference to organisms what this or that gene do: what matters is not whether a particular function is under the control of a specific gene or another, provided that the cell in which some of these genes are expressed is able to perform that particular function. Moreover, the genetic circuitry controlling the development of a trait is frequently redundant, and this circumstance can drastically reduce the liability of a particular gene for a particular function. Arendt (2005), discussing nervous system evolution, went on distinguishing between different levels of comparison involving, respectively, gene functions, expression patterns, and cell type molecular fingerprints.

Nielsen and Martinez (2003) proposed a new term, *homocracy*, to designate organs or structures which are organized through the expression of identical patterning genes, irrespective of whether these structures, as such, can be regarded as homologous. A related concept stressing the conservation throughout phylogeny of genetic networks underlying the production of eventually diverging organs has been suggested by Shubin et al. (2009) under the evocative but controversial name of deep homology.

A virtually opposite concept of homology has been lately suggested by Wagner (2007), who regards the homology of morphological characters as rooted in the historical continuity of gene regulatory networks to which he refers as to 'character identity networks' that enable the execution of a character-specific developmental programme.

3 Sameness Across Developmental Time

In applying the concept of homology to the study of developmental processes we meet, indeed, new and perhaps unexpected problems. On which basis or under which perspective can we say that developmental stage X of animal A is homologous, and therefore directly comparable with stage Y of animal B? Of course, between X and Y, there may be large similarities that invite to call both of them a gastrula, a larva, or a pupa. But these terms have often a purely descriptive value and their current use does not necessarily express a reasonable hypothesis of homology between the ontogenetic stages of different animals.

This is the case, for example, of the term larva. Raff (1999) relied on a hypothesis of homology between the larvae of related species of sea urchins to discuss radical evolutionary changes in the early development of those animals. However, do we imply homology when we use this identical term for organisms as diverse as the trochophore of a polychaete, the caterpillar of a butterfly and the tadpole of a frog? According to Hanken (1999), even the homology between the tadpole of a frog and the tadpole of a newt is far from granted. On the other hand, if those are larvae anyway, why not to call a newly born kangaroo also a larva?

Of course, it is not easy to make a sensible periodization of development into objectively defined and meaningfully comparable stages (Minelli et al. 2006a). The main difficulty is caused perhaps by **heterochrony** (e.g., McNamara 1986; McKinney 1988; Raff and Wray 1989; McKinney et al. 1990; Raff et al. 1990; McKinney and McNamara 1991; Reilly et al. 1997; Slack and Ruvkun 1997; Hart and Wray 1999), that is, by the different times, or different speeds with which the different parts of the body are formed during the development of the two organisms under comparison.

In a sense, it is possible to argue that every organ (or, rather, every body part developing with a certain degree of autonomy from the rest of the animal) goes through its own succession of stages. If so, during its development the animal behaves like a mosaic of parts, which, compared to those of another animal, can have somewhat different ages – a situation that encourages the adoption of a factorial approach to homology, able to account for the composite nature of developmental processes.

It has been suggested (Sander 1983) that along the ontogeny of all metazoans there is a 'phylotypic' stage that is largely invariant within an entire phylum. This concept has evolved into the so-called hourglass model, proposed by Duboule (1994), whereby the phylotypic stage (e.g. the vertebrate pharyngula) is a necessary developmental stage (and a structural model) comparable to the bottleneck that separates the two halves of an hourglass. The animal can reach this stage through different ontogenetic pathways, mainly dependent on the amount of yolk in the egg. Downstream of the phylotypic stadium, ontogenetic trajectories diverge again, with a gradual accumulation of differences eventually culminating in the adults.

As a consequence, phylotypic stages would be privileged points of reference against which to study homologies. Even the phylotypic stages, however, are an idealization, as Richardson (1995; also Richardson et al. 1997) has shown convincingly, for the vertebrates at least.

In an attempt to dissect homology relationship through developmental time, Scholtz (2005, 2008) distinguishes between *homology of developmental stages* as morphologically constrained and independently evolving units and *homology of developmental processes* (as evolutionarily modifiable sequences of otherwise comparable developmental stages), regarding both of them as legitimate targets of enquiry and instruments of comparison. Expressing an opposite view, Minelli et al. (2006a) have argued that while a firm subdivision of arthropod development in stages delimited by the moulting cycle is useful for describing ontogeny, this is limiting as a starting point for studying its evolution. Evolutionary change affects the association between different developmental processes, only some of which are paced by the molting cycle. Events occurring but once in life (hatching; first achievement of sexual maturity) are traditionally used to establish boundaries between major units of arthropod developmental time, but these boundaries are quite labile. The presence of embryonic molts, the 'gray zone' of development accompanying hatching (with the frequent delivery of an immature whose qualification as 'free embryo' or ordinary postembryonic stage is arbitrary), and the frequent decoupling of growth and molting suggest a more complex reality, where homology of developmental events and developmental stages are not relations of the kind all-or-nothing, a fact more easily accounted for by a factorial concept of homology.

4 Sameness Across the Body Space

4.1 Serial Homology

The proximal-cause concepts of homology also apply to different structures within the same individual. Homology between repetitive structures of the same individual, like vertebrae, fingers, or petals is currently known as ***serial homology*** (Fig. 4a). The term is generally used irrespectively of the spatial distribution of the repetitive structures, which are not necessarily in a row, as in the case of the segments of an earthworm, but can also occur in a different regular pattern (i.e., radial, as the sectors of a sea urchin) or even be sparsely distributed as are the setae of a fruit fly.

According to Ax (1989) and Wake (1999), we should not apply the notion of homology to the relationships between members of the same series. In their view, homology can be predicated only of equivalent structures in two separate bodies. It is a curious fact, however, that in the days when homology was known as special similarity, the French anatomists of the early nineteenth century used this term just to indicate what would later be called serial homology. Of course, the issue at stake when we establish a comparison between different parts of the same organism is different from a comparison of structures belonging to individuals of different species.

Wake (1999) remarked anyway that the structural similarity between the front and hind legs of tetrapods, a similarity that often goes down to the smallest details, has challenged the interpretative efforts of comparative anatomists since the days of Owen. Shared developmental constraints between the two pairs of limbs remain very strong even in those tetrapods where fore and hind limbs are subjected to divergent selection, as in the case of apes and humans, with the two pairs of limbs eventually becoming the upper and the lower limbs. Rolian et al. (2010) have shown, in our own species as well as in the chimp, that fingers and toes evolve along highly parallel trajectories deceptively suggestive of parallel directional selection. However, this parallel evolution is possibly due to the fact that both pairs of appendages represent a sort of copies of a common model, that is, of the trunk of the same animal;

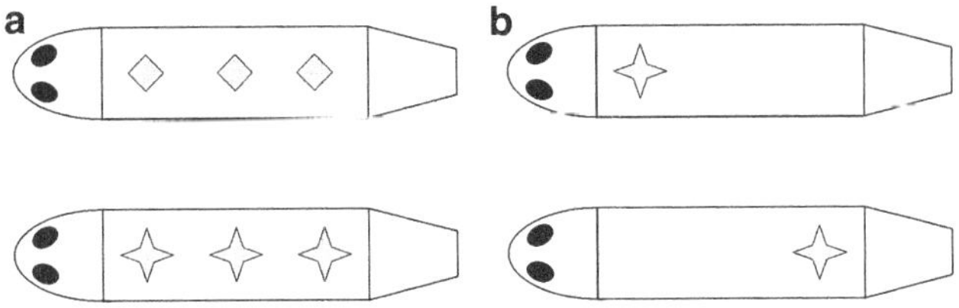

Fig. 4 (**a**) Positional homology. Different, non homologous structures are localized in homologous positions in the body of two species. This drawing also illustrates the concept of serial homology, as within each species there is a serial repetition of the same structure. (**b**) Special homology. The same homologous structure is localized in non-homologous positions within the body of the two species

this is the concept of *paramorphism* (Minelli 2000) according to which the appendages are a sort of duplicates of the main body axis. If so, the serial homology between the anterior and posterior limbs of a vertebrate would derive, by transitivity, from the fact that both are partial homologues of the main body axis.

4.2 Special Homology and Positional Homology

Another source of debate on homology is the distinction between ***positional homology*** and ***special homology*** (Minelli and Schram 1994; Minelli 1998, 2002) (Fig. 4). To illustrate this point, let's consider a homeotic mutant, in which a body part, well built to the finest details, develops at a 'wrong' site i.e., where one would expect to find a different structure. Famous examples are some *Drosophila* mutants, like the one with two pairs of wings, of which the 'wrong' one is the second, borne on the metathorax where wild-type flies carry a pair of halteres, or the one with a pair of legs replacing the antennae. In these cases, ectopic structures (the second pair of wings and the supernumerary feet, respectively) are the positional counterparts of very different structures (halteres and antennae, respectively), but from the standpoint of special homology, they are perfectly comparable to the structures (wings, legs) with the name of which we call them (Wagner 2007). The ability to dissociate the control over the position of a structure from the quality of the same structure was soundly confirmed by experiments that led to the production, in *Drosophila*, of ectopic compound eyes (Halder et al. 1995).

Fruitflies and mice exemplify two animal models of organization so different as to make it very difficult to find homologies between them. For example, *Drosophila* is a representative of the Gastroneuralia, those animals in which the main longitudinal axis of the central nervous system runs ventrally to the gut. The mouse belongs instead to the Notoneuralia, in which the main axis of the central nervous system runs dorsally to the gut. In a sense, we can say that arthropods become comparable to vertebrates as long as you turn them upside down. A pioneer in comparative anatomy as Etienne Geoffroy Saint-Hilaire (1822) did not hesitate to propose this formal exercise, in his attempt to demonstrate the fundamental organizational unit of all animals. No wonder, however, that his attempt was ridiculed (Appel 1987) and relegated to the dark corner of abstract speculation.

Yet the bold comparison of Geoffroy Saint-Hilaire has found support in recent discoveries in the molecular genetics of development. Following a tentative hypothesis formulated by Arendt and Nübler-Jung (1994), DeRobertis and Sasai (1996) explained in detail its possible molecular underpinnings. In vertebrates, the formation of dorsal mesoderm and central nervous system is induced by a region called the organizer. However, the way in which the organizer acts remained uncertain for a long time. Today we know that the *chordin* gene encodes a protein whose presence has the same effect as the organizer. Similar inductive effects are produced by the protein encoded by the *Drosophila* gene *short gastrulation* (*SOG*), which has significant similarities with the nucleotide sequence of *chordin*. However, the expression patterns of these two genes are mutually inverted along the dorsoventral axis. The comparison extends to another pair of genes, whose expression domains are opposite and complementary to *chordin/SOG*. These are *decapentaplegic* in *Drosophila*, which is expressed dorsally and promotes the development of dorsal structures, and its vertebrate counterpart *bone morphogenetic protein 4*, which is expressed on the ventral side and controls the formation of ventral mesoderm.

4.3 Anatomical Frames of Reference

Nineteenth-century comparative anatomists realized that some organs or organ systems are more conservative than others in position and special organization and could be therefore selected as topographical markers with reference to which to recognize homologies, while other organs undergo extensive and frequent changes in shape, development and possibly even position. Thus, insertion points of muscles, blood vessel patterns and, above all, nerve pathways (see Sect. 2.1, Remane's criterion 1; A. Remane 1952, 1963) became the main points of reference which the morphologist called for help, in the not rare cases where clues provided by other structures (for example, the bones in the vertebrates or the sclerites in the exoskeleton of arthropods) were uncertain or contradictory. It is unlikely that the early anatomists had arrived by chance at these guides to the comparative work. More likely, the initial choices had been probably tinged by a somewhat metaphysical and essentialist bias, in the belief that some organs or organ systems are more important than others and therefore more reliable in the search for homologies. No wonder, in particular, that the nervous system was granted a privileged role, since it is only for the brain, the most complex part of the nervous system, that humans can claim a indisputable anatomical superiority over all other animals. Of course, this anthropocentric bias does not detract from the likely interest that patterns of innervation may have in the identification of topographical relationships within a animal's body.

Breidbach and Kutsch (1990) used the neurons that innervate the dorsal longitudinal muscles of adult and juvenile stages of two locusts (*Schistocerca gregaria* and *Locusta migratoria*) and a beetle (*Zophobas morio*) to identify a set of homologies involving the three key dimensions of comparative morphology. These authors were in fact able to identify a set of 11 neurons that are common to all species studied (special homology or historical (= phylogenetic) dimension of the comparison), are repeated in each animal from one segment to another (serial homology) and are also recognizable in the later stages of the insect, regardless of whether it will face (beetle) or not (grasshoppers) a dramatic metamorphosis (ontogeny).

However, other studies have shown that innervation might be subject to change dependent on use and age, changes ranging from simple addition or elimination of synapses up to a whole reorganization of the nervous system during metamorphosis (Hallam and Jin 1998). The question is, rather, what are the real reasons of this biological stability – certainly not absolute, but higher than that of many other systems – the nervous system seems to boast.

According to many indications, a large proportion of genes expressed at early stages of embryonic development, which are involved in defining the overall architecture of the most diverse Metazoa, and the different structures of their bodies, would have originally had a role in specifying the structure of the nervous system. Secondary, and more recent, would be their responsibility in other choices, including segmentation of the longitudinal axis of the animal. If so, in a historical sense the neural organization would be the first aspect of an animal's complex organization to whose service a network of genes would have evolved, capable of producing a highly ordered structure in a reliable way.

5 Homology at Work

Homology is not only a central concept that organizes biological knowledge. It is also a basic conceptual tool for a number of applications in biological research.

5.1 *Homology in Phylogenetic Inference*

Distinguishing between plesiomorphies and apomorphies was the first, critically important step in Hennig's program aimed at providing biological systematics with a solid foundation allowing a methodologically sound reconstruction of phylogenetic relationships, the central problem of what eventually emerged as the phylogenetic, or cladistic, systematics. In cladistic methods of phylogenetic inference, shared plesiomorphies (*symplesiomorphies*) among a set of taxa within a group of interest (*ingroup*) do not convey any information about their phylogenetic relationships, as these are interpreted as an homologous trait inherited (in its primitive form) from an ancestor of the whole ingroup, and are therefore neutral with respect to any phylogenetic hypothesis within it. On the contrary, shared apomorphies (*synapomorphies*) among a subset of taxa of the ingroup support all the phylogenetic hypotheses that see the taxa sharing the character in its innovative form as more closely related to each other than to any other taxa of the ingroup. In other terms, the reconstruction of phylogenetic relationships can only be based on synapomorphies. Methods have been thus developed for distinguishing plesiomorphies from apomorphies, a problem to which phylogeneticists technically refer as to the problem of identifying character polarity. This analysis is independent from the specific natures of the characters to be analyzed, whether e.g. morphological or molecular, but only depends on the distribution of their alternative states among the taxa to be compared. An accessible illustration of these concepts and methods, well suitable as a reference work for educational purposes, is provided by the introductory chapter of Lecointre and Le Guyader (2006), while a classic reference book is Felsenstein (2004). A classic reference to the role of homology in the context of cladistics is Patterson (1982), while more recent discussions of homology in the wider context of phylogenetic inference are found in Sober (1988, 2008), Brower (2000), Stevens (2000), Reif (2004), Williams and Humphries (2004), Richter (2005), Williams and Ebach (2008), Nixon and Carpenter (2011).

In the context of phylogenetic analysis, de Pinna (1991) distinguished between ***primary homology***, i.e., the assumption of a hypothesis of homology, and ***secondary homology***, i.e. the legitimation of such a statement. Primary homologies are the characters that we code in the data matrices for phylogenetic reconstruction. These are characters that have passed the most severe observational tests of similarity and topological correspondence. When the matrix of putative homologues is analyzed and a cladogram is obtained, the primary homologies that survived the test, i.e. the characters that have a single origin on the cladogram, are then elevated to secondary homologies, while the primary homologies that have multiple origins are declassed to homoplasies (Edgecombe 2008).

5.2 Homology and Evolutionary Novelties

A classic problem in evolutionary biology is the origin of evolutionary novelties (Müller and Wagner 2003). As a first approximation, we can define an evolutionary novelty as a trait suddenly emerging in the course of the evolutionary history of a particular group of organisms, a trait that was not present, even in a different form, in earlier representatives of this group. Classic examples of evolutionary novelties are insect wings, the echolocation system (sonar) of bats, the lantern of fireflies, the flower of angiosperms and the digestive enzymes of insectivorous plants.

In the literature there is a degree of inaccuracy in the use of the term 'evolutionary novelty'. This obtains slightly different meanings in different contexts, while alternative terms (evolutionary novelty, invention, or innovation) do sometimes refer to the same concept (Minelli and Fusco 2005; Brigandt and Love 2010). At the root of this confusion is the fact that some authors care for distinguishing between the emergence of a new trait during the evolution of a lineage and its possible evolutionary success through a species radiation, while other authors do not. Not all the novelties lead to a phyletic radiation, and an innovation may prove to be a key character in a radiation long after its first appearance, once other boundary conditions have changed (e.g., the appearance of feathers in theropod dinosaurs and the subsequent radiation of birds). In general, the success of an innovation depends on the environmental context in which this is actually tested. Beyond the problems of terminology, the two concepts are certainly independent, in principle at least.

With respect to the subject of this chapter, the most relevant aspect is the emergence of novelties, regardless of their potential contribution to a lineage's evolutionary success (Müller and Wagner 1991; Hall 2005), which is also a special focus of evolutionary developmental biology (see Love this volume). An evolutionary novelty is thus *a trait that has no obvious homology with any other trait in another organism or the same organism, and whose origin can not be easily traced back to a modification of a body structure existing in an ancestral species.*

However, evolution does not produce novelties from scratch. Evolution 'operates' on what already exists, and thus we expect that at a sufficiently accurate analysis, each novelty will reveal some evidence of these changes, thus appearing as a more or less complex mixture of conserved and novel elements (Moczek 2008; Hall and Kerney 2012). This applies to large-scale transformations that we can trace through evolutionary history, such as the emergence of a new body architecture, but also to changes at smaller scale, limited to the appearance of a new feature in the context of an unchanged body organization. Thus, no novelty is totally new and distinguishing what is new from what is preserved is not necessarily easy or straightforward. The new components (or those preserved) can be structural elements, metabolic pathways, properties of development, genetic regulatory factors, or other. It is therefore difficult to establish where homology ends and novelty begins, if establishing that boundary makes sense at all.

A nice example of the evolution of a new structure is offered by horned scarab beetles. Several thousand species of beetles have rigid, non-articulated 'horns' on the head and/or the prothorax. These are used as weapons in male-male competition to access to the females, the latter generally having less developed horns or no horns at all. Armin Moczek and his collaborators have studied the genetic basis of the development of these structures in a few species of *Onthophagus*, a scarab group which, with 2,400 species, is one of the largest genera in the animal kingdom. These researchers found (Moczek and Rose 2009) that the development of these horns shares many properties with the development of the 'true' insect appendages such as legs and antennae. During pre-pupal and pupal stages, groups of cells that will form the horns of the adult undergo significant proliferation followed by a more or less intense phase of modeling and (often) remodeling, depending on the species and sex. During these stages, horn development is regulated through the expression of *Distal-less* (*Dll*), *dachshund* (*dac*) and *homothorax* (*hth*), three genes otherwise involved in the specification of the proximal-distal axis of insect legs. Two of these genes (*Dll* and *hth*) are expressed in the horns in the same relative positions in which they are expressed in the legs of *Onthophagus*, as well as in those of all other insects studied thus far (Moczek and Nagy 2005), whereas *dac* is expressed all along the axis of the horns, overlapping with the domains of expression of two other genes, suggesting that it does not have a specific role in the development of these structures.

Thus, beetle horns, while representing an evolutionary novelty, are not totally new. And beetle legs and horns, while sharing the same underlying genetic networks, so to be qualified as homologous in terms of genetic control, are historically non-homologous. A factorial concept of homology finds in the study of evolutionary novelties its most obvious application.

5.3 Homology and Nomenclature

With the exclusion, perhaps, of biological terms that are common also in everyday language, like 'head' or 'leg', the use of the same name for two structures, or two features, in two different organisms is easily taken as an implicit declaration of homology (see Edgecombe 2008). Attempts to avoid this over-interpretation can generate an over-proliferation of morphological and anatomical terminology (Fusco 2008). For instance, as a heritage of a specific, and now surpassed, view of animal phylogeny and the evolution of segmented body architectures, there is a variety of names with which serially homologous elements of the trunk of different animals are indicated: 'segments' for annelids and arthropods, 'somites' for vertebrates, 'rings' for rotifers, 'zonites' for kinorhynchs, 'proglottids' for cestodes etc. (Minelli and Fusco 2004).

Undesirable interferences between name choice and homology investigation are found also in gene nomenclature. Ferrier (2008) has convincingly argued that a sensible labeling and classification of developmental control genes, on the basis of their phylogeny, are essential to any research program in evolutionary developmental biology and evolutionary genomics, since it is crucial that the structure, expression and function of orthologous genes are being correctly compared between taxa. This is particularly true for the homeobox genes, for which there are confusing and conflicting names and classifications that bias investigations and understanding of their evolution and their role in the evolution of animal development.

5.4 Homology and the Reconstruction of Ancestors

The genes that first arouse strong illusions of having finally found an objective basis of homology are the *Hox* genes. Based on the presence in many different animals of the same set of *Hox* genes, each of which, in all species studied, has a substantially equivalent expression domain along the anterior-posterior body axis, but different from that of other *Hox* genes, Slack et al. (1993) introduced the important concept of the *zootype*. According to these authors, the origin of the basic body architecture of the Metazoa (or at least the origin of the bilaterian animals) would be associated with the appearance of a first group of *Hox* genes able to specify the main axis of the body, with its "hot spots" at which the basic structures such as the brain, the genital openings etc. will be expressed in precise antero-posterior sequence.

However, following initial enthusiasm (e.g., Akam 1989; Holland et al. 1993; Sondergaard 1993; Tabin and Laufer 1993) more recent authors have become much more cautious about the possibility of using patterns of *Hox* genes expression as safe markers of positional homology, not to mention special homology (see for example Müller and Wagner 1996; Akam 1998a, b; Schierwater and Kuhn 1998; Galis 1999; Holland and Holland 1999). Indeed, when we compare insects to vertebrates, it is easier to find good examples of homeobox genes (including genes other than those of the *Hox* class) involved in the control of non-homologous rather than homologous structures (Galis 1996). For example, the gene *Brachyury (T)* is involved in forming the notochord in vertebrates, while its counterpart in *Drosophila* is involved in producing the terminal intestine.

The discovery that homologies at the level of developmental genes can be traced back to very old ancestors has fuelled an often indiscriminate use of these data to make inferences about the organization of the most recent common ancestor of distantly related groups, including very ancient lineages, for which the fossil record is generally poor or nonexistent. The inference is based on this simple inductive reasoning: since all the extant members of a certain group of organisms possess genes x, y, and z, involved in the development of organs X, Y and Z, respectively, then the most recent common ancestor of this group also had to possess traits X, Y and Z. For example, because all animals have homologous genes for the development of eyes (one for all, the gene *eyeless/Pax6*), then the most recent common ancestor of all Bilateria must have had at least some rudimentary form of photoreceptors, and animal groups without eyes (e.g., echinoderms) must have lost them secondarily (Gehring 2002).

Inferences of this kind are often in obvious conflict with assumptions based on other lines of reasoning. For example, on the basis of the expected distribution of homologous developmental genes in their living descendants, the most recent common ancestor of all Bilateria (often referred to as Urbilateria), probably a tiny inhabitant of the seabed in the Precambrian, would have presented an organizational plan that some have judged 'illogical'. In addition to being equipped with antero-posterior and dorso-ventral polarity, two features on which there is the largest agreement among the students of animal evolution, it would also have had a segmental organization, photoreceptor organs, a heart (or a pulsating vessel), an haemocele, a skeleton, a brain and cephalic sensory appendages. Many of these features obviously would not sit well within the body architecture of a small benthic marine organism, as these traits prove to be useful only at larger body size (Minelli 2003).

Homologies, often at the level of genes, between distantly related organisms are sometimes labeled *deep homologies*, to express the idea of their coalescence in the far past, or 'deep time'. However, the discovery of an homologous gene involved in the development of two traits in two very distantly related organisms does not automatically transform a supposed homoplasy into a homology. Rather, this invites to a 'deeper analysis' of the historical relationship between the two traits, once again, we suggest, through a factorial approach (Fig. 1d).

For instance, the simplistic inferences about Urbilateria briefly discussed above do not take adequately into account the phenomenon of gene co-option, i.e. the recruitment of an already existent gene into a different regulative gene network so that it acquires a novel function. The fact that the same gene, essential for photoreception, is involved in the development of the eyes in all the taxa that have these structures does not necessarily entail that eyes are structurally homologous, i.e., derived from the eye possessed by these animals' most recent common ancestor. The same gene might have been involved independently, over and over again, in different lineages where structurally non-homologous eyes are formed. In addition, genes may change function during evolution. This also explains the apparent paradox of finding genes whose products feature in a prominent role in a particular evolutionary transition to recur within the genome of the sister clade, which suggests that these genes were already present in an evolutionary stage prior to the transition itself. For example, the genome of the unicellular eukaryote *Monosiga brevicollis*, which belongs to a group, the choanoflagellates, believed to be the sister clade of animals, contains genes that encode cell adhesion proteins (cadherins, integrins and lectins) that are considered the key to the evolution of multicellularity in animals. In choanoflagellates the genes encoding these proteins are obviously not involved in the development of a multicellular soma, but in different, oldest functions, such as feeding or reproduction (King et al. 2008).

As emphasized by the developmental biologists Hejnol and Martindale (2008) "There are no such things as 'segmentation', 'eye', 'heart' or 'limb' genes. [...] There are just molecules that can bind to DNA or interact with receptors, phosphorylate other molecules, etc." The association between specific gene products and functions of the body is itself a product of evolution.

6 Educational Suggestions

As a central concept in comparative biology, homology is listed among the inescapable topics of most courses in biology, from basic to advanced levels. However, more often than not, the subject is presented through examples, while a deep analysis of the concept is considered of historical interest only. A closer examination of the notion complemented by practical or conceptual exercises is generally restricted to very specific classes on methods of phylogenetic inference, although a few papers with the character of a tutorial and accompanied by exercises on the more inclusive subject of 'tree thinking' have been recently published (e.g., Baum et al. 2005). Also, an analysis of the concept of homology is relevant in science education in any discussion about the challenge to evolution represented by creationism and intelligent design arguments (see Brigandt this volume).

However, homology can also offer an interesting subject of study per se, providing a special opportunity to inspect in depth the logic behind a fundamental concept in biology.

A first question is how an object can remain itself, preserving its identity, while at the same time changing across time or space. This is a classic problem of sameness in logic, as often exemplified by the *Theseus' ship paradox*, that raises the question of whether a ship which was repeatedly restored by replacing all its original parts one after the other nonetheless remained the same ship. It is interesting to note that among the solutions offered in modern logic there is the distinction between 'objects' and their 'properties', something that has a very strict analogy with the concept of 'character' and 'character state' in comparative biology. In natural language the same verb, to be, is used to express the properties of an object and its identity, but logicians distinguish a 'to be of predication' from a 'to be of identity' and these translate into different symbols in logic formalism (Priest 2000).

A second question relates to the vagueness of the homology concept. In classic logic, the problem of definition fuzziness is often exemplified by the *sorite's paradox*, where in a heap of sand, removing the grains of sand one at a time there is no defined point at which the heap changes from a heap to a non-heap, producing the paradoxical conclusion that also one grain of sand (or even no grains at all, or a negative number of grains) must also form a heap. The question is obviously amplified if one accepts a factorial concept of homology. Homology is a perfect test-bench for a deep reflection on a wide class of objects the student of biology will necessarily face, that is objects with a fuzzy definition. Homology is in good company with 'species', 'sexuality', 'heritability', 'novelty', 'evolvability' etc. Handling of objects with no clear-cut definition or with context-dependent definition requires special care and specific reasoning tools. Nonetheless, inference can be quite rigorous once fuzziness has been recognized and accounted for. Homology is obviously a good subject for class discussion on these matters.

Here we suggest a few exercises to be developed in class under the guidance of a teacher or autonomously tackled by the student:

- Given a set of definitions of homology, find a number of organism features that, depending on the definition, would qualify as cases of either homology or homoplasy (e.g., the historical and proximal-cause definitions of homology and the mouths of an annelid and of a sea urchin, which are historically homologous, as both derive from the mouth of their last common ancestor, but are non-homologous in terms of developmental origin).
- Given a set of structures in different organisms, lay out different meanings of sameness in order to produce different patterns of homologous and homoplastic relationships between the same structures (e.g., the legs of a fly, the legs of a bug, the chelicera of a spider, the tube feet of a sea urchin: all structures basically share the same genetic cassette, the first three historically derive from the appendages of an ancestral arthropod, the first two historically derive from the legs of an ancestral insect).
- Given a set of structures in different organisms that are credited of some degree of homology, carry out a factorial analysis, by identifying the different components of their sameness (e.g., segments in a centipede and in a polychete worm, evaluating their possible historical derivation from a common ancestor, developmental origin, genetic control).

- Find examples of failure for each of Remane's homology criteria taken individually (e.g., for 'position', the anal pore of leeches is historically homologous to the anal pore in the other clitellate annelids, however, in leeches it opens dorsally rather than posteriorly).
- Discuss the question of identity (sameness) in the light of Theseus' ship paradox.
- Discuss the question of partial homology in the light of the sorite's paradox.

Appendixes

Appendix 1

Grammatical tagging and cross-relations of key terms

analogy	Noun. A relation of non-sameness (see main text) between two or more similar traits or states of a trait (a pre-Darwinian concept)
analogous	Adjective. Of a (state of a) trait in analogy relationship with s.e.
analogue/analog	Noun. A (state of a) trait in analogy relationship with s.e.
apomorphic	Adjective. Of a (state of a) trait in a set of homologues that is derived, i.e. in different condition with respect to that in a reference ancestor
apomorphy	Noun. An apomorphic homologue
autapomorphy	Noun. An apomorphy exhibited by one terminal taxon
synapomorphy	Noun. An apomorphy shared by members of a taxon
homology	Noun. A relation of sameness (see main text) between two or more traits or states of a trait
homologous	Adjective. Of a (state of a) trait in homology relationship with s.e.
homologue/homolog	Noun. A (state of a) trait in homology relationship with s.e.
homoplasy	Noun. 1. A relation of non-sameness (see main text) between two or more similar traits or states of a trait. 2. A homoplastic (state of a) trait
homoplastic/ homoplasious	Adjective. Of a (state of a) trait in homoplasy relationship with s.e
orthology	Noun. A type of homology relation (see main text) between two or more traits or states of a trait
orthologous	Adjective. Of a (state of a) trait in orthology relationship with s.e
orthologue/ortholog	Noun. A (state of a) trait in orthology relationship with s.e.
paralogy	Noun. A type of homology relation (see main text) between two or more traits or states of a trait
paralogous	Adjective. Of a (state of a) trait in paralogy relationship with s.e.
paralogue/paralog	Noun. A (state of a) trait in paralogy relationship with s.e.
plesiomorphic	Adjective. Of a (state of a) trait in a set of homologues that is primitive, i.e. in the same condition of that in a reference ancestor
plesiomorphy	Noun. A plesiomorphic homologue
symplesiomorphy	Noun. A plesiomorphy shared by members of a taxon

(continued)

Appendixes (continued)

xenology	Noun. A type of homology relation (see main text) between two or more traits or states of a trait
xenologous	Adjective. Of a (state of a) trait in xenology relationship with s.e.
xenologue/xenolog	Noun. A (state of a) trait in xenology relationship with s.e.

Appendix 2

Non mutually exclusive classifications of homology

- Classification 1
  - historical (evolutionary)
  - non-historical
    - idealistic (pre-Darwinian)
    - proximal-cause (e.g., developmental)
- Classification 2
  - all-or-nothing
  - degree (partial)
    - quantitative
      - one-dimensional (a scalar, e.g., percentage)
      - multidimensional (a vector)
    - qualitative (factorial)
- Classification 3
  - structural similarity (e.g., DNA sequences)
  - sameness
    - in space (body)
      - serial (e.g., annelid segments)
      - radial (e.g., echinoderm sectors)
      - sparse (e.g., arthropod setae)
    - in time
      - developmental
      - evolutionary

Appendix 3

A classification of similarity

similar (or somehow comparable)

- homoplastic
  - by convergence
  - by parallelism
- homologous
  - orthologous
    - apomorphic
    - plesiomorphic
  - paralogous
  - xenologous

References

Abouheif, E. 1997. Developmental genetics and homology: A hierarchical approach. *Trends in Ecology and Evolution* 12: 405–408.
Abouheif, E. 1999. Establishing homology criteria for regulatory gene networks: Prospects and challenges. In *Homology* (Novartis Foundation symposium 222), ed. G.R. Bock and G. Cardew, 207–225. Chichester: Wiley.
Akam, M. 1989. Hox and HOM: Homologous gene clusters in insects and vertebrates. *Cell* 57: 347–349.
Akam, M. 1998a. Hox genes, homeosis and the evolution of segment identity: No need for hopeless monsters. *The International Journal of Developmental Biology* 42: 445–451.
Akam, M. 1998b. Hox genes: From master genes to micromanagers. *Current Biology* 8: R676–R678.
Akam, M. 1999. [Discussion to Abouheif (1999), op. cit., pp. 224–225]
Appel, T.A. 1987. *The Cuvier-Geoffroy debate: French biology in the decades before Darwin*. New York/Oxford: Oxford University Press.
Arendt, D. 2005. Genes and homology in nervous system evolution: Comparing gene functions, expression patterns, and cell type molecular fingerprints. *Theory in Biosciences* 124: 185–197.
Arendt, D., and K. Nübler-Jung. 1994. Inversion of dorsoventral axis? *Nature* 371: 26.
Atz, J.W. 1970. The application of the idea of homology to behavior. In *Development and evolution of behavior*, ed. L.R. Aronson, E. Tobach, D.S. Lehrman, and J.S. Rosenblatt, 53–74. San Francisco: Freeman.
Ax, P. 1989. Homologie in der Biologie – ein Relationsbegriff im Vergleich von Arten. *Zoologische Beiträge (Berlin) N.F.* 32: 487–496.
Baum, D.A., S.D. Smith, and S.S.S. Donovan. 2005. The tree-thinking challenge. *Science* 310: 979–980; Supporting Online Material: www.sciencemag.org/cgi/content/full/310/5750/979/DC1. (Tree-thinking quizzes I and II).
Bock, W.J. 1974. Philosophical foundations of classical evolutionary classification. *Systematic Zoology* 22: 375–392.
Bock, W.J. 1989. The homology concept: Its philosophical foundation and practical methodology. *Zoologische Beiträge (Berlin) N.F.* 32: 327–353.

Bolker, J., and R.A. Raff. 1996. Developmental genetics and traditional homology. *BioEssays* 18: 489–494.
Bookstein, F.L. 1994. Can biometrical shape be a homologous character? In *Homology: The hierarchical basis of comparative biology*, ed. B.K. Hall, 197–227. San Diego/London: Academic Press.
Boyden, A.A. 1935. Genetics and homology. *The Quarterly Review of Biology* 10: 448–451.
Boyden, A.A. 1943. Homology and analogy: A century after the definitions of "homologue" and "analogue" of Richard Owen. *The Quarterly Review of Biology* 18: 228–241.
Boyden, A.A. 1947. Homology and analogy. A critical review of the meanings and implications of these concepts in biology. *American Midland Naturalist* 37: 648–669.
Breidbach, O. 2003. Post-Haeckelian comparative biology – Adolf Naef's idealistic morphology. *Theory in Biosciences* 122: 174–193.
Breidbach, O., and W. Kutsch. 1990. Structural homology of identified motoneurons in larval and adult stages of hemi- and holometabolous insects. *The Journal of Comparative Neurology* 297: 392–409.
Brigandt, I. 2002. Homology and the origin of correspondence. *Biology and Philosophy* 17: 389–407.
Brigandt, I. 2003. Homology in comparative, molecular, and evolutionary developmental biology: The radiation of a concept. *Journal of Experimental Zoology Part B: Molecular and Developmental Evolution* 299B: 9–17.
Brigandt, I. 2006. Homology and heterochrony: The evolutionary embryologist Gavin Rylands de Beer (1899–1972). *Journal of Experimental Zoology Part B: Molecular and Developmental Evolution* 306B: 317–328.
Brigandt, I. 2007. Typology now: Homology and developmental constraints explain evolvability. *Biology and Philosophy* 22: 709–725.
Brigandt, I., and P.E. Griffiths. 2007. The importance of homology for biology and philosophy. *Biology and Philosophy* 22: 633–641.
Brigandt, I., and A.C. Love. 2010. Evolutionary novelty and the evo-devo synthesis: Field notes. *Evolutionary Biology* 37: 93–99.
Brower, A.V.Z. 2000. Homology and the inference of systematic relationships: Some historical and philosophical perspectives. In *Homology and systematics: Coding characters for phylogenetic analysis*, ed. R.W. Scotland and T. Pennington, 10–21. London/New York: Taylor & Francis.
Burke, A.C., and A. Feduccia. 1997. Developmental patterns and the identification of homologies in the avian hand. *Science* 278: 666–668.
Butler, A.B., and W.M. Saidel. 2000. Defining sameness: Historical, biological, and generative homology. *BioEssays* 22: 846–853.
Darwin, C. 1859. *On the origin of species by means of natural selection, or the preservation of favoured races in the struggle for life*. London: Murray.
de Beer, G.R. 1958. *Embryos and ancestors*, 3rd ed. Oxford: Clarendon Press.
de Beer, G.R. 1971. *Homology: An unsolved problem*. London: Oxford University Press.
de Pinna, M.C.C. 1991. Concepts and tests of homology in the cladistic paradigm. *Cladistics* 7: 367–394.
DeRobertis, E.M., and Y. Sasai. 1996. A common plan for dorsoventral patterning in Bilateria. *Nature* 380: 37–40.
Dickinson, W.J. 1995. Molecules and morphology: Where's the homology? *Trends in Genetics* 11: 119–121.
Donoghue, M.J. 1992. Homology. In *Key words in evolutionary biology*, ed. E. Fox Keller and E.A. Lloyd, 170–179. Cambridge, MA: Harvard University Press.
Duboule, D. 1994. Temporal colinearity and the phylotypic progression: A basis for the stability of a vertebrate Bauplan and the evolution of morphologies through heterochrony. *Development* 1994(Supplement): 135–142.
Edgecombe, G.D. 2008. Anatomical nomenclature: Homology, standardization and datasets. In *Updating the Linnaean heritage: Names as tools for thinking about animals and plants*, Zootaxa 1950, ed. A. Minelli, L. Bonato, and G. Fusco, 87–95. Auckland: Magnolia Press.
Ereshefsky, M. 2007. Psychological categories as homologies: Lessons from ethology. *Biology and Philosophy* 22: 659–674.

Ereshefsky, M. 2010. Homology: Integrating phylogeny and development. *Biological Theory* 4: 225–229.

Ereshefsky, M. 2012. Homology thinking. *Biology and Philosophy* 27: 381–400.

Felsenstein, J. 2004. *Inferring phylogenies*. Sunderland: Sinauer Associates.

Ferrier, D.E.K. 2008. When is a Hox gene not a Hox gene? The importance of gene nomenclature. In *Evolving pathways. Key themes in evolutionary developmental biology*, ed. A. Minelli and G. Fusco, 175–193. Cambridge: Cambridge University Press.

Fitch, W.M. 1970. Distinguishing homologous from analogous proteins. *Systematic Zoology* 19: 99–113.

Fusco, G. 2008. Morphological nomenclature, between patterns and processes: Segments and segmentation as a paradigmatic case. In *Updating the Linnaean heritage: Names as tools for thinking about animals and plants*, Zootaxa 1950, ed. A. Minelli, L. Bonato, and G. Fusco, 96–102. Auckland: Magnolia Press.

Galis, F. 1996. The evolution of insects and vertebrates: Homeobox genes and homology. *Trends in Ecology and Evolution* 11: 402–403.

Galis, F. 1999. On the homology of structures and Hox genes: The vertebral column. In *Homology* (Novartis Foundation symposium 222), ed. G.R. Bock and G. Cardew, 80–94. Chichester: Wiley.

Gans, C. 1974. *Biomechanics: An approach to vertebrate biology*. Philadelphia: Lippincott.

Gehring, W.J. 2002. The genetic control of eye development and its implications for the evolution of the various eye-types. *The International Journal of Developmental Biology* 46: 65–73.

Geoffroy Saint-Hilaire, E. 1807. Considérations sur les pièces de la tête osseuse des animaux vertébrés, et particulièrement sur celles du crâne des oiseaux. *Annales du Muséum d'histoire naturelle Paris* 10: 342–365.

Geoffroy Saint-Hilaire, E. 1822. Considérations générales sur la vertèbre. *Mémoires du Muséum d'histoire naturelle Paris* 9: 89–119.

Ghiselin, M.T. 1976. The nomenclature of correspondence: A new look at "homology" and "analogy". In *Evolution, brain, and behavior: Persistent problems*, ed. R.B. Masterton, W. Hodos, and H. Jerison, 129–142. Hillsdale: Lawrence Erlbaum Association Publications.

Ghiselin, M.T. 2005. Homology as a relation of correspondence between parts of individuals. *Theory in Biosciences* 124: 91–103.

Ghiselin, M.T. 2006. The failure of morphology to contribute to the modern synthesis. *Theory in Biosciences* 124: 309–316.

Gilbert, S.F., and J.A. Bolker. 2001. Homologies of process and modular elements of embryonic construction. In *The character concept in evolutionary biology*, ed. G.P. Wagner, 437–456. San Diego: Academic Press.

Goodwin, B. 1993. Homology and a generative theory of biological form. *Acta Biotheoretica* 41: 305–314.

Greene, H.W. 1999. Natural history and behavioural homology. In *Homology* (Novartis Foundation symposium 222), ed. G.R. Bock and G. Cardew, 173–188. Chichester: Wiley.

Griffiths, P.E. 2006. Function, homology, and character individuation. *Philosophy of Science* 73: 1–25.

Grygier, M.J. 1994. Developmental patterns and hypotheses of homology in the antennules of thecostracan nauplius larvae (Crustacea). *Acta Zoologica* 75: 219–234.

Halder, G., P. Callaerts, and W.J. Gehring. 1995. Induction of ectopic eyes by targeted expression of the *eyeless* gene in *Drosophila*. *Science* 267: 1788–1792.

Hall, B.K. (ed.). 1994. *Homology: The hierarchical basis of comparative biology*. San Diego/London: Academic Press.

Hall, B.K. 1995. Homology and embryonic development. *Evolutionary Biology* 28: 1–37.

Hall, B.K. 1998. Germ layers and the germ-layer theory revisited: Primary and secondary germ layers, neural crest as a fourth germ layer, homology, and demise of the germ-layer theory. *Evolutionary Biology* 30: 121–186.

Hall, B.K. 2003. Descent with modification: The unity underlying homology and homoplasy as seen through an analysis of development and evolution. *Biological Reviews* 78: 409–433.

Hall, B.K. 2005. Consideration of the neural crest and its skeletal derivatives in the context of novelty/innovation. *Journal of Experimental Zoology Part B: Molecular and Developmental Evolution* 304B: 548–557.

Hall, B.K., and R. Kerney. 2012. Levels of biological organization and the origin of novelty. *Journal of Experimental Zoology Part B: Molecular and Developmental Evolution* 318: 428–437.

Hallam, S.J., and Jin Yishi. 1998. lin-14 regulates the timing of synaptic remodelling in *Caenorhabditis elegans*. *Nature* 395: 78–82.

Hanken, J. 1986. Developmental evidence for amphibian origins. *Evolutionary Biology* 20: 389–417.

Hanken, J. 1999. Larvae in amphibian development and evolution. In *The origin and evolution of larval forms*, ed. B.K. Hall and M.H. Wake, 61–108. San Diego/London/Boston/New York/Sydney/Tokyo/Toronto: Academic Press.

Hart, M.W., and G.A. Wray. 1999. Heterochrony. In *The origin and evolution of larval forms*, ed. B.K. Hall and M.H. Wake, 159–165. San Diego/London/Boston/New York/Sydney/Tokyo/Toronto: Academic Press.

Haszprunar, G. 1992. The types of homology and their significance for evolutionary biology and phylogenetics. *Journal of Evolutionary Biology* 5: 13–24.

Hejnol, A., and M.Q. Martindale. 2008. Acoel development supports a simple planula-like urbilaterian. *Philosophical Transactions of the Royal Society of London. Series B, Biological Sciences* 363: 1493–1501.

Hendler, G. 1978. Development of *Amphioplus abditus* (Verrill) (Echinodermata: Ophiuroidea). II. Description and discussion of ophiuroid skeletal ontogenies and homologies. *The Biological Bulletin* 154: 79–95.

Hendler, G. 1988. Ophiuroid skeleton ontogeny reveals homologies among skeletal plates of adults: A study of *Amphiura filiformis*, *Amphiura stimpsonii* and *Ophiophragmus filograneus* (Echinodermata). *The Biological Bulletin* 174: 20–29.

Hennig, W. 1966. *Phylogenetic systematics*. Urbana: University of Illinois Press.

Henry, J.J., and R.A. Raff. 1990. Evolutionary change in the process of dorsoventral axis determination in the direct developing sea urchin, *Heliocidaris erythrogramma*. *Developmental Biology* 141: 55–69.

Hinchliffe, J.R., and M.K. Hecht. 1984. Homology of the bird wing skeleton. Embryological versus paleontological evidence. *Evolutionary Biology* 18: 21–39.

Hinchliffe, J.R., and E.I. Vorobyeva. 1999. Developmental basis of limb homology in urodeles: Heterochronic evidence from the primitive hynobiid family. In *Homology* (Novartis Foundation symposium 222), ed. G.R. Bock and G. Cardew, 95–109. Chichester: Wiley.

Hodos, W. 1976. The concept of homology and the evolution of behavior. In *Evolution, brain, and behavior: Persistent problems*, ed. R.B. Masterton, W. Hodos, and H. Jerison, 153–167. Hillsdale: Lawrence Erlbaum Association Publications.

Holland, N.D., and L.Z. Holland. 1999. Amphioxus and the utility of molecular genetic data for hypothesizing body part homologies between distantly related animals. *American Zoologist* 39: 630–640.

Holland, N.D., L.Z. Holland, Y. Honma, and T. Fujii. 1993. *Engrailed* expression during development of a lamprey, *Lampetra japonica*: A possible clue to homologies between agnathan and gnathostome muscles of the mandibular arch. *Development Growth and Differentiation* 35: 153–160.

Holland, P.W.H. 1999. The effect of gene duplication on homology. In *Homology* (Novartis Foundation symposium 222), ed. G.R. Bock and G. Cardew, 226–242. Chichester: Wiley.

Hoßfeld, U., and L. Olsson. 2005. The history of the homology concept and the "phylogenetisches symposium". *Theory in Biosciences* 124: 243–253.

Inglis, W.G. 1966. The observational basis of homology. *Systematic Zoology* 15: 219–228.

Jardine, N. 1967. The concept of homology in biology. *The British Journal for the Philosophy of Science* 18: 125–139.

Jardine, N. 1969. The observational and theoretical components of homology: A study based on the morphology of the dermal skull-roofs of rhipidistian fishes. *Biological Journal of the Linnean Society* 1: 327–361.

King, N., M.J. Westbrook, S.L. Young, A. Kuo, M. Abedin, J. Chapman, S. Fairclough, U. Hellsten, Y. Isogai, I. Letunic, M. Marr, D. Pincus, N. Putnam, A. Rokas, K.J. Wright, R. Zuzow, W. Dirks, M. Good, D. Goodstein, D. Lemons, W. Li, J.B. Lyons, A. Morris, S. Nichols, D.J. Richter, A. Salamoi, J.G.I. Sequencing, P. Bork, W.A. Lim, G. Manning, W.T. Miller, W. McGinnis, H. Shapiro, R. Tjian, I.V. Grigoriev, and D. Rokhsar. 2008. The genome of the choanoflagellate *Monosiga brevicollis* and the origin of metazoans. *Nature* 451: 783–788.

Kleisner, K. 2007. The formation of the theory of homology in biological sciences. *Acta Biotheoretica* 55: 317–340.

Kosswig, G. 1961. Über sogenannte homologe Gene. *Zoologischer Anzeiger* 166: 333–356.

Laubichler, M.D. 2000. Homology in development and the development of the homology concept. *American Zoologist* 40: 777–788.

Lecointre, G., and H. Le Guyader. 2006. *The tree of life: A phylogenetic classification*. Cambridge, MA: Belknap Press of Harvard University Press.

Liem, K.F., and D.W. Wake. 1985. Morphology: Current approaches and concepts. In *Functional vertebrate morphology*, ed. M. Hildebrand, D.M. Bramble, K.F. Liem, and D.W. Wake, 366–377. Cambridge, MA: Harvard University Press.

Love, A.C. 2007. Functional homology and homology of function: Biological concepts and philosophical consequences. *Biology and Philosophy* 22: 691–708.

Love, A.C., and R.A. Raff. 2006. Larval ectoderm, organizational homology, and the origins of evolutionary novelty. *Journal of Experimental Zoology Part B: Molecular and Developmental Evolution* 306B: 18–34.

Mayr, W. 1969. *Principles of systematic zoology*. New York: McGraw-Hill.

McKinney, M.L. 1988. *Heterochrony in evolution. A multidisciplinary approach*. New York/ London: Plenum Press.

McKinney, M.L., and K.J. McNamara. 1991. *Heterochrony. The evolution of ontogeny*. New York/ London: Plenum Press.

McKinney, M.L., K.J. Mc Namara, and L.G. Zachos. 1990. Heterochronic hierarchies: Application and theory in evolution. *Historical Biology* 3: 269–287.

McKitrick, M.C. 1994. On homology and the ontological relationship of parts. *Systematic Biology* 43: 1–10.

McNamara, K.J. 1986. A guide to the nomenclature of heterochrony. *Journal of Paleontology* 60: 4–13.

Meyer, A. 1998. We are devo-evo. *Trends in Genetics* 14: 482–483.

Michaux, B. 1989. Homology: A question of form or a product of genealogy? *Rivista di Biologia* 82: 217–246.

Mindell, D.P., and A. Meyer. 2001. Homology evolving. *Trends in Ecology and Evolution* 16: 434–440.

Minelli, A. 1996. Some thoughts on homology, 150 years after Owen's definition. *Memorie della Società italiana di Scienze naturali e del Museo civico di Storia naturale di Milano* 27(1): 71–79.

Minelli, A. 1998. Molecules, developmental modules and phenotypes: A combinatorial approach to homology. *Molecular Phylogenetics and Evolution* 9: 340–347.

Minelli, A. 2000. Limbs and tail as evolutionarily diverging duplicates of the main body axis. *Evolution and Development* 2: 157–165.

Minelli, A. 2002. Homology, limbs and genitalia. *Evolution and Development* 4: 127–132.

Minelli, A. 2003. *The development of animal form: Ontogeny, morphology, and evolution*. Cambridge: Cambridge University Press.

Minelli, A., C. Brena, G. Deflorian, D. Maruzzo, and G. Fusco. 2006a. From embryo to adult – Beyond the conventional periodization of arthropod development. *Development Genes and Evolution* 216: 373–383.

Minelli, A., and G. Fusco. 1995. Body segmentation and segment differentiation: The scope for heterochronic change. In *Evolutionary change and heterochrony*, ed. K.J. McNamara, 49–63. London: Wiley.

Minelli, A., and G. Fusco. 2004. Evo-devo perspectives on segmentation: Model organisms, and beyond. *Trends in Ecology and Evolution* 19: 423–429.

Minelli, A., and G. Fusco. 2005. Conserved versus innovative features in animal body organization. *Journal of Experimental Zoology Part B: Molecular and Developmental Evolution* 304B: 520–525.

Minelli, A., E. Negrisolo, and G. Fusco. 2006b. Reconstructing animal phylogeny in the light of evolutionary developmental biology. In *Reconstructing the tree of life: Taxonomy and systematics of species rich taxa* (Systematics Association special series volume 72), ed. T.R. Hodkinson, J.A.N. Parnell, and S. Waldren, 177–190. Boca Raton: Taylor & Francis/CRC Press.

Minelli, A., and B. Peruffo. 1991. Developmental pathways, homology and homonomy in metameric animals. *Journal of Evolutionary Biology* 4: 429–445.

Minelli, A., and F.R. Schram. 1994. Owen revisited: A reappraisal of morphology in evolutionary biology. *Bijdragen tot de Dierkunde* 64: 65–74.

Moczek, A.P. 2008. On the origins of novelty in development and evolution. *BioEssays* 30: 432–447.

Moczek, A.P., and L.M. Nagy. 2005. Diverse developmental mechanisms contribute to different levels of diversity in horned beetles. *Evolution and Development* 7: 175–185.

Moczek, A.P., and D.J. Rose. 2009. Differential recruitment of limb patterning genes during development and diversification of beetle horns. *Proceedings of the National Academy of Sciences of the United States of America* 106: 8992–8997.

Müller, G.B. 2001. Homologie und Analogie: die vergleichende Grundlage von Morphologie und Ethologie. In *Konrad Lorenz und seine verhaltensbiologischen Konzepte aus heutiger Sicht*, ed. K. Kotrschal, G.B. Müller, and H. Winkler, 127–137. Fürth: Filander Verlag.

Müller, G.B. 2003. Homology: The evolution of morphological organization. In *Origination of organismal form: Beyond the gene in developmental and evolutionary biology*, ed. G.B. Müller and S.A. Newman, 51–69. Cambridge, MA: MIT Press.

Müller, G.B., and S.A. Newman. 1999. Generation, integration, autonomy: Three steps in the evolution of homology. In *Homology* (Novartis Foundation symposium 222), ed. G.R. Bock and G. Cardew, 65–79. Chichester: Wiley.

Müller, G.B., and G.P. Wagner. 1991. Novelty in evolution: Restructuring the concept. *Annual Review of Ecology and Systematics* 22: 229–256.

Müller, G.B., and G.P. Wagner. 1996. Homology, Hox genes, and developmental integration. *American Zoologist* 36: 4–13.

Müller, G.B., and G.P. Wagner. 2003. Innovation. In *Keywords and concepts in evolutionary developmental biology*, ed. B.K. Hall and W.M. Olson, 218–227. Cambridge, MA: Harvard University Press.

Nielsen, C., and P. Martinez. 2003. Patterns of gene expression: Homology or homocracy? *Development Genes and Evolution* 213: 149–154.

Nixon, K.C., and J.M. Carpenter. 2011. On homology. *Cladistics* 27: 1–10.

Osche, G. 1973. Das Homologisieren als eine grundlegende Methode der Phylogenetik. *Aufsätze und Reden der Senckenbergischen naturforschenden Gesellschaft* 24: 155–165.

Owen, R. 1843. *Lectures on the comparative anatomy and physiology of the invertebrate animals, delivered at the Royal College of Surgeons*. London: Longman Brown Green and Longmans.

Panchen, A.L. 1992. *Classification, evolution, and the nature of biology*. Cambridge: Cambridge University Press.

Panchen, A.L. 1994. Richard Owen and the concept of homology. In *Homology: The hierarchical basis of comparative biology*, ed. B.K. Hall, 21–62. San Diego/London: Academic Press.

Panchen, A.L. 1999. Homology – History of a concept. In *Homology* (Novartis Foundation symposium 222), ed. G.R. Bock and G. Cardew, 5–23. Chichester: Wiley.

Patterson, C. 1982. Morphological characters and homology. In *Problems of phylogenetic reconstruction* (The Systematics Association special volume no. 21), ed. K.A. Joysey and A.E. Friday, 21–74. London: Academic.

Patterson, C. 1988. Homology in classical and molecular biology. *Molecular Biology and Evolution* 5: 603–625.

Peters, K. 1922. Zur Geschichte und Kritik des Begriffs "Homologie" und seine Anwendung in der Embryologie. *Biologisches Zentralblatt* 42: 308–327.

Pigliucci, M. 2001. Characters and environments. In *The character concept in evolutionary biology*, ed. G.P. Wagner, 363–388. San Diego: Academic Press.
Priest, G. 2000. *Logic: A very short introduction*. Oxford: Oxford University Press.
Raff, R.A. 1999. Larval homology and radical evolutionary changes in early development. In *Homology* (Novartis Foundation symposium 222), ed. G.R. Bock and G. Cardew, 110–124. Chichester: Wiley.
Raff, R.A., B.A. Parr, A.L. Parks, and G.A. Wray. 1990. Heterochrony and other mechanisms of radical evolutionary change in early development. In *Evolutionary innovations*, ed. M.H. Nitecki, 71–98. Chicago: University of Chicago Press.
Raff, R.A., and G.A. Wray. 1989. Heterochrony: Developmental mechanisms and evolutionary results. *Journal of Evolutionary Biology* 2: 409–434.
Ramsey, G., and A.S. Peterson. 2012. Sameness in biology. *Philosophy of Science* 79: 255–275.
Reeck, G.R.C., C. De Haen, D.C. Doolittle, W.M. Fitch, R.E. Dickerson, P. Chambon, A. McLachlan, E. Margoliash, T.H. Jukes, and E. Zuckerkandl. 1987. "Homology" in proteins and nucleic acids: A terminology muddle and a way out of it. *Cell* 50: 667.
Reif, W.E. 2004. Problematic issues of cladistics: 3. Homology and apomorphy; ontology versus operationalism. *Neues Jahrbuch für Geologie und Paläontologie, Abhandlungen* 231: 395–422.
Reilly, S.M., E.O. Wiley, and D.T. Meinhardt. 1997. An integrative approach to heterochrony: The distinction between interspecific and intraspecific phenomena. *Biological Journal of the Linnean Society* 60: 119–143.
Remane, A. 1952. *Die Grundlagen des natürlichen Systems, der vergleichenden Anatomie und der Phylogenetik*. Leipzig: Akademische Verlagsgesellschaft Geest und Portig.
Remane, A. 1955. Morphologie als Homologienforschung. *Verhandlungen der Deutschen zoologischen Gesellschaft* 1954: 159–183.
Remane, A. 1963. Über die Homologisierungsmöglichkeiten bei Verbindungsstrukturen (Muskeln, Blutgefäßen, Nerven) und Hohlräumen. *Zoologischer Anzeiger* 166: 481–489.
Remane, J. 1989. Die Entwicklung des Homologie-Begriffs seit Adolf Remane. *Zoologische Beiträge (Berlin) N.F.* 32: 497–503.
Reuter, R. 1994. The gene *serpent* has homeotic properties and specifies endoderm versus ectoderm within the *Drosophila* gut. *Development* 120: 1123–1135.
Richardson, M.K. 1995. Heterochrony and the phylotypic period. *Developmental Biology* 172: 412–421.
Richardson, M.K., J. Hanken, M.L. Gooneratne, C. Pieau, A. Raynaud, L. Selwood, and G.M. Wright. 1997. There is no highly conserved embryonic stage in the verytebrates: Implications for current theories of evolution and development. *Anatomy and Embryology* 196: 91–106.
Richter, S. 2005. Homologies in phylogenetic analyses – Concept and tests. *Theory in Biosciences* 124: 105–120.
Riedl, R. 1980. Homologien; ihre Grunde und Erkenntnisgrunde. *Verhandlungen der Deutschen zoologischen Gesellschaft* 73: 164–176.
Rieppel, O.C. 1988. *Fundamentals of comparative biology*. Basel/Boston/Berlin: Birkhäuser.
Rieppel, O. 1992. Homology and logical fallacy. *Journal of Molecular Biology* 5: 701–715.
Rieppel, O. 1993a. Studies on skeleton formation in reptiles. II. *Chamaeleo hoehnelii* (Squamata: Chamaeleoninae), with comments on the homology of carpal and tarsal bones. *Herpetologica* 49: 66–78.
Rieppel, O. 1993b. Studies on skeleton formation in reptiles. IV. The homology of the reptilian (amniote) astragalus revisited. *Journal of Vertebrate Paleontology* 13: 31–47.
Rieppel, O. 2005. Modules, kinds, and homology. *Journal of Experimental Zoology Part B: Molecular and Developmental Evolution* 304B: 18–27.
Rolian, C., D.E. Lieberman, and B. Hallgrímsson. 2010. The coevolution of human hands and feet. *Evolution* 64: 1558–1568.
Roth, V.L. 1984. On homology. *Biological Journal of the Linnean Society* 22: 13–29.
Roth, V.L. 1988. The biological basis of homology. In *Ontogeny and systematics*, ed. C.J. Humphries, 1–26. London/New York: British Museum (Natural History)/Columbia University Press.

Roth, V.L. 1991. Homology and hierarchies: Problems solved and unresolved. *Journal of Evolutionary Biology* 4: 167–194.
Rutishauser, R., and P. Moline. 2005. Evo-devo and the search for homology ("sameness") in biological systems. *Theory in Biosciences* 124: 213–241.
Sander, K. 1983. The evolution of patterning mechanisms: Gleanings from insect embryogenesis. In *Development and evolution*, ed. B.C. Goodwin, N. Holder, and C.C. Wylie, 137–159. Cambridge: Cambridge University Press.
Sander, K. 1989. Zum Geleit: Homologie und Ontogenese. *Zoologische Beiträge (Berlin) N.F.* 32: 323–326.
Sattler, R. 1994. Homology, homeosis, and process morphology in plants. In *Homology: The hierarchical basis of comparative biology*, ed. B.K. Hall, 423–475. San Diego: Academic Press.
Schierwater, B., and K. Kuhn. 1998. Homology of Hox genes and the zootype concept in early metazoan evolution. *Molecular Phylogenetics and Evolution* 9: 375–381.
Schlosser, G. 2002. Modularity and the units of evolution. *Theory in Biosciences* 121: 1–80.
Schlosser, G., and G.P. Wagner (eds.). 2004. *Modularity in development and evolution*. Chicago/London: University of Chicago Press.
Schmitt, M. 1989. Das Homologie-Konzept in Morphologie und Phylogenetik. *Zoologische Beiträge (Berlin) N.F.* 32: 505–512.
Schmitt, M. 1995. The homology concept – Still alive. In *The nervous systems of invertebrates. An evolutionary and comparative approach*, ed. O. Breidbach and O. Kutsch, 425–438. Basel: Birkhäuser.
Scholtz, G. 2005. Homology and ontogeny: Pattern and process in comparative developmental biology. *Theory in Biosciences* 124: 121–143.
Scholtz, G. 2008. On comparisons and causes in evolutionary developmental biology. In *Evolving pathways*, ed. A. Minelli and G. Fusco, 144–159. Cambridge: Cambridge University Press.
Scholtz, G. 2010. Deconstructing morphology. *Acta Zoologica (Stockholm)* 91: 44–63.
Serres, E. 1827. Recherches d'anatomie trascendante sur les lois de l'organogénie appliquées à l'anatomie pathologique. *Annales de sciences naturelles Paris* 11: 47–70.
Shubin, N., and D. Wake. 1996. Phylogeny, variation, and morphological integration. *American Zoologist* 36: 51–60.
Shubin, N., C. Tabin, and S. Carroll. 2009. Deep homology and the origins of evolutionary novelty. *Nature* 457: 818–823.
Slack, F., and G. Ruvkun. 1997. Temporal pattern formation by heterochronic genes. *Annual Review of Genetics* 31: 611–634.
Slack, J.M.W., P.W.H. Holland, and C.F. Graham. 1993. The zootype and the phylotypic stage. *Nature* 361: 490–492.
Sluys, R. 1996. The notion of homology in current comparative biology. *Journal of Zoological Systematics and Evolutionary Research* 34: 145–152.
Sober, E. 1988. *Reconstructing the past: Parsimony, evolution, and inference*. Cambridge, MA: MIT Press.
Sober, E. 2008. *Evidence and evolution: The logic behind the science*. Cambridge: Cambridge University Press.
Sommer, R.J. 2008. Homology and the hierarchy of biological systems. *BioEssays* 30: 653–658.
Sommer, R.J., and D. Tautz. 1994. Expression patterns of *twist* and *snail* in *Tribolium* (Coleoptera) suggest a homologous formation of mesoderm in long and short germ band insects. *Developmental Genetics* 15: 32–37.
Sondergaard, L. 1993. Homology between the mammalian liver and the *Drosophila* fat body. *Trends in Genetics* 9: 193.
Spemann, H. 1915. Zur Geschichte und Kritik des Begriffes der Homologie. In *Die Kulture der Gegenwart, 3, IV/1, Allgemeine Biologie*, ed. P. Hinneberg, 63–86. Leipzig: Teubner.
Stevens, P.F. 2000. On characters and character states: Do overlapping and non-overlapping variation, morphology and molecules all yield data of the same value? In *Homology and systematics: Coding characters for phylogenetic analysis*, ed. R.W. Scotland and R.T. Pennington, 81–105. New York: Taylor & Francis.

Striedter, G.F., and R.G. Northcutt. 1991. Biological hierarchies and the concept of homology. *Brain Behavior and Evolution* 38: 177–189.
Sudhaus, W. 1980. Problembereiche der Homologienforschung. *Verhandlungen der Deutschen zoologischen Gesellschaft* 73: 177–187.
Szucsich, N., and C.S. Wirkner. 2007. Homology: A synthetic concept of evolutionary robustness of patterns. *Zoologica Scripta* 36: 281–289.
Tabin, C., and E. Laufer. 1993. Hox genes and serial homology. *Nature* 361: 692–693.
Van Valen, L. 1982. Homology and causes. *Systematic Botany* 173: 305–312.
Voigt, W. 1973. *Homologie und Typus in der Biologie*. Jena: Fischer.
Wagner, G.P. 1989a. The biological homology concept. *Annual Review of Ecology and Systematics* 20: 51–69.
Wagner, G.P. 1989b. The origin of morphological characters and the biological basis of homology. *Evolution* 43: 1157–1171.
Wagner, G.P. 1994. Homology and the mechanisms of development. In *Homology: The hierarchical basis of comparative biology*, ed. B.K. Hall, 273–299. New York: Academic Press.
Wagner, G.P. 1996. Homologues, natural kinds and the evolution of modularity. *American Zoologist* 36: 36–43.
Wagner, G.P. 1999. A research programme for testing the biological homology concept. In *Homology* (Novartis Foundation symposium 222), ed. G.R. Bock and G. Cardew, 125–152. Chichester: Wiley.
Wagner, G.P. 2007. The developmental genetics of homology. *Nature Reviews Genetics* 8: 473–479.
Wagner, G.P., and L. Altenberg. 1996. Perspective: Complex adaptations and the evolution of evolvability. *Evolution* 50: 967–976.
Wagner, G.P., and P.F. Stadler. 2003. Quasi-independence, homology and the unity of type: A topological theory of characters. *Journal of Theoretical Biology* 220: 505–527.
Wake, D.B. 1999. Homoplasy, homology and the problem of 'sameness' in biology. In *Homology* (Novartis Foundation symposium 222), ed. G.R. Bock and G. Cardew, 24–46. Chichester: Wiley.
Wenzel, J.W. 1992. Behavioral homology and phylogeny. *Annual Review of Ecology and Systematics* 23: 361–381.
Williams, T.A. 1999. Morphogenesis and homology in arthropod limbs. *American Zoologist* 39: 664–675.
Williams, D.M. 2004. Homology and homologues, cladistics and phenetics: 150 years of progress. In *Milestones in systematics*, ed. D.M. Williams and P.L. Forey, 191–224. Boca Raton: CRC Press.
Williams, D.M., and M.C. Ebach. 2008. *Foundations of systematics and biogeography*. New York: Springer.
Williams, D.M., and C.J. Humphries. 2004. Homology and character evolution. In *Deep morphology: Toward a renaissance of morphology in plant systematics*, ed. T. Stuessy, E. Hörandl, and V. Mayer, 119–130. Königstein: Koeltz.
Wilson, E.B. 1894. The embryological criterion of homology. In *Biological lectures delivered at the Marine Biological Laboratory Wood's Hole*, 101–124. Boston: Ginn & Co.
Wray, G.A. 1999. Evolutionary dissociations between homologous genes and homologous structures. In *Homology* (Novartis Foundation symposium 222), ed. G.R. Bock and G. Cardew, 189–206. Chichester: Wiley.
Wray, G.A., and E. Abouheif. 1998. When is homology not homology? *Current Opinion in Genetics and Development* 8: 675–680.
Wray, G.A., and R.A. Raff. 1990. Novel origins of lineage founder cells in the direct-developing sea urchin *Heliocidaris erythrogramma*. *Developmental Biology* 141: 41–54.
Young, B.A. 1993. On the necessity of an archetypal concept in morphology with special reference to the concepts of "structure" and "homology". *Biology and Philosophy* 8: 225–248.
Young, R.L., and G.P. Wagner. 2011. Why ontogenetic homology criteria can be misleading: Lessons from digit identity transformations. *Journal of Experimental Zoology Part B: Molecular and Developmental Evolution* 316B: 165–170.

Teaching Evolutionary Developmental Biology: Concepts, Problems, and Controversy

Alan C. Love

1 Images of Science: Problems as Organizers of Inquiry

One of the most enduring images of science is due to Karl Popper: *falsification*. Instead of hypotheses receiving incremental confirmation by induction (e.g., via individual observations), Popper famously argued that hypotheses cannot be confirmed, only falsified (Popper 2002 [1963]). A reason for the endurance of this image is a natural resonance with scientific practice. Many biologists see themselves engaged in a form of reasoning that seemingly corresponds to Popperian falsification, and subsequent developments of these ideas for statistical hypothesis testing have sharpened this comparison (Mayo 1996). But Popper also recognized other dimensions of scientific practice, and one of these is germane for students of biology who are contemplating whether to pursue a career in the life sciences, as well as for a scientifically literate citizenry that will seek diverse career options.

> Try to learn what people are discussing *nowadays* in science. Find out where the difficulties arise, and *take an interest in disagreements*. These are the questions that you should take up. In other words, *you should study the problem situation of the day* (Popper 2002 [1963], p. 129, emphasis mine).

The stress on "nowadays" and "of the day" is enshrined explicitly in the *National Science Education Standards*: "Science content increases and changes, and a teacher's understanding [...] must keep pace" (NRC 1996, p. 57). Although it is difficult to disagree with the recommendation to incorporate recent scientific developments into the curriculum, it is a daunting task for educators. The life sciences contain

A.C. Love (✉)
Department of Philosophy, Minnesota Center for Philosophy of Science,
University of Minnesota, 831 Heller Hall, 271 19th Ave. S.,
Minneapolis, MN 55455, USA
e-mail: aclove@umn.edu

K. Kampourakis (ed.), *The Philosophy of Biology: A Companion for Educators*, History, Philosophy and Theory of the Life Sciences 1, DOI 10.1007/978-94-007-6537-5_16,

diverse and heterogeneous disciplines, and the time available to fit the spectrum of new findings into a school day has not increased.

Popper's emphasis on disagreements and problem situations may be even more daunting. The call to "teach the controversy" is laced with intelligent design machinations (Scott and Branch 2003; Sarkar 2007; see also Brigandt this volume), and understanding the problem situations in contemporary biology is a task that goes beyond conveying recent discoveries. And yet focusing on problem situations also exhibits a natural resonance with biological practice, as Popper was aware: "we are not students of some subject matter but students of problems" (Popper 2002 [1963], p. 88). Although the significance of problems as *initiators* of scientific inquiry is woven into the fabric of science education, there is little discussion of problems as *ongoing guides to* or *organizers of* inquiry.[1] The trajectory of research rides on *erotetic*[2] rails: "from problems to problems—to problems of ever increasing depth" (Popper 2002 [1963], p. 301). For problems to have depth, they must have structure, and it is this erotetic structure that plays an organizing role in biological practice. Because of the organizing role of problems, it is critical to incorporate this image of science into multiple levels of life science instruction (Love 2013a), especially when addressing the issue of scientific disagreement. To do so, we need to turn our attention to "what people are discussing nowadays in science."

2 Some Evo-Devo Concepts, Problems, and "Controversies"

Evolutionary developmental biology (Evo-devo) is a vibrant area of contemporary life science and is finding its way into teaching curricula at a variety of different instructional levels. From *Hox* genes to the origin of turtle shells, teaching how development evolves and how development structures the evolution of organismal traits is central to biology education and is encouraged by abundant and accessible presentations (Arthur 2004, 2011; Bateson and Gluckman 2011; Carroll 2005; Kirschner and Gerhart 2005; Minelli 2009; Shubin 2008; Stern 2011). This two-fold elucidation—how development evolves and how development structures evolution—is a helpful starting point for thinking about Evo-devo's loose conglomeration of research programs (Raff 2000; Müller 2007).

1. The *evolution of development* (how development evolves): inquiry into the patterns and processes of how ontogeny varies and changes over time.
2. The *developmental basis of evolution* (how development structures the evolution of organismal traits): inquiry into the causal impact of ontogenetic processes on evolutionary trajectories.

[1] Theories or hypotheses are assumed to guide inquiry: "Hypotheses are widely used in science for choosing what data to pay attention to and what additional data to seek" (AAAS 2009, ch. 1).

[2] "Pertaining to questioning": derived from the Greek noun *erotisis*, which means "a question."

Although much of what gets emphasized in popular (and professional) presentations pertains to the comparative developmental genetics of metazoans (Carroll 2005; De Robertis 2008), where the focus is on conserved genetic regulatory networks and signaling pathways underlying developmental processes (commonly collected under the label ‘genetic toolkit'), the two-fold characterization of Evo-devo encompasses a wider range of disciplinary approaches. For example, because the developmental genetics is *comparative*, phylogenetic systematics plays a critical role in drawing evolutionary inferences (Telford and Budd 2003). Historically, disciplines such as morphology and paleontology were the loci of concerns about both the evolution of development and the developmental basis of evolution (Love 2003, 2007).[3] In contemporary research, paleontology, comparative embryology and morphology, experimental investigations of epigenetic dynamics at different levels of organization, and computational or simulation oriented inquiry are all pertinent (Müller 2007; Raff 2007).

Much of twentieth century evolutionary biology concentrated on adult phenotypes, whether morphological or behavioral.[4] The structural features and adaptive significance of different developmental pathways, such as larval stages or the environmental induction of different traits from one genotype (*polyphenisms*), were relegated to the background of evolutionary theorizing. As a consequence, a bias emerged in model organisms upon which evolutionary biology forged its theoretical commitments. Animals exhibiting complex life histories (e.g., marine invertebrates) were neglected for vertebrates and arthropods without complicated metamorphic events (e.g., a change in basic body plan symmetry), which displayed adult phenotypes that could be measured quantitatively (Love 2009). Life history theory redressed some of this neglect from a population biology standpoint (Stearns 1992), but its abstract population-based explanatory framework of resource investment strategies and parent-offspring conflict is in sharp contrast to one involving the concrete, mechanistic details of development. Many Evo-devo researchers conceptualize evolutionary change in terms of changes in gene regulation during ontogeny, with a special emphasis on *cis*-regulatory elements (Davidson 2006). Revisiting substantive questions, such as the evolutionary origin of larvae (Raff 2008), requires model systems that exhibit the relevant kind of variation in life history for gathering molecular and embryological data to test evolutionary hypotheses in the framework of explicit phylogenies (Raff et al. 2003).

Even though Evo-devo biologists are engaged in hypothesis testing, generating a theory is not central to most investigations of the evolutionary significance of

[3]Claims of a recent, developmental genetic genesis for Evo-devo should be treated suspiciously. For example, "Evo-devo began in the pre-genomic era when genetic studies in *Drosophila* and gene cloning in *Xenopus* revealed that the *Hox* genes that control the anterior-posterior (A-P) axis were unexpectedly conserved" (De Robertis 2008, p. 186).

[4]These adult phenotypes were primarily exemplified in multicellular animals (metazoans), as well as some plants. Microbial phenotypes, whether morphological or behavioral, were largely neglected (see Duncan et al. this volume).

developmental mechanisms. This is at odds with the image of science articulated for teaching students.[5] There is no *Evo-devo theory* to complement *life history theory*; the research is not organized by an overarching theoretical structure. The evolution of development and the significance of development for evolution are routinely explored within the framework of a family of recurring concepts that includes constraints, modularity, evolvability, and novelty, among others (Arthur 2002). Instead of simply serving to categorize phenomena, these concepts play roles in marking out core research problems that represent properties relevant to understanding the evolution of development and the developmental basis of evolution (Brigandt and Love 2010, 2012b). One classic discussion centers on *constraints* or biases on the production of phenotypic variation due to characteristic features of developmental processes (Maynard Smith et al. 1985).[6] The main point of contention at first seemed to be whether these somehow retarded the operation of natural selection (e.g., leading to sustained stasis in the fossil record) or facilitated some evolutionary trajectories over others, thereby diminishing the power of adaptive explanations of phenotypes. For example, the order of formation in amphibian digit development can explain the evolutionary pattern of digital reduction in these lineages (Alberch and Gale 1985): frogs experiencing hind limb digital reduction lost pre-axial digits ('big toes') because they form last during ontogeny; salamanders experiencing hind limb digital reduction lost post-axial digits ('pinky toes') because they form last during ontogeny (Fig. 1).

This disagreement appeared to revolve around the relative explanatory power of natural selection (from evolutionary genetics) versus developmental constraints (from Evo-devo). But the disagreement actually turned on a terminological ambiguity (Amundson 1994). Many evolutionary biologists interpreted constraints through the lens of the problem of adaptation (non-optimal phenotypes are constrained), whereas Evo-devo researchers looked at constraints as an aspect of the problem of variation and how development underlies its distribution (independent of adaptive value). This was not a situation of rival explanations—a hypothesis testing perspective on scientific controversy—but rather a case of distinct explanatory endeavors. The problems representing these differences (adaptation and variation) are oriented around different features of evolution: explaining the process of evolutionary change from one adult phenotype to another via population processes such as natural selection, which sorts phenotypes, alters allele frequencies, and yields adaptive outcomes; versus, explaining the process of evolutionary change from one ontogeny to another via developmental processes such as morphogenesis, which can be altered

[5]"Theories compete for acceptance;" science is construed as "the testing, revising, and occasional discarding of theories" (AAAS 2009, ch. 1).

[6]The terminology of constraints connotes negativity or prevention but developmental constraints sometimes provide positive evolutionary opportunities (Gould 2002). As a result, some authors prefer "bias" as a more general designator, with "constraints" being one species of the genus (Arthur 2004). Here I do not distinguish between these different connotations and use constraint and bias interchangeably.

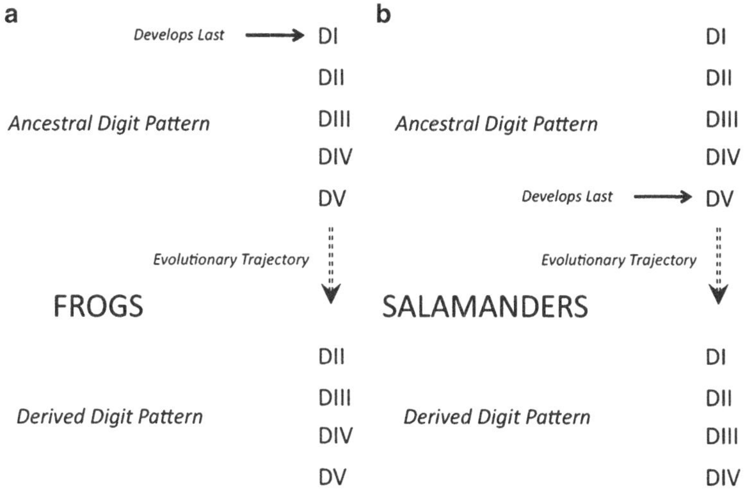

Fig. 1 Digital reduction trends in frogs and salamanders. A simplified, schematic representation of how the order of formation in amphibian digit development explains the evolutionary pattern of digital reduction in these two lineages (Alberch and Gale 1985). (**a**) Frogs experiencing hind limb digital reduction lost pre-axial digits ('big toes') because they form last during ontogeny. (**b**) Salamanders experiencing hind limb digital reduction lost post-axial digits ('pinky toes') because they form last during ontogeny

in different ways to generate novel morphologies (Amundson 2005; see also Potochnik, this volume). Recent treatments of constraint recognize these differences (Arthur 2011).

Another key concept in Evo-devo is *modularity* (Bolker 2000), which represents questions related to our two-fold elucidation of Evo-devo: (a) how do modules originate or dissolve during evolution (the evolution of development)? and, (b) how does modularity affect evolution (the developmental basis of evolution)? Modularity, in the sense of discrete sets of pleiotropic interactions during development, is pervasive (Wagner and Zhang 2011), and this architecture may channel evolutionary change in lineages. Modules, in the sense of quasi-autonomous parts, are ubiquitous in biological systems at different hierarchical levels (Kuratani 2009), whether genes, cells, tissues, organs, or anatomy, and this type of organization has evolved repeatedly. The question of how modularity affects evolution leads us to a third concept: *evolvability* (i.e., the capacity to generate heritable, selectable phenotypic variation). Modularity appears to be a key property that underlies evolvability because modules can exhibit variation and be independently modified without disrupting other features that are critical for an organism's survival and reproduction (Kirschner and Gerhart 1998).[7]

[7]Others properties underlying evolvability include the versatility of cell components, weak regulatory linkages, and exploratory behavior (see Kirschner and Gerhart 2005).

One controversy surrounding evolvability is whether the conserved cellular machinery that produce properties like modularity during development facilitates links between random genetic mutation and phenotypic variation so that viable character assemblages are more likely to emerge (Gerhart and Kirschner 2007). This seems to contradict the standard conception that mutations are random with respect to their effects on viability. But this may not be a conflict between two different theoretical claims—a theory testing perspective on scientific controversy. As with constraint, a more fruitful interpretation is in terms of different problems. The claim about random genetic mutation is meant to apply in the context of the problem of adaptation; adaptations do not arise because genetic mutation is biased toward viability. The claim about facilitated phenotypic variation is meant to apply in the context of the problem of variation; the capacity to generate phenotypic variation is biased due to aspects of organismal development, such as developmental plasticity, which make possible coordinated changes among different traits (e.g., the innervation and vascularization of an appendage). Random genetic variation and facilitated phenotypic variation are consistent claims that emerge from work on different problems (adaptation and variation). Again, this is not a situation of rivalry, and attending to how problems organize (and not just initiate) research around different evolutionary questions can diffuse the controversy.

3 A Genuine Controversy

The diffusion of controversy in the cases of constraint and evolvability by appeal to the biological practice of working on different problems (see above, Sect. 2) might leave the impression that Evo-devo doesn't harbor genuine controversies. Not true. One source of controversy[8] is nestled within the perspective of separate problems requiring distinct approaches in evolutionary biology. Both traditional population biologists and Evo-devo proponents have claimed that their approaches, and hence their problems, are more fundamental than the other. On the side of the former, the claim has been made forcefully:

> The litmus test for any evolutionary hypothesis must be its consistency with fundamental population genetic principles [...] population genetics provides an essential framework for understanding how evolution occurs (Lynch 2007, p. 8598).

On the side of the latter, the claim has been made specifically with respect to the teaching of evolution: Evo-devo approaches better illustrate how evolutionary change can occur (Gilbert 2003), or emphasize the most central features of the evolutionary process.

---

[8]Other controversies that might have been explored include disagreements about whether most molecular change during evolution occurs in *cis*-regulatory regions of the genome that control gene expression or within protein coding regions of the genome (see Hoekstra and Coyne 2007).

> Millions of biology students have been taught the view (from population genetics) that 'evolution is change in gene frequencies.' [...] This view forces the explanation toward mathematics and abstract descriptions of genes, and away from butterflies and zebras [...] The evolution of form is the main drama of life's story, both as found in the fossil record and in the diversity of living species. So, let's teach that story. Instead of 'change in gene frequencies,' let's try 'evolution of form is change in development' (Carroll 2005, p. 294).

Simply detailing how variation and adaptation are distinct problems (see above, Sect. 2) will not diffuse this controversy. It raises thorny issues about how knowledge is structured, and whether some form of reductionism is warranted (Brigandt and Love 2012a). Should one set of approaches or discipline be considered the "main drama" or more fundamental than the others (e.g., population genetics)? Must all of the problems and the approaches necessary to address them be unified under a single explanatory framework? Should we think of some concepts as *most* central to Evo-devo, such as gene regulatory networks (Laubichler 2009) or evolvability (Hendrikse et al. 2007), or should Evo-devo and evolutionary biology be conceptualized as differentially organized depending on the problems in view (Love 2008a, 2013b; Brigandt 2010)? Does Evo-devo challenge the dominant explanatory perspective of the Modern Synthesis (Laubichler 2010), or is it simply complementary (Minelli 2010)? These issues cannot be settled quickly and are indicators of genuine controversy in Evo-devo, a big controversy about how knowledge is organized. Appeals to science as an activity of theory confirmation or hypothesis testing will be inadequate to resolve it since problems are involved in organizing inquiry, not just initiating it. Erotetic units guide biological practice, but we have yet to see how, and this is essential to finding resources for comprehending this genuine controversy over knowledge organization and whether population genetics is more fundamental than Evo-devo (or *vice versa*). One tactic for approaching this more manageably is to zero in on an issue that has been intimately related to the controversy: the origin of evolutionary novelty.

4 Controversy over Explaining the Origin of Novelty

The concept of evolutionary novelty, which represents another Evo-devo research problem (Love 2008a), provides an opportunity to see a controversy of a different kind. Instead of being defined on an axis of different problems (e.g., adaptation and variation), this Evo-devo controversy arises within the context of one problem: explaining the origin of evolutionary novelties. Novelties have been defined as morphological traits that are not homologous to features in an ancestral lineage and designate developmental variation that is not experimentally accessible in extant species meant to represent ancestral lineages (Müller and Newman 2005; see also Minelli and Fusco, this volume). How does qualitatively new phenotypic variation originate at particular phylogenetic junctures? One notable example is the origin of feathers (Prum and Brush 2002); we do not observe and cannot experimentally induce feather-like structures in existing "reptile" species with scales. The study of novelty is a defining aspect of Evo-devo because it concerns the developmental generation of phenotypic variation, not its adaptive spread through a population.

> Evolutionary innovations are outside the scope of any current research program. Through its contribution to the solution of that question, [Evo-devo] genuinely expands the explanatory range of evolutionary theory (Wagner et al. 2000, p. 822).

> It is essential to include developmental mechanisms in the explanation of evolutionary innovations. [...] This is also the reason why developmental evolution makes an indispensable contribution to evolutionary biology (Wagner 2000, p. 97).

As expected given the stress on genetic regulatory evolution in Evo-devo, most explanations of novelty emphasize developmental genetics.

> The evolution of new morphological features is due predominantly to modifications of spatial patterns of gene expression (Gompel et al. 2005, p. 481).

> Ancient regulatory circuits provide a substrate from which novel structures can develop [...] new structures need not arise from scratch, genetically speaking, but can evolve by deploying regulatory circuits that were first established in early animals (Shubin et al. 2009, pp. 818, 822).

But this developmental genetic explanation of the origin of novelty has a rival: the generic physical explanation. Some argue that early in evolution generic properties of cells and tissues (e.g., self-organization and geometry) and physical mechanisms of soft condensed materials (e.g., diffusion and viscoelasticity) interacted with environmental forces to yield basic metazoan morphologies with minimal developmental genetic machinery (Newman 1994; Newman et al. 2006; Newman and Bhat 2009). This 'pre-Mendelian' world was replaced through an evolutionary process in which morphologies became stabilized by developmental genetic mechanisms through genetic assimilation, yielding the 'Mendelian' world that we now observe.[9] The disagreement between the two explanations is stark:

> (i) <u>Developmental genetic explanation</u>: "novelty requires the evolution a new gene regulatory network" (Wagner and Lynch 2010, R50); "evolutionary change in animal form cannot be explained except in terms of change in gene regulatory network architecture" (Davidson 2006, p. 29).

> (ii) <u>Generic physical explanation</u>: "epigenetic mechanisms, rather than genetic changes, are the major sources of morphological novelty in evolution" (Newman et al. 2006, p. 290).

We appear to be on traditional ground; two competing hypotheses is the archetype of scientific controversy. But this may not be the only interpretation. Recall that in Sect. 3 we identified the controversy as epistemological (how knowledge is organized or structured), not empirical. Following Popper's advice—"you should study the problem situation of the day"—we need to scrutinize the anatomy of the problem of evolutionary novelty to determine whether this controversy is solely empirical, or whether it might also contain epistemological elements.

---

[9]Thus, 'Mendelian' refers to standard transmission and developmental genetic processes we find in contemporary organisms. 'Pre-Mendelian' signifies that these standard processes were not yet in place even though phenotypes were being generated through environmental forces interacting with soft condensed materials according to physical principles.

5 Explaining the Origin of Novelty: Problem Structure

Although it is true that complex scientific problems are not structured logically or "well-defined" (Osbeck et al. 2011, ch. 3), this does not mean that they are unstructured and their role in guiding inquiry suggests otherwise. But what is this structure and how does it organize research? Popper argued that as scientists, "we are not students of some subject matter but students of problems" (Popper 2002 [1963], p. 88). This is the role of problems in initiating inquiry, and one often noted when teaching biology. By observing a pattern in the history of science, Popper also saw a role for problems to guide inquiry: "from problems to problems—to problems of ever increasing depth" (p. 301).

For a problem to have 'increasing depth,' it cannot be equated with a standard interrogative. Biological problems—such as how cells differentiate or how evolutionary novelties originate—are not single questions similar to interrogatives like "who broke the window?" They constitute an agenda, a list of things that need to be addressed, and concern many different but interrelated questions. To keep this in mind, we can designate them explicitly as *problem agendas* (Love 2008a). They are indicative of long-term investigative programs and require contributions from more than one disciplinary approach. The interrelations among the questions are not haphazard but constitute an anatomy or problem structure ('depth'), which provides an organizational framework for coordinating inquiry among different disciplinary approaches.

Problem agenda structure has at least three dimensions. First, problem agendas have a *history*, and the discussions and debates surrounding their historical pathways yield structural relations (Hattiangadi 1978, 1979). Second, problem agendas are epistemologically *heterogeneous* in the sense of containing different kinds of questions (Laudan 1977), among which there are specific relationships. Third, structure can be found in the *hierarchical* relationships among questions in problem agendas, such as definable arrays of questions that can be thought of as parts to the whole (Nickles 1981). These three dimensions of problem agenda structure can be observed concretely in the case of explaining evolutionary novelty:

- History: To the extent that twentieth century neo-Darwinian evolutionary theory recognized the evolution of novelty as a problem at all, it assumed that population genetic processes were explanatorily sufficient. Marginalized disciplines, including morphology and paleontology, maintained an active interest in the problem (Love 2007) and the explosion of experimental tools from developmental genetics over the past two decades has facilitated a revisiting of these questions in many disciplines simultaneously (Love and Raff 2003). This historical controversy structures the problem agenda through debate about its component questions and what disciplines are needed to address them.
- Heterogeneity: The existence of different types of questions in the problem agenda requires distinct intellectual contributions to answer them. Empirical questions ('what regulatory genes control axis specification?') are answered differently than theoretical questions ('how is pleiotropy represented in a mathematical model?'); pattern questions ('what are the phylogenetic junctures for understanding the origin of segmentation?') are answered differently from

process questions ('how can changes in *cis*-regulatory binding sites contribute to heterotopy?[10]'); questions about the cellular level of organization are answered differently from questions about anatomy.

- Hierarchy: Different question components and strands of historical debate stand in systematic relations of abstraction and generalization (Love 2008b). Questions that are more abstract ('how is variation generated?' or 'how can complex traits overcome developmental constraints?') are higher up in the hierarchy of the problem structure than others ('how is gene regulatory network variation generated?' or 'how can appendicular skeletal traits overcome developmental constraints due to pleiotropy?'). Questions that are more general ('how do novelties originate in metazoans?') are higher in the hierarchy of problem structure than others (how do novelties originate in mammals?).

We can fill out these three dimensions of structure by noting that they come with associated *criteria of explanatory adequacy*. This provides a template for how explanatory contributions are coordinated, so that reflecting on the problem agenda anatomy makes clear how an interdisciplinary explanatory framework can be generated (Brigandt 2010; Love 2008a):

1. Historical controversy highlights the need for different disciplinary contributors to answer distinct and previously neglected questions in the problem agenda of evolutionary novelty, including phylogeny and paleontology (to reconstruct character polarity, ancestral character states, and transitional stages), morphology (to determine the compositional identity of a feature), and development (to detail the genotype-phenotype mapping relations for how variation emerges during ontogeny). Additionally, the emphasis on the adaptive modification of traits in neo-Darwinian population biology (see Depew, this volume; Forber, this volume) led to a neglect of questions about the origin of structure. The stress on explaining the origin of new morphological units corresponds to an attempted correction of this functional bias and indicates that any adequate explanatory framework for the origin of new characters must address both morphology (form) and function.
2. Different disciplinary approaches and methods will be required to address the heterogeneous types of questions in the problem agenda of novelty. Adequate explanatory frameworks must exhibit sufficient complexity and balance: complexity to match the heterogeneous questions in the problem agenda with corresponding answers, and balance to handle empirical and theoretical questions, not neglect pattern questions for process questions, and deal with lower levels of

[10]Differences ('hetero') in development that contribute to evolutionary change can be classified according to the kind of difference in view: (i) *heterochrony*: differences in the timing of developmental events; (ii) *heterotopy*: differences in the spatial location of developmental events; (iii) *heterotypy*: differences in the type of developmental event, such as cavitation versus invagination; and, (iv) *heterometry*: differences in the amount of activity in developmental events, such as the up-regulation of gene expression (Arthur 2002).

organization as well as higher levels. The focus of one discipline on some questions rather than others creates a fruitful division of labor and organizes different lines of investigation in terms of the kinds of questions they tackle.

3. Hierarchical structure in the problem agenda marks out criteria of adequacy with respect to abstraction and generality. The demand of abstraction requires that necessary disciplinary contributions have been made, such as the generation of variation being investigated using methods from quantitative genetics, developmental genetics, epigenetics, and phenotypic plasticity at different levels of organization. Since more concrete questions involve distinct biological processes ('how is gene regulatory network variation generated?' versus 'how is epigenetic variation generated?'), the ability to offer an explanatory framework at the desired level of abstraction requires multiple methodological approaches. The demand of generality requires that diverse characters in different clades are investigated using many methods, and that appropriate proxies for extinct taxa are utilized in experimental research (Metscher and Ahlberg 1999). Successful explanatory proposals for particular novelties must be evaluated with respect to their applicability to others. Since more specific questions involve clade-level differences, appropriately diverse taxa must be studied and the results judiciously compared. Because the precise phylogenetic pattern leading up to a novelty (character transformations at particular junctures) must be settled prior to assessing the developmental mechanisms that contributed to an evolutionary transition, the architecture of the problem agenda not only requires different approaches (paleontology, phylogeny, developmental biology) but also shows how contributions from different approaches articulate. Thus, the hierarchical structure of a problem agenda provides a scaffold upon which to insert the relevant disciplinary contributions.

We now have a detailed picture of how problems organize research and guide inquiry. This problem structure makes explicit why multiple disciplinary contributors are needed to address complex scientific problems like the origin of evolutionary novelty: "problems may cut right across the borders of any subject matter or discipline" (Popper 2002 [1963], p. 88). This organizing architecture of problems is not in conflict with an image of science focused on hypothesis testing or theory confirmation, but it does lay bare why so much of contemporary life science investigation is interdisciplinary. And because the anatomy of problem agendas underlies the research practices of biologists, it must be communicated to students of biology, especially in the context of scientific controversy.

Returning to the Evo-devo controversy between developmental genetic and generic physical explanations of the origin of novelty, there is no doubt that some of it is empirical (i.e., about the way the world is). But we now have resources for picking out previously invisible epistemological elements. Consider the hierarchical dimension of problem agenda structure. The controversy pertains to the mechanisms that generate variation during development. The more abstract question ('how is variation generated?') is higher up in the hierarchy of problem structure than its subunits ('how is gene regulatory network variation generated?' or 'how is epigenetic variation due to generic physical mechanisms generated?'). This means there is disagreement about the way in which answers to more concrete component questions are fed into an answer at the more abstract level. These disagreements are not just empirical but involve disciplinary biases about what causes are significant (i.e., there is an epistemological element of the controversy). In particular, many developmental biologists interested in evolution have neglected the potential role of physical mechanisms in ontogeny.[11] There also is substantial agreement embedded within this mixture of empirical and epistemological disagreement. Both approaches concur that their explanations must be framed by an explicit phylogenetic hypothesis (a different disciplinary contributor) and that the problem itself is genuine (an element from the historical dimension).

Turning to generalization, the primary focus of the generic physical explanation is the early history of metazoans because this is the time when the 'pre-Mendelian' world exists (according to this viewpoint). There is agreement with the developmental genetic explanation that the clade of interest is metazoans (and not bacteria, for example), in part because multicellular animals exhibit complex developmental processes that can be subject to evolutionary change. But there is disagreement about whether developmental variation has always been produced in the same way, i.e., whether you can generalize from extant experimental inquiry to past evolutionary events. This is an epistemological disagreement about actualism[12] (causes now operating explain past events). The generic physical explanation is committed to the ontogeny of early metazoans operating differently than it does today: "ancient organisms undoubtedly exhibited less genetic redundancy and metabolic integration and homeostasis than modern organisms [...] ancient metazoa were even more developmentally plastic than modern ones" (Newman et al. 2006, p. 290). As a result, simulation methods become more relevant for testing the generic physical explanation since contemporary developmental genetic experiments can only be executed on organisms *as they are today*. Therefore, the hierarchical structure of the problem agenda shows that the empirical disagreements are modulated by epistemological agreements and disagreements about relationships among component questions, assumptions underlying these questions, and the methods required to answer them.

It should be noted that in providing an additional layer of interpretation for the controversy between developmental genetic and generic physical explanations,

[11]"The current preeminence of the molecular genetic approach to biology, in which living systems are conceptualized as networks of interacting genes and proteins, may have obscured this inevitable link between physics and biology in the mind of scientists" (Mulder 2008, p. 1643); "there has been a renewed appreciation of the fact that to understand morphogenesis in three dimensions, it is necessary to combine molecular insights (genes and morphogens) with knowledge of physical processes (transport, deformation and flow) generated by growing tissues" (Savin et al. 2011, p. 57).

[12]Uniformitarianism is a stronger principle that actualism. The former combines actualism and a commitment to extant causes operating with the same intensity throughout history.

we also have generated resources for addressing the larger controversy over fundamentality identified in Sect. 3. The anatomy of problem agendas strongly suggests that different disciplines play explanatory roles to which they are well suited but that no one approach is fundamental *per se*. The "fundamentality" of one particular perspective producing the most empirically adequate explanations is rejected, *even when one hypothesis is favored*: "paleontologists, comparative anatomists, developmental biologists, and molecular geneticists are all contributing data aimed at clarifying the *genetic* basis for novel structures like heads, tails, and limbs" (Freeman 2002, pp. 475–476, emphasis mine). The three dimensions of structure and their concomitant criteria of explanatory adequacy facilitate a more transparent picture of what intellectual contributions are needed for an adequate explanation. Different novelties at different levels of organization may require different explanatory ingredients in different combinations. Successful multidisciplinary coordination with respect to different problems is likely to involve different integrative relations across fields (Brigandt 2010), and therefore we can relinquish the aim of securing a single set of fundamental relations between population genetics and Evo-devo. Viewing science in terms of the erotetic organization that governs its reasoning practices gives us a new perspective on controversy, especially its epistemological elements.

6 Teaching the (Controversial) Elephant of Science

The main result of the preceding discussion is that we need to teach more than one image of science if we are going to adequately convey its diverse aspects. If we only conceptualize scientific reasoning in terms of theory construction and hypothesis testing, we run the risk of miscommunicating the practice of science to our students and misdiagnosing the nature of controversies. This is dangerous because it presents a biased picture of the sciences to students and potentially discourages them from participating. A prophylactic against this possibility is to teach how structured problems organize research and guide inquiry (Love 2013a), not just initiate it. I have illustrated the value of this strategy through specific concepts in Evo-devo (constraints, modularity, evolvability, and novelty). A closer look at attempts to explain the origin of novelty displayed how problem agendas demand a synthesis of methodologies from different biological disciplines to generate empirically adequate explanations. This illuminates the interdisciplinarity of biological practice in a way that is often ignored when science is taught only from the perspective of theory or hypothesis. And the *National Science Education Standards* [*NSES*] encourage the teaching of interdisciplinarity, whether in terms of scientific inquiry generating knowledge via "different types of investigation" (p. 176) or the analysis of alternative explanations and models (pp. 148, 175). The present perspective also fleshes out key claims made in the *NSES*, such as "having students reflect on the concepts that guide the inquiry" (p. 174), because there is now a detailed picture of how concepts—as representative of problem agendas—actually structure inquiry.

At the outset we observed that a challenge to keeping up with the current problem structure in biology is that the life sciences contain diverse and heterogeneous disciplines. Evo-devo is just one example where the structure of problems informs how scientific inquiry is and should be organized. A recent National Research Council study on the role of theory in twenty-first century biology acknowledged this directly in the traditional folk tale about the blind men and the elephant (NRC 2008, ch. 10). As is familiar, each man accesses some feature of the elephant and forms a judgment of the whole in terms of the particular part, subject to his idiosyncratic tactile exploration: the side (like a wall), the tusk (like a spear), the trunk (like a snake), the knee (like a tree), the ear (like a fan), and the tail (like a rope).

The moral is straightforward,[13] but can be given a novel epistemological interpretation in light of our discussion of problem structure. Problems, like the elephant, have structure due to history, heterogeneity, and hierarchy. Teaching this problem structure in the context of Evo-devo—and other areas of biology—assists students in learning about how different data, methods, and theoretical assumptions are brought to bear on complex biological phenomena. This structure is not esoteric (i.e., only comprehensible to a small group of specialists), but can be vague or implicit. Reliance on philosophical reconstruction and explication serves the teacher in this endeavor, as shown in the case of evolutionary novelty. We need to investigate phenomena in the world (like the elephant) with a combination of theories, experiments, methods, and observations, but we need to explore different but complementary images of scientific investigation as well, such as hypothesis testing, modeling, and problem structure (Fig. 2). One immediate advantage of teaching the erotetic image of science is an expanded understanding of controversy.

Eugenie Scott and Glenn Branch provided five conditions for when it is appropriate to teach scientific controversy (Scott and Branch 2003, pp. 499–500).

1. The controversy ought to be of interest to students.
2. The controversy ought to be primarily scientific, rather than primarily moral, social or religious.
3. The resources for each side of the controversy ought to be comparable in availability.
4. The resources for each side of the controversy ought to be comparable in quality.
5. The controversy ought to be understandable by the students.

Students are fickle about their interests, and whether they find Evo-devo topics intriguing is difficult to ascertain *a priori*. The controversies discussed herein are clearly scientific, and the resources in the controversy are comparable in quality and availability, even though the number of advocates on each side is not equal in all cases (e.g., there are fewer advocates of the generic physical explanation of novelty

[13] This conception, minus the metaphorical pachyderm, is present in the *NSES*: "The natural [...] world is complex; it is too large and complicated to investigate and comprehend all at once. Students and scientists learn to define small portions for the convenience of investigation" (p. 116).

Fig. 2 A multi-faceted image of scientific reasoning. Just as the blind men studying the elephant mistake the part for the whole, so also our image of science can be subject to similar biases. Thus, we can think of teaching epistemological aspects of science as another version of the elephant and recognize that the emphasis on particular features, such as problem structure, illuminate specific and distinctive aspects of scientific reasoning, such as interdisciplinarity. Focusing on problem structure also sheds light on genuine scientific controversies, revealing differences between their empirical and epistemological elements (Source: Holton and Curry 1914, illustrator unknown; public domain, http://commons.wikimedia.org/wiki/File:Blind_men_and_elephant.png)

than the developmental genetic explanation). The controversies are not simple but also are not beyond the reach of high school and college age students. Therefore, there is a *prima facie* case for teaching these controversies.

If we scrutinize these conditions in light of our discussion of problem structure and Evo-devo controversy, then it becomes clear that we must distinguish between teaching empirical and epistemological aspects of scientific controversies. To do so reveals how controversies are often complex mixtures of epistemological agreements and disagreements, and even sometimes stem from terminological ambiguity, rather than a competition between mutually exclusive hypotheses. Teaching empirical controversy is not the same as teaching epistemological controversy; students need to know the difference and be able to distinguish them in exemplars from scientific reasoning, such as the controversy between developmental genetic and generic physical explanations of evolutionary novelty.

When this pedagogical strategy is executed in the context of biological problem agendas that demand interdisciplinary explanations, it fulfills training goals for instructors ("to make conceptual connections within and across science disciplines"; *NSES*, p. 59), and students: "No matter how the curriculum is organized, it should provide students with opportunities to become aware of the great range of scientific disciplines that exist" (AAAS 2009, ch. 1). It also foregrounds the Nature of Science considerations that are at the heart of teaching biology, including alternative explanations and methods across disciplinary approaches, the roles of consilience and collaboration, variability in experimental practices, and different standards for what counts as evidence in interdisciplinary situations (Allchin 2003, 2011). This meshes well with the *NSES* emphasis on distinct evidential standards in different disciplinary contexts, which implies that the evaluation of explanations involves more than empirical confirmation. The *Benchmarks for Scientific Literacy* paint a congruent picture:

> Scientific investigations may take many different forms [and]...are conducted for different reasons [...] There are different traditions in science about what is investigated and how [...] disciplines differ from one another in what is studied, techniques used, and outcomes sought [...] many problems are studied by scientists using information and skills from many disciplines (AAAS 2009, ch. 1).

To construe scientific reasoning solely in terms of one image (e.g., hypothetico-deductive methodology; Lawson 2003) is to mistake one part of the elephant for the whole. An image of science that highlights how problem structure organizes research—depicted here in a snapshot from Evo-devo—can increase our apprecia-tion of the complex nature of scientific inquiry. We need a multi-faceted image of scientific reasoning; it's time to grasp the different parts of the elephant simultaneously.

Acknowledgments I am grateful to Kostas Kampourakis for the invitation to contribute this chapter. Many of the ideas set forth here emerged in collaboration with my colleague Ingo Brigandt and I acknowledge my debt to him in working out details related to a problem-oriented conception of scientific inquiry. Wallace Arthur and Kostas Kampourakis provided helpful comments and critical feedback on an earlier version of the manuscript.

References

AAAS. 2009. *Benchmarks for scientific literacy*. AAAS Project 2061. New York: Oxford University Press. http://www.project2061.org/publications/bsl/online/.

Alberch, P., and E.A. Gale. 1985. A developmental analysis of an evolutionary trend: Digital reduction in amphibians. *Evolution* 39: 8–23.

Allchin, D. 2003. Lawson's shoehorn, or should the philosophy of science be rated "X"? *Science & Education* 12: 315–329.

Allchin, D. 2011. Evaluating knowledge of the nature of (whole) science. *Science & Education* 95(3): 518–542.

Amundson, R. 1994. Two concepts of constraint: Adaptationism and the challenge from developmental biology. *Philosophy of Science* 61: 556–578.

Amundson, R. 2005. *The changing role of the embryo in evolutionary thought: Roots of Evo-Devo*. New York: Cambridge University Press.

Arthur, W. 2002. The emerging conceptual framework of evolutionary developmental biology. *Nature* 415: 757–764.

Arthur, W. 2004. *Biased embryos and evolution*. New York: Cambridge University Press.

Arthur, W. 2011. *Evolution: A developmental approach*. Oxford: Wiley-Blackwell.

Bateson, P., and P. Gluckman. 2011. *Plasticity, robustness, development and evolution*. Cambridge, NY: Cambridge University Press.

Bolker, J.A. 2000. Modularity in development and why it matters to Evo-devo. *American Zoologist* 40: 770–776.

Brigandt, I. 2010. Beyond reduction and pluralism: Toward an epistemology of explanatory integration in biology. *Erkenntnis* 73: 295–311.
Brigandt, I., and A.C. Love. 2010. Evolutionary novelty and the Evo-devo synthesis: Field notes. *Evolutionary Biology* 37: 93–99.
Brigandt, I., and A.C. Love, 2012a. Reductionism in biology. In *The Stanford encyclopedia of philosophy*, ed. E.N. Zalta, http://plato.stanford.edu/entries/reduction-biology/
Brigandt, I., and A.C. Love. 2012b. Conceptualizing evolutionary novelty: Moving beyond definitional debates. *Journal of Experimental Zoology. Part B, Molecular and Developmental Evolution* 318: 417–427.
Carroll, S.B. 2005. *Endless forms most beautiful: The new science of Evo-devo*. New York: W.W. Norton.
Davidson, E.H. 2006. *The regulatory genome: Gene regulatory networks in development and evolution*. San Diego: Academic.
De Robertis, E.M. 2008. Evo-devo: Variations on ancestral themes. *Cell* 132: 185–195.
Freeman, S. 2002. *Biological science*. Upper Saddle River: Prentice Hall.
Gerhart, J., and M. Kirschner. 2007. The theory of facilitated variation. *Proceedings of the National Academy of Sciences of the United States of America* 104: 8582–8589.
Gilbert, S.F. 2003. Opening Darwin's black box: Teaching evolution through developmental genetics. *Nature Reviews Genetics* 4: 735–741.
Gompel, N., B. Prud'homme, P.J. Wittkopp, V.A. Kassner, and S.B. Carroll. 2005. Chance caught on the wing: *Cis*-regulatory evolution and the origin of pigment patterns in *Drosophila*. *Nature* 433: 481–487.
Gould, S.J. 2002. *The structure of evolutionary theory*. Cambridge, MA: The Belknap Press of Harvard University Press.
Hattiangadi, J.N. 1978. The structure of problems, part I. *Philosophy of the Social Sciences* 8: 345–365.
Hattiangadi, J.N. 1979. The structure of problems, part II. *Philosophy of the Social Sciences* 9: 49–76.
Hendrikse, J.L., T.E. Parsons, and B. Hallgrímmson. 2007. Evolvability as the proper focus of evolutionary developmental biology. *Evolution & Development* 9: 393–401.
Hoekstra, H.E., and J.A. Coyne. 2007. The locus of evolution: Evo-devo and the genetics of adaptation. *Evolution* 61: 995–1016.
Holton, Martha Adelaide, and Charles Madison Curry. 1914. Blind men and an elephant. In *Holton-Curry readers*, 108. Chicago: Rand McNally & Co.
Kirschner, M., and J. Gerhart. 1998. Evolvability. *Proceedings of the National Academy of Sciences of the United States of America* 95: 8420–8427.
Kirschner, M.W., and J.C. Gerhart. 2005. *The plausibility of life: Resolving Darwin's dilemma*. New Haven/London: Yale University Press.
Kuratani, S. 2009. Modularity, comparative embryology and evo-devo: Developmental dissection of evolving body plans. *Developmental Biology* 332: 61–69.
Laubichler, M.D. 2009. Form and function in Evo devo: Historical and conceptual reflections. In *Form and function in developmental evolution*, ed. M.D. Laubichler and J. Maienschein, 10–46. New York: Cambridge University Press.
Laubichler, M.D. 2010. Evolutionary developmental biology offers a significant challenge to the neo-Darwinian paradigm. In *Contemporary debates in philosophy of biology*, ed. F.J. Ayala and R. Arp, 199–212. Malden: Wiley-Blackwell.
Laudan, L. 1977. *Progress and its problems: Towards a theory of scientific growth*. Berkeley/Los Angeles: University of California Press.
Lawson, A. 2003. Allchin's shoehorn, or why science is hypothetico-deductive. *Science & Education* 12: 331–337.
Love, A.C. 2003. Evolutionary morphology, innovation, and the synthesis of evolutionary and developmental biology. *Biology and Philosophy* 18: 309–345.

Love, A.C. 2007. Morphological and paleontological perspectives for a history of Evo-devo. In *From embryology to Evo-devo: A history of developmental evolution*, ed. M. Laubichler and J. Maienschein, 267–307. Cambridge, MA: MIT Press.

Love, A.C. 2008a. Explaining evolutionary innovation and novelty: Criteria of explanatory adequacy and epistemological prerequisites. *Philosophy of Science* 75: 874–886.

Love, A.C. 2008b. From philosophy to science (to natural philosophy): Evolutionary developmental perspectives. *The Quarterly Review of Biology* 83: 65–76.

Love, A.C. 2009. Marine invertebrates, model organisms, and the modern synthesis: Epistemic values, evo-devo, and exclusion. *Theory in Biosciences* 128: 19–42.

Love, A.C. 2013a. Interdisciplinary lessons for the teaching of biology from the practice of Evo-devo. *Science & Education* 22(2): 255–278.

Love, A.C. 2013b. Theory is as theory does: Scientific practice and theory structure in biology. *Biological Theory*. doi:10.1007/s13752-012-0046-2.

Love, A.C., and R.A. Raff. 2003. Knowing your ancestors: Themes in the history of Evo-devo. *Evolution & Development* 5: 327–330.

Lynch, M. 2007. The frailty of adaptive hypotheses for the origins of organismal complexity. *Proceedings of the National Academy of Sciences of the United States of America* 104: 8597–8604.

Maynard Smith, J., R. Burian, S. Kauffman, P. Alberch, J. Campbell, B. Goodwin, R. Lande, D. Raup, and L. Wolpert. 1985. Developmental constraints and evolution. *The Quarterly Review of Biology* 60: 265 287.

Mayo, D. 1996. *Error and the growth of experimental knowledge*. Chicago: University of Chicago Press.

Metscher, B.D., and P.E. Ahlberg. 1999. Zebrafish in context: Uses of a laboratory model in comparative studies. *Developmental Biology* 210: 1–14.

Minelli, A. 2009. *Forms of becoming: The evolutionary biology of development*. Princeton: Princeton University Press.

Minelli, A. 2010. Evolutionary developmental biology does not offer a significant challenge to the neo-Darwinian paradigm. In *Contemporary debates in philosophy of biology*, ed. F.J. Ayala and R. Arp, 213–226. Malden: Wiley-Blackwell.

Mulder, B. 2008. On growth and force. *Science* 322: 1643–1644.

Müller, G.B. 2007. Evo-devo: Extending the evolutionary synthesis. *Nature Reviews Genetics* 8: 943–949.

Müller, G.B., and S.A. Newman. 2005. The innovation triad: An EvoDevo agenda. *Journal of Experimental Zoology. Part B, Molecular and Developmental Evolution* 304B: 487–503.

Newman, S.A. 1994. Generic physical mechanisms of tissue morphogenesis: A common basis for development and evolution. *Journal of Evolutionary Biology* 7: 467–488.

Newman, S.A., and R. Bhat. 2009. Dynamical patterning modules: A "pattern language" for development and evolution of multicellular form. *International Journal of Developmental Biology* 53: 693–705.

Newman, S.A., G. Forgacs, and G.B. Müller. 2006. Before programs: The physical origination of multicellular forms. *International Journal of Developmental Biology* 50: 289–299.

Nickles, T. 1981. What is a problem that we may solve it? *Synthese* 47: 85–118.

NRC. 1996. *National science education standards*. Washington, DC: National Academy Press (National Research Council).

NRC, 2008. *The role of theory in advancing 21st-century biology: Catalyzing transformative research*. Washington, DC: Committee on Defining and Advancing the Conceptual Basis of Biological Sciences in the 21st Century: National Research Council, National Academies of Science.

Osbeck, L.M., N.J. Nersessian, K.R. Malone, and W.C. Newstetter. 2011. *Science as psychology: Sense-making and identity in science practice*. New York: Cambridge University Press.

Popper, K. 2002 [1963]. *Conjectures and refutations: The growth of scientific knowledge*. London/New York: Routledge.

Prum, R.O., and A.H. Brush. 2002. The evolutionary origin and diversification of feathers. *The Quarterly Review of Biology* 77: 261–295.
Raff, R.A. 2000. Evo-Devo: The evolution of a new discipline. *Nature Reviews Genetics* 1: 74–79.
Raff, R.A. 2007. Written in stone: Fossils, genes, and evo-devo. *Nature Reviews Genetics* 8: 911–920.
Raff, R.A. 2008. Origins of the other metazoan body plans: The evolution of larval forms. *Philosophical Transactions of the Royal Society of London. Series B, Biological Sciences* 363: 1473–1479.
Raff, E.C., E.M. Popodi, J.S. Kauffman, B.J. Sly, F.R. Turner, V.B. Morris, and R.A. Raff. 2003. Regulatory punctuated equilibrium and convergence in the evolution of developmental pathways in direct-developing sea urchins. *Evolution & Development* 5: 478–493.
Sarkar, S. 2007. *Doubting Darwin? Creationist designs on evolution*. Malden: Blackwell Publishing.
Savin, T., N.A. Kurpios, A.E. Shyer, P. Florescu, H. Liang, L. Mahadevan, and C. Tabin. 2011. On the growth and form of the gut. *Nature* 476: 57–62.
Scott, E.C., and G. Branch. 2003. Evolution: What's wrong with 'teaching the controversy'. *Trends in Ecology & Evolution* 18: 499–502.
Shubin, N.H. 2008. *Your inner fish: A journey into the 3.5-billion-year history of the human body*. New York: Vintage Books (Random House).
Shubin, N., C. Tabin, and S. Carroll. 2009. Deep homology and the origins of evolutionary novelty. *Nature* 457: 818–823.
Stearns, S.C. 1992. *The evolution of life histories*. New York: Oxford University Press.
Stern, D.L. 2011. *Evolution, development, and the predictable genome*. Greenwood Village: Roberts and Company Publishers.
Telford, M.J., and G.E. Budd. 2003. The place of phylogeny and cladistics in *Evo-Devo* research. *International Journal of Developmental Biology* 47: 479–490.
Wagner, G.P. 2000. What is the promise of developmental evolution? Part I: Why is developmental biology necessary to explain evolutionary innovations? *Journal of Experimental Zoology. Part B, Molecular and Developmental Evolution* 288: 95–98.
Wagner, G.P., and V.J. Lynch. 2010. Evolutionary novelties. *Current Biology* 20: R48–R52.
Wagner, G., and J. Zhang. 2011. The pleiotropic structure of the genotype-phenotype map: The evolvability of complex organisms. *Nature Reviews Genetics* 12: 204–213.
Wagner, G.P., C.-H. Chiu, and M. Laubichler. 2000. Developmental evolution as a mechanistic science: The inference from developmental mechanisms to evolutionary processes. *American Zoologist* 40: 819–831.

Philosophical Issues in Ecology

James Justus

1 Introducing Philosophy of Ecology

Dobzhansky's (1964) sweeping generalization, "nothing in biology makes sense except in the light of evolution," provocatively captures the centrality of evolutionary theory in biological science (see also Dobzhansky 1973). But his rally call is also grievously partial. Ecology casts the same indispensable light in biology, particularly on evolution. Although the term 'ecology' was not coined until 1866 (Haeckel 1866), ecological insight was an integral part of early evolutionary thinking, it is at the core of Darwin's theory, and it will be crucial to future theorizing about how evolution has shaped the biological world. Evolutionary theory's central concepts—e.g. adaptation and natural selection—and its compelling accounts of biological phenomena—e.g. the transmutation of species and fit between organisms and environments—all reflect an ecological perspective.

Apart from its contribution to evolutionary theory, ecology also endeavors to explain significant portions of the living world directly. It is, for example, canonically characterized as the study of interactions between organisms and the environment. Its explanatory scope therefore includes not only these interactions but also the distributions and abundances of species they produce throughout the globe. Given these ambitious goals, the sophisticated experimental tests and mathematical theories developed to achieve them deserve much more attention from philosophers of science. And despite an ecological perspective underpinning much of evolutionary theory, even philosophers of biology have paid relatively little attention to ecology, largely due to disciplinary inertia. Similarly, the typical undergraduate exposure to biology for pre-medicine and non-biological majors often consists

---

J. Justus (✉)
Department of Philosophy, History and Philosophy of Science Program,
Florida State University, Tallahassee, FL 32306, USA
e-mail: jjustus@fsu.edu

K. Kampourakis (ed.), *The Philosophy of Biology: A Companion for Educators*, History, Philosophy and Theory of the Life Sciences 1, DOI 10.1007/978-94-007-6537-5_17,

largely of evolutionary theory, at the expense of ecology. This has recently changed and philosophical and pedagogical interest is growing in the conceptually rich questions ecologists study.

Although the structure of the sub-discipline continues to evolve, several broad areas of interest have emerged:

1. conceptual issues in the history of ecology (Justus 2008a; Eliot 2011a);
2. characterizing problematically unclear ecological concepts, especially 'biodiversity' and 'stability' (Sarkar 2005; Justus 2008b; Maclaurin and Sterelny 2008; Justus 2011);
3. whether there are distinctively ecological laws (Mikkelson 2003; Lange 2005; Eliot 2011b);
4. reduction in ecological science and the reality of biological communities (Sterelny 2006; Odenbaugh 2007);
5. the role of mathematical modeling in ecology (Ginzburg and Colyvan 2004; Justus 2008b);
6. the management of uncertainty in ecological inference (Regan et al. 2002; Weisberg and Reisman 2008);
7. the relationships between evolutionary theory and ecology, and conservation science and ecology (Cooper 2003; Linquist 2008; Millstein forthcoming).

Beyond a narrow focus on ecology, some of these areas offer novel insights into standard topics in general philosophy of science, such as: emergence and reduction; the nature of laws of nature (see Lange, this volume); conceptual content and concept determination (see Depew, this volume; Forber, this volume); the status and function of models in science; and, the status and function of values in sciences (see Millstein, this volume; Plutynski, this volume; Gannett, this volume).

Others areas involve topics unique to ecology, and on which philosophers can make valuable contributions to scientific practice (e.g. [1] and [4]–[6] from above). Each area, in turn, covers numerous specific issues. With respect to (4), for example, some ecologists and philosophers of science have recently proposed an analogy between Newtonian mechanics and ecosystem dynamics (Ginzburg and Colyvan 2004). Although the status and epistemic utility of this analogy remains controversial, this work suggests that a close parallel should exist between modeling strategies in physics and ecology. But other analyses counter this parallel. For example, Hubbel's (2001) unified neutral theory of biodiversity primarily derives from theories developed within biology proper: R. MacArthur and Wilson's (1967) theory of island biogeography and Kimura's (1983) neutral theory of molecular evolution. And one concept of stability appropriated from physics and often employed in ecological modeling, Lyapunov stability, seems unable to capture the ecological phenomena it is intended to represent (Justus 2008b). Analyses of this unresolved issue shed light on the different role models may have in biology and physics in general. With respect to (1), to cite another prominent example, there are several concepts besides 'biodiversity' and 'stability' central to ecological science and in need of conceptual clarification, including 'carrying capacity,' 'community,' 'complexity,' 'disturbance,' 'ecosystem,' 'habitat,' 'keystone species,' 'niche,' 'population,' and many others. Like most concepts in developing sciences, fully adequate definitions of these and other ecological concepts have not yet been formulated. These and other issues provide rich conceptual grist for philosophers of ecology.

As these examples illustrate, ecology concerns a diverse conceptual terrain and an interesting set of theoretical and methodological issues, thus far largely unexplored by philosophers of science and underappreciated in liberal arts education. An exhaustive survey is impossible, but the following sections describe *some* of the main contours of the newly emerging field of philosophy of ecology. Section 2 describes how an ecological perspective shaped Darwin's theory, particularly the niche concept and the idea that there is a "balance of nature." Section 3 considers the debate between individualists and holists such as Fredric Clements about the character of biological communities as well as metaphysical issues about their reality. Section 4 surveys the perennial controversy about the nature of laws of nature that has recently emerged within ecology. Section 5 attempts to clarify a central, but also problematic concept: ecological stability. Section 6 briefly concludes that ecology holds largely untapped riches for philosophy of science, and should have a more central role in the teaching of philosophy of biology.

2 Ecology and Evolution: The Niche Concept and the "Balance of Nature"

Although ecology only emerged as a distinct biological science in the late nineteenth – early twentieth century, ecological ideas indispensably shaped Darwin's theory. Perhaps the most theoretically fertile issue at the intersection of ecology and evolution—and the key driver of evolutionary dynamics—is the adaptive fit between organisms and their environments (see Forber, this volume for the concept of adaptation).

Darwin's brilliance was to recognize how the austere conditions environments impose on organisms yield a selective mechanism of evolutionary change. Malthus identified the presumed biological predicament: populations grow geometrically but food supplies at best increase arithmetically. For Darwin, this pinpointed the inescapable struggle for limited resources confronting all organisms. In such a struggle, some heritable variations are favored, others prove detrimental. Ecology provides the relevant scientific window into the struggle underlying such natural selection: "ecology is the study of all those complex interrelations referred to by Darwin as the conditions of the struggle for existence" (Ernst Hackel in 1869 as quoted in Stauffer 1957).

The struggle is not just between organisms and the abiotic environment. Darwin frequently emphasized that species realize different functional roles in ecological systems, which he often labeled 'places' and later ecologists termed 'niches' (Worster 1994). Generalizing what counts as a species' environment to recognize the significant selective impact of intra and interspecific interactions between organisms—interactions that may change in form and intensity over time—was one of the key insights the niche concept facilitated. As realizers of particular functional roles in an ecosystem, organisms face more than just a static environment composed of a suite of abiotic factors such as precipitation, temperature, nutrient availability, etc. Their niche is also the product of intraspecific interactions, and shaped by relationships with other species occupying different niches in the overall dynamics of the ecological system, which are also evolving. Species' efficiency in utilizing and expanding their niche, and the nature of relationships between inhabitants of different niches, would then explain why some species succeed and others fail in the struggle for existence. Natural selection then becomes a form of niche dynamics. These interactions and interspecific relationships could also, in turn, produce reproductive barriers that would catalyze the transmutation of species, even in the absence of geographic isolation (Darwin 1859, p. 103). Put in contemporary terms, the niche concept gave Darwin resources to explain speciation even in the absence of geographic barriers, so-called sympatric speciation (Coyne and Orr 2004).

Besides helping explain the origin of species, the niche concept also played a role in Darwin's account of how competition shapes ecological systems and his conception of a "balance of nature." Darwin, with most of his scientific contemporaries, was committed to the idea of such a balance:

> Battle within battle must ever be recurring with varying success; and yet in the long-run the forces are so nicely balanced, that the face of nature remains uniform for long periods of time, though assuredly the merest trifle would often give the victory to one organic being over another (Darwin 1859, p. 73).

The commitment to some type of balance was a staple of the schools of natural philosophy from which biology emerged, long before the term 'ecology' was even coined (Egerton 1973). Darwin and other early ecologists continued this tradition by attempting to derive the existence of a "natural balance" in biological populations from organismic metaphors and analogies with physical systems, although the analogical and metaphorical content often differed (see Kingsland 1995). For example, the ecologist Frederic Clements (1916) is best known for claiming to find functional integration within biological communities that resembled the physiological integration within individual organisms, and which justified conceptualizing communities as a kind of superorganism with analogous homeostatic properties (see §3). But Darwin (1859, pp. 115–116) employed the same metaphor several decades before,[1] with a much less problematic aim:

> The advantage of diversification in the inhabitants of the same region is, in fact, the same as that of the physiological division of labor in the organs of the same individual body [...] No physiologist doubts that a stomach by being adapted to digest vegetable matter alone, or flesh alone, draws more nutriment from these substances. So in the general economy of any land, the more widely and perfectly the animals and plants diversified for different habits of life, so will a greater number of individuals be capable of there supporting themselves. A set of animals, with their organization but little diversified, could hardly compete with a set more perfectly diversified in structure.

Although this conclusion plausibly holds for communities in relatively constant environments and thereby provides a plausible explanation of the greater species diversity found in the tropics than in more environmentally turbulent temperate

[1] See Kohn (2009) for further analysis of Darwin's conceptualization of this metaphor.

regions (see Rosenzweig 1992), later ecologists would show that specialization often constitutes a handicap in fluctuating environments that favor adaptable generalists (e.g. Pianka 2000, Ch. 8).

There were two threads to Darwin's view on the character of this putative balance, particularly the causal forces responsible for it. Most scientists before Darwin did not fully appreciate the extent to which inter *and intra*specific competition shaped communities (Bowler 1976). A balance of nature was considered the result of a predetermined harmony that competition would only undermine. Darwin's balance was undergirded by a much more realistic dynamics. Interspecific competition constrains the populations comprising biological communities by limiting organisms' access to the resources they need to metabolize and ultimately reproduce. This curtails populations' geometric tendency to increase. Other forms of interspecific interaction have similar consequences. Predators and parasites, for instance, inhibit prey and host populations. Intraspecific competition produces the same inhibitory effect within a species, and it can inhibit other species through interspecific relationships. For example, intraspecific competition among prey limits predator populations.

But, as Darwin was well aware, these inhibitory relationships do not alone account for the kind of dynamic balance ostensibly exhibited in the natural world. The problem was the differential power and scope of intra and interspecific competition. Intraspecific competition is fully general: it arguably occurs in all biological populations (but see Cooper 2003, Ch. 3). But its power to restrain population growth is governed by the availability of resources. When resources are plentiful, little check on growth occurs. On the other hand, interspecific competition (predation, parasitism, etc.) can suppress population growth more effectively than intraspecific dynamics in such cases, but it is not universal: not all species seem to be connected in inhibitory interspecific relations. Thus, although intraspecific competition would limit all populations when resources were scarce and interspecific interactions would sometimes suppress growth further, if these were the only checks on populations, it seems that many species would exhibit unrealistic rates of growth for unrealistic periods of time.

For Darwin, the potential problem stemmed from under-appreciating a second important thread in his concept of a balance of nature, the vastly complicated and intricately complementary set of ecological interdependencies between species: "how infinitely complex and close-fitting are the mutual relations of all organic beings to each other and to their physical conditions of life" (Darwin 1859, p. 80). Although most species do not interact directly, Darwin believed they do indirectly through chains of intermediaries. The result is a "web of complex relations" (Darwin 1859, p. 73) in which species are highly ecologically connected. A specific species' position in the web indicates what other species curb or enhance its growth. Darwin described examples of several such food webs, perhaps the most well-known (and engaging) being the ecologically serpentine relation between a clover species (*Trifolium pratense*) and the common cat (Darwin 1859, pp. 73–74). Not all parts of this web and other complex sets of ecological relationships in nature exemplify an antagonistic struggle for survival. Some are beneficial, such as mutualisms benefiting both species. But through those relationships the population suppressing effects of competitive and predatory struggles are propagated throughout the web.

Unlike previous accounts that assumed a static, providentially predetermined pattern or structure, Darwin's web-based balance of nature concept was rooted in the struggle between individual organisms to survive and reproduce. Species were balanced at their current population levels through a complex array of checks and balances finely honed by natural selection. Darwin emphasized that the exact character of the balance could change as species evolved, so in this sense the niche structure of a community was not fixed. But note that even this kind of balance requires an equilibrium assumption: population levels at a given time reflect the homeostatic processes of a biological community at a point of equilibrium. Although this assumption has been supplanted with a recognition that non-equilibrium models with complex dynamics such as chaos, limit cycles, and so-called strange attractors may best represent many types of ecological systems (DeAngelis and Waterhouse 1987), the idea that there is, and perhaps must be, a balance of nature persists. Section 5 considers the contemporary account of this balance—characterized intuitively, but plausibly by Darwin and early ecologists—with the concept of ecological stability.

As the above discussion illustrates, the niche concept seems to be a fundamental abstraction in ecological theorizing, essential to ensuring its generality (Leibold 1995). For example, general accounts of the similar structure of ecosystems composed of different species are only possible, it seems, if a shared underlying niche structure generates the similarity. Grasslands in the central plains of North America and Africa share a similar structure and exhibit similar dynamics because they instantiate roughly the same system of niches, albeit with different species. This is only one of many seemingly indispensable functions of the niche concept. Appeals to niche structure seem to provide the only explanation of convergent evolution, character displacement, as well as evolutionary convergence of ecosystems: remarkably similar biological communities emerging over geologic time scales (e.g. past communities with saber tooth tigers as apex predators and present communities with *Panthera* and *Canis* species functioning similarly) (see Sterelny and Griffiths 1999, Ch. 11).

But despite the significant work the niche concept is employed to do within current ecological theory, the somewhat opaque nature of the concept is worth highlighting. Its content, for example, has evolved significantly. Joseph Grinnell (1917), one of the first to use the term 'niche' in an ecological context, construed niches as portions of habitat in which species persist and reproduce. Abiotic environmental factors, vegetation, and food *supporting animal species* were the primary focus. So construed, plants fell outside the scope of applicability of the niche concept. In *Animal Ecology*, Charles Elton (1927) retained the focus on animals, but shifted perspective to the causal role of species within a broader biological community. Rather than environmental factors undergirding species, a species' niche was then its constellation of causal impacts on other species, its "way of making a living" in a biological community. This is the sense of 'niche' that seems to underlie Darwin's account of the balance of nature (see above). Hutchinson (1957) later returned to Grinnell's perspective, while broadening it to include all species, with his notion of a fundamental niche: a multi-dimensional hyper-volume, each dimension representing a relevant environmental factor (e.g. temperature, precipitation, etc.) and within which a species *can* persist indefinitely. Note the modality. The realized niche is the portion of the fundamental niche a species actually occupies, which competition and other interspecific interactions can often make much smaller.

This fluid conceptual landscape renders different, sometimes incompatible grounds for ecological theorizing. For example, although Grinnell and Hutchinson's conceptions emphasize environmental factors, Hutchinson relativized niches to species: niches are defined by persistence properties of species (Griesemer 1992). Grinnell's account is not similarly relativized. Niches are independently identifiable units of the environment. As such, niches can be occupied by species or be vacant, an idea Hutchinson's concept does not permit.[2] This stronger concept is committed to ecosystems possessing a niche structure irrespective of the species they contain, and is needed to underwrite judgments that an ecosystem is "saturated" with species or that an invasive species is successful because it has occupied a previously vacant niche (Lawton 1984). But these claims and the niche concept underlying them have received significant criticism (Colwell 1992). First, and perhaps more importantly, the growing literature on niche construction challenges the claim there typically is a species-independent niche structure to be found (see Odling-Smee et al. 2003). The recent charge of niche constructionists such as Richard Lewontin that many niches are made and are not simply found seems to make the standard approach inapplicable. If organisms can modify their environments and thereby their niches to increase fitness, it is no longer clear that the niche has explanatory priority. What explains ecosystem structure, convergent evolution, character displacement and the like is no longer an extant niche structure that specific biological systems realize or that imposes a selection regime producing convergence and displacement. Rather, a locus of explanatory force resides within organisms that do the niche constructing. Second, and relatedly, some have argued that clear, defensible criteria for delineating such niches have not been formulated (Herbold and Moyle 1986). One general problem is that some of the modalities involved in these and other niche concepts seem intractable, for example, evaluating whether a species *would* occupy a reputedly vacant niche, or whether a species *could* persist indefinitely in this region of environmental parameter space. Third, "neutral" theories of ecosystem structure, particularly Stephen Hubbell's (2001) "Unified Neutral Theory of Biodiversity," pose an intriguing alternative to the putative indispensability of niche thinking. By emphasizing the role of dispersal limitation, sampling effects, and stochasticity within a cohesive model of community dynamics, neutralists have formulated a cogent alternative to approaches based on niche structure, at least for plants.

Beyond its role in contemporary ecological theory, these worries about the niche concept also raise concerns about perhaps the strongest candidate for the status of a distinctively ecological law: species with identical niches cannot coexist. Before scrutinizing that claim's nomological status in Sect. 4, the next section considers the philosophically intricate issue of the reality of biological communities.

---

[2]Elton's (1927) niche concept seems to also allow vacant niches, but with a very different sense. Vacant Eltonian niches would be unfilled nodes in the causal nexus of interactions between species.

3 Are Biological Communities Real?

At a minimum, a biological community is a set of populations of different species. Usually, the species are taken to interact in *some* way to *some* degree. Beyond these platitudes, controversy emerges. The difficult question is whether communities are "something more" than the individual organisms of different species comprising them. If they aren't, presumably they possess no independent existence. If they are, an account is needed of: (i) this "something more" *and* (ii) how it confers independent existence. Absent either, realist aspirations are frustrated.

Assessments of (i) primarily focus on the nature and intensity of interactions between species comprising candidate communities. This requires a careful dissection of the causal structure of these interactions, their dynamics, species distribution patterns, and what they reveal about how groups of species might be assembled into communities. Appreciating that this is a conceptual *and empirical* issue is essential. Assessing (ii) involves delving into the murky depths of metaphysics, principally to determine whether the additional causal structure these ecological assemblages possess actually "cuts nature at its joints," a proverbial criterion for ontological credibility according to most scientific realists. Ecological science seems to offer little or no new insights regarding the ontological question of whether causal novelty confers independent existence at issue in (ii), and most philosophical analysis has thus far concerned (i).

Different positions on (i) fall on a spectrum. At one extreme is the view that communities are simply aggregations of species at a particular location and time, and that their relationships with the abiotic environment, not other species, largely determines their co-occurrence. On this view, communities as distinct ecological units are no more real than a *collection* of knick-knacks on a mantel, as opposed to the knick-knacks themselves, is real. The contingency of co-occurrence and lack of significant interaction are considered marks of the unreal. At the other extreme is the view that communities are tightly causally integrated units that exhibit a degree of functional cohesion, similar to individual organisms. On this view, communities are as real as individual organisms that possess these attributes.

A long-standing debate about the mechanism(s) of ecological succession in early twentieth century ecology helped catalyze this question about the nature of biological communities. On one side were "holists" such as Fredric Clements with his climax account of succession. For Clements, the specific constellation of abiotic factors in a given locale—cloud cover, elevation, precipitation, soil type, temperature, etc.—yields a deterministic sequence of succession stages that usually culminates, if undisturbed, in a final climax community. Different constellations usually produce different sequences and climax communities: grassland, mangrove, marsh, pine forest, and so on. If the seemingly mechanistic determinism were not controversial enough, Clements added that climax communities were "superorganisms" with the teleology and functional integration among constituent species that that term suggests. On the other side of the debate were "individualists" such as H. A. Gleason, H. G. Andrewartha, and others who discerned much less structure to succession, less cohesion in reputed biological communities, and generally more contingency underlying ecological patterns.[3]

The extreme positions associated with this debate have largely been abandoned as implausible, but the theoretical contrasts it drew continue to reverberate throughout contemporary ecology. For example, although both sides of the dispute considered themselves respectable empiricists, one influential criticism of the climax, "superorganism" theory was that it departed significantly and unjustifiably from what observations of succession, species distributions, and dynamics in supposed biological communities actually indicated. Robert Whittaker made this kind of criticism based on very influential analyses of plant distributions along environmental gradients, such as elevation. Plant species distributions along environmental gradients seem to overlap continuously and significantly, and do not form discrete identifiable boundaries (see Whittaker 1956). But, the argument goes, communities are only real if they have such distinct boundaries. So they aren't.

The key claim being challenged is that species distributions display converging boundaries that communities must possess. Recently, Odenbaugh (2007) gave three responses to this argument. First, Odenbaugh rightly points out that Whittaker's results are inflated in claims such as: "Whittaker found, each species behaved totally independently [...] there is no such thing, really, as a pine forest, or a mixed hardwood forest or a tall-grass prairie or a tundra," (Budiansky 1995, p. 86) and, "There are no discrete communities of plants. The reality is endless blending," (Colinvaux 1979, p. 72). These audacious claims simply assume Whittaker's results can be extrapolated to all plants. But it is worth noting that both of these assertions occur not in top-tier scientific journals, but in works of popular science where epistemically unwarranted flights of rhetorical fancy are less constrained. More measured assessment suggests Whittaker's results constitute strong *evidence* against the existence of the needed boundaries.

Odenbaugh's additional responses address this more reasonable interpretation. He claims that two implicit assumptions seem to underlie Whittaker's analysis:

> Interactions among species should be similar at *all* points along environmental continua. Thus, if two or more species interact in a certain way at a point, then if they interact at other points it is in the same way. [...] Groups of species should be associated at *all* points on a gradient if interdependence is to be accepted. Thus, if two or more species interact at a point on a gradient, then they interact at *all* points on that gradient. (2007, p. 635; italics added)

Odenbaugh correctly notes that these assumptions are false, but it seems implausible that Whittaker or other respected ecologists would be foolish enough to endorse such categorical claims. Exceptionless patterns are a rarity in ecology, and biology generally. One should no more accept the conditional that species interact at *all* points on an environmental gradient if they interact at one, than one should accept that species do not interact at all points if they fail to at one. The same goes

[3]See Eliot (2011a) for a historically engaging and conceptually rich account of this debate, one that locates the divergence more in methodological disagreements about how ecological research should be conducted than contrasting ontological commitments.

for the functional form of the interaction. The degree of continuity of species interactions across environmental gradients is obviously an empirical issue. But all that is required for Whittaker's results to constitute evidence of a lack of boundaries is that interactions persist and remain similar along sufficient portions of the environmental gradient sampled. What counts as 'sufficient' depends on the patterns Whittaker found and the species involved: the higher the turnover in species distributions, the smaller the portion generally required for sufficiency. Intraspecific variability exists, but organisms comprise a species in large part because they share a very similar ecological profile. One would expect, although data and sophisticated statistical analyses could only reliably confirm, that this similarity would ensure the sufficiency required.

Odenbaugh's last response is that Whittaker himself employed a "community-level" property, niche differentiation between two species, to explain the absence of boundaries in species distributions. This, Odenbaugh suggests, shows Whittaker was himself committed to the existence of communities. If Whittaker's contention was to show that no communities exist and two species interacting shows they do, Whittaker would have indeed blundered. But the worry is that this characterization of the debate construes the goal-posts too weakly. Surely what is at stake with respect to the reality of biological communities—and the target of Whittaker's analysis—is more than whether there are binary community-level properties such as predator–prey interactions between populations, and thus two-species-member "communities." If this were the issue, simply observing a population of lions consuming gazelles or parasitic mistletoes infesting a deciduous forest would be sufficient to resolve the debate. Even if they stop short of endorsing full Clementsian "super-organismal" status, proponents of the reality of biological communities have a more ambitious agenda. They believe there are many communities composed of many more than two species, that maintain some type of homeostasis, and that exhibit other kinds of causal integration that two-species interactions, which are often highly unstable, do not. Odenbaugh's criticism therefore seems to invoke an indefensibly weak burden of proof given what is at issue in this debate.

Against this more exacting standard, Sterelny (2001) gives a positive argument for the reality of communities based on paleoecological data and an analogy with organisms and species. What makes the latter real, according to Sterelny, is their internal regulation. For example, organisms maintain a boundary between themselves and the external environment, and they regulate their internal states against environmental changes. In particular, organisms exclude foreign objects and agents of disease through a complex array of processes. Species also regulate membership through behavioral, physiological, and genetic impediments to reproduction. Through an impressive regimentation of paleoecological data and scientific analyses to understand that data, Sterelny convincingly argues that there have been episodes of "co-ordinated stasis":

> Suites of species, drawn from quite different lineages, appear together quite suddenly in the fossil record. They persist together largely intact. The periods of persistence are evolutionarily significant: often a few million years. These species not only persist together; they do so maintaining both their morphological and their ecological characteristics. The commonest species stay common; the relatively rare stay relatively rare. Few new species migrate

> in, or evolve in place. Few of those in place at the establishment of an association disappear before the association breaks up. In general, *associations* persist, not just the individual taxa that make up those associations. Each seems to end with the association dissolving and many of its component species disappearing, to be replaced by a different but persisting assemblage. So the pattern is one of both evolutionary and ecological stability bounded on each side by a turnover event. (2001, pp. 438–439)

Most of Sterelny's analysis is concerned with scrutinizing proposed explanations of these periods of surprising constancy based on the best theories of community dynamics. None proves successful. It therefore remains a tantalizing prize for future theorizing. It is abundantly clear, however, that individualist approaches such as those associated with Whittaker uncontrovertibly fail to account for this phenomenon. The paleoecological data Sterelny considers therefore constitute strong evidence for the community-level internal regulation that, by analogy with the organismal and species cases, indicates the reality of biological communities. Sterelny does not over-extrapolate the significance of these results, and other paleoecological studies seem to provide countervailing evidence. For example, Davis and Shaw (2001) found that after the glaciers receded, individual tree species dispersed at different rates, in different directions, and from different origins. This seems to confirm the individualist view in which abiotic factors, not biotic interactions, drive species distribution patterns. As with most philosophical questions in ecology with an empirical component, the evidentiary issues involved are far from resolved.

4 Are There Distinctively Ecological Laws?

The philosophy of each special science such as biology, chemistry, economics, psychology, etc. inevitably grapples with whether it has distinctive laws (see Lange, this volume). Ecology is no different. And like the question of whether biological communities are real, the relevant issues involve a complex interplay of conceptual and empirical issues. Many take the possession of natural laws to be a signature mark of a science's objectivity, so the philosophical stakes are also significant.

Prima facie, there are various challenges to the idea that such laws exist: the relative paucity of predictive success in ecology, that ecological models and experimental results lack sufficient generality, that candidate laws are riddled with exceptions, and that ecological systems are too complex to name a handful. But complexity, to take the last first, is surely a surmountable obstacle. It is difficult to imagine a more complicated system than the entire cosmos but no one suggests its complexity is not governed by relativistic and quantum mechanical laws (Ginzburg and Colyvan 2004), or that humans do not continue to uncover its underlying law-governed physical and chemical dynamics. Some philosophers have recently argued that other properties thought to preclude a discipline from trading in laws—poor predictive accuracy and limited generality, not being exceptionless—should be jettisoned, and that ecology indicates why. For example, generalizations sometimes accorded nomic status, such as the latitudinal gradient in species diversity (see Rosenzweig 1992) and the so-called species-area power law (see Connor and McCoy 1979), are nevertheless known to have numerous exceptions. One suggestion is that these exceptions can be subsumed under a suitably crafted *ceteris paribus* clause, just as recognized physical laws, for example ideal gas laws and Coulomb's law of electrostatics, do not hold in some circumstances and thereby lack universality.

Even without fully entering the thicket of difficulties with *ceteris paribus* clauses (see Earman et al. 2002), the general inadequacy of this response is apparent. Absent clear and justifiable standards delimiting the extent of *ceteris paribus* clauses, this move is entirely *ad hoc*. No matter how unmotivated, potential counterexamples can simply be folded into the clause's scope in a facile ploy to preserve nomic status. By this rationale, it seems that *any* true claim can be made faux-nomological with a sufficiently well-chosen and comprehensive *ceteris paribus* clause. The important insight that laws possess a kind of natural necessity that simply true generalizations do not has little traction in this account.

The problems with this approach are compounded by the fact that most candidates for laws in ecology are based on models and theories that are *highly* idealized. That is, they incorporate unrealistic, i.e. *false*, assumptions about the systems they are intended to represent, largely to make model and theory analysis tractable. For example, they ignore some components and interactions of ecological systems, treat interactions as instantaneous and assume that their effects propagate similarly, represent discrete components with continuous variables, describe community structure non-spatially, etc. But these and other unrealistic idealizations make it uncertain whether modeling results demonstrate properties of the represented system or are byproducts of the idealizations. Since it is often unclear what properties are primarily responsible for system dynamics given their complexity, idealizations may significantly mischaracterize their most important features. An enhanced sense of understanding conveyed by an idealized model may therefore fail to be about the system it is intended to represent, thereby misdirecting rather than assisting in the discovery of ecological laws. This difficulty is exacerbated by the short supply of extensive and long-term ecological data required to empirically vet model results, in contrast with the usually highly-confirmed models found in physics and chemistry. An added concern is that mathematical ecologists have often uncritically emulated mathematically sophisticated models of physics to ensure their modeling is mathematically rigorous, but the emulation has sometimes led to serious misrepresentation of biological phenomena (see Justus 2008b). Models are the main conduit through which theorizing occurs in ecology, so high degrees of idealization pose a significant impediment to finding distinctively ecological laws. The same concern holds for other areas of biology, such as population genetics.

There is one class of empirical generalizations that avoid these model-based difficulties, and are also apparently very well confirmed, albeit with significant "scatter" of data around the proposed relationships. These are the so-called macroecological allometries (see Ginzburg and Colyvan 2004, Ch. 2). They include:

1. Kleiber allometry – basal metabolic rate is directly proportional ($\propto$) to a 3/4 power of body mass, i.e. $(\text{body mass})^{3/4}$. First noticed by biologist Max Kleiber, the larger the organism, the greater (at a 3/4 power) its calorie consumption rate at rest.

2. Generation-time allometry – organismal maturation time ∝ (body mass)$^{1/4}$.
3. Fenchel allometry – maximum reproduction rate is inversely proportional to (body mass)$^{1/4}$; first studied by Tom Fenchel.

Other ecological allometries exist, but these are perhaps the most empirically vetted and thus strongest candidates for lawhood. Besides their high degree of confirmation, their broad scope also seems to evince nomological credentials. The Kleiber allometry, for example, has been verified for organisms with masses ranging from elephants to bacteria. For these reasons, Ginzburg and Colyvan (2004, p. 12) suggest these allometries "deserve to be called *laws*."

John Beatty's (1995) evolutionary contingency thesis (ECT)—that all distinctively biological generalizations describe contingent evolutionary outcomes—constitutes a formidable obstacle to this claim. The thesis presents a two-horned dilemma for proposed biological laws. If they are distinctively biological, then they are contingent because the evolutionary processes responsible for their existence are highly contingent. The unguided, largely random nature of genetic mutation, and the fact that natural selection acts with respect to environments that frequently (and contingently) fluctuate are two examples of such contingency. This contingency, Beatty argues, is incompatible with lawhood in the same way that, for example, the contingency of 'There are no 5 m^3 gold cubes' is incompatible with lawhood. This view is not without its detractors (e.g. Sober 1987), but the controversy need not detain us because regardless of whether the allometries fall to the first, they do fall to Beatty's second horn: they are not distinctively *ecological*.[4] Take Kleiber's allometry.[5] One recent credible explanation of the pattern is that it is a consequence of fluid dynamics and the geometric structure of circulatory, respiratory, and vascular systems of animals and plants (West et al. 1997). As such, whether or not this pattern should be accorded nomic status, it certainly does not seem to be a distinctively ecological generalization any more than the fact that all organisms have mass, or that organism body mass tends to scale with body volume are truths of ecology. Although these are truths concerning entities studied in ecology, ecological science seems to contribute nothing nomological.

But there is a famous and distinctively ecological generalization with pretensions to lawhood, the *competitive exclusion principle* (CEP). Although Grinnell (1917) drew upon Darwin's work to arrive at the same kind of exclusionary principle a few decades before (see Hardin 1960), CEP's origination is largely credited to the Russian biologist Georgyi Gause (1934). In a series of brilliant experiments, Gause (1934) studied competitive dynamics in *Paramecium* and yeast species, respectively. In constant ecological conditions (e.g. nutrient levels, water temperature, turbidity, *etc.*) and in the absence of refugia that would mitigate the effects of interspecific competition, one species inevitably outcompeted the other to extinction. On this

---

[4] Marc Lange's (2005) recent analysis of what biological laws could be seems vulnerable to Beatty's first horn (see also Lange this volume).

[5] Ginzburg and Colyvan (2004) note that the Kleiber allometry is the most empirically well supported, and that the generation time and Fenchel allometries are likely based on it.

basis, Gause generalized the CEP: species with identical niches, i.e. two species that would compete for exactly the same resources, cannot coexist. The intuitive appeal of the idea and its apparently exceptionless status across many different biological systems has prompted its honorific designation as an ecological law, Gause's Law.

Apart from the extreme case of exclusion, the degrees and types of niche overlap that permit coexistence has become an important focus of contemporary attempts to explain species distribution patterns and dynamics in biological communities (Abrams 1983). The ecologist Robert MacArthur (1958) was one of the first to rigorously document this kind of phenomenon in his rightfully famous dissertation work on New England warblers. The objective of the study was to determine how so many behaviorally and physiologically similar bird species could coexist in boreal forests, which seemed to contradict the exclusion principle. With such similar properties, it seemed that interspecific competition would be especially strong between the birds and would eventually lead to the extirpation of all but one competitively dominant species. Through meticulous observation MacArthur uncovered the mechanism that eluded the exclusionary outcome: different species bred and fed in distinct spatial parts of coniferous trees and, furthermore, warblers exhibited strong territoriality towards those parts. He also found that nesting times differed across the warbler species. This affected the same minimization of competition as spatial partitioning. These behaviors effectively divided the homogeneous arboreal habitat into disparate sections, thereby partitioning (spatially and temporally) the niche space. This process curtails competition and allows the extant set of warblers to coexist.

Despite these successes, there are reasons to doubt that the CEP constitutes a natural law. First, it is of quite limited scope. For example, it is inapplicable when resources are abundant and species are not competing for them. Nor does it apply in fluctuating environments where niche contours are ephemeral, or changes in the direction of competition occur before exclusion can catalyze extinction. Since most environments nontrivially fluctuate, this is a serious limitation, and it recalls difficulties with *ceteris paribus* described above. CEP also seems to have numerous potential counterexamples. For example, migration into an area can prevent the predicted competitive exclusion. Another apparently recalcitrant counterexample was originally identified by MacArthur's advisor, G. E. Hutchinson (1961): the seemingly inordinate number of plankton species given their seemingly simple, homogeneous niche space. Known as the "paradox of the plankton," this issue remains an active area of contemporary ecological research and is yet to be conclusively resolved (see Tilman et al. 1982).

There is another threat to CEP as law, its empirical status. To appreciate the potential difficulty, first note that the relevant niche concept CEP invokes must be the Grinnellian, environmentally-based notion (see Sect. 2). In his seminal publication, Gause (1934) actually appeals to Elton's (1927) work, but Elton's functional niche concept would make CEP false. Different species often serve veritably identical roles in the causal nexus of interspecific interactions comprising a biological community. Different African grazers that migrate and ruminate together, and different pollinators, are but two plausible examples. In fact, ecological literature has labeled species with similar resource requirements that utilize them in similar ways a 'guild' to capture this commonality. Instead, what permits species with similar causal roles in a biological community to coexist, according to CEP, is their partitioning of the environment. For example, two fish eating water fowl, the great cormorant (*Phalacrocorax carbo*) and the European shag (*P. aristotelis*), exhibit approximately the same causal interactions with other terrestrial species, but they nest in different portions of cliffs (as well as consume fish from different sources, estuaries and harbors vs. the open sea, respectively) (Lack 1954).

But with 'niche' construed environmentally, the empirical content of CEP has been questioned. The problem begins with imprecision. For example, the CEP says nothing about what precise degree of niche differentiation is required to ensure coexistence. It seems that this can only be answered on a case by case, ecosystem by ecosystem, basis, if at all. The poor general guidance sets up a troubling scenario. The reasoning proceeds as follows. Suppose two species coexist but ecologically appear very similar. Then, by CEP, their niches must differ. Their ecological similarity initially suggested similar niches, so at what point does investigation of how the species utilize resources and interact with the environment that reveals no significant difference constitute a counterexample to CEP? Without precise guidance about what degree of niche differentiation coexistence requires, the worry is that the CEP is effectively immune from empirical challenge. For this reason, Pianka (2000, p. 248) calls CEP an untestable hypothesis "of little scientific utility." The claim about utility should be rejected. As Slobodkin (1961) convincingly argued and the history of the science confirms, CEP has played a very useful role in ecological theorizing as a research heuristic. But scientific utility should not be confused for nomic status. The mechanistic world-view was extremely valuable in the development of science, but it was not a law, or true.

5 A Theory of Ecological Stability

With some legitimacy, Arthur (1990, p. 30) cites the balance of nature as ecology's "number one" research priority, about which there is "near unanimity on its importance" (1990, p. 35). This priority is recent. Not until the second half of the twentieth century was the concept of a balance of nature rigorously characterized as ecological stability, and predominantly metaphysical speculations about its cause superseded with scientific hypotheses about its basis. But significant uncertainty and controversy remains about what features of an ecological system's dynamics should be considered its stability and thus no consensus has emerged about how ecological stability should be defined. Instead, ecologists have employed a confusing multitude of different terms to attempt to capture apparent stability properties: 'constancy,' 'persistence,' 'resilience,' 'resistance,' 'robustness,' 'tolerance,' and many more. This, in turn, has resulted in conflicting conclusions about debates concerning the concept based on studies using distinct senses of ecological stability.

One such debate, the stability-diversity debate, has persisted as a (perhaps the) central focus of theoretical ecology for half a century. The debate concerns the deceptively simple question of whether there is a relationship between the diversity of a biological community and its stability. From 1955, when Robert MacArthur initiated the debate, to the early 1970s, the prevailing view among ecologists was that diversity is an important, if not the principal, cause of community stability. Robert May, a physicist turned mathematical ecologist, confounded this view with analyses of mathematical models of communities that seemed to confirm the opposite, that increased diversity jeopardizes stability. The praise May's work received for its mathematical rigor and the criticisms it received for its seeming biological irrelevance thrust the SD debate into the ecological limelight, but subsequent analyses have failed to resolve it.

Different analyses seem to support conflicting claims and indicate an underlying lack of conceptual clarity about ecological stability that this section diagnoses and resolves. Below, a comprehensive account of stability is presented that clarifies the concepts ecologists have used that are defensible, their interrelationships, and their potential relationships with other biological properties. In particular, I argue that the concepts of resistance, resilience, and tolerance jointly provide an adequate definition of ecological stability. Roughly speaking, a community exhibits these concepts to a high degree if it: changes little after being perturbed (resistance); returns rapidly to a reference state or dynamic after being perturbed (resilience); and will return to that reference state or dynamic after most perturbations (tolerance).

Besides providing insights about how problematic scientific concepts should be characterized, it is worth noting that the issues involved in characterizing ecological stability have a potential bearing on biodiversity conservation. It seems that for most senses of stability, more stable communities are better able to withstand environmental disturbances, thereby decreasing the risk of species extinction. Positive feedback between diversity and stability would therefore support conservation efforts to preserve biodiversity. This yields a response to an influential criticism. As part of their argument that ecological theory has failed to provide a sound basis for environmental policy, Shrader-Frechette and McCoy (1993) have criticized that several proposed definitions of ecological stability are incompatible and that the concept is itself "conceptually confused" or "inconsistent." The account of ecological stability below answers this criticism.

Stability attributions must be made with respect to two evaluative benchmarks. The first is a system description (M) that specifies how the system and its dynamics are represented.[6] The second is a specified reference state or dynamic (R) of that system against which stability is assessed. In most ecological modeling, M is a mathematical model in which:

1. variables represent system parts, such as species of a community;
2. parameters represent factors that influence variables but are (usually) uninfluenced by them, such as solar radiation input into a community; and

[6]In the following, 'ecological stability' designates stability of a biological community unless otherwise specified, though most of the discussion also applies to the stability of a biological population or an ecosystem.

3. model equations describe system dynamics, such as interactions among species and the effect that environmental factors have on them.

M therefore delineates the boundary between what constitutes the system, and what is external to it. Relativizing stability evaluations to *M* is a generalization of Pimm's (1984) relativization of stability to a "variable of interest" because stability is assessed with respect to (1)–(3) rather than a subset of (1).

The specification of *M* partially dictates how *R* should be characterized, and *vice versa*. A biological community, for instance, is usually described as a composition of populations of different species. *R* must therefore reference these populations in some way. For example, *R* is often characterized in terms of the "normal" population sizes of each species. Since ecological modeling in the late 1960s and 1970s was dominated by the development of mathematically tractable equilibrium models (DeAngelis and Waterhouse 1987), "normal" population sizes were often assumed to be those at equilibrium, *i.e.* constant population sizes the community exhibits unless perturbed. This is not the only possible reference specification, however. A community may be judged stable, for instance, with respect to a reference *dynamic* the populations exhibit. Common examples are a limit cycle—a closed path *C* that corresponds to a periodic solution of a set of differential equations and towards which other paths asymptotically approach—or a more complicated attractor dynamic (see Kot 2001, Ch. 8). Ecological stability can also be assessed with respect to some specified range of tolerated fluctuation. *R* may also be characterized solely in terms of the presence of certain species.[7] Only extinction would constitute departure from this reference state.

The details of *M* and *R* are crucial because different system descriptions—e.g. representing systems with different variables or representing their dynamics with different functions—may exhibit different stability properties or exhibit them to varying degrees relative to different specifications of *R*. Specifying *R* as a particular species composition vs. specifying *R* as an equilibrium, for instance, can yield different stability results. Similarly, different *M* can produce different assessments of a system's stability properties. Describing a system with difference versus differential equations is one example (May 1974). He showed, for instance, that the logistic *difference* equation:

$$N_{t+1} = (1+r)N_t - \frac{r}{K}N_t^2; \quad (1)$$

where t is time; r is the intrinsic growth rate; K is the carrying capacity; and N is the population size, exhibits dramatically different behavior than the corresponding logistic *differential* equation. For $r > 0$, the logistic differential equation has an asymptotically Lyapunov stable equilibrium $N^* = K$. This is also an asymptotically Lyapunov stable equilibrium of the logistic difference equation, but only for

[7] To illustrate the partial dependence of M on R, notice that the species referred to in R must be part of the system description M.

$0 < r < 2$. For $2 < r < 2.526$ the system exhibits a 2-period limit cycle. As r increases beyond 2.526 a 4-period limit cycle emerges, and the system exhibits chaotic behavior for $r > 2.692$. Thus, although the logistic differential and difference equations appear to describe very similar dynamics, the seemingly inconsequential choice of representing time as a discrete or continuous variable has a substantial effect on evaluating stability properties of the system.

Details of *M* and *R* are also important because they may specify the spatial and temporal scales at which the system is being analyzed, which can affect stability assessments. Systems with low resistance but high resilience, for example, fluctuate dramatically in response to perturbation but return rapidly to their reference state *R*. Low resistance is detectable at fine-grained temporal scales, but systems may appear highly resistant at coarser scales because their quick return to *R* prevents detection of fluctuation. Similarly, significant fluctuations in spatially small areas may contribute to relatively constant total population sizes maintained through immigration and emigration in larger regions.

Once (and only once) *M* and *R* are specified, the stability properties of a system can be determined. These properties fall into two general categories, depending on whether they refer to how systems respond to perturbation (relative to *R*) or refer to system properties independent of perturbation response. A perturbation of an ecological system is any discrete event that disrupts system structure, changes available resources, or changes the physical environment (Krebs 2001). Typical examples are flood, fire, and drought. Perturbations are represented in mathematical models of communities by externally induced temporary changes to variables that represent populations, to parameters that represent environmental factors, and/or to model structure. Many, perhaps most, real-world perturbations of communities should be represented by changes to both variables and parameters. A severe flood, for instance, eradicates individual organisms and changes several environmental factors affecting populations. In the following, let P_v, P_p, and P_{vp} designate perturbations that change only variables, change only parameters, and those that change both, respectively.

Perturbations may cause other changes, such as alteration of the functional form of species interactions, that are not adequately represented by changes to variable or parameter values of typical community models, but which should be included in a comprehensive assessment of community's stability. Since these perturbations change community dynamics, they change *M*. How the altered community responds to these (and subsequent) perturbations must then be assessed against the new description of the community's dynamics as long as those dynamics remain altered. Although a completely adequate assessment of the ecological stability of a community requires consideration of all such changes caused by perturbations, most ecological modeling focuses on changes to variable and parameter values.

There are four plausible adequacy conditions for an account of ecological stability:

(A1) the ecological stability of a biological community depends upon how it responds to perturbation ([A2]–[A4] specify the form of the required dependency);

(A2) of two communities *A* and *B*, the more ecologically stable community is the one that would exhibit less change if subject to a given perturbation *P*;
(A3) if *A* and *B* are in a pre-perturbation reference state or dynamic *R*, the more ecologically stable community is the one that would most rapidly return to *R* if subject to *P*; and,
(A4) if *A* and *B* are in *R*, the more ecologically stable community is the one that can withstand stronger perturbations and still return to *R*.

Before considering these conditions in detail, a few remarks help clarify their general basis. First, (A2)–(A4) only place *comparative* constraints on the concept of ecological stability and therefore require only a rank ordering of the stability of biological communities, rather than a particular quantitative valuation. The reason for requiring only comparative constraints is that quantitative valuation of ecological stability depend upon the system description (*M*) and reference state or dynamic (*R*) specified for a community, both of which may vary. Second, conditions (A2) and (A3) order the stability of communities based on their behavior following a *particular* perturbation *P*. As adequacy conditions, they therefore do not require a measure of the strength of perturbations. This reflects the difficulties facing the formulation of a measure of perturbation strength (see below), although such a measure is needed to evaluate the resistance of communities when only their responses to perturbations of different strength are known. If a quantitative measure of perturbation strength for different types of perturbation were available, two further *non-comparative* adequacy conditions could be formulated:

(A2′) a highly stable biological community should change little following weak perturbations;
(A3′) a highly stable biological community should rapidly return to its reference state or dynamic following weak perturbations.

In contrast, condition (A4) does require a measure of perturbation strength.

Condition (A1) captures the idea that a community's behavior is a reliable indicator of its ecological stability only if the behavior reflects how perturbation changes the community. If unperturbed, a community may exhibit great constancy throughout some period, for instance, as assessed by a lack of fluctuations in the biomasses of species in the community. It may be, however, that if it had been even weakly perturbed, it would have changed dramatically. Constancy of this community surely does not indicate ecological stability when it would have changed substantially if perturbed slightly. Similarly, variability of a community does not necessarily indicate lack of ecological stability if it is the result of severe perturbations, perturbations that would cause greater fluctuations or even extinctions in less stable communities.

The reason for (A2) is that more stable communities should be less affected by perturbations than less stable ones. Communities that can withstand severe drought with little change, for instance, are intuitively more stable than those modified dramatically. The justification for (A3) is that more stable communities should more rapidly return to *R* following perturbations than less stable ones. This adequacy condition captures the idea that lake communities that return to R quickly after an incident of thermal pollution, for instance, are more stable than those with slower return rates following similar incidents. The ground for the last condition, A4, is that communities that can sustain stronger perturbations than others and still return to R should be judged more stable.

Three concepts—resistance, resilience, and tolerance—represent the properties required of ecological stability by (A2′)–(A4). **Resistance** is inversely correlated with the degree a system changes relative to R following a perturbation (P_v, P_p, or P_{vp}). Since perturbations vary in magnitude, resistance must be assessed against perturbation strength. Large changes after weak perturbations indicate low resistance; small changes after strong perturbations indicate high resistance. Resistance is thus inversely proportional to perturbation sensitivity.

Depending on M and R, changes in communities can be evaluated in different ways, each of which corresponds to a different measure of resistance. Community resistance is typically measured by changes in species *abundances* following perturbation. It could, however, be measured by changes in species *composition* following perturbation, or in some other way. Pimm's (1979) concept of species deletion stability, for instance, measures resistance by the number of subsequent extinctions in a community after one species is eradicated.

A simple example illustrates the contextual import of M and R in assessing resistance. Consider the classical Lotka-Volterra model of a one-predator, one-prey community:

$$\frac{dx_y(t)}{dt} = ax_y(t) - \alpha x_y(t)x_d(t); \quad (1a)$$

$$\frac{dx_d(t)}{dt} = -bx_d(t) + \beta x_d(t)x_y(t); \quad (1b)$$

where x_d and x_y represent predator and prey populations; a represents prey birth rate; b represents predator death rate; and $\alpha, \beta > 0$ in the second term of each equation represent the effect of prey individuals on predator individuals and *vice versa*. Equations (1a and 1b) are the description of the system, M. There is one non-trivial equilibrium, $x_d^* = \frac{b}{\beta}$ and $x_y^* = \frac{a}{\alpha}$, which is usually specified as the reference state, R.

For this M and R, resistance to a Pv perturbation that eradicates, say, half of x_y can be measured by how far x_d deviates from x_d^*. If M were different, the perturbation could obviously have a different effect on x_d. If x_d and x_y were competitors, for instance, x_d would increase rather than decrease after this perturbation. Similarly, if R were different, assessments of resistance may change. If R were the species composition x_d and x_y (i.e. $X_d, X_y > 0$) rather than their equilibrium values, for instance, resistance would be assessed in terms of changes from this composition, i.e. in terms of species extinction. The equilibrium $X_d^* = \frac{b}{\beta}$, $X_y^* = \frac{a}{\alpha}$ is globally stable for this simple community, so only a Pv perturbation strong enough to eradicate one of the species will cause extinction; this community returns to equilibrium after all other Pv perturbations. For communities with many species and more species dynamics, however, a Pv perturbation that eradicates half or less of one species may cause the extinction of that, or other species.

Different types of perturbations, moreover, yield different measures of resistance. Since evaluating resistance requires considering perturbation strength, strengths of different types of perturbations must be comparable for there to be a single measure of resistance for a system. Such comparisons are sometimes straightforward. If one perturbation eradicates half of species x in a community, for instance, another that eradicates 75 % of x is certainly stronger. If another perturbation eradicates 25 % of 3 species or 5 % of 15 species in the community, however, it is unclear how its strength should be ranked against the perturbation that eradicates 75 % of x. What criteria could be used to compare strengths of P_v, P_p, or P_{vp} perturbations, to which systems may show differential sensitivity, is even less clear. Systems that are highly resistant to P_v perturbations may be extremely sensitive to even slight P_p. Comparing the resistance of communities is therefore only unproblematic with respect to perturbations of comparable kind.

Resilience is the rate at which a system returns to R following perturbation (P_v, P_p, or P_{vp}). Like resistance, resilience must be assessed against perturbation strength unless, although unlikely for many types of perturbation, return rate is independent of perturbation strength. Slow return rates after weak perturbations indicate low resilience and rapid rates following strong perturbations indicate high resilience. If return rate does not depend on perturbation strength, however, resilience can be evaluated by the return rate independent of the perturbation strength, although the rate may vary across different types of perturbations. Systems may not return to R after perturbation, especially following severe perturbation, so, unlike resistance, resilience is only assessable for perturbations that do not prevent return to R. Note that resilience and resistance are independent concepts: systems may be drastically changed by weak perturbations (low resistance) but rapidly return to R (high resilience), and *vice versa*.

Resilience is commonly measured as the inverse of the time taken for the effects of perturbation to decay relative to R. For a specific mathematical model, this can be determined analytically or by simulation. For the community described by Eq. (1) above, for instance, resilience to a P_v perturbation that eradicates half of one species could be simply measured by $\frac{1}{\left|t_{eq} - t_p\right|}$ where t_p is the time at which the community is initially perturbed and t_{eq} is the time at which the community reestablishes equilibrium. Resilience to P_v perturbation is determined by the largest real eigenvalue part for systems modeled by linear differential equations if it is negative, and analytic methods have been developed to assess resilience to P_v perturbation for nonlinear models. Empirical measurement of resilience for communities in nature, however, is often thwarted by subsequent perturbations that disrupt return to R. This difficulty can be avoided if subsequent perturbations can be evaded with controlled experiments. If the return rate is independent of perturbation strength, estimation of resilience is also more feasible because only the decay rate of the perturbation effects need be measured before the system is further perturbed; measurement of perturbation strength is not required (Pimm 1984). Like resistance, furthermore, different types of perturbations yield different measures of resilience since return rate to R may depend upon the way in which systems are perturbed. A system may be highly resilient to P_v perturbation and poorly resistant to P_p perturbation, for instance, or more resilient to some P_v or P_p perturbations than others.

Tolerance, or "domain of attraction" stability, is the ability of a system to be perturbed and return to R, regardless of how much it may change and how long its return takes. More precisely, tolerance is positively correlated with the range and strength of perturbations a system can sustain and still return to R. The magnitudes of the strongest perturbations it can sustain determine the contours of this range. Note that tolerance is conceptually independent of resistance and resilience: a system may be severely perturbed and still return to R (high tolerance), even if it changes considerably (low resistance) and its return rate is slow (low resilience), and *vice versa*.

Similar to resistance and resilience, different kinds of perturbations yield different measures of tolerance. Tolerance to P_v perturbations, for instance, is determined by the maximal changes variables can bear and not jeopardize the system's return to R. With respect to P_v perturbations that affect only one species of a community, for instance, tolerance can be simply measured by the proportion of that species that can be eradicated without precluding the community's return to R. If a nontrivial equilibrium of Eq. (1) from above is globally stable, for instance, the community described by the equation is maximally tolerant to P_v perturbations relative to this reference state because the community will return to it after any P_v perturbation that does not eradicate one of the species. Variables of a system may be perturbed, however, in other ways. A P_v perturbation may change all variables, several, or only one; it may change them to the same degree, some variables more severely than others; and so on. How exactly variables are perturbed may affect whether the system returns to R. System tolerance must therefore be evaluated with respect to different types of perturbation. The same goes for assessing tolerance to P_p or P_{vp} perturbations.

Although resistance, resilience, and tolerance do not adequately explicate ecological stability individually, they do so collectively. In fact, they constitute jointly sufficient and separately necessary conditions for ecological stability, notwithstanding Shrader-Frechette and McCoy's (1993, p. 58) claim that such conditions do not exist. Consider sufficiency first. Since these three concepts represent the properties underlying conditions (A2)–(A4) [and (A2′) and (A3′)], communities exhibiting them to a high degree would change little after strong perturbations ([A2]), return to R rapidly if perturbed from it ([A3]), and return to R following almost any perturbation ([A4]). If R is a point equilibrium, moreover, a community exhibiting high resistance, resilience and tolerance will be relatively constant. As such, these three properties certainly capture ecologists' early conceptions of ecological stability, and there seems to be no further requirement of ecological stability that a community exhibiting these properties would lack.

Each concept is also necessary. Highly tolerant and resistant communities, for instance, change little and return to R after most perturbations. In regularly perturbing environments, however, even a highly resistant and tolerant community may be iteratively perturbed to the boundary of its tolerance range and "linger" there if its return rate to *R* is too slow. Subsequent perturbations may then displace it from this range, thereby precluding return to *R*. If this community rapidly returned to *R* after most perturbations (high resilience), it would rarely reach and would not linger at its tolerance boundary. In general, low resilience preserves the effects perturbations have on communities for extended, perhaps indefinite durations, which seems incompatible with ecological stability.

Similar considerations show tolerance and resistance are necessary for ecological stability. A highly resilient and tolerant but weakly resistant community rapidly returns to *R* following almost any perturbation, but changes significantly after even the slightest perturbation, which seems contrary to ecological stability. The dramatic fluctuation such communities would exhibit in negligibly variable environments is the basis for according them low ecological stability. A highly resilient and resistant but weakly tolerant community changes little and rapidly returns to *R* when perturbed within its tolerance range, but even weak perturbations displace it from this range and thereby preclude its return to *R*, which also seems contrary to ecological stability.

Resistance, resilience, and tolerance are independent concepts and thus biological communities may exhibit them to different degrees. Although the necessity of each concept for ecological stability does not strictly entail they are equally important in evaluations of a community's stability, nothing about the pre-theoretic concept of ecological stability seems to suggest otherwise. As a concept composed of resistance, resilience, and tolerance, ecological stability therefore imposes only a partial, not complete, ordering on communities. Moreover, since communities may differentially exhibit resistance, resilience, and/or tolerance for different types of ecological perturbations, each property also imposes only a partial ordering on communities. This twofold partiality entails inferences from analyses of stability-diversity and stability-complexity relationships are limited by the property and type of perturbations analyzed, beside the particular system description (*M*) and reference state or dynamic (*R*) specified.

It is worth pausing over what the framework for ecological stability presented above shows about the general concept. It certainly shows that ecologists have used the term 'stability' to describe several distinct features of community dynamics, although only resistance, resilience, and tolerance adequately define ecological stability. This plurality does not manifest, however, an underlying vagueness, "conceptual incoherence," or "inconsistency" of the concept, as Shrader-Frechette and McCoy (1993, p. 57) suggest in their general critique of basic ecological concepts and ecological theories based on them.

Two claims seem to ground their criticism. First, if 'stability' is used to designate distinct properties, this indicates the concept is itself conceptually vague and thereby flawed. Although terminological ambiguity is certainly undesirable, most ecologists unambiguously used the term to refer to a specific property of a community and accompanied the term with a precise mathematical or empirical operationalization. Since these were in no sense vague, in no sense was ecological stability "vaguely defined" (Shrader-Frechette and McCoy 1993, p. 40). Ecologists quickly appreciated this terminological ambiguity, moreover, and began explicitly distinguishing different senses of ecological stability with different terms (Odenbaugh 2001).

Shrader-Frechette and McCoy's second claim is that, "There is no homogeneous class of processes or relationships that exhibit stability," (1993, p. 58). The underlying assumption seems to be that concepts in general, and ecological stability in particular, must refer to a homogeneous class in order to be conceptually unproblematic. That ecological stability does not, and worse, that ecologists have supposedly attributed inconsistent meanings to it, shows the concept is incoherent, they believe, much like the vexed species concept (1993, p. 57). Shrader-Frechette and McCoy do not offer an argument for this assumption, and it is indefensible as a general claim about what concepts must refer to. Common concepts provide clear counterexamples. The concepts 'sibling,' 'crystal,' and 'field,' for instance, refer to heterogeneous classes, but there is nothing conceptually problematic about them. There is debate about the idea of disjunctive *properties* in work on multiple realization (Kim 1998), but criticisms raised against disjunctive properties do not necessarily apply to disjunctive *concepts*, nor were they intended to. Kim (1998, p. 110) emphasizes this point:

> Qua property, dormativity is heterogeneous and disjunctive, and it lacks the kind of causal homogeneity and projectability that we demand from kinds and properties useful in formulating laws and explanations. But [the concept of] dormativity may well serve important conceptual and epistemic needs, by grouping properties that share features of interest to us in a given context of inquiry.

Even if criticisms of disjunctive properties were sound, it therefore would not follow that the disjunctive concepts such as ecological stability are also problematic. Shrader-Frechette and McCoy's criticism of ecological stability is therefore indefensible. The conceptual and epistemic utility of a concept is enhanced, furthermore, if clear guidelines for its application exist, which the above analysis has attempted to provide.

Moreover, the definitional status of the concepts of ecological stability and species is not analogous. Biologists have proposed plausible, but incompatible competing definitions for the species concept because it is problematically ambiguous (Ereshefsky 2001; Reydon 2013). That resistance, resilience, and tolerance have been referred to under the rubric 'stability,' however, do not show ecological stability is similarly problematically ambiguous because they are conceptually independent and therefore compatible, as different senses of 'species' are not.[8] In addition, most ecologists recognized that there are several components of ecological stability, and individual stability concepts such as resistance, resilience and tolerance, or measures thereof, were rarely proposed as *the* uniquely correct definition of ecological stability. Rather, they were and should be understood as distinct features of ecological stability or ways of measuring it, not competing definitional candidates. Like many

[8]As noted above, different quantitative measures of resistance, resilience, and tolerance may be incompatible. This does not establish, however, that the corresponding concepts are incompatible.

scientific concepts, ecological stability is multifaceted, and the distinct referents ecologists attributed to it accurately reflect this. Conceptual multifacetedness alone does not entail conceptual incoherence or inconsistency.

6 Conclusion

The discussion above confirms philosophy of ecology's proper place within the purview of philosophy of science. It should, and increasingly is, garnering the philosophical attention it should. Before concluding, the timely significance of the relationship between ecology and conservation science, and environmental science more generally, merits special consideration. Although ecology has received less attention from philosophers of science than other areas of the life sciences, this has begun to change as the severity and complexity of environmental problems, and ecology's potential role in helping solving them, has become more apparent. Threats to coral reefs and the ecologically-informed management strategies developed in response are vivid examples. Ecology provides a scientific basis for conservation action, pollution reduction and mitigation, and other environmentalist objectives. Indeed, conservation biology is sometimes conceptualized as a species of applied ecology, and concepts such as the 'balance of nature' have a central place in how biology students conceptualize environmental challenges (Hovardas and Korfiatis 2011; Ergazaki and Ampatzidis 2012). This marshaling of scientific resources to achieve ethical goals raises a multitude of philosophically rich issues.

One is the precise nature of the relationship between facts and values in such applied sciences. It often seems that facts and ethical values are very closely linked in the experimental and statistical methods these sciences employ (see also Gannett, this volume). For example, values seem to determine how uncertainty is managed, particularly whether the available data are sufficiently strong to overcome uncertainty and warrant accepting or rejecting hypotheses (Shrader-Frechette 1990). One of the most direct ways in which ethical and socio-political values bear on ecology (and *vice versa*) is in population viability analyses (PVAs). These are studies, usually model-based, of the dynamics of biological populations and how they would respond to various disturbance and management regimes. Whether the data are sufficient to show a regime adequately ensures a stipulated viability threshold usually requires a trade-off between minimizing type I and type II errors. This in turn seems to require the input of non-epistemic, ethical values. As such, PVAs have a significant bearing on conservation planning and action and seem to essentially incorporate ethical assumptions and considerations. Choices of scientific categories and terms, such as 'carcinogen' and 'endangered,' seem to be similarly infused with ethical considerations. Numerous other examples could be cited. Some philosophers have recently argued this close connection between fact and value constitutes confluence, that facts and values are indelibly intertwined in such ethically-driven sciences (Putnam 2002); others believe this conclusion is overstated (Sober 2007).

A related issue concerns how the notion of "progress" should be conceptualized in sciences such as applied ecology. Innovations that constitute significant progress towards achieving the ethical goals some applied sciences are intended to help achieve—e.g. biodiversity conservation—often involve improvements in data acquisition, statistical analysis, or algorithm efficacy. That these clear advances do not resemble the kind of developments in theory paradigmatically considered scientific progress in other sciences does not mean they do not constitute genuine progress (*cf.* Linquist 2008). A different, broader notion of progress is therefore needed to recognize the role achieving ethical goals has in some applied sciences.

Although ecology underpins much of evolutionary theory and it seeks to understand vast portions of the biological world, the typical undergraduate exposure to biology often consists entirely of developmental biology and evolutionary theory, at the expense of ecological science. Similarly, philosophers of biology have paid relatively little attention to ecology. This inattention is due to disciplinary inertia rather than principled position, so it is ripe for change. This essay contributes to that change by describing some of the philosophically rich issues ecologists study.

Acknowledgements Interactions with students in a Spring 2012 graduate seminar devoted to philosophy of ecology at FSU helped shape and improve this piece. Besides those students, thanks to Kostas Kampourakis, Roberta Millstein, Jay Odenbaugh, and Carl Salk for very helpful comments.

References

Abrams, P. 1983. The theory of limiting similarity. *Annual Review of Ecology and Systematics* 14: 359–376.

Arthur, W. 1990. *The green machine: Ecology and the balance of nature*. Cambridge, MA: Basil Blackwell.

Beatty, J. 1995. The evolutionary contingency thesis. In *Concepts, theories, and rationality in the biological sciences*, ed. G. Wolters and J. Lennox, 49–81. Pittsburg: University of Pittsburg Press.

Bowler, P. 1976. Malthus, Darwin, and the concept of struggle. *Journal of the History of Ideas* 37: 631–650.

Budiansky, S. 1995. *Nature's Keepers: The new science of nature management*. New York: Free Press.

Clements, Fredric E. 1916. *Plant succession, an analysis of the development of vegetation*. Washington, D.C.: Carnegie Institute.

Colinvaux, P. 1979. *Why big fierce animals are rare: An Ecologist's perspective*. Princeton: Princeton University Press.

Colwell, R. 1992. Niche: A bifurcation in the conceptual lineage of the term. In *Key words in evolutionary biology*, ed. E.F. Keller and E.S. Lloyd, 241–248. Cambridge: Harvard University Press.

Connor, E., and E. McCoy. 1979. The statistics and biology of the species-area relationship. *American Naturalist* 113: 791–833.

Cooper, G. 2003. *The science of the struggle for existence: On the foundations of ecology*. Cambridge: Cambridge University Press.

Coyne, J., and H. Allen. 2004. *Speciation*. Sunderland: Sinauer Associates.

Darwin, C. 1859. *On the origin of the species*, 1st ed. Cambridge: Harvard University Press.

Davis, M., and R. Shaw. 2001. Range shifts and adaptive responses to quaternary climate change. *Science* 292: 673–679.
DeAngelis, D., and J. Waterhouse. 1987. Equilibrium and non-equilibrium concepts in ecological models. *Ecological Monographs* 57: 1–21.
Dobzhansky, T. 1964. Biology, molecular and organismic. *American Zoologist* 4: 443–452.
Dobzhansky, T. 1973. Nothing in biology makes sense except in the light of evolution. *The American Biology Teacher* 35: 125–129.
Earman, J., C. Glymour, and S. Mitchell, (Eds). 2002. *Ceteris paribus laws. Erkenntnis* 57.
Egerton, F.N. 1973. Changing concepts of the balance of nature. *Quarterly Review of Biology* 48: 322–350.
Eliot, C. 2011a. The legend of order and chaos: Communities and early community ecology. In *Handbook of the philosophy of ecology*, ed. K. de Laplante, B. Brown, and K. Peacocke, 49–108. Haarlem: Elsevier.
Eliot, C. 2011b. Competition theory and channeling explanation. *Philosophy and Theory in Biology* 3: 1–16.
Elton, C. [1927] 2001. *Animal ecology*. Chicago: University of Chicago Press.
Ereshefsky, M. 2001. *The poverty of the Linnaean Hierarchy*. Cambridge, MA: Cambridge University Press.
Ergazaki, M., and G. Ampatzidis. 2012. Students' reasoning about the future of disturbed or protected ecosystems & the idea of the 'balance of nature'. *Research in Science Education* 42: 511–530.
Gause, G.F. 1934. *The struggle for existence*. Baltimore: Williams and Wilkins.
Ginzburg, L.R., and M. Colyvan. 2004. *Ecological orbits: How planets move and populations grow*. New York: Oxford University Press.
Griesemer, J. 1992. Niche: Historical perspectives. In *Key words in evolutionary biology*, ed. E.F. Keller and E.S. Lloyd, 231–240. Cambridge: Harvard University Press.
Grinnell, J. 1917. The niche-relationships of the California Thrasher. *Auk* 34: 427–433.
Haeckel, E. 1866. *General morphology of organisms*. Berlin: Reimer.
Hardin, G. 1960. The competitive exclusion principle. *Science* 131: 1292–1297.
Herbold, B., and P. Moyle. 1986. Introduced species and vacant niches. *The American Naturalist* 128: 751–760.
Hovardas, T., and K. Korfiatis. 2011. Towards a critical re-appraisal of ecology education: Scheduling an educational intervention to revisit the 'balance of nature' metaphor. *Science & Education* 20: 1039–1053.
Hubbell, S. 2001. *The unified neutral theory of biodiversity and biogeography*. Princeton: Princeton University Press.
Hutchinson, G.E. 1957. Concluding remarks. *Cold Spring Harbor Symposia on Quantitative Biology* 22: 415–427.
Hutchinson, G.E. 1961. The paradox of the plankton. *American Naturalist* 95: 137–145.
Justus, J. 2008a. Complexity, diversity, stability. In *A companion to the philosophy of biology*, ed. S. Sarkar and A. Plutynski, 321–350. Malden: Blackwell.
Justus, J. 2008b. Ecological and Lyapunov stability. *Philosophy of Science* 75: 421–436.
Justus, J. 2011. A case study in concept determination: Ecological diversity. In *Handbook of the philosophy of ecology*, ed. K. de Laplante, B. Brown, and K. Peacock. New York: Elsevier Press.
Kim, J. 1998. *Mind in a physical world: An essay on the mind-body problem and mental causation*. Cambridge, MA: MIT Press.
Kimura, M. 1983. *The neutral theory of molecular evolution*. Cambridge/New York: Cambridge University Press.
Kingsland, S. 1995. *Modeling nature: Episodes in the history of population ecology*, 2nd ed. Chicago: University of Chicago Press.
Kohn, D. 2009. Darwin's Keystone: The principle of divergence. In *Cambridge companion to the "origin of species"*, ed. M. Ruse and R. Richards, 87–108. Cambridge: Cambridge University Press.

Kot, M. 2001. *Elements of mathematical ecology*. Cambridge/New York: Cambridge University Press.
Krebs, C. 2001. *Ecology*. New York: Benjamin Cummings.
Lack, D. 1954. *The natural regulation of animal numbers*. Oxford: Oxford University Press.
Lange, M. 2005. Ecological laws: What would they be and why would they matter? *Oikos* 110: 394–403.
Lawton, J. 1984. Non-competitive populations, non-convergent communities, and vacant niches: The herbivores of bracken. In *Ecological communities: Conceptual issues and the evidence*, ed. D. Strong, D. Simberloff, L. Abele, and A. Thistle, 67–101. Princeton: Princeton University Press.
Leibold, M.A. 1995. The niche concept revisited: Mechanistic models and community context. *Ecology* 76: 1371–1382.
Linquist, S. 2008. But is it progress? On the alleged advances of conservation biology over ecology. *Biology and Philosophy* 23: 529–544.
MacArthur, R. 1958. Population ecology of some warblers of northeastern coniferous forests. *Ecology* 39: 599–619.
MacArthur, R., and E.O. Wilson. 1967. *The theory of island biogeography*. Princeton: Princeton University Press.
Maclaurin, J., and K. Sterelny. 2008. *What is biodiversity?* Chicago: The University of Chicago Press.
May, R. 1974. *Stability and complexity in model ecosystems*, 2nd ed. Princeton: Princeton University Press.
Mikkelson, G. 2003. Ecological kinds and ecological laws. *Philosophy of Science* 70: 1390–1400.
Millstein, R. forthcoming. Exploring the status of population genetics: The role of ecology. *Biological theory: Integrating development, evolution and cognition*. Special issue on The meaning of 'theory' in biology.
Odenbaugh, J. 2001. Ecological stability, model building, and environmental policy: A reply to some of the pessimism. *Philosophy of Science (Proceedings)* 68: S493–S505.
Odenbaugh, J. 2007. Seeing the forest and the trees. *Philosophy of Science* 74: 628–641.
Odling-Smee, F., K. Laland, and M. Feldman. 2003. *Niche construction: The neglected process in evolution*. Princeton: Princeton University Press.
Pianka, E. 2000. *Evolutionary ecology*. New York: Addison Wesley Longman.
Pimm, S. 1979. Complexity and stability: Another look at MacArthur's original hypothesis. *Oikos* 33: 351–357.
Pimm, S. 1984. The complexity and stability of ecosystems. *Ecology* 61: 219–225.
Putnam, H. 2002. *The collapse of the fact/value dichotomy and other essays*. Cambridge: Harvard University Press.
Regan, H., M. Colyvan, and M. Burgman. 2002. A taxonomy and treatment of uncertainty for ecology and conservation biology. *Ecological Applications* 12: 618–628.
Reydon, T. 2013. Classifying life, reconstructing history and teaching diversity: Philosophical issues in the teaching of biological systematics and biodiversity. *Science & Education* 22: 189–220.
Rosenzweig, M. 1992. Species diversity gradients: We know more and less than we thought. *Journal of Mammalogy* 73: 715–730.
Sarkar, S. 2005. *Biodiversity and environmental philosophy: An introduction*. Cambridge: Cambridge University Press.
Shrader-Frechette, K. 1990. Island biogeography, species-area curves, and statistical errors: Applied biology and scientific rationality. *PSA* 1990: S447–S456.
Shrader-Frechette, K.S., and E.D. McCoy. 1993. *Method in ecology*. Cambridge: Cambridge University Press.
Slobodkin, L.B. 1961. *Growth and regulation of animal populations*. New York: Holt, Rinehart and Winston.
Sober, E. 1987. Does 'fitness' fit the facts. *Journal of Philosophy* 84: 220–223.

Sober, E. 2007. Evidence and value freedom. In *Value-free science – Ideal or illusion?* ed. H. Kinkaid, J. Dupré, and A. Wylie, 109–119. Oxford: Oxford University Press.
Stauffer, R. 1957. Haeckel, Darwin, and ecology. *The Quarterly Review of Biology* 32: 138–144.
Sterelny, K. 2001. The reality of ecological assemblages: A palaeo-ecological puzzle. *Biology and Philosophy* 16: 437–461.
Sterelny, K. 2006. Local ecological communities. *Philosophy of Science* 73: 215–231.
Sterelny, K., and P. Griffiths. 1999. *Sex and death*, 253–280. Chicago: University of Chicago Press.
Tilman, D., S. Kilham, and P. Kilham. 1982. Phytoplankton community ecology: The role of limiting nutrients. *Annual Review of Ecology and Systematics* 13: 349–372.
Weisberg, M., and K. Reisman. 2008. The Robust Volterra principle. *Philosophy of Science* 75: 106–131.
West, G., J. Brown, and B. Enquist. 1997. A general model for the origin of allometric scaling laws in biology. *Science* 276: 122–126.
Whittaker, R. 1956. Vegetation of the Great Smoky Mountains. *Ecological Monographs* 26: 1–80.
Worster, D. 1994. *Nature's economy: A history of ecological ideas*. New York: Cambridge University Press.

Small Things, Big Consequences: Microbiological Perspectives on Biology

Michael J. Duncan, Pierrick Bourrat, Jennifer DeBerardinis, and Maureen A. O'Malley

1 Introduction

Microbes are the most numerous, diverse and ancient of the many life forms on our planet. They are also central to all life and its maintenance. The science of these organisms, microbiology, is the science of how microorganisms function, interact and evolve, in the context of causally influencing all other life forms. So far, microbiology has attracted barely any attention from philosophers of biology, and we outline the many reasons for remedying this state of affairs. We begin this chapter with a general argument for a philosophy of microbiology that recognizes the biological and evolutionary importance of the microbial world, and starts with a basic understanding of what microbes and microbiology are.

1.1 The Entities Studied by Microbiology

'Microbe' is a general colloquial term that covers quite a range of microscopic biological phenomena. It includes all unicellular forms of life and commonly encompasses viruses, despite the fact that these entities have no cells of their own and have to use cellular organisms for reproduction. For this reason, even though

---

Michael J. Duncan, Pierrick Bourrat and Jennifer DeBerardinis contributed equally to this chapter.

M.J. Duncan • P. Bourrat • M.A. O'Malley (✉)
Department of Philosophy, University of Sydney,
Quadrangle A14, Sydney, NSW 2006, Australia
e-mail: maureen.omalley@sydney.edu.au

J. DeBerardinis
Sydney Centre for the Foundations of Science, University of Sydney,
Quadrangle A14, Sydney, NSW 2006, Australia

K. Kampourakis (ed.), *The Philosophy of Biology: A Companion for Educators*, History, Philosophy and Theory of the Life Sciences 1, DOI 10.1007/978-94-007-6537-5_18,

they can be said to evolve, viruses are not usually thought of as living organisms (see Sect. 2 for debates on this topic). Unicellular microbes form three major groups, and these groups are now thought of as representative of the three major 'domains' of life: Bacteria, Archaea, Eukarya. Two of these microbial groups are bacteria and archaea, which together are called prokaryotes. Prokaryotes are unicellular life forms whose intracellular structure is organized without easily recognized compartmentalized functions, such as the nucleus (however, more and more compartments have been recognized in prokaryote cells, so they cannot be thought of as chemicals bounded by membranes). Bacteria and archaea used to be thought of only as bacteria, but fine-grained molecular and cell-biological work in the 1970s showed them to be very different physiologically and genetically. Their genomic content, cell walls, membranes, and replication, transcription and translation machinery all establish these two groups as very distinct life forms even if morphologically they are not obviously distinguished (Cavicchioli 2007).

The third group of unicellular life forms are found in the broad domain or super kingdom of eukaryotes, as are all multicellular organisms (as traditionally understood – cf. Sect. 3). Eukaryotes thus comprise all the unicellular and multicellular life forms that have the most well recognized compartmentalized processes in each cell. Eukaryotic microbes include protists, another colloquially named group, which are distinct from the other groups of eukaryotes (plants, animal and fungi), but which include unicellular algae (microbes once placed in the plant kingdom) and sometimes unicellular fungi (e.g., yeast).

Because 'microbe' means microscopic biological entity (versus 'macrobe'), informal discussions sometimes include very small multicellular organisms such as rotifers (mostly asexual animals found in aquatic environments). Not many microbiologists would do that, however (preferring the distinctions of 'microfauna' or 'meiofauna'), even though they and other biologists are perfectly willing to incorporate visible aggregations of single microbial cells such as filaments and moulds under that heading (it could be argued that organisms such as ourselves would then qualify as microbes, because we too are visible aggregations of single cells, organized in a particular way, as are filaments and other visible, organized aggregations of microorganisms). Although there are many troublesome issues in the nomenclature and taxonomy of microbes, our chapter will not focus on these but instead try to get at some of the deeper underlying issues of biological organization and its evolution.

1.2 Methods and Perspectives in Microbiology

Even a sketchy understanding of the development of microbiology as a science gives good reasons for philosophers to care about microbes. Microbiology has a long history, from the 1660s at least, which is when microscopy permitted close-up microbial observation and experiment. As well as accumulating diverse observations of miniature living things, the microscope provoked reflections on the very nature of microbial and other microscopic life (Hooke 1665; Leeuwenhoek 1694). These reflections led very naturally to profound questions about how life was generated, spontaneously or otherwise (Farley 1974). The more microscopic organisms were observed, the more pressing became questions about their specificity, stability and classifiability. The emergence of pure culture techniques in the late nineteenth century was a tremendous breakthrough for microbiology, since it enabled the stable identification of specific effects of different microbial taxa and thus their recognition as 'species' (Cohn 1875, in Brock 1961; Amsterdamska 1987; Bulloch 1938; Drews 2000; Gradmann 2000). Although in many respects, microbes were problem organisms for classification from Linnaeus's time onwards (Ratcliff 2009), these problems did not stop proposals first in the nineteenth century and then in the mid-twentieth that microbes would make ideal model organisms for all biological study, and especially for genetic and biochemical analyses (e.g., Beijerinck 1900–1901; Theunissen 1996; Bateson 1907, in Summers 1991; Kluyver and Donker 1926, in Friedmann 2004; Demerec 1946; Lederberg 1987). This is ultimately how microbes became the platform for the great golden era of molecular biology in the mid-twentieth century, in which viruses and bacteria as well as some eukaryotic microbes became not just models of life but highly effective laboratory tools.

The centrality of microbes to molecular biology has been even more pronounced in the contemporary genomic era, in which first viruses and then prokaryotes had their whole genomes sequenced and analyzed. Now, with a molecular emphasis on systems (see Braillard, this volume), microbes are once again a major focus of general biology. This is not just because of the tractability of microbes, especially prokaryotes, but also because of the presumed unity of life that assumes common characteristics of life in all organisms, no matter how many cells constitute that life or how many compartments exist in any cell. While there are some important differences between microbial and multicellular life forms, especially in regard to evolutionary processes and patterns, as well as development (see Love, this volume), there is still a strong tendency in biology to emphasize shared properties as the basis of life and evolvability.

1.3 Why Should Philosophy of Biology Include Microbiology?

We will take it for granted that philosophy of biology is concerned with life and the sciences that study it. Whatever the interests of any biologist or philosopher of biology, even if they have specific reasons to focus on the non-microbial aspects of animals, plants and fungi, microbes will be part of those specific biological phenomena and their environments. One reason for this claim is that most biodiversity on Earth is and always has been microbial, despite the fact that animals, fungi and plants are the most visible forms of life. The traditional unit of biodiversity is species, and although the concept of species is even more problematic in microbes than anywhere else

(Ereshefsky 2010), microbiologists still use these taxonomic units. Even after taking into account the many problems of sampling and scale, there are more microbial or even just prokaryotic taxa than there are multicellular taxa (Fierer and Lennon 2011).

On a strictly quantitative basis of entity counts, microbes outnumber all other life forms combined, even if the prolific viruses are not counted (Whitman et al. 1998; Suttle 2007). More than half the living biomass on the planet is prokaryotic (excluding the structural material supporting many plants), despite the much tinier size of prokaryote cells. Every environment on, in or around the planet is occupied by microbial life, whether we are examining the stratosphere or the deepest parts of the planet accessed by human technology (Nee 2004; Wainwright et al. 2004; Newman and Banfield 2002; Pedersen 2000). All cells host a variety of microbes, even prokaryote cells, which are themselves occupied by numerous viruses known as phages (with a few exceptions; see Willner et al. 2011).

But the most important form of biodiversity is metabolic. Microbes can do everything plants, animals and fungi do, and have many unique metabolic tricks up their sleeves. They can use organic and inorganic energy sources, respire aerobically or anaerobically, and fix their own or use already fixed carbon. Prokaryotes and other microbes can combine these metabolic strategies and switch from one to another; they can consort metabolically with other organisms and use or produce substrates from or for these metabolic partners (Madigan et al. 2008; Southam et al. 2007).

Microorganisms are not only diverse in their own right but are fundamental to the maintenance of plant, animal and fungal biodiversity. The planetary chemistry of life is regulated by microbial metabolisms interacting with the Earth's geochemistry (Falkowski et al. 2008; Newman and Banfield 2002; Dietrich et al. 2006). Most of the biogeochemical transformations necessary for life depend on interconnected metabolic pathways in microbes (Falkowski et al. 2008; Strom 2008). Carbon, oxygen and nitrogen cycles are essential for life on our planet, and microbes are major players in every one of these cycles and several others (Ingraham 2010; Kasting 2005; Newman and Banfield 2002). This is what we mean when we say that microbes form the basis of all fundamental life processes and are thus the basis of all the biodiversity philosophers and biologists normally talk about.

Nor is this a recent development. Microbes have dominated all the past eras of the evolving Earth. From the origins of life until now, life has mostly been microbial. Whatever happens in the next great extinction event, many microbes are likely to survive it and continue evolving. If life is ever found anywhere else in the universe, it is most likely to be microbial (Gould 1994) or in the less probable case of multicellular life, to have arisen from microbes. It is well accepted that on our planet, microbes have been the evolutionary bases of all non-microbial life. Eukaryotic life arose from unicellular life; multicellular life began on the basis of eukaryotic microbes (Buss 1987; Bonner 1998; Maynard Smith and Szathmáry 1995; King 2004). All subsequent innovations, such as those of metazoan body plans and flowering plants, have their origins in a microbial context.

The dependence of multicellular life on unicellular life forms is ongoing in biology today. There is a vast array of symbioses that operate at every level of life. Symbiotic relationships, which may be mutualistic, commensal or parasitic, include endosymbioses (within cells), arrangements on the outside of cells (such as 'dancing Yeti crabs' that cultivate microorganisms as food by waving their furry claws over hydrothermal vents – see Thurber et al. 2011), and extensive partnerships between diverse groups of microorganisms (Moya et al. 2008). Every symbiosis involves microbes, even when the main symbionts are multicellular organisms. In humans, for example, our health, development and survival depend in a variety of ways on our microbial symbionts, many of which are parasites but which nevertheless interact intensively with us on both physiological and evolutionary timescales such that our 'nature' is shaped by our symbionts. Metagenomics, the molecular study of the interacting levels of biological organization constituting these collectives, is greatly advancing a more integrated view of biological entities and processes (O'Malley and Dupré 2009).

Our point is that the living world is saturated by microbes and their effects. They may be invisible individually, but collectively they constitute the greatest biological force on the planet. Because of this deep reliance and interdependence of all life on microbes, biologists and philosophers of biology have no choice but to consider microorganisms at least occasionally; far more if they are aiming at a deep embracing view of biology and the phenomena and processes of which it consists. We argue that using a microbiological perspective to analyse concepts central to biology—life, biological individuality and levels of selection—greatly enhances biological discourse by challenging the assumptions on which these concepts are based. Furthermore, demonstrating the connections between these notions reinforces the centrality of microbiological thinking to not only biology itself but also its philosophy and education.

2 Conceptions of Life in Classifying Viruses as Living Things

Biology is the study of life and living things. The problem is that life itself is difficult to identify and define (see Cleland and Zerella, this volume), despite the intuitive difference between a giraffe and a rock. Some microbes challenge any common-sense notion of living and non-living. Indeed, whether a virus is alive has remained a puzzling question since their discovery over a century ago, and scientific evidence amassed since then has only complicated the matter. Even more dramatic but less obvious is the fact that the so-called 'virus debate' has complicated the very objective of defining life. The way scientists and philosophers engage in discussions about the living status of viruses underscores disparate ways of answering the question "What is life?". Answers are given as either uncovering a natural kind or creating a useful heuristic for the categorization of biological entities. Because these two approaches to defining life are often conflated in arguments over the living status of viruses, approaching life from a microbiological perspective illuminates the nature of life debate in a unique way that disentangles natural kind and heuristic definitions.

2.1 *The Living Status of Viruses*

Microscopic entities that seem to straddle the intersection of the living and the non-living, viruses challenge the common-sense distinction that seems so obvious when classifying much of macrobial life. In the last several years, popular science magazines, professional virologists and philosophers have engaged in a debate that fuses questions about the living status of viruses and the definition of life in general (e.g. Forterre 2010; Moreira and López-García 2009; Owen 2008). In some ways, the rhetoric suggests that the debaters take life to be a genuine natural kind. In doing so, they follow in the spirit of definition of life debates in general, which often involve disputing proposed definitions by offering counter-examples (Cleland and Chyba 2002). The classification of living things, seen in this way, should be more than a heuristic tool for determining what sorts of entities biologists should study; rather, it marks out a real, unique group whose interactions with the world are profoundly different from their non-living counterparts. With a realist conception of life as a natural kind, the equally contentious auxiliary debate over the living status of viruses naturally follows. After all, if life is a natural grouping, viruses are either living or non-living, rather than living or non-living *by virtue* of a particular definition of life. Although virologists seem to want to draw the first conclusion, their arguments in fact only support the second.

Some definitions of life seem immediately amenable to viruses, such as that promulgated by Joyce (1995, p.140) and adopted by the USA's National Aeronautics and Space Administration (NASA): 'a self-sustained chemical system capable of undergoing Darwinian evolution'. Though a controversial objection, this definition may fail to capture early life if it did not undergo Darwinian evolution (Cleland and Chyba 2002). Further, definitions that merely require replication and evolution, without including a chemical system requirement, problematically count computer viruses among the living. Even if we are willing to accept this minimal definition, real viruses don't always fit the bill, argue Moreira and López-García (2009), for they are evolved *by* cells rather than evolved independently. Other definitions of life, such as Lwoff's (1967) widely accepted definition of life as cellular, exclude viruses with even greater certainty.

Yet, the recent discovery of a new group of viruses, disturbingly reminiscent of their living microbial counterparts, has challenged the classification of viruses as non-living and reignited debates about the definition of life. The so-called mimivirus, isolated in 1992 but only identified as a virus in 2003, shares critical features of living cellular organisms not previously known to be present in viruses (Claverie and Abergel 2009). Even the properties for which it was named—'mimi' being a reference to its mimicry of bacteria—illustrate the similarities between this virus and cellular life, and are why it took so long to be recognized as a virus rather than a parasitic bacterium (Raoult 2005). At 800 nm across it is much larger than most known viruses, and its 1.2 million base-pair genome encodes 1,260 genes, seven of which are common to all cellular life: eukaryotes, bacteria, and archaea. The mimivirus can even become 'ill' when infected by a smaller virus, thereby moving virologist

Jean-Michel Claverie to exclaim that, 'The fact that it gets sick makes it more alive' (2008, p. 677, in Pearson 2008). In light of these findings, it seems that scientists must either concede that viruses are living organisms or commit to the view that non-living things can become infected.

2.2 Heuristic Definitions of Life

Whether or not these positions involve classifying viruses as living or non-living, both sides of the debate invoke scientific findings about viruses as evidence for their inclusion in either category. These arguments show that viruses are alive with respect to a given definition of life, with 'life' functioning as a heuristic that groups entities for study. That the definition of life is concurrently treated as a genuine natural kind, however, renders this ontological deflationism problematic. This is clear when virologists and others use scientific evidence to classify viruses as living or non-living while simultaneously recasting definitions of life to be either more inclusive or exclusive of them, thereby engaging in self-defeating debates. Forterre (2010), for example, uses the fact that viruses are alive with respect to a particular definition as evidence that viruses are alive *simpliciter* (belonging to a natural kind). From this he reasons (circularly) that we should rewrite the definition of life to be more inclusive of viruses.

Raoult and Forterre (2008) call for the classification of viruses as living things on the basis of their similarities to organisms we currently consider living, thereby suggesting that viruses are alive with respect to our currently accepted definition of what a living organism is. Viruses should be considered living, they argue, because they possess the characteristics that other terrestrial life does: they are made of the same macromolecules and have co-evolved with the three domains of life. Hedge et al. (2009, in Ruiz-Saenz and Rodas 2010, p. 89) summarize similar evidence that suggests viruses are alive because they share features with organisms already considered alive: 'viruses [...] follow Darwin's theory of "survival of the fittest", acquire mutations and evolve to sustain a new environment.' That certain viruses also encode proteins shared by all three living domains and can become infected like their living counterparts—findings made possible because of the discovery of the mimivirus—supports the idea that viruses are in fact living entities.

Elsewhere, Forterre (2010) justifies his inclusion of viruses among the living by suggesting they are even consistent with what seems an impossibly exclusive definition of life: life as purely cellular. In his description, viruses normally form viral 'factories' in order to copy their genomes and produce virions that are dispersed and infect other hosts. Though often confused with the virus itself, according to Forterre, the virion is the part of the viral life cycle in which it is inactive outside of the cell in the form of a protective coat (capsid) encasing genetic material. Because mimivirus factories are nearly the size of the nucleus of their amoebic hosts, scientists have acknowledged that viral factories are more complex than once thought. Claverie (2010) has even proposed that the viral factory corresponds to the real organism, such that viruses virtually transform their cellular host into a viral factory. Once the cellular machinery has been co-opted into replicating viral DNA and producing virions, the cell is actually effectively a virus, Forterre (2010) contends. Even through the lens of the cellular definition of life, then, viruses are living things.

In counter-arguments that viruses are not genuine instances of life, life is also treated as a heuristic that usefully groups biological entities for study without identifying a natural kind. In their paper 'Ten reasons to exclude viruses from the tree of life', Moreira and López-García (2009, p. 307) immediately deflate the ontological status of the definition of life, claiming that whether viruses are alive is a matter of 'inference and logic starting from any given definition of life.' They proceed by outlining the ways in which the characteristics of viruses are not consistent with any potentially viable definitions of life—even those that rely on self-replication and evolution rather than metabolism. Viruses, they point out, do not evolve by themselves but rather rely on their cellular hosts to do so. In this sense, these microscopic entities are not merely non-living (i.e., are not part of a natural grouping of living things): they are non-living *in virtue* of particular conceptions of life. The upshot is that we might construct a definition of life that would be inclusive of viruses, but such a definition would likely be too generous in its inclusivity, offering membership to things such as computer viruses that few would be comfortable to consider living. This liberalism in membership would undermine the usefulness of life as a heuristic that indicates what sorts of entities biologists should study.

2.3 Life as a Natural Kind

Although new scientific evidence about viruses has only suggested they are alive with respect to particular definitions of life, and counterarguments similarly claim the opposite, Forterre (2010) treats the definition of life as picking out a natural kind. After all, if it is enough that viruses are considered living in light of the cellular definition of life, Forterre's job is done. Raoult and Forterre (2008), however, continue to argue that the life-likeness of viruses is grounds for a new, more inclusive, definition of life. On their view, capsid-encoding organisms (viruses) complement the other three domains of life, which are collectively to be called 'ribosome-encoding organisms' (all cells contain ribosomes, which translate transcribed DNA (mRNA) into proteins). Fully defined, capsid-encoding organisms 'are composed of proteins and nucleic acids, self-assemble into a nucleocapsid and use a ribosome-encoding organism to complete its life cycle' (Raoult and Forterre 2008, p. 314).

Forterre's argument is one example of a problematic rhetorical framework also exemplified in Benner's (2010) project of defining life ideally. On Benner's view, an ideal definition of life serves as a standard by which we may evaluate whether all potential organisms, including those we might encounter on distant planets, are living or non-living. In his words, the definition must be inclusive of everything that has features we 'value' in life. Such a definition relies on life as a heuristic:

a convenient categorization to mark out groups of organisms with relevantly similar features. Our current definition is likely to be imperfect, Benner concedes, and will require updating as we encounter new entities that we would like to consider living but which do not fit our definition. As soon as we believe this new form of life is possible—and therefore worth investigating in the same way we do other living things—we will consider it living. The features that most researchers value, however, are generated by heuristic definitions of life; we think that cellular organisms are alive, for example, because of preconceived cellular definitions of life. Thus, those valuable features cannot be used as the basis on which to construct a definition that aims to pick out a natural kind.

Overall, attempts to classify the living status of viruses highlight a conflation of natural kind and heuristic approaches to articulating definitions of life. A microbiological perspective is therefore mandatory for disentangling ways of answering the question "What is life?". These issues of categorization and ontology are emphasized even more when we step away from the broad categorization of life to the entities that exhibit biological individuality.

3 What are Biological Individuals in Light of Microbiology?

Despite the fact that biological individuals (or 'organisms') are one of the primary units of study in the life sciences, surprisingly little attention has been given to explicating precisely what they are. This is probably because the identification of biological individuals is taken to be unproblematic (even given the well known problems of defining life). However, closer consideration of the facts—once again those pertaining to the microbial world—reveals that biological individuality is far from straightforward. In this section we show how microbiology has changed the way biological individuals are viewed.

For most people, animals are the most clear-cut biological individuals. It is certainly uncontroversial that human beings, dogs, birds and so on are organisms. Yet more than 90 % of cells in any 'human' body are microbial (Savage 1977) and the same is true for all other animals. Many of these symbiotic microorganisms play a vital role in digestion, immune response and health in general. Does this mean that these beneficial symbionts are part of the animal body? One might think not because microbial cells are genetically, developmentally, and reproductively distinguishable from animal cells. However, matters get more complicated when the phenomenon of endosymbiosis is considered. It is now well established that mitochondria in eukaryotes, chloroplasts in plants, and perhaps other organelles (substructures within a cell which perform a specific function such as energy production) were originally bacteria that became incorporated into the larger cell (Sagan 1967; Margulis 1970; Archibald 2011; van der Giezen 2011). The emergence of eukaryotes, and of photosynthesizing eukaryotes, occurred when one prokaryotic organism assimilated another one by a form of cellular ingestion ('endocytosis', a major innovation in nutrient acquisition), and the fused cells became a single unit of inheritance and selection. Mitochondria and chloroplasts are the descendants of these engulfed prokaryotic individuals. Like bacteria, mitochondria reproduce by splitting into two, but this process is regulated in various ways by the host cell (Osteryoung and Nunnari 2003). Mitochondria also lack some of their original genome, which is instead located in the nucleus of the host cell. Because of this reliance on the host cell, mitochondria are not considered to be individual organisms (see Sect. 4.3), although there is no scientific doubt that they once were. Nevertheless, this is clearly a matter of degree, as they still carry many of their own genes and reproduce semi-independently.

Discussion of symbionts and endosymbionts highlights a number of dimensions implicit in the notion of biological individuality. Some of those discussed by philosophers are very specific, and include germ-soma separation (reproduction specialist cells versus survival-and-growth cells), policing mechanisms (for punishing cheating and rewarding cooperation, as well as keeping foreign entities at bay), spatial boundaries (demarcating the limits of the individual), co-dispersal (coupled reproduction and spread of the components of the individual) and being bearers of adaptations (Clarke 2010). Three more general dimensions have been discussed in detail by philosopher Jack Wilson (1999). The first of these is *genetic individuality*. In this dimension, organisms are distinguished from one another by their genetic makeup, with a biological individual being made up of more or less genetically identical cells. Genetic homogeneity is not enough, however, to distinguish parts of an individual from the whole, so a genetic individual needs to be genetically homogenous *and* genetically unique (Santelices 1999). The cells in a human heart, for example, are genetically homogenous but not unique because there are other genetically identical cells elsewhere in the same body. Another dimension of biological individuals is developmental. *Developmental individuals* are groups of cells that have developed from a single cell or small group of cells. Host organisms and their symbionts usually have different developmental trajectories despite the fact that host organisms often pass on symbionts to their offspring. The third feature of biological individuals identified by Wilson (1999) is functional integration. *Functional individuals* have parts that are strongly causally connected to one another but not with parts outside the individual. It is important to note that functional integration, just like developmental unity and genetic homogeneity, comes in degrees. There appears to be no sharp line demarcating any of these properties, thus making the identification of anything like a natural kind difficult.

For paradigmatic organisms, genetic, developmental and functional individuality appear at first glance to be neatly aligned. A typical animal is made up of functionally interdependent, genetically unique and homogenous cells that develop from a single fertilized egg. Of course there are exceptions to this, such as identical (monozygotic) twins, which form a single genetic individual but two functional individuals, and either one or two developmental individuals depending on which starting point is selected (i.e., before or after the fertilized egg splits). The importance of considering microbiology does not simply derive from the fact that it furnishes us with more exceptions like this. Instead microbiology demonstrates that alignment of Wilson's three dimensions is the exception rather than the rule (Dupré and

O'Malley 2009). In almost all cases, including paradigmatic biological individuals, the first two criteria are in tension with the third. This is because symbiotic communities, which are by definition genetically heterogeneous, are often functionally integrated wholes. Microbiological research has shown that biological individuality is more complicated than it might have seemed to be.

One of many examples of this is the glassy-winged sharpshooter (*Homalodisca vitripennis*), which is a kind of leafhopper that feeds solely on the sap of woody plants. Because this sap is low in nutrients, the sharpshooter depends on two bacteria, *Baumannia cicadellinicola* and *Sulcia muelleri*, to convert it into vitamins, amino acids and cofactors (Wu et al. 2006). The bacteria in turn rely on the host and one another for their survival, making all three highly co-dependent. Thus the bacteria function very much as parts of the sharpshooter's metabolic system despite being genetically and developmentally distinguishable. Such co-dependence is the norm amongst related insects such as aphids, and allows them to exploit unclaimed ecological niches (Wu et al. 2006). Besides this metabolic role bacteria are also known to play an important role in invertebrate reproduction. Intracellular bacteria *Wolbachia* kill or feminize male organisms, such as *Drosophila*, induce parthenogenesis, and block successful mating between infected males and uninfected females (Werren et al. 2008). *Wolbachia* therefore function as part of the reproductive system of infected invertebrates. This kind of functional integration of symbiotic microorganisms with their hosts is also found in many other animals as well as plants (Barrow et al. 2008).

Microbiology also provides important insights into intra-specific relationships, or relationships between genetically related organisms. It is now widely believed that multicellular organisms evolved from highly cooperative groups of conspecific unicellular organisms (Buss 1987; Maynard Smith and Szathmáry 1995). This emphasis on cooperation is a common theme in much recent work on biological individuality (e.g., Folse and Roughgarden 2010; Michod and Roze 2001; Queller and Strassmann 2009). Research into microbial sociality is seen by many as vital for understanding how multicellular organisms evolved, and also how they function. In fact, many microorganisms are highly social, making them ideal candidates for studying the evolution of cooperation and individuality (Crespi 2001; Shapiro 1998; Velicer 2003). For instance, members of the 'social' slime mould, *Dictyostelium discoideum*, begin their lives as individual amoebae that aggregate together when food is scarce to form a slug that can move faster and traverse environments the individual amoeba cannot. Eventually they form a fruiting body consisting of a sterile stalk on top of which are fertile spores. The stalk is made up of dead amoebae, which give up their lives for the colony (Bonner 2009). Because these latter stages of the *Dictyostelium* life-cycle resemble a single organism in many ways and require a great deal of cooperation (often between genetically heterogeneous amoebae) it is an ideal case study.

This research on cooperation and sociality provides ways of investigating functional integration within an evolutionary framework. Instead of saying that parts of an organism are causally interdependent, the focus becomes the alignment of fitness of these parts and group-level adaptations (e.g. Folse and

Roughgarden 2010; Strassmann and Queller 2010). The more interdependent the parts of an individual are, the more they will function as a single entity and share a common evolutionary fate. Placing biological individuals within a microbiological framework, therefore, is crucial for understanding both what living things are and how they evolve.

From a non-evolutionary perspective, there are several different kinds of equally important biological individuals, such as developmental, genetic, and functional individuals. However, from an evolutionary perspective (more specifically, a Darwinian perspective – see Sect. 4),there is scope to argue that genetic and developmental individuality, as well as criteria such as germ-soma separation, policing mechanisms and spatial boundaries, are all secondary to cooperation or functional integration. That is, these other factors are all means by which high levels of cooperation are achieved and maintained rather than essential conditions (Michod and Roze 2001; Queller and Strassmann 2009). They are adaptations for cooperation and functional integration.

For instance, genetic homogeneity in a group of cells can facilitate cooperation due to kin selection. An organism can pass on its genes either by reproducing itself or by increasing the chances of genetically similar organisms (i.e., kin) reproducing. Cooperative behaviour should thus be more likely amongst kin than amongst non-kin, all other things being equal. In turn, germ-soma separation and policing mechanisms can both be understood as mechanisms by which genetic homogeneity is maintained. As for spatial boundaries, Sterelny and Griffiths (1999) argue that the presence of a physical boundary isolates the individual cell or group of cells, promoting and maintaining functional integration.

From a Darwinian perspective, therefore, most of the characteristics associated with paradigmatic biological individuals are only contingent: our mistake has been to treat them as defining properties. Nonetheless, the presence of these characteristics on this view is often symptomatic of biological individuality. Do symbiotic groups show any such symptoms? In the case of microbe-macrobe symbioses it seems that they often do. Frank (1996) argues that host organisms often control symbiont reproduction by inducing a germ-soma separation in the symbiont. Microbes that live inside host cells (endosymbionts) share the cell wall as a spatial boundary, which reinforces their relationship with the host and encourages long-term interaction. And host immune systems act as policing mechanisms against free-riding or pathogenic microorganisms while tolerating commensalists and mutualists. These phenomena suggest to us that many microbe-macrobe symbioses, such as the glassy-winged sharpshooter and its endosymbionts, constitute biological individuals or at least crucial units of study.

Whatever position anyone might favour in this discussion, it is clear that advances in microbiology have drastically changed the way biological individuality is viewed, and will likely continue to do so. The genetic, developmental, and functional dimensions of biological individuality are not typically aligned as has traditionally been assumed. Furthermore, research into the relationships between these dimensions, involving particularly the study of microbe sociality, is at the forefront of both scientific and philosophical thought on this topic.

4 Can Microbes Help to Understand and Solve Some of the Issues in the Debate About Levels of Selection?

Our discussion of biological individuality suggests that no real consensus about what an organism is can be achieved within a non-evolutionary framework. The situation is similar to the concept of species (e.g., Ereshefsky 2010). Each concept has several working definitions in different disciplines, without complete agreement on what constitutes a species or a biological individual. We proposed, however, that in the latter case the situation might be clarified through a Darwinian analysis acknowledging a microbiological perspective. Indeed, we have seen that new biological individuals can be formed during symbioses. Although these new (functional) individuals are not paradigmatic biological individuals (i.e., multicellular organisms), they nevertheless cannot be disqualified as viable biological individuals existing at a higher level than the symbionts on their own. Extending the notion of biological individuality to levels other than the organism level is precisely what is at stake in the levels of selection debate.[1]

Without a microbiological perspective, this debate would probably have remained stuck at the same stage it had reached at the end of the 1960s, with the organism level being seen as the only level of selection ('individual selectionism'). A better understanding of microbial evolution revealed that what was thought about the primacy of the organism level was insufficient for a more general understanding of evolution by natural selection. A large body of empirical work on microbial evolution showed that group selection was an important force in evolution, contrary to what the individual selectionists of the late 1960s believed. More recent work on major transitions in individuality shows that what were thought to be 'true' individuals by individual selectionists in fact originated from groups of cells. This means that the argument for the individual organism as the sole level of selection is at least inadequate and at worst perniciously misleading. A microbiological perspective on evolution is therefore necessary for anyone who wants to appreciate the subtleties in the levels of selection debate.

4.1 The Levels of Selection Debate in a Nutshell

Prior to the 1960s, groups of multicellular organisms, such as mammals or birds, were usually uncritically considered to function harmoniously for the good of the group or the species. Under this view, which D.S. Wilson and E.O. Wilson (2007)

[1]For clarity's sake, we emphasize here that our claim about biological individuals existing at the group level is distinct from the claim made by Ghiselin (1974) and Hull (1976) that species are *individuals*. The individuals discussed in our paper are functional individuals, organisms, or units of selection, whereas the individuals Ghiselin and Hull care about are merely spatiotemporally localized entities (Wilson and Sober 1989).

label retrospectively 'naïve group selectionism', groups or species were the only true units of selection or 'Darwinian individuals'. However, this idea was forcefully debunked by Williams in his 1966 book *Adaptation and Natural Selection*. Although he recognized in principle the possibility of group selection, he stressed its theoretical implausibility in light of the models of that time, and the lack of empirical evidence for group selection as an important evolutionary force. Moreover at the time that Williams wrote his book, new theories were emerging as alternatives to group selection: inclusive fitness theory and the closely related kin selection theory (Hamilton 1964), evolutionary game theory (Axelrod 1984; Maynard Smith 1982) and selfish gene theory, also known as the 'gene's eye view' (Dawkins 1976, 1982).

By the beginning of the 1970s, the concept of group selection had been renounced by all respectable evolutionary biologists. The most popular argument for the organism as the primary level of selection was that selection at the level of the group would always be disrupted by selection at lower levels. For example, within a group of altruistic individuals sacrificing their interests by helping others, a selfish mutant will have higher fitness than an altruist. This type of reasoning leads most evolutionary biologists even today to think of the organismal individual as the sole significant Darwinian individual. Wilson and Wilson (2007) argue convincingly that this individualistic evolutionary perspective, which always refers to the paradigmatic biological perspective of multicellular organisms, is misguided. They advocate 'neo-group-selectionism'. The difference between naïve group selectionism and neo-group-selectionism is that the latter is placed within a general theory of multilevel selection, which recognizes that all the different levels of organization can be relevant levels of selection, with Darwinian individuals nested within one another.

Wilson and Wilson propose three main reasons in favour of neo-group-selectionism and for each reason a microbiological perspective is essential. The first is that new theoretical models, especially agent-based models (unavailable in the 1960s) clearly and plausibly demonstrate how group selection can occur. Furthermore, some models of virulence show that under realistic assumptions virulence is expected to decrease over time (Bull 1994). The decrease of virulence can hardly be explained *without* a multilevel perspective. Although it is in the interests of the individual pathogen to use as many resources as possible in order to spread (which would kill its host), it is not in the interest of the group of host-pathogens, since the population of hosts is not infinite and killing all of them would result in the conjoint extinction of the pathogens. Decreasing the level of the virulence is hence adaptive at the group level.

Second, many empirical studies show that group selection is an important evolutionary force, or at least that evolution can be interpreted legitimately from a group-selectionist perspective. Much of the best evidence for this claim comes from microbial evolution. A recent exemplar is Rainey and Rainey's work (2003) on the 'wrinkly spreader (WS)' strain (sub-species) of the bacterium, *Pseudomonas fluorescens*. This strain produces a mat on the surface of a liquid medium. Although this mat is costly to produce and strains of *P. fluorescens* that do not contribute to its production (cheaters) have the highest fitness within groups of mixed strains, the WS strain is maintained in the overall population by inter-group selection: groups in which the WS strain is present do better than others. More recently, Rainey and Kerr (2010) have argued that WS cheaters form the germline of the proto-multicellular organisms constituted by the mat. This represents a new microbiological hypothesis about the origin of multicellular organisms, which (as we will see in Sect. 4.3) is an important topic within the levels of selection debate. Another microbiological example of group selection comes from Kerr et al. (2006),who present an experimental setting of a metapopulation (a population of populations) in which the T4 phage can adopt two strategies for killing its host, the bacterium *Escherichia coli*, depending on the dynamics of migration between the different populations. If the migration is unrestricted, 'rapacious' strains of phages out-compete 'prudent' strains, which kill the host more slowly and thus allow bacterial populations to reproduce more before being infected. However, when migration is limited and matched by actual migration rates between populations, the more prudent strains out-compete the rapacious ones. This study is an empirical confirmation of the models of virulence discussed above.

The third piece of evidence that Wilson and Wilson cite in support of neo-group-selectionism is that the three main theories presented as alternatives to group selection are in fact perfectly consistent with it. Inclusive fitness theory is nowadays seen as formally equivalent to group selection (Okasha 2006; Wilson and Wilson 2007; West et al. 2007). The two theories are fully translatable, and this is recognized by many neo-group selectionists *and* individual selectionists (however, cf. van Veelen et al. 2011; Nowak et al. 2010). In addition, the models used in evolutionary game theory implicitly invoke groups, because when n individuals interact they form a group of size n, and during these interactions different strategies can be selected in different groups. Although microbiology is not directly implicated in these theoretical considerations, the development of these two theories would certainly have had a different structure if they had taken microbial groups into account. The group-level evolution of microbes can readily be studied due to their size and rapid generation time (unlike animals, for example).

Finally, say Wilson and Wilson, selfish gene theory is neither an alternative theory to group selection, nor a theory demonstrating the prevalence of individual level selection since it confuses mere 'bookkeeping' with causality (Gould and Lloyd 1999; Okasha 2006). In other words, while it is true that any evolutionary process will have an outcome at the level of genes, it does not follow that genes causally explain this outcome or that the individual organism level should be privileged (Okasha 2006). This is especially true when biological individuality is itself a blurry concept, and becomes even more apparent when we examine major evolutionary transitions in such individuality.

4.2 Microbiology at the Heart of the Major Transitions in Individuality

The levels of selection debate went through a dramatic shift under the influence of Buss and his 1987 book *The Evolution of Individuality*, and Maynard-Smith and Szathmáry and their *The Major Transitions in Evolution* (1995). These accounts show that modern multicellular organisms are the evolutionary product of cooperation amongst groups of microorganisms in which conflicts had been resolved (although see Clarke 2011 for a different account on non-resolution of conflicts in plants). Under this view, individual level selection could be understood as a form of group selection: selection between groups of cells. An individual selectionist could argue that whether we call these cells 'an individual group of cells' or 'an individual' is purely a semantic matter, and that their core argument is that this level is the only important one in evolution. However, this position misses a crucial aspect of the transition from uni- to multicellular organisms, as well as other evolutionary transitions such as the origins of life. The fact that multicellular organisms arose from unicellular organisms means necessarily that group selection was the most important force in evolution during this transition. It was so important that it created what we call today the 'individual level'—a level that did not exist before.

This transition into individuality makes it clear that an exclusively individualist position on selection is untenable. Arguing against group selection as an important force in evolution nowadays (which is mostly what the levels of selection debate is about), commits the individual selectionist to hold that evolution is only about relatively recent multicellular organisms. Moreover, some extant organisms seriously challenge the very concept of multicellularity. This is the case for the slime mould already mentioned, *Dictyostelium,* which is part of the time a solitary amoeba and part of the time a collection of amoebae that reproduces like a paradigmatic multicellular organism (i.e. some of the amoebae become 'somatic cells' while others become 'germ cells' and form a fruiting body; for more details see Bonner 2009). Myxobacteria are a similar example in the prokaryotic world (Shimkets 1990).

Arguing against group selection also commits individual selectionists to the denial of questions about origins of life as an important topic in evolutionary biology. While it is legitimate for a science not to be interested in the question of origins, it is at least a paradoxical decision in a discipline called '*evolutionary* biology', especially when 'origin' means the origin of evolutionary innovations. As Buss (1987, p. 20) notes, individuality is itself an innovation that needs to be explained. An organismal perspective remains deaf to the question of the origin of multicellular organisms from groups of microorganisms. Finally, arguing against group selection and, more generally, multiple levels of selection commits one to ignoring the creation of new individuals from two or more genetically distinct actors. We have already mentioned endosymbiosis theory (Margulis 1970), widely accepted nowadays, which proposes that the eukaryotic cell is a chimera constituted by what were initially distinct individuals. From their endosymbiosis onwards, natural selection acted on this group of organisms as a whole, causing not only the evolutionary future of these organisms to be intertwined but also their increasing biological dependence.

Overall, we suggest that microbiology reveals that a commitment to a single level of selection, namely the traditional organismal level, is too narrow both in light of genuine examples of group selection (many of which are microbial), and because of the problematization of the very concept of the individual. Microbiology is clearly playing a crucial role in the levels of selection debate, which has been a central topic of discussion in evolutionary biology for several decades. Evolutionary analyses of microbial systems have allowed biologists to test new hypotheses, given them new conceptual tools, and helped them clarify the questions at stake in the debate.

5 Philosophy of Microbiology and Biology Education

Each of our three case studies illustrates the contributions of a microbiological perspective to biology and philosophy of biology debates. While our discussions above about defining life, biological individuality and levels of selection speak for themselves, we want to emphasize as our conclusion that there are numerous ways in which a philosophical understanding of microbiology can contribute to education in biology. The first reason is an obvious one: additional information about microbes can be incorporated into a broader biological understanding. This would not necessarily be philosophical, but the philosophical issues we have spelled out could act as an incentive for macrobial biologists to learn a little about microbiology. From a more explicitly philosophical perspective, debates, problems, and unresolved issues in macrobiology can be critically assessed in light of microbiology. Most generally of all, philosophy of biology, by including microbiology, could widen the scope of biology and conceptions of how science is practised.

While microbiology contributes to major philosophical questions that are relevant to all philosophical and historical efforts to understand biology, the discipline of microbiology itself can benefit from an appreciation of its underlying philosophical dimensions. Even though microbiologists have long advocated a broad integrative view of microbiology education (Handelsman 2002), its integration can extend to biology more generally. It is not uncommon for biologists to ignore microbiology, especially its ecological and evolutionary aspects, and our cases above show clearly why this would be a problem.

Not only does the philosophy of microbiology generate a more inclusive representation of life, but it also tests many standard biological assumptions about the study of life. One of them is in regard to the debate about reductionism, which has been an important topic in philosophy of biology, but is now shifting towards closer scrutiny of multilevel integrative explanations (Brigandt and Love 2012; Mitchell 2009).We have shown from a microbiological point of view that methodologically, even if the macroorganism is the primary object of study, it must be understood both in a microbial context *and* in terms of a microbial decomposition. If we want to understand global warming, for example, microbes have to be brought into a broad-brush picture of atmospheric gases, temperature trends and ocean acidification (Singh et al. 2010; Zehr et al. 2011). But at the same time, individual genomes, pathways, and organismal interactions have to be understood at the level of the single cell in order to obtain more precise information and capacities for intervention. Major social health problems, such as obesity, can only be understood when the effects of microbial communities on the generation of adipose tissues in traditional multicellular organisms (such as humans and rodents) are comprehended evolutionarily and ontogenetically (O'Malley and Stotz 2011). Being able to slide up and down ecological, evolutionary and organismal scales via microbial analysis is something we think our sections above show clearly. As a consequence, traditional concerns about reductionism, whether scientific or philosophical, become much less pressing. The epistemological issue that is brought to the forefront by microbiological analysis is therefore not so much reductionism as the integration and importance of multiple levels of analysis.

Philosophy of microbiology also has very important ontological questions to discuss that are highly relevant to all of biology. In this chapter, we have shown through our examination of life, biological individuality and levels of selection how the macrobe-microbe distinction is not ontologically meaningful, even though there can be many good practical reasons to focus on either microbial or non-microbial life. All organisms, no matter their size, are biological individuals of some sort, and yet it is primarily by including microbes in any discussion of the ontology of such individuality that the deep issues in such ontological attributions become clear. The same is true of the Darwinian individual, which we have shown here to be problematized by microbiology not just in a critical but a highly constructive way: once the traditional organism is not assumed as the focus of analysis, there is room to develop a much better understanding of group and multilevel selection. The modern synthesis of evolution, while it has undeniably revolutionized evolutionary biology, is largely the result of an amicrobial picture of the world. Because of the exclusion of microbial processes and patterns (including several we have not discussed here, such as how microbes share genetic resources), traditional forms of evolutionary theory cannot encompass major evolutionary questions, including those about the origins of life and major evolutionary transitions.

For many major philosophy of science discussions, such as those about natural kinds, pluralism and multilevel explanation, philosophy of microbiology has important contributions to make and can be part of a broader agenda of introducing philosophy of science to biology students. This is not for imperialistic disciplinary reasons, but for the simple reason that biologists need to draw out the most profound implications of the science they do. This does not always require discussion of microbes, of course, but it does mean that broad, abstract claims about 'living things', biodiversity and evolution should be examined in light of microbial life as well as in reference to visible life forms. We have only indicated the depth and scope of these issues in regard to a limited set of examples (life, biological individuality, levels of selection), and there is a great deal of work that could be taken up by anyone stimulated by these illustrations.

Overall, we see the philosophy of microbiology working as a corrective to a tradition of *not* thinking about microbes, whether that is in philosophy or biology. Thinking 'microbiologically' has provided valuable new insights into many traditional biology debates, whether over the definition of life, the concept of an individual or levels of selection. On top of this, a philosophically motivated microbiology education is vital for ensuring that the same shallow assumptions do not continue to be guiding principles. Instead of revisiting existing debates from a microbiological perspective (as we have done here), we are optimistic that future developments in biology can start from a more inclusive conception of biology. Biologists may sometimes find it useful to emphasize differences between microbial and macrobial life, but whatever is going on in the biological world, there is no way it can be totally abstracted from a microbial context, and any general biological claim needs testing against the microbial world.

Acknowledgements MJD is supported by an Australian Postgraduate Award. PB acknowledges the support of the Australian Research Council (DP0878650) and an International Postgraduate Research Scholarship from the University of Sydney. JDB is supported by the Australian-American Fulbright Commission. MAO's research is funded by an ARC Future Fellowship.

References

Amsterdamska, O. 1987. Medical and biological constraints: Early research on variation in bacteriology. *Social Studies of Science* 17: 657–687.
Archibald, J.M. 2011. Origin of eukaryotic cells: 40 years on. *Symbiosis* 54: 69–86.
Axelrod, R. 1984. *The evolution of cooperation*. New York: Basic.
Barrow, J.R., M.E. Lucero, I. Reyes-Vera, and K.M. Havstad. 2008. Do symbiotic microbes have a role in plant evolution, performance and response to stress? *Communicative and Integrative Biology* 1(1): 69–73.
Beijerinck, M.W. 1900–1901. On different forms of hereditary variation in microbes. *Proc K Ned Akad Wet* 3: 352–365. www.dwc.knaw.nl/DL/publications/PU00014384.pdf. Accessed 19 Dec 2011.
Benner, S.A. 2010. Defining life. *Astrobiology* 10(10): 1021–1030.
Bonner, J.T. 1998. The origins of multicellularity. *Integrative Biology* 1: 27–36.
Bonner, J.T. 2009. *The social amoebae: The biology of cellular slime molds*. Princeton: Princeton University Press.
Brigandt, I., and A.C. Love. 2012. Reductionism in biology. *Stanford Encyclopedia of Philosophy*, ed. N. Zalta. http://plato.stanford.edu/entries/reduction-biology/.
Brock, T.D. 1961. *Milestones in microbiology*. Englewood Cliffs: Prentice-Hall.
Bull, J.J. 1994. Virulence. *Evolution* 48(5): 1423–1437.
Bulloch, W. 1938. *The history of bacteriology*. Oxford: Oxford University Press.
Buss, L.W. 1987. *The evolution of individuality*. Princeton: Princeton University Press.
Cavicchioli, R. (ed.). 2007. *Archaea: Molecular and cellular biology*. Washington, DC: ASM.
Clarke, E. 2010. The problem of biological individuality. *Biological Theory* 5: 312–325.
Clarke, E. 2011. Plant individuality and multilevel selection theory. In *Major transitions in evolution revisited*, ed. B. Calcott and K. Sterelny, 227–250. Cambridge, MA: MIT Press.
Claverie, J. 2010. Mimivirus: The emerging paradox of quasi-autonomous viruses. *Trends in Genetics* 26(10): 431–437.
Claverie, J.-M., and C. Abergel. 2009. Mimivirus and its virophage. *Annual Review of Genetics* 43: 49–66.
Cleland, C.E., and C.F. Chyba. 2002. Defining 'life'. *Origins of Life and Evolution of the Biosphere* 32: 387–393.
Crespi, B.J. 2001. The evolution of social behavior in microorganisms. *Trends in Ecology & Evolution* 16(4): 178–183.
Dawkins, R. 1976. *The selfish gene*. Oxford: Oxford University Press.
Dawkins, R. 1982. *The extended phenotype: The long reach of the gene*. New York: Oxford University Press.
Demerec, M. 1946. *Minutes of the 23rd annual meeting of the Long Island Biological Association*, July 30th. Available on request from Cold Spring Harbor Library Archives. http://library.cshl.edu/archives/. Accessed 19 Dec 2011.

Dietrich, L.E.P., M.M. Tice, and D.K. Newman. 2006. The co-evolution of life and Earth. *Current Biology* 16: R395–R400.

Drews, G. 2000. The roots of microbiology and the influence of Ferdinand Cohn on microbiology of the 19th century. *FEMS Microbiology Reviews* 24: 225–249.

Dupré, J., and M.A. O'Malley. 2009. Varieties of living things: Life at the intersection of lineage and metabolism. *Philosophy and Theory in Biology* 1. http://hdl.handle.net/2027/spo.6959004.0001.003. Accessed 19 Dec 2011.

Ereshefsky, M. 2010. Microbiology and the species problem. *Biology and Philosophy* 25: 553–568.

Falkowski, P.G., T. Fenchel, and E.F. DeLong. 2008. The microbial engines that drive Earth's biogeochemical cycles. *Science* 320: 1034–1039.

Farley, J. 1974. *The spontaneous generation controversy from Descartes to Oparin*. Baltimore: Johns Hopkins.

Fierer, N., and J.T. Lennon. 2011. The generation and maintenance of diversity in microbial communities. *American Journal of Botany* 98: 439–448.

Folse, H.J., and J. Roughgarden. 2010. What is an individual organism? A multilevel selection perspective. *The Quarterly Review of Biology* 85(4): 447–472.

Forterre, P. 2010. Defining life: The virus viewpoint. *Origins of Life and Evolution of the Biosphere* 40(2): 151–160.

Frank, S.A. 1996. Host control of symbiont transmission: The separation of symbionts into germ and soma. *American Naturalist* 148(6): 1113–1124.

Friedmann, H. 2004. From 'Butyribacterium' to '*E. coli*': An essay on unity in biochemistry. *Perspectives in Biology and Medicine* 47: 47–66.

Ghiselin, M.T. 1974. A radical solution to the species problem. *Systematic Zoology* 23: 536–544.

Gould, S.J. 1994. The evolution of life on Earth. *Scientific American* 271: 84–91.

Gould, S.J., and E.A. Lloyd. 1999. Individuality and adaptation across levels of selection: How shall we name and generalize the unit of Darwinism? *Proceedings of the National Academy of Sciences of the United States of America* 96(21): 11904–11909.

Gradmann, C. 2000. Isolation, contamination, and pure culture: Monomorphism and polymorphism of pathogenic microorganisms as research problem 1860–1880. *Perspectives on Science* 9: 147–172.

Hamilton, W.D. 1964. The genetical evolution of social behaviour. I. *Journal of Theoretical Biology* 7(1): 1–16.

Handelsman, J. 2002. Microbiology as a change agent in science education. *ASM News* 68: 163–167.http://newsarchive.asm.org/apr02/feature1.asp. Accessed 19 Dec 2011.

Hooke, R. 1665. *Micrographia: Or some physiological descriptions of minute bodies made by magnifying glasses with observations and inquiries thereupon*. London: Martyn & Allestry. http://www.gutenberg.org/ebooks/15491.

Hull, D.L. 1976. Are species really individuals? *Systematic Zoology* 25: 174–191.

Ingraham, J.L. 2010. *March of the microbes: Sighting the unseen*. Cambridge: Belknap.

Joyce, G.F. 1995. The RNA world: Life before DNA and protein. In *Extra-terrestrials: Where are they?* ed. B. Zuckerman and M.H. Hart. Cambridge, MA: Cambridge University Press.

Kasting, J.F. 2005. Methane and climate during the Precambrian era. *Precambrian Research* 137: 119–129.

Kerr, B., C. Neuhauser, B.J.M. Bohannan, and A.M. Dean. 2006. Local migration promotes competitive restraint in a host–pathogen 'tragedy of the commons'. *Nature* 442: 75–78.

King, N. 2004. The unicellular ancestry of animal development. *Developmental Cell* 7: 313–325.

Lederberg, J. 1987. Genetic recombination in bacteria: A discovery account. *Annual Review of Genetics* 21: 23–46.

Lwoff, A. 1967. Principles of classification and nomenclature of viruses. *Nature* 215(5096): 13–14.

Madigan, M.T., J.M. Martinko, P.V. Dunlap, and D.P. Clark. 2008. *Brock biology of microorganisms*, 12th ed. San Francisco: Pearson/Benjamin Cummings.

Margulis, L. 1970. *Origin of eukaryotic cells*. Yale: Yale University Press.

Maynard Smith, J. 1982. *Evolution and the theory of games*. New York: Cambridge University Press.

Maynard Smith, J., and E. Szathmáry. 1995. *The major transitions in evolution*. New York: Freeman.

Michod, R.E., and D. Roze. 2001. Cooperation and conflict in the evolution of multicellularity. *Heredity* 86: 1–7.

Mitchell, S.D. 2009. *Unsimple truths: Science, complexity, and policy*. Chicago: University of Chicago Press.

Moreira, D., and P. López-García. 2009. Ten reasons to exclude viruses form the tree of life. *Nature Reviews Microbiology* 7: 306–311.

Moya, A., J. Peretó, R. Gil, and A. Latorre. 2008. Learning how to live together: Genomic insights into prokaryote-animal symbioses. *Nature Reviews Genetics* 9: 218–229.

Nee, S. 2004. More than meets the eye. *Nature* 429: 804–5.

Newman, D.K., and J.F. Banfield. 2002. Geomicrobiology: How molecular scale interactions underpin biogeochemical systems. *Science* 296: 1071–1077.

Nowak, M.A., C.E. Tarnita, and E.O. Wilson. 2010. The evolution of eusociality. *Nature* 466: 1057–1062.

Okasha, S. 2006. *Evolution and the levels of selection*. New York: Oxford University Press.

O'Malley, M.A., and J. Dupré. 2009. Philosophical themes in metagenomics. In *Metagenomics: Theory, methods and applications*, ed. D. Marco. Norwich: Horizon.

O'Malley, M.A., and K. Stotz. 2011. Intervention, integration and translation in obesity research: Genetic, developmental and metaorganismal approaches. *Philosophy, Ethics, and Humanities in Medicine* 6: 2. doi:10.1186/1747-5341-6-2.

Osteryoung, K.W., and J. Nunnari. 2003. The division of endosymbiotic organelles. *Science* 302: 1698–1704.

Owen, J. 2008. Virus-infecting virus fuels definition of life debate. *National Geographic News*. http://news.nationalgeographic.com/news/2008/08/080822-giant-virus-html. Accessed 19 Dec 2011.

Pearson, H. 2008. 'Virophage' suggests viruses are alive. *Nature* 454: 677.

Pedersen, K. 2000. Exploration of deep intraterrestrial microbial life: Current perspectives. *FEMS Microbiology Letters* 185: 9–16.

Queller, D.C., and J.E. Strassmann. 2009. Beyond society: The evolution of organismality. *Philosophical Transaction of the Royal Society of London. Series B: Biological Sciences* 364: 3143–3155.

Rainey, P.B., and B. Kerr. 2010. Cheats as first propagules: A new hypothesis for the evolution of individuality during the transition from single cells to multicellularity. *Bioessays* 32(10): 872–880.

Rainey, P.B., and K. Rainey. 2003. Evolution of cooperation and conflict in experimental bacterial populations. *Nature* 425: 72–74.

Raoult, D. 2005. The journey from *Rickettsia* to Mimivirus. *ASM News* 71: 278–284.

Raoult, D., and P. Forterre. 2008. Redefining viruses: Lessons from mimivirus. *Nature Reviews Microbiology* 6(4): 315–319.

Ratcliff, M.J. 2009. *The quest for the invisible: Microscopy in the enlightenment*. Farnham: Ashgate.

Ruiz-Saenz, J., and J.D. Rodas. 2010. Viruses, virophages, and their living nature. *Acta Virologica* 54: 85–90.

Sagan, L. 1967. On the origin of mitosing cells. *Journal of Theoretical Biology* 14(3): 225–274.

Santelices, B. 1999. How many kinds of individual are there? *Trends in Ecology & Evolution* 14: 152–155.

Savage, D.C. 1977. Microbial ecology of the gastrointestinal tract. *Annual Review of Microbiology* 31: 107–133.

Shapiro, J. 1998. Thinking about bacterial populations as multicellular organisms. *Annual Review of Microbiology* 52: 81–104.

Shimkets, L.J. 1990. Social and developmental biology of the myxobacteria. *Microbiology and Molecular Biology Reviews* 54: 473–501.

Singh, B.K., R.D. Bardgett, P. Smith, and D.S. Reay. 2010. Microorganisms and climate change: Terrestrial feedbacks and mitigation options. *Nature Reviews Microbiology* 8: 779–790.

Southam, G., L.J. Rothschild, and F. Westall. 2007. The geology and habitability of terrestrial planets: Fundamental requirements for life. *Space Science Review* 129: 7–34.

Sterelny, K., and P.E. Griffiths. 1999. *Sex and death: An introduction to philosophy of biology*. Chicago: University of Chicago Press.

Strassmann, E., and D.C. Queller. 2010. The social organism: Congresses, parties, and committees. *Evolution* 64(3): 605–616.
Strom, S.L. 2008. Microbial ecology of ocean biogeochemistry: A community perspective. *Science* 320: 1043–1045.
Summers, W.C. 1991. From culture as organism to organism as cell: Historical origins of bacterial genetics. *Journal of the History of Biology* 24: 171–190.
Suttle, C.A. 2007. Marine viruses – Major players in the global ecosystem. *Nature Reviews Microbiology* 5: 801–812.
Theunissen, B. 1996. The beginnings of the 'Delft tradition' revisited: Martinus W. Beijerinck and the genetics of microorganisms. *Journal of the History of Biology* 29: 197–228.
Thurber, A.R., W.J. Jones, and K. Schnabel. 2011. Dancing for food in the deep sea: Bacterial farming by a new species of yeti crab. *PLoS One* 6(11): e26243.
van der Giezen, M. 2011. Mitochondria and the rise of eukaryotes. *BioScience* 61: 594–601.
van Leeuwenhoek [Leuwenhock], A. 1694. Containing the history of the generation of an insect, by him called, the wolf. With observations on insects bred in rain-water, in apples, cheese, etc. *Philosophical Transactions of the Royal Society of London* 18 :194–200.
van Veelen, M., J. Garcia, M.W. Sabelis, and M. Egas. 2011. Group selection and inclusive fitness are *not* equivalent; The Price equation vs. models and statistics. *Journal of Theoretical Biology*. doi:10.1016/j.jtbi.2011.07.025.
Velicer, G.J. 2003. Social strife in the microbial world. *Trends in Microbiology* 11(7): 330–337.
Wainwright, M., N.C. Wickramasinghe, J.V. Narlikar, P. Rajaratnam, and J. Perkins. 2004. Confirmation of the presence of viable but non-culturable bacteria in the stratosphere. *International Journal of Astrobiology* 3: 13–15.
Werren, J.H., L. Baldo, and M.E. Clark. 2008. *Wolbachia*: Master manipulators of invertebrate biology. *Nature Reviews Microbiology* 6(10): 741–751.
West, S., A. Griffin, and A. Gardner. 2007. Social semantics: Altruism, cooperation, mutualism, strong reciprocity and group selection. *Journal of Evolutionary Biology* 20(2): 415–432.
Whitman, W.B., D.C. Coleman, and W.J. Wiebe. 1998. Prokaryotes: The unseen majority. *Proceedings of the National Academy of Sciences USA* 95: 6578–6583.
Williams, G.C. 1966. *Adaptation and natural selection: A critique of some current evolutionary thought*. Princeton: Princeton University Press.
Willner, D., M.R. Haynes, M. Furlan, N. Hanson, B. Kirby, Y.W. Lim, P.B. Rainey, R. Schmieder, M. Youle, D. Conrad, and F. Rohwer. 2011. Case studies of the spatial heterogeneity of DNA viruses in the cystic fibrosis lung. *American Journal of Respiratory Cell and Molecular Biology*. doi:10.1165/rcmb.2011-0253OC.
Wilson, J. 1999. *Biological individuality: The identity and persistence of living entities*. Cambridge, MA: Cambridge University Press.
Wilson, D.S., and E. Sober. 1989. Reviving the superorganism. *Journal of Theoretical Biology* 136: 337–356.
Wilson, D.S., and E.O. Wilson. 2007. Rethinking the theoretical foundation of sociobiology. *The Quarterly Review of Biology* 82(4): 327–348.
Wu, D., S.C. Daugherty, S.E. van Aken, G.H. Pai, K.L. Watkins, H. Khouri, et al. 2006. Metabolic complementarity and genomics of the dual bacterial symbiosis of sharpshooters. *PLoS Biology* 4(6): 1079–1092.
Zehr, J.P., J. Robidart, and C. Chris Scholin. 2011. Marine microoganisms, biogeochemical cycles, and global climate change. *Microbe*. http://www.microbemagazine.org/index.php/04-2011-home/3280-marine-microorganisms-biogeochemical-cycles-and-global-climate-change. Accessed 19 Dec 2011.

Essentialism in Biology

John S. Wilkins

1 Introduction

When we teach students about evolution, we often set up narratives that are, to put it mildly, triumphalist. This does not aid either the reception of evolution nor understanding the historical contexts in which these ideas of science were developed, and it can set up a false dichotomy between older ideas often embedded in religion and culture and modern science. In this chapter, I shall attempt to bring some clarity to an often-abused term – "essentialism" – in the context of scientific thinking and in particular of biology. It is a term that has real rhetorical power. To be accused of essentialism is to be, variously, an adherent of an outmoded and dangerous metaphysics, to be antiscientific, anti-Darwinian, anti-women, racist, nationalist, anti-LGBT, and very probably some kind of political regressive. Like many other terms of that kind, it is almost entirely defined by its opponents, and has little generic meaning beyond expressing the disapprobation of those opponents, and relegating those who are said to hold the ideas to the outer darkness.

In recent years the term "essentialism" has been much employed by biologists and philosophers of biology, and to a lesser extent psychologists and historians of science. The general claim of what I shall call *scientific essentialism* is that natural kinds must have modally necessary shared properties that nothing else does. A variety of this is *biological [or taxic] essentialism*, in which it is thought, wrongly as I argue, that pre-evolutionary and anti-evolutionary scientists held an essentialistic metaphysics in which evolution was prohibited by sharply divided taxic kinds between which there were "bridgeless gaps". There may be scientific essential kinds in some sciences; I do

J.S. Wilkins (✉)
Department of Philosophy, University of Sydney, Sydney, NSW, Australia

Historical and Philosophical Studies, University of Melbourne,
Parkville 3010, Melbourne, Australia
e-mail: john@wilkins.id.au

K. Kampourakis (ed.), *The Philosophy of Biology: A Companion for Educators*, History, Philosophy and Theory of the Life Sciences 1, DOI 10.1007/978-94-007-6537-5_19,

not think that biology (and other sciences that are special or historical, like geology or psychology) has ever really appealed to them for taxic kinds. When essences have been employed by biologists it has been in a non-modal, non-"Aristotelian" manner. I scare quote "Aristotelian" because the kind of taxic essentialism being attacked was not Aristotle's, and it was never really scientific essentialism.

The understanding of essentialist claims and counterclaims goes a long way to uncovering the tensions and issues in modern biology and the philosophy of biology, and at the same time uncovers how we have generated some of the framing narratives of our time. Teaching the history and philosophy of essentialism would be of great use to students coming to a nuanced and useful understanding of science, of biology, of evolution, and of philosophy.

2 Essentialism and Evolution

2.1 *The Origins of Essentialism*

There are many narratives told about evolution. One of the most widely told is the Essentialism Story, replayed in textbook, popular storytelling and philosophy alike (Hull 1965a; Sober 1980, 1994; Wilson 1999; Okasha 2002; Walsh 2006). It goes like this: Before Darwin, biologists were constrained by essentialist thinking, and were committed to species being natural kinds composed of essential characters shared by every member of the species. This meant that either a species had to evolve in a discontinuous fashion (*saltatively*) where the parents of the first member of the new species were members of the ancestral species, or that evolution was logically impossible. In the narrative, Darwin changed all this by adopting a kind of nominalism,[1] in which every member of a species, and every species, was a unique object, and no species had members that shared characters that all members exhibited and which no other species did. In the place of the traditional metaphysics of essentialism, Darwin developed a view in which species were *populations* (Mayr 1982, 1988, 1991; Hull 1973; Sober 1980). Michael Ghiselin and David Hull developed an individualistic view of species, in which species themselves were Darwinian individuals, particulars not classes (Ghiselin 1974; Hull 1976). The Individuality Thesis consisted of three not entirely connected claims: one, that kinds in biology were not universals but historical objects; two, that as individuals they were causally cohesive and acted as systems (usually populations in respect to species); and three, that they presented themselves to observation with unique sets of observable properties. Metaphysically, however, it is the claim that species are historical individuals, like "The United States of America" or "the blues", that was most influential. A historical individual is something or group that has a beginning, and end,

[1]Nominalism in metaphysics is the view that only individual things exist, and no universal kinds. See below.

and is spatially located (Zalta 1988). This is the new metaphysics of evolution. Anything else is "outmoded metaphysics" (as a review of a colleagues' paper called it). If you aren't with the new evolutionary metaphysics, you aren't modern.

Only, it isn't historically the case. There is little evidence that *anyone* was what I call a "biological [or taxic] essentialist" (Wilkins 2009b, 2010). It is true that writers often talked about the essences of life, of organs, and so forth, but they never accepted that species had to have what we now call *jointly necessary and severally sufficient conditions*, or that members of a species or any other taxon would bear such essential properties. The alarm was first sounded by Paul Farber (1976), and more recently historian of systematics Polly Winsor made the same argument (Winsor 2003, 2006a, b), as have others (Amundson 2005; Richards 2010). So, when did the story arise? Winsor thinks it was based on the ideas of Arthur J. Cain, taken up and disseminated by Mayr, Hull and thence many philosophers and biologists. Hull was influenced directly and personally by Popper, whose graduate seminar he had taken in the early 1960s, resulting in the famous paper "The Effect of Essentialism on Taxonomy – Two Thousand Years of Stasis" which Popper took it on himself to submit without Hull's knowledge (Hull 1965a: personal communication). Popper had defined and criticized "methodological essentialism" in his book, *The Open Society and Its Enemies* (Popper 1945), in the first volume on Plato as the founder of ideas that led to the then-threatening views we call fascism:

> I use the name *methodological essentialism* to characterise the view, held by Plato and many of his followers, that it is the task of pure knowledge or science to discover and to describe the true nature of things, i.e. their hidden reality or essence. It was Plato's peculiar belief that the essence of sensible things can be found in their primogenitors or Forms. But many of the later methodological essentialists, for instance, Aristotle, did not altogether follow him in this, although they all agreed with him in determining the task of pure knowledge as the discovery of the hidden nature or Form or essence of things. All these methodological essentialists also agreed with Plato in maintaining that these essences may be discovered and discerned with the help of intellectual intuition; that every essence has a name proper to it, the name after which the sensible things are called; and that it may be described in words. And a description of the essence of a thing they all called a definition. According to methodological essentialism, there can be three ways of knowing a thing: 'I mean that we can know its unchanging reality or essence; and that we can know the definition of the essence; and that we can know its name. Accordingly, two questions may be formulated about any real thing [...]: A person may give the name and ask for the definition; or he may give the definition and ask for the name.' (p. 25f)

What Popper is critiquing here is sometimes called *rationalism*: that we can know the natures of things through reflection and reasoning, doing science-by-definition (SBD). He contrasts it to

> *methodological nominalism* [which] aims at describing how a thing behaves, and especially, whether there are any regularities in its behaviour. (p. 26)

Popper's view was widely known and influenced many scientists and philosopher of science, especially when his *Logik des Forschung* was translated as *Logic of Scientific Discovery* (Popper 1959), although one thing it lacked was a theory of discovery. Hull's paper set the tone, and clearly established the notion that Aristotle was the author of essentialist thinking, whereas Popper and before him Dewey

(1997/orig. 1908) had suggested it was Plato, with which G. G. Simpson, the palaeontologist and one of the major authors of the Modern Synthesis agreed. Hull gave a longer historical summary in his *Science as a Process* (1988), and Ernst Mayr, in his widely read *The Growth of Biological Thought* (1982), constantly interpreted, sometimes aggressively selecting sources, the history of biology in terms of essentialism. Clearly one of the influences was Popper, via Hull, through to Mayr (who cited Popper's definition on page 864). But Mayr himself gave only a general, and non-philosophical, account of Aristotelian essentialism:

> … a limited number of fixed and unchanging forms, *eide* (as Plato called them) or *essences* as they were called by the Thomists in the Middle Ages. (p. 38)

Essentia preceded Thomas by a comfortable margin; at the very least his teacher Albertus Magnus used the term frequently, and the term is used, seemingly in the usual sense, in Quintilian's *Institutio Oratoria* Book 2, 14.2 (c100CE). But the issue here is where *modern* definitions of essentialism come from. Oddly the *term* "essentialism" has no great philosophical history itself. Apart from its use in education (essentialism is the claim there are essential things that must be taught, what we now call the canon), it was used shortly after Popper in a philosophical sense in a paper on aesthetics (Gallie 1948). These are the two earliest versions I can locate in English. A Google Ngram for "essentialism" places the rise of the term in the late 1930s, far too late for it to have been a label used to describe anything pre-Darwinian. Similar patterns occur for variants and different capitalizations.[2] Although Google Ngrams are a somewhat unreliable source of frequency of uses, this pattern is repeated in German and French, where it often applies to existentialist philosophical discussions.[3] The term first gets used – apart from a small spike around 1900 – beginning in 1939. Some of this is in the logic literature, where it gets discussed in questions of modality (Parsons 1969; Wiggins 1974; Hooker 1976), until it becomes more widely used in philosophical literature, and it really picks up after Mayr's book in 1982.

But the source of the standard definition, and the one that ties it to Aristotle, seems to be one of the most widely read and cited philosophy papers of the century: Quine's "Two Dogmas of Empiricism" (Quine 1951; reprinted in Quine 1953a). Quine is attacking a particular theory of meaning:

> The Aristotelian notion of essence was the forerunner, no doubt, of the modern notion of intension or meaning. For Aristotle it was essential in men to be rational, accidental to be two-legged. But there is an important difference between this attitude and the doctrine of meaning. From the latter point of view it may indeed be conceded (if only for the sake of argument) that rationality is involved in the meaning of the word 'man' while two-leggedness is not; but two-leggedness may at the same time be viewed as involved in the meaning of 'biped' while rationality is not. Thus from the point of view of the doctrine of meaning it makes no sense to say of the actual individual, who is at once a man and a biped, that his rationality is essential and his two-leggedness accidental or vice versa. Things had

[2] http://books.google.com/ngrams/

[3] In the mid-nineteenth century, it is used in a philosophical context and also a medical context in German, on occasion, but not in our sense.

> essences, for Aristotle, but only linguistic forms have meanings. Meaning is what essence becomes when it is divorced from the object of reference and wedded to the word.

A commentator (White 1972) noted that it is unremarkable that Quine did not cite any text of Aristotle in support of this interpretation, since it is only tenuously connected to anything Aristotle wrote. Quine later gave a more technical definition (Quine 1953b):

> … Aristotelian essentialism […] is the doctrine that some of the attributes of a thing (quite independently of the language in which the thing is referred to, if at all) may be essential to the thing and others accidental. E.g., a man, or talking animal, or featherless biped (for they are all the same *things*), is essentially rational and accidentally two-legged and talkative, not merely qua man but qua itself. (p. 173f)

This introduces modal necessity (the "necessary" part of the *necessary and sufficient conditions* definition). What is interesting is that this seems to be the very first use of "Aristotelian essentialism", and while that's just a phrase, not much else marries scientific essentialism with Aristotle. It looks like one of the major preoccupations of modern philosophy of science is no older than the early 1950s. A Google Ngram for the phrase "Aristotelian essentialism" and cognate terms shows that the phrase did not exist in English until the early 1950s. It is clear that Aristotle was not seen to be a *scientific* essentialist before Quine's essay, even had Quine thought that he was (which he didn't). I suspect that this interpretation was inadvertent, and Quine's status as a philosopher led others to think that this *en passant* comment was historically and generally correct, when in fact scientific essentialism was not the kind of essentialism Aristotle actually held (Charles 2002; Matthews 1990). He thought essences were, as Quine noted, about words, not objects: "I want to claim here that Aristotle's grasp of modal notions, and of the use of modal operators, is such that he could not clearly express the Quinian distinction between essential and non-essential attributes of a sensible particular". (White, p. 60; White's argument is subtle, and has to do with the role sensible particulars play in Aristotle's metaphysics and epistemology, that is not relevant here.)

In conclusion, the notion of a scientific Aristotelian essentialism is a mistake based on a casual reading of various philosophers, including (as I detail in my 2009b) Dewey, logic texts, and Popper, but the particular widespread error of ascribing it to Aristotle appears to be based on Quine's passing comment.

2.2 Darwinism and the Essentialist Story

The hardening of the idea of pre-Darwinian essentialism was due to Hull's essay. In it, Hull appeals to Popper's usage, and a discussion by Michael Scriven (1959) about the distinction between "normic" and "analytic" criteria, the former being something like a typical example of a kind, and the latter a defined set of characteristics of a kind. But what is most interesting is that Popper's attack in the *Open Society* (1945), and Scriven's here, are discussing what we might call the assumption that we can define terms in an essentialistic or analytic fashion, and *thereby know something*. Popper's attack is centered on the idea, long held in philosophy, that one can gain knowledge by definition: I call this "science-by-definition" (SBD). Aristotle in his *logical* works did practice a form of SBD, and Plato clearly did, although the famous "carve nature at its joints" comment (*Phaedrus* 265d–266a) applied to *justice* and not any "natural" kind in the modern sense. But the knowledge Aristotle thought he gained from analytic characters, as Scriven might put it, was of a different kind to the knowledge gained by empirical observation and experiment, which is what he typically applies in the natural history works we might call science. When Hull equated logical analytic criteria with criteria in taxonomy, he changed the argument substantially, for it is unclear whether any naturalist ever proceeded by definitional analysis.[4] For instance, Linnaeus, whose system has been deprecated and described in this way (Enç 1975; Ereshefsky 1999, 2000), did not. His was an empirical classification based upon types, and it served a largely diagnostic role. Linnaeus himself knew it was a conventional system, and largely artificial, and he certainly did not intend it to be in some fashion fixist or essentialistic. Unfortunately, the diagnostic criteria in the Linnaean scheme were called the "essential characters" in the English translation (*character essentialis*), which has misled many modern commentators. They would better be called "diagnostic characters". Linnaeus' thoughts on the matter are clear enough:

> If the essential characters of all genera had been discovered, the recognition of plants would turn out to be very easy, and many would undervalue the natural characters, to their own loss. But they must understand that, without regard for the natural character, no one will turn out to be a sound botanist; for when new genera are discovered, the botanist will always be in doubt if [he] neglects the natural character. Anyone who thinks that he understands botany from the essential character and disregards the natural one is therefore deceiving and deceived; for the essential character cannot fail to be deceptive in quite a number of cases. The natural character is the foundation of the genera of plants, and no one has ever made a proper judgement about a genus without its help; and so it is and always will be the absolute foundation of the understanding of plants. (*Philosophia botanica* 1751, quoted in Winsor 2006b, p. 5)

The "natural character" here is something like the key causal properties as identified by a skilled and trained botanist. Linnaeus referred to the "unique idea" (*unica idea*) that was, in effect, a set of characters that distinguished genera.[5] For example, he used the fructative apparatus (calyx, corolla, stamen, pistil, pericarp, seed, and receptacle) on four analytic dimensions as the potential natural character of genera (Atran 1990, p. 174). This is not an essentialism of the constitutive *or* the definitional kind, but merely a diagnostic essentialism. Even more interesting is that whether or not Linnaeus was a diagnostic essentialist in the sense that he defined the taxa analytically, which I doubt, the practice thereafter was to treat these taxonomic definitions as identifying a *type* taxon, around which classifications were arranged.

[4]A possible exception is Louis Agassiz, but I think his practice and his theoretical argument in Agassiz (1859) are not necessarily all that deeply connected. He was an excellent observer (Winsor 1979). What scientists say they are doing, and what they actually do, are often distinct.

[5]I am indebted to Larissa Vasiliyeva for bringing this to my attention, through an advance copy of her forthcoming paper in Botanica Pacifica with Steven Stephenson (2012).

The "type species" of a genus was the "most typical" form of it, and as Whewell noted of this approach

> These lessons are of the highest value with regard to all employments [*sic*] of the human mind; for the mode in which words in common use acquire their meaning, approaches far more nearly to the *Method of Type* than to the method of definition. (Whewell 1840, vol 2, pp. 517–519)

And

> So long as a plant, in its most essential parts, is more like a rose than anything else, it is a rose. (p. 520)

As Whewell notes, the method of classifying by type is more common and a better account of taxonomies in biology. Winsor (2003) calls this the *method of exemplars*; either term will serve. Nevertheless, on this misunderstanding of Linnaean taxonomy a whole story was erected, and Linnaeus became, along with Aristotle and to a lesser extent Plato, the whipping boy of bad taxonomy and systematics (see Hull 1988, chapter 3 "Up from Aristotle").

In fact, the primary use of the essentialism story has been to attack opposing systematic techniques and philosophies. Ernst Mayr used it to attack cladism as being "typological", which he wrongly treated as synonymous with "essentialism". Pheneticists attacked cladistics in a similar fashion. Process cladists attacked pattern cladists as being "creationists" and "typologists" because they failed to include process based, or historical, classifications in their phylogenetic trees.[6] And non-neo-Darwinians were often attacked in the same fashion, particularly those who applied, following D'Arcy Thompson (1917), formal analyses to development and evolution. The use of the very term "form" became an identifier for essentialistic issues. Much of this is too recent to be easily neutrally discussed (cf. Winsor 2006a; Levit and Meister 2006; Love 2009).

2.3 *Transformation and Variation*

The irony, then, appears to be that if scientific essentialism, especially in biology, has ever been promoted, it looks to be a very modern invention, and not something that has preceded Darwinian thinking at all. I suspect that it arose in reaction to

[6]It is widely accepted that there are three kinds of classification philosophies in modern biology. One is called "phenetics", and it relies on mathematically measuring similarities of arbitrarily chosen traits. It was replaced in most instances by "cladistics", which draws treelike diagrams to represent relations based on shared or unique homologies. Process cladists think that these treelike diagrams (cladograms) represent the history of the evolution of the taxa, while pattern cladists think they are merely statements of relationship that might have been evolved in any number of historical pathways. The third view is misleadingly called "evolutionary systematics" (misleading because none of the other views are unevolutionary). It holds that classification is both genealogical (tracing treelike pathways in evolution) and "grade-based", in which groups are put together on the basis of evolutionary novelties like flight or skeletal structures. These novelties represent grades of organization or evolution. For that reason it is sometimes called "gradism".

Darwinian, and more particularly Mendelian genetic versions of Darwinian, thinking, together with ruminations concerning the philosophy of language and in particular of the reference of kind terms in the 1950s and on. Darwin is not the latest metaphysical view of biology at all. This is why biological essentialism has become popular of late: it is a somewhat revisionary response to Darwin himself.

Ernst Mayr, whose work is so influential on the popular narrative, argued that evolutionary thinking shifted from the transformational, in which entire species changed into new species as Lamarck thought, to the variational, in which parts of species (varieties or populations) changed to form new species but leaving the rest of the ancestral species unchanged (Mayr 1991, 1992). This has been taken up by others (Kronfeldner 2007; Shtulman 2006; Shtulman and Schulz 2008).

Why did Mayr propose this, among many, distinctions? It is an extension of his prior use of the terms "population thinking" and "essentialism". Populational accounts of species imply that they vary. Transformational accounts such as Lamarck's (or the neo-Lamarckians still active when Mayr was a student in Berlin) tended to have an instantaneous essentialism; a species was whatever the essential traits were at a moment. As a champion of the so-called "allopatric", or geographical isolation, view of species formation, Mayr naturally preferred to stress the role variation played in species formation, and therefore in the individual natures of species and of the evolutionary process (Wilkins 2007), and so the contrasting views tended to become one big error in his mind (Chung 2003). Thus, transformational views of evolution, typology, and essentialism are all examples of a bad metaphysics misleading science before (and even after) Darwin arrived to set things straight.

3 What Is Essentialism?

3.1 The Meaning of "Essentialism"

The word "essentialism", like its root word "essence", does not refer to a single notion or view, but a group of them, which are not all closely related but which form a family of ideas that resemble each other somewhat (Stone 2004). *Essence* itself has been long held to be contentious. A well known and widely-used dictionary of philosophy at the turn of the twentieth century defined it the traditional way:

> Essence [Lat. *essentia*, from *esse*, to be]: Ger. *Wesen*; Fr. *essence*; Ital. *essenza*. The constant and necessary nature of a thing as contrasted with its accidents. [...] Aristotle uses the word for (1) the form, (2) the matter or substratum, (3) the concrete being, the individual. [...] But the scholastics defined the word more precisely in contrast with substance: essence is the nature of the individual thing, substance is the indeterminate substratum, which, united to the form, makes up the individual thing. Descartes follows the scholastic usage, but since his time the word essence has usually had the same meaning as substance. Kant defines essence as determined by an idea; hence it may be false to reality, while the nature of a thing is actually experienced and cannot be false. (Baldwin 1901, Vol. I)

The traditional logical definition appealed to the properties that marked out a class of things from other things (*differentia)*:

> Whatever term can be affirmed of several things must express either their *whole essence*, which is called the *Species*; or a *part* of their essence (*viz.* either the material part, which is called the *Genus*, or the *formal* and *distinguishing part*, which is called *Differentia*, or in common discourse, *characteristic*) or something *joined to the essence*; whether *necessarily* (*i.e.* to the *whole* species, or, in other words, *universally*, to every individual of it), which is called a *Property*; or *contingently* (*i.e.* to some individuals only of the species), which is an *Accident*. (Whately 1875)

However, the idea had been depreciated somewhat by the loose and often inchoate uses by idealists and those in the Hegelian tradition, leading one nineteenth century philosopher to write somewhat more succinctly than the dictionary that succeeded him:

> Essence, (*essentia*, from *esse*, to be,) "the very being of anything, whereby it is what it is." Locke. It is an ancient scholastic word, which cannot be really defined, and should be banished from use. (Jevons 1870, p. 335)

The quote from Locke comes from the *Essay Concerning Human Understanding*, III.3.15 (Locke 1997 [1690]). Locke, of course, propounded a nominalistic view, in which essences were known only as the meanings of words, and any real essences (physical natures) were forever hidden from our apprehension. Curiously he did not deny that there *were* real essences. Because the term is used in so many ways, it pays us to try to distinguish these different senses, which is ironic, since one of the original and core doctrines of essentialism is that terms have clear meanings. However, I have reconstructed these different senses rather than drawing them directly from the clear usage of the various technical communities that employ the word. Any actual use is likely to apply more than one of these senses, even when those senses in some way conflict with the claims being made about essence. One must not presume that the same word means the same thing even when the same author is using it in the same work. Often, indeed almost universally, people elide from one meaning to another unaware they have even done so, and this has caused no end of confusion in the literature. Susan Gelman and Lawrence Hirschfeld put it this way:

> When we co-taught a graduate seminar in 1996 on essentialism, and read sources from ancient Greek philosophers to postmodern feminist theorists, we were overwhelmed by the scope, richness, and variety in arguments about essentialism. We read authors who treated essences as a property of the real world, others who treated essences as an inevitable product of the human mind, and still others who treated essences as a historical construction imposed on people for political ends. (Gelman and Hirschfeld 1999, p. 404)

Table 1 presents the varieties of essentialism in the modern (post-1940) literature.

A view can be scientifically essentialist without thereby committing the advocate to psychological or human essentialism. It may be that there is a covariance between these views, but I think that it is usually one way: if one is justifying some human essentialistic view, like a racial realism, one often will appeal to taxic, scientific or even metaphysical essentialisms, but it does not thereby mean that these other forms imply a social essentialism.

Table 1 Varieties of essentialism as presented in the modern (post-1940) psychological and philosophical literature

Type of essentialism	Nature, examples, and references
Psychological [folk]	Imputing to objects an internal persistent nature on the basis of superficial appearances (Medin and Ortony 1989; Medin et al. 2000; Gelman 2003). For example, children think of animate objects as having some internal essence that moves them
Human [historical and social]	Imputing to sociocultural groups a shared persistent set of properties of each member of the group. Examples, gender (Heyes 2000), nations (White 1965), ethnicities (Gil-White 2001), races (Sesardic 2010) and medicine (Jensen 1984; Pickering in press)
Logical [semantic, linguistic]	Imputing to terms an invariant and unique meaning. Examples: The Aristotelian/scholastic tradition, Cicero. Criticized influentially by Popper (Popper 1957, Vol 1: Plato, chapter 3). Example: strict definitions of general terms like "life" or "human"
Metaphysical [Aristotelian essentialism, universalism, Platonism]	The claim that there are universals that are facts about the world (Aaron 1952; Quine 1951, 1953b). Examples: colors, numbers, shapes. The opposite of a universal is a *particular*, such as this color, that shape, or the number of people in this room
Scientific [natural kind]	The claim that scientific laws refer to objects that have invariant objects and properties (Ellis 2001, 2002). For example, "mass", or "charge" in physics; innate or species traits in biology
Biological [taxic]	The claim that all members of taxonomic objects in biology (species and higher, subspecies and lower) have invariant properties (Devitt 2008, 2010; Hull 1965a, 1984; Rieppel 2010; Sober 1980; Walsh 2006; Wilkins 2010, 2013). Examples: Linnaean "essential characters" that define a species or genus; the genome of a species

3.2 Kinds of Essences

There are basically three general forms of essentialism available for each type: *constitutive*, *diagnostic* and *definitional*. *Constitutive* essentialism is the view that some class of objects are what they are because they all possess invariant properties. *Diagnostic* essentialism is the view that classes of objects are recognizable because all members share some salient properties. *Definitional* essentialism is the view that kinds have severally necessary and jointly sufficient defining properties. Although Hull listed roughly these three tenets as "essentialistic tenets of typology",[7] these are not all the same, or even necessarily related, ideas, and it is not the case that these views must travel together as Hull insisted. However, it is hard to keep them apart. Even those who study one kind of essentialism, such as the psychological

[7] "The three essentialistic tenets of typology are (1) the ontological assertion that Forms exist, (2) the methodological assertion that the task of taxonomy as a science is to discern the essences of species, and (3) the logical assertion concerning definition" (Hull 1965b, p. 317).

Table 2 A taxonomy of essentialisms as found in the literature

	Constitutive	Diagnostic	Definitional
Physical	X	X	X
Biological	X	X	?
Psychological	X	X	?
Human	X	X	?
Logical	—	X	X
Metaphysical	X	—	X

Dashes indicate the inapplicability of that kind of essentialism to that category or domain, and queries indicate uncertainty as to whether that kind of essentialism has ever actually been imputed to that domain

kind, will elide from one sense to another. For example Susan Gelman and Lawrence Hirschfeld write

> … essentialism […] has a long history of links to other domains [than folkbiology] and indeed much of the evidence for essentialism comes from outside the domain of folkbiology. People appear to attribute hidden essences to social categories such as race, gender and personality […] Racial, gender and personality "essences" may be analogical extensions from a folk biological notion […], but race, gender and personality are not themselves biological categories. Similarly, claims of essentialism in language extend to words such as proper names […] Given these controversies, the present chapter examines the evidence for essentialism and addresses whether essentialism is plausibly a core component of folkbiology; whether it is an untutored belief, universal, and/or biologically specific. (Gelman and Hirschfeld 1999, p. 403f)

Gelman and Hirschfeld have a different taxonomy of essentialisms than the one I present here. They distinguish between the sortal (definitional essentialism), the causal (constitutive essentialism) and the ideal (metaphysical essentialism), and identify four kinds of each essentialism. For our purposes, dividing representational (semantic or psychosocial) essentialism into three disparate types is not helpful. Instead I prefer the following taxonomy, and the kinds of essentialism asserted in the literature are marked with an "X" (Table 2).

It is not clear that the philosophical essentialism of Kripke (1980) and Putnam (1975), who are primarily concerned with philosophical questions about the metaphysical implications of the reference of *terms* (as indeed Aristotle and many other philosophers throughout history were), has anything whatsoever to do with the notions of "essence" employed in folkbiology or race theory.[8] However, a surprising number of discussions of biological essentialism appeal to the structure of water (H_2O versus XYZ, which is Putnam's example, derived from Mill's discussion in the *Logic*[9]) or of the elements, like Gold, rather than biological examples of

[8]Contra Hull (1976, p.179n. 4).

[9]Putnam had argued in his 1975 that the meaning of kind terms did not depend on reference to the constituents of instances of that kind, by a "Twin Earth" thought experiment, in which everything was the same as on our Earth except that "water" denoted a substance XYZ not H_2O. The point was that such general meanings of terms were established by a set of macrolevel properties, not the microlevel ones. Mill's discussion (III.vi.1) of the nature of water is the ancestor of modern theories of emergence, which are only tangential to our topic.

essential kinds, and when philosophers do discuss biological examples, they often use vernacular terms like "swan" (which one? There are 6–7 species of swans, some of which carry black coloration) or "tiger" (which has nine extant or extinct subspecies, each with distinct colorations), basing their arguments upon superficial characters like pelt or plumage.

Sometimes, however, the logical and semantic sense is directly applied to kinds other than these, such as the definition given by Jensen of medical essentialism:

> … terms referring to entities have to be defined by specifying a conjunction of characteristics, each of which are necessary, and which together are sufficient for the use of the term. (Jensen 1984, p. 63)

The appeal to this definition of biological essentialism is common among the biologists themselves, and particular among those who discuss taxonomy and systematics. For example, herpetologist and systematist Kevin de Queiroz cites Karl Popper explicitly, appealing to the notions of *methodological essentialism* and *methodological nominalism* Popper introduced (de Queiroz 1992, 1994).

4 Historical Considerations

4.1 Aristotelian Essentialism

The term *essence* was not actually used by Aristotle, but by the late classical and medieval followers, or so they thought, of Aristotle. It is a Latin word, and Aristotle expressed himself in Greek, and the words (not word) he used – *to ti ên einai* and variations – mean, in a literal translation "what it is to be [that thing]". In this sense it is relatively harmless – even the most nominalistic of thinkers believes there are properties, causes or components that makes something what it is, but Aristotle appeared to make more of this than a simple passing phrase. He introduced the notion of an "accident" (*sumbebêkos*), a property that a thing has which, if changed, would not make it something else. For example, a white bird remains a bird if it changes into another color, so whiteness is not "essential" to being a bird. Those properties that a thing has that if changed *would* make it not a bird, like feathers and a beak, are "essential".

A famous and apocryphal story in Diogenes Laërtius' *Lives and Opinions of Eminent Philosophers* tells of the cynic Diogenes of Sinope challenging Plato's definition of the essential characters of Man as a "featherless biped" by bringing a plucked chicken to his next talk, whereupon Plato redefined Man as a featherless biped with broad nails (Book VI.20). Aristotle, whether he knew this story or not, took steps to avoid this ad hocery in two ways: first by excluding *privative* definitions – in terms of what something is *not*, the *diairesis* of the Academy and Plato[10];

[10]According to the Platonist view, classification had to proceed by dichotomous, or binary, division, hence "diairesis" or "splitting into two". They achieved this by defining things as being some property, or not being it. Aristotle, on the other hand, allowed for groups to be subdivided into many subsets, all of which had to have their own positive definienda (see Wilkins 2009b).

and second by seeking the truly essential necessary properties.[11] Based on his "three souls" account in *De Anima*, where living things essentially have three motive forces – nutrition and growth (life), sensory capacities and motion (animal), and reason – Aristotle was able to define humans as the living animal that reasons. Everything else was accidental. Aristotle's use of essentialism is ironic in some ways. In his discussions of the *what-it-is-to-be*, the examples he used were of *predicates*, which is to say, *terms*. So far as I can tell, he did not develop a *taxic* essentialism, although of course he did divide organisms into functional kinds, like animals that live in water, or fly, or have limbs (cf. Atran 1990; Nelson and Platnick 1981).

The neo-Platonists, and in particular Porphyry, conflated Plato's *diairesis* and Aristotle's logical division and developed what came, much later, to be known as the *Arbor Porphyriana*, or Porphyry's Tree. On this logical structure, one began with Aristotle's *Summum Genus* (most general kind), Being, and divided it into subordinate genera (species of the higher genus), such as Corporeal Being, and its negations (e.g., Incorporeal Being), and then subdivided *those* into subordinate genera, and so forth until one attained a species which had under it, or within it, only individuals. Each lower branch was quite literally more specific, and had *propria* (non-accidental properties) that differentiated them out of the higher genus. Combined with the Aristotelian scale of nature implicit in the *De Anima* and elsewhere, this led to the production of the *scala naturae* or great chain of being. However, the great chain, most popular from the sixteenth century, was not itself essentialistic either, as the scala (Latin for "ladder") was gradualistic. There were no sharp discontinuities in most of the post-medieval versions of either the Tree or the Chain, except at the attainment of reason (Man) and of divinity (God).

4.2 Scholastic Essentialism

One sense of "Aristotelian" that might be in play here is some version of medieval scholasticism. It is well known that Aristotle was "rediscovered" in the twelfth century following Michael Scot's and Willem van Moerbeke's translations of his works from Arabic, and that during the twelfth and thirteenth centuries a revitalization of science and philosophy was inspired by these works, culminating in Thomas Aquinas' theology. The term "essentia" was in use by these authors, especially Aquinas and Albertus Magnus, at this time. However, the mere use of the terms subsequently is not sufficient to make some author an essentialist, since the ubiquity of Thomas' ideas among philosophers and scientists meant the terminology was shared by nominalists (those who think only particular things exist, and that general terms are simply verbal conventions) and universalists (those who think that general properties are real facts about the world, and not merely about words, or *nomina*)

[11] E.g., Metaphysics 1022a22, Categories X, Posterior Analytics, I.4; on necessary properties see Metaphysics Z.4, Topics 102a3, Posterior Analytics, 73a34-5 cf. Cohen (2009).

alike. A nominalist who thinks only particulars exist might still talk about essences in order to deny their reality beyond the words.

Moreover, much of the modern focus is on the use of these ideas by modern thinkers, especially the neo-scholastics who developed from Pope Leo XIII's encyclical letter "Aeterni Patris" in 1879, which recommended Thomas as the philosopher of the Church. Initially this did not result in objections to evolution based upon static or essentialistic doctrines, but instead Catholic thinkers almost universally, from Mivart on, objected to the lack of teleology in Darwin's theory (Artigas et al. 2006; Paul 1979; on Darwin and teleology see Lennox and Kampourakis, this volume). In the early twentieth century, in reaction to Modernism, neo-scholasticism morphed into the neo-Thomism of Etiénne Gilson and Jacques Maritain (Gilson 1964, 1984; Maritain 1955), and around the turn of the century claims that logical essentialism prohibited Darwinian evolution began to surface, following objections raised around the turn of the century (Clarke 1895; Wasmann 1910). It is probably not coincidental that modern special creationism arose in the period just following this development.[12]

5 Philosophical Considerations

5.1 Classes, Types and Family Resemblances

In considering the philosophical arguments over biological essentialism, several distinctions must be made for clarity. One is the distinction between *type* and *essence*. Typologies are roughly phenomenological groupings, that is collections of phenomena based on similarity metrics that are held to be salient. Essential classes, on the other hand are very often held to require *intensional properties*. Intension is the notion of the meaning or definition of a kind term or general term, and is contrasted to *extension*, or the physical spatial extent of the class (consider "being Australian" and "being born in Australia"; people born in Australia can be accused of not being Australian). The use of terms like *class*, *set*, *kind* (especially *natural kind*, see below) and *taxon*, are often thought, at least by philosophers of biology, to entail that they and all their members have an intensional property set. Members of types, on the other hand, can have only some of the property set used to define or identify them.

The rise of the essentialism story coincided with an increasing interest in Wittgenstein at the end of the 1950s, and discussions of the notion of a *family*

[12]Contrary to the received opinion, special creationism as an alternative to evolutionary science is a fairly modern development. First proposed by George Macready Price, a Seventh Day Adventist, in the first two decades of the twentieth century, special "scientific" creationism was introduced onto the wider stage of American discourse in the 1960s. The period Price was writing was one of great turmoil in evolutionary opinion (Numbers 2006).

resemblance predicate. One particular paper was cited by Hull (1965b): Douglas Gasking's on "Clusters" (Gasking 1960). Gasking noted there was a distinction between sets and classes[13] which Hull took up, and he discussed how groups might be formed using a clustering notion similar to and based upon Wittgenstein's famous "predicate" (1968, §§66–67), in which "we see a complicated network of similarities overlapping and criss-crossing: sometimes overall similarities, sometimes similarities of detail." Again, in Wittgenstein's discussion, this is about *terms* (Campbell 1965), although some have applied it to the predicate *species* (Pigliucci 2003; Pigliucci and Kaplan 2006). Hull's discussion of essentialism considers the case where the taxa in question have a set of necessary and sufficient defining properties, a conjunction like $a \wedge b \wedge c \wedge d$, versus the case in which the taxon must only have most of these properties, forming an extended disjunct: $(a \wedge b \wedge c) \vee (a \wedge c \wedge d) \vee \ldots$ for all combinations that obtain. At the time Hull wrote, the so-called "numerical taxonomy" (which later came to be called "phenetics" as discussed above) of Sokal and Sneath (1963) was heavily discussed amongst systematists, and Hull effectively argued that their view was the philosophically mandated view (although he later switched to a cladistic account of systematics). Hull's view was initially derived from Beckner's earlier work (1959) in which Wittgenstein's family resemblance predicate was first applied to biological taxa. Beckner distinguished an *E-definition* (effective definition) from a *W-definition* (well-defined definition) in biology, and considered a cluster definition to be E-defined. About the same time, Douglas Gasking defined a *chain group* as one in which the relationship is one of reflexive similarity and noted that:

> Likewise a field naturalist who has learnt, by long experience, to recognise on sight members of a diversified 'polytypic species' does not normally think of the species as a group of forms serially related to a certain focal form. He thinks of it as a chain-cluster of forms which is not essentially defined in terms of any particular one of them. (p. 13)

Note the final sentence: on Gasking's account, the field naturalist does *not* think that a polytypic species must be essentialistically defined. Contrary to Hull's

---

[13]Gasking (p5f) made the following comment about biological species:

"For our next example consider the symmetrical and non-transitive relation crossable with, defined as follows: Two local populations of plants or animals are said to be 'crossable' if they interbreed freely in nature, or would do so but for geographical or ecological barriers. 2 (It is a matter of biological fact that this relation is non-transitive. There do occur in nature series of populations where A is crossable with B, B with C, and C with D, but where A is not crossable with D.) On the basis of this non-transitive relation we can define the transitive relation serially crossable with. In terms of this, taking a local population as focus, we can define the chain-group as all those populations that are serially crossable with this population. In so doing we define a 'biological species' 3 – for between any two populations belonging to the same biological species there holds the chain-group relation serially crossable with."

Gasking's distinction showed Hull that simply grouping things, in this case living things, into sets did not imply all the logical relations that usually are drawn from talking about classes, such as transitivity. Given Gasking's previous comment that sets do not become (p.1) but are timeless, he clearly thinks that to be a species is a time-indexed relation; one shares the property of being the same species at a particular time *t*. This obviously raises a problem for species evolution, even if he permits them to be clusters.

three tenets (the ontological, methodological, and definitional commitments of essentialism), neither the metaphysical claim that Forms really exist nor the view that species must be defined essentialistically is true of the naturalist, either before or after Darwin (or Darwin himself). Only the claim that there must be some persistent causal process which makes a species a species is correct, and this is typological rather than essentialism. Gasking himself refers to this as a *concomitant of* relations (p. 14). Essentialism was a dead issue before it was defined and then attacked in biology.

However, in the second sense, which I have called the *constitutive* sense, of biological essentialism, not only are all biologists before *and* after Darwin (including Darwin himself) essentialists, but it is also the mission statement of taxonomy and systematics, developmental and evolutionary biology, genetics and ecology, to discover what these constitutive properties are. Something must cause species to be species, either individually and severally, or universally. No matter what causes all species to evolve, or if every species has its own cause, that is the constitutive "essence" of that species that taxonomists seek to explore and determine. Diagnostic and definitional essentialisms are about words and identification; constitutive essentialism is about what happens in the mind-independent world.

5.2 Natural Kinds[14]

Several "essentialist" accounts have been proposed in recent times for biological taxa. Two specific proposals of note are Richard Boyd's *homeostatic property cluster account* (HPC; Boyd 1999) and Paul Griffith's *historical essence account* (Griffiths 1999), and recently a more general account by Michael Devitt, *intrinsic biological essentialism* (IBE, Devitt 2008, 2010). Boyd's account is that some kinds have a shared causal mechanism that causes members of the kind to cluster about a stable point, which he calls homeostatic properties. Like Hull and Ghiselin's Individuals, these homeostatic property cluster kinds do not play a role in laws of nature, but the causal mechanisms are instead an outcome of such laws. Griffiths has argued that biological taxa share a *historical* property, that of having a common origin. This is clearly an extrinsic or relational property, as Griffiths observes. Devitt's IBE is a full-blown "Aristotelian" essentialism: taxa have some *intrinsic* shared properties such as developmental mechanisms or genetic mechanisms (to exclude Griffiths-style extrinsic properties). Where Griffiths' essentialism is relational, depending on how the species and all its members relate to a past event, and so to an ancestral species, Devitt's is internal to the species and its members; such as the developmental and genetic properties of the organisms that constitute it.

How do these new essentialisms affect the older claims made, such as the Individuality Thesis, that taxa are historical objects? This has to do with the

[14]Part of this section was previously published as section 7 of Wilkins (2013).

employment of the notion of a natural kind in modern philosophy of science. The term "natural kind" in the essentialism debate in the philosophy of biology can mean "a category of natural objects", or "a class defined by some physical properties" or "intrinsic property set", and so on. There is not a lot of agreement on what the term implies outside the philosophy of biology either, and this is not the place to cover the issue (see Bird 2009; Bird and Tobin 2009; Anderson 1994; Ben Yami 2001; Boyd 1991; Cordry 2004; Dupré 2002; Hacking 1990; Kathrin 2008; LaPorte 2004; Peterson 1999; Quine 1969; Riggs 1996; Sankey 1997; Sterelny 1983; Wilkerson 1988; Witmer and Sarnecki 1998). Griffiths and Boyd-style biological kinds are not in the philosophical sense "natural kinds", but as Boyd notes, if *they* aren't natural kinds then there's something wrong with the notion of a natural kind in philosophy, and I think this is right.

To resolve this, we may distinguish between three kinds of kinds: *type-kinds*, of the sort that Whewell propounded and which is, I think, the most common notion employed in natural history; *definitional class-kinds*, such as is correctly ascribed to the logical tradition deriving from Aristotle; and *property-based class-kinds*, in which every member of the kind must have the same unique set of properties. I have argued that Mill introduced these property-based kinds of kinds from chemistry and mineralogy (Wilkins 2013). Type-kinds are exemplars, types that one uses as a central or "typical" hook on which to hang a group.

There are several criteria held to be necessary for a natural kind in science: the kind must be an *actual* kind, it must be *natural* (that is, not arbitrary or artificial), it must be *required by the scientific discipline* covering the domain under investigation, and the kinds must participate in *laws* of that domain, or at least generalizations within it. Moreover, natural kinds are typically supposed to license inductive inferences across the domain. Taxa are supposed, under the essentialism story, to do all these things, but cannot, which is why the notion of taxa as kinds has to be abandoned in favor of the Individuality Thesis. Under this story, it is clear that individualism is supposed to be a kind of relative nominalism; the view that for this class of objects, at any rate, only particulars exist. The claim made by Ghiselin and Hull was that species are not class-kinds. To be sure, they also thought that before scientific evolutionary theory the received view was that species were class-kinds, which I reject, but that doesn't affect their argument that species are in fact particulars – spatiotemporally restricted contiguous objects that are unique in evolutionary history. Whether or not anyone thought species were class-kinds at all, the argument that they are not does not depend upon eliminating the view that species have definitional or property criteria. It depends upon the positive argument that species that evolve have beginnings and endings.

Unfortunately, later versions of the Individuality Thesis committed it also to the view that species were functionally coherent individuals, relying on a version of the "biological" species concept. Ghiselin attached several conditions: "individual" means, in addition to the metaphysical sense of "particular" or "instance", that the thing is integrated into a functional whole, like a pistol, and that it is observationally discriminable (Ghiselin 1997). However, a species can be a metaphysical particular without being functionally integrated. To illustrate this, consider a particular of loose objects like a bag of jellybeans. Being a bag of things, it is a metaphysical particular (nothing else in time or space is that bag, which is what makes a particular a particular). But the bag of jellybeans need not function as an integrated whole the way a watch would. The watch is both a particular and a functional system. Likewise a species might be a functionally integrated population through interbreeding or cooperation, but another species might not (if asexual, for example, or if populations were in permanent isolation). In both cases, however, on the Individuality Thesis, these are metaphysical particulars. Another formulation is to say that to some degree, individual organisms within a species are, as Templeton put it, demographically replaceable (Templeton 1989; Wilkins 2007, 2010). Since functional integration usually requires *differentiation* of parts rather than homogeneity, a species need not be an integrated individual. In other words, species can be individuals and yet form kinds, because to a first approximation individual organisms are indiscernibly different, at least ecologically. The indiscernibility of members is a key characteristic of a kind. At any rate, species-as-metaphysical-particulars, and as historical objects, remains untouched by the distinction between type-kinds, class-kinds, and clade-kinds.

Essentialisms of a non-taxonomic but explanatory kind have been offered, a recent example being Dennis Walsh's "adaptive essentialism" approach (Walsh 2006). Here, and in Devitt's intrinsic biological essentialism, the emphasis is on the developmental "natures" of the organisms. This is a rather benign form of essentialism – that there are underlying causal processes – including but not restricted to genes, parental investment, ecological niches, constructed niches, social inheritance and the like, that make a typical member of the species, well, typical – is not at issue. What it cannot be is a Millian class kindship, which is what Devitt's version requires. This sense of *essence* need merely be a type-kind: the essence of the species is just the developmentally typical lifecycle (which is in fact a tautology, as the species would not be a species if it lacked at least one typical developmental lifecycle, however that might play out as reaction norms in different environments).[15] Consequently, Boyd's homeostatic property cluster (HPC) account doesn't provide a malignant essence either. A HPC is a kind maintained by causal mechanisms, that cause properties to cluster together, but which are not jointly necessary. HPC kinds are causal versions of Wittgenstein's family resemblances. However, one might say that HPCs, being causally maintained, can only apply to populations that are in constant (enough) causal contact, which supports the "metapopulational" account of Kevin de Queiroz (2007); higher taxa above the metapopulation level cannot be maintained by HPC kinds. If a particular species does not comprise a metapopulation, then it cannot be a HPC kind, and clades that are not in causal contact (say, because they are isolated temporally or biogeographically) cannot be HPC kinds either.

[15] Some have proposed "cryptic species" or "pseudospecies" for taxa that lack their own distinguishing properties. I think that if they truly lacked all unique properties, they would not even be distinct species; even if we do not know the causes of differentiation, the organisms certainly do, in the sense that they react physically when the right properties exist and not otherwise.

Finally, in Paul Griffiths' historical essence account, species and other taxa have essences in virtue of a shared genealogy, which is consistent with, for example, the genealogical concordance view of species of Avise and Ball (1990). This "essence" is not really an intrinsic essence, but one of genealogical relations between individuals and populations distributed over time. As such it is quite consistent with Whewellian type-kinds in a cladistic manner, because it is identified by conserved developmental homologies, and these need not be conserved in a logical essence, with all and only those properties that class all and only members of the taxon. Instead the developmental mechanisms that conserve these relations, like those of the HPC, are a notional type around which the variations accrue over time.

So biological kinds are best thought of as exemplary types, rather than the arid classes of "Aristotelian essentialism", and as Whewell and Jevons (1878) and many other nineteenth century authors thought before Mill, types and kinds in natural history are classified, and exist, as clusters around exemplars. There never were nor is there any need for there to be biological essences in that sense.

6 Educational Considerations

What does all this mean in terms of teaching biology? One of the most important is that both disciplines are historical themselves, and rely heavily on sociological and cultural context if we are to make sense of them. Biologists do not work in some isolated cultural vacuum, and their ideas about not only what they are studying, but how they view their competitors and predecessors is often based on a kind of triumphalism in which there are the Good Guys who got us to the state of blessed enlightenment we now enjoy (if, for example, you happen to accept the right speciation theory or taxonomic methodology) and those who are the Bad Guys, who don't think what We think and so are retarding progress and knowledge with pre-modern metaphysics and epistemologies. One doesn't have to read far to find these comments made; Hull's (1965) paper and Mayr's *Growth of Biological Thought* (1982) are two classic examples. Science is a human activity, and students need to understand that even the best authorities fall prey to the temptation to be Whiggists trumpeting the modern and denigrating alternative views. Since the arrival of a new theory or result doesn't render past researchers stupid or blind, why should we paint them as fools? Neither the history of a scientific discipline nor the history of a concept will necessarily move from foolish old to clever new science. History is not like development of an organism; there are no predetermined sequences (Wilkins 2009a), and to think that there is we might call the *developmentalist fallacy* in history.[16] Science does not recapitulate cognitive ontogeny, nor do students need to learn science by

[16]Not unrelated to the identically named problem in social history (Dussel 1993), in which the linear idea of history always moves from simple or immature to complex or mature. An example of a developmentalist fallacy can be found in Piagetian "genetic epistemology", which is often take to represent a historical process in individual development.

recapitulating the historical development of the science itself. In teaching the history of science, we are providing context and guarding against simple-minded triumphalism, but a historical narrative is not necessarily a curriculum.

Another educational point to make is the problem of *polysemy*, the fact that even technical words can have multiple loosely connected meanings. That this is more than a point about semantics is clear from the ways in which various inferences are made by appealing to first one, then another disconnected sense of "essentialism". The confusion of logical essentialism with biological essentialism, and both with psychological essentialism is a case in point. Many discussions of race and cognition would be improved by clearly distinguishing these distinct meanings. We would be less inclined to posit racial classifications if we could disentangle ourselves from the idea that race implies property-based class-kinds. Likewise we would be less inclined to think of essentialism as mandated by psychological dispositions if we could clearly separate the idea there is a persistent underlying cause for things being differentiated from diagnostic or definitional essentialism.

Finally, we might attend to some logical fallacies, and in particular, the fallacies of composition and division. The aggregate or average properties of the members of a group do not give the properties of the group, and vice versa. A species like *Homo sapiens* can be two legged and rational without every member of the group being two legged or rational. There can be diagnostic properties for any taxonomic group that doesn't commit us to thinking they are only real groups if every member has them. As obvious as this point is, it is often overlooked by scientists, and even occasionally by philosophers.

It is time to abandon the notion of essentialism and call each kind of conception its own name – psychological essentialism, for example, might better be called "inherentism", philosophical logical essentialism, "definitionalism", and scientific essentialism "elementalism" or some such. That will stop many of the ambiguities and their consequent errors; along with the developmentalist fallacy and the tendency to vilify those whose views do not match some modern consensus, which are often reached by political rather than empirical or theoretical means. Science progresses because there are alternatives, and many alternative views thought long dead can revive and even motivate fruitful research. Considering the influence of now-peripheral and deviant views on science, like neo-Platonism and alchemy on early modern science, formalism in biology, the revival of preformationism in modern genetics, and so on, to denigrate something just because it is old is to commit a terrible and costly mistake. Of course science progresses, and many empirical ideas and avenues are forever closed to us, but philosophical ideas in science deserve to be treated with respect, for one never knows when one will arise and help us out of a hole we have fallen in.

References

Aaron, R.I. 1952. *The theory of universals*. Cambridge: Cambridge University Press.

Agassiz, Louis. 1859. *An essay on classification*. London: Longman, Brown, Green, Longmans and Roberts and Trubner.

Amundson, Ron. 2005. *The changing rule of the embryo in evolutionary biology: Structure and synthesis*, Cambridge studies in philosophy and biology. New York: Cambridge University Press.
Anderson, Erik. 1994. Kant, natural kind terms, and scientific essentialism. *History of Philosophy Quarterly* 11(4): 355–373.
Artigas, Mariano, Thomas F. Glick, and Rafael A. Martínez. 2006. *Negotiating Darwin: The Vatican confronts evolution, 1877–1902*, Medicine, science, and religion in historical context. Baltimore: Johns Hopkins University Press.
Atran, Scott. 1990. *The cognitive foundations of natural history*. New York: Cambridge University Press.
Avise, J.C., and R.M. Ball Jr. 1990. Principles of genealogical concordance in species concepts and biological taxonomy. In *Oxford surveys in evolutionary biology*, ed. D. Futuyma and J. Atonovics, 45–67. Oxford: Oxford University Press.
Baldwin, James Mark (ed.). 1901. *Dictionary of philosophy and psychology including many of the principal conceptions of ethics, logic, aesthetics, philosophy of religion, mental pathology, anthropology, biology, neurology, physiology, economics, political and social philosophy, philology, physical science, and education and giving a terminology in English, French, German, and Italian*. 3 vols. New York/London: Macmillan.
Beckner, Morton. 1959. *The biological way of thought*. New York: Columbia University Press.
Ben Yami, Hanoch. 2001. The semantics of kind terms. *Philosophical Studies* 102(2): 155–184.
Bird, Alexander. 2009. In *Routledge companion to metaphysics*, ed. R.L. Poidevin, P. Simons, A. McGonigal, and R. Cameron, 497–506. Abingdon: Routledge.
Bird, Alexander, and Emma Tobin. 2009. Natural kinds. In *The Stanford encyclopedia of philosophy*, ed. E.N. Zalta. http://plato.stanford.edu/archives/spr2009/entries/natural-kinds/
Boyd, Richard. 1991. Realism, anti-foundationalism and the enthusiasm for natural kinds. *Philosophical Studies* 61: 127–148.
Boyd, Richard. 1999. Homeostasis, species, and higher taxa. In *Species, new interdisciplinary essays*, ed. R. Wilson, 141–186. Cambridge, MA: Bradford/MIT Press.
Campbell, Keith. 1965. Family resemblance predicates. *American Philosophical Quarterly* 2(3): 238–244.
Charles, David. 2002. *Aristotle on meaning and essence*. Oxford: Oxford University Press.
Chung, Carl. 2003. On the origin of the typological/population distinction in Ernst Mayr's changing views of species, 1942–1959. *Studies in History and Philosophy of Biological and Biomedical Sciences* 34: 277–296.
Clarke, Richard F. 1895. *Logic*, Manuals of catholic philosophy, 3rd ed. London: Longmans, Green.
Cohen, S. Marc. 2009. Aristotle's metaphysics. In *The Stanford encyclopedia of philosophy*, ed. E.N. Zalta. http://plato.stanford.edu/archives/spr2009/entries/aristotle-metaphysics/
Cordry, Ben S. 2004. Necessity and rigidly designating kind terms. *Philosophical Studies* 119(3): 243–264.
de Queiroz, Kevin. 1992. Phylogenetic definitions and taxonomic philosophy. *Biology and Philosophy* 7(3): 295–313.
de Queiroz, Kevin. 1994. Replacement of an essentialistic perspective on taxonomic definitions as exemplified by the definition of "Mammalia". *Systematic Biology* 43(4): 497–510.
de Queiroz, Kevin. 2007. Species concepts and species delimitation. *Systematic Biology* 56(6): 879–886.
Devitt, Michael. 2008. Resurrecting biological essentialism. *Philosophy of Science* 75(3): 344–382.
Devitt, Michael. 2010. Species have (partly) intrinsic essences. *Philosophy of Science* 77(5): 648–661.
Dewey, John. 1997. *The influence of Darwin on philosophy and other essays, great books in philosophy*. Amherst: Prometheus Books.
Dupré, John. 2002. Is 'Natural Kind' a natural kind term? *The Monist* 85(1): 29–49.
Dussel, Enrique. 1993. Eurocentrism and modernity (Introduction to the Frankfurt lectures). *Boundary* 220(3): 65–76.

Ellis, Brian David. 2001. *Scientific essentialism*, Cambridge studies in philosophy. Cambridge: Cambridge University Press.
Ellis, Brian David. 2002. *The philosophy of nature: A guide to the new essentialism*. Chesham: Acumen.
Enç, Berent. 1975. Necessary properties and Linnaean essentialism. *Canadian Journal of Philosophy* 5(1): 83–102.
Ereshefsky, Marc. 1999. Species and the Linnean hierarchy. In *Species, new interdisciplinary essays*, ed. R.A. Wilson, 285–305. Cambridge, MA: Bradford/MIT Press.
Ereshefsky, Marc. 2000. *The poverty of Linnaean hierarchy: A philosophical study of biological taxonomy*. Cambridge/New York: Cambridge University Press.
Farber, Paul Lawrence. 1976. The type-concept in zoology during the first half of the nineteenth century. *Journal of the History of Biology* 9(1): 93–119.
Gallie, W.B. 1948. IV. The function of philosophical æsthetics. *Mind* LVII (227): 302–321.
Gasking, Douglas. 1960. Clusters. *Australasian Review of Psychology* 38: 13–18.
Gelman, Susan A. 2003. *The essential child: Origins of essentialism in everyday thought*, Oxford series in cognitive development. Oxford/New York: Oxford University Press.
Gelman, Susan A., and Lawrence A. Hirschfeld. 1999. How biological is essentialism? In *Folkbiology*, ed. D.L. Medin and S. Atran, 403–446. Cambridge, MA: MIT Press.
Ghiselin, Michael T. 1974. A radical solution to the species problem. *Systematic Zoology* 23: 536–544.
Ghiselin, Michael T. 1997. *Metaphysics and the origin of species*. Albany: State University of New York Press.
Gilson, Etienne. 1964. *The spirit of Thomism*. New York: Harper & Row.
Gilson, Etienne. 1984. *From Aristotle to Darwin and back again: A journey in final causality, species, and evolution*. Trans. J. Lyon. Notre Dame: University of Notre Dame Press.
Gil-White, Francisco. 2001. Are ethnic groups biological "species" to the human brain? Essentialism in our cognition of some social categories. *Current Anthropology: A World Journal of the Human Sciences* 42(4): 515–554.
Griffiths, Paul E. 1999. Squaring the circle: Natural kinds with historical essences. In *Species, new interdisciplinary essays*, ed. R.A. Wilson, 209–228. Cambridge, MA: Bradford/MIT Press.
Hacking, Ian. 1990. Natural kinds. In *Perspectives on quine*, ed. R.B. Barrett and R.F. Gibson. Cambridge: Blackwell.
Heyes, Cressida J. 2000. *Line drawings: Defining women through feminist practice*. Ithaca: Cornell University Press.
Hooker, John. 1976. Brody on essentialism. *Philosophical Studies* 29(4): 263.
Hull, David L. 1965a. The effect of essentialism on taxonomy: Two thousand years of stasis. *The British Journal for the Philosophy of Science* 15: 314–326, 316:311–318.
Hull, David L. 1973. A populational approach to scientific change. *Science* 182: 1121–1124.
Hull, David L. 1976. Are species really individuals? *Systematic Zoology* 25: 174–191.
Hull, David L. 1984. Can Kripke alone save essentialism? A reply to Kitts. *Systematic Zoology* 33: 110–112.
Hull, David L. 1988. *Science as a process: An evolutionary account of the social and conceptual development of science*. Chicago: University of Chicago Press.
Jensen, U.J. 1984. A critique of essentialism in medicine. In *Health, disease and causal explanations in medicine*, ed. L. Nordenfelt and B.I.B. Lindahl, 63–73. Dordrecht: D. Reidel Publishing.
Jevons, William Stanley. 1870. *Elementary lessons in logic: Deductive and inductive: With copious questions and examples, and a vocabulary of logical terms*, Macmillan's school class books. London: Macmillan.
Jevons, William Stanley. 1878. *The principles of science: A treatise on logic and scientific method*, 2nd ed. London: Macmillan. Original edition, 1873.
Kathrin, Koslicki. 2008. Natural kinds and natural kind terms. *Philosophy Compass* 3(4): 789–802.
Kripke, Saul A. 1980. *Naming and necessity*. Cambridge, MA: Harvard University Press.

Kronfeldner, Maria. 2007. Is cultural evolution Lamarckian? *Biology and Philosophy* 22(4): 493–512.
LaPorte, Joseph. 2004. *Natural kinds and conceptual change*. Cambridge: Cambridge University Press.
Levit, Georgy S., and Kay Meister. 2006. The history of essentialism vs. Ernst Mayr's "Essentialism Story": A case study of German idealistic morphology. *Theory in Biosciences* 124: 281–307.
Locke, John. 1997 [1690]. An essay concerning human understanding. In *Penguin classics, classics series, penguin philosophy*, ed. R.S. Woolhouse. London: Penguin.
Love, Alan. 2009. Typology reconfigured: From the metaphysics of essentialism to the epistemology of representation. *Acta Biotheoretica* 57(1): 51–75.
Maritain, Jacques. 1955. *Bergsonian philosophy and Thomism*. New York: Philosophical Library.
Matthews, Gareth B. 1990. Aristotelian essentialism. *Philosophy and Phenomenological Research* 50: 251–262.
Mayr, Ernst. 1982. *The growth of biological thought: Diversity, evolution, and inheritance*. Cambridge, MA: The Belknap Press of Harvard University Press.
Mayr, Ernst. 1988. *Toward a new philosophy of biology: Observations of an evolutionist*. Cambridge, MA: The Belknap Press of Harvard University Press.
Mayr, Ernst. 1991. *One long argument: Charles Darwin and the genesis of modern evolutionary thought*. Cambridge, MA: Harvard University Press.
Mayr, Ernst. 1992. Speciational evolution or punctuated equilibria. In *The dynamics of evolution*, ed. A. Somit and S. Peterson, 21–48. New York: Cornell University Press.
Medin, Douglas L., and Andrew Ortony. 1989. Psychological essentialism. In *Similarity and analogical reasoning*, ed. S. Vosniadou and A. Ortony. New York: Cambridge University Press.
Medin, Douglas L., Elizabeth B. Lynch, and Karen O. Solomon. 2000. Are there kinds of concepts? *Annual Review of Psychology* 51(1): 121–147.
Nelson, Gareth J., and Norman I. Platnick. 1981. *Systematics and biogeography: Cladistics and vicariance*. New York: Columbia University Press.
Numbers, Ronald L. 2006. *The creationists: From scientific creationism to intelligent design*. Expanded ed. Cambridge, MA/London: Harvard University Press.
Okasha, Samir. 2002. Darwinian metaphysics: Species and the question of essentialism. *Synthese: An International Journal for Epistemology Methodology and Philosophy of Science* 131(2): 191–213.
Parsons, Terence. 1969. Essentialism and quantified modal logic. *Philosophical Review* 78(1): 35–52.
Paul, Harry W. 1979. *The edge of contingency: French Catholic reaction to scientific change from Darwin to Duhem*. Gainesville: University Presses of Florida/A University of Florida Book.
Peterson, Philip L. 1999. The meanings of natural kind terms. *Philosophia* 27: 137–176.
Pickering, Neil. in press. Extending disorder: Essentialism, family resemblance and secondary sense. *Medicine, Health Care and Philosophy*.
Pigliucci, Massimo. 2003. Species as family resemblance concepts: The (dis-)solution of the species problem? *Bioessays* 25(6): 596–602.
Pigliucci, Massimo, and Jonathon Kaplan. 2006. *Making sense of evolution: The conceptual foundations of evolutionary biology*. Chicago/London: University of Chicago Press.
Popper, Karl R. 1945. *The open society and its enemies*. 2 vols. London: G. Routledge & Sons.
Popper, Karl R. 1957. *The open society and its enemies*, 3rd ed. London: Routledge and Kegan Paul.
Popper, Karl R. 1959. *The logic of scientific discovery*. Trans. K. Popper, J. Freed and L. Freed. London: Hutchinson.
Putnam, Hilary. 1975. *Mind, language, and reality*, His philosophical papers, vol. 2. Cambridge/New York: Cambridge University Press.
Quine, Willard Van Orman. 1951. Two dogmas of empiricism. *Philosophical Review* 60: 20–43.
Quine, Willard Van Orman. 1953a. *From a logical point of view: 9 logico-philosophical essays*. Cambridge, MA: Harvard University Press.

Quine, Willard Van Orman. 1953b. Three grades of modal involvement. In *Actes du XIème Congrès International de Philosophie* XIV: Volume complémentaire et communications du Colloque de Logique, 65–81, Republished in 1966. The ways of paradox, and other essays. New York: Random, 158–176.
Quine, Willard Van Orman. 1969. Natural kinds. In *Essays in honour of Carl G. Hempel: A tribute on the occasion of his sixty-fifth birthday*, ed. N. Rescher, 5–27. Dordrecht: Springer.
Richards, Richard A. 2010. *The species problem: A philosophical analysis*, Cambridge studies in philosophy and biology. Cambridge: Cambridge University Press.
Rieppel, Olivier. 2010. New essentialism in biology. *Philosophy of Science* 77(5): 662–673.
Riggs, Peter J. 1996. *Natural kinds, laws of nature and scientific methodology*, Australasian studies in history and philosophy of science, vol. 12. Dordrecht/Boston: Kluwer Academic.
Sankey, Howard. 1997. Induction and natural kinds. *Principia* 1(2): 239–254.
Scriven, Michael. 1959. The logic of criteria. *The Journal of Philosophy* 56(22): 857–868.
Sesardic, Neven. 2010. Race: A social destruction of a biological concept. *Biology and Philosophy* 25(2): 143–162.
Shtulman, Andrew. 2006. Qualitative differences between naïve and scientific theories of evolution. *Cognitive Psychology* 52(2): 170–194.
Shtulman, Andrew, and Laura Schulz. 2008. The relation between essentialist beliefs and evolutionary reasoning. *Cognitive Science: A Multidisciplinary Journal* 32(6): 1049–1062.
Sober, Elliott. 1980. Evolution, population thinking, and essentialism. *Philosophy of Science* 47: 350–383.
Sober, Elliott. 1994. *From a biological point of view: Essays in evolutionary philosophy*, Cambridge studies in philosophy and biology. Cambridge/New York: Cambridge University Press.
Sokal, Robert R., and P.H.A. Sneath. 1963. *Principles of numerical taxonomy*, A series of books in biology. San Francisco: W. H. Freeman.
Sterelny, Kim. 1983. Natural kind terms. *Pacific Philosophical Quarterly* 64: 110–125.
Stone, Alison. 2004. Essentialism and anti-essentialism in feminist philosophy. *Journal of Moral Philosophy* 1(2): 135–153.
Templeton, Alan R. 1989. The meaning of species and speciation: A genetic perspective. In *Speciation and its consequences*, ed. D. Otte and J. Endler, 3–27. Sunderland: Sinauer.
Thompson, D'Arcy Wentworth. 1917. *On growth and form*. Cambridge: Cambridge University Press.
Vasilyeva, Larissa N., and Steven L. Stephenson. 2012. The hierarchy and combinatorial space of characters in evolutionary systematics. *Botanica Pacifica. A Journal of Plant Science and Conservation* 1(1): 21–30.
Walsh, Denis. 2006. Evolutionary essentialism. *The British Journal for the Philosophy of Science* 57(2): 425–448.
Wasmann, Erich. 1910. *Modern biology and the theory of evolution*. Trans. A.M. Buchanan. 3rd ed. London: Kegan Paul, Trench, Trübner (Original edition, 1906).
Whately, Richard. 1875. *Elements of logic*. Ninth (octavo) ed. London: Longmans, Green & Co (Original edition, 1826).
Whewell, W. 1840. *The philosophy of the inductive sciences: Founded upon their history*. vol. 2. London: John W. Parker.
White, Morton Gabriel. 1965. *Foundations of historical knowledge*. New York: Harper & Row.
White, Nicholas P. 1972. Origins of Aristotle's essentialism. *The Review of Metaphysics* 26(1): 57–85.
Wiggins, David. 1974. Essentialism, continuity, and identity. *Synthese* 28(3–4): 321.
Wilkerson, T.E. 1988. Natural kinds. *Philosophy* 63: 29–42.
Wilkins, John S. 2007. The concept and causes of microbial species. *Studies in History and Philosophy of the Life Sciences* 28(3): 389–408.
Wilkins, John S. 2009a. Darwin. In *A companion to the philosophy of history and historiography*, ed. A. Tucker, 405–415. Chichester: Wiley-Blackwell.
Wilkins, John S. 2009b. *Species: A history of the idea*, Species and systematics. Berkeley: University of California Press.

Wilkins, John S. 2010. What is a species? Essences and generation. *Theory in Biosciences* 129: 141–148.
Wilkins, John S. 2013. Biological essentialism and the tidal change of natural kinds. *Science & Education* 22(2): 221–240.
Wilson, Robert A. 1999. Realism, essence, and kind: Resuscitating species essentialism? In *Species, new interdisciplinary essays*, ed. R.A. Wilson, 187–208. Cambridge, MA: Bradford/ MIT Press.
Winsor, Mary Pickard. 1979. Louis Agassiz and the species question. *Studies in History of Biology* 3: 89–117.
Winsor, Mary Pickard. 2003. Non-essentialist methods in pre-Darwinian taxonomy. *Biology and Philosophy* 18: 387–400.
Winsor, Mary Pickard. 2006a. The creation of the essentialism story: An exercise in metahistory. *History and Philosophy of the Life Sciences* 28: 149–174.
Winsor, Mary Pickard. 2006b. Linnaeus' biology was not essentialist. *Annals of the Missouri Botanical Garden* 93(1): 2–7.
Witmer, D.Gene, and John Sarnecki. 1998. Is natural kindness a natural kind? *Philosophical Studies* 90(3): 245–264.
Wittgenstein, Ludwig. 1968. *Philosophical investigations*. Trans. G.E.M. Anscombe. Repr. of [3rd ed.] English text, with index. ed. Oxford: Basil Blackwell.
Zalta, Edward N. 1988. *Abstract objects: An introduction to axiomatic metaphysics*. Dordrecht: Reidel.

Biological Teleology: The Need for History

James G. Lennox and Kostas Kampourakis

1 Introduction

For the purposes of this chapter, we characterize a teleological explanation as one in which some property, process or entity is explained by appealing to a particular result or consequence that it may bring about.[1] For example, when you return from a visit to the market someone might ask you "Why did you go to the market?" If you answer, "In order to buy milk", you are explaining the process of going to the market by pointing to something that was a consequence of that process, buying milk. Or, someone might ask you "Why do hawks and owls have sharp, hooked beaks and talons?" If you answer "Because those sharp talons and the hooked beaks allow them to capture and eat their prey", you are explaining these traits by reference to the (valuable) consequences for the organism of having those traits.

Teleological explanations have played a central role throughout the history of the life sciences. Having an accurate understanding of that history is of the first importance for biology educators. Most children enter the classroom already prepared

[1] We are thus endorsing the explication of teleological explanation defended by Larry Wright (in Wright 1976). Wright argues that teleological explanations of both goal directed behavior and functions have the same logical form, which he terms 'consequence etiology'. He argues that teleological explanations are a form of causal explanation, and that it is a mistake to explicate functions by appealing to past selection. In all these respects his view is in stark contrast to that of Millikan's defense of "proper functions" (1984), as Millikan was well aware, and with Neander (1991). For a detailed critique of Millikan's and Neander's approach, see Lennox (2010).

J.G. Lennox (✉)
Department of History and Philosophy of Science,
University of Pittsburgh, Pittsburgh, PA, USA
e-mail: jglennox@pitt.edu

K. Kampourakis
Secretariat of Educational Research and Development, Geitonas School,
Vari Attikis, 16602, Greece
e-mail: konstantinos.kampourakis@gmail.com; kamp@geitonas-school.gr

K. Kampourakis (ed.), *The Philosophy of Biology: A Companion for Educators*, History, Philosophy and Theory of the Life Sciences 1, DOI 10.1007/978-94-007-6537-5_20,

to explain the traits of living things in teleological terms—there is a reason why Rudyard Kipling's Just So Stories are appealing to them! A young person's questions about why animals have the many strange and fascinating characteristics they do are often answered by appealing to the adaptive value of those characteristics, as if identifying the adaptive value of a characteristic explains why the animal has it. It is only much later, if ever, that our better students begin to puzzle about how that could possibly be, and about what sort of evidence would support such an explanation (Keil 1992; Kelemen 1999a, b, c; Southerland et al. 2001; Evans 2008; Kampourakis and Zogza 2008; Kampourakis et al. 2012a, b; Kelemen 2012; Kampourakis 2013).

Biological textbooks often explain adaptations by reference to natural selection in language that sounds suspiciously teleological, e.g. 'That color pattern is present in the males of that population of fish because it increases their attractiveness to female mates without increasing their visibility to predators.'[2]

Moreover, explanations that at least appear to be teleological are not restricted to the observable, phenotypic adaptations of vertebrate behavior. Notice the explanatory structure implicit in the following quotation from Albert Lehninger's *Bioenergetics: The Molecular Basis of Biological Energy Transformations* (Lehninger 1971, p. 110).

> Thus photo-induced cyclic electron flow has *a real and important purpose*, namely, *to* transform the light energy absorbed by chlorophyll molecules in the chloroplast into phosphate bond energy. (emphasis added)

A common response to passages like this is to say that their use of term such as 'purpose' is not evidence of a philosophical commitment to teleology. We will address that response directly later, but we first want to discuss the context of the above quotation. That context shows the importance of teleological inquiry for research purposes, even at the molecular level.

The above passage occurs during Lehninger's explication of photosynthesis, and specifically during his explanation of the conversion of the photo-electric energy absorbed by chlorophyll into chemical energy. He briefly recounts the research that lead to the discovery of *non-cyclic* electron flow, which results in the reduction of one form of NADP to another, which in turn is crucial in converting carbon dioxide to sugar. But by the 1950s a series of experiments had established a related process that was puzzling, because it did not have any obvious value for plants. To quote Lehninger again:

> …the question now arises: If illumination of chloroplasts causes electrons to flow out of the excited chlorophyll molecule, around the chain, and then back to the chlorophyll again, thus restoring the normal number of electrons in the latter, *what has been accomplished* by cyclic electron flow? What can be *the purpose* of this process? (Lehninger 1971, p. 109)

[2]Based on the iconic study of balanced mate and predator selection in the wild of John Endler carried out over many years on populations of Caribbean guppies. Cf. Endler (1983, 1989). Compare the following summary comments from Grant and Grant (1989): "Is it a matter of chance who survives to breed and who does not, who reproduces once and who reproduces many times? Or do some birds succeed because they are better equipped than others to exploit the environment and avoid its hazards?"

These questions are, of course, just those that the first quotation is intended to answer. And it is these questions that drove the research that came up with that answer in the first place. For the problem is that we have a very complicated biochemical process going on within photosynthesizing plants that performs no apparent function for them. It is not just that there is no *product* of the process; it is that energy is expended simply to return the system to its initial state. Such a result drives further research *only* if it is assumed that expenditure of energy by an organism occurs *because it serves some end or goal*. Now let us return to the worry that this is merely the convenient, expository language of a textbook, intended simply to help the reader understand a difficult biochemical process.

Let us take this thought seriously. What language, we might then ask, should we hardheaded, twenty-first century scientists substitute for the language of purpose in the above quotations? It will not be the language of mechanism, since the mechanisms are described, in uncompromisingly mechanistic terms, in the very pages where questions are raised about the *purposes* of those mechanisms. It was only after researchers had a very clear idea of the mechanisms involved in each step of the process that they became puzzled about their purpose. We will begin by seeing the origins of a basic fault line created in Ancient Greece in the fourth century BCE when Aristotle decides, in deliberate opposition to Plato, to defend a *natural* teleology, free of the idea that the natural world is the creation of a divine, rational being of some sort, with a plan for his creation. We will argue that the *philosophical* debate over teleological explanation in natural science during the Scientific Revolution was primarily between those who, under Platonic influence, defended a theistic, creationist teleology and those who, for a wide variety of reasons, opposed the use of any sort of teleology in natural science; while the effective *scientific use* of teleological explanation was bearing fruit in the disciplines of anatomy, physiology and medicine, where the natural teleology championed by Aristotle was alive and well. The theistic teleology of that era formed the basis for 'natural theology', the idea that the scientific search for purpose and design in nature was justified because it would assure us of God's goodness and his plan for the Creation. This, of course, was the approach to the study of nature that Darwin admired in the writings of Rev. William Paley which he studied while at Cambridge University, and which his theory of evolution by natural selection was to challenge. And yet, as we shall see, Darwin refused to throw the teleological baby out with the theological bathwater—Darwin happily accepted the praise of those who saw his theory as one that provided naturalistic, teleological explanations for organic adaptations.[3]

Understanding teleology is important for biology education as it is a major conceptual obstacle for understanding biology in general and evolution in particular. Many studies have documented that secondary students, and undergraduates, tend to provide teleological explanations for the origin of biological features. However, it should be noted that researchers have not always explicitly distinguished teleological

[3]Interestingly, Darwin was both praised and blamed by his contemporaries both for promoting and for eliminating teleology (Beatty 1990).

explanation from other types, and although it was found that students tended to explain the origin of biological features in terms of goals or needs, their preconceptions were characterized as Lamarckian, a characterization that actually masks their teleological nature (Kampourakis and Zogza 2007). For example during the pre-test of one study 35 out of 50 secondary students provided teleological explanations for evolution. In particular, these students explained the appearance of traits as the consequence of the animals' requirements for survival, although many of the explanations contained the words *need* or *necessary* (Settlage 1994). In a study with college students, it was found that most of them understood evolution as a process in which species respond to environmental conditions by changing gradually over time and they attributed changes in traits to need-driven adaptive processes (Bishop and Anderson 1990). In another study with college students it was found that their knowledge about evolution before instruction was limited and mixed, and the most common preconceptions were related to teleology (Jensen and Finley 1996). In a very interesting study with students of various ages that explored the patterns that characterize students' explanations of biological phenomena, teleological explanations were the most prevalent category in all grade levels. However, it seemed that the older were the students, the less were the teleological explanations given (Southerland et al. 2001). Finally, in a recent study of 14–15 year old students' intuitive explanations of evolution, it was found that in most cases teleological explanations predominated, with the end or goal being the survival of the species. It was found that in general students had the tendency to provide explanations referring to some purpose or plan when they did not have adequate information (Kampourakis and Zogza 2008).[4]

Instruction about evolution usually takes place for the first time in secondary settings. Secondary evolution instruction might be more effective if children's evolution-related preconceptions (like teleological ones) were diagnosed and addressed during elementary school. Rather than being an undesirable stage in their conceptual development, children's preconceptions are a necessary step (Carey 2000). If teachers neglect them, or are unaware of them because they are not always expressed, children's misunderstandings may become deeper as they grow up, despite their science content knowledge. On the other hand, if children's teleological intuitions were effectively addressed in elementary school, evolution instruction in secondary settings might be more effective. Drawing on the historical analysis of teleological explanations of Sect. 2 and a review of literature from conceptual development research on children's teleological explanations in Sect. 3, in Sect. 4 we propose questions for further research that should be undertaken in order to better understand children's intuitive teleological explanations.

---

[4]It should be noted that teleological explanations are not restricted to biology, but may be given for chemical phenomena as students may think that the behavior of a system is driven by intrinsic purposes (Talanquer 2007). High school students may consider that atoms react in order to form molecules because they need to achieve a full outer shell as a sufficient explanation for chemical reactions (Taber and Watts 1996). Similarly, in the case of physics high school students may believe they can predict which of the objects would be hotter than the others by thinking on the basis of their use, and not based on the properties of their structure (Harrison et al. 1999).

2 Teleology in the History and Philosophy of Science

2.1 Rational Design or Natural Function: The Argument Begins

Plato's dialogue *Phaedo* is best known because its scene is set around Socrates' decision not to accept exile from Athens as his punishment for impiety, but to accept execution—and to console his friends by mounting an argument for the immortality of the soul. In the midst of his discussion of that topic, however, Socrates recounts that in his early youth he was attracted to the investigation of nature, but was disappointed by most of what he read. He then came across Anaxagoras' argument that Nous was responsible for all things, and that gave him great hope:

> For I never supposed that someone who said these things [facts about nature] to be ordered by Intelligence would offer any other cause for them than that these things are best just as they are. (97a7–b1)

Here we see a teleological pattern of explanation that rests on what we will refer to as the *Socratic Assumption*.

Socratic Assumption: If a natural object has a property due to an intelligent agent, then it has that property because it is best that it have it.

On this assumption, Intelligent agency is not merely *purposeful*, but *normative* in nature. In his later dialogues *Timaeus* and *Laws*, Plato explicitly starts by assuming the truth of the antecedent of this assumption. It is of great importance, however, to note that the Socratic Assumption does not underwrite the 'argument from design'. The Socratic Assumption underwrites inferences about a product, assuming it was produced by an intelligent agent. But it does not underwrite the inference that if you come across something that is complex, useful, or good, you can infer it was produced by an intelligent agent. In *Laws* X, in opposition to those who appeal to necessity and chance to explain natural phenomena, the Athenian insists on the priority of "judgment and foresight, wisdom, art and law" to "hard and soft, heavy and light":

> …and the great and primary works and actions just because they are primary, would be those of art; while nature and things that are by nature—which they incorrectly name in this way—would be later, being governed by art and intelligence. (892b5–8)

And in the *Timaeus*, the normative implications of art and intelligence governing and having priority over nature are made explicit:

> For the God, wishing all things to be good and nothing to be bad in so far as possible, took over everything which was visible—not at rest but moving in a discordant and disorderly manner—and led it from disorder to order, judging this to be in all respects better. (*Timaeus* 30a2–5)[5]

[5]The Timaeus is the best known and historically most important statement by Plato of the view that the natural world is the product of an intelligent agent acting to achieve what is best, but it is actually quite pervasive in the middle and later dialogues. Compare: Republic VII 530a6; X 596c4; Laws X 889–906; Philebus 26e5; Sophist 262b5–c4.

In doing so, however, the Demiurge achieves the good by placing soul and intelligence in the natural world of changing things. The materials in nature are cooperative causes that are persuaded by intelligence to act and interact in order to achieve the good (cf. 46c–e, 76c–d).[6]

> [God] made use of causes of this sort as subservient, while he himself contrived the good in all things that come to be. We must accordingly distinguish two kinds of causal account, the necessary and the divine. (68e4–7)
>
> We must speak of both kinds of causes but separate those which, with intelligence, are craftsmen of fine and good things, from those which in the absence of foresight, produce their sundry effects at random and without order. (46e3–6)[7]

In Plato's *Timaeus* we catch a glimpse of one, historically influential view about teleology. The natural world is ultimately the product of a divine intelligence acting to produce the best possible natural world, according to some divine standard of goodness. In Plato's case that standard is clearly mathematical. For the Divine Craftsman's first task is to fashion a harmonically proportioned world soul, which "is invisible and partakes of reason and harmony, and, being made by the best of intellectual and everlasting beings, becomes the best of generated things" (36e6–37a2). His next task is to insure that the four material elements are composed of geometric solids, which are in turn composed of triangles (cf. 53a–56c). This task is described at the outset as God fashioning the elements "by form and number" (53b) and at its conclusion in the same terms:

> And the ratios of their numbers, motions and other properties, everywhere God, as far as necessity allowed or gave consent, has exactly perfected and harmonized in due proportion. (56c3–7)

Notice again, however, God is far from omnipotent—necessity must give consent. The Demiurge imposes as much order on the world of 'becoming' as possible.

Aristotle too defends the view that to understand fully the natural world requires an appeal to teleology, and he further believes that those features of the natural world that are present for the sake of some end, or serve some function are good, according to an appropriate standard. But there the similarities end, and are massively outweighed by the dissimilarities. For Aristotle, it is almost exclusively in the context of explaining why animals (including human beings) have the parts they do, behave as they do, or develop as they do, that teleology is deployed. Animals develop as they do for the sake of becoming complete living organisms,[8] have the

[6]For a detailed account of Plato's teleology, see Lennox (1985), Johansen (2004); note 33 of Lennox (1985) provides a complete list of the passages that offer teleological explanations (see also Lennox 2001).

[7]Compare: "Reason overruled Necessity by persuading her to guide the greatest part of things that become towards the best [...] If, then, we are really to tell how it came into being on this principle, we must bring in also the wandering cause—in what manner its nature is to cause motion." (47e5ff; Johansen trans.).

[8]Three different interpretations of Aristotle's natural teleology can be found in the contributions of John Cooper, David Balme and Allan Gotthelf to Gotthelf and Lennox 1987.

parts that they do in order to perform the activities they need to perform in order to live, and behave as they do to achieve the goals they need to achieve in order to survive.

We can explore this radically naturalistic approach to teleology conveniently within his small work devoted to explaining the differences in the way animals move from place to place, *De incessu animalium* (*IA*). *IA*'s first chapter surveys the questions it is to answer—e.g. "why blooded animals are moved by means of four contact points while the bloodless by more, and generally why some animals are footless, some two-footed, some four-footed and some many footed" (704a11–13). He concludes this introduction with the following words:

> Concerning all these questions and any others that are akin to them, we must study the causes. For *that* (*hoti men*) these things happen to be this way is clear from our natural history; but as to the reason why (*dioti de*), that we must now investigate. (704b8–10)

A systematic study of all the different forms of animal locomotion, and what other differences those differences are correlated with, is the task of a preliminary investigation which Aristotle calls *historia*, from which our word 'history' derives. In *IA* he is assuming we have that information, and are now laying out the results of an investigation into causes. That investigation begins by specifying three *principles* (ἀρχαί)[9] which we are accustomed to use often as posits or *suppositions* (ὑποθέμενοι) in natural *inquiry* (τὴν μέθοδον τὴν φυσικήν). The first of these three principles echoes down through the centuries:

> One of these principles is that nature does nothing in vain, but among the possibilities always does what is best for the being of each kind of animal, so that, if it is *better* in a certain way [for an animal's being], that is also how it is according to nature. (704b14–17)

On the question of what grounds teleological explanation in natural science, then, the differences between Plato and Aristotle are stark. In Aristotle's statement of this principle of natural teleology, there is no hint of the Socratic Assumption that to explain something as present because it is best presupposes rational agency. Moreover, the standard by which one judges that what nature has produced is best is the being of each kind of animal, not some rational agent's plan. Finally, what is better is not determined by reference to a non-natural standard, such as transcendent mathematical forms—what is better is what is in accordance with nature, presumably what contributes to or is part of the nature of the animal in question. This essentially eliminates the possibility that there is some overall good for the natural world. What is good for each different kind of animal will be determined by an examination of the requirements of its life. This becomes explicit in a later passage where this principle is once again invoked:

> The absence of feet in the snakes is both because nature produces nothing in vain, but in all cases [produces] with a view to what is best among the possibilities for each [animal],

---

[9]The other two principles are that there are three axes of directionality (up/down, front/back, left/right) and that all locomotion depends on push/pull mechanics. (704b18–705a2)

> preserving the distinctive being of each thing and its 'being what it is' (διασώζουσαν ἑκάστου τὴν ἰδίαν οὐσίαν καὶ τὸ τί ἦν αὐτῷ εἶναι), and also because of what was said previously, that none of the blooded animals is able to move by means of more than four contact points. (*IA* 8. 708a9–13)

Here, in the context of explaining why snakes lack limbs for motion on land, Aristotle spells out what he means when he says that the nature of the animal produces what is best for it among the possibilities—what is best is what preserves the distinctive being and essence of that particular kind of animal. In this case, it is Aristotle's view that snakes are among the general class of blooded animals (which corresponds closely to what we would refer to as vertebrates), and a universal feature of that kind is that they move by means of only four contact points—typically, four limbs. However, because of the distinctively long, slender nature of snakes, it would not be best for them to have limbs, given that they are limited to four. So instead, it is best that they move in a particular undulating manner.[10]

Plato and Aristotle, then, lay down two distinct defenses of teleological explanation, one that rests on an assumption of intelligent design and one that argues for certain natural processes being inherently goal-directed. In Platonic teleological explanations whatever is explained in teleological terms is the product of rational design and it was made in order to fulfill the intentions of its creator. In this case, the cause of the existence of an object and/or its features is the intention of its creator to fulfill a particular purpose. In contrast, in Aristotelian teleological explanations whatever is explained in teleological terms exists because it is useful to its possessor. In this case no appeal to design and to the intentions of a rational designer is necessary.[11] The cause of the existence of a particular feature is its usefulness to its possessor. It is important to *underscore* these differences since they are often downplayed. One particularly clear example of an attempt to do so is David Sedley's claim, in his *Creationism and its Critics in Antiquity*, that (and we quote) "Aristotle's teleological worldview [i]s a reasoned modification of Plato's creationism" (Sedley 2007, p. 167). That is an odd claim, especially in light of Sedley's reminder, two

[10]The details of Aristotle's account of the locomotion of snakes is not important for our purposes, but we will mention that he does not think they violate the four contact point rule, since they move by bending in two places in an S configuration and move by alternately moving one side of one bend and the opposite side of the other.

[11]A referee made the surprising assertion that "the cited primary texts seem to contain Aristotle's musings on rationality constraints on optimal design by an anthropomorphized 'Nature' to which full intentionality and agency as a designer seem to be attributed." This is astounding. It will be noticed that Aristotle never mentions rationality, design, intentionality–nor does he capitalize nature. The reference of course is to the formal nature (i.e. its living capacities) of the animal (see Lennox 1997). This is simply a confession by the referee that he cannot imagine someone holding the views that living things have goal directed natures that act for ends unless he also believes that this nature is a rational designing agent. A pithy statement of Aristotle's view, from his Physics II. 8 is: "Since the nature of a thing is twofold, on the one hand as its matter and on the other as its form, and since this latter is a goal, and the materials are for the sake of this, this formal nature would be the cause in the sense of that for the sake of which." (199a30–32). Material features of living things are present for the sake of their form. Formal natures in that sense ground teleological explanations for the parts of animals, i.e. their material natures.

pages later, that Aristotle does not believe the world was created, but believes it to be eternal.

Aristotle's deployment of teleological explanation in biology is quite circumscribed and thoroughly naturalistic. He takes it as an empirically basic fact that living things are naturally organized to be well suited to their ways of life, and have formal natures that act for the sake of their coming to be (development) and their being (self-maintenance) so organized. But he also thinks that biological investigation that is not aimed at answering teleological questions is headed in the wrong direction at the outset—teleology thus plays a role in guiding research as well as in scientific explanation, as the following passage from his essay *On Respiration* makes clear.

> The main reason why they [previous natural philosophers] do not speak well about these things is, on the one hand, that they lack experience of the internal parts, and on the other hand, that they do not grasp that nature always acts for the sake of something; for had they inquired for the sake of what respiration belongs to animals, and had they investigated this in the presence of the parts, i.e. gills and lungs, they would quickly have discovered the cause. (*On Respiration* 3, 471b24–29)

The Ancient world thus provides the Renaissance with two very different models of teleological explanation.[12] In the next section, looking toward Darwinism and evolutionary biology, we will argue that, for better or worse, it was the Platonic, not the Aristotelian model that came to dominate, in the form that came to be known as natural theology.

2.2 *Teleology in the (Early) Modern Era*

It is generally acknowledged that Medieval philosophy was dominated by attempts to align either Platonic or Aristotelian philosophical principles with theological principles found in the Islamic and Judeo-Christian religious traditions. It is often assumed, however, as we noted in the introduction, that the dominant position of the seventeenth century was the *rejection* of teleology, and that the adoption of a mechanical understanding of the universe stood in opposition to teleology. Both assumptions are questionable.

Here we will argue that the seventeenth century saw a debate over teleology that was fought primarily on *epistemological* grounds, with a significant role being played by views about our *epistemic access* to God's purposes for his creation. Opposing attitudes toward teleology as an acceptable mode of explanation in natural philosophy were expressed, by René Descartes and Robert Boyle, for example; and that opposition depends to a great extent on different views about our epistemic access to divine intentions.

[12] It can quite fairly be complained that we are leaving the great Greek physician Galen of Pergamum, out of the discussion. And while there are subtle differences in his defense of teleology compared with either Plato or Aristotle, a case can be made that those differences are a result of him attempting to synthesize their approaches to teleology (Hankinson 2008).

We can see this clearly in a passage in *The Wisdom of God Manifested in the Works of Creation*, written by John Ray.[13] After attacking two 'atheistical' approaches to nature (besides modern followers of Epicurus and Lucretius, he includes, interestingly, one form of Aristotelianism, on the grounds that it rejects the idea that the cosmos was created), Ray turns to Descartes. Citing passages in the *Meditations*, *Principles* and letters to Gassendi, he argues that:

> Mons. Des Cartes and his Followers [...] endeavour to disarm us of this decretory Weapon [i.e. final causes]. [...] And this they do, First, By excluding and banishing all Consideration of final Causes from Natural Philosophy, upon Pretence, that they are all and every one in particular *undiscoverable by us; and that it is Rashness and Arrogance in us to think we can find and be partakers of his Counsels.* (Ray 1691, p. 38, emphasis added)

He then claims that this 'false and evil Consequence' has been adequately dealt with by "that excellent Person, Mr. Boyle, in his *Disquisition about the final Causes of Natural Things*",[14] from which he borrows a long quote, discussing the exquisite way in which the eye has been designed for seeing, fine tuned in each species for its particular needs. During this passage, Boyle comments:

> It must needs be highly absurd and unreasonable to affirm, either that it was not Designed at all for this Use, *or that it is impossible for Man to know whether it was or not.* (quoted in Ray 1691, p. 40, emphasis added)

Ray, then, provides a good deal of evidence in favor of the view that the dispute among mechanical philosophers about final causes in nature is to a significant extent about the transparency of God's purposes to us.

Robert Boyle's place in this debate is critical, since it shows that a committed mechanical philosopher had no problem at all integrating that worldview with teleology. Indeed, he argues for the compatibility at the outset, as part of his argument against Descartes.[15] And he is fond of using mechanical analogies to support claims about the necessity of appealing to a Rational Agent in account for the organs of animals, a style of argument that becomes the stock and trade of Natural Theology to the present day.

> And though I keep by me some curious ones, yet I never saw an inanimate production of nature, or as they speak, of chance, whose contrivance was comparable to the meanest limb of the despicablest animal; and there is incomparably more art expressed in the structure of a dog's foot than in that of the famous clock of Strasbourg. (*Disquisition*, sec. II)[16]

---

[13]The first edition was published in London in 1691. This work is often considered the founding document of the tradition that comes to be labeled natural theology, though it is heavily dependent in both style and argument on Boyle's Disquisition on the Final Causes of Natural Things (London 1688), to which Ray here refers.

[14]Indeed, the entire first section of Boyle's Disquisition is devoted to criticizing the position of Cartesians and Epicureans on teleology.

[15]Disquisition (Complete Works V, p. 397–98).

[16]This is a reference to the second such clock, completed in 1574, which was replaced by a third in the 19th century. Similarly: "The Situations of the Coelestial Bodies, do not afford by far so clear and cogent Arguments, of the Wisdom and Design of the Author of the World as do the bodies of animals and Plants, [...] the Eye of the Fly is (at least as far as appears to us) a more curious piece of Workmanship, than the Body of the Sun." Disquisition, p. 404. For a full discussion of Boyle's preference for evidence drawn from living nature, see Lennox (1983).

Boyle's teleology is in the Platonic tradition, then, both because he embraces teleology in nature as a consequence of God's goodness, and because he conceives of God as a craftsman.

> The provident Δημιουργὸς wisely suited the fabric of the parts to the uses, that were to be made of them: as a mechanic employs another contrivance of his wheels, pinions, etc., when he is to grind corn with a mill. (*Disquisition* 409; cf., p. 414)

Boyle and Ray, then, both defend the validity of teleological explanation in the study of living things especially, and ground such explanations on the assumption that adaptive design of complex organs is evidence of a rational designer.

Many prominent thinkers in the seventeenth century, however, opposed the appeal to final causes in natural philosophy. Best known, perhaps, is the pronouncement in Descartes' *Principles of Philosophy*, the target of Ray's and Boyle's responses:

> When dealing with natural things we will, then, never derive any explanation from the purposes which God or nature may have had in view when creating them <and we shall entirely banish from our philosophy the search for final causes>. For we shall not be so arrogant as to suppose that we can share in God's plans. (*Principles* I. 28; compare III. 2)[17]

This seems quite clearly an epistemological proscription on the appeal to final causes. Notice that initially Descartes allows for purposes either of God or nature; however, the explanation for this proscription is that we should not think we are privy to God's plans, and that shows that the references to nature's purposes must rest on the assumption of a nature operating according to divine ordination.

On quite different grounds, Spinoza, in the Appendix to book one of the *Ethics*, rejects appeal to final causes equally forcefully:

> There is no need to show at length, that nature has no particular goal in view, and that final causes are mere human figments. This, I think, is already evident enough, both from the causes and foundations on which I have shown such prejudice to be based, and also from [...] all those propositions in which I have shown, that everything in nature proceeds from a sort of necessity, and with the utmost perfection. However, I will add a few remarks, in order to overthrow this doctrine of a final cause utterly. (Spinoza 1677[18])

Here, the rejection is grounded in metaphysics. Nature (or God; for Spinoza they are equivalent) operates of necessity as it must—we imagine things happening for this or that purpose, but we delude ourselves. Interestingly, as one can see in the above quotation, from Spinoza's standpoint, perfection is the enemy of design.

To take another example, Francis Bacon appears to reject appeal to final causes on pragmatic grounds. After making reference to the four Aristotelian causes (Material, Efficient, Formal and Final) in his *Novum Organum*, he notes that "of these the Final is a long way from being useful; in fact it actually distorts the sciences except in the case of human actions" (*Novum Organum* II. p. 2); and in I. (p. 48) he comments, in the same vein, that the human intellect, "in struggling towards that which is further off [...] falls back upon that which is more nigh at hand; namely on final causes, which have a relation clearly to the nature of man rather than to the nature of the universe." He appears to see the explanatory power of appeal to a goal

[17]The material in angle brackets was added for a later edition.

[18]http://frank.mtsu.edu/~rbombard/RB/Spinoza/ethica1.html#Appendix

as deriving from rational agency; when such appeals are made in natural philosophy, they are inappropriate and typically delude us into thinking we have achieved understanding and are thus roadblocks to knowledge.

One final example will show how varied are the opponents of teleology. Thomas Hobbes, best known as author of *Leviathan*, claims to reduce final cause to efficient cause: "A final cause has no place but in such things as have sense and will; and this also I shall prove hereafter to be an efficient cause." (*Elements of Philosophy* X., p. 7) Unlike Bacon, who thinks final causes are legitimately appealed to in the sphere of human action, Hobbes insists that even there, the real causal agency is the will, and that is a special form of efficient cause.

Not all the Continental Rationalists were opposed to teleology, of course. Gottfried Wilhelm Leibniz was a powerful voice in favor of teleological reasoning in natural philosophy. In the *Monadology* (1714) he argues that souls follow what he refers to as "laws of final causes", and he gives primacy to final causes at the 'monadic' level. As with Boyle, he sees no conflict between mechanism and teleology: monads are, as he puts it, 'divine machines' or 'machines of nature' which are "still machines in their smallest parts ad infinitum. Such is the difference between nature and art, that is to say between Divine art and ours." (*Monadology*, sec. 64) That move from 'nature' to 'Divine art' echoes, perhaps consciously, a central theme of Plato's *Laws* X. On his view, there is a pre-established harmony between the laws governing organic bodies and those governing their souls:

> Souls act in accordance with the laws of final causes through their desires, purposes and means. Bodies act in accordance with the laws of efficient causes or of motion. The two realms, that of efficient causes and that of final causes, are in harmony, each with the other. (*Monadology*, sec. 79)[19]

Leibniz's endorsement of the heuristic value to science of asking teleological questions is also highly reminiscent of Robert Boyle's *Disquisition about Final Causes*. Leibniz writes, for example, that:

> …final causes can sometimes also be introduced to great effect in particular problems in physics—not only so that we can better admire the most beautiful works of the supreme Creator, but also sometimes in order to find out things which by consideration only of efficient causes would be less obvious, or only hypothetical. (SD 24/FW 164)[20]

In the end, then, while, like many in the Platonic tradition, Leibniz is happy to adopt Aristotle's teleological language, his view of two interpenetrating kingdoms of corporeal nature, the realm of power and the realm of wisdom, one realm explained mechanically, the other "architectonically […] or by final causes" has much more in common with the teleology of Plato's *Timaeus* than that of Aristotle's *Parts of Animals*. When it comes to teleology in the seventeenth century, Plato's divine craftsman is much in presence.

[19]Quoted from: Gottfried Wilhelm Leibniz, Discourse on Metaphysics, Correspondence with Arnauld, Monadology, La Salle: Open Court, 1902 (originally published in 1714).

[20]For a good discussion of this aspect of Leibniz's thoughts on teleology, see McDonough (2009).

Nevertheless, much of the most interesting discussion and *use* of teleological thinking in the early modern period is to be found in those who are working in the fields of medicine and comparative anatomy. Even when Boyle, Ray and Leibniz are philosophically reflecting on Final Causes, they are doing so with one eye firmly fixed on the use of teleological reasoning in the practices of anatomy, physiology and medicine all around them. In this respect, William Harvey should be a central character in any narrative about teleology in the seventeenth century, for a number of important reasons. First, all of his published work is, and was seen by him to be, a contribution to natural philosophy, not medicine—a study of the movement of the heart and blood across *all* the animals with hearts; a study of generation in *all* the animals, oviparous and viviparous. Second, he is developing the radical Aristotelian tradition emerging from Padua in the late sixteenth century, where he studied under Fabricius. He was proudly, outspokenly an Aristotelian; his *Exercitationes de Generatione Animalium*—in the preface to which he declared that "Aristotle is my leader, Fabricius my guide" —owes far more to Aristotle's work of the same name than to any contemporary work. It is permeated by the view of animal generation as inherently goal directed, guided by the nutritive generative soul, whose vehicle is the heat in the blood, for the sake of the production of another animal one in form with its parents. And while there are occasional references to a divinity, teleology resides in formal *natures* acting for the sake of ends, and such explanations are grounded in a thoroughgoing *empirical investigation* of generation across a wide range of animals. Harvey's great work on animal generation was widely admired by people with widely divergent attitudes to teleology—both Boyle and Descartes, for example. It demonstrates that a form of teleology rooted in the Aristotelian tradition was also alive and well in the seventeenth century. I will, then, leave the last word of this section to Dr. Harvey:

> For if you carefully consider Nature's works, you will find that not one of them is done to no purpose, but that all of them are ordained to some end and for the sake of some good. (*Ex. Gen. An.*, Ex. 41, 187–8 [Whitteridge trans.])

2.3 Teleology (Just) Before Darwin

In pre-Darwinian times, the Platonic and the Aristotelian teleological approaches to the study of nature found two important proponents: William Paley and Georges Cuvier, respectively. William Paley believed that the complexity and perfection of nature were powerful arguments for the existence and attributes of God. This worldview is widely known as natural theology. Paley famously made an argument by analogy of organisms as watches and of God as a watchmaker, according to which a complex structure, such as a watch, could not have emerged accidentally but required the existence of a rational designer, in this case of a watchmaker (Paley 2006 [1802], pp. 7–8). By comparing a stone and a watch in terms of complexity

Paley came to two major conclusions: (1) the more complex a structure was, the more powerful was the evidence of the existence of a designer and (2) the more complex the design was, the more competent was the designer. Then, based on these conclusions, he argued that organisms were much more complex than any artifact such as a watch and consequently they demanded an even more competent designer than a human watchmaker: God. It should be noted that Paley was drawn into the field of apologetics, as he tried to use rational arguments and data from the study of nature to confront skeptical approaches to Christian religion (see Ruse 2003; Thomson 2005; McGrath 2011).

There are striking similarities between Plato's *Demiurge* and Paley's *Creator*. Perhaps the most important one is that the Divine Creator not only created nature and organisms, but also worked within limitations which he was eventually able to overcome. Throughout his book Paley explains the existence of several biological features in teleological terms, as if they were mechanisms intentionally designed to perform a particular role. For example:

> In order to exclude excess of light, when it is excessive, and to render objects visible under obscurer degrees of it, when no more can be had, the hole or aperture in the eye, through which the light enters, *is so formed, as to contract or dilate itself for the purpose* of admitting a greater or less number of rays at the same time (Paley 2006 [1802], pp. 18–19, emphasis added).

> The tendon of the trochlear muscle of the eye, to the end that it may draw in the line required, is passed through a cartilaginous ring, at which it is reverted, exactly in the same manner as a rope in a ship is carried over a block or round a stay, *in order to make it pull in the direction which is wanted*. All this, as we have said, is mechanical; and is as accessible to inspection, as capable of being ascertained, as the mechanism of the automation in the Strand (Paley 2006 [1802], p. 48, emphasis added)

Paley thinks that the eye alone is sufficient to support the conclusion that an intelligent Creator is necessary in order to explain the existence of contrivances in nature. Even if nothing else was observed to exhibit complexity, or if everything else exhibited disorder and chaos, this single example would still be enough for the inference to a Designer (Paley 2006 [1802], p. 45).

The most important proponent of Aristotelian teleology in the time before Darwin was Georges Cuvier, who had been influenced by the works of Aristotle (Taquet 2006, pp. 179–184). One of his major contributions was the synthesis of paleontology, classification and comparative anatomy on the basis of a teleological view of nature according to which function determined form. The founding principle of Cuvier's biology was the "conditions for existence" (*conditions d' existence*). According to this principle the Creator, having taken into account the conditions that were necessary for the survival and the reproduction of an organism, had created the organs necessary for this (Ruse 2003, pp. 60–64). Cuvier's teleology was accommodated by naturalists both in France and abroad, especially in England because it was in accordance with natural theology. It should be noted though that Cuvier's approach was very different from Paley's (Appel 1987, pp. 55–57). Although Cuvier believed in the existence of a Creator,[21] he did not put emphasis on the confirmation of His existence but on the explanation of the features that organisms possessed on the basis of their function.

> Today comparative anatomy has reached such a point of perfection that, after inspecting a single bone, one can often determine the class, and sometimes even the genus of the animal to which it belonged [...]This assertion will not seem at all astonishing if one recalls that in the living state all the bones are assembled in a kind of framework *[charpente]*; that the place occupied by each is easy to recognize; and that by the number and position of their articulating facets one can judge the number and direction of bones that were attached to them. This is because the number, direction, and shape of the bones that compose each part of an animal's body are always in necessary relation to all the other parts, in such a way that – up to a point – one can infer the whole from any one of them, and vice versa. For example: if an animal's teeth are such as they must be, *in order for it to* nourish itself with flesh, we can be sure without further examination that the whole system of its digestive organs is appropriate for that kind of food; and that its whole skeleton and locomotive organs, and even its sense organs, are arranged in such a way *as to* make it skillful at pursuing and catching its prey. For these relations are the necessary conditions of[22] existence of the animal; if things were not so, it would not be able to subsist (translated in Rudwick 1997, p. 36, emphases added).

According to Cuvier, an organism cannot exist if it does not fulfill the conditions which are necessary for its existence. It should be noted that the conditions that Cuvier refers to have to do with the organism only and not with the environment in which it lives. Thus, the various parts of the organism must be coordinated so that its survival is made possible. A crucial consequence of Cuvier's approach was that evolution is not only empirically false, but also conceptually impossible. If the structure of organisms is arranged in teleological terms, then the transition from one form to another will be impossible because any transitional form will not be appropriate either for the way of life of the ancestors, or for that of the descendants. The parts of transitional forms will fail to fulfill the conditions for existence of either ancestors or descendents and will thus die out. Consequently, there can be no transitional forms and no link between different groups which are necessarily distinct from each other (Ruse 2000, 2002; Reiss 2005, 2009).

---

[21]For example, Cuvier described humans as "the last and most perfect work of the Creator" (see Reiss 2009, p. 93).

[22]Rudwick (1997) translates "conditions d'existence" as conditions *of* existence whereas Reiss (2009) suggests that conditions *for* existence is more appropriate. The reason for this is that it gives the term less ambiguity as it can have two distinct meanings: "The first of these meanings, and that of Cuvier, is that of the necessary conditions for the existence of an organism. These conditions are a characteristic of the organism. For an animal obtaining enough food is a condition of existence in this sense, where 'enough' is obviously relative to the particular organism in question. The second possible meaning is that of the environmental conditions, or circumstances, in which an organism exists. The types of other organisms present in the environment of an animal are conditions of existence in this sense. The reason I have preferred to translate the phrase as 'conditions for existence' is that this latter meaning, entirely different from Cuvier's (Russell 1916, p. 34), is thereby excluded." (Reiss 2005, p. 261; Reiss 2009, pp. 17–19).

2.4 Darwin and a New Teleology

In one of Charles Darwin's private notebooks on the transmutation of species, he writes: "'Natura nihil agit frustra'[23] as Sir Thomas Browne says 'is the only indisputable axiom in Philosophy'." Browne was a seventeenth century Baconian physician who was friends with William Harvey—and our survey of the disputations over teleology in that century cast doubt on Browne's assertion about its 'indisputability'! But this quotation in Darwin's Species Notebooks shows that the question of natural teleology had not disappeared by the nineteenth century.

One way to think about the question of teleology in a Darwinian context is to ask the following, more specific, question: Are explanations of adaptations[24] by appeal to natural selection *teleological* explanations? We think it is fair to say that many philosophers of biology would answer affirmatively, while most practicing evolutionary biologists would answer in the negative. Part of the reason for this discrepancy is simply a matter of terminology: many of those answering in the negative would, we suspect, answer affirmatively to a question like the following: 'Is it appropriate to explain the presence of a trait in a population by appealing to its value in enhancing fitness?' But to explain the presence of a trait by appealing to its value consequence for its possessors is to offer what a number of philosophers of biology would refer to as a teleological explanation of that trait (Ayala 1970; Brandon 1981; Ariew 2007; Walsh 2008).

But lying behind that discrepancy are, we believe, deeper historical and philosophical issues that are complicated and significant for understanding contemporary misunderstandings about evolutionary biology. We have spent a good part of this chapter on Teleology clearing up some historical misunderstandings, and we will continue by making use of these historical results to clarify the sense in which an entirely naturalistic understanding of natural selection is nevertheless robustly teleological.

In 1861 Charles Darwin presented the results of his research on sexual dimorphism in the genus *Primula* to the Botanical section of the Linnean Society. In the published version, he wrote:

> The meaning or use of the existence in *Primula* of the two forms in about equal numbers, with their pollen adapted for reciprocal union, is tolerably plain; namely, to favour the intercrossing of distinct individuals. With plants *there are innumerable contrivances for this end; and no one will understand the final cause of the structure of many flowers without attending to this point.* (Barrett 1977, 2, p. 59; emphasis added)

Darwin had earlier written a well-received study of the 'contrivances' found in Orchids to promote fertilization by insects (Darwin 1862). That work was much admired by Asa Gray, a self-taught botanist who in 1842 had been designated the Fisher Professor of Natural History at Harvard College. Darwin and Gray began corresponding on botanical topics in 1855, and in 1857 Darwin revealed to Gray, a

[23]"Nature does nothing in vain" is the usual translation. It is, as we have seen, a phrase that originates, as far as we know, with Aristotle.

[24]Two distinct definitions of adaptation exist in the literature: a historical one and an ahistorical one. For details and discussion about the teaching of adaptation see Kampourakis (2013).

reform Presbyterian, that he was working on a book that would present a new theory of species transformation—and was pleasantly surprised by Gray's cautiously positive reaction. Emboldened, later that year Darwin sent Gray a brief sketch of his theory. This sketch was then incorporated into the material Darwin presented, along with Alfred Russell Wallace's 'On the tendency of species to depart indefinitely from their type,' to the Linnean Society in 1858—Darwin's first public presentation of his theory of natural selection (for details see Browne 2002, pp. 37–42).

In a later appreciation of Darwin, Gray makes direct reference to the overtly teleological character of the botanical work published after the *Origin*:

> A propos Darwin's botanical papers, which furnish excellent illustrations of it, let us recognize Darwin's great service to natural science in bringing back to it Teleology; so that, instead of Morphology vs. Teleology, we have Morphology wedded to Teleology. (Gray 1874, 80)[25]

In response, Darwin underscores his agreement with Gray's characterization of his theory as teleological.

> What you say about Teleology pleases me especially and I do not think anyone else has ever noticed the point. [Correspondence on line: 9483][26]

Nevertheless, in corresponding with others, Gray shows himself aware that he and Darwin grounded their teleology in very different ways. In a letter to Alphonse de Candolle in 1863, for example, Gray admits he is well aware that Darwin does not accept the inference from the presence of ends in nature to a designer:

> Under my hearty congratulations of Darwin for his striking contributions to teleology, there is vein of petite malice, from my knowing well that he rejects the idea of design, while all the while he is bringing out the neatest illustrations of it! Did time allow, I should like to write at large upon these enticing topics (Gray Jane Loring, 1893, p. 498)

And Darwin is equally aware that there is a deeper disagreement behind their common endorsement of teleological reasoning. In an 1861 letter to Gray Darwin reports to him on Sir John Herschel's first public response to the *Origin*, in a new edition of his *Physical Geography*.

> ...[he] agrees to certain limited extent; but puts in a caution on design, so much like yours that I suspect it is borrowed.—I have been led to think more on this subject of late, & grieve to say that I come to differ more from you. It is not that designed variation makes, as it seems to me, my Deity "Natural Selection" superfluous; but rather from studying lately domestic variations & seeing what an enormous field of undesigned variability there is ready for natural selection to appropriate for any purpose useful to each creature.—(*Correspondence* vol. 9 (1994), 162: 3176)[27]

---

[25] We find similar language already in his 1862 review of Darwin's On Contrivances: Gray (1862, pp. 428–429), where he applauds Darwin for having "brought back teleological considerations into botany".

[26] In his reply to this note, Gray reminds Darwin that he had been stressing Darwin's teleology from the publication of the Origin onward. We have noted the Darwin online number of this letter—the published correspondence currently only goes up to 1868.

[27] Compare, from a couple of months later: "...seeing what Fanciers have done by selecting individual differences in the nasal bones of Pigeons, I must think that it is illogical to suppose that the variations, which natural selection preserves for the good of any being, have been designed." (Correspondence vol. 9 (1994), 267: 3256).

This exchange reveals deep disagreement over whether a significant element of chance, namely with respect to the sources of variation, is compatible with a teleological account of adaptive modification. As with a number of Darwin's closest scientific allies, Gray saw natural laws as laws of 'intermediate causes,' instituted, and perhaps maintained, by God. Insofar as he could interpret the operation of these laws in this way, he was prepared to endorse Darwin's theory—biological adaptation was achieved by divinely instituted laws of nature. Darwin suggests, in his autobiographical remarks about his changing religious views, that when he wrote the *Origin* he shared this view, and the frontispiece quotes from Bacon and Whewell are further evidence for this.[28] But that requires that the production of variation also be due to divinely instituted natural law. And by that Gray does not simply mean there must be deterministic laws of variation; he means that there must be design in the production of variation. The primary meaning of 'chance' for Gray is 'undesigned' or 'useless'.

Darwin's choice of 'contrivance' for the title of this work[29] is intended to highlight one key feature that distinguishes his theory from a Lamarckian one: variations arise for reasons unrelated to an organism's adaptive needs—they 'chance to occur', as he often puts it. Adaptation results from the differential survival and preservation of those variations that are advantageous. Near the close of *Various Contrivances* (Darwin 1862), Darwin gives the following example of a case where a very simple 'optimal design' solution to an adaptive problem appears to be passed over in favor of a more complicated, sub-optimal, solution, a species of the genus *Malaxis* (for further discussion of this example and passage, see Beatty 2006, pp. 634–635). He supposes that at one point in the past its ovarium was oriented so that the labellum hung downward, but at a certain point in its history it became advantageous to have the labellum in the more typical, upward position.

> ...this change, it is obvious, might be simply effected by the continued selection of varieties which had their ovarium a little less twisted; but if the plant only afforded varieties with the ovarium more twisted, the same end could be attained by their selection until the flower had turned completely round on its axis: this seems to have occurred with the Malaxis, for the labellum has acquired its present upward position, and the ovarium is twisted to excess. (Darwin 1862, pp. 349–50)[30]

Asa Gray, in a good-natured response, saw clearly what Darwin was about.

> Of course we believers in real design make the most of your "frank" and natural terms, "contrivance, purpose," etc., and pooh-pooh your endeavors to resolve such contrivances into necessary results of certain physical processes, and make fun of the race between long noses and long nectaries! (*Correspondence* vol. 11 (1999), 253, p. 4056)

---

[28] Works vol. 29, 1989, p. 123.

[29] The 1862 edition was titled "On the various contrivances by which British and foreign orchids are fertilized by insects, and the good effects of intercrossing". It was shortened to The Various Contrivances by which Orchids are Fertilized by Insects in the second edition of 1877.

[30] There are a number of differences in the 1877 edition that tend to put more stress on the number of variations. This would support a view suggested by John Beatty in correspondence, that in 1862 Darwin is inclined to think that, at least in nature, variation is more limited and restricted than he does later.

Darwin hits on a craft analogy to make the role of chance in his 'two step' mechanism clear. He first expresses it in a letter to Gray in August of 1863, while he is working on *Variations in Animals and Plants under* Domestication (1868):

> In my present book [*Variations*] I have been comparing variation to the shapes of stones fallen from a cliff, & natural or artificial selection to the architect,[31] but I cannot work a metaphor like you do.—That seems a very pretty case of the orchid with prominence on labellum. (*Correspondence* 11 (1999), 581: 4262)

In the last two pages of *Variations* he mines this analogy for all it is worth and reveals that the lengthy correspondence with Gray has helped him to differentiate two notions of chance that are not clearly distinguished in the *Origin*. Darwin first expands the analogy, imagining rock fragments of various shapes and sizes accumulating, as a consequence of erosion, at the base of a precipice. An architect then selects those with shapes and sizes, best suited to play various roles in a building he is erecting.[32] These rock fragments were not *designed* for these roles—they are *selected* for them. He goes on: "...the fragments of stone [...] bear to the edifice built by him the same relation which fluctuating variations[...] bear to the varied and admirable structures ultimately acquired by their modified descendants." (Darwin 1868 Vol. II, p. 430) He goes on to argue that ignorance of the cause of each variation doesn't detract from the explanatory power of selection. "[I]t would be unreasonable," to claim "that nothing had been made clear [...] because the precise cause of the shape of each fragment could not be told." (Darwin 1868 Vol. II, pp. 430–431) He then distinguishes two senses of what is 'accidental'. In one sense the shapes of the fragments are not accidental: "...the shape of each depends on a long sequence of events, all obeying natural laws...". "But," he goes on, "in regard to the use to which the fragments may be put, their shape may be strictly said to be accidental." (Darwin 1868 Vol. II, p. 431)

Darwin is explicitly accounting for the contrivances of Orchids as a consequence of chance variation and the 'paramount power of selection'.

> If we assume that each particular variation was from the beginning of all time preordained, the plasticity of organisation, which leads to many injurious deviations of structure, as well as that redundant power of reproduction which inevitably leads to a struggle for existence, and, as a consequence, to the natural selection or survival of the fittest, must appear to us superfluous laws of nature. (Darwin 1868, Vol. II, p. 432)

Darwin, then, appears to think of himself as providing teleological explanations, but without any backing from theology, and in conjunction with a view of the sources of variation being 'accidental' with respect to the organism's well being. The critical question for those interested in the philosophical question of whether selection explanations are in some significant sense teleological, is whether there is a way in which Darwin can pull this off. In a 1993 paper,[33] one of us (JGL) defended

---

[31] The architect metaphor was a metaphor used by Darwin to explain the origin of chance variation (see Beatty 2010).

[32] Darwin had actually introduced the metaphor earlier, in volume I, p. 395.

[33] Lennox (1993); cf. Ghiselin (1994), Lennox (1994), and Gotthelf (1999) for further discussion of the issues at stake.

a reading of Darwin's explanatory method according to which he can do so. Here we present, in a slightly modified form, the explanatory method he attributed to Darwin at that time.

As we have seen, Darwin became very interested in 'contrivances' in plants that encouraged cross-fertilization and discouraged self-fertilization, recognizing in this a fruitful mechanism for the production of new variation in populations. This was a major focus of his book on insect pollination in Orchids, but as we've seen, it was the sole topic of his paper on dimorphism in *Primula* (in Barrett 1977, 2., pp. 45–63). This paper is entirely focused on presenting the results of Darwin's search for the final cause of this dimorphism in *Primula* (Primoses and Cowslips). He explicitly characterized that work as the search for its final cause. We earlier highlighted the robust teleological language in this passage. The 'end' in question is that of promoting crosses between distinct plants. But the teleological investigation Darwin is involved in is more fundamental than that: he is interested in knowing what value is achieved by mechanisms that promote intercrossing and discouraging self-fertilization. Based on a careful analysis of the argument in that paper, one of us (JGL) was able to abstract the following argument schema (Lennox 1993):

1. Dimorphism is present in *Primula veris*. [Variation of interest V is present in Organism of interest O.]
2. Dimorphism has the effect of increasing heteromorphic crosses and decreasing homomorphic fertilization. [V has a certain Effect E.]
3. Heteromorphic crosses are more fertile and produce more vigorous offspring than homomorphic fertilizations. [E is advantageous to O.]
4. Natural selection would thus favor increased dimorphism in *Primula veris*. [Therefore V in O would be selectively favored]

Conclusion: Dimorphism is present in *Primula veris because* it promotes intercrossing. [Therefore V is present in O *because* of E.]

Darwin, without a blink, refers to the promotion of intercrossing as the "Final Cause" of the dimorphic condition of *Primula*. Is this merely a careless mode of expression, or does the above reasoning reveal a legitimate sense in which the reproductive consequences of sexual dimorphism are the *cause* of its presence in *Primula?* It is unlikely that Darwin would have used such a loaded expression unreflectively; and indeed there is a clear sense in which Darwin has identified the Final Cause of the trait in question. The various environmental "checks" to population expansion, which Darwin thinks of as the principal mechanisms promoting adaptive evolutionary change, bias reproductive frequencies on the basis of whether the consequences of particular variations are advantageous or disadvantageous to their possessor's living to sexual maturity and reproducing. If a variation functions, in a particular environment, to increase its relative frequency in subsequent generations, that variation is selectively favored *for* that function.

Darwin's explanation thus has the form of what has come to be termed, following Larry Wright's analysis of teleology, *consequence etiology*. But it is also clear that there is a significant *value* component to Darwin's understanding of teleology. Those traits are selectively favored which provide a relative *advantage* (to adopt Darwin's language) to the organisms that have them.[34] Thus, the explanation for the presence of a trait is based on its contribution to the survival of its possessors. If an organ is maintained through the survival of its possessors during the process of natural selection as a consequence of its contribution to their survival, it can be suggested that the cause of the presence of this organ is the fact that it has this particular contribution. In other words, the explanation for the presence of the organ is based on the consequence of its existence, which is its contribution to the survival of its possessors (for a discussion of consequence etiologies see Depew 2008). Consequently, the explanation for the presence of a particular feature is teleological in form but does not rely on supernatural final ends or transcendental purposes. The presence of a trait is explained by its value to the organism which is the basis of its preservation by natural selection.

For Darwin, the selective agent is "nature", i.e. factors in an organism's environment (predation, drought, disease, etc.) that present challenges to the organism's surviving and reproducing, challenges that some meet better than others. It is in a common sense way *external, but the external agent is not a rational, deliberating, designing agent.* So as far as causation is concerned the agent is external but not rational. On the other hand, from the standpoint of the "teleological" question (for whose sake?), Darwin is very consistent and explicit that *selection acts for the good of the organism*:

> Man selects only for his own good; Nature only *for that of the being which she tends*. Every selected character is fully exercised by her. (Darwin 1859, p. 83; emphasis added)

> It may be said that natural selection is daily and hourly scrutinising, throughout the world, every variation, even the slightest; rejecting that which is bad, *preserving and adding up all that is good;* silently and insensibly *working [...] at the improvement of each organic being in relation to its organic and inorganic conditions of life.* (Darwin 1859, p. 84; emphases added)

> It is, however, far more necessary to bear in mind that there are many unknown laws of correlation of growth, which, when one part of the organization is modified through variation, *and the modifications are accumulated by natural selection for the good of the being*, will cause other modifications, often of the most unexpected nature. (Darwin 1859, pp. 85–86; emphasis added)

It is not an external agent's good (e.g. as it is with a breeder's desire to have a horse with stronger legs or sheep with more wool) that determines selective preference, but what is advantageous for the organism given the environmental hazards it must deal with. Thus, selection acts for the sake of preserving the best adapted organisms (that is, it acts for the good of the organism, not the agent).

---

[34]The same referee who wanted to find rationality, intentionality and design in Aristotelian natures also claimed he couldn't find, in any of the evidence presented here, Darwin endorsing teleology! If the above evidence isn't sufficient, there is much more in Lennox (1993), Beatty (1990, 2006), and Lennox (2010); but in our view, the above evidence is quite sufficient.

3 Teleology in Conceptual Development Research

3.1 Teleological Explanations: Design or Natural Selection?

An investigation of the historical development of the life sciences has shown us, first of all, that from the beginning there were two distinct forms of teleological explanation: one depends on the idea of the natural world as a product of a supernatural creator whose work is modelled on human craftsmanship and can be seen in a tradition stretching from Plato through Robert Boyle, John Ray and William Paley to today's defenders of Intelligent Design; the other depends on the idea that living processes are naturally goal-oriented and that living things have the features they do because those features support or enhance their lives, a view which originates in Aristotle and is found in various guises throughout the history of biology, including the selection-based teleology found in Darwin and articulated by a number of contemporary biologists and philosophers of biology. Recognizing this distinction should have a positive impact both on research on the conceptual development of children and on biological pedagogy: such research should not fail to recognize these two distinct kinds of teleology, and simply assume that the teleological explanations for natural phenomena of young children rest upon assumptions about organisms being designed by a craftsman designer; neither should the adoption of a teleological perspective on living things be something children should be encouraged to "grow out of". A careful reading of the many philosophically and biologically sophisticated defenders of teleology throughout the history of biology shows why this is the case.

Let's compare these two types of teleological explanations with an example. The shape and the size of the wings of airplanes can be explained in terms of their function: *airplanes have wings in order to fly*. No airplane can fly unless it has wings of the appropriate shape and size (other aircrafts, like helicopters, also fly but in a different way). Of course, possessing wings is a necessary but not sufficient condition for flying. Airplanes can fly as long as they also have engines which are powerful enough to make them take off. A similar explanation can be given for the wings of birds: *birds have wings in order to fly*. No bird can fly unless it has wings of the appropriate shape and size. Again, possessing wings is a necessary but not sufficient condition for flying. Birds can fly as long as several preconditions are fulfilled. For example, ostriches have wings, but they are relatively heavy and thus unable to fly.

Despite the apparent similarities, there is a major difference between these two teleological explanations: birds are organisms whereas airplanes are artifacts. In both cases, we have teleological explanations because we have consequence etiologies. An effect of having wings in both cases is the capacity for flight; and in both cases, providing the capacity for flight is the reason why they have wings (or, if you prefer, both have wings because wings provide the capacity for flight). Even more, in both cases, one can say wings were selected because they are organs for flying. What is different is that in the artifact case, a conscious agent, who has studied the requirements of flight, consciously selects the design that would be most appropriate; in the biological case, the agent is unconscious and it is natural selection: in a population where flight is advantageous, those organisms with variations providing increased ability for flight will tend to survive better and tend to leave more offspring.

Table 1 Major similarities and differences among organisms, artifacts and non-living natural objects

	Organisms	Artifacts	Non-living natural objects
Alive	Yes	No	No
Designed	No	Yes	No
Goal-directed	Yes	Yes	No

Consequently, teleological explanations for artifacts are based on design: they presuppose an intentional agent external to the object who is responsible for this design. Accordingly, teleological explanations for the features of organisms are based on natural selection: features that contribute to the organism's survival are selected because of this contribution which is a consequence of their existence. Teleology is natural and appropriate for the features of organisms when the explanandum is the outcome of natural selection, without appeal to intentional design.[35] To sum up, there are two types of teleological explanations: teleological explanations for artifacts which are based on design and teleological explanations for organisms which are based on natural selection.

3.2 Organisms, Artifacts and Non-living Natural Objects: When Are Teleological Explanations Legitimate?

Although one cannot easily provide a definition of what an organism is, because this requires a definition for life itself (see Cleland and Zerella this volume), nevertheless there are a number of familiar properties which are characteristic of organisms such as the fact that they consist of cells, undergo development, reproduce, digest food, respire, react to stimuli from their environment etc. As already mentioned, artifacts are by definition objects which were designed and made for a purpose. It should be noted that not all objects made by humans are artifacts: a pile of papers on a desk is human-made but it is not an artifact, unless the person who made it had a particular intention in mind (e.g. made it for artistic purposes). Finally, some objects that exist in nature, such as clouds and rocks, are non-living like artifacts and non-designed like organisms. The similarities and differences among organisms, artifacts and non-living natural objects are summarized in Table 1 (for a relevant discussion see Keil 1989, chapter 3).[36]

[35]Organisms exhibit some goal-directedness that has nothing to do with design. This is due to the fact that organisms are self-organized, goal-directed systems, in which the goal in this case is a stable state which organisms can achieve thanks to their homeostatic properties (Walsh 2008)

[36]A referee of this chapter raised a concern about where in this table domesticated organisms or genetically modified ones would belong, suggesting that they should be considered as artifacts because they are intentionally modified by humans for some purpose. Human intervention in the case of artificial selection of domesticated animals differs enormously from genetic modification in the laboratory (resulting from example in the production of transgenic organisms). Humans may also use rocks as they are to create a path in a river or modify them extensively to produce objects of art. However, most natural entities (both organisms and non-living natural objects) are not modified by humans and most importantly come to existence in nature without any human intervention. Therefore, for our purposes here we will consider those cases in which humans modify organisms or non-living natural objects as exceptions.

Organisms are living entities that come into existence through natural processes and are self-organized and goal-directed. Organisms exhibit functions and properties (e.g. homeostatic ones) that serve particular goals (e.g. their well being) (Walsh 2008). Teleological explanations based on design are entirely inappropriate for organisms because they presuppose the existence of an external intentional agent who has designed organisms for some purpose. Organisms do not seem to be designed at all as they may possess fundamental useless features and rudimentary organs which are the relics of an evolutionary past. Organisms evolve through natural processes which may maintain useful features, selected for their effects, but also useless ones which simply happened not to be eliminated by natural selection. Thus, birds have wings because they have evolved by natural selection. Wings currently contribute to the survival of many extant birds, like eagles and hawks, which can fly. However, other birds like penguins and ostriches have wings despite the fact that they do not fly. In the case of penguins wings have been co-opted for another use (swimming) whereas in the case of ostriches wings may be rather useless (at least with regard to their original function). The wings of ostriches and penguins are derived from the common ancestor of birds through evolution and still exist either because they are useful (albeit in a different way) or because they have not been eliminated yet (but they may be in the future).[37] Teleological explanations based on design make no sense at all for organisms; one cannot claim that birds' wings were designed for flying because one cannot explain why a rational designer created all birds with wings although some of them use them differently or do not use them at all. The use of teleological explanations based on design for organisms may be due to the fact that they are perceived as artifacts, created for some purpose, a stance called creationism.[38] Organisms do not seem to be intelligently designed, as they possess rudimentary organs and fundamentally useless features (Williams 2001 [1996]; Avise 2010). Contrary to artifacts, organisms are the products of historical, evolutionary processes the outcomes of which are contingent and unpredictable (Beatty 2006).[39]

[37]Those that are useful are probably under selection pressure (a penguin's "wing" is about as much like a fin as a wing can be) and those that are not probably are too (an ostrich's wings are reduced—it still "flaps" them when it runs, they just are sufficient for lift).

[38]It should be noted that many evolutionists, such as Daniel Dennett and Richard Dawkins, use the language of 'design' for organisms, arguing that it is a legitimate metaphorical extension from the artifact domain. The metaphorical use of design for organisms may be legitimate in a discussion among scholars who understand evolution correctly. However, it can be problematic in the case of non-experts who may interpret design literally (intelligent design) and not metaphorically (natural design) due to their intuitions of purpose and design in nature.

[39]Artifact production has an evolutionary history as well, with contingencies and unpredictabilities. Why do cars in one part of the world have steering wheels on the left and in other places on the right? Why do they stay that way, even when it would be ideal to have them all the same? However, this is cultural, not biological evolution; design has a place in the evolution of artifacts because they are made for some purpose that is evident even when they have changed significantly. Organisms in contrast are more "plastic" than artifacts, and despite developmental constraints or biases, they may evolve to something very different than the original.

Artifacts are non-living objects which are created by humans and which exhibit properties (e.g. shapes) intentionally designed to serve particular purposes (whatever humans wanted to use them for e.g. knives have particular shapes in order to be used for cutting). Consequently, only teleological explanations based on design are appropriate for artifacts. Teleological explanations based on natural selection are entirely inappropriate for artifacts because their properties have been determined by external, conscious, intentional agents. If an artifact is selected over others, this is a process of artificial selection because it is consciously done to fulfill the aim of the human agent who selects. There is nothing like natural selection taking place, acting for the sake of the artifact itself. Artifacts are designed and created to serve the purposes of their designers. A knife has sharp edges which are useful to humans (e.g. in order to cut other objects) and is in no way useful to the knife itself (e.g. to protect itself). Thus, artifacts do not have any homeostatic properties because, not being alive, they do not have mechanisms to support such properties. And even if they do e.g. in cases of machines that have built-in computers to adjust their function, these are systems designed by humans for their own purposes. The use of teleological explanations based on selection for their own sake may be due to the fact that they may be perceived as alive, a stance called animism (Carey 1985, pp. 15–40). Artifacts are not alive because they do not consist of cells and do not exhibit all the characteristic properties of life that organisms do.

Finally, non-living natural objects like clouds and rocks are, like organisms, the outcomes of natural processes and, like artifacts, they are not alive. However, contrary to the wings of birds and the sharp edges of knives which can be explained in teleological terms, non-living natural objects can only have effects and not purposes. A pointy rock was neither designed, nor selected for its shape e.g. to protect itself from animals or to help them scratch their back. A mountain may be a cause of rain, because it causes uplifting of clouds which results in rain, however one cannot claim that mountains exist in order to cause rain. In short, non-living natural objects are not designed, are not alive, and are not goal-directed systems. Therefore teleological explanations, either based on design or based on natural selection, are not appropriate for non-living natural objects.

3.3 Children's Intuitive Teleological Explanations

Conceptual development research suggests that there are distinct stages in children's perceptions of evolution-related phenomena. It seems that 5–7 year olds tend to think that kinds of animals are eternal and unchanging, and that they cannot undergo radical changes during their lifetime. On the other hand, 8–9 year olds are more likely to accept developmental and intra-species variation but are less likely than older children to accept common descent. They also realize that species are not eternal. In addition, they tend to think that each kind of organism was specially created. Finally, 10–12 year olds are more likely to accept the notion that one species may have descended from an entirely different one, and eventually to accept the idea of common descent for animals, but not always for humans (Evans 2008, pp. 281–282).

Biological thinking emerges as a distinct domain of theorizing during childhood, although a disagreement exists on how autonomous it is (Carey 1985; Keil 1992; Springer 1999). Conceptual change in children's biological thinking seems to consist of spontaneous conceptual change, which takes place during development due to maturity and independently of any instruction, and instruction-based conceptual change, that is due to particular teaching interventions that aim at promoting conceptual change. It is difficult to distinguish between these two types of conceptual change during elementary school, and whatever is happening is probably the outcome of both maturity and schooling (Inagaki and Hatano 2002, pp. 153–154). However, in the case of intuitive teleology such a distinction may be possible, not only because elementary science instruction usually does not explicitly address children's design-based teleological intuitions and preconceptions, but also because children are often implicitly driven to an enhancement of these intuitions. For example, many popular wildlife and nature programs seem to present evolution using design-teleology, because they often describe organisms as perfectly designed in order to survive (Aldridge and Dingwall 2003). Consequently, children are not expected to undergo any spontaneous conceptual change from design-teleology to selection-teleology,[40] and even adults tend to intuitively provide design-based teleological explanations as well (e.g. Kelemen and Rosset 2009). Evidence from conceptual development research suggests that children tend to provide teleological explanations for the features of organisms and artifacts from very early in childhood (3–4 years old). Children may tend to provide teleological explanations for the features of organisms and artifacts because of their particular properties (Keil 1992, 1994, 1995). Or they may tend to explain everything in teleological terms, as they may provide teleological explanations for the features of non-living natural objects, as well (Kelemen 1999a, b, c).

According to Keil, teleological explanations are mostly given for organisms but also for artifacts. Despite that, children are able to distinguish organisms from artifacts due to two main differences between them. First, the properties of artifacts serve the purposes of the intentional agents who use them, whereas the properties of organisms serve the organisms themselves. For example, roses have thorns to keep animals from getting at them whereas barbed wire has barbs to prevent animals from accessing something that is valuable to humans. The second main difference is that organisms are perceived to have clearer essences than artifacts. This may be due to the way in which causal relations link properties together and explain their presence in organisms (Keil 1994, pp. 248–249). Keil has suggested that when they consider nature, children provide teleological explanations only for the features and properties of organisms, and not for those of non-living natural objects,

[40]Conceptual change related to teleology (i.e. from design-teleology to selection-teleology) can be understood in different ways. One way is to describe it as a change in the concept of adaptation. If children intuitively form a concept of adaptation, it is one that is based on consciousness and design. Thus, conceptual change from design-teleology to selection-teleology is actually the change of the concept of adaptation from a state that is the outcome of conscious design to the state that is the outcome of natural selection. This is why reference to evolutionary history is important in teaching about adaptation (Kampourakis 2013).

such as mountains. In a study, pre-school and second-grade children were shown either an emerald or a plant and were asked to choose between two explanations for their green color: a teleological explanation (e.g. they are green because it helps more of them to survive) and a physical explanation (e.g. they are green because tiny parts mix together to give them a green color). Both pre-school and second-grade children preferred teleological explanations for plants and physical explanations for emeralds (Keil 1992, pp. 129–130). In short, Keil has argued that children tend to provide teleological explanations mainly for the features and properties of organisms and artifacts, but not for those of non-living natural objects. Hereafter, we will refer to this view, according to which teleology is used for organisms and artifacts but not for non-living natural objects as *discriminative teleology*.[41]

Kelemen has suggested that from early in life children are aware that intentional agents make objects and use them in particular ways. Consequently, from the age of 3–4 years old children understand that agents use objects in order to achieve their own goals. This early awareness of intentional use of objects might influence children's explanations, particularly as most of the objects around them are artifacts, making them perceive all kinds of entities as designed and made for some purpose. Such experiences may eventually contribute to their tendency, in the absence of other alternative explanations, to provide teleological explanations for all kinds of entities, as if they were artifacts (Kelemen 1999a). In one study, 4- and 5-year-old children explained in teleological terms not only the properties of clocks and pockets, but also of non-living natural objects, such as mountains (for climbing) and clouds (for raining) as well as of organisms, such as babies (for loving) and of animals (for walking around). They did so, although they had been explicitly given the option of replying that these properties did not exist for some purpose (Kelemen 1999b). In another study it was found that 1st, 2nd and 4th grade children explained both animal properties, such as long necks, and non-biological natural properties, such as pointy rocks, in teleological terms. These results support the view that children tend to think about both living and non-living entities as possessing properties for particular purposes (Kelemen 1999c). Hereafter, we will refer to this view, according to which teleology is used for organisms, artifacts, and non-living natural objects as *non-discriminative teleology*.

More recent studies have not made clear whether children's teleology is discriminative or non-discriminative. For example, it has been suggested that 5- and 6-year old children think that organisms, artifacts and non-living natural objects have functions, although they perceive function as more characteristic of artifacts

---

[41] In previously published research by one of us (KK) the adjective used to describe this type of teleology was "selective". There, the term selective teleology simply referred to the fact that children may provide teleological explanations selectively for organisms and artifacts but not for non-living natural objects (e.g. Kampourakis and Zogza 2008; Kampourakis et al. 2012a, b). The adjective "selective" as used there had nothing to do with the process of selection. However, given the philosophical focus of this chapter and the extensive discussion of natural selection, we have used the adjective "discriminative" instead. Thus, although not consistent with the terminology used in previously published work, in this chapter we restrict the use of the adjective "selective" to references to natural selection.

compared to natural entities (DiYanni and Kelemen 2005). In contrast, it has been concluded that pre-school children clearly distinguish between organisms and artifacts, as they identify artifacts in terms of functions and animals in terms of appropriate biological characteristics (Greif et al. 2006).

However, it might be more important to understand whether children consider the parts or properties of organisms, artifacts and natural objects as *designed* for some purpose or not. Thus, teleological explanations based on design (where the part or property explained was intentionally designed for some purpose) can be distinguished from teleological explanations based on natural selection (where the part or property explained emerged without design and is maintained by natural selection because it is advantageous). In this case, a teleological explanation based on design for the existence of barbs would suggest that humans created them for some purpose, i.e. to protect something that is valuable for them, whereas a teleological explanation based on natural selection for the existence of thorns would refer to the advantageous consequences of thorns for roses. The preceding discussion and distinction between teleological explanations based on design and those based on natural selection shows that the often made assumption that teleological explanations are by definition design explanations is wrong. Depending on the explanandum (e.g. organism or artifact feature), the explanans relies on a consequence etiology but not necessarily on design.

4 Questions for Further Research

Whether children provide teleological explanations for organisms and artifacts only or for non-living natural objects as well, may not be that important for biology education. A more in-depth analysis of children's teleological explanations is necessary, and in order to make appropriate comparisons, children's teleological explanations could be classified as based on design or based on natural selection as we have suggested above. Perhaps what matters is not whether children intuitively provide teleological explanations for some but not for other entities, but what kind of teleological explanations these are. As we have shown, teleological explanations based on natural selection differ in some important aspects from teleological explanations based on design. Consequently, conclusions from conceptual development research that children provide teleological explanations for organisms and artifacts are not very informative if there is no clear indication of the form that these teleological explanations take. The conclusions from Keil's studies seem to suggest that children perceive organisms and artifacts differently, without necessarily attributing their features to design. In contrast, the conclusions from Kelemen's studies seem to suggest that children provide design-based explanations for all kinds of entities.

In what follows, we propose questions and directions for further research. It would be interesting to understand in more detail children's teleological explanations. This might be achieved by designing research to investigate whether children

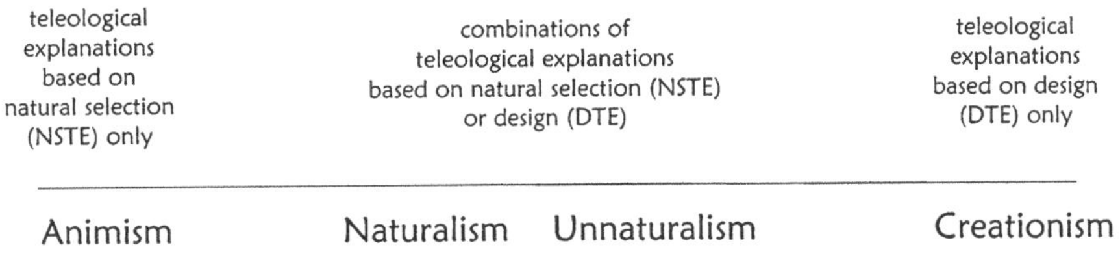

Fig. 1 Teleology in the animism/creationism continuum

are able to distinguish between different forms of teleological explanations and whether they could be differentiated into those who typically appeal to design or to selection. It would also be interesting to see in how many different ways children approach teleology, at what ages, whether they adopt a form of teleological explanations up to some age and then some of them undergo some kind of conceptual shift to adopt another form. For example, if children were found to provide teleological explanations based on design for artifacts and teleological explanations based on natural selection for organisms then they might be able to properly identify their characteristics that call for teleological explanations. This is a conceptual scheme, compatible with evolutionary theory, which could be called *Natural.* If children were found to provide teleological explanations based on natural selection for both organisms and artifacts, this might be evidence that they perceive both of them as alive. Thinking of artifacts as alive is described as animism and thus this scheme can be called *Animist.* If children were found to provide teleological explanations based on design for both organisms and artifacts it could be evidence that they perceive both of them as created by an intelligent designer. Thinking of organisms as created is described as creationism and thus this scheme can be called *Creationist.* Finally, if children were found to provide teleological explanations based on design for organisms and teleological explanations based on natural selection for artifacts, and either of them for non-living natural objects, the respective scheme can be called *Unnatural.*

What kind of explanations children would provide for artifacts would not affect distinctions such as those described above. Thus, the discriminative/non-discriminative distinction may not be that important. Animism and creationism are schemes that may exist independently of whether children's teleology is discriminative or non-discriminative, i.e. independent of the kind of teleological explanations children provide for non-living natural objects, for which teleological explanations are illegitimate anyway. Whether children's teleology is discriminative or non-discriminative may make a difference on the type of instruction that is required (see Kampourakis et al. 2012a, b for evidence of discriminative teleology in 2nd grade students and a possible shift toward this from a non-discriminative teleology in kindergarten). The resulting framework presented in Fig. 1 presents two widely known conceptual schemes, animism and creationism, as cases of teleology. This is just an example of what might be revealed after a careful and detailed study of the form of children's intuitive teleological explanations.

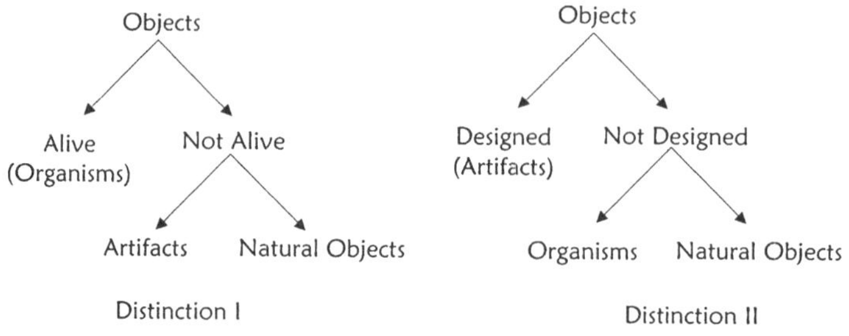

Fig. 2 Two major distinctions on which instruction addressing children's animist (*distinction I*) and creationist (*distinction II*) explanations

Ever since Jean Piaget, animism has been known as a distinctive characteristic of children's conception of the world. According to Piaget, children tend to attribute characteristics of animals (mostly humans) to non-living objects, and to make predictions or provide explanations about them based on their knowledge about animals (again about humans mostly) (Piaget 1960 [1929], pp. 196–206). Susan Carey has suggested that children's animism was the result of their lack of specialized knowledge about biology. Consequently, being more familiar with humans, they extended their knowledge about them to non-living objects as well (Carey 1985, pp. 39–40). Creationism has also been found to be a prevalent conceptual scheme, regardless of the religiosity of children's background (Evans 2001; Kelemen 2003). Then, perhaps children tend to perceive nature as an intentionally designed artifact (Kelemen 2004). Children provide purpose-based explanations of nature also endorse the existence of a creator agent in a manner that might be informed by their understanding of artifacts (Kelemen and DiYanni 2005).

Once children's teleological explanations are classified according to the framework presented in Fig. 1, instruction should focus on dealing with animistic and creationistic conceptual schemes, or in particular with addressing children's teleological explanations. We suggest that two types of distinctions should be presented to children either simultaneously or consecutively (Fig. 2). The first distinction would be between living and non-living entities. Only organisms are alive, whereas artifacts and non-living natural objects are not. By explaining that neither artifacts, nor non-living natural objects evolve by natural selection the way organisms do,[42] children might realize why teleological explanations based on natural selection are only appropriate for organisms. Accordingly, the second distinction would be between designed and not designed entities. Only artifacts are designed, whereas organisms and non-living natural objects are not. By explaining that both organisms and non-living natural objects have fundamental flaws in their structure or have

[42] Populations sharing common descent undergo processes of change in terms of their genetic makeup and developmental trajectories (or as Darwin described in short: "descent with modification").

central parts that are useless contrary to artifacts, children might realize why teleological explanations based on design are only appropriate for artifacts.

There is a lot of interesting research on children conceptual development and science education. What philosophy of science can contribute is a clarification of concepts and of their content. In the case of teleology, history is important because it shows how different teleological approaches have been adopted in the study of nature in the course of history, as well as that not all of them assumed an external, conscious designer. There are different kinds of teleological explanations, as we have shown here. They are similar in their structure but they are very different on their causal grounds. In both cases, a characteristic or object exists because of its consequences. However, in the case of design teleology this characteristic or object is itself a consequence of someone's intentions. However, when it comes to organisms a characteristic may exist because of its consequences, being itself a consequence of a natural, unintentional process: natural selection. Thus, it is not wrong to say that humans have hearts in order to pump blood or that eagles have wings in order to fly, as long as one is aware that these characteristics do not exist because they were intentionally designed *in order to* fulfil some ultimate purpose but *because* they are favored by selection as they confer an advantage and contribute to the survival and reproduction of their bearers.

References

Aldridge, M., and R. Dingwall. 2003. Teleology on television?: Implicit models of evolution in broadcast wildlife and nature programmes. *European Journal of Communication* 18(4): 435–453.

Appel, T.A. 1987. *The Cuvier-Geoffroy debate: French biology in the decades before Darwin*. New York/Oxford: Oxford University Press.

Ariew, A. 2007. Teleology. In *Companion to the philosophy of biology*, ed. D. Hull and M. Ruse, 160–181. Cambridge: Cambridge University Press.

Avise, J.C. 2010. *Inside the human genome: A case for non-intelligent design*. Oxford: Oxford University Press.

Ayala, F.J. 1970. Teleological explanations in evolutionary biology. *Philosophy of Science* 37: 1–15.

Barrett, P.H. (ed.). 1977. *The collected papers of Charles Darwin*. Chicago: University of Chicago Press.

Beatty, J. 1990. Teleology and the relationship of biology to the physical sciences in the nineteenth and twentieth centuries. In *Newton's legacy: The origins and influence of Newtonian Science*, ed. F. Durham and R. Purrington. New York: Columbia University Press.

Beatty, J. 2006. Replaying life's tape. *Journal of Philosophy* CIII(7): 336–362.

Beatty, J. 2010. Reconsidering the importance of chance variation. In *Evolution: The extended synthesis*, ed. Gerd Müller and Massimo Pigliucci. Cambridge: MIT Press.

Bishop, B.A., and C.W. Anderson. 1990. Student conceptions of natural selection and its role in evolution. *Journal of Research in Science Teaching* 27: 415–427.

Brandon, R.N. 1981. Biological teleology: Questions and explanations. *Studies in History and Philosophy of Science* 12(2): 91–105.

Browne, J. 2002. *Charles Darwin: Voyaging*. Princeton: Princeton University Press.

Carey, S. 1985. *Conceptual change in childhood*. Cambridge, MA/London: MIT Press.

Carey, S. 2000. Science education as conceptual change. *Journal of Applied Developmental Psychology* 21: 13–19.
Darwin, C. 1859. *On the origin of species by means of natural selection*. London: John Murray.
Darwin, C.R. 1862. *On the various contrivances by which British and foreign orchids are fertilised by insects, and on the good effects of intercrossing*. London: John Murray.
Darwin, C.R. 1868. *The variation of animals and plants under domestication*. London: John Murray.
Depew, D. 2008. Consequence etiology and biological teleology in Aristotle and Darwin. *Studies in History and Philosophy of Biological and Biomedical Sciences* 39: 379–390.
DiYanni, C., and D. Kelemen. 2005. Time to get a new mountain? The role of function in children's conceptions of natural kinds. *Cognition* 97: 325–335.
Endler, J.A. 1983. Natural and sexual selection on color patterns in Poeciliid Fishes. *Environmental Biology of Fishes* 9: 173–190.
Endler, J.A. 1989. Sexual selection and predation risk in Guppies. *Nature* 332: 593–594.
Evans, E.M. 2001. Cognitive and contextual factors in the emergence of diverse belief systems: Creation versus evolution. *Cognitive Psychology* 42: 217–266.
Evans, E.M. 2008. Conceptual change and evolutionary biology: A developmental analysis. In *International handbook of research on conceptual change*, ed. S. Vosniadou, 263–294. New York: Routledge.
Ghiselin, M.T. 1994. Darwin's language may seem teleological, but his thinking is another matter. *Biology and Philosophy* 9: 489–492.
Gotthelf, A. 1999. Darwin on Aristotle. *Journal of the History of Biology* 32: 3–30.
Gotthelf, A., and J.G. Lennox (eds.). 1987. *Philosophical issues in Aristotle's biology*. Cambridge: Cambridge University Press.
Grant, B.R., and P.R. Grant. 1989. *Evolutionary Dynamics of a Natural Population: the Large Cactus Finch of the Galápagos*, 280. Princeton: Princeton University Press.
Gray, A. 1862. 'Discussion: Fertilization of orchids through the agency of insects'. *American Journal of Science, 2nd Series* xxxiv: 420–429.
Gray, A. 1874. Scientific Worthies III: Charles Robert Darwin. *Nature* 10: 79–81.
Gray, Jane Loring. 1893. *The letters of Asa Gray*, 2 vols. New York: Houghton-Mifflin Co.
Greif, M., D. Kemler-Nelson, F.C. Keil, and F. Guiterrez. 2006. What do children want to know about animals and artifacts?: Domain-specific requests for information. *Psychological Science* 17(6): 455–459.
Hankinson, R.J. 2008. The philosophy of nature. In *The Cambridge companion to Galen*, ed. R.J. Hankinson, 210–241. Cambridge: Cambridge University Press.
Harrison, A.G., D.J. Grayson, and D.F. Treagust. 1999. Investigating a first-grade1 student's evolving conceptions of heat and temperature. *Journal of Research in Science Teaching* 36(1): 55–87.
Inagaki, K., and G. Hatano. 2002. *Young children's naive thinking about the biological world*. New York: Psychology Press.
Jensen, M.S., and F.N. Finley. 1996. Changes in students' understanding of evolution resulting from different curricular and instructional strategies. *Journal of Research in Science Teaching* 33(8): 879–900.
Johansen, T.K. 2004. *Plato's natural teleology: A study of the Timaeus-Critias*. Cambridge: Cambridge University Press.
Kampourakis, K. 2013. Teaching about adaptation: Why evolutionary history matters. *Science & Education* 22(2): 173–188.
Kampourakis, K., and V. Zogza. 2007. Students' preconceptions about evolution: How accurate is the characterization as "Lamarckian" when considering the history of evolutionary thought? *Science & Education* 16(3–5): 393–422.
Kampourakis, K., and V. Zogza. 2008. Students' intuitive explanations of the causes of homologies and adaptations. *Science & Education* 17(1): 27–47.
Kampourakis, K., E. Palaiokrassa, M. Papadopoulou, V. Pavlidi, and M. Argyropoulou. 2012a. Children's Intuitive teleology: Shifting the focus of evolution education research. *Evolution Education and Outreach* 5(2): 279–291.

Kampourakis, K., V. Pavlidi, M. Papadopoulou, and E. Palaiokrassa. 2012b. Children's teleological intuitions: What kind of explanations do 7–8 year olds give for the features of organisms, artifacts and natural objects? *Research in Science Education* 42(4): 651–671.
Keil, F.C. 1989. *Concepts, kinds and cognitive development*. Cambridge, MA/London: MIT Press.
Keil, F.C. 1992. The origins of an autonomous biology. In *Modularity and constraints in language and cognition*, Minnesota symposium on child psychology, vol. 25, ed. M.R. Gunnar and M. Maratsos, 103–138. Hillsdale: Erlbaum.
Keil, F.C. 1994. The birth and nurturance concepts by domains: The origins of concepts of living things. In *Mapping the mind: Domain specificity in cognition and culture*, ed. L.A. Hirschfeld and S. Gelman, 234–254. Cambridge: Cambridge University Press.
Keil, F.C. 1995. The growth of causal understanding of natural kinds. In *Causal cognition: A multi-disciplinary debate*, ed. D. Sperber, D. Premack, and A.J. Premack, 234–262. Oxford: Clarendon Press.
Kelemen, D. 1999a. Function, goals and intention: Children's teleological reasoning about objects. *Trends in Cognitive Sciences* 3(12): 461–468.
Kelemen, D. 1999b. The scope of teleological thinking in preschool children. *Cognition* 70: 241–272.
Kelemen, D. 1999c. Why are rocks pointy?: Children's preference for teleological explanations of the natural world. *Developmental Psychology* 35: 1440–1452.
Kelemen, D. 2003. British and American children's preferences for teleo-functional explanations of the natural world. *Cognition* 88: 201–221.
Kelemen, D. 2004. Are children "intuitive theists"?: Reasoning about purpose and design in nature. *Psychological Science* 15(5): 295–301.
Kelemen, D. 2012. Teleological minds: How natural intuitions about agency and purpose influence learning about evolution. In *Evolution challenges: Integrating research and practice in teaching and learning about evolution*, ed. K.S. Rosengren, S. Brem, E.M. Evans, and G. Sinatra, 66–92. Oxford: Oxford University Press.
Kelemen, D., and C. DiYanni. 2005. Intuitions about origins: Purpose and intelligent design in children's reasoning about nature. *Journal of Cognition and Development* 6(1): 3–31.
Kelemen, D., and E. Rosset. 2009. The human function compunction: Teleological explanation in adults. *Cognition* 111: 138–143.
Lehninger, A. 1971. *Bioenergetics: The molecular basis of biological energy transformations*, 2nd ed. Menlo Park: W. A. Benjamin.
Lennox, J.G. 1983. Robert Boyle's defense of teleological inference in experimental science. *Isis* 74: 38–52.
Lennox, J.G. 1985. Plato's unnatural teleology. In *Platonic investigations*, ed. Dominc J. O'Meara, 195–218. Washington, DC: Catholic University of America Press. (Reprinted in Lennox 2001).
Lennox, J.G. 1993. Darwin was a teleologist. *Biology and Philosophy* 8: 409–421.
Lennox, J.G. 1994. Teleology by another name: A reply to Ghiselin. *Biology and Philosophy* 9: 493–495.
Lennox, J.G. 1997. Nature does nothing in vain. In *Aristotelische Biologie: Intentionen, Methoden, Ergebnisse*, ed. W. Kullmann and S. Föllinger, 163–182. Stuttgart: Franz Steiner.
Lennox, J.G. 2001. *Aristotle's philosophy of biology: Studies in the origins of life science*. Cambridge: Cambridge University Press.
Lennox, J.G. 2010. La fonction biologique: Phylogénie d'un Concept'. In *Les fonctions: des Organismes aux Artefacts*, ed. Jean Gayon and Armand de Ricqlès, 17–42. Paris: Presses Universitaires de France.
McDonough, J. 2009. Leibniz on natural teleology and the laws of optics. *Philosophy and Phenomenological Research* 7(3): 505–544.
McGrath, A.E. 2011. Darwinism and the divine: Evolutionary thought and natural theology. Chichester: Wiley-Blackwell.
Millikan, R. 1984. Language, thought, and other biological categories: New foundations for realism. Cambridge, MA: MIT Press.
Neander, K. 1991. Functions as selected effects: The conceptual analyst's defence. *Philosophy of Science* 58: 168–184.

Paley, W. 2006 [1802]. *Natural theology or evidence of the existence and attributes of the Deity, collected from the appearances of nature*. Oxford/New York: Oxford University Press.
Piaget, J. 1960 [1929]. *The child's conception of the world*. Patterson: Littlefield, Adams.
Ray, J. 1691. *The Wisdom of God manifested in the works of creation*. London: William Innys.
Reiss, J.O. 2005. Natural selection and the conditions for existence: Representational vs. conditional teleology in biological explanation. *History and Philosophy of the Life Sciences* 27: 249–280.
Reiss, J.O. 2009. *Not by design: Retiring Darwin's Watchmaker*. Berkeley: University of California Press.
Rudwick, M. 1997. *Georges Cuvier, fossil bones and geological catastrophes: New translations and interpretations of the primary texts*. Chicago: The University of Chicago Press.
Ruse, M. 2000. Teleology: Yesterday, today, and tomorrow? *Studies in History and Philosophy of Biological and Biomedical Sciences* 31(1): 213–232.
Ruse, M. 2002. Evolutionary biology and teleological thinking. In *Functions: New essays in the philosophy of psychology and biology*, ed. A. Ariew, R. Cummins, and M. Perlman, 33–59. Oxford: Oxford University Press.
Ruse, M. 2003. *Darwin and design: Does evolution have a purpose?* Cambridge, MA/London: Harvard University Press.
Russell, E.S. 1916. *Form and function: A contribution to the history of animal morphology*. London: John Murray.
Sedley, D. 2007. *Creationism and its critics in antiquity*. Berkeley: University of California Press.
Settlage, J. 1994. Conceptions of natural selection: A snapshot of the sense-making process. *Journal of Research in Science Teaching* 31(5): 449–457.
Southerland, S.A., E. Abrams, C.L. Cummins, and J. Anselmo. 2001. Understanding students' explanations of biological phenomena: Conceptual frameworks or p-prims? *Science & Education* 85: 328–348.
Spinoza, B. 1677. Ethica, ordine geometrico demonstrata. Leiden.
Springer, K. 1999. How a naive theory of biology is acquired. In *Children's understanding of biology and health*, Cambridge series in perceptual and cognitive development, ed. M. Siegal and C. Peterson. Cambridge: Cambridge University Press.
Taber, K.S., and M. Watts. 1996. The secret life of the chemical bond: Students' anthropomorphic and animistic references to bonding. *International Journal of Science Education* 18(5): 557–568.
Talanquer, V. 2007. Explanations and teleology in chemistry education. *International Journal of Science Education* 29(7): 853–870.
Taquet, P. 2006. *Georges Cuvier: Naissance d'un Genie*. Paris: Odile Jacob.
Thomson, K. 2005. *Before Darwin: Reconciling God and nature*. New Haven/London: Yale University Press.
Walsh, D. 2008. Teleology. In *The Oxford handbook of philosophy of biology*, ed. M. Ruse, 113–137. Oxford: Oxford University Press.
Williams, G.C. 2001 [1996]. *Plan and purpose in nature: The limits of Darwinian evolution*. London: Phoenix.
Wright, Larry. 1976. *Teleological explanations*. Berkeley: University of California Press.

Biology's Functional Perspective: Roles, Advantages and Organization

Arno G. Wouters

1 Introduction

A bird's vocal organ consists of two parts that can produce sounds independently (Greenewalt 1968; Stein 1968). As a result, most birds are able to produce sound in two voices. Songbirds typically don't do so (they produce an unisonous sound either by alternating the two sources, or by simultaneously producing the same note with each source), but most other birds generate sounds in two voices. Why?

Aubin et al. (2000) explain the emperor penguin's use of this two voice system by appealing to its role in solving the problem of how to survive and reproduce in the circumstances in which these penguins live (see also Sturdy and Mooney 2000).

Emperor penguins live in the Antarctic where they breed on sea ice during the harsh arctic winter, a hundred miles away from their feeding grounds. During the breeding season they form couples that raise one young. They do not make nests but carry the egg and young on their feet until the young becomes independent. After laying an egg, the female leaves the breeding grounds for 2–3 months, to hunt for food in the ocean. In the mean time, the male incubates the egg. To keep themselves warm the males huddle together in colonies of thousands of individuals at densities of 10 birds per square meter. When the female returns, she takes over the egg or chick from her mate, who now heads off to the hunting grounds. She feeds the chick by regurgitating fish stored in her stomach. After a few weeks the male returns from the sea with new supplies. Then, the parents take turns in hunting and keeping the chick warm in increasingly shorter cycles until the chick can regulate its own temperature. After that, both parents are busy feeding the chick for several weeks more.

A.G. Wouters (✉)
Department of Philosophy, Erasmus University Rotterdam,
Burg. Oudlaan 50, 3062 PA Rotterdam, The Netherlands
e-mail: wouters@fwb.eur.nl

K. Kampourakis (ed.), *The Philosophy of Biology: A Companion for Educators*, History, Philosophy and Theory of the Life Sciences 1, DOI 10.1007/978-94-007-6537-5_21,

It will be clear that in these circumstances survival and reproduction is only possible if the breeding partners (mate and young) are able to recognize each other in a dense crowd in lack of visual cues: it is night, there are no nests, the vast mass of white ice does not provide reliable landmarks, and all penguins look the same. How have they solved this problem?

In absence of visual cues, auditory signals might do the trick and previous research had shown that emperor penguins indeed recognize their partners by their display call (mates recognize each other, the chicks recognize their parents). The syllables of this call have a temporal structure that is highly stereotyped for an individual and highly variable among individuals. It was also known that the two voice system offers an additional means of individual recognition: the temporal beat pattern generated by the interaction of the two voices.

By recording display calls and playing them back after modification (e.g. after filtering one voice) and by measuring call propagation at different distances in different circumstances (e.g. in the midst of a colony and without intervening bodies), Aubin et al. (2000) show that the two voices are needed for individual recognition in the circumstances in which these penguins live: the pattern of syllables is severely compromised by propagation through a penguin colony, whereas the pattern of beats produced by the interaction of the two voices remains largely unchanged over the distances that penguins normally bridge by calling (up to 7 m).

It will be clear from this short description, that Aubin et al. (2000) explain the emperor penguin's use of the two voice system by describing why it is needed to survive and reproduce in the circumstances in which these penguins live. The study identifies a problem that must be solved if the penguins are to stay alive (the problem of individual recognition), it points out how the two-voice system helps to solve this problem (it is used to produce two-voiced display calls that can be transmitted over the required distances in the circumstances in which the penguins live), shows that the resulting solution is better than alternatives such as visual recognition or unisonous calls and explains why this is the case (visual recognition doesn't work in a monotonous environment and unisonous calls loose their individual characteristics when transmitted over the relevant distances).

This way of explaining an organism's characteristics, by viewing organisms as solutions to the problem of staying alive has bothered biology since its conception as an explanatory science at the end of the eighteenth century. In what sense is staying alive a problem? In what sense are organisms solutions? Can we talk of problems without presupposing the existence of someone who experiences them or foresees them? Can we talk of solutions without assuming that they were brought about or put in place for the purpose to solve or prevent a problem? Aren't we treating organisms as if they are human made artifacts when we apply a functional perspective?

Many attempts to answer these questions start from the assumption that talk of problems and solutions makes sense only in relation to purposes. This assumption is very intuitive and has been taken for granted for a large part of the history of biology. If it is true, it means that the functional approach in biology rests on the teleological assumption that organisms and/or their parts and activities are there for a certain purpose (for a discussion of teleology in biology see Lennox and Kampourakis this volume).

In this chapter I outline a different way to make sense of biology's functional perspective, one that does not assume teleology. More precisely, I explain how it is possible to understand organisms as solutions to the problem of staying alive without assuming that this problem is experienced as such, without assuming an organizer in any sense of that term, without assuming that organisms or their parts and activities have purposes, and without treating organisms as if they are human made artifacts. The key to this non-teleological view is the insight that biology's functional perspective is grounded in the way in which organisms hang together, rather than in the way in which their parts and activities come into existence. Elsewhere, I have shown by means of many examples (Wouters 1995, 1999, 2003, 2005b, 2007) that this non-teleological and non-experiential view of biology's functional perspective is in line with the practice of reasoning in contemporary functional biology. In *Not by Design* (2009), evolutionary biologist John O. Reiss develops an allied account from a different perspective (see also Reiss 2005). My main concern has been to understand explanation in functional biology (see Wouters 1999); Reiss seeks a way to get rid of the teleology implicit in the metaphor of natural selection in evolutionary biology. I wholeheartedly agree with Reiss' intention and solution and highly recommend his book to anyone interested in understanding adaptation, natural selection, function, or teleology.

As I shall explain, functional reasoning in biology is different from what many people intuitively think. A substantial part of a biologist's training consists in learning to think in ways that differ from our everyday ways of thinking about function, adaptation and evolution (just as a large part of a training in physics consists in learning to think in a way that differs from our intuitive physics). This new way of thinking is not taught explicitly; students learn it by example, by solving textbook problems, by struggling with real problems and by talking with each other and their supervisors during laboratory work and field courses. As a result of this implicit character of the way biologists learn functional reasoning, they are often not aware of the way their reasoning differs from our intuitions.

By explicating functional reasoning in biology and the way it differs from our everyday intuitions, philosophy may help biology educators to understand what their students are struggling with and to respond adequately to their problems. Unfortunately, the debate on function in philosophy is motivated more by the concerns of philosophy of mind and language than by a wish to understand biology (McLaughlin 2001), and proceeds all too often by explicating the intuitions of philosophers, rather than by studying what is actually going on in biology (Wouters 2005a). For that reason, I shall not review the philosophical discussion of function but present a philosophical account of functional reasoning in biology that is, I hope, of use in teaching biology (see Wouters 2005a; Garson 2008 for reviews of the debate).

The structure of this chapter is as follows. After this introduction, I explain how to talk about problems and solutions without assuming that someone experiences that problem or aims to solve it (Sect. 2). Central to this account is a definition of organization in terms of critical dependence of a certain ability of a system on the composition, arrangement and timing of its parts and their activities. In the next section (Sect. 3) I explain how the functional stance allows biologists to understand this organization without appeal to the way in which the organization comes about (and, hence, without appeal to teleology). In Sect. 4 I discuss the problem of how it comes about that organisms are organized the way they are. The difficulty to solve this problem was the main motive for the teleological interpretation of the functional perspective in eighteenth and nineteenth century biology. Darwin's insight that the solution to the problem of adaptation must be sought in the history of populations rather than in the interaction between the parts of the organism abolished the need for such an interpretation. Unfortunately, many people (including Darwin himself) failed to understand the relation between adaptation, function, and evolution. This failure is an important source of the misunderstanding that functional explanations explain why a certain trait is or was selected for, as well as for the continuation of teleological interpretations of the functional perspective. In Sect. 5 I explain how functional explanations differ from evolutionary explanations and why it is important not to confuse the two. In Sect. 6 I explain how functional reasoning in biology differs from our everyday intuition that talk of functions assumes reasons or purposes by comparing my account of functional reasoning in biology, with the philosophical theory of function that provides the best account of that intuition: the selected effects theory of function. I conclude this chapter, in Sect. 7 with some tentative suggestions for teaching about function.

2 Function Without Teleology

The key to the non-teleological view of function is a notion of organization derived from William Wimsatt's work (cf. Wimsatt 1986, 1997, 2007 Ch. 7; Craver 2001; Wouters 2005c).[1] On this notion of organization, a system is organized for a certain property or ability if that property or ability is critically dependent not only on the system's material composition, but also on the arrangement of its components and on the order and timing of their activity.

'Critically dependent' means that the system would not have that property or ability if its parts were arranged in a different manner or if their activities occurred in a different order or were differently timed. For example, my computer's current ability to process the text on its screen is critically dependent on the way it is wired: that ability would break down, for example, if there were no dependable relation between the input it receives from the keyboard and the characters that appear on the screen. My computer's weight on the other hand is not critically dependent on its organization: no matter how I rearrange its parts, their combination weighs the same. Similarly the emperor penguins' ability to recognize their partners in the circumstances in which they live is an organized ability because it would break down if, for example, the penguins were not able to produce sounds whose individual characteristics are transmitted through dense masses of moving penguins.

[1] Wimsatt uses the term 'emergence' where Craver and I use 'organization'.

An organized ability breaks down under some rearrangements of the system's parts and their activities, but not necessary under all. My computer's organized ability to process text, for example, would be disturbed if there were no consistent relations between input signals and the characters that appear on the screen, but it doesn't matter whether signal 1 is associated with an 'a' and signal 2 with a 'b', or the other way round. Similarly, the emperor penguins' ability to recognize each other in dense colonies would be disturbed if one of the two sound sources were disabled, but not when the two sources were swapped.

As a result it is possible to talk of degrees of organization of an organized ability: the more rearrangements of the parts of a system and the more changes in the order and timing of their activities would be possible without disturbing the relevant ability, the less organized is that ability. My computer's ability to keep the temperature of its processor beneath the point at which it would stop working is an organized ability (it is dependent on the location and type of the processor, on the presence and location of ventilators, on the location, size and form of holes, and so on). However, because all changes that would disturb the working of the processor would also disturb the computer's capacity to process text and because some rearrangements of the relation between input and screen would disturb the latter capacity but not the former, the computer's ability to process text is more highly organized than its ability to keep the processor's temperature beneath a certain limit.

Because the notion of 'organization' as it is used here is oblivious to both the way in which the organization came into being and the way in which it is maintained (if it is at all maintained), it does not assume the existence of an organizer in any sense of that term. No doubt my computer was designed for certain purposes (which might or might not have included text processing), put together for certain purposes (including, no doubt, the purpose to gain money), and bought for certain purposes (one of which was the purpose to process text). No doubt, the computer's activity is controlled by an organizer (my writing program) that was activated for a certain purpose (my wish to write this text). Although all these factors are in a certain sense the causes of the computer's current organization in that they explain how the computer became organized the way it is, and why it remains so, they are irrelevant with respect to the question whether or not the computer is organized for the ability to process text. The only thing that matters for this issue is the observation that a computer's ability to process text depends not only on the computer's material composition, but also on the way in which its parts are wired and on the order and timing of the activity of those parts. As long as this is the case, the machine in front of me is organized for processing text, even if this ability was brought about for other purposes or for no purpose at all, and even if this ability was not controlled by a program. Similarly, the emperor penguin is, no doubt, a product of evolution by natural selection, but it is not the type of history that allows us to say that the penguin is organized for being alive or that individual recognition is an organized capacity, but the fact that those capacities are critically dependent on the penguin's material composition, the arrangement of its parts and the order and timing of their activities.

The notion of organization as discussed above easily yields notions of problem and solution that do not assume purposes, sentient beings or intelligent beings. Organization was defined in terms of critical dependence on the arrangement of a system's parts and the order and timing of their activity. So if a system is organized for a certain ability this means that not all arrangements of its parts and their activity would result in the system having that ability. Some would and some would not. Apparently there are certain requirements for producing an organized ability. The forms of organization that produce this ability meet the requirements, the others not. Because of this, systems that are organized for a certain ability can be seen as systems that have solved the problem of how to meet the requirements for producing that ability. Hence, saying that a certain system solves a problem is just another way of saying that it produces an organized ability. Because talk of organization does not assume purposes (as I explained above), talk of problems and requirements does neither. Nor does it assume that the problem is experienced by a sentient being or foreseen by an intelligent one or that someone has knowledge of the requirements. The amount of organization needed to produce or maintain an organized ability suffices to talk of problems and requirements. Saying that something solves a problem or that it meets the requirements for solving a problem is just another way of looking at an ability that is critically dependent on an underlying organization.

There can be no doubt that staying alive is a problem in this non-teleological and non-experiential sense of 'problem,' for the ability to stay alive is clearly critically dependent not only on the organism's material composition, but also on the arrangement of its components and on the order and timing of their activity.

Having solved a problem, however, is not enough to warrant a functional perspective. The occurrence of eclipses is dependent on a certain arrangement of three celestial bodies, one that produces light and two that do not. So, the system of our sun, the Earth and its moon can be said to have solved the problem of how to produce eclipses in the non-teleological and non-experiential sense of problem and solution described above. Yet, it makes no sense to understand the sun-moon-Earth system as a solution to the problem of how to produce eclipses. This is because the sun-moon-Earth system would exist even if it didn't solve the problem of how to produce eclipses.

There also are, however, systems whose very existence depends on having solved an organizational problem. Such systems can exist and continue to exist only because of one or more of their organized abilities. Human artifacts are a case in point: they exist only because of their ability to meet the desires of the people who design, produce, buy and use them. Organisms provide another example. Organisms exist far from thermodynamic equilibrium and can exist only because they actively maintain their own organization ('actively' means that they need to invest energy in doing so) (Prigogine and Stengers 1984). In other words, they can exist only because they have solved the problem of how to organize material parts and their interaction in such way that they maintain their own organization.[2]

[2]See Schlosser (1998), McLaughlin (2001), Christensen and Bickhard (2002), Delancey (2006), Mossio et al. (2009) and Saborido et al. (2011) for accounts of function in terms of self-maintenance. My account differs from such accounts in two ways. First, in my account self-maintenance is a condition for being alive, but not for having functions. Second, in my account functional explanations explain why a trait is needed for continued existence of the organism, but not how the trait is maintained in the organism.

In regard to such systems it makes sense to ask why they can exist whereas other ways of organizing their components cannot, in addition to questions about how they came into being and how they produce the properties they have. To address these questions (that is, to understand the continued existence of systems whose existence depends on having solved an organizational problem) these systems need to be viewed as solutions to the problem of their existence. This is where the functional approach comes in.

To explain a system whose existence depends on an organized ability (to explain how it works and why it has the structure it has), we typically employ a kind of organization chart: a sketch or picture that outlines the main components of the system, their function and the way they interact. The chart need not be a visual diagram (although it often is), it can also be sketched by means of words or it might reside in the minds of the producers of the explanation and their audience. The chart provides the basis to explain the system's working and structure.

Let me explain this by means of a simplistic example: an electronic label maker with a small display that previews a line of text (a real example from biology follows in Sect. 3). To explain the working and structure of this label maker, I start by observing that its existence depends on the ability to print labels with the text the user wishes to have on it. Next, I identify its main components: a keyboard (with character keys, a delete key and a print key), a display, a printing unit, an input circuit that connects the keyboard with the display, and an output circuit that connects the display with the printing unit. I then sketch the function of those components. The keyboard allows the user to enter and delete characters and to start the printing. The display presents the text to the user. The printer prints the labels. The input circuit organizes the interaction between the keyboard and the display in such a way that the right character (the character corresponding to the picture on the key) appears at the right place (that is: at the end of the text) on the display. The output circuit organizes the connection between the display and the printer in such way that the right graphics (the figures corresponding to the characters on the display) are printed in the right order (the figure corresponding to the rightmost character at the display should be printed first).

In general a function is a position in an organization. The role of a component in the organization of a system whose very existence is an organizational problem might be called an 'essential' function. The functions of the *parts* and *activities* of artifacts in the organization of their ability to meet our expectations (such as the keyboard's function to enter text) are 'artifact functions'.[3] The functions of the parts and activities of organisms in enabling their continued existence are 'biological functions' or 'biological roles'.

This sketch of the components, their functions and the way they are connected, is used to explain (bottom up) how the label maker works, that is how it is able to meet the conditions for its existence (namely the ability to print a label with the text wanted by the user). When the user presses a character key, an electric signal enters

[3]The artifacts themselves do not have artifact functions. I think they have social functions, but that is a topic for another paper.

the input circuit causing a new character to appear on the display. Because of the organization of that circuit the character that appears on the display corresponds to the picture on the key. As a result, the user can enter text by pressing keys and use the screen to check whether the text is the one she wants to print. When the user presses the delete key an electric signal enters the circuit that causes the disappearance of the rightmost character on the display. Because of this and because the user can also enter new characters by pressing character keys, the user can make sure that the text on the display is the text she wishes to have on the label. When the user presses the print key an electric signal triggers the output circuit to print a series of characters. Because of the way in which that circuit is organized the printed characters correspond to the characters on the display. Because the rightmost character is printed first, the text on the label is the same as the text on the display. Because, as I have already explained, the user can also make sure that the text on the display is the text she wishes to have on the label, this means that the user can make sure that the text printed on the label is the text she wishes to have on that label.

The organization chart is also used to explain (top down) why the label maker has the structure it has by relating why that structure is useful to meet the conditions for its existence. Why does the label maker have a keyboard? Because, in order to make sure that the label maker prints the text the user wishes it to print, the user needs to have a way to tell the device what to print. Why does a certain key press always have the same result? Because the user would not be able to enter the text she wishes to print if there were no dependable relation between what is entered, what is displayed and what is printed. Why are the character on a certain key, the associated character on the display and the associated character on the printer the same? Because it would be more difficult for the user to enter the text she wishes if the displayed character would be a different from the entered one, and nearly impossible to check whether the printer will print what she wishes if the printed characters would differ from the ones displayed.[4] Why is the text on the display printed rightmost first? Because otherwise the text on the display must be read in the opposite direction of the text on the label, which would make it difficult for the user to understand the text on the display.

The (bottom up) explanations of how a certain system works are called 'mechanistic explanations' (see Bechtel this volume), the (top down) explanations of why the system has the structure it has are 'functional explanations'. Explanations of both kinds are needed to understand systems whose existence depends on an organized ability. The production of eclipses is fully understood if one can explain how they result from the properties of their components and the way they are organized. The relevant organization is fully explained by relating its history. However, if someone would be able to calculate the label maker's behavior on the basis of his knowledge of the wiring, but would not be able to answer the relevant why-questions, he has not understood the label maker's organization. Not even if he would be able to relate how the label maker was put together and why it was put together the way it was.

[4]An experienced typist can use an AZERTY key board on a computer configured for a QWERTY one, but I doubt that it will be possible to process text with a computer that prints a Q where there is an A on the screen, a Z where there is a W, and so on.

In this section, I explained how it is possible to make sense of function talk and functional explanation without assuming teleology or sentience. Functions can be defined as positions in an organization. Organization can be defined in terms of dependence on the composition, arrangement and timing of the relevant system, its parts and their activities. The functional approach is appropriate with respect to systems whose very existence depends on an organized ability.

I have illustrated this with the example of a simple label maker. In order to understand label makers and other artifacts we must, of course, appeal to human expectations, experiences, and purposes. After all, the existence of label makers and artifacts depends on such attitudes. However, the justification for a functional approach to artifacts lays in their character as systems whose existence depends on an organized ability, not in their ability to meet human purposes, nor in the role of human attitudes in their production. Human purposes enter into the functional approach to artifacts because the relevant ability is the ability to meet such purposes, but they are not what justifies understanding them as solutions to a problem.

As I have indicated above, the continued existence of organisms depends on their organized ability to maintain their own organization. For this reason, they should be understood as solutions to the problem of how to stay alive. In the next section I discuss how functional biologists do so.

3 The Functional Approach to Living Organisms

For a start, note that Aubin et al. (2000) are concerned with the two types of explanation characteristic of a functional approach. They explain *how* emperor penguins locate their partners (by means of temporal patterns produced by the interaction of two voices) and they explain *why* those penguins locate their partners in the way they do it (because of the lack of visual cues in their environment and the dense crowds in which the emperor penguins live, visual recognition, recognition by characteristic audio frequencies, or recognition of temporal patterns produced by one voice wouldn't do the job, but recognition of temporal patterns produced by two voices works excellent in these circumstances).

Such simultaneous investigation of how-questions and why-questions is characteristic of a lot of research in organismic biology. The how-questions are answered by describing features of the system that produce the relevant abilities, activities or characteristic (the emperor penguin's ability to recognize their partners is brought about by the interaction of two voices that produces a temporal pattern of beats characteristic of the individual). The why-questions are answered by describing features of the organism, its environment or its way of life due to which the characteristics to be explained are advantageous to the organisms that have it. Biologists typically call this type of answers to how-questions 'mechanistic' or 'causal' explanations and this type of answers to why-questions 'functional' or 'ecological' explanations.

Central to both kinds of explanation is the biological role of the two-voice system. Aubin et al. (2000) research indicates that this system is used to produce a sound with a temporal beat pattern by means of which the penguins recognize their partners.

Such ascriptions of biological roles are the handle by means of which functional biologists understand organization. In the course of history, they have developed a standard way of dividing an organism into a hierarchy of component systems that each has a role in bringing about the organism's ability to live the life it lives. The body of multicellular animals, for example,[5] consists of organ systems such as the respiratory system, the circulatory system, the digestive system, and the musculoskeletal system. The digestive system converts food into nutrients (such as glucose). The respiratory system takes up oxygen from the outside and releases carbon dioxide into the environment. The circulatory system transports oxygen from the respiratory system and nutrients from the digestive system to the organs (including the musculoskeletal system) and carbon dioxide from the muscles to the respiratory system. The musculoskeletal system converts oxygen and glucose from the circulatory system into mechanical energy and movement.

The ability of the different organ systems to perform their biological role, is, in turn, explained by reference to the roles that the different parts of those systems play in bringing about that ability. The circulatory system, for example, roughly consists of a transport medium, the blood, that carries the oxygen and carbon dioxide, arteries and veins that contain the blood and channel it in certain directions, capillaries that facilitate the exchange of gases and nutrients between the organs and the blood, and a heart that pumps the blood around.

The working of these parts (e.g. blood, arteries, veins, capillaries, and heart), in their turn, is explained by describing the roles of their components in bringing about these parts' abilities to perform their biological role, and so on, until a level is reached at which the relevant subsystems can be explained in terms of the physical and chemical characteristics of the molecules that make up that subsystem (Cummins 1975; Craver 2001).

In other words, functional biologists are working on big organization charts of the different ways in which organisms manage to keep themselves alive. Such organization charts provide a unifying framework that enables biologists to relate detailed studies of specific mechanisms at different levels to the project to explain how organisms are able to stay alive.

Returning to my example: by identifying individual recognition as a biological role of the two-voice system, the researchers situate that system in the organization chart of the emperor penguin's ability to stay alive. This gives them a handle both to explain how these penguins are able to find each other in the circumstances in which they live (by means of a characteristic pattern of beats produced by the interaction of two sound sources) and to study the question why these penguins use a two-voice system.

In the remainder of this section I take a closer look at the way in which this second question is addressed.

[5]The functional perspective is equally applicable to other kinds of organisms such as plants and bacteria. Because I am more familiar with zoology than with plant biology and microbiology, most of my examples are from zoology. I apologize for this unfortunate bias.

Thanks to the insight that the two-voice system is used for individual recognition, the original question of why emperor penguins use the two-voice system can be rephrased as a question about different ways to perform the biological role of individual recognition. In addition, it becomes clear what alternatives should be considered. The question can now be rephrased as something like 'why do emperor penguins recognize their partners by using the two voice system rather than topographical cues, visual characteristics, characteristic frequency specters, or temporal patterns produced by one voice?'

To answer this rephrased question Aubin et al. (2000) point to the settings in which individual recognition is performed: the penguins live in dense colonies in a monotonous environment. In this setting, two-voiced sounds can be used for individual recognition, while the alternatives can't.

Note that by pointing to these settings it is not explained how or why the two-voice system *came into being*, but rather how and why that system is useful (advantageous or needed) to solve the problem of being alive, useful in the sense that the penguins' ability to survive and reproduce would decrease if the two-voice system were replaced by one of the alternatives.

The fact that penguins live in dense colonies is relevant to the two-voice system because in this setting the organism's ability to survive and reproduce would decrease if the two-voice system were replaced by one of the alternatives, whereas this replacement would have negative or no effect on the organisms' capacity to survive and reproduce in other circumstances. A penguin's ability to live in dense colonies in a monotonous environment is, in other words, *functionally dependent* on its ability to produce two-voice sounds.

More generally, functional explanations in biology answer the question why certain organisms (for example emperor penguins) have a certain trait (the trait to be explained – the use of a two voice system for individual recognition) rather than some specific alternatives (such as the use of visual cues) by pointing out that the trait to be explained is advantageous to those organisms because some of their other traits (the explaining traits – they breed in dense colonies in a monotonous environment) are functionally dependent on the trait to be explained. Functionally dependent in the sense that the ability to maintain the living state of an organism with the explaining traits would diminish if the trait to be explained were replaced by an alternative, whereas replacing the trait to be explained would not make much difference or have a negative effect if the organism lacked the explaining traits.

Roughly spoken, a biological advantage is an ability that increases an organism's ability to stay alive, including its chances to survive and reproduce.[6] In the example of

[6]See Canfield (1964, 1965), Ruse (1971), Wimsatt (1972), Bigelow and Pargetter (1987), and Horan (1989) for 'life chances' or 'forward looking' accounts of function, Frankfurt and Poole (1966), Mitchell (1993, 1995) and Millikan (1989a, 1993) for criticism, and Wouters (2003) for a response to some of those criticisms. In my view, life chances accounts are basically right as accounts of biological advantages, but not as accounts of biological roles, and they fail to account for the explanatory force of functional explanation by viewing them as forward looking nomological or causal explanations, rather than as non-causal explanations of what is needed to stay alive.

the penguins, the existence of two independent sound sources is advantageous to emperor penguins because it enables them to produce a sound that is characteristic of the individual who produces it and propagates far enough to be recognized by their partners in the circumstances in which emperor penguins live. Without this ability emperor penguins would not be able to survive and reproduce in those circumstances.

Biologists often speak of 'function', 'survival value', or 'adaptive value' in this context. I prefer the term 'biological advantage' because the other terms are ambiguous or misleading ('function' might also refer to the biological role of a part, activity or ability, or to the activity of a system, 'adaptive value' often refers to evolutionary advantages, and 'survival value' erroneously suggests that survival is the only aspect of being alive – see Wouters 2003).

Although functional explanations explain a trait (e.g. the use of two voices) by appealing to its advantages, they start, as I emphasized, by identifying a relevant biological role (the two-voice system is used in individual recognition). This prior identification of a biological role is crucial to the explanation because of the way in which advantages are related to biological roles. Advantages, by definition, increase an organism's ability to stay alive, so they must be involved in the performance of a biological role, and they are advantageous precisely because they enable the organism to perform this role better than certain alternatives (the two voice system is advantageous because without it the penguins would not be able to recognize each other). So, biological advantages can be seen as abilities involved in the performance of a certain biological role without which the organism would perform that role less well (or not at all).

Advantages are, by nature, comparative: performing a certain biological role (e.g. individual recognition) in a certain way (e.g. by means of a two-voice system) is advantageous (or not) as compared to other ways of performing that role (e.g. by visual means or by means of one voice). Functional explanations compare the life chances of two (or more) kinds of systems: on the one hand the organism as it is and on the other hypothetical systems similar to the organism except that the trait to be explained is replaced by an alternative (or removed). The hypothetical systems with which the organism is compared need not exist or have existed. They even do not need to be able to stay alive. Quite often a comparison is made between a real organism and a hypothetical system that cannot possibly be alive and the point of the comparison is precisely that: to show that it cannot be alive because it lacks an essential ability. The penguin example is a case in point: the researchers show that emperor penguins would not be able to stay alive if they used visual cues, frequency patterns, or temporal patterns of one voice for individual recognition, instead of the beats produced by the two-voice system.

Which ways of performing a certain biological role are advantageous and which are not depends, of course, on the other characteristics of that organism and the environment in which it lives. Using auditory cues rather than visual ones for individual recognition is, for example, advantageous to organisms that have to recognize each other in a monotonous environment, but in environments that provide clear visual landmarks it might be the other way round. A two voice system is advantageous over one voice if the sound has to travel several meters through a mass of moving animals or a turbulent atmosphere, but in other circumstances it might be the other way round.

As I explained above, to understand how a certain biological role is performed it suffices to study the mechanism underlying the performance of that role. However, one cannot understand why a biological role is performed the way it is without an eye on the life of the organism as a whole. This is because the role itself does not determine what counts as a better way of performing it. What counts as 'better' is not intrinsic to the relevant role, but, depends solely on the effects on the organism's ability to stay alive. A certain way of performing a biological role is better than another way if the organism's ability to survive and reproduce is higher when the role is performed in the first way than when it is performed in the second way.

As I indicated above, functional explanations explain why certain organisms have a certain characteristic by pointing out that this characteristic is advantageous to those organisms because some of their other characteristics are functionally dependent on the characteristic to be explained. For example, the two-voice system is advantageous to emperor penguins because their ability to locate their partners in a dense mass of moving penguins is functionally dependent on the ability to produce sounds that maintain their individual characteristics when transmitted through that mass.

Functional dependence is a synchronic relation between two or more traits of an individual organism (see Wouters 2007). This relation is not of a causal nature but a requirement for being alive: organisms whose ability to stay alive depends on their ability to recognize each other by their call in a dense mass of moving organisms cannot exist if their call does not maintain its characteristic features when transmitted through that mass. Of course, there may also exist causal relations between a trait and its dependents. Perhaps, the two-voice system was maintained in the lineage of emperor penguins because variants with a damaged two-voice system were eliminated in the process of natural selection as the result of their display call not being recognized by their partners. Perhaps individual recognition in dense colonies plays a role in the ontogeny of the two voice system – maybe the penguins improve their ability to synchronously produce different sounds when their display call is not successful. However, the relation of functional dependence exists independently of the history of the related traits and independently of the causal relations that might also hold between these traits. It would also exist if the traits arose out of entirely different causal processes, for example if the penguins were put together in a laboratory or tele-transported from Alpha Centauri.

Suppose that, due to some small developmental problem, one or more emperor penguins lose their ability to produce two voiced sounds. These penguins would not be able to survive and reproduce in the circumstances in which their fellows live: the returning female would not find her mate and young, so the male would either starve to death or go for a hunt and let its young die in the cold. Suppose a hard-working scientist would be able to synthesize organisms that are exactly like emperor penguins except that they do not or cannot produce two-voiced sounds. If she would put those homemade organisms in the Antarctic they would have exactly the same problem as the emperor penguins that lost their ability to produce two-voiced sounds. Suppose that somewhere on Alpha Centauri the conditions are such that organisms similar to emperor penguins can recognize each other by means of unisonous sounds and that on that planet there actually evolved organisms similar to our emperor penguins except that their display calls are unisonous. If those organisms would be tele-transported to the Earth's Antarctic they would have exactly the same problems as the dysfunctional emperor penguins and the 'penguins' from the laboratory.

To sum up this section: I explained the functional approach to understanding how organisms are able to stay alive and why they do so the way they do. To answer such questions biologists work together on organization charts that outline the main components of organisms and their role in maintaining the ability to stay alive. By identifying the biological role of the system they study, functional biologists position that part or activity into such an organization chart. This enables them to determine what questions are relevant to understand that system, namely 'how is that biological role performed' and 'why is that role performed in the way it is, rather then in other ways'. To answer the latter kind of question biologists point to traits of the organism that are functionally dependent on the trait to be explained, in the sense that the ability of an organism with those traits to stay alive would diminish if the trait to be explained were replaced by an alternative, whereas replacing the trait to be explained would not make much difference if the organism lacked the explaining traits. Such functional explanations explain why a certain trait is useful to the organisms that have it (why they need that trait for continued existence) but they do not explain how the relevant traits came into existence. This gives rise to the question of how it can be that the different parts are organized the way they need to be. I discuss that problem in the next section.

4 The Problem of Adaptation

The manifold and complex forms of organized life in our world give rise to what is known as 'the problem of adaptation' (for the concept of adaptation see Forber this volume). How come the features of an organism are adapted to each other (how come an organism has the features it needs to have)? How come the emperor penguins have the ability to produce two voiced sounds at the moment they need it? How come the different parts are there when they are needed, have the form and structure they need to have, and perform their activities in the right order and at the right time?

Such questions have been a main concern since the emergence of biology as an explanatory science in the eighteenth century. Pre-Darwinian functionalists, such as Johann Friedrich Blumenbach (1752–1840), Georges Cuvier (1769–1832), Johannes Müller (1801–1858) and Karl Ernst von Baer (1792–1876), tentatively assumed that adaptation results from the way in which the interaction between an organism's parts is organized. They thought that this interaction consists in some kind of material exchange but they had no precise idea about the way in which that exchange takes place, nor about the way in which it results in adaptation and organization.

Most of them were open to the idea that this was something to be discovered by future research, and at the same time sympathetic to philosopher Immanuel Kant's (1790) view that the non-linear arrangement of cause and effect in the living world prevents us from explaining how organization arises out of material interaction (see Lenoir 1982).

However, although it was thought that adaptation arises out of material interaction and it was deemed impossible (in principle or for the moment) to explain how adaptation and organization arise out of that interaction, according to the pre-Darwinian functionalists there is nevertheless a way to explain adaptation and organization, namely by assuming that material interaction is organized *for the purpose of* enabling the organism as a whole. Under this assumption functional explanations (explanations that explain how a certain part or activity meets the needs of the organism) are at the same time explanations of why that part or activity is *generated*, for, on that assumption, fulfilling the needs of the organism is the purpose for which that part or activity is generated.

Darwin's (1859) theory of evolution by natural selection solved the problem of adaptation in a surprising way: it abolished the need to explain adaptation, and, hence, the need for a teleological interpretation of functional explanation (but see Lennox and Kampourakis, this volume). Pre-Darwinian biologists tried to explain the characteristics of organisms and their adaptation entirely as the result of *individual level* processes. As Kitcher (1985) points out, Darwin's *Origin* (1859) shows the power of a new way of explaining, namely by relating the history of *lineages of populations*. This new form of explanation offered a new way to solve the problem of adaptation.

In the Darwinian view, the fact that the organization of individual level interaction (interaction between the parts and activities of the organism and between the organism and the environment) results in an organism capable of staying alive is explained not by an appeal to teleology, but by the assumption that the inherited factors that influence this interaction (such as DNA and other materials transferred via the gametes, learned behavior and the environment in which the organism lives) are already adapted to each other. In other words, the organization and adaptation of an organism result from the coordinating influence of already adapted factors inherited from that organism's parents.

The pre-existing adaptation of inherited factors is, in turn, explained by the theory of evolution by natural selection. Basically, this theory explains how new forms of organized life arise out of existing forms of organized life. Imagine a population of reproducing organisms with a certain form of organization (e.g. gazelle-like herbivores) whose offspring resemble their parents to a certain degree, but there is some variation with respect to one or more characteristics (e.g. some of the offspring have longer necks than others). If this variation is heritable (e.g. progeny of individuals with longer than average necks tend to produce offspring whose neck is on average longer than average) and some of the variants produce more offspring than other variants (e.g. individuals with longer than average necks tend to produce more offspring than average), the characteristics of the organisms in that population will gradually change (the necks will become longer and longer) (see also Depew this volume for different perceptions of this process).

This change might involve many characteristics. Perhaps, in a certain population, both variants with a longer than average neck and variants with longer than average legs produce more offspring than average (for example because both can reach for leaves higher up in the trees). In that case, both the average neck length and the average leg length increase in the course of time. This increase goes faster if variants that have both longer legs and a longer neck are even more productive than variants with one of those traits. If there is an additional advantage in having a combination of two favorable traits (having both a longer neck and longer legs is better than having either a longer neck or longer legs not only because of the additional food source, but also because of a better balance) the change might go pretty fast.

The gradual change in the population's characteristics has an effect on what variants are favored in the selection process. Perhaps there is some variation in the heart's ability to generate pressure. Perhaps, in the original population this variation was unfavorable or neutral. If the neck is longer, more pressure is needed to pump the blood to the head. So, it might well be that when the neck becomes longer, variants with a heart that can generate more than average pressure produce more offspring than variants whose heart generates average or less than average pressure. Similarly, when, due to the increase in neck length, the pressure in the arteries becomes higher, it might well be that variants with a thicker than average wall become favored. Perhaps, taller animals can spot predators at distances at which they could not see them before. This will affect the reproductive effects of variation in abilities related to properties like speed and strength (perhaps speed is less important for survival than before). Perhaps, animals with longer necks and legs are more suspicious, so camouflage becomes more important.

Such an accumulation of a variety of changes can, in the course of time, result in the emergence of new forms of organized life (e.g. giraffe-like herbivores). In this way the complex forms of organization we find today might have arisen out of simple forms of organized life that arose spontaneously in a suitable environment billions of years ago.

So Darwin's theory of evolution by natural selection solves the problem of adaptation by describing how one form of organization can give rise to other forms of organization. In this way it abolishes the need to explain adaptation and, hence, the need to give a teleological twist to the functional perspective. In the course of time functional explanations came to be seen in the way described in Sect. 3, as explaining why certain organisms need certain traits by explaining how certain other traits of that organism functionally depend on the trait to be explained.

Unfortunately, the relation between adaptation, function and natural selection in evolutionary theory is often misunderstood (see Reiss 2009 for an elaborate discussion of this problem). A main source of this misunderstanding is Darwin's metaphor of natural selection. This metaphor depicts nature, natural selection or the organism's environment as an agent who, in analogy to a dove breeder, picks the variants which will be allowed to produce offspring. This easily leads to the mistaken views that natural selection explains how adaptation came about (just as the dove breeder's intervention explains how his stock became adapted to his wishes), and that the parts of organisms have functions because organisms are subject to selection (just as artifacts are often supposed to have functions because they are picked by humans).

As I explained in this section, Darwin's theory solves the problem of adaptation by describing a way in which the different forms of organization we find today might have arisen out of simple forms of organized life, rather than by describing a way in which adaptation came about. Adaptation is just not the kind of thing that can come about in the course of evolution: organisms can exist and continue to exist only if their parts are adapted to each other and to the organism's way of life.

As I explained in Sect. 2, talk of functions of parts of organisms and artifacts is rooted in their character as systems that can exist only because of an organized ability, not because of the way they come into being or are maintained. Biological functions are positions in the organized maintenance of the individual organism, not contributions to the generation or maintenance of a certain type of organisms in the population.

Because of this tendency to confuse function and selection, it is important to be clear about the relation between function and selection and the differences between functional explanations and selection explanations. This is the subject of the next section.

5 Function and Evolution

5.1 Function and Evolutionary History

Traits that arise through and are maintained in the process of evolution by natural selection can be explained by a form of evolutionary explanation called 'adaptation explanation' (see Forber this volume). Adaptation explanations explain the availability, prevalence or relative frequency of a certain trait in a population by specifying why it was selected for, that is by specifying why that trait was favored in the process of natural selection. If in the example in Sect. 4, individuals with longer than average necks produced more offspring than average, because thanks to their longer neck they could browse the higher parts of the trees, the increase and the current neck length are said to occur as the result of selection for the ability to browse the higher parts of the trees. Similarly, if in a certain population a certain trait (e.g. a four chambered heart) did not change during a certain period because, due to failures to produce a certain effect (e.g. failing to pump blood), individuals with a deviant trait state (e.g. a heart in which blood leaks from one chamber into another) got less offspring than average or no offspring at all, the lack of change and the current trait state result from selection for that effect (for pumping blood). If a certain trait evolved due to selection for performing a certain function, that trait is an adaptation for that function (see Sober 1984, pp. 97–102 for this notion of 'selection for' and pp. 196–211 for this notion of 'adaptation for').

The relation between adaptation explanations and functional explanations has been subject to many controversies and confusions.

The influential school of reductionist physiology in mid-century Germany, including Karl Ludwig (1816–1895), Herman von Helmholtz (1821–1894) and

Emil du Bois-Reymond (1818–1869), claimed that Darwin's theory of evolution by natural selection means victory for their view that functional biology should attempt to explain life by filling in the parameters of physical and chemical laws without appeal to purpose or organization. Darwin's theory would show that functional biologists should not be bothered by adaptation, and that, hence, biology can do away with purpose and organization by replacing functional explanation by adaptation explanation.

However, these reductionists did not succeed in their attempt to explain life in terms of physics and chemistry. Ironically, the successful application of physics and chemistry characteristic of contemporary functional biology results from the adoption of the functional perspective described in Sect. 3 in the course of the twentieth century. As I discussed, this perspective combines mechanistic and functional explanations from a functional point of view in order to understand how organisms are able to maintain themselves.

In this new framework, the evolutionary approach complements the functional perspective rather than replaces it: the combination of mechanistic and functional explanations explains how organisms hang together, evolutionary explanations explain how this organization came about in the course of the history of the population, and developmental explanations explain how this organization comes about in the course of individual development.[7]

In his influential 'Cause and Effect in Biology' (1961) the evolutionary biologist Ernst Mayr, one of the architects of the evolutionary synthesis in the 1940s, claims that "the word 'biology' is a label for two largely separate fields which differ greatly in method, *Fragestellung*, and basic concepts" (p. 1501). In his eyes, functional biology is a reductionist science that applies physics and chemistry to answer how-questions about the parts and processes of organisms. Evolutionary biology is a historical science that applies the comparative method to answer why-questions about the differences between organisms in terms of their evolutionary history. Mayr associates these two disciplines with two types of 'causes': proximate and ultimate. Proximate causes are the causes that immediately influence the working of an organism's mechanisms (such as hormonal changes that influence a bird's readiness to migrate and the shift in temperature that cause the bird to start migrating at a certain day); ultimate causes consists in an organism's genetic constitution and the factors that shaped this constitution 'through many thousands of generations of natural selection' (p. 1503).

Mayr emphasizes (p. 1503) that the failure to distinguish these two 'causes' has all too often let to erroneous rejection of adaptation explanations. As an example he mentions Loeb's (1916) rejection of the view that tadpoles grow legs as an adaptation to land life on the ground that even very young tadpoles not yet able to live on

[7]These issues are often presented as 'the four problems of behavioral biology' and typically attributed to Tinbergen (1963). Tinbergen himself, however, emphasizes that 'the questions we ask' are 'the same throughout Biology' (p. 411) and refers to them as 'the four problems of *Biology*' (p. 426). Similar frameworks can be found in other biological disciplines (e.g. Huxley 1942: 40; Dullemeijer 1974: 95). See Dewsbury (1992) and Wouters (2005c) for more extensive discussion.

land develop legs when they are fed pieces of thyroid gland. In Mayr's view the administration of thyroid hormone is a proximate cause that triggers the development of legs pre-programmed in the genes by the ultimate cause: selection for life on land.

Mayr's dichotomous distinction between proximate and ultimate 'causations' has become the received view among evolutionary biologists (but see Uller this volume). However, its association of why-questions with selection history is utterly confused: it ignores functional explanation and it confounds evolution, selection and history.

As Francis (1990), Armstrong (1991), Dewsbury (1992, 1994, 1999) and Wouters (1995, 2005c) point out, why-questions are often answered by functional explanations (which explain why a certain trait is useful) rather than evolutionary ones (which explain why a certain trait evolved) and such explanations are neither causal nor historical in character. Aubin et al. (2000) explanation of the emperor penguin's two-voice system is a case in point. Functional explanations also serve to explain differences between organisms. For example, Aubin et al. (2000) suggest that their functional analysis of the two-voice system also explains why the non-nesting species of penguins (2 species: the emperor penguin and the king penguin) use a two-voice system for individual recognition whereas the nesting species of penguins (14 genera) do not do so: the latter do not need such a system because they can use the nest as a meeting point.

Questions such as 'why do emperor penguins produce two voiced sounds?' are, in other words, ambiguous. They may mean 'why is it useful to emperor penguins to produce two voiced sounds?' (the question addressed by Aubin et al. 2000) or 'why did the two voice system of sound production evolve?' The first type of question is answered by means of functional explanations, the second type by means of adaptational explanations. Functional explanations explain a trait in terms of what is needed to stay alive; adaptation explanations explain a trait in terms of what actually happened in the past. Because of this, functional explanations may compare existing organisms (such as emperor penguins) with hypothetical variants that need not have existed and often cannot plausible exist (e.g. 'organisms' that use visual cues to recognize each other, but are otherwise like emperor penguins), whereas adaptation explanations must compare variants that actually existed in the past. Functional explanations identify traits on which the trait to be explained is functionally dependent; adaptation explanations identify the effects in ancestral populations of the trait to be explained that were causally effective in the maintenance of that trait in that population. Functional explanations are concerned with non-causal relations (namely functional dependencies) at the individual level; adaptation explanations with causal relations at the level of lineages of populations.

Confusion might arise because adaptation explanations often use a kind of functional analysis to make claims about fitness (claims about the number of offspring produced by organisms of a certain type). Such functional analyses might serve to provide evidence for the claim that some variant was fitter than another as well as to explain why that variant was fitter (why variants with the trait to be explained had more offspring than their competitors). An adaptation explanation of the length of the giraffe's neck, for example, would not merely point out that longer necks evolved by selection for browsing the higher parts of the trees, it would also explain why browsing the higher parts of the trees increased their fitness. Perhaps, the higher leaves contained essential nutrients not present in the lower leaves. Perhaps, the population went through a period of food shortage in which individuals with longer than average necks had an advantage because there were more leaves within their reach. The part of an adaptation explanation that explains why a certain trait increased fitness is a kind of functional analysis that compares past variants that actually existed and were actually competing with respect to their reproductive capacity.

Although adaptation explanations are often supported by functional analysis, it would be a serious mistake to take functional explanations for adaptation explanations: claims about selection require evidence about the variants that actually occur in the population, the occurrence of selection, the genetics of the relevant traits, the mating system, the structure of the population, and the order in which the traits evolved (did long necked giraffes evolved out of ancestors with shorter necks or out of ancestors with even longer necks?), in addition to evidence about the fitness increasing effects of the trait (Brandon 1990, pp. 161–176).[8]

To show that the two voice system is useful to emperor penguins because, unlike conceivable alternatives, it allows the breeding partners to recognize each other in the circumstances in which they live, Aubin et al. (2000) provide evidence for the following claims: (1) the two voice system is used for individual recognition, (2) individual recognition plays an essential role in the emperor penguin's way of life, (3) in the circumstances in which emperor penguins live the alternatives would not be able to perform that role.

To argue that the two voice system was maintained by selection for individual recognition it should be evidenced that this effect was actually operative in the population's past and explains the current prevalence. To do so, information is needed about the selection situation as well as the effect of selection on the composition of the population. To provide evidence of selection, we must know which variants actually turned up in the past, show that these variants differ in their ability to produce offspring in the circumstances in which they lived, that these differences in reproductive capacity are due to differences in the ability to recognize partners and mates, and that the variant with a two voice system is the one with the greatest reproductive capability. To work out the effect of selection on the composition of the population (that is to show that selection resulted in the evolution or maintenance of the two voice system) we need information about the genetic make up of the different variants, their frequency distribution and the structure of the population.

To show that the two voice system was not only *maintained* by selection for individual recognition but *evolved as an adaptation* for that function, more detailed information is needed about the evolutionary history, especially about the order in which the relevant traits evolved and the actual circumstances in which they evolved.

---

[8]See Gould and Lewontin (1979), Lewontin (1981), Coddington (1988), Harvey and Pagel (1991), Brooks and McLennan (1991) and Reiss (2009) for extensive criticism of the view that in order to explain the evolution of a trait it suffices to point out its functional advantages.

5.2 *Function and Evolutionary Maintenance*

The second problem with Mayr's association of why-questions and selection is the confusion of evolution with selection history. Ariew (2003) and Thierry (2005) rightly object that there is more to evolution than natural selection: drift, mutation, recombination and gene flow, to name a few. Moreover, as Antonovics (1987), among others, points out, there is more to evolution than history: many evolutionary studies are concerned with the *current maintenance* of certain population characteristics, rather than with past evolution and maintenance.

Behavioral ecologists tend to subdivide Mayr's how-questions and why-questions each into two subtypes, a tradition that probably originates with Alcock (1975) (see Dewsbury 1992). How-questions are divided into questions about, on the one hand, mechanisms and the causes that trigger those mechanisms, and, on the other hand, development. Why-questions are divided into questions about, on the one hand, evolutionary maintenance and, on the other hand, evolutionary history. Within this scheme, questions about function are often erroneously presented as questions about evolutionary maintenance.

The kind of reasonings about evolutionary maintenance that Sober (1983) calls 'equilibrium explanations' might easily be confused with functional explanation. Equilibrium explanations explain the relative frequency of different variants in a population by relating why plausible deviations of that distribution will be selected against (cf. Sober 1983, 1984, pp. 139–142). A famous example is Fisher's (1930) explanation of the main pattern of distribution of males and females among all kinds of sexual reproducing organisms (50 % males, 50 % females) by looking at what would happen if there were more females than males (or the other way round) in a certain population: in such a population individuals of the minority sex produce, on average, more offspring than individuals of the majority sex (after all, it takes two to tango). But this would mean that individuals who produce more than average offspring of the minority sex are selected for. As a result the relative frequency of individuals of the minority sex would increase until there is no longer a minority sex. In a similar way, the prevalence of a certain trait state (e.g. a heart that pumps at a certain rate) in a certain population might be explained by pointing out that in the relevant circumstances deviant variants that may plausibly turn up in that population (e.g. individuals with a heart that pumped faster or slower) will get less offspring than average.

As Sober (1983, 1984, pp. 139–142) argues, equilibrium explanations are not of a causal nature. They don't tell us how a certain trait is *actually* maintained in the population, they tell us in abstract terms under which conditions a certain frequency distribution *would be* stable. As Sober puts it: even if it were true that every person with a certain character would commit murder in the circumstances of the crime under investigation, we wouldn't say that the detective who correctly inferred the character of the murderer, but does not know who committed the murder, knows who caused the victim's death. Fisher's model, for example, explains how it is that if (1) with respect to a certain trait (sex) there are two variants (male/female) in a population, (2) an organism of one of these types can reproduce only together with an organism of the other type (sexual reproduction), (3) mating is random within the population, and (4) there are heritable differences in the distribution of the two variants among the offspring of one mating pair, the 50/50 distribution would be stable.

But although equilibrium explanations are not causal, they are not functional either. Functional explanations are concerned with functional dependencies between the traits of an organism, that is with relations at the individual level. Equilibrium explanations are concerned with the way in which fitness depends on population characteristics.

As Reeve and Sherman (1993) point out, many studies of evolutionary maintenance aim to show that the variant that prevails in a certain environment is the one that, among a set of plausible variants, on average produces the highest number of offspring (let's call this 'the fittest variant') in that environment. If this is the case (if the prevailing variant is the fittest), it indicates that the prevailing variant is maintained by natural selection. If the prevailing variant turns out not to be the fittest, it indicates that there are other factors or processes involved in the maintenance of that variant.

One of the methods to support claims about the number of offspring organisms of a certain type produce is what Reeve and Sherman (1993) call 'the teleonomic approach,' which is a kind of functional analysis. In this approach, the efficiency in performing a certain function is used as a fitness indicator: it is assumed that the fitness of an organism increases if the function is performed more efficiently.

As in the case of adaptation explanations (discussed above), the 'teleonomic approach' uses functional analyses to support evolutionary explanations. However, once again, this should not mislead us into thinking that a functional explanation constitutes an evolutionary explanation. To explain evolutionary maintenance evidence should be given for the claims that (1) the fittest variant is the prevailing one in the relevant environment, (2) the variants with respect to which the prevailing variant is the fittest are the variants that can be expected to regularly occur in the population,[9] (3) the environmental context is the one specified in the analysis, (4) the variants are heritable, and (5) the mating system and population structure are as assumed, in addition to evidence for the claim that the prevailing variant is the fittest.

6 Functions, Reasons and Purposes

As I said in the introduction, it is highly intuitive to think that talk of functions, problems and solutions is teleological in character. I also said that the notion of function as used in biology is different from what we might intuitively think and that it might help in teaching to be aware of these differences. In this section, I discuss

[9]If the variants in the set do not occur, the variant is maintained by lack of variation rather than by selection; if there occur variants that are not in the set used for the fitness analysis, that analysis doesn't show that the prevailing traits is the fittest.

how talk of functions, problems and solutions in biology differs from our everyday function talk. I do so by discussing the philosophical theory that in my opinion accounts best for our everyday intuitions, namely the selected effects theory of function. I start by explaining this theory. I then discuss the relevant intuitions: I describe each intuition and explain how it is accounted for by the selected effects theory. Finally, I discuss the extent to which functional reasoning in biology parts with these intuitions.

Central to the philosophical discussion of function is the intuition that functions are a special kind of effect and that function attributions such as 'the heart has the function to pump the blood around' serve to distinguish effects that are functions (e.g. pumping blood) from effects that are side-effects (e.g. heart sounds and pulses) (Hempel 1959). On this assumption, the central aim of a philosophical theory of function is to define or explicate what distinguishes those effects that are functions from those effects that are not.

According to teleological theories of function, functions distinguish themselves from side-effects by being the effects for which the function bearer is generated.[10] On such theories the heart has the function to pump blood because it is generated for pumping blood. Pulses and heart sounds are only side-effects because they aren't effects for which the heart is generated.

The selected effects theory of biological function applies this teleological definition of function to the functions of the parts, activities and traits of organisms. In the light of contemporary evolutionary theory, an organism has the traits it has because of the way in which those traits influenced the number of offspring produced by that organism's ancestors as compared to competitors with other traits. In other words, an organism's traits are there because of *what they did in that organism's ancestors* that caused those ancestors to be favored in the process of natural selection.

So, if functions are defined as the effects for which the function bearer is generated, it is the function of a part or behavior of an organism to produce the effects for which that part or behavior was maintained in the process of natural selection. For example, the heart of certain organisms (say humans) has the selected function to pump blood if and only if those organisms have the type of heart they have (e.g. a hollow muscle steadily beats at a rate of 60–80 beats per minute when at rest and increases this rate up to four times within seconds when needed) because hearts of that type were favored by selection because of the way in which hearts of that type pumped blood in ancestral organisms in ancestral populations. In other words, on the selected effects theory, saying that a heart of a certain type has the function to pump blood is just another way of saying that an adaptation explanation of why hearts of that type are present in the population would appeal to past selection for the heart pumping blood in the way it is done in hearts of that type.

[10]See Wright (1972), Neander (1980, 1983, 1991a, b), Millikan (1984, 1989b), Mitchell (1989, 1995), Brandon (1990), Griffiths (1993), Godfrey-Smith (1994), Buller (1998), Schwartz (1999; 2002), and Garson (2011) for expositions and defenses of teleological theories of function, and Boorse (1976), Kitcher (1993), Amundson and Lauder (1994), Walsh (1996), Davies (2001), Cummins (2002), and Wouters (2003) for criticism.

A key feature of the selected effects theory is the idea that a function is not an effect a part, activity or trait currently produces, tends to produce or can produce, but an effect produced by parts, activities or traits of the same type in ancestral organisms. Due to this feature the selected effects theory meets some everyday intuitions that are difficult to account for if one sees functions as something the function bearer does.

One such intuition is the intuition that functions are the purposes for which or the reasons why the function bearer is what it is and where it is. This intuition easily leads to an association of function with design. For if something is designed by a designer, it is easy to make sense of this intuition: the function is the purpose for which or the reason why the designer made it or put it there.

The selected effects theory of biological function as defined above is teleological in sense that the selected function of a part, activity or trait is defined as 'what that thing (part, activity, trait) *is for*.' The theory explicates this 'what it is for' as 'what it evolved for,' which, in turn, is explicated as 'why it was maintained in the process of selection.'

Note that compared to the pre-Darwinian teleological view of functions as purposes that explain the generation of the function-bearer, selected functions are only weakly teleological: selected functions are what a trait is for, but that 'what it is for' is defined as a past effect produced by traits of the same type, not as the *purpose* of that trait or the *reason* why it is generated.

However, I suspect that the selected effects theory owes much of its attraction to 'what it is for' being tentatively equated with 'for what *purpose* it is there' or with 'for what *reason* it is there.' I shall refer to this intuitive modification as 'the combined selected effects theory of function, purpose and reason.' This combined theory maintains that it is the proper function of a thing to fulfill its purpose, that is the purpose of a thing to do what it is produced for, that what the thing is produced for is determined by the reasons why the thing is produced, and explicates this 'for what purpose' and 'for what reason' as 'why it was maintained in natural selection.' It might seem a bit idiosyncratic to stipulate that it is a thing's natural purpose to do what it was selected for or to see those past effects as evolutionary reasons, but it cannot be denied that this combined theory yields notions of natural purpose and evolutionary reason firmly grounded in evolutionary theory.

The combined selected effects theory of function, purpose and reason depicts evolution by natural selection as the kind of process that bestows purposes to the parts and processes of organisms, without assuming either that evolution put the things there for that purpose or that evolution itself has a purpose. In the same way, it views selected functions as the reasons why evolution produced the function bearer (made it into what it is, put it where it is, or maintained it the way it is) without assuming that evolution somehow knew what it did. In this way, it defines functions in terms of purposes and reasons without assuming a conscious designer.

Another everyday intuition asserts that functions are somehow normative: a function is an effect the function bearer should produce, is supposed to produce or is expected to produce. This intuition is a bit vague but it seems to imply at least that things can have functions they do not or cannot perform: although my pancreas does not and cannot produce insulin, it is function to do so, isn't it?

As any teleological theory of function, the selected effects theory easily accounts for this possibility, for there is nothing incoherent or impossible in a thing failing to produce the effects for which it is there: if it is the function of my pancreas to produce the effects for which it is there (for which my ancestors were favored over their competitors), it is irrelevant to its function whether or not it produces those effects now.

Note, however that selected functions are not normative in the ordinary sense of that term (they are not what the thing should do, is supposed to do or is expected to do), unless it is, in addition, stipulated that a function bearer should produce or can be expected to produce the effects for which it is there.

Another pervasive intuition that leads to difficulties if functions are something the function bearer does is the conviction that function attributions are by themselves explanatory of the presence of the function bearer (in other words that functional explanations are function attributions in answer to a why-question). This intuition was made explicit for the first time by philosopher John Canfield:

> Someone might say, 'Explain the function of thymus', or ask, 'What is the function of the thymus?' or 'Why do animals have a thymus?' When we answer, 'The function of the thymus is [such and such]' we have, it seems plain, given an explanation (Canfield 1964, p. 293)

In combination with the equally pervasive intuition that explanations cite the causes of what happened, it is difficult to see how functions can be explanatory of the presence of the function bearer if a function is something the function bearer does: effects cannot be the cause of the presence of that what causes it. The selected effects theory evades this problem by defining functions as *past* effects. Past effects can be the causes of the presence of the current function bearer, of course, and the selected effects theory defines functions as precisely those effects that were causally effective in the maintenance of the relevant type of function bearer in the population. Problem solved (Salmon 1989; Neander 1991b; Mitchell 1995).

Note, however, that, precisely because the selected effects theory conceives of functions as *past* effects (as effects produced in ancestral organisms), adaptation explanations cannot and do not appeal to effects *as* selected functions. For an organism's fate in the struggle for life depends ultimately on what happens to that very organism, not to what happened to its ancestors. It may matter for selection how well an organism's heart pumps blood, but it does not matter for selection how this ability evolved, and, hence, it doesn't matter for selection whether or not pumping blood is that organism's heart's selected function. Selected functions are, in other words, irrelevant to selection.

Because of this irrelevance, selected functions cannot have an explanatory role. If humans have the type of heart they have as the result of past selection for pumping blood, it follows that the heart has the selected function to pump blood. In that case the heart was selected because of the way it pumped blood, not because it had pumping blood as its selected function. So, although the effects appealed to in an adaptation explanation are, by definition, selected functions, adaptation explanations do not appeal to those effects *as* selected functions (that is, to those effects *being* selected functions). Selected functions may serve to summarize an adaptation explanation, but, unlike the teleological functions of the pre-Darwinian functionalists (and unlike the non-teleological notion of function in current biology), selected functions can't have and don't have an explanatory role.

To sum up, the selected effects theory of biological function, when combined with selected effects theories of purpose, reasons and norms, offers a way to reconcile our everyday intuitions about the teleological and normative character and the explanatory force of functions with contemporary evolutionary theory. It offers a way to view biological functions as the purpose for which or the reason why the function bearer is generated without assuming a designer, but it does not bestow an explanatory role to functions of this kind.

However, as I indicated in Sect. 4 biology drifted away from our everyday intuitions and uses terms like function in an entirely different way. Functions in the biologist's sense are positions in an organization and they do not assume purposes or reasons.

The point of the study of the emperor penguin's two voice system, for example, is not to distinguish functions from side-effects. Rather, the study seeks to understand how the two-voice system is useful in the context of the life of the organism. The function of a part or activity of a certain organism is determined neither by its (actual or potential) effects, nor by its past effects, but by the way in which that organism meets the requirements for being alive and the position of the function bearer in that organization. What function a thing has is entirely determined by its current context (that is, by the organization of the system of which it is a part) not by its future effects, its current properties, or its past effects. A function in the biologist's sense is, in other words, not a special kind of effect, but the way in which a part or activity fits into the organism's organization.

Because function is determined by the organism's form of organization rather than by the characteristics of the function bearer, parts or activities can have functions they do not perform. If an individual's display call is not recognized by its partner (perhaps because of that individual's state or constitution, perhaps because of external circumstances, perhaps because of its partner's condition or because that partner got lost or died) the performance of that call does not contribute to individual recognition and not to the producer's ability to survive and reproduce. Because the emperor penguins' way of meeting the requirements for being alive rests on their ability to recognize each other by auditory means, which in turn rests on the ability to produce two voiced sounds, it is nevertheless the function of that call to enable individual recognition. Without the ability to recognize each other by means of their sounds in dense colonies, the breeding partners would not be able to find each other back after weeks of separation and hence they would not be able to raise offspring in the circumstances in which they do, unless they would adopt a different way of life. So given the emperor penguins' form of organization it is the function of their display calls and of their two voice system to enable individual recognition, independent of whether or not they can and do produce that effect, and independent of whether or not they were selected for that effect.

Although this way of using the term 'function' allows parts and activities to have unperformed functions, the biologist's notion of function is not normative in the ordinary sense of that term: a biological function is something that needs to be done (given the form of organization), not something a thing should do or can be expected to do.

Function attributions have an important explanatory role in biology: often it is because a certain item or activity has a certain role that that item or activity needs to have a certain characteristic, or that it is useful that it has that characteristic. For example, it is because the display call has a role in individual recognition that it needs to be characteristic of the individual and it is because the sound producing organ has a role in individual recognition that it is useful that that organ consists of two parts that can work independently. If the display call had another role (for example, if it would have the function to make the caller more attractive to those who see him, or to warn its neighbors not to come too near) these things wouldn't be useful. Similarly, the average neck length in a certain population of gazelle-like herbivores might increase because the neck has a role in grasping food: if another part such as the trunk (as in elephants) or the forelimbs (as in many apes) has that role, individuals with longer than average necks wouldn't produce more or healthier offspring than individuals with shorter than average necks.

However, it is a misunderstanding to think that attributions of role functions are by themselves explanatory. The discovery that the two voice system has a role in individual recognition tells us how that system is used by the organism, but to understand why that system is useful (that is to explain it functionally) information about its role must be combined with information about the internal and external circumstances in which that role is performed, as well as with an explanation of why in those circumstances that role is better performed by a two voice system than by conceivable alternatives. The answer to the question 'What is the function of the emperor penguin's two-voice system?' is, in other words, *pace* Canfield quoted above, not the same as an answer to the question 'Why do emperor's penguins have a two-voice system?' The answer to this latter question starts with an answer to the first ('Penguins use the two-voice system to recognize each other,' as Aubin et al. (2000) put it in the title), and uses this insight to explain why in the circumstances in which the emperor penguins live 'two voices are better than one' (as Sturdy and Mooney 2000, put it in the title):

> More than 50 years after the first report of the 'two-voice' phenomenon in birds, the results of Aubin et al.'s experiments offer a functional explanation for this phenomenon [...] Aubin et al. describe both a function – mate/chick recognition and location – and an adaptive value'[11] – signal propagation through a noisy and crowded environment – for the 'two voices' of emperor penguin display calls (Sturdy and Mooney 2000: R635).

Intuitively, the purpose of a thing is the reason why that thing was designed, bought or put into use. A reason, in turn, is a desired state or activity to which the production, purchase or use of a thing is thought to contribute. For example, my computer has the purpose to help me to process text, because its assumed capacity to do so was one of the main reasons why I bought it. So, although purposes can be seen as effects to be pursued, what the purpose of a thing is, is determined by a

[11] 'adaptive value' is Study & Moony's term for what I call 'biological advantage'.

certain feature of its causal history: the state or activity for which it was designed, bought, or put into use. Function talk in biology on the other hand is not about an organism's causal history but about the way in which that organism is currently organized. So the biologist's way of talking about functions is quite different from our intuitive way of talking about purposes and does not assume reasons.

To a certain extent, the same holds true for artifact function. Where biological functions can be seen as roles in maintaining an organism's state of being alive, artifact functions can be seen as roles in maintaining a desired organized ability. This means that purposes are relevant to artifact functions because they are *artifact* functions, not because they are *functions* (just as self-maintenance is important to biological functions because they are *biological* functions, not because they are *functions*).

In other words, the similarity between organisms and artifacts that invites a functional stance resides in their organized character rather than in the way in which they are brought about. This means that the functional perspective in biology doesn't treat organisms as if they are designed artifacts. Rather, it treats the state of being alive as critically dependent not only on the organism's material composition but also on the arrangement of its parts and the timing of their activities.[12]

7 Summary and Conclusion

In this chapter I explained how function talk in biology is talk about organization. Functional biologists view organisms as solutions to the problem of how to stay alive and explain their organization by investigating what is needed to stay alive and how those requirements are met. This approach makes no assumption about the way the required organization came into being, and, hence, does not assume purpose, foresight or sentience.

This way of talking about functions differs from our intuitive way of doing so. Most people intuitively associate functions with purposes or reasons and take it for granted that talk of problems and solutions requires foresight and sentience. Intuitively, function attributions are explanatory (they identify the effects for which

[12]Lewens (2004) presents an interesting investigation of the assumed analogy between artifact design and biological function. Like many philosophers, he takes it for granted that function talk in biology is rooted in this analogy. In his view, the main connotations of the notion of function derived from artifact design are the idea that function ascriptions are explanatory and normative and that function attributions distinguish functions from accidents. Lewens argues that there are several ways to construct the analogy between biological and artifact function but that none of them perfectly matches these connotations. He draws the conclusion that the analogy between artifact design and organic evolution is not strong enough to warrant these ideas. Function talk is therefore (according to Lewens) merely a heuristic tool to draw conclusions about likely effects of selection. I agree with Lewens' conclusion about the weakness of the analogy between the *processes* of artifact design and organic evolution. However, I don't see that as a reason to put function talk aside as merely heuristic: in my view, function talk is not rooted in an analogy between these processes, but in the organized character of their *products*.

the function bearer was brought about) and normative (they set the standard for what counts as the right performance).

Biologists, on the other hand, treat functions as positions in an organization rather than as effects for which the function bearer was generated. The biological function of a part or activity is its role in the organization that enables the continued existence of that organism. Functional explanations explain why that role is better performed the way it is performed than in other conceivable ways. They explain why the function bearer needs to be present, needs to have the structure it has, or needs to do what it does, but not how it came to be present, got the structure it has, or generated its activities. What is better is not intrinsic to the role, but depends on that organism's organization and its way of life. Parts and activities of organisms can have functions they do not perform, but function attributions are not normative in the everyday sense of that term: a function in biology is not something the function bearer should perform or is expected to perform.

Research in cognitive development seems to indicate that the pervasive tendency to teleological thinking is an important obstacle to understanding evolution (Kampourakis 2007; Bardapurkar 2008; Galli & Meinardi 2011). In order to prevent the association of adaptation with teleology, Kampourakis (2013) recommends to avoid all talk of adaptation when teaching biology unless an explicit reference is made to selection history. Because natural selection is too difficult to understand for students at the elementary level, he recommends that teachers at that level avoid talk of adaptation, as well as any other linguistic constructions with a teleological load (such as 'birds have wings for flying').

The tendency to intuitively associate function with purpose might make it difficult to avoid talk that is perceived as teleological, for function seems essential to understand and remember the structure and working of biological entities and their parts and organs. It will be very unpractical, perhaps impossible to teach biology without implicit or explicit appeal to function.

It might be more practical to dissociate function from history and, hence, teleology as soon as possible. Perhaps, it is best to teach about function initially without bothering too much about teleological associations. After the students have some experience with functional reasoning the difference between talk about function and organization and talk about purpose and how it came should be introduced explicitly. As this difference is relevant to the parts and activities of both artifacts and organisms, it might help to illustrate this distinction by means of a simple artifact, such as a clothes peg, before explaining its relevance to living organisms.

References

Alcock, John. 1975. *Animal behavior. An evolutionary approach*. Sunderland: Sinauer.

Amundson, Ron, and George V. Lauder. 1994. Function without purpose: The uses of causal role function in evolutionary biology. *Biology and Philosophy* 9: 443–469.

Antonovics, Janis. 1987. The evolutionary dys-synthesis: Which bottles for which wine? *American Naturalist* 129(3): 321–331.

Ariew, André. 2003. Ernst Mayr's ultimate/proximate distinction reconsidered and reconstructed. *Biology and Philosophy* 18(4): 553–565.
Armstrong, Doug P. 1991. Levels of cause and effect as organizing principles for research in animal behaviour. *Canadian Journal of Zoology* 69: 823–829.
Aubin, Thierry, Pierre Jouventin, and Christophe Hildebrand. 2000. Penguins use the two-voice system to recognize each other. *Proceedings of the Royal Society of London Series B* 267(1448): 1081–1087.
Bardapurkar, Abhijeet. 2008. Do students see the "Selection" in organic evolution? *Evolution: Education and Outreach* 1: 299–305.
Bigelow, John, and Robert Pargetter. 1987. Functions. *Journal of Philosophy* 84(4): 181–196.
Boorse, Christopher. 1976. Wright on functions. *Philosophical Review* 85(1): 70–86.
Brandon, Robert N. 1990. *Adaptation and environment*. Princeton: Princeton University Press.
Brooks, Daniel R., and Deborah A. McLennan. 1991. *Phylogeny, ecology and behavior*. Chicago: The University of Chicago Press.
Buller, David J. 1998. Etiological theories of function: A geographical survey. *Biology and Philosophy* 13: 505–527.
Canfield, John V. 1964. Teleological explanation in biology. *The British Journal for the Philosophy of Science* 14(56): 285–295.
Canfield, John V. 1965. Teleological explanation in biology: A reply. *The British Journal for the Philosophy of Science* 15(60): 327–331.
Christensen, Wayne, and Mark H. Bickhard. 2002. The process dynamics of normative function. *The Monist* 85(1): 3–28.
Coddington, Jonathan A. 1988. Cladistic tests of adaptational hypothesis. *Cladistics* 4: 3–22.
Craver, Carl F. 2001. Role functions, mechanisms, and hierarchy. *Philosophy of Science* 68: 53–74.
Cummins, Robert. 1975. Functional analysis. *Journal of Philosophy* 72(20): 741–765.
Cummins, Robert. 2002. Neo-teleology. In *Functions*, ed. André Ariew, Robert Cummins, and Mark Perlman, 157–172. Oxford: Oxford University Press.
Darwin, Charles. 1859. *On the origin of species*. London: John Murray.
Davies, Paul Sheldon. 2001. *Norms of nature. Naturalism and the nature of functions*. Cambridge: MIT Press.
Delancey, Craig. 2006. Ontology and teleofunctions: A defense and revision of the systematic account of teleological explanation. *Synthese* 150(1): 69–98.
Dewsbury, Donald A. 1992. On the problems studied in ethology, comparative psychology, and animal behavior. *Ethology* 92(2): 89–107.
Dewsbury, Donald A. 1994. On the utility of the proximate-ultimate distinction in the study of animal behavior. *Ethology* 96(1): 63–68.
Dewsbury, Donald A. 1999. The proximate and the ultimate: Past, present, and future. *Behavioural Processes* 46(3): 189–199.
Dullemeijer, Peter. 1974. *Concepts and approaches in animal morphology*. Assen: Van Gorcum.
Fisher, Ronald A. 1930. *The genetic theory of natural selection*. Oxford: Clarendon.
Francis, Richard C. 1990. Causes, proximate and ultimate. *Biology and Philosophy* 5(4): 401–415.
Frankfurt, Harry G., and Brian Poole. 1966. Functional analyses in biology. *The British Journal for the Philosophy of Science* 17: 69–72.
Galli, Leonardo González, and Elsa Meinardi. 2011. The role of teleological thinking in learning the Darwinian model of evolution. *Evolution: Education and Outreach* 4: 145–152.
Garson, Justin. 2008. Function and teleology. In *A companion to the philosophy of biology*, ed. Sahotra Sarkar and Anya Plutynski, 525–549. Cambridge: Blackwell.
Garson, Justin. 2011. Selected effects and causal role functions in the brain: The case for an etiological approach to neuroscience. *Biology and Philosophy* 26(4): 547–565.
Godfrey-Smith, Peter. 1994. A modern history theory of functions. *Noûs* 28(3): 344–362.
Gould, Stephen Jay, and Richard C. Lewontin. 1979. The spandrels of San Marco and the Panglossian paradigm: A critique of the adaptationist programme. *Proceedings of the Royal Society of London B* 205: 581–598.

Greenewalt, Crawford H. 1968. *Bird song: Acoustics and physiology*. Washington, TN: Smithonian Institute Press.
Griffiths, Paul E. 1993. Functional analysis and proper functions. *The British Journal for the Philosophy of Science* 44(3): 409–422.
Harvey, Paul H., and Mark D. Pagel. 1991. *The comparative method in evolutionary biology*. Oxford: Oxford University Press.
Hempel, Carl G. 1959. The logic of functional analysis. In *Symposium on sociological theory*, ed. L. Gross, 271–287. New York: Harper and Row.
Horan, Barbara L. 1989. Functional explanations in sociobiology. *Biology and Philosophy* 4: 131–158, 205–228.
Huxley, Julian L. 1942. *Evolution: The modern synthesis*. London: Allen & Unwin.
Kampourakis, Kostas. 2007. Teleology in biology, chemistry and physics education. *Review of Science, Mathematics and ICT Education* 1(2): 81–93.
Kampourakis, Kostas. 2013. Teaching about adaptation: Why evolutionary history matters. *Science & Education* 22(2): 173–188.
Kant, Immanuel. 1790. *Kritik der Urteilskraft*. Berlin: Lagarde und Friederich.
Kitcher, Philip. 1985. Darwin's achievement. In *Reason and rationality in natural science*, ed. Nicholas Rescher, 127–190. Lantham: University Press of America.
Kitcher, Philip. 1993. Function and design. In *Philosophy of science*, ed. Peter A. French, Theodore E. Uehling, and Howard K. Wettstein, 379–397. Notre Dame: University of Notre Dame Press.
Lenoir, Timothy. 1982. *The strategy of life*. Dordrecht: Reidel.
Lewens, Tim. 2004. *Organisms and artifacts: Design in nature and elsewhere*. Cambridge: MIT Press.
Lewontin, Richard C. 1981. The organism as the subject and object of evolution. *Scientia* 118: 65–82.
Loeb, Jacques. 1916. *The organism as a whole*. New York: Putnam.
Mayr, Ernst. 1961. Cause and effect in biology. *Science* 134: 1501–1506.
McLaughlin, Peter. 2001. *What functions explain. Functional explanation and self-reproducing systems*. Cambridge: Cambridge University Press.
Millikan, Ruth Garrett. 1984. *Language, thought, and other biological categories*. Cambridge: MIT Press.
Millikan, Ruth Garrett. 1989a. An ambiguity in the notion "function". *Biology and Philosophy* 4(2): 172–176.
Millikan, Ruth Garrett. 1989b. In defense of proper functions. *Philosophy of Science* 56(2): 288–302.
Millikan, Ruth Garrett. 1993. Propensities, exaptations, and the brain. In *White Queen psychology and other essays for Alice*, 31–50. Cambridge: The MIT Press.
Mitchell, Sandra D. 1989. The causal background of functional explanation. *International Studies in the Philosophy of Science* 3: 213–230.
Mitchell, Sandra D. 1993. Dispositions or etiologies? A comment on Bigelow and Pargetter. *Journal of Philosophy* 90(5): 249–259.
Mitchell, Sandra D. 1995. Function, fitness and disposition. *Biology and Philosophy* 10: 39–54.
Mossio, Matteo, Cristian Saborido, and Alvaro Moreno. 2009. An organizational account of biological functions. *The British Journal for the Philosophy of Science* 60(4): 813–841.
Neander, Karen. 1980. *Teleology in biology*. Paper presented to the AAP conference in 1980.
Neander, Karen. 1983. *Abnormal psychobiology*. Ph.D. thesis, La Trobe University, Melbourne.
Neander, Karen. 1991a. Function as selected effects: The conceptual analyst's defense. *Philosophy of Science* 58(2): 168–184.
Neander, Karen. 1991b. The teleological notion of 'function'. *Australasian Journal of Philosophy* 69(4): 454–468.
Prigogine, Ilya, and Isabelle Stengers. 1984. *Order out of chaos*. London: Heinemann.
Reeve, Hudson Kern, and Paul W. Sherman. 1993. Adaptation and the goals of evolutionary research. *The Quarterly Review of Biology* 68: 1–32.

Reiss, John O. 2005. Natural selection and the conditions for existence: Representational vs. conditional teleology in biological explanation. *History and Philosophy of the Life Sciences* 27(2): 249–280.

Reiss, John O. 2009. *Not by design: Retiring Darwin's watchmaker*. Berkeley: University of California Press.

Ruse, Michael. 1971. Functional statements in biology. *Philosophy of Science* 38: 87–95.

Saborido, Cristian, Matteo Mossio, and Alvaro Moreno. 2011. Biological organization and cross-generation functions. *The British Journal for the Philosophy of Science* 62(3): 583–606.

Salmon, Wesley C. 1989. *Four decades of scientific explanation*. Minneapolis: University of Minnesota Press.

Schlosser, Gerhard. 1998. Self re-production and functionality: A systems-theoretical approach to teleological explanation. *Synthese* 116(3): 303–354.

Schwartz, Peter H. 1999. Proper function and recent selection. *Philosophy of Science* 66: S210–S222.

Schwartz, Peter H. 2002. The continuing usefulness account of proper function. In *Functions*, ed. André Ariew, Robert Cummins, and Mark Perlman, 244–260. Oxford: Oxford University Press.

Sober, Elliott. 1983. Equilibrium explanation. *Philosophical Studies* 43: 201–210.

Sober, Elliott. 1984. *The nature of selection*. Cambridge: The MIT Press.

Stein, Robert Carrington. 1968. Modulation in bird sounds. *The Auk* 85: 229–234.

Sturdy, Christopher B., and Richard Mooney. 2000. Two voices are better than one. *Current Biology* 10(17): R634–R636.

Thierry, Bernard. 2005. Integrating proximate and ultimate causation: Just one more go! *Current Science* 89(7): 1180–1183.

Tinbergen, Niko. 1963. On aims and methods of ethology. *Zeitschrift für Tierpsychologie* 20: 410–433.

Walsh, Denis M. 1996. Fitness and function. *The British Journal for the Philosophy of Science* 47: 553–574.

Wimsatt, William C. 1972. Teleology and the logical structure of function statements. *Studies in History and Philosophy of Science* 3: 1–80.

Wimsatt, William C. 1986. Forms of aggregativity. In *Human nature and natural knowledge*, ed. Donagan Alan, N.Perovich Anthony Jr., and V.Wedin Michael, 259–291. Dordrecht: Reidel.

Wimsatt, William C. 1997. Aggregativity: Reductive heuristics for finding emergence. *Philosophy of Science* 64: S372–S384.

Wimsatt, William C. 2007. *Re-engineering philosophy for limited beings*. Cambridge, MA: Harvard University Press.

Wouters, Arno G. 1995. Viability explanation. *Biology and Philosophy* 10(4): 435–457.

Wouters, Arno G. 1999. Explanation without a cause. Ph.D. thesis, Utrecht University, Utrecht.

Wouters, Arno G. 2003. Four notions of biological function. *Studies in History and Philosophy of Biology and Biomedical Science* 34(4): 633–668.

Wouters, Arno G. 2005a. The function debate in philosophy. *Acta Biotheoretica* 53(2): 123–151.

Wouters, Arno G. 2005b. Functional explanation in biology. In *Cognitive structures in scientific inquiry*, Essays in debate with Theo Kuipers, vol. 2, ed. Festa Roberto, Aliseda Atocha, and Peijnenburg Jeanne, 269–293. Amsterdam: Rodopi.

Wouters, Arno G. 2005c. The functional perspective of organismal biology. In *Current themes in theoretical biology*, ed. Thomas A.C. Reydon and Lia Hemerik, 33–69. Dordrecht: Springer.

Wouters, Arno G. 2007. Design explanation: Determining the constraints on what can be alive. *Erkenntnis* 67(1): 65–80.

Wright, Larry. 1972. Explanation and teleology. *Philosophy of Science* 39(2): 204–218.

Understanding Biological Mechanisms: Using Illustrations from Circadian Rhythm Research

William Bechtel

1 Introduction

In many fields of biology, when investigators seek an explanation for a phenomenon, what they are seeking is an account of the mechanism responsible for it. The search for mechanisms to explain phenomena has played an important role in biology for over two centuries. Twentieth century philosophy of science, however, largely neglected mechanisms; the dominant account of explanation held that explanation involves deriving descriptions of phenomena from statements of laws and initial conditions (Hempel 1965). Noting that laws are seldom averted to in biological explanations (see Lange this volume for laws in biology) but references to mechanisms are ubiquitous, in the last two decades several philosophers of biology have turned their attention to characterizing what biologists take mechanisms to be and the strategies they employ to discover, represent, and evaluate mechanistic explanations (Bechtel and Abrahamsen 2005; Bechtel and Richardson 1993/2010; Machamer et al. 2000). These philosophical analyses of mechanistic explanation can be valuable for educators seeking to present the framework of biological inquiry to students (van Mil et al. 2013).

The distinguishing feature of mechanistic explanation is that scientists seek to explain a phenomenon of interest by identifying the working parts of the responsible mechanism—the parts that perform the various operations that go into producing the phenomenon. Since mechanistic investigation proceeds by decomposing a mechanism into its component parts, which in a straightforward sense are at a lower level of organization (they are necessarily smaller than the mechanism as a whole), it is often characterized as reductionistic. This sense of reduction is rather different

W. Bechtel (✉)
Department of Philosophy and Center for Chronobiology, University of California, 9500 Gilman Drive, La Jolla, San Diego, CA, USA
e-mail: bill@mechanism.ucsd.edu

K. Kampourakis (ed.), *The Philosophy of Biology: A Companion for Educators*, History, Philosophy and Theory of the Life Sciences 1, DOI 10.1007/978-94-007-6537-5_22,

from that often presented in the philosophical literature (Nagel 1961) in which laws characterizing phenomena at one level are derived from those at a lower level and according to which in the end everything might eventually be explained from the lowest-level. For one thing, the parts and operations appealed to in mechanistic explanations may not be characterized in terms of laws. But even more fundamentally, the lower level does not provide all the information needed to account for the phenomenon. The working parts only produce the phenomenon when they are organized and their operations are appropriately orchestrated. Many components of biological mechanisms operate differently when situated within the mechanism than when removed from it (Boogerd et al. 2005).

Knowledge of the manner in which a mechanism is organized and how it affects the parts is additional to the knowledge of the parts that is gleaned from focusing on them treated individually. Moreover, the conditions under which the mechanism works may depend on conditions imposed on the mechanism from its environment. Thus, the reductionistic knowledge provided by mechanistic decomposition requires additional knowledge of the modes of organization realized at higher levels, including levels in which the whole mechanism is just a part, in order to explain a phenomenon. Accordingly, although the techniques for doing so are less developed than those for decomposing mechanisms, mechanistically oriented scientists must also recompose and situate mechanisms in order to account for how they produce the phenomenon. Drawing students' attention to both the valuable aspects of reductionistic decomposition and the need to complement decomposing with recomposing and situating mechanisms can help them develop a more comprehensive understanding of biology (see also Braillard this volume for reductionism and systems biology).

After first further articulating the nature of mechanistic explanation in the next section, I will in subsequent sections discuss the key tasks in developing such explanations—delineating the phenomenon, identifying and decomposing the responsible mechanism, and recomposing and situating the mechanism. To help make the exposition concrete, I will develop as an example throughout the chapter research on circadian rhythms—endogenously controlled oscillations of approximately 24 h in many physiological processes (e.g., basic metabolism and body temperature) and behaviors (e.g., locomotion and cognitive performance). Circadian rhythms are a fascinating phenomenon that readily attracts student interest and research on them provides a reasonably accessible case for introducing students to the intricacies of mechanistic research.

Since circadian rhythms are maintained endogenously, researchers have sought an internal mechanism, a clock, to explain them. Because rhythms are manifest even in single-celled organisms (cyanobacteria and fungi), researchers have assumed that the clock mechanism exists inside individual cells and over the last 20 years an explanatory schema involving a transcription-translation feedback loop (in which a product of gene expression builds up and as it does so inhibits gene expression until it is degraded) has received strong support. In cyanobacteria Nakajima et al. (2005) demonstrated that a system consisting of the KaiA, KaiB, and KaiC proteins together with ATP sufficed to generate circadian oscillations, pointing to a simpler feedback loop involving phosphorylation and dephosphorylation of a crucial protein as sufficient for circadian rhythms. But even in cyanobacteria, the tr anscription-translation loop is assumed to play a fundamental role in normal maintenance of circadian rhythms. In multi-celled organisms the components of the clock mechanism are found and exhibit the appropriate behavior in (nearly) all cells of the organism, but in animals the ability to maintain circadian rhythmicity depends upon the clock mechanism in specific cells in the brain. In mammals, these cells are located in a small structure within the hypothalamus known as the suprachiasmatic nucleus (SCN). As I will discuss below in the context of discussing the process of recomposing a mechanism, there is a complex coordination process whereby individual SCN neurons depend on others to maintain a reliable circadian rhythms.

2 Characterizations of Mechanisms

The ideas scientists deploy to explain nature often originate as metaphors in which they transfer a framework humans have developed in technological applications to understand a system they encounter in nature. The recent attraction of the computer metaphor for understanding the mind and brain is an illuminating example. Once humans had constructed machines that could manipulate symbols (encoded as strings of 0s and 1s), cognitive scientists and neuroscientists began to apply the idea of computation to characterize how information is processed. (Over time, the idea of computation has been extended beyond that which Turing (1936) and Post (1936) initially proposed, and now often seems to involve nothing more than a series of causal processes that transform one representation into another.) This example illustrates a more general practice in which scientists have tried to understand natural processes using ideas developed in the context of humanly constructed machines.[1] Descartes was one of the first to give clear expression to this idea, characterizing

[1]Pigliucci and Boudry (2011) argue that what they call the machine-information family of metaphors has negative consequences in both science education and scientific research (see also Brigandt this volume). In particular, in science education it provides an opening for intelligent design. While the comparison to human made machines does invite appeals to a designer, rejecting the practice of several hundred years in biology of seeking mechanisms to explain phenomena is not a good educational strategy. A better strategy is to consider seriously the sort of mechanisms that are found in living organisms (e.g., ones that build and repair themselves) as well as to emphasize that machines are typically not optimally designed, and this is especially true of biological mechanisms. The origin of biological mechanisms is better explained by evolutionary processes (drift as well as selection) than by appeal to an omniscient intelligent designer. A further part of Pigliucci and Boudry's critique focuses on the use of the information or blue-print metaphor for the relation of genes to biological traits. They appeal to work focusing on biological development to show that the relation between genes and traits is far more complex—organisms and environments figure centrally in explaining how genes are expressed. This is the view of developmental systems theory and one way proponents such as Griffiths and Gray (1994) present their message is by viewing genes as just one part of the complex developmental mechanism responsible for the appearance of traits in organisms in successive generations.

"this Earth and indeed the whole universe as if it were a machine." For Descartes a machine operated as a result of its parts having specific shapes that affected each other's movement brought about by contact action, and so he continued: "I have considered only the various shapes and movements of its parts" (*Principia* IV, p. 188). Descartes extended this idea to the bodies of organisms, proposing to explain all animal functions and behavior (including the same behaviors when they occurred in humans) in terms of the push and pull of component parts. Many investigators followed in Descartes' footsteps, but unlike Descartes, engaged in empirical investigation as they sought to understand biological mechanisms. Moreover, over time they extended the range of basic processes they viewed as appropriate for introducing in mechanistic explanations to include Newtonian attraction between objects, the creation and breaking of chemical bonds, and conduction of electrical currents.

The philosophers of science who have begun to focus on mechanistic explanation have been concerned primarily to provide an account of explanation that fits the practice of biologists. Thus, they have often started with particular examples of explanations found in biology: explanations of basic metabolic processes such as fermentation, of the synthesis of proteins, or of the transmission of chemical signals between neurons. They have then noted that in these cases, scientists decompose the mechanisms into both entities or parts and the activities or operations these perform, and also appeal to the ways in which these components are organized. Thus, Machamer et al. (2000) offer the following characterization of mechanisms: "*Mechanisms are entities and activities organized such that they are productive of regular changes from start or set-up to finish or termination conditions.*" The characterization Bechtel and Abrahamsen (2005) advanced is quite similar: "A mechanism is a structure performing a function in virtue of its component parts, component operations, and their organization. The orchestrated functioning of the mechanism is responsible for one or more phenomena." The differences in terminology (parts vs. entities; operations vs. activities) are not important for purposes here, but a significant point of disagreement concerns Machamer et al.'s contention that organization is sequential "from start or set-up to finish or termination conditions." While they acknowledge that biological organization does sometimes involve forks and cycles, they nonetheless require start and termination conditions in their accounts and characterize the activities in between sequentially. Bechtel and Abrahamsen speak more generally of "orchestrated functioning" and have focused on examples, such as circadian rhythms, in which simultaneous operations in multiple feedback loops are crucial to the functioning of the mechanism. Noting the importance of computational modelling and the use of dynamical systems theory to understand how these mechanisms operate, they have proposed a framework of *dynamic mechanistic explanation* that incorporates dynamical analysis of the recomposed mechanism with the traditional focus on decomposition into parts and operations (Bechtel and Abrahamsen 2011; Bechtel 2011).

Mechanistic explanations seek to characterize the mechanism responsible for a given phenomenon. Sometimes such accounts of mechanisms are presented in linguistic descriptions. But often scientists find it helpful to represent the mechanism visually in diagrams that take advantage of two dimensions to situate parts and use conventions such as shape or color of icons to distinguish types of parts and to indicate their operations. Arrows, with different styles of arrows representing different operations, portray the operations by which particular parts affect others. I will present some diagrams of mechanisms below, but for now I note how such use of diagrams distinguishes mechanistic explanations from nomological ones. Whereas nomological explanation insists on representing all information in propositions, researchers trying to understand mechanisms seek representational formats that support their reasoning about the mechanism. This links to another difference: nomological and mechanistic explanations provide alternative accounts as to how what is offered in explanation relates to the phenomenon being explained. In nomological explanations, logical derivations connect the laws to linguistic descriptions of the phenomena. Diagrams do not figure in logical derivations (although many logicians since Euler and Venn have appealed to diagrams to represent logical relationships), so the process of relating the explanations to the phenomena to be explained must be different on the mechanist account. Instead of drawing inferences, scientists simulate the operation of a mechanism to understand how it could generate the phenomenon. In some cases this can be done by mentally rehearsing each of the operations, often supported by a diagram or mentally imagining the mechanism as it would be presented in a diagram. As they confront more complex mechanisms with multiple simultaneous operations, scientists are increasingly appealing to animations to illustrate how a mechanism generates a phenomenon (McGill 2008).

Two weaknesses of both mental simulations and animations is that one may take the components of a mechanism as being able to do more than in fact they can or one may neglect important consequences of the operations of the components. This is particularly a risk when many operations are occurring simultaneously in the actual mechanism and these operations interact with each other in non-linear ways. Accordingly, researchers often appeal to model systems that they have constructed to emulate the way the parts operate or to computational models in which they characterize operations mathematically, to determine what the actual mechanism will do.

Philosophers of science adopting the nomological framework have traditionally eschewed trying to understand scientific discovery, insisting that the tools of logic could illuminate the evidential support for laws but not the processes by which they were discovered (Reichenbach 1938). (Some cogntive scientists, though, have ventured where philosophers feared to tread and advanced accounts of how such laws could be discovered. See Holland et al. 1986; Langley et al. 1987). When the project of explanation is understood as the discovery of mechanisms, philosophers are in a position to articulate the strategies through which scientists make discoveries. One approach is to focus on the reasoning strategies scientists use as they try to piece information together to develop an account of a mechanism (Bechtel and Richardson 1993/2010; Darden 2006; Darden and Craver 2002). Another is to focus in detail on how, in their experimental work, scientists intervene and manipulate biological mechanisms to elicit information about their parts and operations (Bechtel 2006; Craver 2002, 2007). Less has been done to date on the strategies through which scientists recompose mechanisms, especially in computational models, and use the results either to guide further experiments or revisions in the proposed mechanism (Bechtel and Abrahamsen 2010), but this is a topic ripe for additional philosophical research.

3 Delineating Phenomena

Mechanisms are invoked to explain phenomena and so it is important to specify what phenomena are. While many accounts of explanation have assumed that scientists try to explain observations (data), Bogen and Woodward (1988) convincingly demonstrate that what they in fact seek to explain are phenomena in the world. Although some phenomena only occur once, most are repeatable occurrences for which one can seek to specify the conditions under which they occur (doing so often requires considerable experimental inquiry). Examples of biological phenomena include the growth of plants, the births of mammals, circulation of blood, the metabolism of sugars, fats, and proteins to procure energy, and the conductance of an electrical signal down a nerve. These are what scientists seek to explain, not the data that provide evidence for them. Only when an observation or an experiment produces what is regarded as anomalous data do scientists turn their attention to explaining the data themselves.

An important part of delineating phenomena is to develop appropriate representations of them. Many of the philosophers advancing mechanistic explanation have focused on linguistic descriptions of phenomena (e.g., "proteins are synthesized by constructing strings of amino acids in the order specified in a sequence of DNA"). However, scientists are often interested in explaining much more specific features of phenomena, such as the rates at which a process occurs, and so characterize phenomena in terms of numerical values determined by empirical research. Frequently the numerical data they collect to characterize a phenomenon is presented in tables. However, because what they are generally interested in is the pattern exhibited in the numerical values, researchers must employ other representational techniques that make the pattern apparent. One way of doing this is in terms of equations, such as Weber's (1834) psychophysical law identifying a constant proportion between a just noticeable change in a stimulus (ΔI) and the total quantity of the stimulus(I):

$$\frac{\Delta I}{I} = \mathrm{k}.$$

Although this relation is often called a law, it does not play the role in explanation identified in nomological accounts: it specifies a relation for which an explanation is sought but does not explain its instances. It is important to distinguish between laws that offer explanations (Netwon's force laws were intended to explain the motion of bodies) and those such as Weber's law that characterize phenomena. The latter still require explanation, either in terms of other laws or in terms of mechanisms that explain why the regularities hold. Weber himself proposed a possible mechanism, but a satisfactory explanation has not yet been advanced. Moreover, in the meantime researchers have advanced alternative mathematical representations, such as Stevens' power law, that they claim to better characterize the relation between physical stimuli and how they are perceived. Often it proves difficult to develop an equation to characterize the phenomenon, and researchers instead develop diagrammatic formats (such as illustrated below) to illustrate the pattern for which an explanation is sought.

Philosophers have typically presented the phenomenon to be explained as identified by observations. Even when phenomena are identified through observations (e.g., astronomical observations that were used by Copernicus, Kepler, Galileo, and Newton to characterize the behavior of planets), complex instruments and procedures are frequently required to secure the data from which the phenomenon can be elicited. In addition, in fields such as biology, scientists must intervene in nature to elicit the pattern for which explanation is developed. Even observational techniques, such as microscopic observation, require intervention to prepare specimens for observation through a microscope: water is removed from the specimen and it is chemically modified by stains and fixatives. What one observes in the microscope is the product of these manipulations, which are often quite brutal. An important question that biologists must address is whether they have generated an artifact or have accurately portrayed the phenomenon.[2]

I will illustrate the process of delineating the phenomenon in the case of circadian rhythms. The daily oscillations in some activities of living organisms are easily detectable if someone takes the effort to carry out observations across different periods of a day (the sleep-wake cycle in animals, phosphorescence in *Gonyaulax*, and the folding and unfolding of leaves in some plants). But others require instruments to detect them (e.g., the cycle in body temperature, of cell division in animals, or of gene expression). One of the key features of the phenomenon of circadian rhythms is that they are endogenously generated—they are not simply responses to the oscillations in daylight, temperature, etc., in the environment. To demonstrate that an oscillation in some feature or behavior of an organism meets this condition, researchers must set up conditions in which no likely environmental cue (known as a *Zeitgeber*) is available to set the phase of the oscillation. Thus, when de Mairan (1729) suspected the opening and closing of leaves in the *Mimosa* plant that he had observed was controlled endogenously, he had to resort to raising plants in continuous darkness and show that they continued to fold and unfold their leaves. However, it was always possible that the organism was responding to some other Zeitgeber that was itself responsive to the time of day. A crucial source of evidence in showing that rhythms were endogenously controlled was that when Zeitgebers were removed, they oscillated with a period slightly different from 24 h, which should not be the case if they were responses to cues that were responsive to the day-night cycle of

[2]I have explored the challenge in determining whether evidence is an artifact or reflective of the real phenomenon and the strategies biologists use to address this challenge in the context of modern cell biology in Bechtel (2006). Since typically there is much that is unknown about how the procedures used actually work, researchers rely on such considerations as whether the evidence exhibits a distinct pattern distinguishable from noise, whether it can be at least partially corroborated using other techniques, and whether there are compelling theoretical explanations of the putative phenomenon. Although we commonly think of evidence as more secure and foundational to the explanatory hypotheses advanced, in fact evidence is often just as contested in science and evaluated in conjunction with the explanations offered.

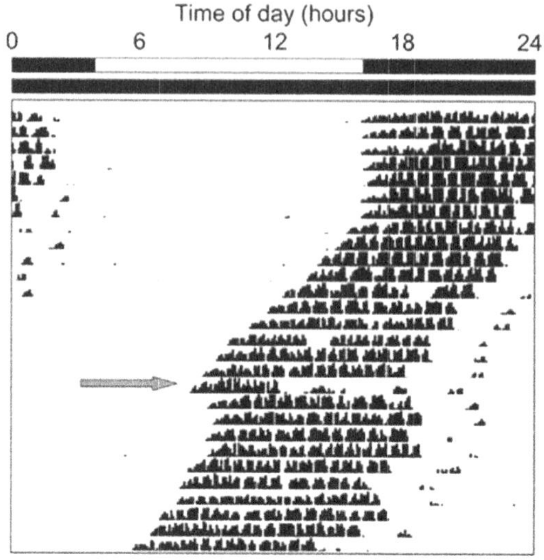

Fig. 1 A basic actogram in which the *top bar* indicates a normal *light–dark* condition for the first 7 days and *constant darkness* for subsequent days. The *grey arrow* identifies the day a light pulse was administered. (From http://www.photosensorybiology.org/id16.html)

our planet (The name *circadian* reflects this: *circa* [about] + *dies* [day]). This required finding and representing patterns of behavior that oscillated with a period somewhat different from 24 h. The actogram format was developed to make this phenomenon manifest in a diagram.

An actogram is typically constructed from an automatic recording of an animal's activity. In perhaps the first actogram, Johnson (1926) devised a rotating disc on which a deflection was recorded every time a mouse moved. More recently, rodent activity is recorded from every rotation of a running wheel provided to the animal. As illustrated in Fig. 1, the time of day is represented across the top and successive days are represented by successive rows. A hash mark or other indicator represents the time when the animal rotated the wheel. Conventionally, the horizontal bars at the top show the periods during which the animal is exposed to light (the top bar shows that the animal was exposed to light from 4 to 16 h during the first 7 days, and subsequently kept in constant darkness). When information is represented in this format and one has learned the conventions that are employed, one can easily distinguish patterns in the animal's behavior. In this case, the animal began its activity somewhat earlier each day, revealing that its endogenous period is somewhat less than 24 h. This period when no Zeitgebers are present is known as free-running. Actograms can also be used to show how various perturbations affect free-running behavior. In this case, a pulse of light was presented 4 h after activity onset on the day indicated by the grey arrow. The actogram shows how this reset the animal's circadian rhythm, inserting a phase delay into what was otherwise a continuing pattern of phase advance due to constant darkness.

Varying the periods of light and darkness to which an animal is exposed is an effective way to explore the features of circadian phenomena. As already suggested in the previous actogram, when animals are exposed to relatively normal light–dark cycles, or to total darkness, they exhibit circadian rhythms. Individual pulses of

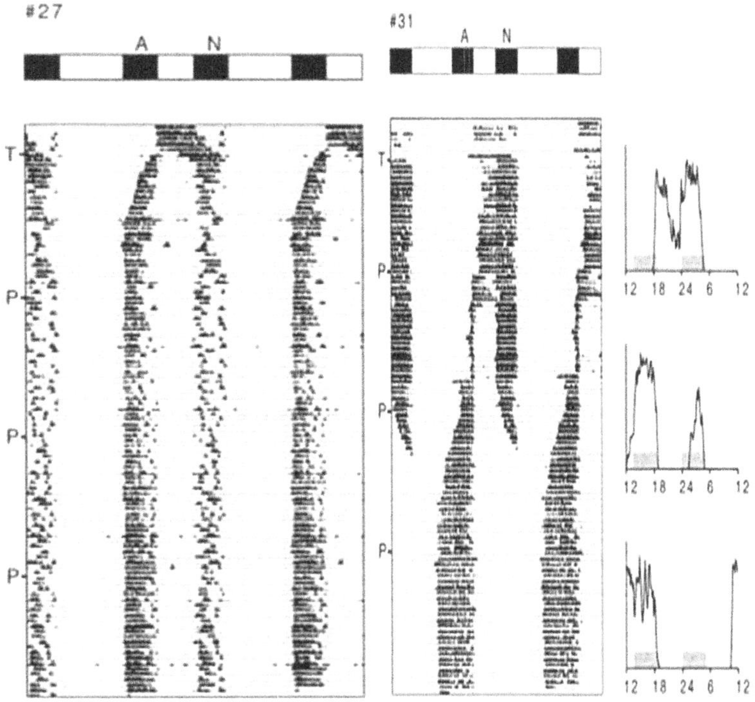

Fig. 2 Two actograms from Gorman (2001) showing the effects on the locomotor behavior of hamsters as a result of exposure to an unusual *light–dark* cycle

light while maintained under constant darkness can reset the circadian clock. But how do animals respond to atypical light–dark cycles? If the light–dark period is either very short (e.g., 19 h) or very long (e.g., 29 h), they typically become arrhythmic, which is manifest in an actogram by scattered bouts of activity on each day. More exotic are conditions under which an animal is exposed to two light and two dark periods each day. Gorman (2001) explored an arrangement in which hamsters were exposed to an arrangement of 9 h of light, 5 h of darkness, 5 h of light, and 5 h of darkness. In the actograms shown in Fig. 2, under these conditions hamsters would often develop split rhythms so that they were active during both dark phases. Sometimes, as illustrated in the actogram on the left, this would occur shortly after being introduced to the unusual lighting conditions (indicated by T on the left margin); in other cases, as illustrated in the actogram on the right, the splitting was delayed and not sustained.

An actogram is a particular diagrammatic format that circadian rhythm researchers have developed to represent a phenomenon of interest to them—changing patterns of behavior under different light–dark conditions. They are not, though, the only format employed—to portray how organisms are affected by light at different phases of their circadian period, researchers have developed phase-response curves that exhibit how much a particular form of light exposure advances or delays the phase

(C. H. Johnson 1999). Different fields of biology focus on different phenomena, and different diagrammatic formats have been developed and found effective for representing the phenomena of interest to them. Two general points should be noted. First, it is biologists themselves who must develop appropriate representational devices. Typically, in a given field the format is developed over time as researchers revise initial attempts until they arrive at a perspicuous format. Second, those using the diagrams must learn the conventions and this often takes time. After one has learned the technique, such diagrams seem transparent—they directly reveal the phenomenon. But students, for example, who have not yet become adept with the technique, fail to understand the phenomenon the diagram is illustrating.

4 Identifying and Decomposing a Mechanism

Once a phenomenon has been characterized,[3] the challenge for researchers is to identify the mechanism and decompose it into its parts and operations. Although often researchers can identify the mechanism before identifying its parts and operations, sometimes the first clue to the mechanism results from identifying one of its parts. Circadian rhythm research illustrates both scenarios. In mammals researchers (Stephan and Zucker 1972; Moore and Eichler 1972; Moore and Lenn 1972) located the mechanism in the SCN both by showing that lesions to the SCN result in animals becoming arrhythmic and by tracing fiber tracts from the retina to the SCN (it was assumed that the clock must have input from the eyes if it were to be synchronized with the local light–dark cycle). In fruit flies, on the other hand, the first link to the mechanism was provided by identifying a gene, *period* (*per*), in which mutations result in flies with short or long endogenous rhythms or which are arrhythmic (Konopka and Benzer 1971). The brain locus of circadian control in flies (a small population of lateral and dorsal neurons) was only identified on the basis of studying the expression patterns of *per* (Helfrich-Förster 1996).

In either identifying the mechanism itself, or one of its parts, the strategies researchers used were much the same—they were seeking some entity whose operation correlates with the phenomenon of interest or that connects with entities to which the phenomenon is related. Perhaps the most common means of showing such a correlation is to ablate or mutate an entity and show a corresponding deficit in the phenomenon of interest. Studies ablating the SCN in mice and demonstrating loss of circadian rhythmicity exemplify this approach as do the mutation studies in

[3]Initial characterizations are often revised in the course of developing an explanation for them as that inquiry may reveal aspects of the phenomena that were not appreciated at the outset. In Bechtel and Richardson (1993/2010) we speak of this as reconstituting the phenomenon. Sometimes the reconstitution is quite major. For example, after 100 years of attempting to explain the phenomenon of animal heat, where such heat was viewed as an energy source that could support animal activity, it was recognized that such heat is actually a waste product and that the phenomenon of interest really involved the synthesis of adenosine triphosphate (ATP).

fruit flies. The case was made even more compelling when subsequent researchers showed that many circadian functions could be recovered when the SCN from another animal was placed in the third ventricle of the brain of the lesioned animal (Ralph et al. 1990). Other ways of generating such a correlation are to stimulate an entity and show an increase in the phenomenon of interest and to measure activity in the entity while the phenomenon of interest is occurring.[4] While these various techniques show that the entity that is manipulated or recorded from is related to the phenomenon, they leave open the question whether on the one hand it is actually the responsible mechanism or a part within it or on the other hand only involved in a related activity. Removing the carburetor from a car or altering its operation affects whether and how the car will generate locomotion, but does not show it to be the mechanism responsible for the conversion of chemical energy to mechanical energy that explains locomotion. Typically, this latter question is only addressed by further developing the account of how the mechanism operates.

If research has successfully identified the mechanism itself, then subsequent research is directed at decomposing it into its parts and operations. In the case of the SCN, the immediate parts are the individual neurons and supporting glia cells. Initially, however, these were passed over as researchers proceeded directly to decomposing individual cells to identify the relevant constituents within them. As this research quickly dovetailed with research on fruit flies that had begun by identifying component parts of a mechanism, I turn first to how decompositional research proceeds once a part has been identified. As the idea of decomposing a mechanism into multiple parts suggests, a given part can only produce the phenomenon if it interacts with other parts. The quest is to identify these other parts and what they do. One approach is to iterate the first strategy, finding other components whose manipulation affects the phenomenon using the same strategies noted above, and once one or more additional components is identified, investigate what operation each performs. Another is to figure out what operation the part identified first performs and advance hypotheses about the other operations that are required to generate the phenomenon.

Research on circadian rhythms illustrates both strategies, but began with considering what operations can be attributed to *per*. Genes have effects when transcribed into mRNA and translated into proteins, and had researchers known what protein was expressed from *per* they might have begun by considering the reactions in which it could participate (proteins are frequently enzymes that catalyze reactions, and one might have considered whether the protein PER catalyzed reactions that relate to circadian function). Initially no known proteins were linked to *per* and it was only with the advent of cloning technology that researchers could begin to characterize the protein in terms of amino acid sequences that partially constituted it. This, however, provided a clue that focused researchers on a different way

[4]As with securing evidence about the phenomenon itself, these techniques involve manipulations, sometimes severe, and raise questions whether the information that is procured is just an artifact of the experimental manipulation.

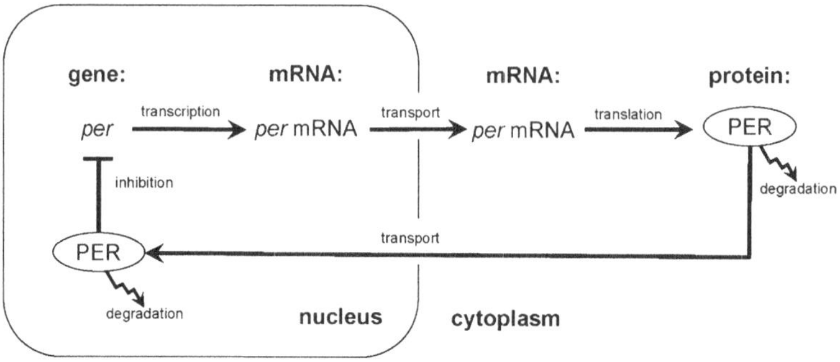

Fig. 3 A representation of the mechanism proposed by Hardin et al. (1990) to explain circadian oscillations in fruit flies

to understand *per*'s contribution. Using cloning techniques, Hardin et al. (1990) determined that both *per* mRNA and PER oscillated with a circadian period, with the mRNA peaking several hours before that of the protein. This led them to suggest a mechanism consisting of three central parts: *per*, *per* mRNA, and PER. On their proposal, each performed a different operation: When uninhibited, *per* was transcribed into its mRNA. The mRNA in turn was transported from the nucleus to the cytoplasm, where it was translated into PER. Finally, PER was continually subject to degradation, but when enough accumulated, it was transported back into the nucleus where it inhibited *per* transcription (Fig. 3). At this point no new PER was produced, and as the existing molecules degraded, inhibition ceased and transcription began again.

While Hardin et al. proposed a mechanism schema that seemed to possess productive continuity, there were in fact numerous gaps. One concerned the manner in which PER degraded. The timing of that operation is crucial to generating an approximately 24 h oscillation. Initially of even greater concern was the question of how PER inhibited transcription of *per*. As the structure of PER was revealed, it seemed to lack a DNA binding region, suggesting the existence of another part which mediated between PER and *per*. Such a part was found when research on the mammalian clock mechanism revealed a gene dubbed *Clock* whose mutants exhibited disrupted circadian rhythms (Vitaterna et al. 1994). The CLOCK protein does possess a DNA binding region and mammalian CLOCK was demonstrated to bind to the promoter region of fruit fly *per*. PER was then viewed as interacting with CLOCK so that it no longer activated the transcription of *per*. Soon thereafter researchers identified both a fruit fly homolog of mammalian *Clock* and three mammalian homologs of *per*, resulting in highly intertwined research on the clock mechanisms in flies and mice (Bechtel 2009). Research into new clock components exploded. Both PER and CLOCK were found to operate as parts of compound molecules known as *heterodimers* (PER with TIM in fruit flies or CRY in mammals and CLOCK with CYCLE in fruit flies and BMAL1 in mammals). Concentrations of CLOCK in fruit flies and BMAL1 in mammals were themselves shown to oscillate, leading to investigations that identified transcription factors that regulated their expression. As well, a number of kinases were identified that figure in the degradation of PER and TIM or CRY. As these parts were identified, researchers also established the immediate operations in which each was involved so that there is now a large catalog of known parts and operations of the circadian clocks of fruit flies and mammals.

Mechanistically oriented biologists have developed a host of tools for identifying component parts of mechanisms and determining what they do. Traditional tools, such as inducing mutations and registering concentrations of mRNAs with Northern blots and proteins with Western blots were extremely time consuming, but newer techniques have greatly simplified the study of individual genes and proteins. Researchers can directly target specific genes to knock-out or to suppress their transcription with tools such as small interfering RNA. And the ability to knock-in genes such as luciferase, an oxidative enzyme that generates light in organisms such as fireflies, has enabled researchers to observe oscillation in tissue-culture preparations through changing luminescence. Moreover, whenever a putative new clock gene is discovered, one can compare its DNA sequence to sequences stored in gene databanks to discover homologues in other organisms whose operations may be easier to assess. The research has provided an enormous wealth of information about parts and operations. Before leaving the topic of decomposition, I should note that the process of decomposition can be iterated—just as researchers decomposed the circadian mechanism into genes and proteins, they could decompose them into nucleotides and amino acids, and decompose them in turn into their component atoms, etc. Some accounts of reduction emphasize the iteration of decomposition down to the most basic entities science has identified at a given time (Bickle 2003). But from the point of view of mechanistic explanation, that is not the goal.

Researchers decompose a mechanism into the parts of the mechanism that explain its behavior. Some researchers might in turn want to explain how the parts work, and then they need to decompose those parts. Recently, for example, researchers have begun to identify how PER inhibits its own transcription, revealing the presence of PSF (polypyrimidine tract–binding protein–associated splicing factor) as a component of the PER complex and determining that it recruits SIN3A to scaffold assembly of a transcription inhibition complex that deacetylates histones in the chromatin of the *per* gene, preventing transcription (Duong et al. 2011). But it is important to note that this mechanism explains a different phenomenon—the inhibition of *per*—not the original phenomenon of maintaining circadian rhythms (in which *per*'s inhibition of PER was a basic operation).

5 Recomposing and Situating the Mechanism

Acquiring the catalog of parts and operations is an important step in developing mechanistic explanations, but until investigators determine how the operations of parts affect other parts (those they operate on) in a coordinated fashion to produce the phenomenon, they have not yet explained the phenomenon. A researcher is no more satisfied with the catalog of parts and operations than you would be if you ordered a new car and it arrived as a collection of parts without even directions for putting them together. Before you have a functioning car, you need to figure out how the parts should be organized and work together to produce a vehicle that one can drive. Determining how the parts are organized and their operations orchestrated in the generation of the phenomenon is what I refer to as *recomposition*. In the course of science, scientists don't wait until they have a complete catalog of parts and operations to try to recompose the mechanism; rather, as they discover parts and operations they try to figure out how they go together to produce the phenomenon. Often the efforts at recomposition reveal the existence of other parts and the operations they perform. Thus, Fig. 3 above already reflects an attempt to recompose the mechanism, and this effort made clear to researches that PER could not directly inhibit *per* transcription and other parts and operations remained to be discovered.

Although one can recompose a mechanism by narrating the envisioned operation of its parts, as I did above for Hardin et al.'s hypothesized mechanism, scientists commonly find it valuable to recompose the mechanism in a diagram. A diagram not only serves to present one's conception of the mechanism, but also supports reasoning about the mechanism. Just as in diagrams of phenomena, researchers can often see patterns in diagrams of mechanisms that would not otherwise be apparent.[5] But there is yet another advantage of working with diagrams—a researcher can manipulate the diagram in the search for an alternative arrangement of operations that may better account for the phenomenon. This sometimes involves identifying gaps in the proposed account of the mechanism that need to be filled in. In this regard, it is interesting to consider how Hardin et al. themselves diagrammed the mechanism they were proposing (Fig. 4). While they clearly presented the idea of a feedback loop, they showed three alternative pathways by which the inhibition might arise (from the protein itself, from a product of a further reaction involving the protein, or from a behavior of the organism resulting from the protein) and two points at which it might effect the process of gene expression (transcription or translation). They also inserted question marks to indicate places where yet additional parts or operations might figure. Such a diagram becomes part of the reasoning processes of scientists as they considered both whether the proposed mechanisms could produce the phenomenon and how one might gain evidence for or against various hypotheses.

A diagram is a static representation that does not reveal how the operations are actually coordinated in the production of the phenomenon. With relatively simple mechanisms, especially ones that operate sequentially, mentally rehearsing the operations, perhaps guided by a narrative text, suffices to show how the mechanism

[5]For example, by comparing diagrams that have been developed for the circadian mechanisms identified in cyanobacteria, fungi, plants, and animals, researchers can readily see that although many of the parts are different, the overall organization is remarkably similar. This is in turn inspiring further inquiry that is revealing that even when the clocks contain different proteins, the domains and motifs that are crucial for the operations they perform are remarkably similar and may well have been evolutionary conserved (Stuart Brody, personal communication, January 2012.)

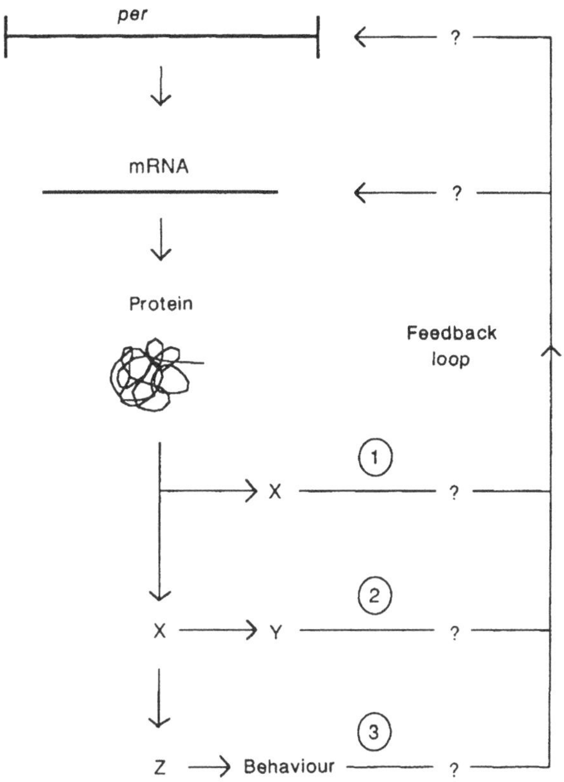

Fig. 4 Hardin et al.'s (1990) diagram of the mechanism they proposed to explain circadian rhythms in fruit flies in which they included *question marks* to indicate alternative places from which feedback might originate and at which it might terminate

will work. If the phenomenon is characterized quantitatively, though, it may be necessary to characterize each operation in equations and show mathematically that the phenomenon results. This becomes even more necessary when the mechanism involves non-sequential organization and the operations are non-linear. Here the ability to imagine the operation of the mechanism becomes highly unreliable Humans are prone to fail to follow one of multiple effects of an operation and are poor at anticipating the results of non-linear interactions.

The circadian example makes this clear. Negative feedback was known to be a mechanism that could generate oscillatory behavior, but not all negative feedback processes generate sustained oscillations: If all the operations are linear, a feedback mechanism will settle into an equilibrium state in which each operation is equilibrated to the others—in the example I have been considering, just enough PER might be synthesized to maintain a constant level of suppression of synthesis to PER. A negative feedback system will only produce sustained endogenous oscillations if there are sufficient non-linear operations in the mechanism. Accordingly, one cannot determine by mental simulation whether a given feedback mechanism will generate sustained oscillations. The only options are either to construct such a mechanism (Elowitz and Leibler 2000) or to represent the operations of the mechanism in

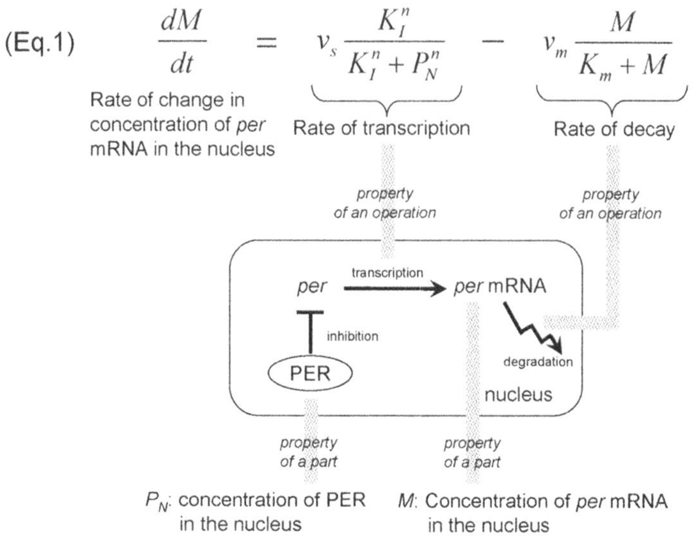

Fig. 5 Equation (1) in Goldbeter's (1995) model shown in relation to the relevant portion of Hardin et al.'s (1990) proposed circadian mechanism

mathematical equations and, through analysis or simulation, determine how the characterized mechanism will behave.[6]

Recognizing the problem of determining whether the mechanism proposed by Hardin et al. would sustain oscillations, Goldbeter (1995) represented an elaborated version of it that included operations that synthesized and degraded each component in a set of five differential equations. Each described the rate of change in the concentration of a substance in the mechanism (*per* mRNA, nuclear PER, etc.). Figure 5 shows how one of these equations describes operations that affect the concentrations of *per* mRNA—the transcription of *per* into *per* mRNA and the hypothesized degradation of *per* mRNA. The equation includes both variables (*M* for the concentration of *per* mRNA; *P* for the concentration of PER in the nucleus) and parameters (K_I and K_m represent the Michaelis constant and v_s and v_m the maximum rates for the two reactions respectively). Modelers choose values for the parameters both with an eye to making the simulation achieve the desired end and to characterizing the actual biological operations—they may speak of the parameters they employ as "biologically plausible."

[6]Neither of these strategies is perfect. Mathematical modeling depends on developing equations that accurately describe the operations in the mechanism. Researchers avoid this problem in synthetic models, letting actual physical components operate as they do. Such models often produce quite surprising results, which then motivate the creations of new computational models in the attempt to explain them. See Danino et al. (2010) who provide a particular compelling example of the productive interaction between synthetic models and computational models in understanding synchronization of oscillations.

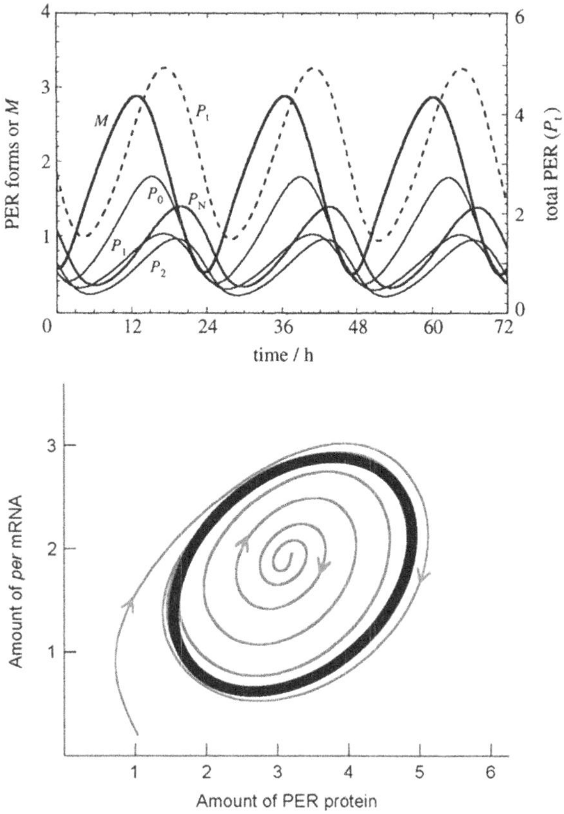

Fig. 6 Two diagrams Goldbeter (1995) used to illustrate the behavior of his computational model of Hardin et al.'s proposed mechanism. In the top diagram the changes in variables in his model (after initial transients) are plotted against time, whereas in the one on the *bottom* successive values (*indicated by arrows*) of two variables are plotted in phase space. This shows the transients as the mechanism approaches the limit cycle (*dark oval*), at which point it oscillates indefinitely

Sometimes the behavior of a mechanism can be deduced directly from the equations and specified parameter values, but typically this is not possible and modelers rely on simulations run on computers. A given simulation begins by assigning initial values to the variables and then iteratively applying the equations to determine values at subsequent times. Typically multiple simulations will be conducted to ascertain the behavior of the mechanism. The results of the simulation are a set of numbers, which modelers then plot in a diagram to decipher whether a pattern results. Goldbeter chose to plot the results in two ways. As shown at the top in Fig. 6, he first plotted time on the abscissa and values of variables on the ordinate, revealing how the values oscillated across time (once the simulation reached a stable oscillation). Often oscillatory patterns, especially before a stable oscillation is achieved, are more perspicuously shown in phase space, as illustrated at the bottom of Fig. 6. Here the values of two variables, the total concentration of the PER protein and *per* mRNA, are indicated on the abscissa and ordinate and time is conveyed only in the succession of locations plotted (the order of successive points is indicated with arrows). In this case the phase space plot makes clear that from different initial values, the behavior will follow a trajectory towards the dark oval, referred to as a *limit cycle* (in this figure two trajectories are shown). If the values of PER and *per* mRNA are on the oval, they will continue to change so as to follow the oval, thus showing that the oscillations will be sustained indefinitely. This analysis illustrates the integration of mechanism and dynamics in dynamic mechanistic explanations.

I noted above that in the years after Goldbeter put forward his model, researchers discovered numerous other components of the intracellular clock mechanism. Since additional components could significantly alter the behavior of the mechanism, disrupting its ability to sustain oscillations, it was necessary to build more complex models to ascertain their behavior. Goldbeter together with Leloup developed models incorporating all the known components of the fruit fly (Leloup and Goldbeter 1998) and the mammalian (Leloup and Goldbeter 2004) clock mechanism. Other modelers have pursued different strategies, developing less detailed models that enable running simulated experiments that, for example, might reveal which components of the mechanism are most crucial for producing the phenomenon (Smolen et al. 2001; Relógio et al. 2011).

So far I have focused on recomposing the mechanism in diagrams and computational models, activities that are important to investigators' attempts to understand how mechanisms will actually function. In these efforts, however, researchers commonly abstract from the larger contexts in which the mechanism actually functions, but these can have important consequences for the functioning of the mechanism. Appreciating them requires not just recomposing the mechanism, but situating it in contexts in which it usually operates. Just as researchers often must iterate the process of decomposition, they often must also iterate the processes of recomposing and situating the mechanism since its operation may be affected not just by other activities in its local environment, but also by activities in the environment of the system of which it is a component. I will briefly indicate three such levels of situating and recomposing required to understand circadian rhythms.

Earlier I noted that in mammals researchers first located the circadian clock in the SCN, a relatively small region of the hypothalamus consisting, in mice, of 16,000–20,000 neurons on each side of the brain. As they pursued the question of how individual neurons in the SCN maintained time, researchers implicitly assumed each SCN neuron operated in the same manner. Welsh et al. (1995), however, discovered that when dispersed in culture both the phase and the period of oscillations varied substantially across neurons. More recent research has shown that many SCN neurons do not maintain rhythmicity when deprived of inputs from their neighbors, and some shift between normal and super-long periods (Meeker et al. 2011). These discoveries make it clear that the oscillations within individual neurons are modulated by the behavior of their neighbors. To understand this behavior researchers are turning more and more to computational modeling as providing the most tractable way of investigating how such a complicated system might operate. These modeling efforts typically begin by using models constructed for the intracellular oscillators and adding terms and equations to them to represent hypothesized interactions between SCN neurons. For example, Vasalou et al. (2009) showed that by assuming the SCN had the structure of a small-world (a network structure in which most connections are local but a few extend long distances) they could capture many details of the phenomenon of SCN behavior. The assumption that the SCN is a small world is highly plausible given recent discoveries of the ubiquity of such organization in biological systems, but before one can assume that the model reveals the behavior of the actual mechanism, further research is required to determine how closely the actual organization of the SCN corresponds to what is proposed in the models.

The fact that lesioning the SCN in mice eliminated circadian behavior and that implanting SCN tissue from a mouse with a clock mutation into the third ventricle of a lesioned mouse would restore some circadian behaviors with the period found in the mutant convinced researchers that the SCN was the master clock (Ralph et al. 1990). One of the consequences of identifying the components of the central clock in the SCN was the discovery that the same genes are also expressed in most cells of the body and in them they also work together as clock mechanisms. Since timing in peripheral tissue was lost with lesions to the SCN, these peripheral clocks were thought to be slaves. This supported viewing the SCN as the largely autonomous locus of the clock mechanism, constrained only by inputs it received from Zeitgebers and sending outputs to other tissues. More recently it has been demonstrated that the clock mechanisms in peripheral tissues do not stop oscillating in the absence of the SCN but rather become desynchronized from each other. This has led to rethinking the relationship of the SCN to this other clocks: the SCN is better thought of as an orchestra conductor than a slave master (Davidson et al. 2004). Circadian oscillators in the liver have been shown to operate semi-independently of the SCN, being entrained to feeding schedules outside the organism's normal feeding times (Damiola et al. 2000). Moreover, there is increasing evidence that metabolic activities, many of which are regulated by the liver, have effects of the SCN (Nakahata et al. 2008). In addition, researchers are increasingly discovering avenues through which clocks in other organs of the body and the processes they regulate feed back on the SCN and affect its behavior, rendering it important to understand how the SCN is situated in the organism (Pezuk et al. 2010).

Finally, organisms with circadian rhythms operate in an environment that on this planet has a 24 h light–dark cycle. It has long been recognized that having an endogenous circadian clock is crucial for organisms to function in their environments as many activities must be performed at appropriate times of day and it often takes several hours for biological systems to prepare to carry out these activities (for example, enzymes need to be synthesized to perform photosynthesis in plants or to digest food in animals). For most organisms the light–dark cycle is externally provided and they must only entrain their endogenous clocks to it. Humans, however, have developed artificial environments in which their exposure to light and conduct of activities is dissociated from the light–dark cycle provided by the Earth's rotation. As a result, our endogenous circadian rhythms are confronted with discordant entrainment signals from environments we have created. The experience of jet lag after travel across multiple time zones makes apparent the disruptions abruptly altered light–dark cycles can have until, over several days, our endogenous clock is re-entrained to the local light–dark cycle. The severe health effects of shift work, which results in our endogenous clock being confronted with a very unnatural light–dark cycle, are increasingly being identified (Maywood et al. 2006; Wang et al. 2011).

Recomposing and situating mechanisms is crucial for understanding their behavior. The result is often a much more holistic understanding of the phenomenon one seeks to explain. It should be noted, though, that appreciation of the whole is often dependent on first developing at least a basic understanding of the composition of the mechanism. Models of mechanisms generated before research reveals actual parts and operations are at best only hypotheses about how the mechanism might operate. Once accounts can be grounded on some understanding of parts and operations, then the task of understanding how the whole mechanism is actually organized and its functioning orchestrated is much more tractable (although certainly not easy). Moreover, once one has an account of how the recomposed mechanism might work, researchers can both evaluate the effects of different contexts on its operation and figure out how these effects are realized in the mechanism.

6 Conclusion

When biologists seek to explain a phenomenon, in many contexts what they are looking for is an account of the mechanism responsible for it. I have described some of the distinctive features of mechanistic research, including its frequent reliance on diagrammatic representations and the strategies for discovering mechanisms. A critical first step is to delineate the phenomenon for which explanation is sought. Although sometimes characterized in verbal descriptions (e.g., maintain circadian oscillations), typically phenomena are characterized quantitatively with diagrams used to represent the pattern elicited from quantitative information. The second step is to identify the responsible mechanism and decompose it into its parts and operations. This process of taking systems part—finding the part that constitutes the mechanism and discovering its components—is what distinguishes mechanistic research. I described how such research led to the discovery of the SCN as the locus of the central clock in mammals and the genes and proteins whose operations figure in maintaining oscillations.

As critical as decomposition is, however, it is equally important to recompose the mechanism so as to understand how the parts operate together. Scientists often present their understanding of the mechanism in diagrams, but to determine what will actually result from the parts performing their operations they turn to simulations. In simple cases, researchers can mentally simulate the mechanism of interest, but increasingly, as research reveals non-sequential organization involving nonlinear operations, biologists appeal to computational models. The result is what I term dynamic mechanistic explanation. I illustrated how a relatively simple model was used to demonstrate that an early hypothesis about how parts interacted might in fact produce circadian oscillations. Beyond recomposing individual mechanisms, researchers often come to recognize ways in which the mechanism is affected by the context in which it operates, and this requires that they situate the mechanism in various environments and assess, often through computational models, how they will affect the mechanism's behavior.

Thinking in terms of mechanisms is quite intuitive, especially in cultures exposed to modestly complex technology. Most of us are familiar with taking mechanisms apart, either to diagnose problems and repair them or just out of curiosity as to how they work. But, as van Mil et al. (2013) discuss, students do not readily extend this perspective to biological mechanisms. Their diagnosis is in part that biology education focuses on the functions of mechanisms, not on how they work. Another part of the explanation is perhaps that our prototypes of machines are devices made out of solid materials (e.g., wood or metal) with relatively clearly delineated parts organized in a fairly straightforward manner. As these characteristics of prototypical machines are violated, as they often are in biology, people are less inclined to adopt a mechanistic perspective.[7] Professional biologists have a several hundred-year history of adopting and enriching the mechanistic perspective (although one finds biologists repudiating the machine metaphor as they focus on the complex organization found in many biological systems). Moreover, they have become accustomed to thinking of components just in terms of the contributions they make to the whole mechanism and then seeking evidence as to their physical constitution.[8] This strategy of starting with a hypothesized functional decomposition requires imagination (and reasoning by analogy to other known mechanisms) and may require cultivation.

While biologists have become enculturated into this extension of mechanical thinking to biology, it may not come naturally to students and the process of developing mechanistic explanations in biology may need to be explicitly articulated. The example of circadian rhythms, as developed in this chapter, drawing as it does on the analogy with manufactured clocks, can provide a helpful entrée for getting students to think mechanistically about biological processes. I have noted the value of diagrams in conveying an understanding of mechanisms. But diagrams themselves may require commentary that explicitly notes the parts and operations shown and helps students learn to rehearse these operations so as to understand how they work

[7]In fact, as machines become more complex and rely on electronic circuitry that is not readily decomposed, we relatively quickly abandon our mechanistic perspective on how they work and settle simply for learning how to control their operation.

[8]van Mil et al. (2013) identify the failure to understand how proteins change conformation as one factor in students' failure to think causally about protein interactions in the cell. While not denying the importance of conveying such an understanding of proteins, I suspect it is not the main problem in thinking causally about cell constituents as biologists themselves came to understand protein actions well before they understood their conformation. Rather, the problem seems more immediate—students are not encouraged to think about how component processes may work together in producing physiological effects.

together to produce the phenomenon. Spending the time needed to work through a diagram and to illustrate how one can use it to think through the workings of a mechanism may help bring students into the culture in which biologists operate.

As I have noted above, discovering or learning how a mechanism works is a reductionist inquiry—it requires decomposing a mechanism into its parts and the operations they perform. But, as I have emphasized, efforts at decomposition need to be complemented by research seeking to recompose a mechanism and to understand how it is situated. Understanding a mechanism requires dexterity in moving down and up between levels of organization. van Mil et al. (2013) note that thinking in terms of levels and moving between them is often difficult for students. Given the challenges that scientists have confronted in moving up and down levels in their inquiries, it is not surprising that students face a challenge. The problem is exacerbated by the fact that nature does not present itself well-delineated in terms of levels—it is only in the process of developing mechanistic explanations that researchers come to recognize both components and the often transient structures into which they are composed. Nonetheless, if students are to develop appropriate sophistication in understanding and developing mechanistic accounts, they need to develop facility in thinking about levels. Working through examples is often the best way to acquire such facility, and the circadian example developed here may prove useful in cultivating such facility.

Acknowledgments I thank Kostas Kampourakis and an anonymous reviewer for helpful comments and suggestions that improved this essay. Research for this chapter was supported both by a grant from the Dean of Arts and Sciences at UCSD and by the National Science Foundation (USA) under Project Number 1127640.

References

Bechtel, W. 2006. *Discovering cell mechanisms: The creation of modern cell biology*. Cambridge: Cambridge University Press.
Bechtel, W. 2009. Generalization and discovery by assuming conserved mechanisms: Cross species research on circadian oscillators. *Philosophy of Science* 76: 762–773.
Bechtel, W. 2011. Mechanism and biological explanation. *Philosophy of Science* 78: 533–557.
Bechtel, W., and A. Abrahamsen. 2005. Explanation: A mechanist alternative. *Studies in History and Philosophy of Biological and Biomedical Sciences* 36: 421–441.
Bechtel, W., and A. Abrahamsen. 2010. Dynamic mechanistic explanation: Computational modeling of circadian rhythms as an exemplar for cognitive science. *Studies in History and Philosophy of Science Part A* 41: 321–333.
Bechtel, W., and A. Abrahamsen. 2011. Complex biological mechanisms: Cyclic, oscillatory, and autonomous. In *Philosophy of complex systems*, Handbook of the philosophy of science, vol. 10, ed. C.A. Hooker, 257–285. New York: Elsevier.
Bechtel, W., and R.C. Richardson. 1993/2010. *Discovering complexity: Decomposition and localization as strategies in scientific research*. Cambridge, MA: MIT Press. 1993 edition published by Princeton University Press.
Bickle, J. 2003. *Philosophy and neuroscience: A ruthlessly reductive account*. Dordrecht: Kluwer.
Bogen, J., and J. Woodward. 1988. Saving the phenomena. *Philosophical Review* 97: 303–352.

Boogerd, F.C., F.J. Bruggeman, R.C. Richardson, A. Stephan, and H.V. Westerhoff. 2005. Emergence and its place in nature: A case study of biochemical networks. *Synthese* 145: 131–164.
Craver, C.F. 2002. Interlevel experiments and multilevel mechanisms in the neuroscience of memory. *Philosophy of Science* 69: S83–S97.
Craver, C.F. 2007. *Explaining the brain: What a science of the mind-brain could be*. New York: Oxford University Press.
Damiola, F., N. Le Minh, N. Preitner, B. Kornmann, F. Fleury-Olela, and U. Schibler. 2000. Restricted feeding uncouples circadian oscillators in peripheral tissues from the central pacemaker in the suprachiasmatic nucleus. *Genes and Development* 14: 2950–2961.
Danino, T., O. Mondragon-Palomino, L. Tsimring, and J. Hasty. 2010. A synchronized quorum of genetic clocks. *Nature* 463: 326–330.
Darden, L. 2006. *Reasoning in biological discoveries: Essays on mechanisms, interfield relations, and anomaly resolution*. Cambridge: Cambridge University Press.
Darden, L., and C. Craver. 2002. Strategies in the interfield discovery of the mechanism of protein synthesis. *Studies in History and Philosophy of Biological and Biomedical Sciences* 33: 1–28.
Davidson, A.J., S. Yamazaki, and M. Menaker. 2004. SCN: Ringmaster of the circadian circus or conductor of the circadian orchestra? In *Molecular clocks and light signalling*, ed. D.J. Chadwick and J.A. Goode, 110–125. Chichester: Wiley.
De Mairan, J.-J.d.O. 1729. Observation Botanique. *Histoire de l'Academie Royale Sciences*, pp. 35–36.
Duong, H.A., M.S. Robles, D. Knutti, and C.J. Weitz. 2011. A molecular mechanism for circadian clock negative feedback. *Science* 332: 1436–1439.
Elowitz, M.B., and S. Leibler. 2000. A synthetic oscillatory network of transcriptional regulators. *Nature* 403: 335–338.
Goldbeter, A. 1995. A model for circadian oscillations in the *Drosophila* Period protein (PER). *Proceedings of the Royal Society of London B: Biological Sciences* 261: 319–324.
Gorman, M.R. 2001. Exotic photoperiods induce and entrain split circadian activity rhythms in hamsters. *Journal of Comparative Physiology A: Sensory Neural and Behavioral Physiology* 187: 793–800.
Griffiths, P.E., and R. Gray. 1994. Developmental systems and evolutionary explanation. *Journal of Philosophy* 91: 277–304.
Hardin, P.E., J.C. Hall, and M. Rosbash. 1990. Feedback of the *Drosophila period* gene product on circadian cycling of its messenger RNA levels. *Nature* 343: 536–540.
Helfrich-Förster, C. 1996. Drosophila rhythms: From brain to behavior. *Seminars in Cell and Developmental Biology* 7: 791–802.
Hempel, C.G. 1965. Aspects of scientific explanation. In *Aspects of scientific explanation and other essays in the philosophy of science*, ed. C.G. Hempel, 331–496. New York: Macmillan.
Holland, J.H., K.J. Holyoak, R.E. Nisbett, and P.R. Thagard. 1986. *Induction: Processes of inference, learning and discovery*. Cambridge, MA: MIT Press.
Johnson, M.S. 1926. Activity and distribution of certain wild mice in relation to biotic communities. *Journal of Mammalogy* 7: 254–277.
Johnson, C.H. 1999. Forty years of PRCs – What have we learned? *Chronobiology International* 16: 711–743.
Konopka, R.J., and S. Benzer. 1971. Clock mutants of *Drosophila melanogaster*. *Proceedings of the National Academy of Sciences USA* 89: 2112–2116.
Langley, P., H.A. Simon, G.L. Bradshaw, and J.M. Zytkow. 1987. *Scientific discovery: Computational explorations of the creative process*. Cambridge: MIT Press.
Leloup, J.-C., and A. Goldbeter. 1998. A model for circadian rhythms in *Drosophila* incorporating the formation of a complex between the PER and TIM proteins. *Journal of Biological Rhythms* 13: 70–87.
Leloup, J.-C., and A. Goldbeter. 2004. Modeling the mammalian circadian clock: Sensitivity analysis and multiplicity of oscillatory mechanisms. *Journal of Theoretical Biology* 230: 541–562.

Machamer, P., L. Darden, and C.F. Craver. 2000. Thinking about mechanisms. *Philosophy of Science* 67: 1–25.
Maywood, E.S., J. O'Neill, G.K.Y. Wong, A.B. Reddy, and M.H. Hastings. 2006. Circadian timing in health and disease. *Progress in Brain Research* 153: 253–269: Elsevier.
McGill, Gl. 2008. Molecular movies ... Coming to a lecture near you. *Cell* 133: 1127–1132.
Meeker, K., R. Harang, A.B. Webb, D.K. Welsh, F.J. Doyle, G. Bonnet, et al. 2011. Wavelet measurement suggests cause of period instability in mammalian circadian neurons. *Journal of Biological Rhythms* 26: 353–362.
Moore, R.Y., and V.B. Eichler. 1972. Loss of a circadian adrenal corticosterone rhythm following suprachiasmatic lesions in the rat. *Brain Research* 42: 201–206.
Moore, R.Y., and N.J. Lenn. 1972. A retinohypothalamic projection in the rat. *The Journal of Comparative Neurology* 146: 1–14.
Nagel, E. 1961. *The structure of science*. New York: Harcourt, Brace.
Nakahata, Y., M. Kaluzova, B. Grimaldi, S. Sahar, J. Hirayama, D. Chen, et al. 2008. The NAD+–dependent deacetylase SIRT1 modulates CLOCK-mediated chromatin remodeling and circadian control. *Cell* 134: 329–340.
Nakajima, M., K. Imai, H. Ito, T. Nishiwaki, Y. Murayama, H. Iwasaki, et al. 2005. Reconstitution of circadian oscillation of cyanobacterial KaiC phosphorylation in vitro. *Science* 308: 414–415.
Pezuk, P., J.A. Mohawk, T. Yoshikawa, M.T. Sellix, and M. Menaker. 2010. Circadian organization is governed by extra-SCN pacemakers. *Journal of Biological Rhythms* 25: 432–441.
Pigliucci, M., and M. Boudry. 2011. Why machine-information metaphors are bad for science and science education. *Science & Education* 20: 453–471.
Post, E.L. 1936. Finite combinatorial processes – Formulation I. *Journal of Symbolic Logic* 1: 103–105.
Ralph, M.R., R.G. Foster, F.C. Davis, and M. Menaker. 1990. Transplanted suprachiasmatic nucleus determines circadian period. *Science* 247: 975–978.
Reichenbach, H. 1938. *Experience and prediction*. Chicago: University of Chicago Press.
Relógio, A., P.O. Westermark, T. Wallach, K. Schellenberg, A. Kramer, and H. Herzel. 2011. Tuning the mammalian circadian clock: Robust synergy of two loops. *PLoS Computational Biology* 7: e1002309.
Smolen, P., D.A. Baxter, and J.H. Byrne. 2001. Modeling circadian oscillations with interlocking positive and negative feedback loops. *Journal of Neuroscience* 21: 6644–6656.
Stephan, F.K., and I. Zucker. 1972. Circadian rhythms in drinking behavior and locomotor activity of rats are eliminated by hypothalamic lesions. *Proceedings of the National Academy of Sciences USA* 69: 1583–1586.
Turing, A. 1936. On computable numbers, with an application to the Entscheidungs problem. *Proceedings of the London Mathematical Society Second Series* 42: 230–265.
van Mil, M., D. Boerwinkel, and A. Waarlo. 2013. Modelling molecular mechanisms: A framework of scientific reasoning to construct molecular-level explanations for cellular behaviour. *Science & Education* 22(1): 93–118.
Vasalou, C., E.D. Herzog, and M.A. Henson. 2009. Small-world network models of intercellular coupling predict enhanced synchronization in the suprachiasmatic nucleus. *Journal of Biological Rhythms* 24: 243–254.
Vitaterna, M.H., D.P. King, A.-M. Chang, J.M. Kornhauser, P.L. Lowrey, J.D. McDonald, et al. 1994. Mutagenesis and mapping of a mouse gene, *Clock*, essential for circadian behavior. *Science* 264: 719–725.
Wang, X.-S., M.E.G. Armstrong, B.J. Cairns, T.J. Key, and R.C. Travis. 2011. Shift work and chronic disease: The epidemiological evidence. *Occupational Medicine* 61: 78–89.
Weber, E.H. 1834. *De pulsu, resorpitione, auditu et tactu: Annotationes anatomicae et physiologicae*. Leipzig: Koehlor.
Welsh, D.K., D.E. Logothetis, M. Meister, and S.M. Reppert. 1995. Individual neurons dissociated from rat suprachiasmatic nucleus express independently phased circadian firing rhythms. *Neuron* 14: 697–706.

Information in the Biological Sciences

Alfredo Marcos and Robert Arp

1 Introduction

> Life, too, is shaped by information. All living creatures are information-processing machines at some level…
>
> Charles Seife, *Decoding the Universe*

Why does information matter in the teaching of biology? How can the biology educator benefit from the philosophy of biology regarding information? These are the two basic questions that we explore in this chapter.

1.1 Information Is Pervasive in Biology

Concerning the first question above, the concept of information is important in the teaching of biology simply because it is integral to the biological sciences themselves. Charles Seife (2007) is correct above in noting that "all living creatures are information-processing machines at some level," and information has been a central concept for contemporary work in the biological sciences (and other sciences), especially since the publication of Claude Shannon and Warren Weaver's, *The Mathematical Theory of Communication*, in 1949 and the discovery of the genetic code around the middle of the twentieth century by Marshall W. Nirenberg and coworkers, who won

A. Marcos (✉)
Department of Philosophy, University of Valladolid, Plaza del Campus, 47011 Valladolid, Spain
e-mail: amarcos@fyl.uva.es

R. Arp
Independent Scholar, 14713 Walmer Street, 66223 Overland Park, KS, USA
e-mail: robertarp320@gmail.com

K. Kampourakis (ed.), *The Philosophy of Biology: A Companion for Educators*, History, Philosophy and Theory of the Life Sciences 1, DOI 10.1007/978-94-007-6537-5_23,

the Nobel Prize in 1968 "for their interpretation of the genetic code and its function in protein synthesis" (NPO 2012). In fact, the pervasiveness of Shannon's information theory—as well as the very terms themselves—becomes evident when one takes a moment to reflect upon just a few of the concepts that are standard in the biomedical sciences, such as genetic *code*, *messenger* RNA, ion *channel*, cell *signaling*, intracellular *communication*, *signal transduction*, pathogen *transmission*, positive *feedback* loop, expressive *noise* minimization, and many others. Biology has developed what we might call an *informational paradigm*. This is a fact.

One may take a positive or a negative view regarding this fact, or even remain indifferent; indeed, all these positions are present in contemporary literature. But this leaves the fact unchanged. And this fact is important, for teaching biology cannot be achieved without a reflective and critical understanding of informational concepts.

There is, moreover, another reason why the concept of information should interest *every* teacher: the educational process itself may be considered an informational relationship existing between and among multiple minds engaged in communicating, processing, and learning.

1.2 Philosophy of Biology and Information

These considerations lead us to the second question posed above: How can the biology educator benefit from the philosophy of biology regarding information? Stated in another way: What does the philosophy of biology contribute regarding the concept of information and its relationship to the biological sciences?

The word *philosophy* comes from two Greek words: *philos* deriving from *philein*, "love," and *sophia* meaning "wisdom." Love here means something like an intense desire for something, while wisdom is arguably a kind of knowledge gained from experience, whether this is practical experience (gained from living life with all of its ups and downs) or theoretical experience (gained from understanding, evaluating, critiquing, and synthesizing ideas, positions, and concepts). Ever the theoretician, the philosopher has always been the person who not only desires to look deeper into some claim, idea, argument, event, or state of affairs by questioning assumptions and challenging status quo thinking, but also attempts to explain and systematize aspects of reality as it is perceived. In Bertrand Russell's (1912/1999) words, which are appropriate given the nature of this book, "Philosophy, like all other studies, aims primarily at knowledge. The knowledge it aims at is the kind of knowledge which gives unity and system to the body of the sciences, and the kind which results from a critical examination of the grounds of our convictions, prejudices, and beliefs" (p. 9).

The word *biology* comes from two Greek words as well: *bios* meaning, "life" and *logos* meaning, "word" or "rational account." Thus, biology is the kind or type of rational account (or science) that studies life, which most of us already know. Whereas *biology* can be characterized as a set of sub-disciplines (the biological or life sciences) under science, the concern of which includes the description, classification, analysis, explanation, prediction, and ultimately control of living things, *philosophy of biology* can be characterized as a sub-discipline of philosophy, the concern of which is the meta-leveled attempt on the part of philosophers, biologists, and other thinkers to understand, evaluate, and critique the methods, foundations, history, and logical structure of biology in relation to other sciences, disciplines, and life endeavors so as to better clarify the nature and purpose of biological science and its practices (see Ayala and Arp 2009; Rosenberg and Arp 2009; Rosenberg and McShea 2007; Ruse 2008; Sober 1993).

Now, the epistemological, computational, linguistic, and logical aspects of information have been dealt with extensively in the philosophical tradition. When the use of informational concepts was extended to biology, philosophers immediately began to react, reflect, ruminate, and even ridicule, so we can expect major contributions from the philosophy of biology.

Specifically, we expect this discipline to help us understand the meaning of the different versions of the concept of information—especially bioinformation—from historical as well as from contemporary perspectives. Philosophy of biology also contributes to clarifying the scope of the use of informational terms in biology, that is, whether they are used metaphorically, in a linguistically instrumental way, or in such a way as to capture the real, objective aspects of living things. If the philosophy of biology can offer no definitive answer to this issue of scope, it can at least make us aware of the problems and ensure that they are clearly posited.

Furthermore, the philosophy of biology also helps the educator understand the complex relationships existing between different concepts that have a great presence in the biological literature. We refer here to the concept of information itself and to others such as *form*, *correlation*, *order*, *organization*, *complexity*, *meaning*, *knowledge*, and *entropy*, to name just a few. The concept of entropy is now used standardly when discussing protein synthesis in cellular functions, for example, and since this kind of entropy—known as *Shannon entropy*—quantifies the expected value of the genetic information contained in the messages delivered between and among various mRNA molecules so that protein synthesis may occur, we can see how a clear understanding of the concepts of information and entropy, as well as their relationship to one another, is crucial for the biology educator if a robust explanation of protein synthesis is to be put forward (Ewens 2010; Collier 2003; Brooks and Wiley 1988; Weber et al. 1988; Wicken 1987).

Philosophers of biology also make contributions to the problem of the *location* of information. For example, we often wonder where hereditary information is to be found. Seemingly in the genes and the configuration of codons and switches (but see Burian and Kampourakis this volume); but there is no doubt that the epigenetic level is also important for the development of the organism (see Uller this volume), as is the cellular cytoplasm, the configuration of tissues, the organism itself as a whole and, in general, the environment. We sometimes speak of information as if it resided exclusively in the genes, but on other occasions we speak of it as if it were present everywhere. In short, all these questions come into play in the teaching of biology, and in all of them, the philosophy of biology can be of help, as we shall see.

Finally, the biologist—*qua* educator—may be interested in the informational aspects of the educational process itself. For this topic, valuable contributions may also be expected from the field of philosophy, especially the philosophy of education, as well as from communication theory, linguistics, psychology, sociology, anthropology, and other related disciplines.

1.3 Outline of the Chapter

The underlying viewpoint in this chapter is that teaching biology cannot be achieved without a reflective and critical understanding of informational concepts. So we begin in Sect. 2 by looking at the different ways in which the concept of information was historically understood up to the 1950s, when it began to make its presence felt in the life sciences. Next, in Sect. 3, we examine the influence that the informational paradigm has had in the different areas of life sciences, such as genetics, cell biology, neurobiology, and ecological studies. Here, we provide several standard examples of information processing in living systems.

Once the apparent pervasiveness of informational terms has been demonstrated through examples from different areas of the life sciences, in Sect. 4 we then examine some of the debates to which this pervasiveness has given rise. In the first place, there is an argument on the advisability of using informational concepts in biology, with some researchers maintaining that informational jargon should be kept out of the life sciences, while others argue that the informational perspective is indispensable for understanding biological phenomena. Second, those authors who accept the informational perspective as legitimate continue to debate about its possible interpretation: for some, informational concepts must be taken as metaphors in biology; for others, they have a merely instrumental use; while still others consider information to be a real and substantial aspect of living things. Finally, there is an argument concerning the very nature of bioinformation, which may be considered as a thing, a property, or a relationship. We think that bioinformation is best understood as a relationship between and/or among entities; for instance, DNA is informational only in relation to a given cellular context, and it is misguiding to locate information in a particular molecule.

In Sect. 5, we turn attention to the relationship between the concept of information and related concepts that are integral to the life sciences, such as entropy, organization, complexity, and knowledge, as well as the problem of the location of information in living systems. In Sect. 6, we offer some final thoughts concerning the philosophy of education in light of information existing as a relational and informing phenomenon. Our hope is that the information we provide *about* information in this chapter will be helpful for biology educators.

2 From Information to Bioinformation: A Historical Overview

The English word *information* derives from the Latin noun *informatio*, which can mean, "representation," "idea," or "explanation." Also, the Latin verb *informo* can mean, "to sketch," "to draw," or "to represent" something as well as "to give shape or form" to something. In ancient times, the term was used in both everyday and learned discourse, as for instance, in the works of Virgil, Cicero, Tertullian, and Augustine of Hippo (Capurro 1978; Floridi 2003, 2011). It was used in different domains: ontological ("to shape something"), epistemological ("to become acquainted through the sensorial or intellectual reception of a form"), pedagogical, and moral ("to instruct," "to form"). But it was not the object of any special philosophical reflection.

During the Middle Ages, the verb *informo* and its derivatives were incorporated into philosophical language from Scholastic discourse. Throughout this period, the verb retained its ontological, epistemological, didactic, and moral connotations as well as its active sense, whereby *informatio* was an action rather than a thing. It referred to the action of shaping and its result. Interestingly enough, the great medieval philosopher and theologian, Thomas Aquinas (1225–1274 CE), used information to refer to the act of shaping/forming and its result when he defined *per modum informationis*, a natural biological process whereby a living thing begins to exist. Also in connection with biological domains, we can refer to Marcus Terentius Varro (116–27 BCE), who describes the development of a fetus as a process of information, whereby it is "shaped" or "informed" (*informatur*) (Capurro and Hjørland 2003).

During the fourteenth and fifteenth centuries, the use of the word *information* spread into. European languages from French. At that point, *investigation*, *education*, and *intelligence* were added to its traditional meanings. However, and perhaps because of the rejection of Scholastic terminology, from then on *information* ceased to be a philosophical term, and others, such as *impression*, *idea*, and *representation* came into play, especially when discussing mental forms of information. Descartes, Locke, Hume, Bacon, Kant, and other modern philosophers did not think of their philosophy in terms of information, and in the few places where we find a word that derives from the term *information*, it came to be understood as an idea or a representation inside the subject's mind. In his dialogue, *Alciphron* (1735/1901) George Berkeley (1685–1753) has Euphranor claim "I love information upon all subjects that come in my way, and especially upon those that are most important" (Dialogue 1, Section 5). Modern *idea-ism*—that is, the preference for philosophizing about ideas rather than things—is clearly related to this change from the view of information as an action to an idea (Collins 1956; Musgrave 1993). Interestingly enough, Thomas Reid (1710–1796), one of the authors who most bitterly criticized the modern *idea-ism*—the "theory of ideas" in his own terms—was also one of the few who used the term *information* profusely. In Reid's *Inquiry into the Human Mind on the Principles of Common Sense* (1764), the term appears no less than 15 times in different contexts and with varied meanings, frequently in connection with the term *knowledge*, and even sometimes in connection with terms such as *input*, *artificial language*, *sign*, *receive*, *perception*, and *channel* (Reid 1764/2001, pp. 48, 53, 61, 64, 103, 117–123).

It was during the nineteenth century that the term *information* grew in importance, to the point of acquiring a crucial place in contemporary culture. It was bound up with the expansion of communication technologies, such as the telegraph, and with the use given to it in military intelligence service (Adriaans and van Benthem 2007; also the papers in Davies and Gregersen 2010). Thus, information acquired a great economic and political value. A 1902 issue of *The Economist*, for example, notes that the telegraph has "taken the place of the Ambassador" whose "business [...] undoubtedly is to collect information" (*The Economist* 1902, p. 1881).

Since then, mathematical theories of communication have been developed that seek to facilitate the transmission of the greatest amount of information at the lowest possible cost, in the shortest possible time, and with the maximum security. After World War II, these developments accelerated thanks to the progress of information technology. The linking of communication and computation, and the growth of their social presence, has done the rest. As a result, the term *information* currently occupies a central place in ordinary speech and in almost all sciences and disciplines, from communications to computer science, statistics to systems theory, and criminology to cytology (Miller 2005; Seife 2007; Floridi 2003, 2007, 2011; Gleick 2011).

2.1 The Shannon Model of Information

The unavoidable locus for the theory of information is the classical work by Claude E. Shannon and Warren Weaver (1949). However, the term *information* does not even appear in its title, *The Mathematical Theory of Communication*. The expression, *theory of information*, probably comes from an article by Ralph Hartley (1928) entitled, "Theory of Information Transmission." Although Shannon focuses attention on communication, we should understand that his theory deals specifically with the communication *of information*. The explanation of this concept given by Warren Weaver is still very useful:

> Information must not be confused with meaning [...] To be sure, this word *information* in communication theory relates not so much to what you *do* say, as to what you *could* say. That is, information is a measure of one's freedom of choice when one selects a message. If one is confronted with a very elementary situation where he has to choose one of two alternative messages, then it is arbitrarily said that the information, associated with this situation, is unity [...] The amount of information is defined, in the simplest cases, to be measured with the logarithm of the number of available choices. (Shannon and Weaver 1949, pp. 8–9)

More specifically, the transmission of information concerns the reduction of statistical uncertainty in the communication between transmitter and receiver (Cover and Thomas 2006; Yeung 2006; West and Turner 2006). In this way, the information of a message is measured by a probabilistic function, $I(m_i) = -\log P(m_i)$, where $I(m_i)$ is the information attributed to a message m_i. In consequence, the amount of information generated by a source of messages is measured by this formula: $H(M) = -\Sigma_i P(m_i) \cdot \log P(m_i)$. This magnitude is also called the *entropy* of a source. The name "entropy" was chosen by Shannon in attention to the formal similarity between this formula and Boltzmann's formula for thermodynamic entropy. We shall return below to this point and its conceptual implications.

Shannon identifies the elements that comprise the communication of information processes. He represents them by means of the diagram in Fig. 1, which is our rendition of it (Shannon and Weaver 1949, p. 34). Shannon's objective was to apply his theory to technical systems of communication, such as a telephone or a telegraph system. For this reason, his diagram includes a transmitter and a receiver.

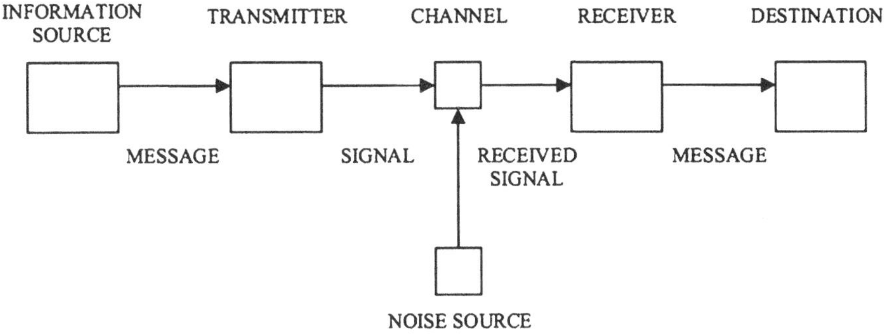

Fig. 1 A rendition of Shannon's diagram

The function of the transmitter is to transform the original message—or instance, a sequence of letters—into a signal suitable for transmission over the channel. Shannon defines a channel as a "pair of wires, a coaxial cable, a band of radio frequencies, a beam of light, etc." (p. 34). For its part, the receiver performs the inverse operation of that which is performed by the transmitter. But, we could devise diagrams with more boxes, depending on the nature of the problems to which we are applying the theory (see, for example, Moles 1972). In Shannon's diagram, the functions of encoding and decoding the message are performed by the transmitter and the receiver, respectively, but we could design new boxes for an encoder and a decoder.

It is possible to construct simpler diagrams with no more than three elements: a source or emitter, a channel, and a receiver. And we can even adopt an abstract interpretation of Shannon's theory free from spatiotemporal connotations. In this regard, Abramson (1963) interprets an information channel as a simple mathematical relationship between the probabilities of two sets of symbols. A channel of information consists only of an incoming alphabet, an outgoing alphabet, and a set of conditional probabilities. For instance, $P(b_j|a_i)$ is the probability of receiving the symbol b_j, if a_i were emitted. Here, a source of information is no longer imagined as a dimensional *box*. It is an abstract entity comprising a set of symbols and their corresponding probabilities (Cover and Thomas 2006; Yeung 2006; West and Turner 2006).

2.2 *Problems to Understand and Overcome*

As Shannon himself warns, there are more problems regarding the concept of information than those that his theory deals with. In order to organize the many informational problems, we can follow the threefold classification suggested by Weaver (Shannon and Weaver 1949, p. 31).

First, there are *technical problems* concerning the maximum amount of information a message can convey. These concern the statistical regularities of the source, such as the internal structure and constraints of the messages, together with the conditions of noise and equivocation of the channel itself. Given these conditions, we ask: "What is the best possible configuration of the message?" That is, which configuration optimizes the balance between length and reliability of the message. Thus, we have problems at a *syntactic* level, of the type dealt with by Claude Shannon's mathematical theory of communication. Let us add that the measure of complexity proposed by Andrey Kolmogorov (1903–1987)—namely, the measure of the computational resources necessary to specify an object, or what has come to be known as the *Kolmogorov complexity*—remains also at the syntactic level (Kolmogorov 1965; Solomonoff 2003; Li and Vitányi 1997; Grünwald and Vitányi 2003; Gilbert and Sarkar 2000).

Second, there exist *semantic problems* that concern the meaning and theoretical truth of the messages, and the correlation between the message and some other thing. Weaver makes it clear that Shannon's theory does not seek to explain problems at this level or at the next one. In the last few decades, several theories have appeared that do deal with semantic aspects of information (Barwise and Seligman 1997; Hanson 1990; Villanueva 1990).

Finally, there are *pragmatic problems* concerning the efficiency of the message to modify the receiver's behavior. Weaver says that, "the *effectiveness problems* are concerned with the success with which the meaning conveyed to the receiver leads to the desired conduct on his part" (p. 5). In biological terms, we find here the functional aspects of information, its ability to affect the receiver's behavior in a functional or adaptive sense.

More recently, Luciano Floridi (2007, 2011) distinguishes between information *as* reality, information *about* reality, and information *for* reality, and it is tempting to correlate these categories with Weaver's levels. On the syntactic level, what we study is information as reality, that is, the properties of the message itself. On the semantic level, we deal with information about reality, or what a message tells us about another part of reality. On the pragmatic level, we observe the capacity of a message to alter reality. This is like saying that we observe the message as information for (making or modifying) reality. A variety of approaches have arisen to address the syntactic, semantic, and pragmatic levels of information (Shannon 1993; Landauer 1996; Winder 2012). However, our main interest here is bioinformation, and so our concern is mainly with pragmatic or functional problems.

3 The Many Faces of Bioinformation[1]

One of the earliest links between information and biology in the twentieth century occurs in August Weismann's 1904 book, *The Evolution Theory* (Weismann 1904; Maynard Smith 2000a). In an important paper over 100 years later,

[1]This section could be read as a kind of phenomenological or descriptive approach to the use of information in some relevant biological contexts. We do not try here even to suggest that the different actual uses of the information concept respond to a unique and coherent meaning of information. Our goal instead is simply to show that the use of informational concepts is pervasive in biology, and illustrate as well as justify this assertion with specific examples of especial relevance.

Artmann (2008) affirms the central role of bioinformation, and his ideas are worth quoting at length:

> The most remarkable property of living systems is their enormous degree of functional organization. Since the middle of the twentieth century, scientists and philosophers who study living complexity have introduced a new concept in the service of explaining biological functionality: the concept of information [...] Let us adduce some of the highly controversial theses that the proponents of biological information theory claim to be true: In molecular *genetics*, a set of rules for transmitting the instructions for the development of any organism has been discovered that is most appropriately described as a genetic code. The main research problem of *developmental biology* is how the decoding of these ontogenetic instructions depends upon changing biochemical contexts. *Neurobiology* cannot make decisive progress before neural codes that are needed for storing, activating, and processing simple features of complex cognitive representations are discovered. *Ethology* is a science of communication since it studies the astonishing variety of information-bearing signals, whose transmission can be observed, for example, in social insects, birds, and primates. Information-theoretical considerations are also of great importance to *evolutionary biology*: macroevolutionary transitions—from co-operative self-replication of macromolecules, to sexual reproduction, to human language—established more and more complex forms of natural information processing. If all these claims prove true, the following answer must be given to the old problem of defining life: *life is matter plus information*. (pp. 22–23, italics added)

Consistent with Artmann's claims, since the 1950s the notion of information has become increasingly important in most fields of biology (see Paton 1992; Gatlin, 1972). It has even been used to define life itself (see Tipler 1995, pp. 124–127; Küppers 1990, 2000, p. 40; Ruiz-Mirazo and Moreno 2011). The biological sciences have adopted a theoretical stance derived from information theory. This perspective holds that all biological processes involve the transfer, processing, or storage of information, and has been referred to as *bioinformational equivalence* in a famous paper by C.I.J.M. Stuart (1985; Burian and Grene 1992, p. 6).

A glance at the current bibliography will suffice to show that, since Stuart's 1985 paper, the use of the concept of information in biology has become widespread (for a historical perspective, see Kull 1999; also Queiroz et al. 2005; Jablonka 2002; Artmann 2008; Collier 2007). In molecular biology, biomolecules are considered to contain information and are the result of informational processes (Holzmüller 1984). In genetics especially, biological thinking is shaped by the idea of information transfer (Brandt 2005; Kjosavik 2007), while in developmental biology and aging, much is said about the expression of information and phenotypic information (Waddington 1968; Oyama 2000; Atlan 1972, p. 96; Peil 1986). In cell biology, tissue biology, zoology, and botany, we study different ways of communicating information with chemical, neuronal, or linguistic bases (Albrecht-Buehler 1990; Marijuan 1991; Stegmann 2005; Pfeifer 2006). In ecology, the concepts of complexity and biodiversity are closely bound up with information through notions of entropy and order (Margalef 1968).

In neurophysiology and endocrinology, the study of communication, storage, and processing of information is central, as are the various electric and chemical codes (Baddeley et al. 2000). The immune system is also researched in terms of knowledge understood as information, both acquired and accumulated (Forrest and

Hofmeyr 2000). Evolution, from the origin of life onward, is thought of as the accumulation of information in macromolecules (Elsasser 1975; Küppers 1990; MacLaurin 1998; Moreno and Ruiz-Mirazo 2002; for information and the origin of life, see Yockey 1977, 1981, 2005). The latest research into the human genome, and the genomes of other organisms, has required the application of powerful methods of computation, classification, and querying of data and information, and this coming together of disciplines has given rise to what is known as *bioinformatics* (see Arp et al. in preparation; Nishikawa 2002).

The concept of information, however, is also central to epistemology and the cognitive sciences and, as several research programmes are attempting to link the cognitive phenomenon with its biological basis, it would be desirable to have one general concept of information that could be applicable to both cognitive and biological contexts. Examples of such programs include evolutionary epistemology along the lines of Lorentz and Wuketits (1983) or Popper (1990), Piagetian epistemology (Piaget 1970), psychobiology (Bond and Siddle 1989), evolutionary psychology (Horan 1992), cognitive ethology (Allen 1992), neural Darwinism (Edelman 1987) and, in general, a widespread current tendency to naturalize epistemology (Giere 1988). An analogy could be drawn between the programs of artificial life, computational science, and the social sciences, where the confluence with biology is evident and the need for a common concept of information is urgent.

Below are a few more-detailed examples of information at work in the biological sciences at various levels. As we hope to demonstrate, many basic life processes—from the molecular foundations of inheritance to the behavior of higher organisms in relation to their environments—are self-organizing processes of storing, replicating, varying, transmitting, receiving, and interpreting information.

3.1 Genetic Information

In general, biologists and other researchers who describe biological phenomena are aligned with Mayr (1996) in his description of organisms as "hierarchically organized systems, operating on the basis of historically acquired programs of information" (Yockey 2005; Terzis and Arp 2011; Gould 2002; Bogdan 1994; Boi 2011).

The "programs of information" part of Mayr's description of organisms above is what is significant for us here. But what does this mean? As most people know, a *gene* is a functional segment of deoxyribonucleic acid (DNA) located at a particular site on a chromosome in the nucleus of all cells (but see Burian and Kampourakis this volume). DNA and ribonucleic acid (RNA) are composed of nucleotides that specify the amino-acid sequences of all the proteins needed to make up the physical characteristics of an organism, much like a cryptogram or code. This genetic code consists of specific sequences of nucleotides that are composed of a sugar (deoxyribose in DNA, ribose in RNA), a phosphate group, and one of four different nitrogen-containing bases, namely, adenine, guanine, cytosine, and thymine in DNA (uracil replaces thymine in RNA). These four bases are like a four-letter alphabet, and triplets of bases form three-letter words or *codons* that comprise the "program of information" which identifies an amino acid or signals a function.

DNA is the template from which RNA copies are made that transmits genetic information concerning an organism's physical and behavioral traits (phenotypic traits) to synthesis sites in the cytoplasm of the cell. mRNA takes this information to ribosomes in a cell where amino acids, and then proteins, are formed according to that information. The proteins are the so-called *building blocks* of life, since they ultimately determine the physical characteristics of organisms (Boi 2011; Carroll 2005).

Two significant processes utilized by researchers that have contributed to, and continue to contribute to, our understanding of the genetic code are *genetic sequencing* and *genetic annotation*. Genetic sequencing refers to the methods and technologies used since the early 1970s to determine the specific order of the bases in a molecule of RNA (adenine, guanine, cytosine, and uracil) or DNA (adenine, guanine, cytosine, and thymine). Walter Fiers and colleagues (1976) published groundbreaking work in RNA sequencing in their *Nature* paper titled, "Complete Nucleotide Sequence of Bacteriophage MS2 RNA: Primary and Secondary Structure of the Replicase Gene." A sequencing-by-separation technique was developed by Frederick Sanger and Alan Coulson (1975) for DNA in 1975, and this "plus and minus" method still acts as the basis for a lot of gene sequencing performed today. Various genetic sequencing methods have been utilized for RNA and DNA since the 1970s, including what is known as *high-throughput sequencing* that can produce millions of sequences at once (Shendure et al. 2004).

Understanding the particular configurations of As, Gs, Cs, and Ts (or Us) in the genetic code is one thing; understanding what processes are initiated, amino acids are identified, or functions are signaled by virtue of these particular configurations is another. Genome annotation—*annotation* read here as "commentary" or "explanation"—refers to the methods and technologies used to identify the locations of genes (as well as the coding regions in a genome) and determine specifically what those genes do. "What are all these genes doing, how do their functions interact, and how may we take advantage of the sequences to advance understanding and cure human disease" (FlyBase 2001). This is the question posed at the beginning of one of the earliest white papers produced by members of FlyBase, a consortium of researchers devoted to annotating the genetic makeup of *Drosophila melanogaster*, a fruit fly. In fact, many organ systems in mammals have well-conserved homologues in *drosophila*, and this species of fruit fly not only was utilized by Thomas Hunt Morgan and his researchers in the early 1900s—in the now famously dubbed "Fly Room"—so as to understand genetic functioning generally (Morgan et al. 1915), but it was also utilized by various groups attempting to annotate the human genome through the Human Genome Project, which was completed in 2003 (HGPI 2012). It is estimated that some 66 % of human disease genes having a clear cognate in *drosophila* (Stein 2001; Reiter et al. 2001; Tweedle et al. 2009).

3.2 ATP, Euglenas, and Information

Cells use energy, and one of the primary functions of the mitochondrion of an animal cell is by using the energy released during the oxidation of sugars to produce a nucleic acid called adenosine triphosphate (ATP). However, this can happen only if there is a line of communication between other organelles of the cell and the mitochondria themselves. ATP acts as the material catalyst of information communicated between mitochondrion and other organelles. When there are low levels of ATP, the mitochondria receive this information and oxidize more sugars; conversely, when sugars are oxidized (this activity, among other activities), the other organelles receive this information and cellular homeostasis can be maintained.

Euglena gracilis is an abundant one-celled microorganism that is a member of the protist kingdom found in freshwater environments; in colloquial terms, it is known as a kind of algae. It is about 10 μm in length and looks like a sperm cell with a more elongated body. It is equipped with a flagellum, eyespot, vacuoles, chloroplasts, mitochondria, plastids, and a cell nucleus. Each one of these components has a function: the flagellum is a whip-like tail that enables the euglena to move around; the eyespot is light/dark sensitive so that the euglena can move toward sunlight, its food source; vacuoles allow for wastes to be disposed; chloroplasts and mitochondria work together to transform sunlight energy through ATP; plastids store the food; the cell nucleus contains a nucleolus that synthesizes and encodes ribosomal RNA, which is important for euglena structure and reproduction (Buetow 1982).

For an organism like the euglena to function effectively in some external environment—basically, live its life in its microbial world—it is necessary that information be exchanged between and among the various subsystems of this system. Food storage in the euglena can be viewed as a subsystem activity, which itself is made up of processes dealing with electron transport and oxygen exchange in photosynthesis. Concerning these processes in the euglena, electrons are transferred from a donor molecule (such as nicotinamide adenine dinucleotide - NADH) to an acceptor molecule (such as O_2) across a membrane, with the resulting H^+ ions used to generate energy in the form of ATP. The information must be exchanged in these processes; otherwise, there would be no storage of food. At the same time, this subsystem works with the subsystems concerning food acquisition and mobility. If information were not being exchanged between the eyespot and the flagellum, then there would be no movement toward sunlight; in turn, there would be no photosynthesis, and then no food storage.

3.3 Action Potentials, Reflex Arcs, and Information

When a neuron produces an action potential (colloquially, when it *fires*), information associated with spiking signals is communicated between that neuron and at least one other neuron. In the language utilized by Shannon (see Fig. 1 above), the axon of one neuron A acts as a *transmitter* and the dendrites of another neuron B, to which the axon of neuron A is connected, acts as a *receiver*. Protein synthesis in neurotransmitter release is the information that is communicated between neurons. Depending on the amount and intensity of the neurotransmitter emitted from the transmitter neuron, the receiver neuron may become excitatory, making it more likely to produce its own action potential. Networks of neurons can fire more quickly when they are used more frequently, as if the information associated with the particular network's firing has been stored. The complex workings of trillions of these connections throughout an animal with a complex nervous system enable it to fight, flee, forage, feast, and the like (Kandel et al. 2000).

A clear illustration of the communication of neuronal information in a systemic fashion is a mammal's muscle coordination in a *reflex arc*. In this activity, information is communicated to and from the spinal cord and a particular muscle group of the body (Kandel et al. 2000; Pelligrino et al. 1996). Consider a situation where a very curious cat decides to jump atop a very hot stove. The intense motion of the molecules from the stovetop is impressed upon the pads of the cat's paws. That motion affects the sensory neurons in the cat's skin, causing them to fire. The sensory neurons send a message to the spinal cord. These *messages* consist of billions of action potentials and neurotransmitter releases, affecting cell after cell that is along the pathway of this particular reflex arc. In an instant, the spinal cord then sends a message back to the muscle groups associated with the cat's legs, diaphragm, and back. In a flash, the cat jumps off the stove, screaming while arching its back.

However, now the cat must coordinate its fall to the ground. This time, information is sent from the visual system to the brain, and then back through the spinal cord to other muscles in the cat's body. All of this information must be integrated by the brain, and motor responses must be orchestrated by the combined effort of brain-body communication of information. The cat narrowly avoids falling into the garbage can placed next to the stove.

3.4 Visual Perception and Information

In their textbook devoted to the principles of neuroscience, Keith Kandel et al. (2000) describe the processes associated with perception in the cerebral cortex using a hierarchical model: "Sensory information is first received and interpreted by the primary sensory areas, then sent to unimodal association areas, and finally to the multimodal sensory areas. At each successive stage of this stream more complex analysis is achieved, culminating eventually, as with vision, for example, in object and pattern recognition in the inferotemporal cortex" (p. 353).

Kandel et al. actually divvy up the hierarchy of sensory systems into four parts, viz., (a) the primary sensory areas, (b) the unimodal areas, (c) the unimodal association areas, and (d) the multimodal association areas.

The primary sensory areas act as the base level, and they refer to the way in which information initially is communicated to the spinal cord and/or brain through one of the five sensory modalities, viz., touch, hearing, taste, smell, and vision. For example, in the visual system the primary sensory area is comprised of the eye, lateral geniculate nucleus, and the primary visual cortex located in the occipital lobe of the brain.

The unimodal areas build upon the data received from some prior particular primary sensory area, and refer to a higher-leveled integration of the data received from one of the primary sensory areas processed in a part of the brain different from that of the primary sensory area. In the visual system, there are two primary unimodal areas that process information concerning *where* an object is and *what* an object is, located along trajectories between the occipital lobe and parietal and temporal regions, respectively.

The unimodal association areas, in turn, refer to an even higher-leveled integration of the data received from two or more unimodal areas. In the visual system, the unimodal association area integrates data about the color, motion, and form of objects, and is located in the occiptotemporal area of the brain.

Finally, the multimodal association areas build upon the data received from the unimodal association areas and, depending upon the sensory modality, process this information in the parietotemporal, parietal, temporal, and/or frontal areas of the brain (also see van Essen et al. 1992).

The result is this: information is exchanged at the various levels of the visual system and between the visual system and the central nervous system and, because of these exchanges, an animal is able to form a coherent picture of an object in its visual field, a visual perception (Crick and Koch 2003; Baddeley et al. 2011; Gray 1999; Singer 1999; Bullot 2011; Arp 2008).

3.5 Environments and Information

Organisms interact with external environments. However, because organisms are hierarchically organized living systems composed of subsystems, processes, and components engaged in various operations, they have their own internal environments as well. An *environment* can be defined as any pressure or force that interacts with, or affects somehow, the organism and its components. We can draw a distinction between the information that is exchanged *within* the organism's environment and the information that is exchanged *between* the external environment and the organism. So, there are really two types of environments, namely, environments that are *internal to* an organism and environments that are *external to* an organism.

Concerning internal environments, for example, the other organelles, nucleus, ATP, water, and various organic molecules act as the environment for a mitochondrion in the eukaryotic cell; other eukaryotic cells, cancerous cells, water, and all kinds of organic molecules and chemical elements act as the environment for a typical eukaryotic cell; a myriad of molecules including hydrogen, carbon, nitrogen, and oxygen surround and exert influence upon organs in a multi-cellular organism's body; a piece of food taken in from the environment external to the organism becomes part of the environment within the organism and, depending on the content, may be digested or expelled.

At the same time, the organism itself is interacting with external environments that are exerting pressures upon, as well as exchanging and communicating information with, the organism. Concerning external environments, we see that organisms are members of species that live in populations. These populations usually co-exist with other populations in communities. Many communities living with their non-living surroundings comprise an ecosystem, and the sum of all ecosystems make up the biosphere of the Earth (see Justus this volume). Other members of a species, different species, and the non-living surroundings of an organism all are considered as parts of the external environment for an organism. The organism constantly experiences environmental pressures, and these pressures can be described in terms of information that is exchanged between the environment and the organism (Brandon 1984, 1992). This kind of information exchange can be witnessed as a result of research accrued and experiments performed by biologists and other thinkers.

It is common knowledge that an organism's survival is dependent upon both genetic and environmental factors. For example, if there is an alteration in a rodent's genetic makeup causing it to have a malformed foot, then it is more likely to be eaten by a hawk out on the open range. However, if the same handicapped rodent lives in a forested area where it can hide under rocks and bushes, it is less likely to become a predator's victim. Also, if an environment happens to be made up of trees having fruit high up on its branches, and it just so happens that a fruit-eating animal's genes "coded" it to have a neck long enough to reach the fruit, then such an animal likely will survive. Conversely, if your animal genes "coded" you to have a short neck, it is unlikely you would survive in such an environment (that is, if the fruit high up in the trees was your only food source). In the words of Tim Berra (1990): "The environment is the selecting agent, and because the environment changes over time and from one region to another, different variants will be selected under different environmental conditions" (p. 8).

Another famous example that illustrates the informational transfer between the environment and an organism has to do with the finches that Darwin (1859/2009) described on the Galapagos Islands during his voyage on *The Beagle*. These finches clearly exhibit *adaptive radiation*, i.e., in the words of Berra (1990): "the evolutionary divergence of members of a single phylogenetic lineage into a variety of ecological roles usually resulting, in a short period of time, in the appearance of several or many new species" (p. 163). Darwin noted several different beak shapes and sizes that apparently were modified in the finches, depending upon the ecological niche the particular bird inhabited. Some finches had massive beaks ideal for crushing their seed food source, others had thinner pointed beaks ideal for probing flowers, still others had curved beaks ideal for picking food out of woody holes. In this situation, the environments in which the various finches inhabited were all different, and the finches with beaks most fit for a particular environment survived to reproduce.

Phenotypic traits are the physiological characteristics or behaviors of organisms that are under genetic control. The genetic information determines what a particular member of a species will look like, how fast it will run, what coloration it will have, how successful it will be at mating, etc. In the finch example, the different beaks represent the variety of phenotypic characteristics under genetic influence. If it just so happened that a certain beak style was effective in gathering food in an environment, then that finch would survive and pass its genes onto its offspring. Soon, that particular niche would be dominated by the beak style that was most advantageous for that environment.

Research has been conducted on animals to determine how the external environment affects the functioning of various systems of the body. One experiment has to do with occluding or removing the eyes of cats, rats, and birds at various stages of development to see if the neural connections of the brain necessary to the visual system either would develop abnormally, or cease to function altogether. These studies indicated that when occluding or removing the eyes, certain neural connections in the brains of these animals would not be made. This resulted in the cessation of certain visual processes, causing the overall subsystem to be under-developed in relation to other animals that have not had their eyes occluded or removed (Shatz 1992; Clayton and Krebs 1994). This example illustrates what happens when information *is not* exchanged between environment and organism.

A final example that demonstrates the information exchange between an organism and its environment has to do with the artificially controlled speciation of the fruit fly, *drosophila*. Experimenters are able to take out, move around, or add genetic sequences in the DNA of the fruit fly, causing radical phenotypic alterations in it to occur such as the deletion of some organ, legs growing where antennae should be, and antennae growing where legs should be. The experimenter's adjustments to the genetic material of the fruit fly are analogous to the radioactive material and other kinds of natural external forces of mutation that alter the genetic codes of fruit fly populations. We find similar monstrosities in fruit flies when we study them in their natural habitats (Duncan et al. 1998). Just as researchers tap into and alter the genetic makeup of fruit flies in controlled experiments, so too, external forces "tap into" and alter the genetic makeup of fruit fly populations in nature. These fruit fly abnormalities are another example of the property of environmental-organismic information exchange found in organisms.

4 A Few Debates Concerning Bioinformation and Bioinformation as a Relation

Despite its application to a broad range of disciplines—including the aforementioned examples in the biological sciences—appealing to the notion of information as an explanatory feature of living systems is a matter of much dispute, which in recent years has arisen over its need and usefulness.

4.1 Bioinformation: Metaphor or Reality?

Some authors consider information a distinctively linguistic phenomenon, so that its application in other fields is purely metaphorical and instrumentally useful. For example, when we were discussing action potentials of neurons above, we noted that protein synthesis in neurotransmitter release is the information that is communicated between neurons. It is possible to render information in the above description as a purely linguistic tool utilized for explanatory purposes—here, one might say, "protein synthesis in neurotransmitter release is not *really* information," and, in fact, we can skip the informational part and go directly to the *real* explanation associated with action potentials, namely, *protein synthesis in neurotransmitter release*. The notion of information is just that—a metaphorical notion—and does some explanatory work in explaining action potentials. But the real entities and processes doing all of the work consist of proteins, neurotransmitters, and the like physicochemical phenomena.

This last point strikes a reductionist tone, and there are many reductionists who argue that the use of information concepts is redundant in fields like biology, which are subject to general laws of matter and energy. Such researchers think that biological phenomena should be explained in mechanical, electromagnetic, chemical, and thermodynamic terms, thus rendering informational conceptions—as well as other conceptions, for example, *function*—superfluous. According to this reductionist perspective, to speak of information in biology would just be an odd way of speaking of correlation and causation (Stuart 1985; Griffiths 2001; Sarkar 1996, 2000; Janich 1992; Kitcher 2001).

Many researchers, however, think that the informational perspective sheds considerable light on biological phenomena, allowing us to understand living things in a way that would be otherwise impossible (Terzis and Arp 2011; Maynard Smith 2000a, b; Queiroz et al. 2005; Godfrey-Smith 2000; Griffiths 2001; Roederer 2005; Avery 2003; Yockey 2005). Proponents of biological information theory argue that many basic life processes include the storing, replicating, varying, transmitting, receiving, and interpreting of *real pieces of information* of various types; and these processes are perhaps irreducible to physical and chemical terms. Stated simply, such researchers are convinced that "there is more to informational talk in biology than mere metaphor" (Sarkar and Plutynski 2008, p. xxi; also Sarkar 1996, 2000, 2005; Griffiths 2001).

There is no doubt that the presence of metaphors in biological texts is ubiquitous, and it is not just a question of informational metaphors (Keller 1995, 2002). Darwin himself was called a "master of metaphor" by Stephen Jay Gould (1989). From "natural selection" to the "immune self" (Tauber 1994), all branches of biology constantly use very diverse metaphors. And all this is not incompatible with a realist reading of biological texts, for metaphors themselves may be interpreted in a realist way (Marcos 1995, 1997, 2010, chapter 10).

The use of information theory as an instrument is very common in biology. As biological systems—from macromolecules to organisms—are very complex, we can use information theory to measure their structural complexity. In John Collier's (2007) words:

> I will compare the use of information as a technology of measurement, which does not imply that there is anything present that might be called 'information' with a stronger usage of information in biology that attributes information to biological systems in a non-instrumental way. This distinction between instrumental and substantive uses of information in biological studies often turns on the notion of information used, so it is important in each case to be clear what is at stake [...] The instrumental usefulness of information technologies does not in itself imply the existence of substantive information. (p. 763)

But this instrumental application of the theories of information is also found outside biology. Any structure, living or otherwise, may be studied from this point of view. Following Collier (2007): "Some of the applications, however, present interesting issues for the philosophy of biology, especially concerning whether the instrumental use of information is sufficient to explain the use of the idea of information by biologists" (p. 767).

In other words, an instrumental interpretation is possible if we do not consider the purely biological, that is, if we consider living beings as mere physico-chemical structures. But, then, what sets living beings apart? "Arguably," Collier (2007) affirms, "to be alive requires this sort of separation of function and the requisite dynamical decoupling between metabolism and replication" (p. 770). So, the mutual reference between metabolism and replication must surely have an informational and functional character (also see Brooks and Wiley 1988; Maynard Smith and Szathmáry 1995). We can shed light on the structure of a gene only by showing its informational connection with a protein. We can say then that the function of a given fragment of DNA is to encode a protein. In an analogous way, we can explain the structure of a protein only by its reference to a vital function. And there are vital functions only when there exists an individual living being. So, living beings distinctively include information. This is the best possible explanation of the usefulness of informational concepts in biology. So, in the opinion of many authors a substantive explanation of bioinformation is required as part of the broader explanation of genetics.

4.2 Bioinformation as a Triadic Relationship

It is probably better to use the term *realist* rather than *substantialist* here. This is because when we speak of a substantialist interpretation of bioinformation, it would seem that we take for granted that bioinformation is a *substance*. From our point of view, bioinformation is a real entity, but not necessarily a substance. This observation leads us to another debate: if we accept that bioinformation is a real entity, what kind of entity is it precisely?

Some authors have viewed information as a thing, third *substance*, or primitive element. Wiener (1961), for example, thinks that information straightforwardly is "information, not matter or energy" (p. 132). Also, information has been seen as a *property* of a thing in terms of form, order, organization, negative entropy

(Brillouin 1962), complexity (Kolmogorov 1965), or diversity (Margalef 1980). Information as a property raises the problem of its location, which is a recurrent difficulty and, as such, one of the major arguments against the bioinformational paradigm. Actually, the problem of information location will be unsolvable unless we abandon this view of information as a simple property. Further, we find information conceptualized as a dyadic (semantic) and a triadic (pragmatic or functional) *relation*, as we hinted at in Sect. 2 above. As Barwise (1986) notes: "But is information relational? Surely so. The basic intuition about the information content C_s of a situation *s* is that it is information *about* something besides *s*" (p. 326; also Dennett 1987; Mackay 1969; Küppers 1990; Queiroz et al. 2005).

On the other hand, information as a thing or basic substance should be the last hypothesis to explore, for the principle of ontological economy implies that, all things being equal, if some other hypothesis works, it is clearly preferable. The other three possibilities could be equated with the three parts of Weaver's classical distinction (1949), which we explored briefly in Sect. 2.

The *technical problems*, which Weaver places at level A, are studied by considering the formal and statistical properties of messages. At this level, we are dealing with information as a property. The *semantic problems*, or level B problems, are concerned with the dyadic relationship between the message and its meaning. The *effectiveness problems*, or problems of level C, imply three elements. Weaver (1949) suggests that they are the message, its meaning, and a change in the receiver's behavior caused by the reception of the message (p. 5). Therefore, problems of level C have a pragmatic aspect, which in biological contexts could be construed as a function. For instance, the change in cell behavior caused by the reception of a genetic message may consist in the accomplishment of a given function such as the synthesis of a determined protein.

In light of the above distinctions, we argue that bioinformation should be conceived as a triadic relationship, i.e., a relation involving three entities. The pragmatic or functional concept of information as a triadic relationship is the concept that best adapts to biological contexts, where functional explanations are very common (Cummins 2002; Millikan 2002; Perlman 2004; Arp 2006; see also Wouters this volume). We consider an explanation for the existence of an organ or a molecule satisfactory only if it includes reference, not only to its structure and material composition, but also to its *function* in the organism. Thinkers cannot seem to get around Trivers' (1985) claim that "even the humblest creature, say, a virus, appears organized to *do* something; it acts as if it is trying to achieve some purpose" (p. 5), or Arnhart's (1998) observation that "although the evolutionary process does not serve goals, the organisms emerging from that process do. Darwin's biology does not deny—rather, it reaffirms—the immanent teleology displayed in the striving of each living being to fulfill its specific ends [...] Reproduction, growth, feeding, healing, courtship, parental care for the young—these and many other activities of organisms are goal-directed" (p. 245). And what has been communicated in this paragraph above comports with the thinking of many biologists and philosophers of biology, including Collier (2007): "The relevant level for specifically biological information is the functional level" (p. 771).

Having said all of this, we can construe information—including bioinformation—in the following triadic, relational way. Information implies a relationship between:

1. a message, m, which may be any event, linguistic, or otherwise;
2. a system of reference, S, which the message informs the receiver about; and
3. a receiver, R.

Let's consider a fragment of mRNA as a concrete example of a message, while its system of reference could be a fragment of a protein. The receiver is a formal scheme residing in a concrete subject (a human being, another living system, a part of a living system, an ecosystem, a computer, or the cytoplasm of a cell for the preceding example). A concrete subject could, of course, use more than one receiver and use them alternately (playing with different "hypotheses") or successively (owing to an evolutionary or individual process of learning). We can also see the receiver as an internal (that is, residing in a concrete subject) predictive model of S, along the lines suggested by Rosen (1985), who characterizes living beings as "anticipatory systems."

A system of alternative messages in one relation can be a system of reference in another relation, and vice versa—and the process could be iterated. A segment of DNA can be a message informing the appropriate part of the cell about the mRNA to be synthesized. The same mRNA, initially part of a system of reference, may later become a message informing the cytoplasm about the synthesis of a certain protein, and so on. As Queiroz et al. (2005) state using Peircean vocabulary, "semiosis entails the installation of chains of triads" (p. 60). This is why a metaphor like "the flow of information" is sometimes useful.

Comparing this triadic view to the classical Shannon one (see Fig. 1), it may seem surprising that the emitter or source is not even mentioned. However, as Millikan (1989) rightly notes: we should "focus on representation *consumption*, rather than representation production" (pp. 283–284). Furthermore, there is often no specific emitter in non-linguistic contexts, like some biological ones, so a concept of bioinformation should not demand the presence of an emitter.

The matter of the channel is more complex because, usually, we have a dimensional image of it. However, it is possible to construe a channel in a more abstract way, as a set of conditional probabilities, along the lines suggested by Abramson (1963). In the same spirit, Barwise and Seligman (1997) note that a channel could be understood, basically, as an objective correlation of any degree between two domains.

Most of the conceptual problems concerning information actually stem from the ellipsis of some element of the informational relation. We often speak about the information of a message with no reference to a receiver or a referential system, although both of them exist implicitly. Bioinformation is always functional, transitive, and pragmatic. The message is always referred to something by a receiver; otherwise it is not a message, just an event (Millikan 1989, p. 286). If messages were not referred to something by a receiver, Griffiths (2001) would be perfectly right to say that "most information talk in biology is a picturesque way to talk about correlation and causation" (p. 400).

However, factors conditioning information are often mistaken for information itself. Such is the case regarding the formal characteristics of the system of reference, and either those of the message or the system to which it belongs. The correlation between the messages and the system the information is about also affects the amount of information involved, but neither this *correlation* nor *form* constitutes the information itself.

The relationship among the three above-mentioned elements (m, R, and S) is informative when it changes the receiver's knowledge of the system of reference. By *knowledge*, we mean the distributions of probabilities of the possible states of the system of reference in the receiver. Knowledge, therefore, should be understood here along the lines suggested by Karl Popper (1990) in a very general way: "Can only animals know? Why not plants? Obviously, in the biological and evolutionary sense in which I speak of knowledge, not only animals and men have expectations and therefore (unconscious) knowledge, but also plants; and, indeed, all organisms [...] Flowering plants know that warmer days are about to arrive [...] according to sensed changes in radiation" (pp. 9, 10, 35). This understanding of knowledge does not necessarily imply consciousness, so the notion is applicable to human as well as to non-human living systems. Consider Rosen's (1985) claim as well: "I cast about for possible biological instances of control of behavior through the utilization of predictive models. To my astonishment I found them everywhere [...] the tree possesses a model, which *anticipates* low temperature on the basis of shortening days" (p. 7).

We can describe information (I) as a relationship between a message (m), a receiver (R), and a system of reference (S). In this relationship can be found the triad formed by a message, receiver, and system of reference where the message alters the receiver's previous knowledge of the system of reference (Dretske 1981, 2007). Moreover, the more probable an alternative is to a receiver, the more information will be received when a message says that a different one has occurred, unless it is a simple contradiction. So, for example, the introduction of a certain genetic message into the cytoplasm increases the probability of the cell carrying out a certain function, for the probabilities of alternative behavior decrease. Now, we can say that the receiver knows—or knows better—how to do something. Again, this understanding of knowledge does not necessarily imply consciousness.

The informational relation, in accordance with our realistic interpretation of bioinformation, may be perfectly objective (see Barwise 1986; Fodor 1986; Denbigh and Denbigh 1985). For example, it is quite objective that a genetic message informs the cytoplasm about synthesizing proteins. But, this does not necessarily mean that the information has been in the world since the beginning, preceding any subject capable of using it, as Dretske (1981) notes. Without cellular machinery there is no connection between DNA and protein. As Moreno and Ruiz-Mirazo (2002) state, the genetic message is, in principle, "decoupled from the dynamical organization of the system" (p. 73).

Information can be measured from the magnitude of its effects, that is, by the changes to the receiver's knowledge of the system of reference (for a measure of this kind, see Marcos 2011a). This is a traditional and standard way of measuring different physical magnitudes. Measuring information—like measuring anything else—requires a subject to do it, and this subject acts according to theoretical grounds. To assess the quantity of information given by a genetic message to the cytoplasm, we need extensive biochemical knowledge, but this does not make the informational relation any less objective.

5 Bioinformation and Related Concepts

The concept of information is usually presented in connection with others that seem to form a constellation. The relationships between them, however, do not usually appear with sufficient clarity, which may make educational tasks difficult. At this point, a philosophical approach may be helpful, one that introduces clarity to the concepts and their mutual relationships. Throughout the chapter, such concepts as *correlation* and *form* have been appearing, and in this section others will make their appearance, such as *entropy*, *order*, *organization*, *complexity*, and *knowledge*. All of them are closely related to the notion of information, but none of them simply identifies itself with it.

5.1 Bioinformation and Thermodynamic Entropy, Order, and Organization

Above, we mentioned that August Weismann correlated information with the biological sciences in his 1904 book, *The Evolution Theory*. However, information first appeared in biology in connection with the concept of physical entropy and its different measures (thermodynamic or statistic) through physicists Rudolf Clausius (1822–1888) and Ludwig Boltzmann (1844–1906), who formulated the measures of entropy. Clausius was the first to introduce the term *entropy* to thermodynamics in 1876, while Boltzmann gave a statistical interpretation to the term. Boltzmann considered that a macrostate of a given system is more entropic in the same measure as it is compatible with a greater number of microstates.

The classic example is that of a gas-filled box. The box has two compartments, right (R) and left (L), connected by a door. The system can be in a macrostate A, in which the temperature in one compartment is relevantly higher than in the other one, or in a macrostate B, with equal temperature in both. According to the kinetic theory of heat, the temperature of each compartment varies depending on the kinetic energy of the particles in it. Thus, if the temperature in R is higher than the temperature in L, this is because the particles in R are on average faster than the particles in L. If there were the same temperature in both compartments, this would be due to a uniform distribution of the fast and the slow particles along the box. Let's call "a microstate" to a concrete distribution of the particles. So, the macrostate A has obviously a lower statistical probability than B because it is compatible with fewer microstates than B, and so B has higher entropy than A.

Boltzmann proposed the following formula for measuring thermodynamic entropy: $S = K \, Ln \, W$, where S is the entropy of a given macrostate of a system, K is Boltzmann's constant, and W is the number of microstates compatible with this given macrostate. As can be easily observed, this equation is similar to the Shannonian formula for informational entropy, $H(S) = -K \sum_i P(s_i) \cdot \log P(s_i)$.

James Maxwell (1831–1879) took the next step with a thought experiment. If we place inside the box a demiurgic being (Maxwell's Demon) that allows the faster particles to pass to one compartment and the slower ones to the other, then the system evolves toward a less entropic state. Apparently, this situation is incompatible with the second law of thermodynamics.

Leo Szilard (1898–1964) found a sound answer to Maxwell's paradox. Maxwell's Demon overcomes the universal tendency to entropy thanks to the information he obtains about the speed of the particles. However, he had to measure the speed by means of whatever physical process, and any measurement process must involve some transaction of energy and increase of entropy. It seems, therefore, that an (inverse) link exists between entropy and information.

Taking inspiration from this idea, Léon Brillouin (1889–1969) developed the concept of *negentropy*, or negative entropy, as equivalent to information (Brillouin 1962). Thinkers such as Tribus et al. (1966) and Layzer (1990) have attempted to equate information with a positive magnitude, the distance from thermodynamic equilibrium (also see Brooks and Wiley 1988; Weber et al. 1988; Marcos 1991a).

The last step prior to the *solidification* of the concept of information in biology was Erwin Schrödinger's (1944) classic, *What is Life? The Physical Aspect of the Living Cell*, where he claims that living things overcome the universal tendency to entropy by exporting entropy to their environment, as Maxwell's Demon does. Thus, a connection was made between thermodynamic order and biological complexity. Schrödinger contributes to the link between biological phenomena and physical entropy, and physical entropy had already been connected with information, so the stage was set for the encounter of information and biology. A slogan for this approach applied to biology could be, "A gain in (physical) entropy means a loss of (biological) information" or even Schrödinger's (1944) own claim that, "life feeds on negative entropy" (p. 70).

From the point of view proposed here, thermodynamic entropy conditions the information that the macrostate of a system can offer about its possible microstates to a receiver equipped with the right physical laws. If the particles of the system act together, the system as a whole is more dynamic. Correspondingly, the movement of the system offers a great deal of information about its elements. If entropy increases, the system is less dynamic and reflects less efficiently the positions and moments of its components. Thus, thermodynamic entropy is linked *specifically* with the information that a macrostate can give about a system's currently accessible microstates. So, the basis for a general measure of information could not be entropy, negentropy, or distance from equilibrium (Marcos 1991a).

Physical entropy is currently linked with (structural) order and (functional) organization, but order and organization are, respectively, relative to a structure and a function. Several types of order or organization may be identified even within the same system. Organization is also relative to a receiver connecting the message and the system of reference. A fragment of DNA is organized for the synthesis of a certain protein only if one knows how the cellular apparatus works. Physical entropy, therefore, should not be considered a general measure of organization; rather, it is a correct approach to *one* type of organization able to render work (Denbigh and Denbigh 1985; Nauta 1972). In biology, organization is always established with regard to a certain function. It is not just a question of structural regularity.

This is why Schrödinger (1944) conjectured, before the discovery of the double helix, that genetic information must be contained in some kind of *aperiodic* crystal. As it is well known, each crystal is formed by the periodic repetition of the same module. Biological macromolecules, such as proteins and DNA, are also modular compounds, but they are not formed by a periodic repetition of a singular module. In this sense, one can speak of them as aperiodic crystals. Schrödinger's book, with its concept of aperiodic crystal, exerted a great influence on physicists and paved the way for many physicists to move into biological studies. One of the most prominent was Francis Crick, who noted that the book was "extremely well written" and made the subject seem "exciting" (Crick 1965). Schrödinger's idea favored also the use of radiographic methods for the study of biological molecules; methods that were first developed for the study of the structure of the crystals. James D. Watson and Francis Crick, the two co-discoverers of the structure of DNA in 1953, used X-ray diffraction data collected by the British crystallographer Rosalind Franklin (Ceccarelli 2001; Ridley 2006).

5.2 Bioinformation and Shannon's Entropy

Some remarks are in order here about the relation between Shannon's entropy and bioinformation. On the one hand, the structure of the system the message belongs to affects the information, but in the opposite way to that of the system of reference. When we try to pass information, we do not want the system to which the message belongs to impose any structural limitations on our communication, or at least we want them kept to a minimum. This is what Shannon calls *entropy* (freedom of choice within a source), and is recommended for a system acting as a symbolic one. A symbolic system is a system whose elements have a symbolic function, such as for instance the genetic system and the language. The word "table" symbolizes a table, as the codon UCG given the correct context symbolizes Serine. This is why in some parts of biological systems—for example, in neuronal, genetic, immunological, and linguistic domains—unities like can be combined in many different ways, for they must be flexible when representing other parts of the systems or external realities.

On the other hand, a higher level of structure or regularity in the system of reference brings about the possibility of transmitting more information about it with a given number of symbols, in line with common sense and philosophical tradition (Eco 1962; Moles 1972; Kolmogorov 1965). For instance, a few words could be enough for describing the full molecular structure of a crystal, but the same number of linguistic symbols could say almost nothing on the molecular structure of a volume of gas.

Consequently, this matter is sometimes shrouded in confusion. It could be seen as a paradox that some authors correlate information with freedom of choice or low structural constraints, as Shannon does; while others, like Eco and Moles, correlate information with structural order, constraints, or regularity. But it is not paradoxical at all, but rather expresses two aspects of information. One aspect is the relative order of the system of reference, while another aspect is that of the symbolic system. Shannon's entropy of the symbolic system correlates positively with information: while regarding the object that the system informs about, it is the case that the greater the order and organization, the greater the amount of information that can be produced by a given sequence of symbols. Finally, another factor limiting the amount of information is the correlation between the structure of the message and that of the referential system. If it is perfect, a maximum amount of information can be transmitted. No greater correlation exists than between a system and itself. In this regard, Shannon's measure is an absolute limit on the amount of information: no more information can be given about a system than is given by the system itself. Therefore, Shannon's measure is often referred to as a measure of possible information.

5.3 *Bioinformation and Complexity*

Another approach to information has appeared more recently, based on the work of Andrey Kolmogorov (1965) and Ray Solomonoff (2003): algorithmic or computational theory. Here, information is viewed as a special kind of complexity. Any sequence describing a text, image, musical composition, etc., may be generated by means of a program and a suitable computer. If the sequence shows any regularity, symmetry, or redundancy, the program could be shorter than the sequence itself. If the sequence is more *complex*, or even random, it will be less susceptible to compression, so the greater the complexity, and the lesser the compressibility. Thus, for instance, under this approach, the "aperiodic crystals" to which Schrödinger refers are more complex than a periodic standard crystal.

But, it must be remembered that information, unlike complexity, is not a property of a single thing, but a relation between at least three entities (as we have mentioned already), so some remarks may be made on the relationship between Kolmogorov's complexity and bioinformation. First, the relationship between information and the complexity of a sequence is not a direct one, that is, complexity cannot be simply equated with information. The need for a long program to generate a sequence does not translate directly into that sequence "having" a great deal of information. It would be counterintuitive, for random sequences would be the most informational ones. Kolmogorov's measure of complexity can distinguish between a crystal and a protein, but a relevant concept of bioinformation must also distinguish between a functional protein and a random peptidic compound.

Second, Kolmogorov's notion of complexity has also been used to calculate the informational content of an individual object as a direct function of the length of the shortest program describing or producing it. Here, we must remember the difference between things and words. When complexity is assessed from the compressibility of a description encoded in a binary sequence, it could normally be referred to a universal Turing machine. The input into such a computer is a binary sequence, as is the output, so the computer cannot relate the description to the object itself. Therefore, a measure of the complexity of sequences is available, but this does not mean that we can calculate the complexity of the object described, because the information that a description gives about an object is always referred to a certain receiver in a concrete subject. For example, a DNA sequence is a good description of the three-dimensional structure of a protein *to certain cytoplasmic machinery*, but it would not make sense to say that it is generally, or for a Turing machine (see Rosen 1985).

Third, there are doubts as to whether natural selection *alone* can explain the increase in complexity throughout evolution (Marcos 1991b, 1992). After all, organisms exist that are very simple but seem perfectly adapted, a classic objection to Darwinism (Bertalanffy 1968). The connection we have established between complexity and information may clarify the issue. Later variants in evolutionary succession may "take into account" those already existing, but not vice versa (Rosen 1985). Once an organism A is settled into its environment, any other organism B will adapt to this environment more effectively if it is equipped to relate informationally with A. This informational asymmetry means that both the environment and organisms become more and more complex, and so maintain their adaptational dynamics throughout evolution. In this regard, complex biological organisms could be indicative of a complex environment, for more information is required to adapt to a complex environment than to exist in a simple one. The existence of living beings that adapt to an environment in which others already exist may ensure the survival of the latter, rather than threaten it, since the environment to which the new system adapts is also that on which they depend. Humanity's acceptance of this idea is not unconnected with the increase in ecological awareness. This remark, of course, has a direct link to the teaching of biology, which should promote ecological awareness.

5.4 Bioinformation and Knowledge

Information is also related to knowledge, which we hinted at in our discussion of mammalian visual perception in Sect. 3. In his highly influential work, *Knowledge and the Flow of Information*, Fred Dretske (1981) defines information as "a commodity that, given the right recipient, is capable of yielding knowledge" (p. 47). So a triadic relation is also needed here: we have the message, the circumstances it informs about, and the "right recipient." Information is therefore related to knowledge in a dual way: it depends on the receiver's previous knowledge, while knowledge is an effect produced by information. So, knowledge itself can be viewed as the property of a subject (edification), or as a dyadic relation between subject and object (correspondence, correlation). It is easy to connect the first notion of knowledge with biology: bioinformation contributes to the construction of living beings themselves (see Devlin 1991).

5.5 Location of Bioinformation

Where is bioinformation? In our opinion, conceiving information as a relation could avoid the (pseudo-)problem of finding the location of bioinformation. It could be (dis)solved by considering information, not as being already present somewhere (in the genes, cytoplasm, proteins, environment, ecosystem, brain, or wherever) but as being established by the *interactive relations* between and among the parts and processes of living systems. This is a very important point for biology educators to consider and may help to clarify many misunderstandings.

The typical textbook presentation of DNA as "encoding" or "including information" make people think of it as the Holy Grail of biology. However, we must remember that DNA on its own codes for nothing; it is informational only in the cellular context. So, in this sense, genetic information is not *wholly* genetic. Further than this—as we have tried to show primarily in Sect. 3—biological information is not an exclusive property of the genes, but exists as a relationship between biological entities of different levels. Especially since the research that has resulted from the Human Genome Project, we have witnessed the development of various *omic* sciences, such as transcriptomics (the study of the set of all RNA molecules produced in one or more cells), metabolomics (the study of metabolites and other products associated with metabolism), and proteomics (the study of the structure and function of proteins). This is another indicator that bioinformation is not a simple property of the genes, but a complex relationship between different biological entities.

The functioning of any living system (or part of a living system) depends on various factors. For example, the three-dimensional structure of a protein depends on DNA, but also on the very machinery of the cell. What the message is and what the receiver is are chosen conventionally but not arbitrarily. A message is usually defined as a small factor of great specificity in relation to a given function and displaying a high potential for variability. The DNA codifying a certain protein possesses these characteristics in relation to the function of synthesizing the protein in question, and the protein in relation to its biological function. In other words, the slightest alteration of the DNA could destroy the structure of the protein, and the slightest change in a protein could destroy its function, as it happens in widely known genetic diseases such as sickle-cell anemia.

Such an effect is unlikely to be the result of a similar change in an environmental factor. But this does not force us to identify the information with a property of the message. The information in a fragment of DNA about a protein obviously depends on its specificity, but only regarding a given receiver (Sattler 1986). Actually, the probability of any given protein arising in a prebiotic environment (Yockey 1977, 1981), even in the presence of a specific DNA, is much smaller than in a cytoplasmic environment. Therefore, information is located neither before nor after the triadic relation. Kampis and Csányi (1991) state: "we have to give up the idea of a complete localization of information" (p. 23; also see Kampis 1990).

On the other hand, any one fragment of DNA may, of course, produce information on more than one function, and not necessarily in the same quantity. For example, attempts could be made to calculate the amount of information in a fragment of DNA in relation to the transportation of oxygen, which is different from the function of producing a particular protein. The difference lies in the fact that the same function can be performed by different proteins or variants of a protein.

Finally, let us deal very briefly with the location of information in living systems according to different hierarchical levels (Collier 2003). An organism can be conceived of as a hierarchically organized living system made up of components that are engaged in processes constituting coordinated subsystems, with the product of these processes and subsystems being a *particularized* homeostasis relative to their operations that contributes to the overall *generalized* homeostasis of the organism.

For all intents and purposes, in the absence of connecting principles, the amount of information obtained by an external observer—for example, a scientist—on a living system at different levels should be considered as amounts of information about different systems. Otherwise, more information would supposedly be derived about a living being from the knowledge of, for example, its atomic state than of its genetic makeup (see Atlan 1972). Information concerning the atomic state is not about the living being *per se*, unless we have theoretical principles connecting atomic states with some functional characteristics. Developing principles of connection between levels is like developing a receiver that allows us to obtain information about one level from another, acting as a message. We know that, given certain principles of connection, one biological level can inform us about another, but we also know that a complete reduction is not viable, for any concrete informational relation is subject to imperfections.

6 Information and Education

6.1 Bioinformatics

Humans seem to be the only species that can produce *information* about information, biological or otherwise. In the last 50 years, the development of new technologies and the massive increase in the use of computers in all areas of human activity have led to a veritable explosion in the amount of data and information (about information) that is produced, used, and in need of management worldwide, constituting a veritable sea of extraordinary depth and breadth. This is especially true in the biological sciences, medical research, and medical practice. In these disciplines, thousands of scientists and clinicians are contributing daily to the accumulation of a massive body of biomedical knowledge and information, which we have hinted at already in our discussions of genome annotation and the newer omic sciences above.

Bioinformatics is now the word used for the categorizing, cataloguing, and coding of this biomedical information with the help of computers. The 11th edition of the *Merriam-Webster Dictionary* (2004) defines *bioinformatics* as "the storage, classification and analysis of biological information using computers" (p. 71), while Baxevanis and Ouellette (2005) define it simply as the "storage, organization, and indexing of biomedical information in computers" (p. 77). The challenge nowadays definitely concerns the ability to collect, categorize, manage, store, process, retrieve, disseminate, mine, and query all of this biomedical data and information appropriately and efficiently by computational means (see Arp et al. in preparation; Mathura and Kangueane 2009; Nishikawa 2002; and the journals *Bioinformatics* and *Bioinformation*). In fact, every science, organization, and business has its own informatics, teeming with data (Beynon-Davies 2003; Taylor and Joudry 2008).

Further, when the biomedical data and information possessed by experts in the various subfields of biology and medicine is organized and stored in interconnected, calibrated, interoperable computer repositories, it is accessible to anyone anywhere in the world, in real time, and could be continuously updated in light of new scientific and medical discoveries. Also, the information contained in these databases could, in principle, be used as the basis for certain kinds of automated reasoning that would independently assist in furthering the goals of scientific research and clinical practice. And one can imagine the ways in which this is immediately beneficial for biomedical research, the curing of diseases, the treatments of patients, the construction of new technologies, the annotating of data, and the general welfare of humankind. Think of a doctor with immediate access to the most current information about all known diseases at the click of a mouse. Or, imagine a single, calibrated, integrated biomedical knowledge base—a kind of Great Bioinformatics Encyclopedia—comprehensive of all biomedical knowledge within one system. The authors of a 2007 *Scientific American* article concerning bioinformatics and the World Wide Web share a similar dream of a database that, when queried, would "give us a single, customized answer to a particular question without our having to search for information or pore through results" (Feigenbaum et al. 2007).

6.2 Human Communication

The organism can be conceptualized as a hierarchical organization whereby levels of operation, in the forms of subsystems and processes, function interdependently with one another in this unified system. In order for all of the functions to take place in this living system, informational relationships must exist on different levels, from the genetic to the social. Genes communicate information, cells communicate information, subsystems and processes communicate information, the environment surrounding the organism communicates information, and we humans are unique in communicating *conceptual* information precisely about these various forms of bio-information (Boeckx and Uriagereka 2011). Conceptual information that exists in the social sphere of human communication and interaction is of particular importance to the biology educator for research reasons, as well as for teaching reasons having to do with conveying concepts concerning biological research, ideas, and principles in books, journals, the classroom, the lab, or on-line.

Some years back, John Tyler Bonner (1980) described culture itself as being rooted in informational terms. According to Bonner, culture is understood as the transfer of information through behavior and especially by virtue of the process of teaching and learning (also see Hintikka 1973; Hintikka and Suppes 1970; Baddeley et al. 2000). Bonner did not limit this process of teaching and learning to human cultures; rather, he extended the concept of culture to other species. We can also speak of cultural information in different living systems, not only in humans (Laland and Galef 2009). For example, one may speak of cultural learning in some aspects of birdsong, and in other forms of animal communication (see Oller and Griebel 2008). We especially find transmission of information in primates, too (Goodall 2000).

In 1976 Richard Dawkins described units of cultural information analogous to genes that he termed *memes*, and such an idea virtually single-handedly spawned an area of study known as *memetics* (Dawkins 1976), that has become significant for biologists, psychologists, sociologists, anthropologists, philosophers, and many other researchers (Blackmore 1999). Like genes, memes can replicate, mutate, compete, and even go extinct. Examples given by Dawkins include fashion, catch phrases, melodies, and various forms of technology. Of course, concepts expressed as theories, hypotheses, ideas, data, arguments, principles, and the like that one would find in a standard discipline like biology exist as straightforward examples of memes, too.

Now, it seems clear that genetic evolution and cultural evolution inform one another in mutual ways. It could be argued that cultural evolution gives continuity to genetic evolution, as would seem to be the case with memes mirroring genes. Within this framework of mutual informing, one could understand the human educational process as a type of memetic informational relationship that prolongs biological evolution, interacts with it, and maintains certain analogies with it. For this reason, when we speak of the transmission of conceptual information between teacher and student in the educational process, we are not talking about something absolutely distinct from the kinds of bioinformation we have described already, such as genetic information, cellular information, visual information, and organismal/environmental information exchange.

Although, as may seem obvious, in the context of education, informational relationships also have their own distinctive features, of which we should like to point out the following. First, as we have argued, the conceptual informational relationship is centered on the receiver. This is never more certain than in the educative process, the locus of which must obviously be the student. Here, lectures from teachers, books, articles, and other educational media carry on the function of messages generators, whose mission is to propitiate changes in the student's knowledge. In our case, the system of reference will be the world of living systems.

This could appear to be an excessively passive view of education, where the student is characterized simply as a receiver; however, this is not correct. Indeed, the cognitive changes are produced *in the* student and *by* the student, by means of the construction and management of different possible receivers. If we see the educational process as an informational relationship, we realize that it depends on messages received, generated by the lectures from teachers and other educational media. It also depends on the activity of the system of reference, that is, in this case, of objective and dynamic biological reality. If nature were not active—in the field, in the laboratory, and in the classroom—we could learn or teach but little biology. The upshot is that education in biology critically depends on the activity of the student, who cannot *learn* without *doing* (Marcos 2011b).

Also, if education itself is considered in informational terms, we underscore one of the historical meanings of information that we described in Sect. 2. Information here is a *formative* relationship—in the moral sense of the term—that forms and informs students, and the teacher as well. The educational process is, indissolubly, a process of information and of formation. But the old idea that it is simply the teacher who forms the disciple is erroneous and incomplete. Formation is the result of an informational relationship in which the teacher and the student take an active part.

When all is said and done, then, our hope is that we not only continue to be students of the biological sciences and philosophy of biology in our own research, but also that in this chapter we have played a bit of the role of teacher for you, the reader, concerning the concept of bioinformation.

References

Abramson, N. 1963. *Information theory and coding*. New York: McGraw-Hill.

Adriaans, P., and J. van Benthem. 2007. Introduction: Information is what information does. In *Philosophy of information*, ed. P. Adriaans and J. van Benthem, 3–28. London: Elsevier.

Albrecht-Buehler, G. 1990. In defense of "nonmolecular" cell biology. *International Review of Cytology* 120: 191–241.

Allen, C. 1992. Mental content and evolutionary explanation. *Biology and Philosophy* 7: 1–12.

Arnhart, L. 1998. *Darwinian natural right*. Binghamton: SUNY Press.

Arp, R. 2006. Evolution and two popular proposals for the definition of function. *Journal for General Philosophy of Science* 37: 2–12.

Arp, R. 2008. *Scenario visualization: An evolutionary account of creative problem solving*. Cambridge, MA: MIT Press.

Arp, R., B. Smith, and A. Spear. in preparation. *Building ontologies with Basic Formal Ontology*. Cambridge, MA: MIT Press.

Artmann, S. 2008. Biological information. In *A companion to the philosophy of biology*, ed. S. Sarkar and A. Plutynski, 22–39. Oxford: Blackwell.

Atlan, H. 1972. *L'Organisation biologique et la theorie de l'information*. Paris: Hermann.

Avery, J. 2003. *Information theory and evolution*. London: World Scientific Publisher.

Ayala, F., and R. Arp (eds.). 2009. *Contemporary debates in philosophy of biology*. Malden: Wiley-Blackwell.

Baddeley, R., P. Hancock, and P. Földiák (eds.). 2000. *Information theory and the brain*. Cambridge: Cambridge University Press. In Terzis, G., and R. Arp (eds.). *Information and living systems*, 289–308. Cambridge, MA: MIT Press.

Baddeley, R., B. Vincent, and D. Attewell. 2011. *Information theory and perception: The role of constraints, and what do we maximize information about?* Cambridge, MA: MIT Press.

Barwise, J. 1986. Information and circumstance. *Notre Dame Journal of Formal Logic* 27: 324–338.

Barwise, J., and J. Seligman. 1997. *Information flow: The logic of distributed systems*. Cambridge: Cambridge University Press.

Baxevanis, A., and B. Ouellette. 2005. *Bioinformatics: A practical guide to the analysis of genes and proteins*. Hoboken: Wiley.

Berkeley, G. 1735/1901. Alciphron, or the minute philosopher. In *The works of George Berkeley*, ed. A. Fraser. Oxford: Clarendon.

Berra, T. 1990. *Evolution and the myth of creationism*. Stanford: Stanford University Press.

Bertalanffy, L. 1968. *General systems theory: Foundations, development, applications*. New York: George Brazilller.

Beynon-Davies, P. 2003. *Information systems: An introduction to informatics in organizations*. New York: Palgrave.

Blackmore, S. 1999. *The meme machine*. Oxford: Oxford University Press.

Boeckx, C., and J. Uriagereka. 2011. Biolinguistics and information. In *Information and living systems*, ed. G. Terzis and R. Arp, 353–370. Cambridge, MA: MIT Press.

Bogdan, R. 1994. *Grounds for cognition: How goal-guided behavior shapes the mind*. Hillsdale: LEA Publishers.

Boi, L. 2011. Plasticity and complexity in biology: Topological organization, regulatory protein networks, and mechanisms of genetic expression. In *Information and living systems*, ed. G. Terzis and R. Arp, 205–250. Cambridge, MA: MIT Press.

Bond, N., and D. Siddle (eds.). 1989. *PsychoBiology: Issues and applications*. Amsterdam: Elsevier.

Bonner, J.T. 1980. *The evolution of culture in animals*. Princeton: Princeton University Press.

Brandon, R. 1984. On the concept of environment in evolutionary biology. *Journal of Philosophy* 81: 613–665.

Brandon, R. 1992. Environment. In *Keywords in evolutionary biology*, ed. E. Keller and E. Lloyd, 81–86. Cambridge, MA: Harvard University Press.

Brandt, C. 2005. Genetic code, text, and scripture: Metaphors and narration in German molecular biology. *Science in Context* 18: 629–648.

Brillouin, L. 1962. *Science and information theory*. London: Academic.

Brooks, D., and E. Wiley. 1988. *Evolution as entropy*. Chicago: University of Chicago Press.

Buetow, D. (ed.). 1982. *The biology of Euglena*. New York: Academic.

Bullot, N. 2011. Attention, information, and epistemic perception. In *Information and living systems*, ed. G. Terzis and R. Arp, 309–352. Cambridge, MA: MIT Press.

Burian, R., and M. Grene. 1992. Editorial introduction. *Synthese* 91: 1–19.

Capurro, R. 1978. *Information: Ein Beitrag zur etymologischen und ideengeschichtlichen Begründung des Informationsbegriffs*. Munich: Saur.

Capurro, R., and B. Hjørland. 2003. The concept of information. *Annual Review of Information Science and Technology* 37: 343–411.

Carroll, S. 2005. *Endless forms most beautiful: The new science of EvoDevo and the making of the animal kingdom*. New York: W.W. Norton & Company.

Ceccarelli, L. 2001. *Shaping science with rhetoric: The cases of Dobzhansky, Schrödinger, and Wilson*. Chicago: University of Chicago Press.

Clayton, N., and J. Krebs. 1994. Hippocampal growth and attrition in birds affected by experience. *Proceedings of the National Academy of Sciences of the United States of America* 91: 7410–7414.

Collier, J. 2003. Hierarchical dynamical information systems with a focus on biology. *Entropy* 5: 100–124.
Collier, J. 2007. Information in biological systems. In *Philosophy of information*, ed. P. Adriaans and J. van Benthem, 763–787. London: Elsevier.
Collins, J. 1956. *A history of modern European philosophy*. New York: Bruce Publishing.
Cover, T., and J. Thomas. 2006. *Elements of information theory*. New York: Wiley.
Crick, F. 1965. Recent research in molecular biology: Introduction. *British Medical Bulletin* 21: 183–186.
Crick, F., and C. Koch. 2003. A new framework for consciousness. *Nature Reviews Neuroscience* 6: 119–126.
Cummins, R. 2002. Neo-teleology. In *Functions: New essays in the philosophy of psychology and biology*, ed. A. Ariew, R. Cummins, and M. Perlman, 157–172. Oxford: Oxford University Press.
Darwin, C. 1859/2009. *The origin of species by natural selection: Or, the preservation of favored races in the struggle for life*. New York: Bantam Books.
Davies, P., and N.H. Gregersen. 2010. *Information and the nature of reality: From physics to metaphysics*. Cambridge: Cambridge University Press.
Dawkins, R. 1976. *The selfish gene*. Oxford: Oxford University Press.
Denbigh, K., and J. Denbigh. 1985. *Entropy in relation to incomplete knowledge*. Cambridge: Cambridge University Press.
Dennett, D. 1987. *The intentional stance*. Cambridge, MA: MIT Press.
Devlin, K. 1991. *Logic and information*. Cambridge: Cambridge University Press.
Dretske, F. 1981. *Knowledge and the flow of information*. Oxford: Blackwell.
Dretske, F. 2007. Epistemology and information. In *Philosophy of information*, ed. P. Adriaans and J. van Benthem, 29–48. London: Elsevier.
Duncan, D., E. Burgess, and I. Duncan. 1998. Control of distal antennal identity and tarsal development in *Drosophila* by *Spineless-aristapedia*: A homolog of the mammalian dioxin receptor. *Genes and Development* 12: 1290–1303.
Eco, U. 1962. *Opera aperta*. Milano: Bompiani.
Edelman, G. 1987. *Neural Darwinism: The theory of neural group selection*. New York: Basic Books.
Elsasser, W. 1975. *Chief abstractions in biology*. New York: American Elsevier.
Ewens, W. (ed.). 2010. Special issue: Entropy in genetics and computational biology. *Entropy* 12: 5–7.
Feigenbaum, L., I. Herman, T. Hongsermeier, E. Neumann, and S. Stephens. 2007. The Semantic Web in action. *Scientific American* 297: 90–97.
Fiers, W., R. Contreras, F. Duerinck, G. Haegeman, D. Iserentant, J. Merregaert, … M. Ysebaert. 1976. Complete nucleotide sequence of bacteriophage MS2 RNA: Primary and secondary structure of the replicase gene. *Nature* 260: 500–507.
Floridi, L. (ed.). 2003. *The Blackwell guide to the philosophy of computing and information*. London: Blackwell.
Floridi, L. 2007. Trends in the philosophy of information. In *Philosophy of information*, ed. P. Adriaans and J. van Benthem, 113–132. London: Elsevier.
Floridi, L. 2011. *The philosophy of information*. Oxford: Oxford University Press.
FlyBase. 2001. *Drosophila white paper*. Retrieved from: http://flybase.org/static_pages/news/whitepapers/DrosBoardWP2001.pdf.
Fodor, J. 1986. Information and association. *Notre Dame Journal of Formal Logic* 27: 307–323.
Forrest, S., and S. Hofmeyr. 2000. Immunology as information processing. In *Design principles for immune systems and other distributed autonomous systems*, ed. L. Segal and I. Cohen, 361–388. New York: Oxford University Press.
Gatlin, L. 1972. *Information theory and the living systems*. New York: Columbia University Press.
Giere, R. 1988. *Explaining science*. Chicago: University of Chicago Press.
Gilbert, S., and S. Sarkar. 2000. Embracing complexity: Organicism for the 21st century. *Developmental Dynamics* 219: 1–9.

Gleick, J. 2011. *The information: A history, a theory, a flood*. London: Harper Collins Publishers.
Godfrey-Smith, P. 2000. Information, arbitrariness, and selection: Comments on Maynard Smith. *Philosophy of Science* 67: 202–207.
Goodall, J. 2000. *Through the window: My thirty years with the chimpanzees of Gombe*. Boston: Mariner Books.
Gould, S.J. 1989. The wheel of fortune and the wedge of progress. *Natural History* 89(3): 14–21.
Gould, S.J. 2002. *The structure of evolutionary theory*. Cambridge, MA: The Belknap Press.
Gray, C. 1999. The temporal correlation hypothesis of visual feature integration: Still alive and well. *Neuron* 24: 31–47.
Griffiths, P. 2001. Genetic information: A metaphor in search of a theory. *Philosophy of Science* 68: 394–403.
Grünwald, P., and P. Vitányi. 2003. Kolmogorov complexity and information theory. *Journal of Logic Language and Information* 12: 497–529.
Hanson, P. (ed.). 1990. *Information, language, and cognition*. Vancouver: University of British Columbia Press.
Hartley, R. 1928. Transmission of information. *Bell System Technical Journal* 7: 535–563.
HGPI (Human Genome Project Information). 2012. Home page. Retrieved from: http://www.ornl.gov/sci/techresources/Human_Genome/elsi/patents.shtml.
Hintikka, J. 1973. *Logic, language games and information*. Oxford: Clarendon.
Hintikka, J., and P. Suppes. 1970. *Information and inference*. Dordrecht: Reidel.
Holzmüller, W. 1984. *Information in biological systems: The role of macromolecules*. Cambridge: Cambridge University Press.
Horan, B. 1992. What price optimality? *Biology and Philosophy* 7: 89–109.
Jablonka, E. 2002. Information: Its interpretation, its inheritance, and its sharing. *Philosophy of Science* 69: 578–605.
Janich, P. 1992. *Grenzen der Naturwissenschaft*. Munich: Beck.
Kampis, G. 1990. *Self-modifying systems in biology and cognitive science*. Oxford: Pergamon Press.
Kampis, G., and V. Csányi. 1991. Life, self-reproduction and information: Beyond the machine metaphor. *Journal of Theoretical Biology* 148: 17–32.
Kandel, E., J. Schwartz, and T. Jessell. 2000. *Principles of neural science*. New York: McGraw-Hill.
Keller, E.F. 1995. *Refiguring life: Metaphors in twentieth-century biology*. New York: Columbia University Press.
Keller, E.F. 2002. *Making sense of life: Explaining biological development with models, metaphors, and machines*. Cambridge, MA: Harvard University Press.
Kitcher, P. 2001. Battling the undead: How (and how not) to resist genetic determinism. In *Thinking about evolution: Historical, philosophical and political perspectives (Festchrifft for Richard Lewontin)*, ed. R. Singh, K. Krimbas, D. Paul, and J. Beatty, 396–414. Cambridge: Cambridge University Press.
Kjosavik, F. 2007. From symbolism to information? Decoding the gene code. *Biology and Philosophy* 22(3): 333–349.
Kolmogorov, A. 1965. Three approaches to the quantitative definition of information. *Problems of Information Transmission* 1: 1–7.
Kull, K. 1999. Biosemiotics in the twentieth century: A view from biology. *Semiotica* 127: 385–414.
Küppers, B.-O. 1990. *Information and the origin of life*. Cambridge, MA: MIT Press.
Küppers, B.-O. 2000. The world of biological complexity: Origin and evolution of life. In *Many worlds: The new universe, extraterrestrial life, and the theological implications*, ed. S.J. Dick, 32–43. Radnor: Templeton Foundation Press.
Laland, K., and B. Galef (eds.). 2009. *The question of animal culture*. Cambridge, MA: Harvard University Press.
Landauer, R. 1996. The physical nature of information. *Physics Letters A* 217: 188–193.

Layzer, D. 1990. *Cosmogenesis: The growth of order in the universe*. New York: Oxford University Press.
Li, M., and P. Vitányi. 1997. *An introduction to Kolmogorov complexity and its applications*. New York: Springer.
Lorenz, K., and F. Wuketits. 1983. *Die evolution des denkens*. Munich: Piper.
Mackay, D. 1969. *Information, mechanism and meaning*. Cambridge, MA: MIT Press.
MacLaurin, J. 1998. Reinventing molecular weismannism: Information in evolution. *Biology and Philosophy* 13: 37–59.
Marcos, A. 1991a. Información y entropía [Information and entropy]. *Arbor* 549: 111–135.
Marcos, A. 1991b. Información y evolución [Information and evolution]. *Contextos* 9: 197–214.
Marcos, A. 1992. *El Papel de la información en Biología* [*The role of information in biology*]. Ph.D. thesis. Barcelona: University of Barcelona.
Marcos, A. 1995. Biología, realismo y metáfora [Biology, realism and metaphor]. *Ágora* 14(1): 77–97.
Marcos, A. 1997. The tension between Aristotle's theories and uses of metaphors. *Studies in History and Philosophy of Science* 28: 123–139.
Marcos, A. 2010. *Ciencia y acción [Science and Action]*. Mexico DF: F.C.E.
Marcos, A. 2011a. Bioinformation as a triadic relation. In *Information and living systems*, ed. G. Terzis and R. Arp, 55–90. Cambridge, MA: MIT Press.
Marcos, A. 2011b. Learning by doing: *Paideia* and *Phronesis* in Aristotle. *Educação (Brazil)* 34: 13–24.
Margalef, R. 1968. *Perspectives in ecological theory*. Chicago: Chicago University Press.
Margalef, R. 1980. *La Biosfera, entre la termodinámica y el juego*. Barcelona: Blume Ediciones.
Marijuan, P. 1991. Enzymes and theoretical biology. *BioSystems* 25: 259–273.
Mathura, V., and P. Kangueane (eds.). 2009. *Bioinformatics. A concept-based introduction*. New York: Springer.
Maynard Smith, J. 2000a. The concept of information in biology. *Philosophy of Science* 67: 177–194.
Maynard Smith, J. 2000b. Reply to commentaries. *Philosophy of Science* 67: 214–218.
Maynard Smith, J., and E. Szathmáry. 1995. *The major transitions in evolution*. Oxford: W.H. Freeman Spektrum.
Mayr, E. 1996. The autonomy of biology: The position of biology among the sciences. *The Quarterly Review of Biology* 71: 98–106.
Miller, K. 2005. *Communication theories: Perspectives, processes, and contexts*. New York: McGraw-Hill.
Millikan, R. 1989. Biosemantics. *The Journal of Philosophy* 86: 281–297.
Millikan, R. 2002. Biofunctions: Two paradigms. In *Functions: New essays in the philosophy of psychology and biology*, ed. A. Ariew, R. Cummins, and M. Perlman, 113–143. Oxford: Oxford University Press.
Moles, A. 1972. *Théorie de l'information et la perception esthétique*. París: Éditions Denoël.
Moreno, A., and K. Ruiz-Mirazo. 2002. Key issues regarding the origin, nature and evolution of complexity in nature: Information as a central concept to understand biological organization. *Emergence* 4: 63–76.
Morgan, T.H., A. Sturtevant, H. Muller, and C. Bridges. 1915. *The mechanism of Mendelian heredity*. New York: Henry Holt.
Musgrave, A. 1993. *Common sense, science and scepticism*. Cambridge: Cambridge University Press.
Nauta, D. 1972. *The meaning of information*. The Hague: Mouton.
Nishikawa, K. 2002. Information concept in biology. *Bioinformatics* 18: 649–651.
NPO (Nobel Prize Organization). 2012. The Nobel Prize in physiology or medicine 1968: Robert W. Holley, H. Gobind Khorana, Marshall W. Nirenberg. *Nobelprize.org*.
Oller, D., and U. Griebel (eds.). 2008. *Evolution of communicative flexibility: Complexity, creativity, and adaptability in human and animal communication*. Cambridge, MA: MIT Press.

Oyama, S. 2000. *The ontogeny of information: Developmental systems and evolution*. Durham: Duke University Press.
Paton, R. 1992. Towards a metaphorical biology. *Biology and Philosophy* 7: 279–294.
Peil, J. 1986. Information theoretical aspects of the aging process. *Archives of Gerontology and Geriatrics* 10: 97–105.
Pelligrino, G., L. Fadiga, L. Fogassi, V. Gallese, and G. Rizzolatti. 1996. Understanding motor events: A neurophysiological study. *Experimental Brain Research* 91: 176–180.
Perlman, M. 2004. The modern philosophical resurrection of teleology. *The Monist* 87: 3–51.
Pfeifer, J. 2006. The use of information theory in biology: Lessons from social insects. *Biological Theory* 1: 317–330.
Piaget, J. 1970. *L'épistemologie génétique*. Paris: PUF.
Popper, K. 1990. Towards an evolutionary theory of knowledge. In *A world of propensities*, ed. K. Popper, 27–51. Bristol: Thoemmes.
Queiroz, J., C. Emmeche, and C. El-Hani. 2005. Information and semiosis in living systems: A semiotic approach. *SEED* 5(1): 60–90.
Reid, T. 1764/2001. In *An inquiry into the human mind on the principles of common sense*, ed. D. Brookes. State College: Pennsylvania State University Press.
Reiter, L., L. Potocki, S. Chien, M. Gribskov, and E. Bier. 2001. A systematic analysis of human disease-associated gene sequences in *Drosophila melanogaster*. *Genome Research* 11: 1114–1125.
Ridley, M. 2006. *Francis Crick: Discoverer of the genetic code*. New York: Harper.
Roederer, J. 2005. *Information and its role in nature*. Berlin: Springer.
Rosen, R. 1985. *Anticipatory systems*. Oxford: Pergamon Press.
Rosenberg, A., and R. Arp (eds.). 2009. *Philosophy of biology: An anthology*. Malden: Wiley-Blackwell.
Rosenberg, A., and D. McShea. 2007. *Philosophy of biology: A contemporary introduction*. London: Routledge.
Ruiz-Mirazo, K., and A. Moreno. 2011. The need for a universal definition of life in twenty-first-century biology. In *Information and living systems*, ed. G. Terzis and R. Arp, 55–90. Cambridge, MA: MIT Press.
Ruse, M. (ed.). 2008. *The Oxford handbook of philosophy of biology*. Oxford: Oxford University Press.
Russell, B. 1912/1999. The problems of philosophy. In *Introduction to philosophy: Classical and contemporary readings*, ed. J. Perry and M. Bratman, 9–12. Oxford: Oxford University Press.
Sanger, F., and A. Coulson. 1975. A rapid method for determining sequences in DNA by primed synthesis with DNA polymerase. *Journal of Molecular Biology* 94: 441–448.
Sarkar, S. 1996. Biological information: A skeptical look at some central dogmas of molecular biology. In *The philosophy and history of molecular biology: New perspectives*, ed. S. Sarkar, 187–232. Dordrecht: Kluwer.
Sarkar, S. 2000. Information in genetics and developmental biology: Comments on Maynard Smith. *Philosophy of Science* 67: 208–213.
Sarkar, S. 2005. *Molecular models of life: Philosophical papers on molecular biology*. Cambridge, MA: MIT Press.
Sarkar, S., and A. Plutynski (eds.). 2008. *A companion to the philosophy of biology*. Oxford: Blackwell.
Sattler, R. 1986. *Biophilosophy*. Berlin: Springer.
Schrödinger, E. 1944. *What is life? The physical aspect of the living cell*. Cambridge: Cambridge University Press.
Seife, C. 2007. *Decoding the universe: How the new science of information is explaining everything in the cosmos, from our brains to black holes*. New York: Penguin.
Shannon, C. 1993. In *Collected papers*, ed. N.J.A. Sloane and A.D. Wyner. New York: IEEE Press.
Shannon, C., and W. Weaver. 1949. *The mathematical theory of communication*. Urbana: University of Illinois Press.
Shatz, C. 1992. The developing brain. *Scientific American* 267: 60–67.

Shendure, J., R. Mitra, C. Varma, and G. Church. 2004. Advanced sequencing technologies: Methods and goals. *Nature Reviews Genetics* 5: 335–344.
Singer, W. 1999. Neuronal synchrony: A versatile code for the definition of relations? *Neuron* 24: 49–65.
Sober, E. 1993. *Philosophy of biology*. Boulder: Westview Press.
Solomonoff, R. 2003. The universal distribution and machine learning. *The Computer Journal* 46: 598–601.
Stegmann, U. 2005. John Maynard Smith's notion of animal signals. *Biology and Philosophy* 20: 1011–1025.
Stein, L. 2001. Genome annotation: From sequence to biology. *Nature Reviews Genetics* 2: 493–503.
Stuart, C. 1985. Bio-informational equivalence. *Journal of Theoretical Biology* 113: 611–636.
Tauber, A. 1994. *The immune self: Theory or metaphor?* Cambridge: Cambridge University Press.
Taylor, A., and D. Joudry. 2008. *The organization of information*. Santa Barbara: Linworth/Libraries Unlimited.
Terzis, G., and R. Arp (eds.). 2011. *Information and living systems*. Cambridge, MA: MIT Press.
The Economist (1902). Journalists and diplomats. December 6, p. 1881.
Tipler, F. 1995. *The physics of immortality*. New York: Anchor Books.
Tribus, M., P. Shannon, and R. Evans. 1966. Why thermodynamics is a logical consequence of information theory. *Journal of the American Institute of Chemical Engineering* 12: 244–248.
Trivers, R. 1985. *Social evolution*. Menlo Park: Benjamin-Cummings.
Tweedle, S., M. Ashburner, K. Falls, P. Leyland, P. McQuilton, S. Marygold, ... and The FlyBase Consortium. 2009. FlyBase: Enhancing drosophila gene ontology annotations. *Nucleic Acids Research* 37: D555–D559.
van Essen, D., C. Anderson, and D. Felleman. 1992. Information processing in the primate visual system: An integrated systems perspective. *Science* 255: 419–423.
Villanueva, E. (ed.). 1990. *Information, semantics & epistemology*. Oxford: Blackwell.
Waddington, C. (ed.). 1968. *Towards a theoretical biology*. Edinburgh: Edinburgh University Press.
Weber, B., D. Depew, and J. Smith (eds.). 1988. *Entropy, information and evolution*. Cambridge, MA: MIT Press.
Weismann, A. 1904. *The evolution theory*. Trans. J. Thomson and M. Thomson. London: Edward Arnold.
West, R., and L. Turner. 2006. *Introducing communication theory: Analysis and application*. New York: McGraw-Hill.
Wicken, J. 1987. *Evolution, information and thermodynamics: Extending the Darwinian program*. Oxford: Oxford University Press.
Wiener, N. 1961. *Cybernetics, or control and communication in animal and machine*. Cambridge, MA: MIT Press.
Winder, R. 2012. *Philosophical aspects of information systems*. London: Taylor and Francis.
Yeung, R. 2006. *A first course in information theory*. New York: Springer.
Yockey, H. 1977. A calculation of the probability of spontaneus biogenesis by information theory. *Journal of Theoretical Biology* 67: 377–398.
Yockey, H. 1981. Self-organization origin of life scenarios and information theory. *Journal of Theoretical Biology* 91: 13–31.
Yockey, H. 2005. *Information theory, evolution, and the origin of life*. Cambridge: Cambridge University Press.

Systems Biology and Education

Pierre-Alain Braillard

1 Introduction

Biology has been transformed in the last two decades in many ways. First, the Human Genome Project (HGP) and genomics[1] in general have marked the advent of a new area of "Big Science", characterized by huge financial, technical and institutional means, at a scale never known before in biology (Roberts et al. 2001; Collins et al. 2003). From a scientific point of view, genomics has opened new strategies of research and new opportunities, but it has also confronted biologists to a "tsunami" of data, which need to be interpreted and analyzed in order to transform them into biological knowledge. This was however only a first step, and even before the completion of the HGP, biology entered a post-genomic phase (Nowak 1995; Hieter and Boguski 1997), also called functional genomics, which aims at developing new experimental and analytical approaches able to elucidate the functions of genomic regions. The most common of these methods are the study of gene expression by micro-arrays, the elucidation of protein-protein and protein-DNA interactions, or systematic gene knock-out with phenotypic screening. Although such data could be obtained in the past through various techniques, functional genomics' goal is to extend them to whole-genome scale (reflected in new expressions, like the various -omics approaches: proteomics, transcriptomics, metabolomics, etc.). These "high-throughput" approaches have revolutionized the way research is conducted and they made biologists dream about "complete" knowledge of the cells and organisms they study. But changes have not been limited to these large-scale experimental methods.

[1]Genomics can be defined as projects aiming at the establishment of complete maps (genetic and DNA sequences) of organisms' genomes (Brent 2000).

P.-A. Braillard (✉)
UFR de Biologie, Cité scientifique-Bâtiment SN3, Université Lille 1, France
e-mail: brailla6@hotmail.com

K. Kampourakis (ed.), *The Philosophy of Biology: A Companion for Educators*, History, Philosophy and Theory of the Life Sciences 1, DOI 10.1007/978-94-007-6537-5_24,

Small-scale research has also been transformed by the use of new modeling approaches based on various mathematical and computational tools. These families of experimental and analytical approaches are commonly referred to as systems biology, which will be the focus of this chapter.[2]

What is exactly systems biology is a difficult and probably futile question. More interesting is the fact that since the expression has become popular 10 years ago, an increasing number of scientists have described themselves as systems biologists. One can of course see financial interests behind such strategies, because systems biology is a kind of "big science" potentially bringing a lot of money. But there is certainly more than that. Part of these changes has no doubt been driven by the technological progresses just mentioned. However, underlying the emergence of systems biology is also a call for a new vision and new explanatory strategies. Many scientists have come to acknowledge that molecular biology needs to become more mathematical and quantitative, like physics and engineering, in order to address the complexity of biological systems. What they point to is the insufficiency of representing and analyzing biological systems with the classical schemes and diagrams used by molecular and cell biologists. The formalism of mathematical and computational models offers a more productive way to gain a broader picture by looking at large networks of interactions, and also to develop a fully dynamical picture of molecular mechanisms. Traditional pictorial models give a rather rough and poor picture of what is going on in a cell and systems biologists insist that a fine description of dynamical processes is necessary to have a good understanding of how biological functions are produced. To achieve this, biology needed help from other disciplines. Since the 1980s bioinformatics had contributed to protein and DNA sequences analysis, but systems biology involves a much broader use of mathematical tools and formal models, from differential equations representing in great detail gene regulatory or signal transduction mechanisms to very coarse-grained models able to represent whole-cell networks.

Whatever systems biology exactly is, most commentators agree that it is clearly transforming biological practices and theories. Like in the early days of molecular biology, we are witnessing an exciting time in biology. These changes have not only a purely scientific dimension, but they also raise more philosophical questions concerning the nature of biological methods, models and knowledge.[3] It is thus also an exciting time for philosophers interested in biological sciences. But all of this has in addition consequences for how biology should be taught, and this is true in several senses. This chapter's aim is to discuss scientific and philosophical issues raised by systems biology and to explore some of their implications for biology education. I will explore potentially fruitful interactions between philosophical analysis of systems biology and issues in biology education at three levels.

The first level concerns explanatory ideals, by which I mean where explanations and causes of phenomena are sought. Since its beginnings molecular

[2]For introductory textbooks, see for example Alberghina and Westerhoff (2007), Alon (2007), Bringmann et al. (2006), Palsson (2006), and for an early general review see Kitano (2002a, b).

[3]For some philosophical analyses on systems biology, see for example (O'Malley and Dupré 2005, Boogerd et al. 2007, Braillard 2010, O'Malley and Soyer 2012).

biology's aim has always been to ground explanations of biological phenomena on molecular (physico-chemical) properties. Genes have come to occupy a central place, being conceived as bearers of genetic information and fundamental causes of phenotypic traits. Despite impressive successes, it has become increasingly evident that focusing on molecular details has important limitations and that gaining a broader (more systemic) view is necessary. Systems biology incarnates this effort, based on recent experimental technologies and mathematical and computational models, and for this reason it is often described as overcoming molecular biology's reductionism.

Philosophers of science have produced in the last decades a rich literature on the nature of molecular and cellular biology's explanatory practices and strategies, on the reductionism defended by molecular biology, on the mechanistic nature of these sciences, and on the status of the gene as a fundamental theoretical and explanatory entity (see for example Bechtel and Richardson 1993; Schaffner 1993; Sarkar 1998, 2007; Keller 2000; Bechtel 2006; Darden 2006; Weber 2005; see also Bechtel this volume; Burian and Kampourakis this volume). Such philosophical analyses must be pursued in order to help us understand the conceptual changes induced by systems biology. The fact that important parts of biology's explanatory framework are being challenged must be reflected in biological education at all levels. It is a good occasion to revise simplistic vision of genes' action or causality in biology.

The second level is related to the needs of education programs to keep up with the rapid pace of scientific progress. Systems biology requires scientists with multiple competences and a good ability for interdisciplinarity. But reaching this goal is far from trivial, because what biologists need is more than "simple" knowledge of new tools that can be applied to ask old questions. On the other hand, what scientists coming from computer sciences, physics or engineering need is more than basic notions in biology, which would allow them to apply their formal skills to a new class of systems. Biology is now facing deep and difficult questions about how to model and analyze biological systems. Many approaches are explored and each scientist has his own intuition about what paths will be more fruitful. To make progress scientists need to think critically about the limits of classical approaches (what cannot be explained by them), and what new ones might or might not bring. The new generation of biologists needs conceptual and intellectual tools to think deeply about new explanatory strategies. The challenges for an adequate education must not be underestimated. By pointing to an area where transfer of concepts and models is both potentially fruitful and problematic I will try to indicate how a philosophical analysis can contribute to this debate.

The third level concerns biology's philosophical foundations. The changes introduced by systems biology go beyond a shift towards a systems view (my first point), they also question the very nature of biological knowledge. Post-genomic biology is sometimes described as large-scale (and partly automated) science, which works mainly by collecting data, analyzing them and building models through a kind of purely inductive method. But this is an oversimplification and a mischaracterization. One of the most exciting parts of systems biology has a somewhat more ambitious project: not only piling up data about molecules and their interactions, but proposing a new vision of life and finding general principles of biological systems.

Such principles can offer biologists a way to find some degree of intelligibility behind the formidable complexity of molecular mechanisms. Philosophers have analyzed in details the nature of theories, models and explanations in biology (see Potochnik this volume), but systems biology is challenging several of these views. It thus offers a formidable opportunity for scientists, teachers and students to think about the nature of biological sciences. I will argue that it is not only important as part as a general intellectual inquiry, but that it can also help to think more clearly about research strategies and methodological issues (my two other points). The way biology is taught should fully incorporate these reflections.

Before starting, I must emphasize that because most of systems biology is quite formal and technical, it is certainly difficult to discuss it in the context of secondary or even undergraduate level. Some of the ideas developed in this chapter will be mainly of interest for graduate education, but I hope it will be clear that there are also broader lessons that can be learned and which are relevant for teaching biology at any level.

2 Towards Systemic Approaches

In this first part, let us see what is systemic in systems biology and what are the consequences that can be of some importance for biology education. I will begin by characterizing molecular biology's reductionism and then discuss some of its limits. This will help us understand systems biology's novelty, and how some of the traditional ideas about analyzing and explaining cells and organisms in terms of genes and molecular mechanisms are being challenged.

2.1 Reductionism in Biology

Molecular biology is usually considered as the paradigm of reductionism in biology. But what is the meaning of reductionism here? The issue of reductionism has been much debated in the philosophy of science in the last decades. During the 1970s, the whole problematic of reductionism was conceived in terms of theory reduction (Hull 1974; Ruse 1971; Schaffner 1969). In this framework, a reduction is successful when a reduced theory can be formally (logically) derived from the reducing theory (with some translations between the terms of the theories). This way to frame the question was directly inspired by the situation in physics, which was the paradigm of science for philosophers as much as for scientists. However, it rapidly appeared that this is not the best way to make sense of reductionism in biology, one of the main reasons being that there are few if any physics-like theories in biology.

Theoretical reductionism has been then gradually replaced by an explanatory reductionism, which deals with relations between particular phenomena instead of theories (Rosenberg 2006). Here the reduction is between different levels of description. More generally, the framework inherited from early philosophers of science (logical empiricists), which focused so much on formal descriptions (axiomatized theories, universal laws) has been replaced by a mechanistic framework, which better accounts for relations between levels (Bechtel and Richardson 1993; Machamer et al. 2000; Glennan 2002; Craver and Darden 2005; see also Bechtel this volume). In this context, a reductive explanation explains by showing how a phenomenon is produced by interactions between its parts (Wimsatt 2000). The question whether mechanistic models are reductive is not simple, but leaving this issue aside we can admit that it corresponds to one sense of reduction commonly used by biologists. It is reductive because biological phenomena (or properties) can be grounded on (and thus explained by) the properties of their parts.

If we look at molecular biology's achievements, we can see that it is reductionist in this sense. It has offered explanations of many biological phenomena in terms of physico-chemical properties of their constituents. The paradigmatic and most famous example is certainly the discovery of the structure of DNA, which was immediately accepted as a reductive explanation of the properties of the gene (most notably its ability to replicate and also to mutate), which had been studied by geneticists but remained wholly mysterious. Similarly, the elucidation of the structure of the hemoglobin protein provided an explanation for many aspects of respiratory function, like the ability to transport oxygen from the lung to tissues and then to release it where it is needed (Judson 1979, chapter 9). Even more impressive was the discovery that a single mutation in this gene can result in a modified conformation and then in an aberrant cell shape, which causes sickle-cell anemia (Ingram 1956). This example is famous because it is one of the first molecular explanations of a disease. This model traces the causal links from the gene, to the protein, to the cell, to the tissue and to the organism (and to population genetics), and it has constituted the paradigm of reductionist explanations. Similar examples could be multiplied, but the message should be clear. Molecular biology's reductionism has mainly consisted in discovering molecular mechanisms involved in the production of biological functions (heredity, metabolism, regulation, etc.).

These early successes naturally lead to think that the best explanations of biological properties are to be found in molecular details. In the 1960s and the 1970s, many biologists turned to molecular approaches to address various problems, from physiology to medicine, and from development to neurobiology. An important aspect of this reductionism is the central place that the gene has come to occupy. As the bearers of genetic information, genes were seen as the fundamental cause of the phenotype. It is well known that an important part of biology and medicine in the last decades was oriented towards the search for "genes for" (Burian and Kampourakis this volume; Moore this volume). The most successful examples are the discovery of oncogenes (Morange 1998, chapter 19) and developmental genes (Nüsslein-Volhard and Wieschaus 1980; Gehring 1999), but virtually every domain has been guided by a similar "hunt for genes". These approaches have been extraordinarily successful, but they have sometimes led to reductionist excesses and a kind of naïve view of genes' role. This is partly due the informational framework adopted by molecular biologists, with its notions of code and genetic program, which has deeply shaped this discipline (see Marcos and Arp this volume). The idea that genes code for, or contain information about phenotypic traits gave rise to a kind of "new preformationism",[4] considering the phenotype as "contained" in the genome (Oyama 2000; Lewontin et al. 1984). This is particularly true of popular conceptions of biology, but scientists have often expressed such a kind of idea and have not always been cautious enough with this metaphor. But the reductionist strategy has more general limits, as we will see now.[5]

2.2 The Limits of Reductionism and the Need for a Systems View

The models and explanations usually developed in molecular and cell biology have several limits that have been acknowledged by some biologists for a long time, but which have become particularly evident with the progress made in the knowledge of molecular mechanisms during the 1990s and also with the advent of genomics projects (especially the HGP). Genomics offered the prospect of establishing a complete list of the genes present in the genome, and indirectly of the proteins encoded by these genes, or in other words a list of the cells' parts. The most enthusiast proponents of the HGP announced that once the genome is sequenced our understanding of biological functions as well as diseases would quickly follow (Guyer and Collins 1993). But even though the HGP was a big success (it was completed ahead of time), what the genome sequence can teach us about our biology was more disappointing (though this was to be expected and was indeed expected by some). First of all, it turned out that there are far less genes than estimated in humans (estimations were about 100,000 genes; recent estimations say around 25,000), in fact not much more than in other organisms. How then can we explain the peculiarities of our organization, for example the complexity of our nervous system? The secret lies of course in the way components interact and produce the complexity.

The central problem was then how to close the gap between sequences and functions, or between the genome and the phenotype? Genomics offers a fundamentally static knowledge (sequences), which tells nothing in itself about how functions emerge from the parts.

The very idea of identifying a list of genes and then explaining biological phenomena in terms of these genes also raise questions because the gene has become an entity rather elusive and difficult to define (Falk 2000; Keller 2000; Griffiths and

[4]The old preformationism being the view holding that organisms develop from miniature versions of themselves, which was widely held during the eighteenth and nineteenth century.

[5]Biologists' views should however not be caricatured. When they use the expression "gene for", what they mean is that different alternative alleles of a gene correspond to differences in forms of a phenotypic trait. But of course this relation depends on other genes and the environment. However, even though all biologists acknowledge this, it remains true that there has been a tendency to view the causal origin of many traits in one or few genes and to consider the rest of the system and the environment as only background conditions, necessary but not explanatory important.

Stotz 2006). The fundamental problem is that the relation between DNA sequences and proteins has become much more complicated than originally thought, as the picture of molecular and cellular regulatory mechanisms became increasingly detailed. Processes like alternative splicing[6] and post-translational modifications, which depend on the cellular context, show that a given sequence can produce many different proteins. Moreover, which sequence is recognized as a gene (i.e. is transcribed) also depends on the context. Thus the gene is not a structural object, but it cannot be easily defined functionally either (see Burian and Kampourakis this volume).

Progress in the understanding of molecular mechanisms has shown that what a gene, a protein or a mechanism does can be highly dependent on the context in which it is embedded. Signal transduction cascades (which transmit extracellular signals from the membrane to the nucleus for example) offer a good illustration of this phenomenon. Depending on the state of the cell (which includes the state of other pathways or the metabolism), the activation of a specific receptor can have very different effects. It is also well known that many genes are involved in different and unrelated cellular processes (for example Pax-6, which is involved in the development of the eye but also in pancreatic cells). Such cases show that a molecular mechanism has limited explanatory power in itself.

More generally, even when new (putative) genes are identified, how should we analyze their contribution to the production of biological functions? The discovery that some genes are implicated in a developmental process or in cancer has only a limited value for understanding how these processes unfold. If we are to explain how functions emerge at the cellular or organismic level, it is necessary to understand how all the components (genes, proteins, RNA, metabolites, etc.) and mechanisms are interconnected and regulate each other. The difficulties are enormous because the influence of the context is not only important at the cellular level, but also at the tissue, organ and organism level (not to mention the environmental level), as we will see.

The challenge is even greater when one realizes that complex dynamical and functional properties can emerge from very simple mechanisms (i.e. composed of few parts). As stressed above, molecular and cell biologists have mostly reasoned by using relatively simple diagrams representing these mechanisms, and such models are useful because it is easy to follow the functioning of these mechanism. In other words, they are cognitively manageable. However, progress in the study of these mechanisms has revealed that a rough and qualitative understanding is not sufficient.

Let us take the famous case of negative feedback, which is omnipresent in biology. It is intuitively easy to see how it can produce homeostasis, but it can also produce oscillations in certain conditions and this is more difficult to assess. Positive feedbacks were usually neglected by biologists because they were thought to produce unstable and chaotic behaviors, but they can also produce bistability (like a switch), which can be biologically significant. When two or more feedback or feed-forward

[6] In this process, exons of pre-mRNA are put together in multiple combinations. With this process, a single gene can produce different mRNAs and hence different proteins.

loops are linked together (which is common in regulatory mechanisms), reasoning with diagrams is not sufficient for predicting dynamical behavior. This is mainly due to non-linear interactions, which cause the relation between causes and effects to become very complex and difficult to predict. This problem is of course worse with larger systems.

To sum up, classical models used in molecular and cell biology have progressively reached their limits because they cannot properly explain the influence of the context in which molecular mechanisms operate and because they poorly represent the dynamical complexity of most regulatory mechanisms through which biological functions emerge. These are some of the reasons why in the last two decades a growing number of biologists have felt the need to develop and use new models and tools to represent and analyze the complexity of biological systems. Let us now see how systems biology takes up these challenges.

The first point to note is the focus on the dynamics of biological processes. Systems biology corresponds to the introduction and generalization of a variety of mathematical and computational models, which are used in order to give a quantitative and precise dynamical description of molecular mechanisms. Several formalisms can be used, but systems of differential equations are certainly the most common. Such models, partly based on experimentally measured parameters, allow a much richer analysis of mechanisms' behavior and hence of their functional properties. The case of the cell cycle is a good example, since many components and their interactions had already been characterized through traditional approaches. Mathematical and computational modeling facilitates the characterization of dynamical and functional properties that cannot be grasped intuitively. It allows rigorously testing alternative hypotheses about components' roles or interactions. What is even more important is that also shed new light on the "logic" of this regulatory process. What I mean here is that by describing a mechanism with dynamical concepts one can reveal new principles of functioning (Tyson et al. 2003). It has for example been hypothesized that the cell cycle progression is driven by shifts between stable states, produced by a combination of regulatory loops (Pomerening et al. 2003). This might explain several essential features of the cell cycle, in particular its directionality (the fact that once a step has been accomplished the system never goes back). Only through dynamical analyses is it possible to frame and test such ideas in precise mathematical terms.

Understanding how properties emerge from the dynamics of the system is only one dimension of systemic explanations. Systems biology is adding a second dimension, which corresponds to the shift of focus from individual components towards large networks. This is the result of both the recognition that mechanistic models must be extended (i.e. a larger context must be taken into account; see above) and the ability given by functional genomics techniques to follow in parallel a very large number of regulatory events (for example changes in gene expression at the whole-genome level). The new "paradigm" is that instead of only focusing on few components, its is necessary to take a broader view in order to gain a better understanding of processes taking place at the whole system level (usually the cell).[7] To use a classical metaphor, we must see the forest, not just the trees.[8]

However, adopting a systems view is not straightforward, for different theoretical and practical reasons. One of the problems is that building detailed mechanistic models at the whole-cell level is infeasible for lack of precise empirical data (reactions constants, concentration, etc.) and computational limits. A virtual cell based on chemical reaction is not achievable in a foreseeable future. For this reason, some scientists argue that bottom-up modeling approaches must be complemented with top-down strategies, whose goal is to offer a coarse-grained picture of molecular processes (Bruggeman and Westerhoff 2007). As Hiroaki Kitano puts it (Kitano 2002a), biologists need to look beyond the details and "zoom-out" in order to gain a better understanding of how systemic properties emerge.

This new explanatory strategy has been largely based on the adoption of a network framework (Albert 2007). Network models have many advantages. First of all, they can easily integrate the flow of data coming from functional genomics. Each component can be represented as a node and each interaction as a link. Since the turn of the century, models of gene regulatory networks, protein-protein interactions networks, metabolic networks, or signaling networks have become widespread in the literature. This kind of representation has opened the way to rigorous mathematical and computational analyses of these systems. I will describe some examples in the next section, but the point I would like to make here is that molecular networks have become a new biological object, whose structural and dynamical properties is now the focus of much interest.

Now, if we come back to the way causality is usually analyzed in molecular biology, we can appreciate the difference introduced by this framework. With network models, the explanatory weight does not lie primarily on the components, but more on the structure of the system (its topology). The dynamics of the whole becomes more explanatory than the parts. In examples such as the cell cycle mentioned above, the fact that some negative and positive feedback or feedforward loops are connected together in a specific way is seen as the most important feature of such mechanisms. Details can (and indeed do) vary from cases to cases, but the structure of the circuit (or network) is central in the emergence of the function.

A noteworthy consequence of this new framework is that causality is much more circular and is in a sense "diluted" in the system's complexity. Indeed, it becomes difficult to point to the causal factor responsible for one phenomenon or event. All the components collectively produce the behavior of interest (again, reductionists have always recognized in principle this fact, but in practice they tended to put all the weight on few causal factors). Consequently, talking about the control of a process becomes also more problematic. It remains of course possible to distinguish the contribution of each component in a mechanism, but when one wants to explain

[7] Few models really try to represent processes at the whole-cell level, but this is where some modeling strategies are aiming at.

[8] Of course, both dimensions are closely linked since the goal is to understand the dynamics of the whole system.

(let's say) how an extracellular signal induces a change in cellular state, it is difficult to point to one or few targets, because a larger network of interaction must be taken into account to understand such a switch. It is important to understand that the very dynamics of the mechanism, which causes a cellular "decision", depends on all these interactions.

One of the upshots of this discussion is that genes have lost some of their explanatory power and must be seen as one type of actor playing a role among many others. Genes are as much controlled as they themselves control cellular mechanisms.

If we come back to the case of cancer, we see that an explanation in terms of one or a few mutations must give way to a much richer analysis in which multiple changes at the molecular, cellular, tissular and organismic levels occur. Many scientists have moved towards the study of the alterations (and deregulation) of important pathways that are involved in the production of pathological behaviors (Kreeger and Lauffenburger 2010). This shift of approach has also an impact on the search of treatments. Some results show that the best strategy is not necessarily to target the mutated or deregulated protein, but another part of the pathway. Another active path is the study of tissular processes involved in cancer (Soto and Sonnenschein 2004).[9]

A third dimension of systems biology's antireductionism is the development of multilevel modeling approaches. We have seen how molecular biology has always been focused on molecular components and mechanisms. It would be a caricature to say that molecular biologists only looked at the molecular level (for example, they often try to correlate molecular process with cellular or other higher-level phenomena), but it is certainly true that they have always been eager to find molecular mechanisms as possible explanations of their phenomena of interest (development, cancer, etc.). This has led to some reductionist excesses, one of the best examples being the studies conducted in the 1970s announcing the discovery of molecules of memory (Morange 1998, chapter 15). Such examples of extreme reductionism are probably exceptional, but there has nevertheless been a tendency to quickly "jump" from the molecular level to the phenotype. Systems biology offers the prospect of a better articulation and integration of the different levels, from the atoms to the organism (and beyond). Denis Noble, who has been working on dynamical models of cardiac cell since the 1960s, has described this kind of pluralistic methodology as "middle-out" (Noble 2006), by which he means that instead of starting from the molecular level up, scientists must start from the level which looks the more appropriate to the phenomenon of interest (and their experimental tools), and then to look both downward (towards the molecular level) and upward (towards the organism). One of the challenges is to understand the reciprocal interactions between levels. The concept of inter-level causation and especially downward causation (when higher-level has causal influence on lower-level) has been considered as highly problematic from a philosophical point of view (Emmeche et al. 2000). This issue would take us too far here, but leaving metaphysical worries aside, I think it is possible to make sense of what these scientists try to understand. In the same way that a protein's properties (for example its diffusion, its enzymatic activity, or its

[9]See also van Leeuwen et al. (2007) and Hornberg et al. (2006).

folding) can be influenced by the cellular context, cellular behavior is also influenced by the context of the tissue or the organism. It has for example become recently evident that mechanical forces induced by tissue activity (e.g. migration during development) can alter cell state and gene expression (Brouzés and Farge 2004). A biologist interested in a process like cell differentiation must look at the same time at molecular mechanisms (protein phosphorylation, binding of a protein on a DNA region, etc.) and at higher-level processes, like the micro-environment in which cells live and the physiological state of the whole organism, and then he must also understand the mutual influences between these levels.

The challenge of integrating such diverse data and approaches is however daunting. Systems biology's methods are of great help here, because it is possible to develop formal models at different levels (for example a kinetic model of chemical reactions, a coarser model of gene regulation, a model which describes cell migration and pattern formation, a model of whole organ function, etc.) and then link these models together so as to analyze and understand how these processes influence each other. Such approaches are called multi-level modeling (see for example Hunter and Borg 2003). They are still in their infancy, but they offer very exciting prospects. They illustrate biologists' efforts to build less reductionist and more pluralistic explanations of biological systems.

2.3 Lessons for Biology Education

So what are the lessons we can learn from this discussion that might be useful for biology education? I think that it has some deep import for the general conceptual framework of biology, which should be reflected in biology teachings at all levels. The classical framework is largely based on the gene as the primary causal agent of cells and organisms' properties, bridging the molecular and phenotypic levels. As I mentioned above, the information metaphor introduced by molecular biologists has profoundly shaped scientific and popular conceptions of what is an organism. The selfish gene metaphor made popular by Richard Dawkins at the end of the 1970s (and still very popular today) has reinforced the view that fundamental biological processes take place at the level of the gene. Even though gene centrist biologists are not naïve genetic determinists (they recognize in principle the influence of the environment), the picture they have created clearly suggests that genes control cellular and organismal events, and that environment (cellular or ecological) mainly plays a passive role in allowing genes to express the information they carry. This is reflected both in the scientific metaphor of the genetic program, which is supposed to control development, and in the popular idea that our genome contains our destiny ("the DNA mystique"; see Nelkin and Lindee 1996; see also Moore this volume).[10] I mentioned

[10] See for example the dream of many that in a near future it will become possible to select one's child intellectual or artistic abilities through preimplantation genetic diagnosis or even genetic engineering.

how this has led to think that explaining a phenomenon consists mainly in identifying the genes controlling it (for example oncogenes or development master control genes).

Some scientists, philosophers and historians have started to critically examine this framework well before systems biology became popular (Lewontin 2002; Keller 2000; Oyama 2000; Griffiths 2001; Moore 2002; Moss 2003; Griffiths and Stotz 2006; Stotz 2006). These authors have shown that the concepts of gene, of genetic information and the whole dichotomy between genes and environment are highly problematic. They have also stressed the possible social and ethical consequences of these ideas (like genetic determinism). I believe that systems biology offers a chance to move beyond this situation by building an alternative framework in which causality is conceived in a less simplistic way and biological complexity is fully recognized. Genes will no doubt remain important in this new picture, but other factors' explanatory role and the richness of multilevel interactions can be better integrated. The new direction biology is taking must be reflected as much as possible in education programs at all levels with the aim of overcoming the oversimplified and distorted views that have been too often derived from genetics and molecular biology.[11]

3 Systems Biology and the Need for Interdisciplinary Approaches

3.1 The Need for the Transfer of Knowledge

In the last 20 years biologists have been increasingly focusing on the dynamical and complex aspects of biological systems. However, this has required the adoption of new models and new concepts.

All the efforts to develop new modeling and analysis methods correspond to the recognition that a new language was needed to describe biological systems. Some scientists realized that other domains, like physics, computer sciences, or engineering could provide biologists with useful languages (Lazebnik 2002). But such transfers between scientific disciplines are never straightforward. What languages are the most appropriate in order to address various biological problems? How to import and use languages that have been developed to describe and analyze systems very different from what biologists study? How to create new languages? When one thinks about biology education, more questions arise: how to teach these languages to students, and more ambitiously, how to teach them to create

[11] A word of caution should be added. Although describing systems biology as antireductionist or organicist certainly captures important aspects of its explanatory strategies, one should avoid any oversimplification. It cannot be denied that systems biology also continues in an important sense molecular biology's fundamental enterprise, which is to uncover molecular mechanisms underlying biological phenomena.

new languages they will need to address and solve original questions? In other words, how can teachers, institutions, and educational programs create the most favorable conditions for the development of this new integrative and interdisciplinary framework?

Such questions have naturally been asked in several circles in the two last decades. Institution like the American Association for the Advancement of Science (http://visionandchange.org/) or the National Research Council (2003) have published reports offering guidelines for the development of new programs, especially at the undergraduate level.[12] Discussing in detail these efforts to adapt biological teaching to rapidly evolving research would bring us too far here. But some important ideas have emerged and I will connect them to philosophical questions raised by the use of physical and engineering approaches to study biological systems.

Most commentators seem to agree that in order to help the next generation of biologists to develop and use new models and concepts, it is not enough to simply teach them more mathematics, physics or computer sciences. Similarly, it is not sufficient to give introductory biology classes to scientists from other fields. This cannot produce a genuine interdisciplinarity. What is seriously needed is a deep reflection on how each particular method or tool might contribute to solve biological problems and what is perhaps even more important to frame new questions. Biologists need to be able to master these tools with enough skills and understanding, so that they can really make the most of them and not only apply them automatically and superficially (Pevzner 2004). Conversely, other scientists must gain a familiarity with biological systems and problems, so they do not apply precipitously their models and concepts.

This necessity becomes more apparent when one realizes that formal tools are not only useful for analyzing data, but also for designing experiments and framing hypotheses. This is especially true in the case of dynamical modeling of mechanisms. Distinguishing between different hypotheses requires a careful design of experiments. Modern technologies allow producing a lot a data rapidly, but accumulating data just because it is possible is obviously not the best strategy. Although some scientists praise "fishing expedition" style of experiments (Weinstein 1998; Golub 2010), which favor an inductive methodology, it is clear that most groups do not follow such a strategy. Starting from some phenomena to be explained, the goal is to build hypotheses, to test them, refine them and build ever more complete and realistic models. This requires going constantly back and forth between experiments and modeling. Formal methods are of invaluable aid here, because they can for example help to determine which measure is the most informative or under what range of conditions a mechanism produces a specific behavior. However, biologists need to have a good understanding of their possibilities and their limits. Biologists can easily misuse these tools and draw biologically unsound conclusions. But they cannot rely only on the skills of physicists or engineers, because those scientists do not possess a sufficient knowledge of the biological questions being asked and

[12] See also Bialek and Botstein (2004), Gross et al. (2004), and Hodgson et al. (2005).

might not be aware of the peculiarities of biological systems (as we will see below). Using new tools is certainly exciting for biologists, partly because they can "see" new things. For example, mathematical models and simulations have revealed the potential roles of dynamical behaviors like bistability or oscillations in many cellular processes (Tyson et al. 2003). But in the same way that the use of new instruments like the microscope often leads to the observation of artifacts, which are difficult to tell apart from real phenomena (Bechtel 2006; this volume), methods of analysis and modeling can also introduce various biases and create things that do not exist (I will discuss below the case of network topological properties). Therefore, it is essential to clearly understand which questions are being asked, which hypotheses are being tested, what bias might be introduced in data analysis, etc. In short, each tool's strengths and limits must be clearly assessed both by biologists and by scientists from other fields collaborating with them.

Keeping this in mind, we can understand why many commentators have stressed the necessity of offering more than introductory classes in formal sciences to biology students. However, it is also clear that no one can master in sufficient depth several disciplines and for this reason it is necessary to create research groups or collaborations between scientists with different backgrounds. But the essential point is that if each specialist has to rely on someone else's skills at some point, the conditions for genuine exchanges of ideas and cross-fertilizations must be created. And this requires understanding several languages, even if one is not fluent in all of them (we will see in the next section that it requires more than that).

The kind of problems discussed here is of course not completely new, since transfers of knowledge from other fields to biology have occurred in the past. A revolution like the birth of molecular biology in the 1940s and 1950s owed much to few talented physicists who managed to open new fields of research. But this story also tells us that nothing was straightforward in this process. Physicists as bright as Delbrück or Crick also explored many paths that led nowhere (which is of course to be expected). The formulation of the coding problem by George Gamow and then the race to crack the genetic code offers a nice illustration of these difficulties (Kay 2000). This problem attracted many outstanding scientists (for example John von Neumann and Herbert Simon) from mathematics and physics who saw that they might apply their knowledge in cryptography and other analytical methods. It turned out that these theoretical efforts were to a large extent vain, and the problem was eventually solved by "simple" biochemical approaches (Nirenberg and Matthaei 1961). However, the adoption of an informational framework played an important role in the emergence of molecular biology, though it is difficult to determine exactly what part it really played (Sarkar 1996).

3.2 Some Examples of Transfer of Knowledge

No doubt systems biology provides examples of both successful and misconceived transfers of knowledge, though it is often too early to make any firm judgment. There are certainly very exciting and promising encounters of biological and other approaches, and I will briefly discuss two of them. In these examples I would like to stress not only the positive side of this research (i.e. how interdisciplinarity has opened new paths, formulated new questions, etc.), but I think we should also look at and think about the difficulties and possible limits of these approaches. Again, my goal is to stress that hybridization of scientific domains needs a lot of precaution and constant vigilance, something that students must be aware of.

The two examples discussed in this section are taken from the kind of "network" approaches mentioned above. I said that using a network framework is the most natural and convenient way to represent all the interaction data produced by functional genomics (for example gene regulatory or protein-protein interactions), but what do networks tell us from a biological point of view?

These examples illustrate two different ways of searching for order and general principles behind the complexity of these networks (Huang 2004). The first might be called top-down, because it focuses on statistical properties of very large networks, whereas the second one is more a bottom-up approach in the sense that it analyzes smaller structures in more detail.

Mathematicians have been interested for a long time in network properties and a whole domain, graph theory, deals with such problems. This domain can be traced back to the work of Euler, but it is mainly from the 1930s and especially in the 1960s that this field has known important developments. In the mathematical language, a graph is a collection of nodes (or vertices) linked by edges. A graph can be oriented, when edges have an orientation, or not. In systems biology, protein-protein interaction networks are represented by undirected graphs, because there is no direction in these interactions (proteins just interact). On the other hand, metabolic or genetic networks are more naturally represented by directed graphs. This is easily understood because regulatory links have a direction. It is important to represent which gene is the regulator and which one is regulated. Various models are thus needed to capture the specificities of each kind of network.

In the 1960s Stuart Kauffman has been one of the first scientists to think that global properties of complex networks might throw new lights on the nature of biological systems (Kauffman 1969, 1971). Because the structure of gene networks was unknown at that time, Kauffman theoretically studied classes of random networks with varying connectivity. He showed that within some range of parameters, complex networks show very remarkable dynamical properties, which he described as "order at the edge of chaos" because they stand between chaotic and frozen behaviors (Kauffman 1993). His central idea is that order can emerge "for free" in such systems, or in other words spontaneously and not only as the result of a long process of fine-tuning by natural selection. This was a very original way to look at biological systems, because Kauffman assumed that some of life's most fundamental properties could be explained without pursuing the reductionist strategy of studying all the molecular details.

When functional genomics began to reveal the structure of molecular networks, it became apparent that their structure differed from Kauffman's models. It turned out that the distribution of links per node does not correspond to randomly constructed networks. The observed distribution corresponds to a class of networks that is called "scale-free". In scale-free networks, most nodes have few connections whereas few nodes are highly connected. These highly connected nodes are called "hubs". This is formally expressed by the following equation: $P(k)=Ak^{-\gamma}$, where P(k) is the probability that a node is connected to a number k of other nodes. This relation is called "power law".[13]

A great excitement has followed a series of papers by the physicist Laszlo Barabasi and other scientists working on scale-free networks (Barabási and Albert 1999; Barabási and Oltvai 2004). They argued that scale-free property is a kind of universal principle of complex networks that can shed light on many of their properties. This remarkable feature was identified in metabolic networks (Jeong et al. 2000; Ravasz et al. 2002), protein-protein interaction networks (Yook et al. 2004), ecological networks (Sole and Montoya 2001) but also outside biology, for example in social networks (Newman 2001), and the Internet (Barabasi and Albert 1999). Barabasi argued that this class of network is characterized by several essential properties, which are biologically relevant. The most important of these properties is robustness. According to Barabasi, thanks to their structure, these networks will generally not suffer serious consequences of internal failures. Imagine that one node is randomly destroyed. Because most nodes have very few connections, it is likely that the problem will not propagate in the whole network. But of course, if a hub is destroyed, then it is highly probable that the whole structure and behavior will be impaired (the network might break down in small sub-networks). Scale-free networks are supposed to be tolerant to random failure, but highly fragile to specific attack (Barabási and Oltvai 2004).

But do biological networks really follow this "law"? And what does it exactly reveal about biological systems? Several commentators have formulated doubts about the interest of this relation, even if it is confirmed.

The first problem is that such an analysis seems too coarse grained to be very informative. What can we learn about biological systems from a law that holds for systems as different as gene networks, the Internet or social networks? Even if we restrict ourselves to biological networks it seems clear the gene networks and metabolic networks are rather different in nature. What exactly does the generality of such features reflect? Biological networks are the result of a process of evolution partly determined by natural selection, which is not the case of social networks. Should we expect that both kinds of networks share the same properties? By looking for such level of abstraction, one runs the risk of missing the peculiarity of biological systems.

Evelyn Fox Keller (2005) has also expressed doubts about the relevance of scale-free property to characterize biological networks. She noted that this feature is relatively frequent in nature and should not cause such surprise. Moreover, she stresses that completely different architectures share this property and that many processes can produce a particular architecture. Keller thinks that the excitement about scale-free networks reflects the contemporary situation of biology. Biologists are more than ever before tempted to use the tools and models developed in physics and many

[13] For a review on complex network models see for example de Silva and Stumpf (2005).

physicists are interested in applying their skills in the study of biological systems. Keller sees such interdisciplinary approaches as naïve.

One prediction of the scale-free hypothesis is that hubs are much more important than nodes with few connections, which means that these networks are generally quite robust but very fragile if a hub is removed. This prediction has been tested experimentally, but results were far from supporting this hypothesis. From such studies, Mahadevan and Palsson conclude that "even though networks of biological entities might have some similar properties, there appear to be fundamental differences in the nature of these networks, investigation of which can lead to valuable insights on their functions" (Mahadevan and Palsson 2005). These authors stress that we must not lose sight of the differences between for example gene networks, which represent regulatory interactions, and metabolic networks, which represent flow of matter and energy. By representing different classes of networks with the same formalism, one ignores what makes them functionally unique.

In other words, such very abstract analyses seem to miss the most essential aspects of biological systems (for a harsh criticism see Lima-Mendez and van Helden 2009). Statistical analyses undeniably offer a new perspective on network structures, but they give a rather rough picture of functional constraints that have probably shaped these networks.[14]

The point I want to make through this example is not that studying the structural properties of large molecular networks is necessary misleading or worthless. It is certainly necessary to gain a broad, if coarse-grained, view on the structure and dynamics of whole cellular systems in order to understand how multiple regulatory mechanisms are integrated and guarantee a smooth and robust functioning in most cases. Representing these mechanisms as networks and then analyzing them mathematically might be a way to reveal fundamental properties of these systems. But this cannot be done in any straightforward way. To assess how informative network models are, we need a careful investigation of what aspects are faithfully represented, what aspects are lost in the modeling and what features are artifacts.

Moreover, despite impressive progress in functional genomic databases, we are still far from being able to model in detail all the interactions taking place in a cell. We have for example a very partial understanding of the role of microRNA in regulation of gene expression (Piro 2011).

To summarize, it is not yet clear exactly what network models can reveal and explain about biological systems. The knowledge developed in past decades by mathematicians and physicists about complex networks is certainly to be appreciated by biologists, but how to apply and modify such modeling approaches is a very delicate question.

A different, but not necessarily incompatible way of finding some order in regulatory networks is to focus on smaller structures, as I will describe with my second

[14]The functional constraints are linked to the selective pressures acting on these systems. Because each type of network plays specific roles in the functioning of the systems (transfer of energy, "treatment" of external signals, regulation of developmental events, etc.) its functional requirements (for example robustness) and its ability to evolve (be "rewired") are different.

Fig. 1 The coherent feed forward loop (*FFL*) with AND gate. When X is activated, it activates Y and Z. In order to activate Z, both X and Y must be present (hence the AND logic). This simple structure has an interesting dynamical property, because it introduces a time delay between X and Z activation. This is due to the time needed for Y to accumulate before it can activate Z. Because only long enough inputs will eventually activate Z, this structure can act as a filter for signals (For more details, see Alon 2007)

example. A recurrent theme in recent research is that large networks are composed of (and thus can be decomposed in) basic structural components called motifs. These motifs have been first identified through top-down approaches. It appeared that some small structures, usually composed of less than five elements, (which are usually genes or proteins connected in a specific way) are found very frequently. The methodology to detect such motifs is to compare a "real" network inferred from functional genomics data with a set of random networks (which have the same numbers of nodes and connections). The patterns of connections that appear more frequently in the real network are defined as network motifs.

What makes this kind of research different from the one discussed above is that these structures are then studied as functional integrated mechanisms that perform distinct dynamical behaviors. Because these behaviors are functionally important in many cellular processes, they have been described as general-purpose devices. This is what confers them a relatively high degree of generality.

Several studies have shown the presence of motifs in organisms such bacteria, yeast and humans (Milo et al. 2002; Shen-Orr et al. 2002). This has lead biologists like Uri Alon to make the hypothesis that "out of the many possible patterns that could appear in the network, only a few are found significantly and are network motifs" (Alon 2006, p. 41). These motifs are for example different kinds of feedback or feed forward loops. Because each interaction can be positive or negative and integration of two signals can follow different logics (AND, OR),[15] even motifs composed of only three nodes are numerous and show very different behavior. Figure 1 represents the coherent feed forward loop, which has received a lot of attention recently (Mangan et al. 2003), because it might play important roles as a regulatory structure.

Feedback loops are hardly new in biology, but biologists have generally been relatively uninterested in comprehensive and rigorous analyses of the dynamical properties of different regulatory structures. On the other hand engineers have developed in the second half of the twentieth century general theories of circuits, which apply to different kind of systems, electric systems being of course the most obvious.

[15]Logic gates process signals, which represent true (1) or false (0). The AND logical gate associates an output 1 if the first and the second input are 1. In the case of OR gate, the output is 1 if either or both inputs are 1.

Systems biologists like Alon think that biological networks are built in a similar fashion, from small circuits that follow a general logic. They see their task as identifying, characterizing, and classifying these motifs based on their structural and functional properties and then showing how they fulfill specific biological functions in different contexts.

Among the general functions that network motifs have been said to perform, we find different "treatments" of signals, like filtering and integration of several signals (Alon 2006). Some motifs can speed-up the response time of a circuit. Other can compute signals so as to guarantee a specific decision given different inputs. They can generate temporal programs of expression, in which genes are activated successively in a defined order. There are also motifs that can produce oscillations, something that is obviously important in biology.

The underlying idea in all these cases is that these specific structures allow performing these functions robustly. The criterion of robustness explains why these motifs are thought to occur at high frequency in many biological systems: they have a selective advantage. The rational for holding that network motifs are general is thus convergent evolution. This means that similarity of structure is not explained by common descent, but by independent appearance (biologists sometimes use the anthropomorphic term "rediscovery", which should of course be used with caution) of the same structure several times during evolution.[16] The emphasis on robustness also makes clear why connections are established with engineering more than with physics here. Engineers have always been interested in finding robust solutions to functional problems and have produced a rich body of theoretical work in this domain. The connection with engineering also explains why many authors refer to these structures in terms of "design principles" (Alon 2006). I hasten to add that the term "design" as used by these biologists has nothing to do with the "Intelligent Design" controversy. It simply points to the fact that some structures solve better functional problems (e.g. animal locomotion like flight, vision, physiological regulation, etc.) than others. Authors like Dennett (1995) have argued that biology cannot dispense itself of such engineering-minded view (which again is fully consistent with evolution by natural selection). Although one can dislike these analogies (Pigliucci and Boudry 2011), they are pervasive in many fields of biology (see Bechtel this volume).

Such analyses look very promising and have undeniably some beauty. But is it really the case that networks obey this kind of logic? How confident can we be in the existence of these general principles? Are they really the products of natural selection? I will not be able to discuss all the arguments in favor or against this hypothesis, but I would like to point to some problems with this application of engineering principles.

Although these analyses look plausible, they are not necessarily true. They must be compared with alternative hypotheses. It turns out that it is possible to construct alternative non-adaptive (neutralist) models (see Dietrich, this volume) for most of

[16] Famous cases of convergent evolution are the eye and the wing.

these features. Instead of viewing motifs as the (nearly optimal) products of natural selection, some evolutionary biologists have argued, using models based on population genetics, that they are in reality the by-products of other evolutionary processes. Michael Lynch is one of the most critical commentators of engineering minded approaches (Lynch 2007a, b). He does not deny that some aspects of biological networks can be explained in adaptive terms, but he urges that we need null-hypotheses to test adaptive models (for adaptation see Forber, this volume). If neutral processes can produce these structures, why should we accept the adaptive story?

Lynch is particularly interested in the case of multicellular organisms, which have relatively small population sizes, a property that increases the effect of genetic drift (for drift see Pigliucci this volume; Dietrich this volume). His conclusion is that "many aspects of complexity at the genomic, molecular and cellular levels in multicellular species are likely to owe their origins to these non-adaptive forces, representing little more than passive outcomes of the unique population genetic environments that are presented by such lineages" (Lynch 2007b, p. 812).

It is thus possible that design principles are merely an illusion created by a hasty and imprudent application of engineering models and concepts. If that were the case, engineering methods would only be misleading in the analysis of biological networks. Of course this is an open question, but what can safely be concluded from these debates is that an evolutionary perspective has to be fully incorporated in these approaches. Only after we have a better understanding of how biological networks can (in principle) evolve and have (actually) evolved, will we be in a position to judge the plausibility of design principles guiding the evolution of these networks. So this means that systems biology needs even more interdisciplinarity.

The questions raised by the application of methods and concepts from physics and engineering in molecular biology are complex and deep, and cannot be fully discussed here. But they nicely illustrate the challenges that systems biologists face. What are the most appropriate tools for understanding the complexity of living systems? What help can biology expect from other disciplines? What approaches are misleading because they precipitously apply methods and modes of reasoning developed in other fields and that do not recognize the particular characteristics of life?

It is of course impossible to provide simple answers to these questions, and as often in science, many paths must be explored and then abandoned. So it would be unwise to posit a priori that one method is superior or that another is worthless. But the point is that scientists must develop a good sense of these problems: to see them clearly, to understand the complexity of these methodological questions, and have them constantly in mind so that they can avoid conceptual blindness. This is a problem that reappears constantly in the development of science. Every time two or more domains try to get closer difficulties appear. Each domain tends to ignore the complexity of other fields and some become imperialistic.

I think that these difficulties should be seriously considered when thinking about how to offer science students the best intellectual tools to contribute to this exciting exploration of biological complexity. By scrutinizing the nature of explanatory models and methodologies, philosophy of biology can no doubt contribute to a deeper awareness of these problems. But its contribution extends further as I will show in the next section.

4 Some Philosophical Reflections on the Nature of Systems Biology

I have briefly described how systems biology has made complexity become a central problem for molecular biology and some of the methodological challenges that have resulted, but we can see deeper consequences at the level of biology's philosophical foundations. Here by philosophical foundations I mean how one apprehends the nature of biology as a science, the goal of which is to describe part of the natural world, and also the nature of its objects, i.e. living systems.

One might wonder how such considerations might be relevant or useful for biological education. I will try to show that they are essential, because seeing more clearly biology's underlying philosophical conceptions is necessary for a better and deeper understanding of the nature of its different approaches, its explanatory strategies, its methods, its goals, and importantly of the difficulties it faces.

Since its beginnings as a science, scientists like Lamarck have held that biology's goal is to uncover the general laws of life. Others have disagreed and insisted that it should avoid such philosophical speculations and content itself with describing "positive facts". This question has been disputed ever since. A different disagreement concerns the relation that holds between biology and physics (and chemistry). The nineteenth century is known for violent dispute between vitalists and mechanists or materialists, about the possibility to explain life in physical terms. Even after the vitalist position had been abandoned the issue of biology's autonomy has been much discussed by philosophically oriented biologists, like Ernst Mayr (1988).

Molecular biology nicely illustrates the importance of such fundamental issues in the constitution of a new science (see for example Morange 1998). It should be clear that it has brought much more than new tools to biology. It has also deeply modified biology's general framework (its philosophical foundations). For example, physicists like Max Delbrück, who initiated the so-called phage group in the 1940s (Morange 1998, chapter 4), have defended the idea that some of the most fundamental questions in biology could be addressed by using very simple experimental systems (physics often proceeds in this way). The use of bacteriophages and techniques like crystallography to dissect life's basic mechanisms illustrates this reductionist ideal. Watson and Crick's model was an important step in the establishment of this new vision of life and of biology.

Molecular biology's emergence cannot be dissociated from its founders' reflections on these broad issues. Many commentators have the feeling that systems biology is in a comparable situation. What is perhaps the most fascinating aspect of systems biology is the search for general principles, which I have described above. I will now come back to this issue and look at it from a more philosophical point of view.

Molecular biology's first major successes have been seen as the discovery of life's most fundamental properties. Watson and Crick said that they had discovered the "secret of life" and Crick's central dogma is indeed a scheme supposed to hold for all forms of life. Everybody knows Jacques Monod's famously dictum "what is true for E. coli is true for the elephant". However, as the knowledge of cell's regulatory mechanisms became more detailed, the general picture turned out to be much more complicated than expected. There were many exceptions to the general principles, but what is more embarrassing its logic seemed difficult to understand.[17]

In the same period (in the 1970s and 1980s), philosophers reflecting on the nature of the life sciences began to argue that the essence of biology is not the discovery of universal laws like physics. They remarked that biologists study historical objects, which are the product of contingent evolution by natural selection, and for this reason even if some general features are to be found (for example nucleic acids are the genetic material), they will only be valid on our planet and cannot be considered to be universal (Smart 1962; Rosenberg 1994). It is true that some scientists have always believed in the possibility to build a strong theoretical biology,[18] however it cannot be denied that the vast majority of biologists do not consider their goal to be the search for very general principles. They are instead engaged in building models (usually mechanistic), which have of course always some degree of generality thanks to shared evolutionary history (conserved genes and mechanisms, homologies), but whose goal is to faithfully describe particular cases, without speculating too much about their generality.

Now it seems that systems biology revives the old dream of finding general theoretical principles in biology. The examples discussed above show that the search for general network principles can take different forms. What both approaches have in common is their fundamental belief that the complexity of biological systems obeys very general (perhaps universal) principles.

The fact that systems biology seriously challenges this scientific and philosophical consensus, according to which biology cannot (or only rarely) formulate universal principles, is something that should not be underestimated, because it touches biology's foundations. This has practical consequences, because it determines what is seen as legitimate scientific goals. If biologists distrust this possibility, they will never search for general principles and will consequently probably never find them (see also Lange this volume). Here we see how scientists' basic views influence their practice and their research agenda. Philosophical scrutiny can help to make these views explicit and then examine them critically.

But this idea has further consequences, which bring us to another aspect of the remarkable transformation of scientific research brought about by the increasingly formalized methods and modeling techniques. Many biologists have noted and

[17]Richard Lewin asked "why is development so illogical?" (Lewin 1984) and other scientists started to describe molecular mechanisms as baroque.

[18]See for example Conrad Waddington's constant efforts in this direction (1968 1969, 1970, 1972), and more recently Stuart Kauffman (1993) or Brian Goodwin (Goodwin and Saunders 1989).

some have complained that this has resulted in a loss of intelligibility and intuition. Part of molecular biology's beauty and appeal comes from its ability to represent some of the most fundamental principles of life with simple diagrams (think of the central dogma or the lac operon). Biologists can visualize them, follow mechanisms' operations, imagine the consequences of different perturbations and can in this way relatively easily imagine experiments to refine their models and test alternative hypotheses. But now that biologist are dealing with quantitative models that show non-linear behaviors and with very large mechanisms composed of hundreds or thousands of components, how can they make sense of all this complexity? Without simplifying principles, the cell appears as a horribly complicated and messy tangle of molecular mechanisms, whose robust and tightly regulated functioning seems impossible to understand. How to explain that with such complexity chaos does not break out?

General principles of organization would of course not diminish in any sense this complexity, but they would allow biologists to find some order that can be expressed in a form cognitively manageable. If no general principles in fact exist, no doubt that biology will continue to progress, but this intellectual discomfort will become larger. This search for general principles thus also concerns biological systems' intelligibility.

5 Conclusion

I think that the goal of any biology curriculum should be to give students a rich and varied scientific culture, which includes philosophical considerations. This makes the study of biology much more interesting, because it shows students that it is more than a sum of facts and tools, which have various applications in technology and medicine. Biology is a particular way to inquire into our world and to try to understand it. It has its specific objects and methods, which are partly different from those of other sciences. Thinking about the way biology can or should describe nature can help us recognize its value as an extraordinary intellectual adventure. In the last decades molecular biology has become so focused on the details of molecular mechanisms that many students and researchers complain that they lose sight of fundamental questions and problems. The development of systems biology is an opportunity for scientists to address such questions. Biology education should draw attention to this aspect of systems biology. It is more exciting to think about systems biology as a way to ask fundamental questions about the nature of biological systems than as an accumulation of empirical data and generation of models by automated methods (it is indeed sometimes described as a purely inductive science).

We have seen examples of such fundamental issues in the first part of this chapter, when I showed how systems approaches have started to challenge many ideas and concepts inherited from molecular biology's framework. Fundamental aspects such as the explanatory role of the gene, the way causality is analyzed in biological systems, and the articulation between different levels (from molecules to organisms and beyond) have already been transformed and this will certainly continue in the near future. The consequences are very deep, not only for science, but also for the general public. Systems biology offers a chance to replace oversimplified and caricatural visions based on excessive reductionist discourses (which have flourished during the Human Genome Project) with a richer and finer general picture.

Another reason for integrating such philosophical discussions in biology education is that it can facilitate the successful building of bridges between disciplines that I have outlined above and which is absolutely necessary if systems biology is to have strong foundations. Scientists must not only understand the methods and concepts developed in other domains, but they must also seriously consider the fundamental differences in their respective scientific frameworks, like their styles of explanation and their underlying ideals, the hidden assumptions they make, and their general "scientific philosophy". Philosophy of biology can contribute to make explicit and clarify these various explanatory projects and research strategies. As I have argued throughout this chapter, this is essential because systems biology is not completely clear about the direction it is taking, as many paths seem to open. These reflections should be seriously integrated in the intellectual formation of future scientists, because it will help systems biology to move forward without misunderstanding between disciplines, without unproductive oppositions between different approaches, without wearing blinders, without being imperialist, and more generally without loosing sight of the necessity but at the same time the difficulties of articulating different approaches to unravel and explain life's complexity. These are the conditions necessary to avoid clashes of scientific cultures (Keller 2007) and guarantee fruitful collaborative work.

Acknowledgments I would like to thank Kostas Kampourakis and an anonymous referee for very useful comments on a previous draft of this chapter.

References

Alberghina, L., and H. Westerhoff (eds.). 2007. *Systems biology, definitions and perspectives*. Berlin: Springer.

Albert, R. 2007. Network inference, analysis, and modeling in systems biology. *The Plant Cell* 19: 3327–3338.

Alon, U. 2006. *An introduction to systems biology: Design principles of biological circuits*. Boca Raton: Chapman & Hall.

Alon, U. 2007. Network motifs: theory and experimental approaches. *Nature Reviews Genetics* 8: 450–461.

Barabási, A.-L., and R. Albert. 1999. Emergence of scaling in random networks. *Science* 286: 509–512.

Barabási, A.-L., and Z.N. Oltvai. 2004. Network biology: Understanding the cell's functional organization. *Nature Reviews Genetics* 5: 101–113.

Bechtel, W. 2006. *Discovering cell mechanisms: The creation of modern cell biology*. Cambridge: Cambridge University Press.

Bechtel, W., and R.C. Richardson. 1993. *Discovering complexity: Decomposition and localization as strategies in scientific research*. Princeton: Princeton University Press.

Bialek, W., and D. Botstein. 2004. Introductory science and mathematics education for 21st-century biologists. *Science* 303: 788–790.
Boogerd, F., F.J. Bruggeman, J.H.S. Hofmeyr, and H.V. Westerhoff. 2007. *Systems biology, philosophical foundations*. Amsterdam/Boston: Elsevier Science.
Braillard, P.A. 2010. Systems biology and the mechanistic framework. *History and Philosophy of the Life Sciences* 32: 43–62.
Brent, R. 2000. Genomic biology. *Cell* 100: 169–183.
Bringmann, P., E. Butcher, G. Parry, and B. Weiss. 2006. *Systems biology: Applications and perspectives*. Berlin: Springer.
Brouzés, E., and E. Farge. 2004. Interplay of mechanical deformation and patterned gene expression in developing embryos. *Current Opinion in Genetics and Development* 14: 367–374.
Bruggeman, F.J., and H.V. Westerhoff. 2007. The nature of systems biology. *Trends in Microbiology* 15: 45–50.
Collins, F.S., M. Morgan, and A. Patrinos. 2003. The Human Genome Project: Lessons from large-scale biology. *Science* 300: 286–290.
Craver, C.F., and L. Darden (ed.). 2005. Special issue: Mechanisms in biology. *Studies in History and Philosophy of Biological and Biomedical Sciences* 36: 233–464.
Darden, L. 2006. *Reasoning in biological discoveries*. New York: Cambridge University Press.
de Silva, E., and M.P.H. Stumpf. 2005. Complex networks and simple models in biology. *Journal of the Royal Society Interface* 2: 419–430.
Dennett, D. 1995. *Darwin's dangerous idea*. New York: Simon & Schuster.
Emmeche, C., S. Koppe, and F. Stjernfelt. 2000. Levels, emergence, and three versions of downward causation. In *Downward causation*, ed. P.B. Andersen, C. Emmeche, N.O. Finnemann, and P.V. Christiansen. Aarhus: University of Aarhus Press.
Falk, R. 2000. The gene – A concept in tension. In *The concept of the gene in development and evolution: Historical and epistemological perspectives*, ed. P. Beurton, R. Falk, and H.J. Rheinberger, 317–348. Cambridge/New York: Cambridge University Press.
Gehring, W.J. 1999. *Master control genes in development and evolution*. New Haven: Yale University Press.
Glennan, S. 2002. Rethinking mechanistic explanation. *Philosophy of Science* (Supplement) 69: S342–S353.
Golub, T.R. 2010. Counterpoint: Data first. *Nature* 464: 679.
Goodwin, B., and P. Saunders (eds.). 1989. *Theoretical biology: Epigenetic and evolutionary order from complex systems*. Edinburgh: Edinburgh University Press.
Griffiths, P.E. 2001. Genetic information: A metaphor in search of a theory. *Philosophy of Science* 68: 394–412.
Griffiths, P.E., and K. Stotz. 2006. Genes in the postgenomic era. *Theoretical Medicine and Bioethics* 27: 499–521.
Gross, L., R. Brent, and R. Hoy. 2004. The interface of mathematics and biology. *Cell Biology Education* 3: 85–92.
Guyer, M.S., and F.S. Collins. 1993. The Human Genome Project and the future of medicine. *Archives of Pediatrics and Adolescent Medicine* 147: 1145–1152.
Hieter, P., and M. Boguski. 1997. Functional genomics: It's all how you read it. *Science* 278: 601.
Hodgson, T., R. Keck, R. Patterson, and D. Maki. 2005. Mathematics in the biology classroom. A model of interdisciplinary education. *Journal of College Science Teaching* 34: 46–50.
Hornberg, J.J., F.J. Bruggeman, H.V. Westerhoff, and J. Lankelma. 2006. Cancer: A systems biology disease. *Biosystems* 83: 81–90.
Huang, S. 2004. Back to the biology in systems biology: What can we learn from biomolecular networks? *Briefings in Functional Genomics and Proteomics* 2: 279–297.
Hull, D.L. 1974. *Philosophy of biological science*. Englewood Cliffs: Prentice-Hall.
Hunter, P.J., and T.K. Borg. 2003. Integration from proteins to organs: The physiome project. *Nature Reviews Molecular Cell Biology* 4: 237–243.
Ingram, V.M. 1956. A specific chemical difference between globins of normal and sickle-cell anemia hemoglobins. *Nature* 178: 792–794.

Jeong, H., B. Tombor, R. Albert, Z.N. Oltval, and A.L. Barabasi. 2000. The large-scale organization of metabolic networks. *Nature* 407: 651–654.
Judson, H.F. 1979. *The eighth day of creation: Makers of the revolution in biology*. New York: Touchstone Books.
Kauffman, S.A. 1969. Metabolic stability and epigenesist in randomly connected nets. *Journal of Theoretical Biology* 22: 437–467.
Kauffman, S.A. 1971. Gene regulation networks: A theory for their global structure and behavior. *Current Topics in Developmental Biology* 6: 145.
Kauffman, S.A. 1993. *The origins of order*. Oxford: Oxford University Press.
Kay, L. 2000. *Who wrote the book of life?* Stanford: Stanford University Press.
Keller, E.F. 2000. *The century of the gene*. Cambridge, MA: Harvard University Press.
Keller, E.F. 2005. Revisiting "scale-free" networks. *BioEssays* 27: 1060–1068.
Keller, E.F. 2007. A clash of two cultures. *Nature* 445: 603.
Kitano, H. 2002a. Looking beyond the details: A rise in system-oriented approaches in genetics and molecular biology. *Current Genetics* 41: 1–10.
Kitano, H. 2002b. Systems biology: A brief overview. *Science* 295: 1662–1664.
Kreeger, P.K., and D.A. Lauffenburger. 2010. Cancer systems biology: A network modeling perspective. *Carcinogenesis* 31: 2–8.
Lazebnik, Y. 2002. Can a biologists fix a radio? – Or what I learned while studying apoptosis. *Cancer Cell* 2: 179–182.
Lewin, R. 1984. Why is development so illogical? *Science* 224: 1327–1329.
Lewontin, R. 2002. *The triple helix: Gene, organism and environment*. Cambridge, MA: MIT Press.
Lewontin, R., S. Rose, and L.J. Kamin. 1984. *Not in our genes: Biology, ideology and human nature*. New York: Random House.
Lima-Mendez, G., and J. van Helden. 2009. The powerful law of the power law and other myths in network biology. *Molecular BioSystems* 5: 1482–1493.
Lynch, M. 2007a. The frailty of adaptive hypotheses for the origins of organismal complexity. *Proceedings of the National Academy of Sciences of the United States of America* 104: 8597–8604.
Lynch, M. 2007b. The evolution of genetic networks by non-adaptive processes. *Nature Review Genetics* 8: 803–813.
Machamer, P., L. Darden, and C. Carver. 2000. Thinking about mechanisms. *Philosophy of Science* 67: 1–25.
Mahadevan, R., and B.O. Palsson. 2005. Properties of metabolic networks: Structure versus function. *Biophysical Journal: Biophysical Letters* 88: L07–L09.
Mangan, S., A. Zaslaver, and U. Alon. 2003. The coherent feed-forward loop serves as a sign-sensitive delay element in transcription networks. *Journal of Molecular Biology* 334: 197–204.
Mayr, E. 1988. *Toward a new philosophy of biology, observations of an evolutionist*. Cambridge, MA: Harvard University Press.
Milo, R., S. Shen-Orr, S. Itzkovitz, N. Kashtan, D. Chklovskii, and U. Alon. 2002. Network motifs: Simple building blocks of complex networks. *Science* 298: 824–827.
Moore, D.S. 2002. *The dependent gene*. New York: Times Books/Henry Holt.
Morange, M. 1998. *A history of molecular biology*. Cambridge, MA: Harvard University Press.
Moss, L. 2003. *What genes can't do*. Cambridge: MIT Press.
National Research Council. 2003. *BIO2010: Transforming undergraduate education for future research biologists*. Washington, DC: National Academies Press.
Nelkin, D., and M.S. Lindee. 1996. *The DNA mystique*. New York: W. H. Freeman.
Newman, M.E.J. 2001. The structure of scientific collaboration networks. *Proceedings of the National Academy of Sciences of the United States of America* 98: 404–409.
Nirenberg, M.W., and H.J. Matthaei. 1961. The dependence of cell- free protein synthesis in *E. coli* upon naturally occurring or synthetic polyribonucleotides. *Proceedings of the National Academy of Sciences of the United States of America* 47: 1588–1602.
Noble, D. 2006. Systems biology and the heart. *BioSystems* 83: 75–80.
Nowak, R. 1995. Entering the postgenome era. *Science* 270: 368–369.

Nüsslein-Volhard, C., and E. Wieschaus. 1980. Mutations affecting segment number and polarity in Drosophila. *Nature* 287: 795–801.
O'Malley, M.A., and J. Dupré. 2005. Fundamental issues in systems biology. *BioEssays* 27: 1270–1276.
O'Malley, M.A., and O.S. Soyer. 2012. The roles of integration in molecular systems biology. *Studies in History and Philosophy of Biological and Biomedical Sciences* 43: 58–68.
Oyama, S. 2000. *The ontogeny of information: Developmental systems and evolution*. Durham: Duke University Press.
Palsson, B.O. 2006. *Systems biology: Properties of reconstructed networks*. Cambridge/New York: Cambridge University Press.
Pevzner, P.A. 2004. Educating biologists in the 21st century: Bioinformatics scientists versus bioinformatics technicians. *Bioinformatics* 20: 2159–2161.
Pigliucci, M., and M. Boudry. 2011. Why machine-information metaphors are bad for science and science education. *Science & Education* 20: 453–471.
Piro, R.M. 2011. Are all genes regulatory genes? *Biology and Philosophy* 26: 595–602.
Pomerening, J.R., E.D. Sontag, and J.E. Ferrell Jr. 2003. Building a cell cycle oscillator: Hysteresis and bistability in the activation of Cdc2. *Nature Cell Biology* 5: 346–351.
Ravasz, E., A.L. Somera, D.A. Mongru, Z.N. Oltvai, and A.L. Barabasi. 2002. Hierarchical organization of modularity in metabolic networks. *Science* 297: 1551–1555.
Roberts, L., R.J. Davenport, E. Pennisi, and E. Marshall. 2001. A history of the Human Genome Project. *Science* 291: 1195.
Rosenberg, A. 1994. *Instrumental biology or the disunity of science*. Chicago: The University of Chicago Press.
Rosenberg, A. 2006. *Darwinian reductionism*. Chicago: The University of Chicago Press.
Ruse, M. 1971. Reduction, replacement, and molecular biology. *Dialectica* 25: 39–72.
Sarkar, S. 1996. Biological information: A sceptical look at some central dogmas of molecular biology. In *The philosophy and history of molecular biology: New perspectives*, ed. S. Sarkar, 187–232. Dordrecht: Kluwer Academic Publishers.
Sarkar, S. 1998. *Genetics and reductionism*. Cambridge: Cambridge University Press.
Sarkar, S. 2007. *Molecular models of life*. Cambridge, MA: MIT Press.
Schaffner, K. 1969. The Watson-Crick model and reductionism. *The British Journal for the Philosophy of Science* 20: 325–348.
Schaffner, K. 1993. *Discovery and explanation in biology and medicine*. Chicago: The University of Chicago Press.
Shen-Orr, S., R. Milo, S. Mangan, and U. Alon. 2002. Network motifs in the transcriptional regulation network of *Escherichia coli*. *Nature Genetics* 31: 64–68.
Smart, J.C.C. 1962. *Philosophy and scientific realism*. London: Routledge and Kegan Paul.
Sole, R.V., and J.M. Montoya. 2001. Complexity and fragility in ecological networks. *Proceedings of the Royal Society London Series B* 268: 2039–2045.
Soto, A.M., and C. Sonnenschein. 2004. The somatic mutation theory of cancer: Growing problems with the paradigm? *Bioessays* 26: 1097–1107.
Stotz, K. 2006. Molecular epigenesis: Distributed specificity as a break in the central dogma. *History and Philosophy of the Life Sciences* 28: 527–544.
Tyson, J.J., K.C. Chen, and B. Novak. 2003. Sniffers, buzzers, toggles and blinkers: Dynamics of regulatory and signaling pathways in the cell. *Current Opinion in Cell Biology* 15: 221–231.
van Leeuwen, I.M.M., C.M. Edwards, M. Ilyas, and H.M. Byrne. 2007. Towards a multiscale model of colorectal cancer. *World Journal of Gastroenterology* 13: 1399–1407.
Waddington, C.H. (ed.). 1968, 1969, 1970, 1972 (4 volumes). *Towards a theoretical biology*. Edinburgh: Edinburgh University Press.
Weber, M. 2005. *The philosophy of experimental biology*. Cambridge: Cambridge University Press.
Weinstein, J.N. 1998. Fishing expeditions. *Science* 282: 628–629.
Wimsatt, W. 2000. Emergence as non-aggregativity and the biases of reductionisms. *Foundations of Science* 5: 269–297.
Yook, S.H., Z.N. Oltvai, and A.L. Barabási. 2004. Functional and topological characterization of protein interaction networks. *Proteomics* 4: 928–942.

Putting Mendel in His Place: How Curriculum Reform in Genetics and Counterfactual History of Science Can Work Together

Annie Jamieson and Gregory Radick

1 Introduction

Classical or transmission genetics is the study of patterns of inheritance: how traits in an organism – and, either implicitly or explicitly, the genes that give rise to those traits – are passed from one generation to the next. For most undergraduate students, their first (and sometimes only) experience of the study of inheritance is bound up with Mendelian principles. Anyone who has studied genetics at high school or first year at university will be very familiar with Mendel's peas: varieties of the garden pea with clear-cut, dichotomous traits such as smooth seeds versus wrinkled, yellow seeds versus green, or tall habit of growth versus short. Explanations of the way these traits are transmitted between generations, especially in the pedagogical context, have traditionally been framed around the concepts of *dominance* and *recessiveness*. For example, when a smooth-seeded plant is crossed with a wrinkled-seeded one, the offspring are all smooth-seeded. Smooth is therefore said to be *dominant* to its *recessive* partner, wrinkled. In the same way, yellow is dominant to green, and tall is dominant to short. Further, the 'gene for' the smooth trait is taken to be dominant to the 'gene for' the wrinkled trait, and so on.

This Mendelian approach in the teaching of genetics has persisted into the twenty-first century despite the increasing recognition in many disciplines – genetics, molecular biology and neuroscience amongst others – that, contrary to the Mendel's-peas picture, genes should not be regarded as the sole causes of organismic ('phenotypic') traits, but rather as elements in a complex network of factors involved in the development of an organism. Of course teaching methods and materials in the sciences always lag behind the sciences themselves, for well-known reasons. In the genetics case, moreover, the start-with-Mendel pedagogic strategy has its defenders

A. Jamieson (✉) • G. Radick
School of Philosophy, Religion and History of Science, University of Leeds, Woodhouse Lane, Leeds LS2 9JT, UK
e-mail: A.K.Jamieson@leeds.ac.uk; G.M.Radick@leeds.ac.uk

K. Kampourakis (ed.), *The Philosophy of Biology: A Companion for Educators*, History, Philosophy and Theory of the Life Sciences 1, DOI 10.1007/978-94-007-6537-5_25,

even among those who appreciate that biology has moved on.[1] A Mendelian conceptual foundation, they will say, gives students precisely the basis they need in order to cope with the messiness of the rest of biology. To stick with Mendel's peas, in the defenders' eyes, is thus to embrace the venerable wisdom of the good teacher who knows that it is always best to start with something simple, and only gradually thereafter to introduce difficulty, complexity, complications, exceptions. Start simply, and your students will proceed with understanding and confidence. Do otherwise, and you risk overwhelming them, leaving them confused and demoralized.

But even the most useful of strategies can be problematic. To a growing number of critics, including geneticists, historians and educators (though these are by no means mutually exclusive groups), Mendel's peas no longer look fit for purpose as a starting point for organizing knowledge of inheritance, because the simplifications they bring with them – and which have made them so attractive to teachers of genetics for so long – may ultimately do more harm than good. The Mendelian concept of dominance in particular has come under suspicion, as generating a highly exaggerated but also deeply engrained sense of the determinative power of genes.[2] Such a stance toward genes can adversely affect students' ability to understand and do genetics in the age of genomics and epigenetics. Just as importantly, it can affect students' perceptions of genetics in wider contexts, including those that bear on important social issues (see Moore, this volume).

These critics do not, it should be emphasized, necessarily have Mendel himself in their sights. In his original papers, he used the term 'dominant' – or rather, its German counterpart – to describe the behaviour of a certain trait in a specific context, not as a defining quality of that trait (Falk 2001; Allchin 2005, p. 440). However, within the modern usage of 'Mendelian', dominance has come to be perceived as a fixed, unvarying quality of a gene variant (an 'allele'), where X is dominant to Y in any and all circumstances, and the 'heterozygote' (having X and Y) will always display the dominant phenotype, which overrides the recessive phenotype. As early as 1900, it was recognized that this view of dominance matches reality only in special cases, and fails utterly to account for phenotypic variability (see Allchin 2002, 2005). This recognition in turn underpinned the introduction, after 1900, of a number of variations on the concept of dominance, including:

- Partial or incomplete dominance, which results in offspring with a range of intermediate phenotypes in the heterozygote. The classic example is the Andalusian Blue fowl, with blue-grey plumage, which is the result of crossing a pure-breeding black with a pure-breeding white bird. The Andalusian Blue is not, itself, pure-breeding because it is inevitably heterozygous.

[1] See Skopek (2011) on the birth and early career of the start-with-Mendel strategy among writers of textbooks in genetics.

[2] Although we shall concentrate on dominance here, other aspects of traditional Mendelian pedagogy have also been criticized. As Kampourakis (2013) has shown, the presentation of the process of science is also flawed, promoting, as it does, a view of Mendel as a lone pioneer, rather than as the author of one contribution to the social activity that is science.

- Co-dominance, as seen in the ABO blood types system, where both parental phenotypes are expressed simultaneously
- Over-dominance, or heterosis, where the heterozygous phenotype is outside the parental range, and can provide a survival advantage with respect to certain characteristics, as seen in the increased resistance to malaria conferred by the sickle cell trait in the heterozygote

Such exceptions to Mendelian patterns of inheritance are collectively characterized as 'non-Mendelian inheritance', including, in addition to the above, pleiotropy (a single gene having effects on multiple traits), expressivity (the variation in the degree of expression of a given trait in individuals with a given genotype), penetrance (the proportion of individuals with a given genotype that express the expected phenotype), phenotypic plasticity (changes in the phenotype of an adult organism in response to its environment) and epigenetic factors (heritable changes in phenotype which do not involve alterations to the nuclear DNA sequence) (see Love, this volume; see Uller, this volume). Even so, this dichotomous categorization – into whatever is Mendelian and whatever is not – tends to reinforce the perception that straightforward Mendelian dominance and recessiveness is the usual or normal case. A similar situation holds with respect to two versus more-than-two alleles, where one gene with two alleles is presented as 'normal' and genes with multiple alleles as the exception.

A look at current university-level textbooks suggests that the critics' concerns are not groundless, though neither should the textbooks be caricatured. A typical example, Peter Russell's *iGenetics: A Mendelian Approach* – commonly recommended as reading for undergraduate courses – begins with the premise that "Mendel's work constituted the foundation of modern genetics" (Russell 2006, p. 2). By page 4, the student is introduced to the concepts of dominant and recessive traits. Although gene/environment interactions, and the idea that genes confer only the potential for the development of a certain trait (rather than determining developmental fate absolutely), are touched upon, they are no more than touched upon, briefly. Thus is the precedence of the Mendelian perspective established at the very beginning, at least among students inclined, in the familiar way, to cling to the most straightforward explanation, and to give less weight to complexities presented as side issues (an impression reinforced, with respect to gene-environment interactions, by their not being fully explained at this stage). In the standard way, the textbook goes on to use the language of a "gene for..." a particular characteristic, and of genes as "controlling"/"being responsible for" phenotypic properties. By such language, arguably, genes are directly connected to the adult phenotype, bypassing the complexities of development. Furthermore, 'dominant' is defined as "describing an allele or phenotype that is expressed in either the homozygous or heterozygous state" and 'recessive' as "an allele or phenotype that is expressed only in the homozygous state" (Russell 2006, p. 733 and p. 742) – a conflation of 'allele' and 'phenotype' that may well encourage students to accept the more readily that dominant alleles mask or overpower recessive ones.

There are attempts to present genetics in a different way. Russell himself also publishes a version of his textbook that is subtitled *A Molecular Approach* (Russell 2010), and that postpones bringing Mendelian genetics into the story until chapter 11. However, this represents merely a reordering of topics, not a different presentation of the subject. Once students travelling the molecular route emerge from the details of DNA, RNA and so on to consider how all of that affects the organism, they encounter exactly the same descriptions and definitions of dominance and recessiveness as do the students approaching the material via Mendel's peas.

In this chapter we wish to propose and explore a more far-reaching option, at once new and old, for reorganizing the genetics curriculum. It is new in that it is not represented in the current debate on that curriculum. But it is old in that it revives a concept of dominance – and a way of thinking about inheritance generally – from more than a century ago, before the Mendelian perspective took off and took over. They are to be found in the work of the very first critic of the Mendelian perspective, W. F. R. Weldon (1860–1906), in particular in an unpublished manuscript *Theory of Inheritance* (1904–1905) where Weldon expressed concerns about the dogmatic nature of Mendelism. In his attempt to achieve a better understanding of inheritance by combining the best of both Mendelism and 'biometry' (the statistical biological studies for which he was best known), Weldon adopted, as we shall see, a contextual interpretation of dominance of the sort that many people now seem to be looking for.

We shall proceed from here as follows. The next section offers a more in-depth review of present-day discussion of dominance and its discontents. In Sect. 3 we will describe the alternative conception of dominance proposed by Weldon in his *Theory of Inheritance*. Section 4 will then consider how Weldon's ideas – and ideas about the genetics that might have been had Weldon and his allies not lost their battle over Mendelism – can be useful now in devising a genetics curriculum suitable for the twenty-first century.

2 The Trouble with Tradition

Concerns about the difficulties of teaching genetics, including the role of misinterpretations of dominance, have been expressed for decades (see, for example, Stewart 1982 and Collins and Stewart 1989). This is not surprising, since the evolution of the concept has been highly complex, and scientific understandings of dominance have been subject to much debate (see Falk 2001 for a history of the concept). The pedagogic problems have been most clearly analysed by Douglas Allchin in a series of papers (see Allchin 2000, 2002, 2005) in which he emphasizes the power of language to influence thinking, and the way that an everyday understanding of dominance – suggesting both power and prevalence – can colour understanding of genetics and genetic issues (see especially Allchin 2005). Common misconceptions based on everyday interpretations of the word include:[3]

- Dominant traits are 'stronger' than recessive ones
- Dominant traits are more likely to be inherited than recessive ones
- Dominant traits completely mask or overpower recessive ones
- Dominant mutations occur more frequently than recessive ones
- Dominant mutations are more likely to survive than recessive ones
- For any trait the 'wild-type' tends to be dominant whereas any mutations will be recessive

Even though more advanced genetics teaching modifies the concept of dominance, the very fact that these modifications are overtly framed as 'non-Mendelian' implies, as noted above, that they are exceptions to the fundamental rule, thus propping up the deterministic interpretation of dominance. Furthermore, since a substantial proportion of students will study only elementary genetics – a single introductory genetics module in a general biology degree, say – they will never encounter these exceptions, thus perpetuating the exaggerations surrounding the subject.

It is not exclusively in the discipline of genetics that the Mendelian concept of dominance is problematic. Consider, for example, how 'dominance' talk can produce misunderstandings about the relationship between genetics and evolution. In the evolutionary context, the commonplace interpretation of 'dominance' can lead students to assume that if the 'fittest' individuals survive, surely they must be the dominant ones. In a similar vein, if an allele for a trait confers increased fitness and thus becomes more prevalent in the population, the linguistic and cultural implications discussed above can lead to the inference that the selected allele must therefore be dominant to its counterpart, or even that its increased occurrence is caused by its dominance over the 'wild-type' allele. These misconceptions can lead to a reduced understanding of both genetics and evolution. We see here too how the forging of an educational link between Darwin and Mendel, whilst clearly attractive, can exacerbate the problem of misunderstandings of genetics and must therefore be handled carefully in a new curriculum (see Bizzo and El-Hani 2009 for a discussion of this problematic relationship in the context of high school biology teaching).

'Dominance' talk in other biological disciplines can also add to the confusion. For example, in behavioural ecology, dominance can imply a power relationship between individuals, whereby one exerts physical or social power over another to gain resources, and the dominated individuals are defeated, give way, or submit to dominant ones. Dominance in this context is not only in accord with our everyday usage but is also presented as an "implicit natural model" (Allchin 2005, p. 431), promoting competition and conflict. If students are studying behavioural ecology and genetics simultaneously, as may well be the case in many modular courses, it would not be surprising if their grasp of the genetic meaning of 'dominance' is somewhat

[3] Allchin (2002), p. 50, citing also Donovan (1997).

unsophisticated and deterministic. Allchin (2005, pp. 441–446) further argues that the reification of dominance as an essential quality has shaped biological practice as well as pedagogy throughout the twentieth century. The perception of dominance as "a heritable property" of a trait or allele, rather than a contextual epiphenomenon, leads readily into a perception of dominance (and by association, recessiveness) as pre-requisites of natural order. In fact, dominance in nature is a special case; the majority of traits do not show simple dominance (Allchin 2005, p. 441 and 433).

Allchin (2005, p. 435) also highlights the inadequacy of even more sophisticated models of dominance, such as the so-called linear model for sickle-cell anaemia. On conventional representations of its genetics, the disease is treated as the dominant phenotype, with its expression being a matter of degree. If you are homozygous for 'normal' hemoglobin, you will not have the disease; if you are homozygous for sickled hemoglobin, you will have the disease; and if you are heterozygous, you are somewhere in between, perhaps suffering some symptoms under certain circumstances. What such a scaling scheme fails to capture, however, is the equally legitimate sense in which the heterozygous phenotype – sickle trait, as it is known – is dominant, since, for well-understood reasons, it is the heterozygous condition that confers resistance to malarial infection.

Another major problem arising from the traditional method of teaching genetics is that of oversimplification. The simple patterns of inheritance shown in Mendel's peas – smooth vs. wrinkled, yellow vs. green, and so on – encourage the view that one gene controls one trait. As Lewis (2011) argues, however, such a view is contrary to the understanding which has emerged through more recent research in genomics, according to which single-gene characteristics are very rare. Instead of focussing on these uncommon single-gene traits and diseases in our teaching, and thereby potentially promoting a hard-line determinism, we need, say Lewis and other critics, to move towards a genomic approach which encompasses the interactions of the entire genome with both the internal and external environments, and which acknowledges the complex nature of the vast majority of traits (see also Burian and Kampourakis, this volume).

There are also concerns about the impact on public understanding of science (Allchin 2000, 2005, p. 430). Increasingly, people are required to make personal and political decisions that touch upon genetics – decisions about pre- or post-natal testing, or about social policy regarding the results – and it is essential that we should have a well-educated population to cope. The problem does not first manifest in higher education but is rooted much earlier. As Mills-Shaw et al. (2008) have shown, misconceptions about a range of genetic issues begin when education about genetics begins, in high school or earlier, including misunderstandings of patterns of inheritance, the genetic basis of disease and genetic determinism.[4] This is not surprising when we consider that genetic determinism is a "general phenomenon" in school textbooks, as recently argued by Gericke et al. (2012). In their study

[4]For more on the increasingly inadequate public understanding of genetic issues, see Condit et al. (1998), Smerecnik (2010), Condit (2011), and Lewis (2011).

of 38 high-school textbooks from six countries, these authors show that due, in large part, to oversimplification in the way that the nature of the gene is presented in these books, students are in effect encouraged to develop a very reductionist, deterministic view of genetics. Mills-Shaw and her co-workers highlight a further area of concern, noted too more recently by Lewis (2012, citing Nowgen 2011), which is that secondary-level teachers do not always feel that they have the necessary knowledge and expertise to teach this subject properly. The potential problems of inadequately educated teachers in this area are explored in an international study by Castéra and Clément (2012) who show that the level of teachers' training and understanding in genetics affects not only the grip of their students on genetic facts and concepts but the attitudes that the students acquire to gender, ethnicity and other socially sensitive topics.

Taken together, the above constitute compelling reasons to improve the presentation of inheritance in both school and university curricula. It is in everyone's interests for school teachers to feel better able to teach genetics (and genuinely to be so), for those students who do not move on to higher education to have a better grounding in important genetic issues, and for those students who do move on to be better prepared for the transition from school to university biology studies. The result would be a better-informed citizenry, more capable of making decisions involving genetics, and with a more subtle attitude to human differences, genetic or otherwise.

One way forward, proposed by Allchin (2005, p. 436) would be to abandon the terminology of 'dominance' altogether and to teach genetics without it. An early emphasis on what Allchin describes as the 'haplophenotype' – that is, consideration of the expression of each allele separately – would remove the need for the dominant/recessive dichotomy (see further discussion in Sect. 4). Other suggestions of how to alleviate the problems by devising new curricula include the development of an "inverted" curriculum, where complex traits are addressed first, before later moving on to the more simple patterns seen in dichotomous traits such as smooth and wrinkled, which are then presented as exceptional examples of discrete variation, rather than as the norm (Dougherty 2009). In this way, students are discouraged from relying on the (over)simplified situation seen in Mendelism as traditionally taught. This approach is not entirely new; J.B.S. Haldane and Julian Huxley, in their 1927 *Animal Biology,* aimed at school boys, began their account of Mendelism with the example of the Blue Andalusian fowl, in which, as noted above, the hybrid of black and white strains shows an intermediate colour. Only after this did they briefly refer to the idea of dominance, to demonstrate that there are *some* instances where one allele can "mask the appearance" of the other in the first hybrid generation.

It is nevertheless striking how persistent old Mendeldian habits of thinking and talking have remained, even among those aiming to update the genetics curriculum. Guilfoile (1997), for example, provides an informative analysis of the molecular basis of the round/wrinkled trait in peas, presenting this as a classical example of a dominant/recessive trait, molecular understanding of which can help students to integrate classical and molecular genetics. His approach certainly contributes to resolving that particular problem. But it does so at the price of overstating the overlap between Mendelian and molecular descriptions of the phenotype. At the molecular level, it turns out, there are not two things, a gene 'for' roundness and a gene 'for' wrinkledness. There is, as far as DNA is concerned, mainly just one thing: DNA encoding an enzyme that coverts sugar into starch. Depending on the number of functional copies of that sort of DNA in a given pea plant, the seeds on that plant will have different quantities of the enzyme, hence different quantities of starch, hence – for reasons to do with the effects on water absorption – different seed shapes. Guilfoile draws no attention to it, but the molecular account he supplies should lead us to expect to find that real pea seeds are not either round or wrinkled, as per textbook Mendelism, but instead show every gradation, from extreme wrinkledness to full roundness. That expectation would no doubt only get stronger with supplementary attention to how other genes in the pea genome, ambient temperature and pressure, mineral content in the absorbed water, and so forth also affect seed shape. And indeed, there are many degrees of wrinkledness (and of other traits) in real pea seeds, as – we shall see shortly – was clearly demonstrated by Weldon in 1902 (Weldon 1902a, b).

The point can be stated more generally. The further we explore the molecular basis of traits, the more unwieldy the simple concept of dominance and recessiveness becomes, since we become less and less able to identify dichotomous phenotypes. As our knowledge increases, through the Human Genome Project and other efforts, we increasingly recognize that the vast majority of human genetic traits are multifactorial, involving the interaction between different genes within the genome and between the genome and the internal and external environments. Mendel's classic examples are idealized models which, on the one hand, can help students to grasp basic concepts but, on the other, can promote deep-seated misconceptions which can interfere with the ability of students to engage with the full complexity of inheritance.

We need to teach genetics in such a way as to leave students in no doubt that dominance, as Allchin (2005, p. 437) writes, "... is not a property inherent in any isolated allele, but rather varies with context." And this is precisely what Weldon, more than a century ago, went to great lengths to demonstrate by analogy, in development and regeneration, as the next section will recount.

3 The Recovery of an Alternative View of Dominance (And How It Got Lost)

It helps, in understanding Weldon's views and how and why he came to hold them, to place him in relation to another English biologist of the same generation, William Bateson. Born in 1860 and 1861 respectively, they met as undergraduates at Cambridge University, where they studied zoology in the heyday of evolutionary 'morphology', and so undertook extensive study and research in embryology. They became friends, though there was always an instability in the relationship, in that Weldon, a year older, seems always to have outshone Bateson, in one respect or another – and this was an instability that increased in the 1890s as their careers progressed, and Weldon developed a habit of criticizing Bateson in print whenever he thought Bateson's work merited it. By 1900, they appeared to be heading for very different, and unequal, professional futures; for where Weldon had become the Linacre Professor of Zoology at Oxford, Bateson was, in professional terms, hanging on by his fingertips, with a relatively low-status position at a Cambridge college (though both by this time were sufficiently distinguished to be members of the Royal Society). But 1900 saw changes that brought about quite a reversal of fortune for the both of them.

1900 is famous for biologists and for historians of biology as the year in which Mendel's paper, which was published in 1866, was 'rediscovered'. The word has to go in quotation marks in part because the paper was never completely forgotten among specialists. What happened in 1900 was that the paper suddenly became a talking point throughout European botany. Bateson and Weldon responded in very different ways to this development. Between 1900 and 1902, Bateson became increasingly persuaded that Mendel's paper represented a new foundation for a truly scientific science of inheritance, which would be experimentally precise as well as quantitative. Weldon, by contrast, came to think that any attempt to put Mendel's work at the centre of the understanding of inheritance was wrongheaded, and indeed a huge backward step for biology. In 1902, he published a critique of the Mendelian perspective in the journal *Biometrika* (Weldon 1902a). Within the historiography of biology, the tradition, right up to the present, has been to treat Weldon's critique as wilfully obstructive and deeply confused (Schwartz 2008, ch. 7). It was neither of these things; and an effort to reinhabit Weldon's point of view is well worth making.

We can usefully attend first of all to two photographic plates that accompanied his 1902 article, and with which he aimed to show readers that Mendel's laws of inheritance do not seem to work even for peas. Weldon took it upon himself to collect hybrid pea varieties and to study them, conveying some of his results photographically to emphasize his point. The top of his first plate shows a line of peas; the leftmost peas are green and the rightmost peas are yellow. But in between, the peas range from greenish yellow to yellowish green. It appears that nature actually presents a continuum of colours. Certainly no one else's peas seemed to look the way Mendel reported his peas as looking. We see a similar situation with wrinkledness, the subject of Weldon's second plate, which shows peas ranging from smooth to wrinkled, in gradually increasing degrees. Another thing you would never imagine with a Mendelian mindset, Weldon suggests, is that, sometimes, descendant pea varieties in their wrinkledness recall not their immediate ancestors but, as his evidence shows, their more distant ancestors.

What is happening here? It may look like nitpicking. In the textbooks, Mendel is celebrated precisely for the brilliant methods that allowed him to cut through all of the complexity that nature presents and find an underlying simplicity and order. His stroke of genius was his insight that for his hybridization experiments to yield clean results he had to purify his starting materials, which took him years – years spent ensuring that his white-flowering pea plants only ever gave rise to white-flowering pea plants, and that his purple-flowering pea plants only ever gave rise to purple-flowering pea plants, and so on. It is hard to view all of that effort with anything other than reverence. One way into salutary irreverence is to recall an old philosophy-of-science joke. It's night-time. A man is walking down the street, and he sees a second man down on his knees, searching the ground around him:

First man: "What are you doing?"
Second man: "Oh, I dropped my keys on the other side of the street."
First man: "So why are you looking for them over here?"
Second man: "Well, the light is better here."

As Weldon saw it, performing experiments à la Mendel in the name of understanding inheritance is an exercise in looking where the light is better, rather than where the keys are – the keys that unlock the most profound mysteries about inheritance. It suits *us*, the investigators, to eliminate all of the variability that creates complexity, and thus to generate patterns which are simple, so much so that they can be treated with simple combinatorial mathematics. But there is no reason, in Weldon's view, for regarding what is produced thereby as somehow basic, foundational, fundamental. It is just arbitrary to do that. Furthermore, in taking that step – in treating that arbitrary order as God-given – we cease to take a serious interest in variability and complexity and the lessons they might hold about how real inheritance (as distinct from an artificially engineered and arbitrarily simplified version of it) really works.

Weldon himself sought illustrations not in the world of jokes but in the history of science. When he tried to engage his audience – notably students on a summer course at Oxford in 1905 – he recalled an episode from the recent history of physics and chemistry. The middle years of the 1890s saw a new column added to the Periodic Table: the noble gases, starting with argon. As Weldon told the story, this development resulted from William Ramsey and Lord Rayleigh being unwilling to let themselves off the hook when confronted with a discrepancy in their data, unwilling to fictionalize it away – in this case, a difference of hundredths of a gram between nitrogen collected from nitrogen-bearing compounds and nitrogen isolated from the atmosphere, after all the other then-known gases had been removed from a sample. Atmospheric nitrogen was just a little bit heavier – again, just hundredths of a gram – but not forgetting about that difference, indeed taking it very seriously, worrying about it and at it, led to the discovery of new elements (and, in 1904, a Nobel Prize). That, in Weldon's view, is great science; it depends on not turning a blind eye to the complexity you experience, and not idealizing it into conceptual oblivion.

In drawing the reader's attention to all the natural variability whose existence is ignored in Mendel's exposition, Weldon was trying to suggest to biologists that it was absolutely vital – when it came to identifying new patterns, when it came to avoiding misleading organizing concepts – to bear in mind exactly what nature presents them with. And he drove the point home in the paper with an example now strongly associated with Mendelism: eye colour in humans. We now live in a world where educated people think they know that dark-coloured eyes are dominant and light-coloured eyes recessive, with family lineages showing classic Mendelian patterns. A light-eyed child from dark-eyed parents? The parents must be heterozygous for eye colour. A dark-eyed child from light-eyed parents? Well, that cannot happen, unless… So widespread is the latter pattern, and the related chain of Mendelian reasoning (and suspicions of adultery), that one can now easily find on the web popular-science columns reassuring parents that yes, light-eyed parents *can* have dark-eyed children! Because, unlike what educated people tend to have learned, eye colour is now known to be the result of interaction between many genes, whose collective effects can result in any of the possible colours we see in humans.[5]

Here is Weldon in 1902 (our emphasis):

> It would almost certainly be possible, by selecting cases of marriage between men and women of appropriate ancestry, to demonstrate for their families a law of dominance of dark over light eye-colour, *or of light over dark. Such a law might be as valid for the families of selected ancestry as Mendel's laws are for his peas and for other peas of probably similar ancestral history, but it would fail when applied to dark and light-eyed parents in general – that is, to parents of any ancestry who happen to possess eyes of given colour.* (1902a, p. 242)

In other words, the experimental investigator could justify either 'law' – dark dominant over light or light dominant over dark – depending on the starting materials chosen (and excluded). And one remembers in coming across this passage that one of Weldon's own greatest scientific heroes, Francis Galton, had in his collection a tin box of glass eyes, whose iris colours divided up not into two kinds, light and dark, but into a spectrum comprising 16 different kinds.

With the publication of Weldon's 1902 critique, the "biometrician-Mendelian" debate, as it has subsequently become known, was launched. What followed constitutes one of the most ferocious controversies in the whole of the history of science (on this debate see MacKenzie and Barnes 1974; Olby 1988; Kim 1994). A more recent name, 'the Mendel wars', captures the tenor. For Bateson's part, he was very energetic and successful at recruiting clever allies to his cause, such as his Cambridge colleague Reginald Punnett, famous to genetics students the world over for the Punnett square, a diagrammatic way of keeping track of the various outcomes of crosses and their probabilities. He also published the first textbook on Mendelian inheritance, *Mendel's Principles of Heredity: A Defence*. It came out in the late spring of 1902 and was intended as a rebuttal – a very rude rebuttal – both to Weldon's attack and to Weldon personally. As the book went on into successive editions, the direct attack on Weldon dropped away. But what was left became a template which has moulded genetics textbooks right through to the present: Mendel's experiments show us the first step; the rest is extension; and when the extensions do not work in a straightforward way there follows allowance for exceptions and exculpatory explanations. But it all starts with Mendel and his peas. For Bateson, the peas were important in more than just a theoretical sense. He always emphasized the utility of Mendelism, and was actively engaged in marketing the new Mendelian genetics ("genetics" was Bateson's own coinage) to farmers and animal and plant breeders (Charnley and Radick 2013).

From the time of Francis Bacon, our culture has come to expect that from the finding of true principles there will flow useful techniques; and Bateson accordingly

[5]For a review of the complex, polygenic nature of the inheritance of eye colour, see Sturm and Frudakis (2004); for discussion of the reasons why determinist explanations of patterns of inheritance are attractive, see Moore (2008).

represented the Mendelian perspective as giving to the breeder at last the kind of power the chemist had to plug qualities in and out. Bateson promised in his 1902 book that the breeder would no longer have to trudge benightedly along the paths of tradition, because breeding would have a new, scientific basis. Experimental plots were set up at Cambridge and elsewhere aiming precisely to show the power of Mendelian views in generating new and commercially attractive plant and animal varieties. And in extolling the usefulness of Mendelism, Bateson did not shy away from stressing its usefulness to eugenics – the breeding of better people through science. His was an era, of course, in which more or less everyone with scientific interests was supportive of eugenics, and Bateson was no exception. Even so, Mendelism was seized upon by eugenicists as especially apt for their cause, suggesting as it did that quite complex traits could be governed by single genes and, by extension, would therefore be amenable to selective breeding (see, for instance, the inheritance-of-eye-colour genealogical charts that adorned walls in Nazi Germany).

What about Weldon? Weldon was different. He spent the years from 1902 to 1906 trying to develop an alternative science of inheritance. He knew what he was against: the placing of Mendel's experiments at the centre. But what was he for? He set down his ideas most fully in a remarkable manuscript entitled *Theory of Inheritance* which, though never completed, nevertheless serves to suggest vividly where Weldon was going (and, for anyone who knows the standard secondary literature on Weldon, where he was going comes as an immense surprise).[6] Two features of Weldon's vision are especially relevant for present purposes. One – and this is unsurprising to anyone familiar with Weldon's biometrical allegiance – is the commitment to statistical description. Statistics are essential, Weldon thought, because it is only with statistical language that biologists can describe biological variability precisely. And again, he thought that the really momentous insights in science, the great leaps forward, depend on keeping visible all of the data in all of their complexity – and statistics is the means.

A second more surprising feature is his emphasis on experimental embryology. Successor to the morphological embryology in which Weldon trained, experimental embryology was one of the premier sciences of the late nineteenth century; and a significant proportion of the manuscript reviews experiments involving artificially induced regeneration, as when the parts of an individual *Stentor* (a protozoan) cut into three each went on to develop into a fully formed individual. The lesson that Weldon drew from all of this extraordinarily detailed experimental work, by himself and others – which, fascinatingly, he regarded as merely confirming the theoretical insights of Galton decades before – was that dominance is not something permanently associated with a biological character. Rather, what is expressed by a biological tissue depends essentially on the tissues surrounding it. Expression is fundamentally context dependent. One would never guess that the middle region of the *Stentor* had the capacity to become an entire *Stentor* until the tissues surrounding that region are removed experimentally. Similarly, higher up the animal scale, a cell that would normally always

[6]The authors are currently working on an annotated edition of Weldon's manuscript.

develop into an anterior structure will, when relocated to the posterior of the animal during a certain stage of development, develop as a posterior structure. And lower down, at the level of the hereditary factors – Weldon favoured Weismann's term "determinants" – one and the same character can be dominant (i.e. expressed), or recessive (i.e. not expressed), or falling between those categories, depending on the company kept. Weldon's is a view of dominance and recessiveness emerging from *interaction*: interaction among the hereditary determinants, interaction in turn with their environments, biological and physico-chemical.[7]

Before we go on to suggest how this recovered Weldonian perspective might prove fruitful for reorganizing the genetics curriculum in the age of genomics, and so remedying some of the shortcomings surveyed in the previous section, we offer a few quotations from the manuscript, not least because Weldon's clarity on these issues is so striking (our emphases throughout):

> The group of properties which normally becomes dominant in the case of any particular unit is determined by its position relatively to the other units which make up the body; for we have seen that any group of units, forming the cut surface of a divided *Stentor*, can be made to exhibit dominance of any given group of properties, so that it can be made to produce any given set of organs, if we remove certain adjacent portions of the body, so as to leave the units in question in a suitable relation of position with regard to the next. (Weldon, ch2a, p. 6)

> If we disturb the normal relation of these tissue elements to their neighbours, as we do by removing them from the body, *we can render dominant properties which were previously recessive, and vice versa*. (Weldon, ch2a, p. 21)

> Whatever may happen during the process of inheritance, it is clear that during the life and growth of single individuals such as those we have examined, *Mendel's conception of dominance as a property permanently belonging to the determinants of certain characters, wherever these are in the presence of certain others, is altogether inadequate*. The tissues of these animals are neither "pure" in the sense that they contain only determinants representing a single character or group of characters, nor constant in the sense that some of the determinants they contain are of necessity dominant over the others: on the contrary, *each tissue can be shown by experiment to be in the condition indicated by Galton's hypothesis, behaving as if it contained determinant elements which represent a large series of characters, any one or more of which can become either dominant or recessive, according to circumstances*. Experiments of the kind we have described show that one factor which determines the dominance or latency of characters in a tissue is the relation between the tissue itself and other parts of the body, and apparently in many cases, as in reversed grafts of the Hydra, an essential factor is rather relative position than any of the more complicated relations connected with nutritive or other processes. (Weldon, ch3, pp. 18–19)

In the spring of 1906, this debate over Mendelism – but above all, as we have seen, over the concept of dominance that ought to dominate in the science of inheritance – more or less came to an end with the unexpected death of Weldon at the age of 46 (from pneumonia brought on, some said, by overwork and stress). Weldon's manuscript was never finished and never published, and the field was left open for Bateson to develop his programme to the pitch of textbook glory that it later achieved.

[7]On heredity-environment interactions in Galton's work and Mendel's, see Radick (2011).

4 What if? Remaking the Curriculum Along Weldonian Lines

A question that naturally arises in cases where someone who might have changed things dies or is otherwise incapacitated before finishing the job is to ask: what if? Or, more precisely, what might have been? What might biology have looked liked had Weldon lived longer and – to give this speculation a little spice – been more successful than he actually was at recruiting clever young allies and at marketing his point of view to agriculturalists and doctors? Could there have been a successful Weldonian science of inheritance? What would it have looked like? And would it have been anything like as successful as the science of inheritance that we actually have?

We are persuaded that these sorts of what-if or counterfactual questions are very important, for all of us, but for historians and philosophers of science especially. They are important historically because if we do not pose, and try to answer, these questions, we cannot weigh the significance of past events; and to that extent we cannot really explain why things happened as they did and not otherwise. Did Weldon's death matter for the history of biology? Not much, if Mendelian genetics was destined to emerge in one form or another and eventually settle down to something like its current form; it matters a great deal, however, if that was *not* the inevitable outcome. Counterfactual history-of-science questions are also important philosophically because at stake in confronting them is our sense of what scientific knowledge is, and in particular what makes it worthy of the esteem in which that knowledge is held. Underpinning the widespread notion that scientific knowledge represents human reason at its most objective is the further notion that scientific knowledge is independent of local historical conditions. One group of investigators may, for contingent reasons, come to hold one view as true; another equally competent group may, for different contingent reasons, come to hold another, incompatible view. But in the end, if science is working well, all views will converge on the truth. But will they really? Why should we think so?[8] There are puzzles here aplenty for philosophers, however historically inclined. And for scientists too. Suppose, for argument's sake, that a distinctively Weldonian science of inheritance could have developed out of Weldon's manuscript. And suppose that that science might have yielded insights into inheritance that our science now lacks because of the particular pathway it followed. What might we be missing?

An obvious way to find out, it seems to us, is to try to create the Weldonian science that never was. One can imagine doing this on various scales, from the grand to the modest. At a modest scale, a promising focus is precisely the problems in current genetics pedagogy discussed above. Again, the textbook tradition has been remarkably conservative. From Bateson's day to the present, students start with Mendel's peas; and all of the complex information about inheritance accumulated by the biological sciences in the intervening century is treated as an add-on, tucked

[8] On counterfactuals and the history of science in general, see Radick 2008; on the case for a counterfactual history of genetics, with Weldon as a focus, see Radick 2005.

in along the way. And so we find ourselves in a situation where the points that Weldon was making about the interactive nature of hereditary factors with each other and with their environments are now widely acknowledged, even taken for granted, among biologists, and yet the textbooks remain, by and large, organized around Mendelian phenomena which – and this is their traditional attraction – can be made sense of without considering environmental interactions of any kind.

Does it have to be like that? What are the alternatives? Is it possible to imagine a genetics pedagogy which, in a Weldonian spirit, set off differently? Suppose, instead of starting with Mendel and his peas, we start with something genuinely representative of how genes function in bodies,[9] and even better, representative of the sort of genetic causation that may actually matter to the lives of a large number of people – say, the contribution that genes make to the condition of a human heart. Here is a case of interaction on a massive scale: lots of genes interacting with one another, and in complex ways with exercise and diet (including ingestion of pollutants from air and water), and the whole collocation changing over time as the individual person matures. What would happen if you set beginning students to thinking about cases like that in the first instance; and if, as you developed your instruction, week by week, chapter by chapter, you foregrounded gene-environment interaction as pervasive and primary, not secondary and selectively present? And suppose further that the students, having followed the curriculum for several weeks and assiduously worked their way through the chapters of this imagined textbook, eventually – let us say, around chapter 8 in the textbook (roughly where they meet gene-interactions now) – they get to Mendel and his peas. But they see the pea case as a special case. These patterns do arise; but they arise only under special conditions, notably when humans have engineered artificially purified lineages into being, by deliberately excluding unwanted variability (and note here how reliant genetics textbooks are on *domesticated* plants and animals, rather than their wild and genetically untidy counterparts; the specific strains of fruit flies, zebra fish and mice, for example, used in the experiments described in textbooks do not – indeed, often could not – exist outside the laboratory).

What would those students be like? One prospect is that they might have a more sceptical attitude than is generally the case when it comes to the notion, familiar enough outside biologists' circles, that genes are 'super' causes – 'genetic determinism', in the philosophers' jargon. They might be that bit more prepared to insist, when confronted with claims for the discovery of a gene for a particular trait, that they be told about the range of environments, genetic and otherwise, in play in the course of the investigations, because they will know, in their bones as it were, that any such claims are implicitly founded on observations made within such a (limited) range.[10]

---

[9]The strategy outlined here has affinities with that advocated in Dougherty 2009. Of course, the very concept of the 'gene' is problematic; see Burian and Kampourakis, this volume, for discussion of an alternative way to conceptualize the material of heredity.

[10]On the problems of genetic determinism see Lewontin (1993); see also Moore, this volume.

It is instructive, in the light of that possibility, to look back with renewed attention at how determinism leaks into genetics textbooks even as their authors attempt to disabuse students of determinist attitudes. Consider the following, witty back-of-the-chapter problem from a textbook already discussed, Russell's *iGenetics: A Mendelian Approach*:

> After a few years of marriage, a woman comes to believe that, among all of the reasonable relatives in her and her husband's families, her husband, her mother-in-law, and her father have so many similarities in their unreasonableness that they must share a mutation. A friend taking a genetics course assures her that it is unlikely that this trait has a genetic basis and that, even if it did, all of her children would be reasonable. Diagram and analyze the relevant pedigree to evaluate whether the friend's advice is accurate (Russell 2006, p. 38).

The friend scoffs at the possibility that something as complex as unreasonableness could have "a genetic basis". But notice what happens when the friend – and, in the friend's footsteps, the student attempting to solve the problem – decides to suppose, in a hypothetical spirit, that unreasonableness does have a genetic basis, and furthermore that the nature of that basis can be disclosed through pedigree analysis. There is, of course, a textbook-sanctioned right answer to the question; and it can be reached only by supposing that there is a gene for unreasonableness, and that it behaves exactly as a gene for seed colour or flower colour in peas is supposed, by Mendelian textbooks, to behave. But consider again that phrase, "genetic basis." It suggests, of course, a trait caused by genes interacting with environments. On a certain view, one might with equal justification describe the trait as having an "environmental basis" – but nothing so described would tempt the student toward the just-like-Mendel's-peas problem-solving paradigm. The approved answer to this question, by the way, is that the friend's advice is wrong: the woman's children would have a 50 % chance of being unreasonable whether the gene for unreasonableness is dominant or recessive.[11]

Where current genetics pedagogy remains still so much the product of its Mendelian history, the alternative pedagogy we have outlined is rooted in the Weldonian history that never happened (but might have). Another statement from Weldon's *Theory of Inheritance* manuscript can serve as a credo:

> Since the character of any organ depends, not only upon a specific something transmitted to it through the germ-cells out of which it was developed, that is to say upon something inherited, but also on two sets of conditions external to the organ itself, namely its relation to the parts of the body to which it belongs, and its relation to the environment in which that body exists, we may say that every character of every animal is both 'inherited' and 'acquired'. (ch. 5, p. 24, Jamieson transcription).

Recently, a small group at the University of Leeds, including ourselves and a distinguished genetics education specialist, Dr Jenny Lewis (whose work we have already mentioned), has come to be in a position to translate this vision into reality. Over the

[11]On the critique of "genetic basis" talk see Kitcher (1997, ch. 11, esp. p. 251); on the concept of heritability, and why the nature/nurture dichotomy no longer holds explanatory power in biology, see Moore, this volume.

course of a 2-year project now underway, we aim, in pilot-study form, to devise a curriculum for genetics that will, in Allchin's phrase, "dissolve dominance" in order to give students a genetics education fit for purpose in the twenty-first century.

The challenges before us are not to be underestimated. In closing let us mention three. An obvious one is to find a way to combine a simple, pragmatic language, accessible to the non-specialist, with content that is still sufficiently detailed and accurate to avoid the current problems with misconceptions and misunderstandings. One promising strategy here is Allchin's suggested emphasis on haplosufficiency. To teach with that concept in mind would be to convey to students the message that many "nonfunctioning" alleles produce something. It might be a 'faulty' product; but it is nevertheless present; and every allele product has some action (albeit imperceptible at the phenotypic level in some cases). Furthermore, for some physiological processes, one normal product is sufficient for normal function/phenotype, whereas others need two fully functioning alleles, and still others will have some intermediate state for the heterozygote. As we noted above in our comments on the molecular genetics of both sickle-cell anemia and seed shape in peas, such an approach will help students to avoid the presence/absence implications of classical Mendelian analysis.

A second challenge is, in a Weldonian spirit, to keep the focus on development, and the extent to which organisms develop as they do thanks to a set of complexly interacting factors, 'inheritance' being only one of these, and in itself subject to the vagaries of time and place as much as everything else biological. Consider, for these purposes, the pedgagic potential of maternal inheritance and maternal effect, both of which display 'non-Mendelian' inheritance in the phenotype. Maternal inheritance is the sum of the transmissible characteristics carried on the mitochondrial genome, and hence entirely matrilineal and unaffected by any paternal contribution. Maternal effect is the effect of the products of maternal genes present in the cytoplasm of the egg – genes which control fundamental early developmental processes, such as body axis formation, before the zygotic genome begins to be expressed. Critical early events in development are thus under the influence of a different genome than that of the zygote. Development also provides the opportunity to emphasize the influence of environmental effects. Neural tube closure, for example, is dependent on a number of known genes and also on a variety of environmental factors, some well understood – folic acid, for example – others still quite mysterious, such as socioeconomic group (Gilbert 2010, pp. 340–341). Such examples, introduced very early in the curriculum, could do much to encourage in students an appreciation for the multiple factors involved in development, rather than genetic determinism.

A third, quite different but no less important challenge is to build in scope for problem solving. Students on a traditional genetics course are expected to be able to predict the outcome of specified crosses, and to analyse pedigrees so as to establish patterns of transmission of traits. Manifestly, the dominance/recessiveness dichotomy is very powerful in this context, underpinning a set of problem-solving techniques which, over the century-plus during which Mendelian pedagogy has been honed, have become partnered to problem sets well-calibrated to instilling mastery of those techniques and so to distinguishing degrees of mastery in the students. Any alternative curriculum must either ensure that it provides commensurately ample scope for prediction and analysis or find appropriate surrogates. Perhaps, in our own attempt at such a curriculum, it will be sufficient to introduce dominance/recessiveness at a later stage, along with a sample of the standard problem sets, having made sure the students are firmly grounded in gene-environment interactionism and the contingent nature of development, so that they clearly grasp that the concept of dominance is merely a tool and not a fact of biology.

Acknowledgments We are grateful to the British Academy for the Small Research Grant that funded our work on Weldon's manuscript; to UCL Special Collections for permission to quote from that manuscript; to the Uses and Abuses of Biology programme at The Faraday Institute for Science & Religion, St Edmund's College, Cambridge, for funding the ongoing pedagogic project described above; to Andrew Cuming, Faculty of Biological Sciences, University of Leeds, for much helpful discussion; and to Kostas Kampourakis for his exemplary editorial support. An earlier version of Sects. 3 and 4 appeared as part of Greg Radick's inaugural lecture at the University of Leeds, "Scientific Inheritance: How History Matters for the Sciences," 16 May 2012 (available at http://lutube.leeds.ac.uk/clasew/videos/8153).

References

Allchin, Douglas. 2000. Mending Mendelism. *The American Biology Teacher* 62(9): 633–639.

Allchin, Douglas. 2002. Dissolving dominance. In *Mutating concepts, evolving disciplines: Genetics, medicine and society*, ed. Lisa Parker and Rachel Ankeny, 43–61. Dordrecht: Kluwer.

Allchin, Douglas. 2005. The dilemma of dominance. *Biology and Philosophy* 20: 427–451.

Bateson, William. 1902. *Mendel's principles of heredity: A defence. With a translation of Mendel's original papers on hybridisation*. Cambridge: Cambridge University Press.

Bizzo, Nelio, and Charbel N. El-Hani. 2009. Darwin and Mendel: Evolution and genetics. *Journal of Biological Education* 43(3): 108–114.

Castéra, Jérémy, and Pierre Clément. 2012. Teachers' conceptions about the genetic determinism of human behaviour: A survey in 23 countries. *Science & Education* [online] (2012). doi: 10.1007/s11191-012-949-0. Accessed 18 Sept 2012.

Charnley, Berris, and Gregory Radick. 2013. Intellectual property, plant breeding and the making of Mendelian genetics. *Studies in History and Philosophy of Science*. [online] (2013). doi: http://dx.doi.org/10.1016/j.shpsa.2012.11.004. Accessed 25 Mar 2013.

Collins, Angelo, and James H. Stewart. 1989. The knowledge structure of Mendelian genetics. *The American Biology Teacher* 51(3): 143–149.

Condit, Celeste M. 2011. When do people deploy genetic determinism? A review pointing to the need for multi-factorial theories of public utilization of scientific discourses. *Sociology Compass* 5(7): 618–635.

Condit, Celeste M., Nneka Ofulue, and Kristine M. Sheedy. 1998. Determinism and mass-media portrayals of genetics. *American Journal of Human Genetics* 62: 979–984.

Donovan, M.P. 1997. The vocabulary of biology and the problem of semantics. *Journal of College Science Teaching* 26: 381–383.

Dougherty, Michael J. 2009. Closing the gap: Inverting the genetics curriculum to ensure an informed public. *American Journal of Human Genetics* 85: 6–12.

Falk, Raphael. 2001. The rise and fall of dominance. *Biology and Philosophy* 16(3): 285–323.

Gericke, Niklas M., Mariana Hagberg, Vanessa Carvalho dos Santos, Leyla Mariana Joaquim, and Charbel N. El-Hani. 2012. Conceptual variation or incoherence? Textbook discourse on genes in six countries. *Science & Education* [online] (2012). doi: 10.1007/s11191-012-9499-8. Accessed 18 Sept 2012.

Gilbert, Scott F. 2010. *Developmental biology*, 9th ed. Sunderland: Sinauer Associates.
Guilfoile, Patrick. 1997. Wrinkled peas & white-eyed fruit flies: The molecular basis of two classical genetic traits. *The American Biology Teacher* 59(2): 92–95.
Haldane, J.B.S., and Julian Huxley. 1927. *Animal biology*. Oxford: Oxford University Press at the Clarendon Press.
Kampourakis, K. 2013. Mendel and the path to genetics: Portraying science as a social process. *Science & Education* 22(2–3): 293–324.
Kim, Kyung-Man. 1994. *Explaining scientific consensus: The case of Mendelian genetics*. New York: Guilford Press.
Kitcher, Philip. 1997. *The lives to come: The genetic revolution and human possibilities*. London: Allen Lane/Penguin.
Lewis, Jenny. 2011. Genetics and genomics: Teaching secondary biology. In *Teaching secondary biology*, ed. M. Reiss, 173–214. London: Hodder Education.
Lewis, Jenny. 2012. Young people's understandings of gene technology – From Flavr Savr tomatoes to stem cell therapy. *Science & Education* [online] (2012). doi: 10.1007/s11191-012-9523-z. Accessed 18 Sept 2012.
Lewontin, Richard. 1993. *The doctrine of DNA: Biology as ideology*. London: Penguin Books Ltd.
MacKenzie, D.A., and S.B. Barnes. 1974. *Biometrician versus Mendelian: A controversy and its explanation*, Kölner Zeitschrift für Soziologie special edition 18, 165–196.
Mills-Shaw, R. Kenna, Katie Van Horne, Hubert Zhang, and Joann Boughman. 2008. Essay contest reveals misconceptions of high school students in genetics content. *Genetics* 178: 1157–1168.
Moore, David S. 2008. Espousing interactions and fielding reactions: Addressing laypeople's beliefs about genetic determinism. *Philosophical Psychology* 21(3): 331–348.
Nowgen. 2011. *Genomics in schools: An interim report from the Nowgen Schools Genomics Programme*. Manchester: Biomedical Research Centre.
Olby, Robert. 1988. The dimensions of scientific controversy: The biometric-Mendelian debate. *British Journal for the History of Science* 22: 299–320.
Radick, Gregory. 2005. Other histories, other biologies. In *Philosophy, biology and life*, Supplement to philosophy, Royal Institute of Philosophy Supplement, vol. 56, ed. O'Hear Anthony, 21–47. Cambridge: Cambridge University Press.
Radick, Gregory. 2008. Why what if? *Isis* 99: 547–551.
Radick, Gregory. 2011. Physics in the Galtonian sciences of heredity. *Studies in History and Philosophy of Biological and Biomedical Sciences* 42: 129–138.
Russell, Peter J. 2006. *iGenetics: A Mendelian approach*. San Francisco/London: Pearson/Benjamin Cummings.
Russell, Peter J. 2010. *iGenetics: A molecular approach*, 3rd ed. San Francisco/London: Benjamin Cummings/Pearson Education.
Schwartz, James. 2008. *In pursuit of the gene: From Darwin to DNA*. Cambridge, MA: Harvard University Press.
Skopek, Jeffrey K. 2011. Principles, exemplars, and uses of history in early 20th century genetics. *Studies in History and Philosophy of Biological and Biomedical Sciences* 42: 210–225.
Smerecnik, Chris M.R. 2010. Lay responses to health messages about the genetic risk factors for salt sensitivity: Do mass media genetic health messages result in genetic determinism? *Psychology, Health & Medicine* 15(4): 386–393.
Stewart, James H. 1982. Difficulties experienced by high school students when learning basic Mendelian genetics. *American Biology Teacher* 44(2): 80–84, 89.
Sturm, Richard A., and Tony N. Frudakis. 2004. Eye colour: Portals into pigmentation genes and ancestry. *Trends in Genetics* 20(8): 327–332.
Weldon, W.F.R. 1902a. Mendel's laws of alternative inheritance in peas. *Biometrika* 1(2): 228–254.
Weldon, W.F.R. 1902b. On the ambiguity of Mendel's categories. *Biometrika* 2(1): 44–55.
Weldon, W.F.R. 1904–1905. *Theory of inheritance*. Ref 264/2D, Papers of Karl Pearson. Special collections, University College London, London.

Against "Genes For": Could an Inclusive Concept of Genetic Material Effectively Replace Gene Concepts?

Richard M. Burian and Kostas Kampourakis

1 Introduction

In an earlier paper one of us (RMB) argued that genetics is committed to several fundamental questions, involving the localization, nature, and structure of genes, their physiology (what sorts of entities they are, what material they are made of if they are made of matter at all, what molecules they interact with, how their expression and interactions are regulated), how they influence organismal development, how they affect evolution, and how they are altered in the course of evolution (Burian 2000). According to the fundamental argument of that paper Mendelian genetics was, from the very beginning, committed to several distinct research programs that could be conveniently classed under three headings. These dealt with gene function, gene localization and composition of genes, and the functional organization of genes (meaning the functional pieces of which they were built and the relations of genes and parts of genes to one another). The case studies sketched in that paper dealt with work completed before 1940 and belong to a period described as Mendelian genetics (see also Burian 2013; Kampourakis 2013; Jamieson and Radick this volume).

We argue in this chapter that recent developments show that scientists dealing with different problems or working in different disciplines have distinct concepts of genes, but that the discrepancies in their usage (about the boundaries of genes, their precise

R.M. Burian (✉)
Department of Philosophy, Virginia Polytechnic Institute
and State University, 220 Major Williams, Blacksburg, VA 24061, USA
e-mail: rmburian@vt.edu

K. Kampourakis
Secretariat of Educational Research and Development, Geitonas School,
Vari Attikis, 16602, Greece
e-mail: konstantinos.kampourakis@gmail.com; kamp@geitonas-school.gr

K. Kampourakis (ed.), *The Philosophy of Biology: A Companion for Educators*, History, Philosophy and Theory of the Life Sciences 1, DOI 10.1007/978-94-007-6537-5_26,

localization, and the sorts of biological roles and functions that can be assigned to them) can be readily understood by understanding the disparate roles played by the genetic material (DNA and sometimes RNA) in different contexts and on different scales. We argue quite generally that in courses for students who are not planning to make professional use of genetics, rather than starting with genes it is more helpful to teach about the ways in which geneticists handle the fundamental questions about the structure of the genetic material. That is the key to understanding the differences in the claims that scientists and lay people make about what genes are and what they do. The ways in which the genetic material behaves (and the products that it yields) in different biological contexts support conflicting claims about what functions a given portion of the genetic material may have and about what its effective structure is in different contexts. Understanding this also clarifies how geneticists and other biologists test and correct their views in the light of new evidence and go about gathering and evaluating evidence which is perhaps more important than to teach specifics of highly developed models of the gene[1] or of gene action. It also helps to explain why there are continuing disagreements about what, exactly, to count as a gene and about the powers of genes. We do not believe that talk about genes is wholly dispensable, but that understanding the behavior of the genetic material is the fundamental basis for understanding the terminology involved and the continuing disputes about the nature of genes and the extent to which they "control" the traits of organisms.

By the second decade of the twentieth century, when Mendelian genetics was well established, its adherents had embarked on significant research programs. These programs fall, more-or-less, into three groups that survived and continued into the molecular era.

1. ***Understanding gene function***: Insofar as Mendelian genes were defined, they were defined by use of regularities concerning the inheritance of phenotypic traits. Early Mendelian research sought to characterize genes in terms of their functions and/or consequences, together with the patterns of inheritance that they exhibited. This is the source of descriptions of 'genes for X' (e.g., eye color, height, or amount of sugar in the kernels of corn, for modifying the action of other genes, etc.). Thus, seeking to understand gene function or gene action was the core of one research program.
2. ***Determining gene composition and localization***: Insofar as Mendelians were materialists regarding genes (some were right from the start, some weren't until after the Watson-Crick structure of DNA was published when most became

[1]Etymologically, the term 'gene' originated in the (Hippocratic) idea of Pangenesis, advanced by Darwin (1868). Darwin thought that gemmules from all parts of the body are transmitted to the reproductive organs and, from there, to the next generation. Hugo de Vries suggested that pangenesis did not involve transportation of gemmules between cells; rather all specification of information was intracellular (de Vries 1910/1889). He called the hereditary elements 'pangens', occasionally written 'pangenes' in English. It is from this that the term 'Gene' was suggested by Wilhelm Johannsen.: 'it appears simplest to isolate the last syllable, gene, which alone is of interest to us [...] The word gene is completely free from any hypotheses'. (Keller 2000, p. 2, quoting from Johannsen 1909).

materialists), one needed to know WHAT a gene is, meaning 'what are genes made of?'. This was also associated with the question 'where (within the cell) are the genes located?', a key question that led to the success of the chromosomal theory. (This question has not totally disappeared in molecular genetics; it is still crucial in seeking to locate regulatory genes and to delimit the boundaries of protein-encoding genes.)

3. ***Understanding gene structure***: A final group of research programs, intertwined with (but partly independent of) the second group concerned gene structure. In particular, especially in molecular genetics, it sought to understand how the features of genes correlated with the phenotypes they produced, that is how gene structure shed light on gene action or function.

Notice that the question about what genes are made of does not necessarily answer the question of gene structure. The structure question asks not only what genes are made of but also how they can store and transmit some kind of information (for information in biology see Marcos and Arp, this volume) and the ways in which they can (and cannot) determine the traits of organisms. Once the question of how genes store information (and what sort of information they store) was solved, the structure question became more prominent and then was greatly amplified by the discoveries of split genes, promoters, enhancers,[2] and other sorts of 'elements' that modify the likelihood of readout, the stopping point of readout, the speed of readout, the combinations of genetic material actually read out, etc. Again, the question about where genes are did not end with the molecular era. The questions of where regulatory genes are located and how to delimit the boundaries of genes, including protein encoding genes, are still open.[3]

One way of characterizing the switch from Mendelian to molecular genetics is that with molecular genetics one *could* (though one did not have to) switch from working 'down' from the phenotype to the gene to working 'up' or 'out' from the gene (or the genetic material) to the phenotype. The fact that working this way is sometimes called 'reverse genetics' shows something about the limitations that genetics used to face, but no longer does thanks to molecularization.[4] One limitation

[2]Split genes are DNA sequences which consist of two kinds of sections: (a) those called exons, which are transcribed to corresponding RNA sections that are in turn translated into protein and (b) those called introns which are transcribed to RNA sections that are then excised and not translated; promoters are specific DNA sequences to which RNA polymerase binds; enhancers are specific DNA sequences to which proteins bind which facilitate the binding of RNA polymerase to the promoter (enhancers increase the transcription of genes – other DNA sequences have the opposite effect and are called silencers).

[3]We can generally distinguish between protein encoding genes, i.e. genes which are implicated in the synthesis of a particular protein molecule which is directly related to some trait or phenotype, and regulatory genes which are implicated in the synthesis of a particular RNA or protein molecule which in turn affects the expression of other genes.

[4]Working the other way round, i.e. from phenotype to gene or genetic material is described as forward genetics. In this case, one attempts to relate an observed phenotype to a DNA sequence. In reverse genetics, DNA sequences are usually altered in order to see which phenotypes are affected and in what way.

of molecular genetics is that the phenotypes that a gene – or genetic material – can deliver are all (more-or-less) molecular. In an important sense, there is no such thing as 'the gene for red eye' in drosophila – several to (probably) a few hundred genes are involved, including those encoding information required for producing the relevant red and the brown pigments, but also all those required to process the pigments in a coordinated way in just the right sets of cells for those pigments to yield eye color. Naming conventions are not so hard for immediate products, but are quite difficult for complex phenotypes. In Mendelian days, genes were "the factor that makes a difference to X" (where X names a phenotypic trait) but in the molecular world they are chunks of DNA (or, exceptionally, RNA) that (in some normative sense) "normally" encode certain specific products (or types of products?) or cause certain kinds of changes in what is read out, or cause the reading out to proceed at a different rate, etc. And one chunk of DNA might belong to one (or more) distinct genes, not only thanks to frameshift[5] in cases of distinct readouts, but also because inside the introns for one gene there are regulatory or even protein-encoding genes, some of which affect the readout of the gene within which they are embedded or of other distinct genes.

These considerations yield two points that greatly affect the morals that should be drawn for education. First, all three sorts of programs are required to fill in an adequate account of the gene concept. Second, the findings about the molecular structure and functions of the genetic material point to different directions – directions that may, in the end, break the concept of the gene into pieces or leave us with a much altered concept of the gene. Thus, genes are *not* contiguous pieces of genetic material yielding one (and only one) product, they may be made of different kinds of material (RNA or DNA), they need not be on (standard eukaryotic or prokaryotic) chromosomes, they need not be the least unit of function (they often encode separable functional domains[6]), etc. Empirical findings have produced (sometimes nasty) surprises and have caused major reevaluations of previous claims about how to delimit genes. There is no end in sight to the process of obtaining findings that cause scientists to revise their account(s) of gene identity, to alter the ways in which they delimit genes, and to revise what they consider to be 'necessarily' true of genes. Tensions between the findings of the different programs continue to turn up and the problem of determining the relative importance of their findings for the 'proper' delimitation of genes is not likely to go away in the near future.

Recently, much new attention has been given to these issues (e.g., Beurton et al. 2000; Dietrich 2000; Griffiths and Neumann-Held 1999; Kay 2000; Keller 2000;

[5] Frameshift refers to the fact that different proteins can be produced by different readouts of the same DNA sequence. The reading frames are shifted by one or two nucleotides and thus yield entirely different amino acid sequences over the length of the genetic material in which their readouts overlap.

[6] Functional domains are segments of a protein (often encoded in single exons) that play a particular well-defined role in different contexts, e.g., attaching the protein to a membrane or facilitating the interaction of the protein with a specific signaling molecule. The fact that many proteins include exons encoding distinct functional domains and that those domains are separated by introns facilitates the evolutionary process by allowing the modular swapping, addition, or subtraction of pieces performing particular functions or subfunctions.

Morange 1996, 2000, 2001; Moss 2001, 2003; Neumann-Held 1999, 2001; Portin 2002; Sarkar 1998; Snyder and Gerstein 2003; Waters 2000). In this chapter we describe various gene concepts proposed since the early twentieth century and the relevant problems in accurately defining what a gene is. Our conclusions are based on contemporary findings arising from the impact of evolutionary, developmental, genetic and medical research on the delimitation of genes and on the consequences of gene expression, plus some issues concerning public communication. We conclude that the most appropriate way of describing current genetic findings severely limits and circumscribes the use of locutions that enhance intuitive notions of genetic determinism. On the basis of these considerations, we suggest that the more inclusive concept of "genetic material" should replace the notion of the "gene" in general education about the findings of molecular genetics and allied disciplines and that it can do so effectively.[7]

2 The Gene Concept of Mendelian Genetics

The main elements of the classical chromosomal theory of the gene were fairly well established with the publication of *The Mechanism of Mendelian Heredity* by T.H. Morgan and his coworkers in 1915 (Morgan et al. 1915). According to this theory, the term 'gene' refers to a segment of a chromosome which, when activated or deactivated, performs a certain function or has a characteristic effect. But how much of a chromosome? And what functions or effects? Much of the effort that went into mapping genes may be viewed as an attempt to answer the first question; much labor was expended on the determination of which part of which chromosome contained which genes. In the process, certain criteria were developed for telling one gene from another. According to one of these, if two mutations affecting the same phenotypic trait – say two eye color mutations – could be separated by recombination, then they belonged to separate genes; if they could not be so separated, then they belonged to the same gene, and they were counted as alternative forms (alleles) of the same genetic locus.[8]

This way of individuating genes was proposed by Sturtevant (Sturtevant 1913a, b), who suggested that two closely linked eye color mutations (called 'white' and 'eosin') that Morgan and Bridges had been unable to separate in an experiment using 150,000 flies (Morgan and Bridges 1913) should be considered to be two alternative abnormal alleles of a single gene at a specified locus on the X chromosome. Now the more closely two genes are linked, the more difficult it is to separate them by recombination, and the larger the number of flies that must be used to execute the test. Thus, it should be no surprise that such claims are sometimes wrong and

[7]Some sections of this chapter draw in part on chapters 7 and 9 of Burian (2005).

[8]In sexual reproduction, recombination (exchange of segments of some specific length at matching loci by means of a mechanism called 'crossing over') occurs between homologous chromosomes which pair during meiosis, yielding chromosomes that have partially maternal and partially paternal DNA.

that it was established many years later, in this very case, that one can separate the two genes in question if one performs a truly gigantic recombination experiment (cf. Carlson 1966, p. 64; Kitcher 1982, p. 351).[9]

Consider the problem this creates when one asks what is referred to by subsequent uses of such terms as 'the gene for white eyes' or 'the eosin locus'. If one conforms to the usage established on the basis of Sturtevant's results, one refers to that portion of the chromosome that contains both the white and the eosin genes. But if one is working with the recombination criterion for theoretical purposes, one may refer, instead, to the smaller portion of the chromosome containing one, but not both of these genes. This is to say that two rather different segments of the chromosome belong to the reference potential (a term introduced by Kitcher 1978) of these phrases. Very often it makes no difference which portion of the chromosome one refers to, as they are, after all, virtually inseparable by ordinary techniques. But occasionally it may matter whether one purpose or the other dominates one's usage – conformity to established usage in order to accomplish coreference with other scientists or correct application of the criteria separating genes from one another. For a long time, the ambiguity was inescapably built into the mode of reference which was available in discussing these genes.

Indeed, at various stages in the history of genetics, it became a theoretical and practical necessity to distinguish between different gene concepts each of which picked out different segments of the chromosome or employed different criteria of identity for genes. For example, in the 1950s Seymour Benzer pointed out that many geneticists had assumed that the smallest unit of mutation with a distinct functional effect coincided with the smallest unit of recombination – and he performed some elegant experiments that showed that this claim is false (Benzer 1955, 1956, 1957). As a result, in some circumstances it became necessary to choose between the unit of function (which, for reasons that need not concern us, Benzer called the *cistron*), the unit of mutation (which he called the *muton*), and the unit of recombination (which he called the *recon*). This particular result showed that there had been hidden openness in the reference potential of the term (and the concept) 'gene' and that, in some arguments, though not in general, it was necessary to divide the reference of that term (concept) according to the separable modes of individuating genes.

The actual history is, of course, much richer than we have let on here, particularly when one pursues the story into the present, where one encounters transposable control elements, parasitic ("selfish") DNA, split genes with separately movable subunits, and so on. Thus, there are at least four ways in which the reference of a particular use of the term 'gene', or one of its cognates, might be specified (compare the discussion in Kitcher 1982, p. 342 ff.). Which one of these is relevant will depend on the dominant intention of the scientist and the context of the discussion.

[9]Carlson (1966, chap. 8) discusses the conceptual importance of Sturtevant's analysis which provided the key step in recognizing that mutation often involves alteration rather than loss of genes.

One such intention is conformity to conventional usage. Taking Sturtevant's early experimental results for granted, conformist usage would refer to the same segment of the X chromosome whether one spoke of the white or the eosin locus. Another, sometimes conflicting, intention is accuracy in the application of the extant criteria for identifying the relevant kinds or individuating the individuals of those kinds. When accuracy is the dominant intention, 'white' and 'eosin' refer to different segments of the chromosome. From this perspective, Sturtevant's 'mistake' expanded the reference potential of the term 'gene' by adding a compound chromosomal segment to the items potentially referred to by that term. In some, but only a few, contexts it proved terribly important to take the resultant long-unrecognized ambiguity of reference into account in order to understand the actual use of the relevant terms and to reconcile conflicts between competing descriptions of the outcomes of experiments. What is at stake here is the precise roles that one's theoretical presuppositions and accepted experimental results play in fixing the reference of one's terms. Although this discussion has not provided a general resolution of that difficult problem, it has given some indication of the proper apparatus to employ in carrying out case by case analyses.

The Benzer case illustrates another way in which reference may be fixed: once an ambiguity (such as that between 'cistron' and 'recon') becomes troublesome, it is sometimes necessary to stipulate as clearly as possible which of the available options one is taking as a way of specifying the reference of one's terms. Even at the risk of total failure to refer – which might happen if one's analysis is mistaken – one fixes one's reference to all and only those things which fit a certain theoretical description. The result is clarity, and when clarity is the dominant intention, reference is fixed by the relevant description. The sense of a term is determined by a description, and reference depends on whether or not anything, in fact, fits that description. Finally, one may operate with a dominant intention, which Kitcher (1978) calls naturalism, to wit, the intention to refer to the relevant effective natural kind occurring or operating in a certain situation or in a certain class of cases. It seems that one must have recourse to naturalism over and above conformity, accuracy, and clarity in order to put forth a successful account of the grounds on which Mendel, Bateson, Morgan, Benzer, and all the rest may be construed as employing concepts referring to the same thing – the gene.

3 Mendelian and Molecular Genetics

A considerable amount of laborious but fascinating experimental work during the period described as classical genetics, resulted in significant revision of Mendelian genetics. The 'pure' Mendelian concept of the gene, which was 'atheoretical' in the sense that it made no specific commitments about 'what a gene is' other than that it determined specific hereditary traits inherited in a specific pattern, was replaced by a series of improved successors which can be grouped under the label *transmission genetics.*[10] These successors were committed to the locations of genes and gradually became committed to restricted accounts of the material of which genes were composed – roughly the protein or nucleoprotein (or some portion thereof) contained at the locus on a chromosome within which the gene was located. This extended process, both in its theoretical and its empirical aspects helped prepare the way for the advent of molecular genetics. For present purposes, we may mark that advent of molecular genetics by the identification of DNA (and RNA) as the genetic material and the publication of the justly famous solution of the principal structure of DNA (Watson and Crick 1953).

It is useful to comment briefly about the relationship between Mendelian and molecular genetics. As the reference of the term 'gene' became more tightly specified during the development of transmission genetics, in a large range of central cases the concept of the gene became that of a minimal chromosomal segment (or perhaps some compound or material within that segment) performing a certain function or causing a certain effect. The relevant effect was known as the (primary) phenotype of the gene and was essential to the identification of the particular gene in question. Not surprisingly, a major part of the history of the gene, not addressed here (see Burian 2000, 2013), concerns the interplay between what one counts as genes and how one restricts or identifies the phenotypes which can be used to specify individual genes. But when all this is said and done, a great variety of phenotypes can legitimately be used to single out genes. In this context, the reference of the concept of the gene depended on the range of phenotypes investigated. Thus, geneticists interested in improving breeds of plants and animals identified genes with effects on desirable traits (such as adult weight for meat animals and flower shape for garden plants) that could not be biochemically characterized. Such genes were not acceptable to biochemical geneticists, who required a definite identification of the biochemical differences between different gene products before they admitted differences, even if they were inherited, to count as the effects of gene differences. In contrast, evolutionary geneticists came to accept changes in the nucleotide sequence as changes in genes even when they had no other phenotypic effect. These mutations, called 'neutral mutations' came to play a major role in the development of evolutionary genetic theory (see Dietrich this volume); among other things, nucleotide sequences changes that do not alter other phenotypes (and thus do not affect fitness) help to provide a 'molecular clock'.[11]

Let us expand this last point. Thanks to the advances made in molecular genetics, it is now possible to examine changes in the DNA (mutations) fairly directly. In some cases, at least, it is also possible to track the effects of those changes rather exactly. It is now well known that some changes in the DNA are silent. That is, they have no effect on any other aspect of the structure, the development, or the composition

[10] It is not possible to review the major advancements of that period here but the interested reader may refer to Carlson (1966).

[11] Not all parts of the genome turn over at a uniform rate, either within an organism or between organisms. For this reason, the calibration of molecular clocks is tricky and imperfect, but with care they have proved to be very powerful analytical tools.

of the organism. Effectively, such changes in the genetic material do not amount to changes in the function of any gene, though, when suitably located, within a locus identified with a gene, they do constitute changes in the structure or composition of the relevant gene. Other changes in the DNA do, of course, result in changes in other features of the organism, but some of them do so in ways which, arguably, are of no importance to its structure, development, or function. For example, some so-called point mutations result in the substitution of one amino acid for another in some particular protein manufactured in accordance with the information contained in the gene in question. Many such substitutions have very drastic effects. But some of them, so far as can be told, do not significantly alter the way the protein folds and do not alter its biological activity or function. In such cases, there are strong reasons for tolerating in perpetuity important ambiguities regarding the referents of the concept of the gene or regarding which concept of the gene is deployed in context.

The reason for this is that phenotypes at different levels are of concern for different purposes. Consider, for example, medical genetics. If one is concerned with phenylketonuria (PKU) and allied metabolic disorders, the phenotypes one deals with will range from gross morphological and behavioral traits down to what turns out to be the heart of the matter – enzyme structure and function (Burian 1981–1982, pp. 55–59; Paul 1995). For medical purposes, both silent changes in the DNA and those changes with no effect on enzyme function often are not counted as mutations, i.e., as relevant changes in the relevant gene. Even though these changes occur within that segment of DNA which constitutes the gene of interest, because they have no relevant functional effects, the gene counts as unchanged. The reason for this is clear: the concept of the gene is coordinate with the concept of the phenotype. And the phenotype of concern is not defined biochemically at the level of DNA, but (if it is defined biochemically at all) at the level of protein or via some functional attributes consequent on the biochemistry of the relevant proteins.[12]

It is important to recognize that there are legitimately different interests that lead us to deal with different sorts of phenotypes. Evolutionists, for example, may be interested in the rate of amino acid substitutions in proteins or of nucleotide substitutions in DNA. That is, the phenotypes they are concerned with might be defined by amino acid or even nucleotide sequences, not protein function. Accordingly, their definitions of the phenotype and of the gene may be discordant with those of the medical geneticist. And it is not a matter of right or wrong, but simply a matter of legitimately different interests and explanatory aims. There are large and important specialized sub-communities in biological research with legitimately different interests, which lead them to deal with legitimately different phenotypes. As the examples introduced in the last few paragraphs show, there are serious cases in which there is no question but that those differing phenotypes correspond to different concepts of the gene and different criteria for individuating genes. Especially

[12]There are other changes within the boundaries of genes that may count as mutations in medical genetics, namely changes in regulatory segments of the gene that alter whether, when, where, or with what intensity the gene is expressed, or, in some cases, which exons encoded by the gene are transcribed and translated.

important is that a certain stretch of genetic material may belong to distinct genes, depending on which gene concepts are employed and on the ways in which that genetic material is utilized in the cells in which it is found. Significant examples are provided by overlapping genes (e.g., those rare cases, found mainly in prokaryotes, in which different proteins are produced by reading out sequences that are 'frame-shifted'). Again, this time mainly in eukaryotes, there are other cases in which the genetic material is read in opposite directions, with some area of overlap (Tycowski et al. 1996).[13]

Work in molecular genetics may well show that some contemporary attempts at establishing gene concepts are ill-founded. Indeed, we believe that there are clear cases of this sort, for example in sociobiology (cf. Burian 1981–1982), but also, much more generally, cases in which certain gene concepts will simply have to be abandoned in light of some of the findings of molecular genetics. But molecular genetics is compatible with several well-founded gene concepts in spite of their discordance. There is a fact of the matter about the nucleotide sequence and the structure of DNA, but there is no single fact of the matter about what the gene is, or about which gene some genetic material that has multiple uses belongs to. Even though their concepts are discordant, the community of evolutionists concerned with the evolution of protein sequence and the community of medical geneticists working on metabolic disorders are both employing perfectly legitimate concepts of the gene. This provides strong, concrete support for the claim that the concept of the gene is open rather than closed with respect to both its reference potential and its reference.

A dangling thread provides a moral for biologists to consider. Stadler (1954) distinguished between the "operational" concept of the gene and the various "hypothetical" concepts of the gene. Stadler is right that proper use of an operational concept can ensure conformity and protect against the pernicious effects of certain theoretical errors. But, as the example of white and eosin genes shows, operational criteria (here, specifically for the individuation of genes) are themselves theory-laden and quite often erroneous. Furthermore, there is no single operational concept (or set of operational criteria) for the gene. In the end, as the brief discussion of molecular genetics in the last few paragraphs suggests, the best arbiter we have of the legitimacy of both operational and hypothetical concepts of the gene comes from molecular analysis. The latter, in turn, cannot be extricated from what Stadler would have considered a hypothetical concept, namely that of the structure of the DNA molecule. It follows that genetic concepts (and theoretical concepts generally) are inescapably open in the ways we have been describing.

[13]For many other puzzling cases in which geneticists employ concepts that yield discordant answers regarding whether a gene is present or how genes should be delimited or enumerated, see Stotz et al. (2004) and Griffiths and Stotz (2006).

4 Gene Concepts

Broadly speaking, there are two kinds of gene concepts. In this section we will offer a modest account of each. Both kinds are legitimate and understanding their interplay is crucial for understanding the history of genetics and a number of current issues in genetics. The first kind of concept makes sense of the conceptual continuities in the history of genetics, but yields concepts that are too generic or schematic to specify adequately what is referred to by '*the*' gene concept and allied concepts. Without such generic or schematic concept(s) of the gene, there would be no such discipline as genetics. However, without supplementation by more specific gene concepts, the schematic concepts do not suffice for specifying the referent of the term 'gene' – indeed, they do not specify well enough what genes are to ensure that the term refers successfully at all. In less philosophical language, these schematic concepts are impotent to specify exactly what we are talking about when we talk about genes.

The second kind of gene concept, in contrast, yields specific gene concepts, but does so at the price of conceptual discontinuity. If one restricts oneself to the series of discontinuous gene concepts, the findings of molecular genetics favor abandoning a univocal and specific concept of the gene altogether in favor of a pair of concepts – the concept of genetic material plus that of the expression of genetic information. This conceptual change allows molecular genetics to bypass the problem of discontinuity, currently solved by the use of schematic gene concepts. It also solves several other problems. As some other scholars have argued, the information content of the genetic material is extremely dependent on the cellular or subcellular context in which it is expressed.[14] This provides one of the rationales for suggesting that molecular biologists could abandon specific concepts of the gene, deploying, instead, concepts focusing on the continuous genetic material and the controls governing what is still called gene expression.

4.1 Schematic (i.e., Referentially Indefinite) Gene Concepts

Any science that seeks to locate hidden causes of some spatio-temporally delimited class of phenomena must use indefinite descriptions. These are descriptions that leave the exact referent of a term open. An example would be a Mendelian description like '*the factor, whatever it is, in the germ cells of these peas that causes them to produce plants that are much shorter than the tall plants produced from peas from the same pod*'. Such specifications are indefinite in not specifying what the causal factor in question is or even what category or sort of thing or process the factor is. Indefinite descriptions can genuinely refer to entities, as does the example

[14] The papers by Falk, Fogle, Gifford, Gilbert, Holmes, Rheinberger, and Schwartz in Beurton et al. (2000) are particularly relevant to this point.

we just gave when used in the right circumstances, but they can also be associated with seriously false descriptions or commitments. This is illustrated by the commitment, common before the middle of the twentieth century, that Mendelian factors (or genes) are composed of proteins. Mendelian genetics, taken strictly (i.e., without commitment to the localization of genes on chromosomes), used gene concepts based on very open-ended indefinite descriptions of exactly the form illustrated above.[15]

We call concepts like that of a gene thus understood *referentially indefinite causal (or functional) concepts*. In particular, the identification of a gene illustrated above is indefinite, but is accomplished in terms of a two part functional description. The first part specifies a difference in the phenotype of the organism bearing a gene (tall vs. short); the second requires a pattern of transmission of the factor(s) responsible for the change. One can distinguish different genes affecting, say, a plant's height or its flower color by their behavior in breeding experiments, by whether or not they 'Mendelize' or follow some recognizable variant of classic Mendelian patterns of inheritance (e.g., 3:1 or 9:3:3:1). Transmission genetics adds a third constraint on identifying genes, namely, their localization on a chromosome.

Here is a schematic formulation of a referentially indefinite functional gene concept: A gene for trait x is any stably inherited factor that causes an organism [or certain cells of the organism], given the rest of what it has in common with conspecifics, to have the potential for manifesting x, where x will (or can be made to) appear under the appropriate developmental plus environmental circumstances.[16] Distinct genes for x may exist and may be discriminated from each other either by specific differences in the phenotypes they cause or by demonstrating that they can be inherited independently of each other. This scheme instantiates Stadler's (1954)

[15]Johannsen's attempt at an atheoretical definition (Johannsen 1909) illustrates the point precisely. In Carlson's translation (Carlson 1966, pp. 20–22): "The word 'gene' is completely free from any hypotheses; it expresses only the evident fact that, in any case, many characteristics of the organism are specified in the gametes by means of special conditions, foundations, and determiners which are present in unique, separate, and thereby independent ways – in short, precisely what we wish to call genes." Genes are thus the differences, whatever they may be, between gametes that cause organisms to have the potential for revealing patent, independently-heritable, traits. Darden (1991) amplifies this point usefully in firmly separating Mendelian genetics as developed after the 'rediscovery' of 1900 from the chromosomal theory developed by the Morgan group and others.

[16]It is important to note that as we develop an account of the relevant causal chains, we may come to adjust what we count as a trait or, at least, what we count as a trait caused in a particular, stably inherited, manner. Think, for instance, of the multiplication of distinct disease entities, e.g., some of the cancers formerly believed to be a single disease, as we have learned to distinguish different underlying ways in which, e.g., the regulatory apparatus of certain types of cells can be disrupted so as to yield phenotypically similar outcomes. It is also important to recognize that the schematic definition may require specification in a great variety of ways. Thus the specification of 'modifier genes' (i.e., genes that have the function of altering the expression or function of other genes) and 'regulatory genes' may be relative to a specific gene or control pathway carried by some, but not all, conspecifics affecting their manifestation of the relevant traits affected by the modified gene. Again, transmission genetics requires the specification of a chromosomal location for the gene over and above the rest of its Mendelian characterization.

'operational gene concepts', indefinitely described. Two points are involved: first, for a long time there were competing theories about the material constitution of particular operationally defined genes, between which no decision was possible. Second, breeding procedures allowed workers to distinguish between distinct genes with otherwise identical phenotypic effects.

Such concepts need not imply any direct claims about what genes *are*, e.g., what they are made of; in general they do not specify the material or structure of the gene(s) in question and even in the best cases they do not, by themselves, pin down their full structure. Without independent knowledge of gene structure or composition, then, these concepts do not provide a fully adequate way of individuating genes. (That is why, in the absence of knowledge about the material composition of genes, Stadler was so pessimistic about our ability to resolve questions about 'the hypothetical gene'.) If adequate information about structure or composition of genes is not built into the gene concept or if it is not determined on independent grounds, it is not possible to count genes in a stably satisfactory way. This helps make sense of the fact that the chromosome theory – or something like it – was flatly needed to complement or complete Mendelian genetics. And it helps explain part of what is accomplished by the specification of genes as composed of DNA and RNA. Once such additional information is built into the concept of the gene, the theoretical presuppositions of gene concepts are radically strengthened – and, for most of the history of genetics, the presuppositions involved have been substantially false.[17]

One can view the history of genetics as involving, among other things, a series of attempts to obtain experimentally and conceptually sound ways of filling in indefinite descriptions of genes of the sort suggested above. What *should* count as a gene, given the indefinite starting point, depends on the specific traits or functions examined and the patterns of inheritance that they exhibit. It also depends on larger commitments, such as the means we employ to determine that something (e.g. a particular sequence of nucleotides), in context, is causally responsible for the trait differences in question. It depends, further, on the restrictions we place *in context* on the ascription of causal responsibility. In the century or so with which we are concerned, it has been at various times stoutly affirmed and stoutly denied that in order to count as a gene an entity had to be on, or to be a part of, a chromosome, or composed of protein, or composed of nucleic acid, and so on. In general, there is no adequate way of telling when such claims were intended as conceptual and when they were intended as factual claims. For this and other reasons, to make sense of the history of genetics we need to understand that *when such commitments had conceptual force, there was always a pathway of retreat open.* The underlying concepts to which people retreated when necessary were referentially indefinite functional concepts.

[17]This claim is, of course, contentious, but we believe it is correct. Consider the sorts of substantially false conceptual commitments that have commonly been made: genes are discrete particles, genes are composed of proteins, they are located only on chromosomes, they are linearly contiguous, they are non-overlapping, etc. Note that our claim that such commitments of detail have been built into gene concepts and are substantially false, does not imply that genetics is based on fundamental mistakes.

It should be clear that indefinite descriptions of genes, even when conjoined with massive sets of experimental results, are not sufficient to specify exactly what terms like 'gene' or 'gene for x' and their cognates refer to. One thing that is often meant by a (or 'the') theory of the gene is the theory-based specification of what it is that goes into individuating genes beyond the indefinite descriptions plus sheer experimental findings. A great deal is involved here. Among the things that should be included are abstract principles for the delimitation of causes, the delimitation of the biological functions to be examined (cf. visible phenotypes vs. behaviors vs. protein structure), and commitments about the material composition, structure, or localization of genes that constrain the concept of a gene and the possible referents of that concept. To understand the historical continuities that make genetics into a discipline and give geneticists a series of problematics on which to work, it is necessary to recognize this role of referentially indefinite concepts, but also to recognize that referentially definite concepts (or, at least, referentially more definite concepts) are requisite for specifying *what genes are* and what is needed to develop means of testing the principal claims made about them – claims about how to individuate them, how they act, and so on. The need to answer such questions has had considerable impact on the character of theory in genetics. Indeed, the failure to develop globally satisfactory definite descriptions of genes is part of what moves us to suggest the need for conceptual reform in molecular biology.

4.2 Definite Gene Concepts

More specific concepts of the gene, though they may still allow further specification, are committal, at least to some degree, about the structure or the localization of genes. What is typically required is a mixed mode of identification in terms of both structure and function. When such definite concepts embody false presuppositions they may, if taken literally, turn out not to refer to anything (e.g., when they make the mistaken commitment that genes are composed of proteins) or they may apply to a subclass of the entities currently considered to be genes in molecular biology (as do those gene concepts that require genes to be composed of DNA, which miss the genes of RNA viruses and several other relatively obscure entities that utilize RNA as their genetic material).

It is always possible to retreat to a less definite description of genes and to constrain successful use of the terms in question so that they must refer to a causal factor contributing to the occurrence of a well specified phenomenon. Of course, in principle, they might then end up referring to an integron (see Rheinberger 2000), and not DNA or RNA as such at all. Thus, it is (nearly) always possible to retreat from false presuppositions so that it is clear that the claims of scientists who employed those presuppositions made good sense (see Burian 2005 chap. 7; see also Burian et al. 1996; Kitcher 1978, 1982). But it is also true that one must specify the substrates out of which genes are built and the structures that can count as relevant causes (and thus deserve to be identified as genes) in order to individuate genes among the thicket of factors contributing to the relevant functional state. Note that for this class of gene concepts the choice of a phenotype is crucial in determining what counts as a gene; when the phenotype is an amino acid sequence, genes will be individuated differently than when the phenotype is something like the suppression of the expression of certain other genes. And it will continue to be the case that biologists with different interests will seek genes for phenotypes of different sorts. Thus, one cannot escape the recognition that there are sharp discontinuities in the history of genetics – discontinuities that cannot be bridged directly ('genes must be composed of protein' vs. 'genes must be composed of nucleic acids'). Nonetheless, such differences can be bridged via a retreat to less definite descriptions.

Once this point is granted, it is clear that the findings of molecular biology, some of which we allude to briefly in the next section, are readily interpreted as calling into question whether genes are particulate without preventing those of us who deny that they are from referring to the same things that our forefathers in Morgan's and Bateson's groups did when they used terminology committed to particulate genes and dynamic equilibria respectively. Indeed, given our treatment of concepts, the findings of molecular biology allow us to deny that the terminology of genes is well-defined and that it picks out a well-delimited group of entities. Given the range of functions for which we seek genes, one may even doubt whether all the gene-like causes are restricted to nucleic acids (cf. prions). But let us set that issue aside so that we may deal with the question whether we have a good way of settling which parts of which DNA and RNA molecules ought to be considered to be genes in light of contemporary knowledge. To this question there seems to be no systematically satisfactory answer. The best answer in a given case depends on our purposes and on the schemes of classification we employ, both of the functions that may be caused genetically and of nucleic acid molecules and their parts.

5 Continuities in the Genetic Material or Why It Is Impossible to Structurally Individuate Genes

Within rather broad limits, we are free to use terminology as we choose. We should, of course, be clear about our usage in order to avoid the confusion that results from using preempted terms in ways that conflict with common usage. The term 'gene' in molecular biology is a genuine accordion term – its expansion and contraction have caused a great deal of semantic quibbling. But the arguments involved are sometimes substantive, for they turn on the inclusion or exclusion of a number of genetic functions performed by nucleic acids that do not fit any of the standard structural constraints on genes. Underlying the different terminologies are serious disagreements about the status of parts of nucleic acid molecules that behave or are treated in different ways in different cellular contexts and at different phases of ontogeny. Here, for example, is one of the broadest gene definitions (specifically, of eukaryotic genes) in the literature:

> We define a [eukaryotic] gene as a combination of DNA segments that together comprise an expressible unit, a unit that results in the formation of a specific functional gene product that may be either an RNA molecule or a polypeptide. The DNA segments that define the gene include the following:
>
> 1. The transcription unit refers to the contiguous stretch of DNA that encodes the sequence in the primary transcript; this includes (a) the coding sequence of either the mature RNA or protein product, (b) the introns, and (c) the 5′ leader and 3′ trailer sequences that appear in mature mRNAs as well as the spacer sequences that are removed during the processing of primary transcripts of RNA coding genes.
> 2. The minimal sequences needed to initiate correct transcription (the promoter) and to create the proper 3′ terminus of the mature RNA.
> 3. The sequence elements that regulate the rate of transcription initiation: this includes sequences responsible for the inducibility and repression of transcription and the cell, tissue, and temporal specificity of transcription. These regions are so varied in their structure, position, and function as to defy a simple inclusive name. Among them are enhancers and silencers, sequences that influence transcription initiation from a distance irrespective of their orientation relative to the transcription start site (Singer and Berg 1991, pp. 461–462, see also pp. 435 ff. and 457 ff.).[18]

This definition includes a great deal that others would exclude. A more orthodox definition, like that of Goodenough and Levine (1974, p. 291), would restrict the gene to those nucleotides which, "when transcribed, will produce a biologically active nucleic acid," thus excluding promoter sites, enhancers, silencers, introns, and the like. But no matter: on either definition most eukaryotic genes are discontinuous stretches of continuous DNA, since introns are excised from biologically active protein-encoding RNAs. Worse yet, in many eukaryotes and quite a few prokaryotes, chain termination is dependent on physiological circumstances and/or is developmentally regulated. This means that the size of a gene – or what parts of the DNA of a multigene family function as genes rather than counting as pseudogenes – depends on physiological circumstances or developmental stage. Even worse are the cases in which RNA is edited (i.e., systematically altered by specific biological processes after it has been transcribed from a DNA source) or DNA encoding immune system proteins is systematically 'shuffled' during development in different cell lines, thus making a greater variety of immune proteins than were originally encoded in the zygote. Such shuffling of the genetic material means that the original genetic contents of a zygote (i.e., a fertilized egg) are not preserved in certain somatic cell lineages. The dynamism of the genome is of great importance for the definitional and conceptual issues that belong at the heart of this chapter.[19]

It might be thought that this argument can easily be dismissed as a trivial semantic argument about how we should define terms, rather than as an argument bearing on

[18]Chapters 11 and 12 in Burian (2005) contain illustrations that will help the reader unfamiliar with the technical terminology to understand Singer and Berg's text.

[19]For reviews providing some details and amplifying on the importance of such issues, see Fogle (2000) and Portin (2002).

how we should think about genes in light of the findings of molecular biology. The next argument, however, which focuses on protein-encoding genes, shows that the issues just raised are not merely semantic in this pejorative sense, but have significant impact on our interpretation of the history of genetics and impinge on how biologists and lay people should be thinking at this point.

The argument concerns the continuity of the genetic material: It yields an important intermediate conclusion: An examination of intrinsic features of RNA or DNA is not sufficient to delimit precisely which parts of these molecules should count as protein-encoding genes because of the context dependence of the "read-out" produced from a sequence of nucleotides.[20] It takes an enormous amount of biological machinery for genes to be expressed; exactly which parts of the genome are processed depends on specific settings and structure of that machinery. Again, a huge number of processing steps affect the times and places at which informational molecules yield products as well as exactly what products they yield. It was known as early as 1987 that the translational apparatus alone requires some 200 macromolecules (Freifelder 1987, p. 367)! Corresponding to the richness and variability of the mechanisms involved, is the richness of the alternative results (even at the molecular level) when a given stretch of nucleic acid is transcribed or enters into an interaction of some sort. The answer to the question *which stretches of nucleic acid should count as genes* depends not only on the functions and the sequence of nucleotides that we have chosen to examine, but also on the particular machinery present in particular cells or compartments within cells, for that is what determines which parts of the signal remain intact and are contiguously read out and what the molecular results of the network of interactions involved turns out to be.

As is generally known, there is cellular machinery that determines which stretches of DNA are accessible to RNA polymerases and where it is that the RNA polymerases get stopped or knocked off the DNA (both dependent, for a given stretch of DNA, on physiological conditions), and how the resulting RNA is processed – immediately in prokaryotes and before it can get through the nuclear membrane in eukaryotes. It is worth recalling at this point, that in eukaryotes, most genes are processed in such a way that the material corresponding to introns is snipped out of the RNA molecule before the transcript gets through the nuclear membrane. At least occasionally, some of the material thus snipped out is, in turn, translated to yield a functional polypeptide or is functional in some other way (Tycowski et al. 1996; Coelho et al. 2002), so that it is natural to talk of one gene embedded inside another.[21] There may still be further

[20] This is also the reason for which the amino acid sequence cannot be determined (or determined up to permutations) by an examination of the structure of DNA or mRNA molecules alone. In different cellular contexts (nucleus vs. mitochondria, some species of organisms vs. others), there are sometimes some regular differences in the transfer RNAs. Thus, in a few cases, the same codon in different contexts codes for a different amino acid or for a stop signal instead of an amino acid.

[21] A brief technical description of such a case is given by Singer and Berg (1991, pp. 705–706) for introns in the mitochondria of yeast.

post-transcriptional processing of mRNA,[22] and, at that, what precise polypeptide sequence the RNA yields is still a function of the tRNAs in the relevant cytoplasmic location. Further, post-translational processing of proteins is, at least in some cases, critical to whether or not the product that results in fact enters into a final product that plays a functional role.

Perhaps a schematic example will make the point clearer. Consider an ORF,[23] located by appropriate molecular techniques. Does the ORF mark the beginning of, or even delimit, a gene? The answer, insofar as there is one, depends on the physiological context, the alternative splicing and readout controls present in the relevant cell compartment (for the stop signals are different in mitochondria than in the nucleus), the tRNAs present in the immediate context and so on and on. Often enough, a single ORF begins a transcript that contains multiple genes.[24] Our conclusion is that even when one works at the molecular level, what counts as a gene is thoroughly context dependent.

An important effort to take context into account is Lenny Moss's *What Genes Can't Do* (Moss 2003).[25] Moss distinguishes sharply between two sorts of gene concepts, labeled *gene-P* and *gene-D*. The label *gene-P* is meant to capture the connection between preformationism and genes that determine a phenotype; thus, a gene-P is defined as a gene *for* a phenotype (i.e., one that is identified by its causal link to that phenotype) (Moss 2003, p. 45). In contrast, a *gene-D* (the 'D' indicates that the gene is interpreted as a developmental resource) is defined by its molecular sequence (i.e., intrinsically, without reference to what it produces). Moss rightly insists that a nucleotide sequence may enter into many different interactions and may be processed so that the products it yields have many different structures that occur in many different tissues. Similar things may be said for non-coding (regulatory) nucleotide sequences and the reactions that they affect. Accordingly, it is simply incorrect to identify molecular sequences in terms of particular effects. No gene-D is properly understood as a gene for X, where X stands for a single phenotype or a function; the effects of a gene-D depend on the biological context and (often) on the history of the organism. Hence, the effects of a gene-D are "*indeterminate* with respect to phenotype" (Moss 2003, p. 45).

---

[22] Alternative splicing, i.e. the production of different mature mRNAs from the same primary RNA transcript through differential excision of introns, is just one of many relevant post-transcriptional phenomena that are relevant here. Gilbert (2000) and Singer and Berg (1991, pp. 578) provide helpful accounts of alternative splicing and other technicalities discussed below. This phenomenon again demonstrates the impossibility of employing the intrinsic features of the DNA or RNA alone to determine which stretches of a DNA or RNA molecule produce "biologically active RNA." For further explanation of many of the issues discussed below, see Burian 2005 chap. 12.

[23] The abbreviation stands for "Open reading frame," which is the name for the nucleotide sequence that signals (in many but not all contexts) a place at which to initiate the readout of DNA.

[24] Chapter 5 of (Gilbert 2000), which covers differential gene expression, includes useful reviews of differential RNA processing (pp. 130–133) and of (contextually variable) translational and post-translational controls of the end products of the expression of nucleotides sequences (pp. 134–136).

[25] Waters (1990, 1994, 2000) offers a contrasting approach to this problem.

This point about nucleotide sequences and the indirectness of their relationship to phenotypes is entirely correct. But we are skeptical of Moss's deployment of the terminology of genes-D. The problem is how one delimits one gene-D from another. Not all nucleotide sequences should count as genes. Some short nucleotide sequences are repeated millions of times within the genome. Should each arbitrary length of such a sequence count as a distinct gene? For good reasons, even when one is working at the molecular level, it is often desirable to identify distinct nucleotide sequences as instances of the same gene – e.g., in numerous contexts in which the relation between a gene and amino acid sequences is at stake, synonymous substitutions are counted as alterations that do not change the identity of the gene, even at the molecular level. Moss would probably consider this a confusion of gene-P interpretations of the gene with gene-D interpretations of the gene. We consider it evidence that *even at the molecular level, functional criteria of delimitation are built into gene concepts*. The issues here obviously ramify far beyond this immediate, partly linguistic, partly conceptual point. Moss's insistence that we take seriously the idea of a sequence-defined or sequence-delimited concept of the gene is salutary. The issue is over the need to restrict sequence-based definitions with further (functional) criteria in order to save the gene concept from picking out any and all arbitrary sequences. In either case, the result is that *the context-dependence of the effects of nucleotide sequences entails that what a sequence-defined gene does cannot be understood except by placing it in the context of the higher order organization of the particular organisms or subcellular units in which it is located and in the particular environments in which those organisms live.*[26] This argument provides a synopsis of the one strand of support for the claim that the science of genetics has argued itself out of the most stringent versions of reductionism.

6 From the Reductionism of Genes to the Complexities of the Genetic Material

We have not yet given a working definition of the genetic material. It is now incumbent on us to do so. Genetic material is any material that provides the information utilized in constructing (other) materials within the same cell or organism with specific biological functions. In contemporary genetics and molecular biology, the use of 'information' in this context is very special and widely misunderstood. Information in this special sense is always *sequence information* for constructing sequences in new (potentially information-bearing) molecules; so far as is now known, the constructed sequences are either sequences

[26]It should be noted that (to the best of current knowledge) sequences of nucleotides in plasmids, viruses, mitochondria, and plastids do not replicate or reproduce outside of the laboratory except in cellular contexts. Thus, for practical purposes, the only contexts in which these entities have functional genes are when they are in some sort of cellular context.

of nucleotides in a nucleic acid or sequences of amino acids in a polypeptide in accordance with the (contextually specific) genetic code. In principle, other materials might have similar information-bearing functions, but the only known materials with such functions are the nucleic acids DNA and RNA. This special sense of information was first proposed by Francis Crick (1958).[27]

The key point is that sequence information goes from nucleic acids to proteins. Proteins do alter nucleic acids, e.g., by, by annealing nucleic acids, cutting pieces out of them, or providing machinery for occasional substitutions of one nucleotide for another, but proteins do *not*, as such, contain or provide sequence information for determining sequences of nucleotides or sequences of other amino acids. Thus, proteins can cause alterations of nucleotide sequences, but they do not contain information for constructing specific sequences. If one understands 'information' as sequence information, it becomes clear (and remains correct) that genetics has captured an extraordinary feature of nucleic acids that is not matched by proteins. This justifies the distinction between hereditary traits that are genetic (i.e., specified by genetic information) and hereditary features that are not genetic (i.e., that are specified in other ways). But it also restricts the phenotypes that count as genetic and justifies the claim that there are also a variety of forms of non-genetic or extra-genetic inheritance, i.e., of epigenetic inheritance. Thus the pigment molecules that produce the red color of drosophila eyes are specified genetically, but that it is the *eyes* that are red is specified by developmental controls that are (in part, at least) epigenetic, for those controls determine when and where the two relevant pigments are distributed and in what proportions. The current technical definition of epigenetic inheritance is (regular, lawlike or mechanistically explained) inheritance of specific states or changes of state that do not depend only on nucleotide sequences or changes of nucleotide sequence. Cellular inheritance and organismal inheritance of methylation of nucleotides or histones, or of chromosome conformation (e.g., via histone modifications) are the easiest examples of epigenetic inheritance, but more contentious examples include behaviors of mothers (for example, grooming of rat pups that causes heritable methylation that, in turn, causes many inherited effects, including increased likelihood of grooming behavior) (on this topic, see Uller this volume; Jablonka and Lamb 2005; Jablonka and Raz 2009).

[27]This is Crick's version of the 'Central Dogma of Genetics'. Watson's, version was different: he interpreted the Central Dogma as claiming that information flows from DNA to RNA to protein, and not backward. Watson's formulation was especially influential thanks to the importance of his textbook *The Molecular Biology of the Gene* (Watson 1965), but it was mistaken and caused a lot of the resistance to reverse transcription (see Strasser 2006; Morange 2008; Olby 1972, 1975). Unfortunately, although in practice geneticists often use the term information in accordance with Crick's account, they often present Watson's account of the Central Dogma when they discuss it, a situation that has caused much misunderstanding, even by geneticists, of their discourse about genetic information. The confusion has been exacerbated by confusion between claims about sequence information with claims about information, in some more general sense, e.g., as it is used in cybernetics or information theory.

Not all of the genetic material is (or should be) counted as belonging to specific genes. Accordingly, conceptually speaking, what counts in classifying some genetic material as belonging to a gene depends on the genetic material in question having an effect on what is counted as a phenotype in at least some circumstances. What one may choose as a phenotype, however, is somewhat constrained by what we learn about the genetic material. Factually speaking, the delimitation of genes at the molecular level depends on the entire system for processing DNA and RNA, the translation of processed RNA into protein or into regulatory products, and also the post-transcriptional processing of those products and the post-translational processing of proteins. As a result, the task of delimiting genes contains an inextricable mixture of conceptual and factual elements. To be sure, the 'lowest' level', i.e., the molecular level, though it is most distant from naive observation, brings the argument closer to a context-fixed factual basis than the others. But the price for this is that one must deal with the interactions of all of the relevant macromolecules and regulatory elements within their physiological setting to tease out the more narrowly delimited specific definitions of genes and gene functions. This has the consequence that precise definitions of genes must be abandoned, for there are simply too many kinds of genes, delimited in too many ways for a single characterization to work. Taken in combination, these arguments provide powerful support for the principal contention of this chapter, namely that when we reach full molecular detail we are better off to place careful limits on specific gene concepts.

Since the 1980s, with the advent of genomics, high throughput databases, and the many other technological and experimental advances fostered by the Human Genome project, serious work in molecular phylogeny and comparative and technical studies at the molecular level have brought about a revolution at a foundational level of our understanding of genes, genetics, and genomes. Molecular and bioinformatic tools have enforced reorganization of our knowledge and what we used to consider as solidly established findings about genes became contextually limited or approximate truths. This revolution is largely quiet; although a lot of the details are familiar, they have seemed fairly particular and the large-scale changes that they will almost surely bring in their wake have remained largely undigested and have not yet been assimilated into wider public consciousness. This revolution is ignored in the medical world (at least as understood by the larger public) to the extent that the Holy Grail that is (all too often) sought there is "the gene for". In fact, what is typical, and what quite a few researchers have cottoned to, is that researchers seek to identify key steps in various physiological process that are controlled by some product of some gene in rather particular contexts. Worse yet, it is also widely recognized that in most interesting cases, there are several networks of various sorts (gene networks, protein networks, physiological process networks, and networks that have nodes of all these sorts of entities) that intersect in controlling or contributing to the disease or processes of medical interest (Goh et al. 2007).

Most eukaryotic genes do not have very well defined boundaries. If one looks at the standard definitions of a protein encoding gene, what one gets back is a mixed bag that amounts to this: what counts as a gene is the largest unit that corresponds to a member of a family of proteins (such as one of the myosins), and can be read out in various ways, differentially in different tissues, to yield different members of that family. In general, this is NOT the largest unit that can be read out from the same start site since about 0.5 % of readouts do not end at the standard stop signals but contain material from two, three, four, or more conventionally delimited genes, so care must be applied in delimiting what one counts as the same family of proteins, and hence genes.[28] For example, some definitions require overlap of at least two exons for belonging to the same gene in cases of multiple exon readouts from the same start site (a condition that is violated by some genes that have lots of short exons with complex combinatorics, contributing to medium sized proteins that biochemists consider to belong to the same gene family). Furthermore, as soon as one goes beyond protein encoding genes to try to take account of active sites that include such widely scattered entities as promoters, enhancers, silencers and other regulatory units that need to interact to create some compound proteins and have sometimes been considered to be part of protein encoding genes, one loses contiguity and other similar criteria that were retained by such definitions as the one we just provided. And if one is asking for gene counts, how many regulatory genes are there? There is no stock answer, as there is thanks to the convention that we just cited for explaining how only ~20,000 genes can yield the more than 200,000 proteins in our bodies.

Consequently, genetics education must aim to accommodate effectively and accurately current knowledge, advancements and practices. Perhaps we should move toward a process rather than a material entity account of genes to try to cope with the complex developments that this yields. But it is clear that no neat single definition will work and that authority in developing adequate answers as to what we do and should count as genes is distributed among experts from a variety of different disciplines who ask key questions and are armed with close knowledge of cases in which we hold genes responsible for various outcomes or states of affairs or processes. To get across the excitement of all the material on the forefront AND the need to have command of an enormous range of experimental facts AND the need to bend to shared and distributed authority is a hugely important job, of major importance for education and the public understanding of genetics. To overcome such problems we propose an instrumental concept of genetic material in the next section, a concept that could replace the various gene concepts in substantial parts of our textbooks and in the classroom (see Keller 2010, p. 77, for a similar suggestion for replacing talk of genes with talk of DNA).

[28] A recent study (Djebali et al. 2012) found that a shocking 85 % of 492 protein-encoding transcripts for human chromosomes 21 and 22 were chimeric in the sense that they contained transcripts from more than one gene, using the standard boundary definitions for genes. However, as the authors warn, the technical tools involved may yield a significant number of false positives and the proportion of these transcripts that are translated and actually yield proteins is not yet known. These authors, like some others, suggest that the appropriate functional units that should be investigated are *transcripts* rather than *genes*, a stance that would be justified if, as they argue is likely, a substantial proportion of the chimeric transcripts they examined are functional.

7 Towards an Inclusive Concept of Genetic Material to Replace the Concept of Gene in Genetics Education

People learn about genetics in formal (school), informal (science museums), and non-formal (mass media) ways. One of the aims of formal and informal science education is to educate scientifically literate citizens. One can distinguish between two types of science/scientific literacy (Roberts 2007). The first refers to issues within science and it is related to the content of science taught in classrooms. In the case of genetics this should be knowledge about DNA, genes, chromosomes, patterns of inheritance etc. The second is related to questions that students may encounter as citizens, e.g. about the implications of scientific knowledge for society. In the case of genetics, this should be knowledge about e.g. the ethical questions related to genetic testing or to disclosing genomic information about individuals. Thus, future citizens, literate about science, should have a sufficient level of updated and accurate knowledge about the content of science in order to be able to make informed decisions about socio-ethical issues.[29]

For instance, in order to make an informed decision about whether a couple, both of whom are heterozygotes for β-thalassemia, should go through preimplantation genetic diagnosis in order to ensure that their children will not have the disease, they should be aware that they would have to go through an in vitro fertilization procedure and that some healthy and potentially viable embryos might not be eventually transferred to the mother. They should also know that, in case they carry different defective alleles of the β-gene (or, better, of the DNA sequence that is implicated in the production of β-globin peptide chains), their child would be a compound heterozygote who might or might not suffer from the disease. To achieve this, people need to realize the enormous complexity of development, as well as that phenotypes are not simply "controlled" by genes.

However, this is not currently the case. It seems that the contemporary presentation of genetics in schools eventually teaches students that there are genes that "control" or "code for" individual properties. Important phenomena such as epistasis, pleiotropy, plasticity, epigenetics, gene regulation, gene overlap, alternative splicing, antisense reading, etc. (Barnes and Dupré 2008; Stern 2011) are overlooked or at best treated as exceptions. The contemporary presentation of genetics in biology textbooks does not take into account the reality and complexities of development, as a recent study has revealed (Gericke et al. 2012). Most interestingly, in a recent study of teachers' conceptions of genetic determinism in several countries, it was found that even biology teachers may hold strong views of genetic determinism (Castera and Clement 2012). The conclusions from these two studies should alert textbook authors, curriculum developers and science educators about the prevalence of outdated models that enhance mistaken notions of genetic determinism. If these

[29] Scientists should, and must, have a role in communicating contemporary knowledge about genetics to the public (see Reydon et al. 2012 for a relevant proposal, which also includes some of the arguments made here).

models are to remain in textbooks and if teachers are not sufficiently familiar with contemporary knowledge of genetics and development, it should be no surprise that people embrace a strong view of genetic determinism (Moore 2008, see also Moore this volume) or that students' writings reveal important misconceptions (Mills et al. 2008; Dougherty 2009).

One possible cause of this problem is the fact that Mendelian genetics is still what most people are taught at school. This is problematic in various ways (see Jamieson and Radick, this volume). Of course, Mendelian genetics still is a valuable heuristic tool and a useful starting point for teaching genetics. Indeed, the description of alleles that control specific characteristics is comprehensible and even middle school students can easily perform simple crosses using so-called Punnett squares. However, if genetics education does not also accommodate recent knowledge about genetics, students will not be able to understand the contemporary issues. With the increasing availability of direct-to-consumer genetic tests for several types of disease, it is important to provide students with the tools to understand what these tests can and cannot reveal. Perhaps the most crucial issue is to help them understand that the intuitive idea of genetic determinism is simply wrong. We hope to have shown not only that the idea of "genes for" is misleading, but also that genes, as such, are not generally distinct units, except when the context is adequately specified (which is seldom the case), with respect to particular phenotypes!

Perhaps the most crucial, neglected component for understanding genetics is that of development. People often do not realize that genes can do nothing outside their cellular contexts and that even evolution proceeds not primarily due to changes in protein coding genes but rather due to changes in regulatory sequences that control the expression of these genes (Stern 2011; Bateson and Gluckman 2011). Genetics education should make clear that the contribution of genes cannot strictly be distinguished from the contribution of their cellular and external environment. Although genes make a partial contribution to a final outcome, they can do nothing on their own. Consequently, only comparisons are possible. To illustrate this, Keller (2010) uses the metaphor of a drummer and his/her drums. There is no point in asking whether the sound produced is more due to the drummer or due to the drums. What would make sense would only be to compare two drummers playing with the same drums, or the same drummer playing with different drums. It is only then that distinguishing between the contributions of the drummer and the drums would make sense. Similarly, distinguishing between the contribution of someone's genes and someone's environment – food, lifestyle, etc. – generally makes sense when comparing genetic differences in persons with highly similar environments, or environmental differences for persons with highly similar genetic makeup.

This is not what one finds even in otherwise excellent textbooks. In a recently published biology textbook (Walpole et al. 2011), the definition of gene given is the following: "A gene is a particular section of a DNA strand that, when transcribed and translated, forms a specific polypeptide" (p. 67). In the glossary of the same book gene is defined as: "a heritable factor that controls a specific characteristic" (p. 586). This is an excellent example of a referentially indefinite gene concept. The two definitions are not entirely consistent with each other. The definition in the main text of the book is a definite one, which is explicit about the composition (DNA) and the function (forming a polypeptide) of genes. In contrast, the definition in the glossary is an indefinite one that is not explicit about the composition (the factor could be any kind of molecule) or the function (the characteristic is certainly a phenotypic one but it could be either an enzyme or a macroscopic feature such as eye color). The definition in the glossary is thus a less definite one as it does not identify genes with DNA or the synthesis of a particular peptide. As such, it includes both epigenetic and genetic causes of heredity and it ties genes to functions in a different way than the first definition. For example, it would include as a gene a stretch of DNA that makes a regulatory RNA that blocks translation of the message for a key protein, thus regulating the functions of that protein. Since the key to this sort of control is not a protein, the main definition would not acknowledge this sort of gene. Note that an acetylated histone that causes the conformation of a chromosomal region to make certain DNA inaccessible and thus prevents a key gene from being expressed in an embryo, usually considered an epigene or an epigenetic mark on the histone, would count as a gene on the glossary definition. A third definition, set out in a box next to main text, seems to be an attempt to encompass both these definitions, but it rather makes things more complicated: "Gene [is] a heritable factor that controls a specific characteristic, or a section of DNA that codes for the formation of a polypeptide" (p. 68).

How is the gene concept used in the book? Definite descriptions seem to predominate: "Hemophilia is a condition in which the blood of an affected person does not clot normally. It is a sex-linked condition because the genes controlling the production of the blood-clotting protein factor VIII are on the X chromosome." (p. 82). Genes are composed of DNA and control particular characteristics, in this case the production of a protein that is involved in blood clotting. What is worse, the book gives the impression that genes are all powerful. Here is an example: "The fertilised egg of any organism contains all the information needed for developing that single cell into a complex organism consisting of many different types of cell. This information is all within the genes, inherited from the maternal and paternal DNA as fine threads called chromosomes" (p. 16). This definition refers to the robustness of development but is absolutely blind to developmental plasticity. It gives the impression that literally all information about development is included in genes and ignores the fact that information is not, as such, a property of DNA. Information is a kind of relationship between DNA and the translational machinery of the cell as influenced by relationships between cells and by environmental factors.

Another textbook (Sadava et al. 2011) poses similar problems. 'Gene' is defined in the glossary as: "A unit of heredity. Used here as the unit of genetic function which carries the information for a single polypeptide or RNA" (p.G-12). Although the idea of information is identified with the particular unit, the relational character of the gene-as-information is not reflected in the definition of the gene as a unit. Furthermore, the concept of the gene is not identified with DNA or any other molecule. However, this is not the case in the main text of the book where the concept of gene is more definite and actually identified with DNA: "Genes are specific segments of

DNA encoding the information the cell uses to make proteins" (p. 6); "The sequences of DNA that encode specific proteins are transcribed into RNA and are called genes" (p. 64); "Genes are now known to be regions of the DNA molecules in chromosomes. More specifically, a gene is a sequence of DNA that resides at a particular site on a chromosome, called a locus (plural loci). Genes are expressed in the phenotype mostly as proteins with particular functions, such as enzymes" (p. 242). These characterizations do not recognize the context-dependence of the boundaries that are read out to make proteins. Nor do they acknowledge that there are RNA genes or that genes may be located on plasmids and other non-chromosomal molecules. Information is once more not presented as a relation.

We suggest that, given the analyses of the previous sections, these definitions of genes are problematic. Therefore, we propose that genetics education should utilize the wider and more inclusive concept of the genetic material rather than the concept of the "gene for". One could then base the discussion on the evolution of genetic material, the interaction between the genetic material and its intracellular or extracellular context, and the expression of genetic material to produce RNA, proteins or other molecules, and introduce the distinction between genetic and epigenetic inheritance. Biology education ought not focus on DNA and genes and then make a leap to organisms and their phenotypes, overlooking the developmental processes that produce them. There is more to biology than nucleotide sequences, as there is more to language than letter sequences. All cells in an organism contain the same genes (up to mutations acquired during the organism's lifetime[30]), but their expression is differentiated according to their environment and the regulatory apparatus in the cells of the organism. Epistatic and pleiotropic interactions also influence the phenotype. Thus, it is important for biology education to make clear that development is a complex process in which DNA is an important, but not the only, factor.

Based on all the above, we propose that the concept of the gene could be replaced by an instrumental concept of genetic material, explicitly linked to development. The resulting presentation would be more inclusive and more accurate, and could bypass the difficulties raised by the indefinite or functional and definite or structural gene concepts proposed so far as. The proposed concept:

- refers to particular macromolecules (DNA, but also RNA) which are related to the expression and inheritance of traits
- does not refer to particular functions since all functional parts of the genetic material may be implicated in various phenomena and phenotypes

[30]In an important footnote, Peter Godfrey-Smith sketches a calculation that shows that after 40 cell divisions any two cells of an adult human would, on average, differ by about 144 point mutations. Most of these mutations would be in non-coding regions and have no phenotypic effects. Nonetheless, it is important to recognize that the casual assumption that most human cells within a human body are genetically identical is simply false (Godfrey-Smith 2009, fn. 9, pp. 82–83). Additionally, within a human body there are about ten times more bacterial and archaeal cells than human cells, some of which are required for proper development (Gilbert and Epel 2009), further undermining simplistic accounts of the genetic uniformity of the cells within our bodies that are essential for our normal functioning.

- does not refer to contiguous DNA sequences because functional units may encompass different parts of the genome

In addition, the proposed concept takes into account the three distinct research programs of Mendelian and molecular genetics, and provides a narrower description of their aims:

1. ***Structure***: The concept of genetic material refers to molecules which are by definition informational. Remember here that 'information' means sequence information and that it is misguiding to locate information in a particular molecule, DNA for instance, except in relation to a given cellular and extracellular context (see Marcos and Arp, this volume). Any molecule with similar informational relationships can be considered to be genetic material, but as far as is now known, only DNA and RNA qualify.
2. ***Function***: Instead of trying to characterize genes in terms of their functions and/or consequences, which was the source of descriptions of "genes for" (e.g., genes for eye color or height, for modifying the action of other genes, for altruism, etc.), we encourage recognition of the multifunctionality of the genetic material, both when one has isolated particular portions of that material and as a whole. This point applies to any kind of information-bearing nucleic acid that directly affects or is implicated at some phenotype at the molecular, cellular or organismal level and provides a convenient way to take into account the contextual dependence of the functions assigned to the nodes of the complex genetic networks that are related to several types of disease.
3. ***Composition and localization***: Instead of seeking to provide precise boundaries for regulatory and protein-encoding genes, we recommend careful examination of the multiple ways in which informational nucleic acids, wherever they are located, relate to other molecules. This procedure reveals the polyfunctionality of the genetic material and the fluidity of its functional boundaries.

To sum up, we propose that the concept of "gene" be used only heuristically in educational books and materials and, for many purposes, that it be replaced by the concept of genetic material which is more inclusive in terms of composition (as it clearly includes RNA as well as DNA), that the localization of genetic material be determined by its sequence-informational function (which takes into account potential multiple effects in multiple contexts) and that the structure of the genetic material be treated as fluid (since whether a molecule, or part of a molecule, is informational or not depends on its interactions and not solely on its molecular structure).

In this sense, we might replace the definition of gene as unit of heredity or a section of DNA which controls a particular polypeptide or phenotypic feature with a definition of genetic material like the following:

> Genetic material: any nucleic acid [composition] in the cell [localization] that interacts with other cellular components and transmits a specific message determining the sequence of other molecules [structure] and thus results in particular, but often quite variable, outcomes inside or outside the cell [function]. These nucleic acids are (usually) reliably copied and maintained from generation to generation, preserving their structure and resulting in the

same functions in similar environments (robustness), though with a range of variation in functions and consequences that depends on cellular and environmental conditions (plasticity). The functions of particular portions of the genetic material may affect or be implicated in cellular processes with local (cellular) or extended (organismal and even environmental) impact; this allows the assignment of fitnesses to particular differences in the genetic material.

Put more simply:

Genetic material: any nucleic acid with the propensity to be inherited and to interact with other cellular components as a source of sequence information, eventually affecting or being implicated in cellular processes with local or extended impact.

This definition is more accurate and inclusive than the typical definitions of the gene. It allows a clear distinction between genetic and epigenetic inheritance, which is not feasible with many standard textbook definitions of the gene. It would free textbooks from referring to "gene(s) for" particular characteristics or diseases but would allow them to refer to particular parts of the genetic material that, in identified contexts, interact with each other and with other cellular components to affect the production of molecular, cellular, or organismic characteristics or to increase the susceptibility of affected individuals to acquire certain traits or diseases.

Let us illustrate why this conception of genetic material is more accurate than the traditional conception of genes. Beta-thalassemia is considered a monogenic disease because various specific mutations at a single region of chromosome 11 affect the production of β-globin molecules. Hemoglobin is produced by the formation of a molecule containing two β-globin and two α-globin molecules. The more defective the β-globin allele, the fewer β-globin molecules are produced or the less well the β-globin molecules trap oxygen when complexed into hemoglobin and consequently the worse the disease is. It is not easy to define the precise functional effects because there are single mutations that can bring about the disease even in heterozygotes, while homozygotes for other mutations have less severe effects. Thus, alleles at the β-hemoglobin locus are evaluated not just by their molecularly specific effects but also by their functional contribution to defects measured by their relevance to health. Familial hypercholesterolemia is also considered a genetic disease, and it is due to a mutation that affects the structure of the LDL-receptor in the liver. However, people who possess the mutation may have milder problems if they follow a proper diet and if they regularly take medication (e.g. statins). In some cases, the problem may totally disappear, so again the mutation is not by itself sufficient to cause anything at the phenotypic level (or if it causes something, it can eventually be reversed). Finally, things are even more complex in cancer which generally is a genetic but not an inherited disease. Somatic mutations, epigenetic changes (sometimes called epimutations in recent literature), and the environment all have crucial influence on most kinds of cancer.

We argue that the concept of the genetic material that affects or is implicated in these situations is more appropriate than any standard concept of the gene as it can be applied in all of these cases and makes appropriate allowance for various degrees of environmental influence (in the wider sense). Our proposal is based on the importance of cellular, organismal, and environmental influences on the expression (i.e., the expressed or delivered informational content) of the genetic material. Because of the impact of these contextual factors on the information derived from the genetic material, indefinite descriptions of the interplay between the genetic material and the cellular machineries should replace gene concepts. Of course such descriptions can be extremely specific in those cases in which we know what is happening in sufficient detail.

If the aim is to educate scientifically literate citizens, then we should teach non-experts not only what genetics is about but also refrain from enhancing such intuitive conceptions as that of "genes for". Accordingly, we recommend encouraging non-experts to employ indefinite descriptions based on the influences that the genetic material has on (perhaps multiple) characteristics, including, of course, the particular salient characteristics that were formerly used in identifying "genes for" particular traits, while discouraging the genetic determinism that would be reinforced by the idea of "genes for".

References

Barnes, B., and J. Dupré. 2008. *Genomes and what to make of them*. Chicago/London: University of Chicago Press.

Bateson, P., and P. Gluckman. 2011. *Plasticity, robustness, and evolution*. Cambridge: Cambridge University Press.

Benzer, S. 1955. Fine structure of a genetic region in bacteriophage. *Proceedings of the National Academy of Sciences USA* 41: 344–354.

Benzer, S. 1956. Genetic fine structure and its relation to the DNA molecule. *Brookhaven Symposia in Biology* 8: 3–16.

Benzer, S. 1957. The elementary units of heredity. In *A symposium on the chemical basis of heredity*, ed. W.D. McElroy and B. Glass, 70–93. Baltimore: Johns Hopkins University Press.

Beurton, P., R. Falk, and H.-J. Rheinberger (eds.). 2000. *The concept of the gene in development and evolution: Historical and epistemological perspectives*. Cambridge/New York: Cambridge University Press.

Burian, R.M. 1981–1982. Human sociobiology and genetic determinism. *Philosophical Forum* 13: 40–66.

Burian, R.M. 2000. On the internal dynamics of Mendelian genetics. *Comptes rendus de l'Académie des Sciences, Paris. Série III, Sciences de la Vie/Life Sciences* 323(12): 1127–1137.

Burian, R.M. 2005. *The epistemology of development, evolution, and genetics*. New York/Cambridge: Cambridge University Press.

Burian, R. 2013. On gene concepts and teaching genetics – Episodes from classical genetics. *Science & Education* 22(2): 325–344.

Burian, R.M., R.C. Richardson, and W.J. Van der Steen. 1996. Against generality: Meaning in genetics and philosophy. *Studies in History and Philosophy of Science* 27: 1–29.

Carlson, E.A. 1966. *The gene: A critical history*. Philadelphia/London: W. B. Saunders.

Castera, J., and P. Clement. 2012. Teachers' conceptions about genetic determinism of human behaviour: A survey in 23 countries. *Science & Education* (online first article doi:10.1007/s11191-012-9494-0).

Coelho, P.S.R., A.C. Bryan, A. Kumar, G.S. Shadel, and M. Snyder. 2002. A novel mitochondrial protein, Tar1p, is encoded on the antisense strand of the nuclear 25S rDNA. *Genes & Development* 16: 2755–2760.

Crick, F.H.C. 1958. On protein synthesis. *Symposia of the Society for Experimental Biology* 12: The Biological Replication of Macromolecules: 138–163.
Darden, L. 1991. *Theory change in science: Strategies from mendelian genetics*. New York: Oxford University Press.
Darwin, C.R. 1868. *The variation of animals and plants under domestication, 2 vols*. London: John Murray.
de Vries, H. 1910 [1889]. *Intracellular pangenesis*. Chicago: The Open Court Publishing Co. www.esp.org
Dietrich, M.R. 2000. The problem of the gene. *Comptes rendus de l Académie des Sciences. Série III, Sciences de la Vie* 323(12): 1139–1146.
Djebali, S., et al. 2012. Evidence for transcript networks composed of chimeric RNAs in human cells. *PLoS One* 7(1): e28213.
Dougherty, M.J. 2009. Closing the gap: Inverting the genetics curriculum to ensure an informed public. *American Journal of Human Genetics* 85(1): 6–12.
Fogle, T. 2000. The dissolution of protein coding genes in molecular biology. In *The concept of the gene in development and evolution: Historical and epistemological perspectives*, ed. P. Beurton, R. Falk, and H.-J. Rheinberger, 3–25. Cambridge/New York: Cambridge University Press.
Freifelder, D. 1987. *Molecular biology*. Boston: Jones and Bartlett.
Gericke, N., M. Hagberg, V. Carvalho Santos, L.M. Joaquim, and C. El-Hani. 2012. Conceptual variation or incoherence? Textbook discourse on denes in six countries. *Science & Education* (online first article doi:10.1007/s11191-012-9499-8).
Gilbert, S.F. 2000. *Developmental biology*. Sunderland: Sinauer.
Gilbert, S.F., and D. Epel. 2009. *Ecological developmental biology: Integrating epigenetics, medicine, and evolution*. Sunderland: Sinauer Associates.
Godfrey-Smith, P. 2009. *Darwinian populations and natural selection*. Oxford/New York: Oxford University Press.
Goh, K.I., M.E. Cusick, D. Valle, B. Childs, M. Vidal, and A.-L. Barabasi. 2007. The human disease network. *Proceedings of the National Academy of Sciences USA* 104(21): 8685–8690.
Goodenough, U., and R.P. Levine. 1974. *Genetics*. New York: Holt, Rinehart, and Winston.
Griffiths, P.E., and E.M. Neumann-Held. 1999. The many faces of the gene. *BioScience* 49: 656–662.
Griffiths, P.E., and K.C. Stotz. 2006. Genes in the postgenomic era. *Theoretical Medicine and Bioethics* 27: 499–521.
Jablonka, E., and M.J. Lamb. 2005. *Evolution in four dimensions: Genetic, epigenetic, behavioral, and symbolic variation in the history of life*. Cambridge, MA: MIT Press.
Jablonka, E., and G. Raz. 2009. Transgenerational epigenetic inheritance: Prevalence, mechanisms, and implications for the study of heredity and evolution. *The Quarterly Review of Biology* 84(2): 131–176.
Johannsen, W. 1909. *Elemente der Exakten Erblichkeitslehre*. Jena: G. Fischer.
Kampourakis, K. 2013. Mendel and the path to genetics: Portraying science as a social process. *Science & Education* 22(2): 173–188.
Kay, L.E. 2000. *Who wrote the book of life: A history of the genetic code*. Stanford: Stanford University Press.
Keller, E.F. 2000. *The century of the gene*. Cambridge, MA: Harvard University Press.
Keller, E.F. 2010. *The mirage of a space between nature and nurture*. Durham: Duke University Press.
Kitcher, P. 1978. Theories, theorists and theoretical change. *Philosophical Review* 87: 519–547.
Kitcher, P. 1982. Genes. *The British Journal for the Philosophy of Science* 33: 337–359.
Mills Shaw, K.R., K. Van Horne, H. Zhang, and J. Boughman. 2008. Essay contest reveals misconceptions of high school students in genetics content. *Genetics* 178(3): 1157–1168.
Moore, D.S. 2008. Espousing interactions and fielding reactions: Addressing laypeople's beliefs about genetic determinism. *Philosophical Psychology* 21(3): 331–348.

Morange, M. 1996. Construction of the developmental gene concept. The crucial years: 1960–1980. *Biologisches Zentralblatt* 115: 132–138.
Morange, M. 2000. Gene function. *Comptes rendus de l'Académie des Sciences. Série III, Sciences de la Vie* 323(12): 1147–1153.
Morange, M. 2001. *The misunderstood gene*. Trans. M. Cobb. Cambridge, MA: Harvard University Press.
Morange, M. 2008. What history tells us XIII. Fifty years of the central dogma. *Journal of Biosciences* 33(2): 171–175.
Morgan, T.H., and C.B. Bridges. 1913. Dilution effects and bicolorism in certain eye colors of Drosophila. *The Journal of Experimental Zoology* 15: 429–466.
Morgan, T.H., A.H. Sturtevant, H.J. Muller, and C.B. Bridges. 1915. *The mechanism of Mendelian heredity*. New York: Henry Holt and Co.
Moss, L. 2001. Deconstructing the gene and reconstructing molecular developmental systems. In *Cycles of contingency: Developmental systems and evolution*, ed. S. Oyama, P.E. Griffith, and R.D. Gray, 85–97. Cambridge, MA: MIT Press.
Moss, L. 2003. *What genes can't do*. Cambridge, MA: MIT.
Neumann-Held, E.M. 1999. The gene is dead – Long live the gene! Conceptualizing genes the constructionist way. In *Sociobiology and bioeconomics: The theory of evolution in biological and economic theory*, ed. P. Kosowski, 105–137. Berlin/New York: Springer.
Neumann-Held, E.M. 2001. Let's talk about genes: The process molecular gene concept and its context. In *Cycles of contingency: Developmental systems and evolution*, ed. S. Oyama, P.E. Griffith, and R.D. Gray, 69–84. Cambridge, MA: MIT Press.
Olby, R.C. 1972. Francis Crick, DNA, and the central dogma. In *The twentieth-century sciences*, ed. G. Holton, 227–280. New York: W. W. Norton.
Olby, R.C. 1975. The protein version of the central dogma. *Genetics* 79(Suppl): 3–14.
Paul, D.B. 1995. Toward a realistic assessment of PKU screening. In *PSA: Proceedings of the biennial meeting of the philosophy of science association, 1994*, vol. 2, ed. A. Fine, M. Forbes, and L. Wessels, 322–328. East Lansing: Philosophy of Science Association.
Portin, P. 2002. Historical development of the concept of the gene. *The Journal of Medicine and Philosophy* 27: 257–286.
Reydon, T.R., K. Kampourakis, and G.P. Patrinos. 2012. Genetics, genomics and society: The responsibilities of scientists for science communication and education. *Personalized Medicine* 9(6): 633–643.
Rheinberger, H.-J. 2000. Gene concepts: Fragments from the perspective of molecular biology. In *The concept of the gene in development and evolution: Historical and epistemological perspectives*, ed. P. Beurton, R. Falk, and H.-J. Rheinberger, 219–239. Cambridge/New York: Cambridge University Press.
Roberts, D.A. 2007. Scientific literacy/science literacy. In *Handbook of research on science education*, ed. S.K. Abell and N.G. Lederman, 729–780. Mahwah: Erlbaum.
Sadava, D., D.M. Hillis, H.C. Heller, and M. Berenbaum. 2011. *Life: The science of biology*. Gordonsville: WH Freeman Publishers.
Sarkar, S. 1998. *Genetics and reductionism*. New York/Cambridge: Cambridge University Press.
Singer, M., and P. Berg. 1991. *Genes and genomes: A changing perspective*. Mill Valley: University Science.
Snyder, M., and M. Gerstein. 2003. Defining genes in the genomics era. *Science* 300: 258–260.
Stadler, L.J. 1954. The gene. *Science* 120: 811–819.
Stern, D.L. 2011. *Evolution, development, and the predictable genome*. Greenwood Village: Roberts & Company Publishers.
Stotz, K.C., P.E. Griffiths, and R. Knight. 2004. How biologists conceptualize genes: An empirical study. *Studies in History and Philosophy of Biological and Biomedical Sciences* 35: 647–673.
Strasser, B.J. 2006. A world in one dimension: Linus Pauling, Francis Crick and the central dogma of molecular biology. *History and Philosophy of the Life Sciences* 28: 491–512.
Sturtevant, A.H. 1913a. A third group of linked genes in *Drosophila ampelophila*. *Science* 37: 990–992.

Sturtevant, A.H. 1913b. The linear arrangement of six sex-linked factors in drosophila, as shown by their mode of association. *The Journal of Experimental Zoology* 14: 43–59.

Tycowski, K.T., S. Mei-Di, and J.A. Steltz. 1996. A mammalian gene with introns instead of exons generating stable RNA products. *Nature* 379: 464–466.

Walpole, B., A. Merson-Davies, and L. Dann. 2011. *Biology for the IB diploma coursebook*. Cambridge: Cambridge University Press.

Waters, C.K. 1990. Why the anti-reductionist consensus won't survive: The case of classical Mendelian genetics. In *PSA 1990*, vol. 1, ed. A. Fine, M. Forbes, and L. Wessels, 125–139. East Lansing: Philosophy of Science Association.

Waters, C.K. 1994. Genes made molecular. *Philosophy of Science* 61: 163–185.

Waters, C.K. 2000. Molecules made biological. *Revue Internationale de Philosophie* 54: 539–564.

Watson, J.D. 1965. *Molecular biology of the gene*. New York: Benjamin.

Watson, J.D., and F.H.C. Crick. 1953. Molecular structure of nucleic acids. *Nature* 171: 737–738.

Current Thinking About Nature and Nurture

David S. Moore

> A devil, a born devil, on whose nature Nurture can never stick
>
> William Shakespeare (1611 a.d., *The Tempest*)

1 Introduction

Curious people typically wonder at some point in their lives whether they might have been different if they had had different experiences while growing up. It is clear to all of us from casual observation that some of our characteristics are affected by our experiences; children growing up in Calais, France typically speak French, while children growing up just across the English Channel in Dover, England typically speak English, reflecting these children's exposure to French and English, respectively. In contrast, some of our characteristics are not obviously affected by our experiences at all; children often have facial features like their biological parents' facial features, regardless of whether or not they are adopted at birth. Likewise, some of our normal characteristics, such as five fingers on each hand, are present at birth, contributing to the impression that experiences play no role in the development of these traits. Such observations lead us to think that certain aspects of our behavioral characteristics, too—for example, a person's intelligence or personality—might not be affected by experience. But despite the intuitive appeal of such a perspective, empirical and theoretical investigations have now made it clear that this way of thinking misrepresents the development of both our biological and psychological traits (Bateson and Gluckman 2011; Blumberg 2005; Gottlieb 2007; Jablonka and Lamb 2005; Lewkowicz 2011; Lewontin 2000; Lickliter 2008;

D.S. Moore (✉)
Pitzer College, 1050 N. Mills Avenue, Claremont 91711, CA, USA

Claremont Graduate University, Claremont, CA, USA
e-mail: dmoore@pitzer.edu

K. Kampourakis (ed.), *The Philosophy of Biology: A Companion for Educators*, History, Philosophy and Theory of the Life Sciences 1, DOI 10.1007/978-94-007-6537-5_27,

Meaney 2010; Moore 2008a; Noble 2006; Oyama 2000; Robert 2004). In fact, all of our characteristics are influenced by both biological and experiential factors.

The idea that some characteristics are caused by experiences whereas others are inborn has a long history, dating back at least to William Shakespeare's early seventeenth century work in the humanities and to Sir Francis Galton's late nineteenth century work in the sciences. As the first scientist to juxtapose the words Nature and Nurture (Plomin 1994), Galton defined Nurture as consisting of "every influence from without that affects [a person] after his birth... [including] food, clothing, education, or tradition [...] all these and similar influences whether known or unknown" (Galton 1874, p. 12). In contrast, he used the word Nature to refer to the causes of traits that appear uninfluenced by experience. In large part because he was Charles Darwin's half cousin, Galton was interested in the transmission of characteristics across generations (Kevles 1995), and as one of the first individuals to investigate how experiences and heritages influence people's characteristics, the path he blazed strongly influenced modern conceptions. In particular, he believed that a sharp distinction between Nature and Nurture was justifiable (Gottlieb 1992). Galton's proposition that Nature and Nurture can be considered as dichotomous factors that contribute independently to our traits led directly to the modern characterization of Nature and Nurture as oppositional, as implied by the word 'versus' in the stock phrase *Nature versus Nurture.* Although Galton's conceptualization was ultimately unable to withstand close scrutiny, Nature and Nurture continue to be presented in some quarters as contrasting influences on development.

Galton's erroneous view has implications that go far beyond academic debates about biology. Having established the notion of "eugenics" based on his ideas about Nature and Nurture, Galton advocated policies wherein governments would "rank people by ability and authorize more children to the higher- than to the lower-ranking unions... [while the unworthy would] be comfortably segregated in monasteries and convents, where they would be unable to propagate their kind" (Kevles 1995, p. 4). The emergence of these kinds of ideas in the early twentieth century ultimately led to forced sterilizations in the United States and to genocide in Nazi Germany. As was appropriate, the rejection of eugenics after World War II did not entail the rejection of Galton's broader framework for the study of human characteristics; if Nature and Nurture really were oppositional factors influencing human development, people would simply have to come to terms with any implications of this reality, even if they found such implications politically distasteful. As it happens, scientists now know that Nature and Nurture collaborate to make us what we are (Moore 2002), but one of the lessons of the tragedies of the early twentieth century is this: our beliefs about these issues have important influences on our behaviors in both the public and private domains.

Molecular biology is a relatively arcane science, but to the extent that discoveries in this field bear on questions of Nature and Nurture, they are likely to have implications for our political and personal actions. For example, if the public generally believed that obesity can be avoided with a vegan diet, their reaction to skyrocketing rates of obesity would likely be different than if they believed some people have genes that cause them to gain weight over time no matter what they eat. Of course, molecular biologists understand that individual genes never single-handedly cause characteristics like obesity—or any other phenotypes for that matter (Noble 2006; Stotz 2006)—but some molecular biologists sometimes speak and write in ways that can confuse readers about this point. And regardless, the public does not get most of their information about genes directly from molecular biologists. Instead, they often receive information like the account in an article on the *Newsroom* website of the University of California, Los Angeles (Wheeler 2010), which reported that geneticists have made:

> the startling discovery that nearly half of all people in the U.S. with European ancestry carry a variant of the fat mass and obesity associated (FTO) gene, which causes them to gain weight – from three to seven pounds, on average – but worse, puts them at risk for obesity… [and that the same gene] is also carried by roughly one-quarter of U.S. Hispanics, 15 percent of African Americans and 15 percent of Asian Americans.

Those uneducated in molecular biology could be forgiven for concluding—mistakenly!—that if a prestigious university like UCLA is reporting on the discovery of an "obesity gene" that *causes* weight gain and that is "carried by more than a third of the U.S. population," the obesity epidemic currently plaguing the U.S. need not be a reflection of the high-calorie diets and sedentary lifestyles typical of contemporary Americans. Such a conclusion could easily lead an obese person to attribute their condition to their genes and thereby rationalize continuing gluttony. Similar arguments could be made about people's beliefs in genes that determine IQ, which could lead to voting against the use of tax revenues for supporting public schools; why, some might argue, should we spend money on the education of children who might be "biologically" unable to learn?

Our beliefs about genetic and environmental contributions to people's characteristics influence what we do. For this reason, there is significant value in biology teachers being able to impart to their students an accurate understanding of how Nature and Nurture interact to produce our biological and psychological characteristics.

2 Cultural Lag

Among those who have considered the issue in great detail, thinking about Nature and Nurture has not changed significantly in the past few decades. Certainly by the turn of the millennium, it was already clear that construing Nature and Nurture as discretely different influences on development was an obsolete way of approaching questions about the origins of biological and psychological characteristics (Moore 2002). In fact, 10 years ago, the biologist Sir Patrick Bateson chose the title "The corpse of a wearisome debate," for his review of Steven Pinker's (2002) book *The blank slate: The modern denial of human nature*. From his review, it is clear that Bateson already believed in 2002 that books like Pinker's are not a valuable contribution to our understanding of "human nature." Nonetheless, as is evident from the recent publication (or re-issuing) of books such as *The mirage of a space between*

Nature and Nurture (Keller 2010), *The Nurture assumption: Why children turn out the way they do* (Harris 2009), or *Kids: How biology and culture shape the way we raise young children* (Small 2011), theorists continue to write about "the Nature versus Nurture debate" and publishers continue to believe there are people interested in reading about it. One sensible question we can ask is: why?

One reason this "debate" continues to generate interest is captured by the words "cultural lag," which Bateson (2002) used to refer to the fact that some people remain unaware of theoretical advances in a field long after the new way of thinking has become canonical in that field. Because of cultural lag in some quarters, reiteration of the essential interdependence of Nature and Nurture can still be merited, which is why a book like *The mirage of a space between Nature and Nurture* (Keller 2010) continues to be a valuable contribution to the literature on this topic. However, the recalcitrant persistence of Galton's outmoded perspective is not merely a function of passive cultural lag but rather is, in some cases, actively maintained. For example, in *The blank slate,* Pinker argued that "another book on nature and nurture" (Pinker 2002, p. vii) was warranted, *not* because of how important it is to debunk the simplistic Nature-versus-Nurture idea, but because of his perceived need to defend the idea that certain characteristics—for instance, intelligence (Herrnstein and Murray 1994) and rape (Thornhill and Palmer 2000)—are influenced by biology. In writing such a book, Pinker succumbed to the temptation to "pour scorn [...] on those people suffering from cultural lag" (Bateson 2002, p. 2212), namely those people who continue to cling to the indefensible idea that some human characteristics are completely *un*influenced by biology. But in so doing, Pinker (perhaps inadvertently) perpetuated the beliefs that Nature and Nurture are separable and that they are independently measurable influences on our characteristics. Thus, although a nuanced understanding of how genetic and non-genetic factors *really* interact has obviated the Nature-Nurture debate, the debate lives on because some writers preserve it (whether they intend to or not). Books like *The blank slate* encourage a false understanding of the determination of our characteristics, by claiming that even if Nature and Nurture typically interact in complex ways, "in some cases, an extreme environmentalist explanation is correct ... [whereas in] other cases [...] an extreme hereditarian explanation is correct" (Pinker 2002, p. viii). In fact, neither of these extreme views is ever correct, and claims to contrary themselves reflect a form of cultural lag.

So, there are multiple forms of cultural lag, all of which need to be addressed by writers who can reiterate what has been accepted for decades in some corners of the biological and social/behavioral sciences (Beach 1955; Blumberg 2005; Gottlieb 1997; Johnston 1987; Lehrman 1953; Lewontin 1983). To those who would argue that Nature is more powerful than Nurture in determining our characteristics (i.e., cultural lag dating to Galton in the nineteenth century), the case must be made that Nature and Nurture are equally influential during development. To those who would argue that Nurture is more powerful than Nature (i.e., cultural lag dating to the 1950s, when behaviorists held sway in American psychology), the same case must be made. To those who would argue that Nature-Nurture interactionism "might turn out to be wrong" (Pinker 2002, p. viii)—a form of cultural lag dating only to the early twenty-first century, but which is nonetheless significant—the case must be made that Nature and Nurture are now known to *always* interact during development. To those who would argue that it is a reasonable goal to attempt to measure *how much* Nature and Nurture each contribute to the development of particular characteristics (e.g., Plomin 1994), the case must be made that this question does not actually make sense once we acknowledge that Nature and Nurture are both *essential* to the development of those characteristics (a point considered in more detail in the next section). Once these various forms of cultural lag have been addressed, scientists can turn their attention to the truly consequential question of *how* Nature and Nurture interact in the production of particular characteristics. That is, rather than spending time answering nonsensical questions about how *much* Nature or Nurture influences the development of a characteristic, the question that should be driving our research programs and that should be situated at the center of our life sciences curricula is: *how is it* that genetic factors, proteins, cells, organs, organisms, populations of individuals, cultural factors, and other aspects of an organism's environment co-act to produce the organism's traits (i.e., phenotypes) in development?

3 Definitions and Conceptual Problems

Making the case that Nature and Nurture are both always essential—and therefore equally important—contributors to development requires clear definitions of these words. Early in the scientific consideration of Nature and Nurture, Galton adopted a decidedly vague definition of Nurture (cited previously), and considered everything else to be Nature. More than a century later, after biologists elaborated their understandings of molecular (i.e., genetic) contributions to inheritance, things became clearer; in the latest edition of their textbook *Behavioral Genetics*, Plomin et al. (2008) effectively defined Nurture as "environment" and Nature as "genetics" (p. 2). Because Galton was primarily concerned with the extent to which characteristics could be inherited and thereby run in biological families, it makes sense that his intellectual heirs—quantitative behavioral geneticists like Plomin and colleagues—would define Nature as "genetics;" after all, biologists for the past 100 years have generally believed that only DNA—the genetic material—is transmitted from generation to generation (Jablonka and Lamb 2005). Numerous theorists have recently argued that this belief reflects an unhelpfully narrow understanding of inheritance, and that a convincing case can be made that non-genetic factors can be inherited from our ancestors too, albeit via different mechanisms than those responsible for transmitting genetic factors (Carey 2011; Gottlieb 1992; Griffiths and Gray 1994; Harper 2005; Jablonka and Lamb 2005; Johnston 2010; Laland et al. 2001; Lickliter and Honeycutt 2010; Moore 2013; Uller, this volume). But regardless, if we accept the definition of Nature as "genetics" and Nurture as "environment," two problems with Galton's foundational conceptualization of the Nature/ Nurture issue immediately become apparent.

First, there are a number of biological components that lie between genes and environments, and although these components occupy levels at which crucial phenotype-building interactions occur (Johnston and Edwards 2002), they are typically ignored in Galton-style behavioral genetics investigations. For those who have not studied biology, it can be easy to forget that genes can be considered to be physical structures with specific spatial locations and that they operate, therefore, within specific contexts (see Burian and Kampourakis, this volume). Genes can be thought of as analogous in some ways to the smallest elements in a set of nesting Russian *matryoshka* dolls; our genes constitute parts of our chromosomes, which are located within the nuclei of most of our cells, which constitute our organs, which are surrounded by hormones, fluids, and other organs, all of which are located within our bodies. Because genes and the environment outside of the body are both able to influence the states (or existence) of the various bodily components that lie *between* the genes and the environment (Gottlieb 1991a, 2007; Lickliter and Honeycutt 2010), it follows that an understanding of trait development that references only Nature and Nurture—and not these other in-between levels of biological systems—must be an incomplete understanding. In fact, a gene does what it does in part because of molecules present in its local environment (i.e., inside the nucleus of a cell). The simplistic idea that genes and environments are independent contributors to trait development fails to capture the complex reality that one gene's products can constitute the "environment" of another gene, and that environmental factors (e.g., a specific nutrient, a specific person, an altered light cycle, etc.) can have their effects on a trait by influencing biological factors that lie between genes and environments (e.g., hormones, epigenetic marks, neurons, etc.). When one considers the space between an animal's genes and its environment, it becomes rather more difficult to define Nature and Nurture in a way that clearly distinguishes between them (see Bateson and Gluckman 2011, for additional examples that strengthen this argument).

A second, related point arises when Nature is defined strictly as "genetics." Galton famously claimed that "when nature and nurture compete for supremacy on equal terms [...] the former proves the stronger" (Galton 1874, p. 12), but this claim becomes utterly inconceivable when we define Nature as "genetics." Although modern behavioral geneticists, too, sometimes imply that genetic factors can be "stronger" than environmental factors in the development of some traits (e.g., see Deater-Deckard et al. 2006; Yamagata et al. 2006), the fact is that genetic factors, when isolated from their cellular and broader contexts, are inert (Noble 2006; Keller 2010); independently of other factors, genes *per se* have no "strength" at all. Instead, genetic and environmental factors *collaborate* to build traits (Moore 2002; Lewkowicz 2011), and when two or more factors are both *required* to produce an outcome, none of the factors can be more important—stronger—than any other. By analogy, consider the internal combustion engine under the hood of most automobiles. Such engines require fuel and an ignition spark to operate normally, and the absence of either of these components renders the engine non-functional. Just as it makes no sense to ask if the gasoline or the spark has the "stronger" effect on the functioning of the engine, it makes no sense to conceive of Nurture and genetics as factors that "compete for supremacy" with one another (Moore 2011). Of course, different observers in different contexts might have reasons for choosing to focus on one factor over another, but it would be a mistake to believe that either factor ever actually has a *stronger* influence than the other on an outcome in a given situation. In their natural contexts, genes are essential contributors to processes that require essential non-genetic contributors as well.

4 Heritability and Its Weaknesses

Modern quantitative behavioral geneticists understand what Galton did not, namely that "the environment plays a crucial role at each step" (Plomin et al. 2008, p. 305) in the development of our psychological/behavioral characteristics. Nonetheless, a research method Galton pioneered to tease apart Nature and Nurture—studies of identical and fraternal twins—provided the data for 5,000 articles on behavioral genetics published between 2001 and 2006. Thus, even though modern behavioral geneticists understand that genetic and environmental factors always both play vital roles in trait development—which necessarily means that neither can ever be more important than the other—they continue to rely on a century-old technique that Galton devised specifically to "appraise [Nature's and Nurture's] relative importance" to the appearance of traits (Galton 1907, p. 131). Moreover, in their empirical research reports, modern behavioral geneticists write about statistical "heritability estimates," which are the primary product of twin studies, in ways that make it seem as if it is possible to measure the relative importance of Nature and Nurture. To give one of many possible recent examples, the authors of a twin study on impulsivity in adolescence concluded that their calculated heritability estimates were "consistent with estimates from [...] past studies, suggesting that impulsivity is influenced around 40–45 % by genetic factors" (Niv et al. 2012). Such a claim would imply to many readers that an accurate measurement has been made of the relative strength-of-influence of genetic factors on impulsivity. But although numbers like these suggest that traits can be more influenced by genetic or by non-genetic factors, it is actually not possible to apportion causation of traits to such factors in this way.

A reasonable question to ask, then, is why our modern research literature is littered with what appear to be estimates of the relative importance of Nature and Nurture to trait development when the facts of molecular biology clearly indicate that both factors are always indispensible, and that therefore, it is never possible to evaluate which is the more important factor. The answer to this question likely has to do with the fact that the products of twin studies—heritability statistics—are notoriously misleading, in that they *appear* to reflect the relative importance of genetics in trait development even though they really do not (Block 1995; Keller 2010; Moore 2006, 2008a, 2013). Rather than revealing anything about the extent of genetic influence on trait development, these statistics (e.g., the 40–45 % reported by Niv et al. 2012) actually reflect the extent to which *variation in a trait* across a population can be "accounted for" by variation in genes across that population.

At first glance, a factor that accounts for the variation in a trait seems like it must be the cause of the trait, but in fact there are crucial differences between *causing* a trait and *accounting for variation* in that trait. Quantitative behavioral geneticists use an approach that can reveal *statistical* interactions that account for variation, but these kinds of interactions are very different from the "causal-mechanical" interactions (Griffiths and Tabery 2008, p. 341) known to characterize the developmental process itself. For this reason, even if a twin study of a characteristic reveals *no* statistical interaction between genetic and environmental factors, it is still the case that the development of the characteristic in individuals is *caused by mechanical interactions* between such factors (see Griffiths and Tabery 2008, for additional consideration of these two very different meanings of the word "interaction"). Because heritability statistics are about accounting for variation and not about causation, they do not actually reflect the strength of influence of genes on the development of a trait, even if it seems like they do. Moreover, it is not clear that there are interconnections between accounts of trait variation across a population and explanations of trait development in individuals (Moore 2008b), so the heritability estimates generated by twin studies do not even necessarily point the way toward genetic factors that might warrant further study (see Block 1995, for additional consideration of these issues).

These are not novel points. For example, nearly 40 years ago, Lewontin (1974) pointed out that it is possible for variation in genetic factors to account for a high percentage—even 100 %—of the variation in a trait in a population, but that this does not mean genetic influences on that trait are any "stronger" than non-genetic influences. The development of a trait with a heritability of .80 (or even 1.0) can be influenced by environmental factors just as much as can the development of a trait with a heritability of .05 (Moore 2006, 2013). Of course, quantitative behavioral geneticists (e.g., Plomin 1990) understand this distinction between what heritability statistics can do (account for variability) and cannot do (explain the cause of a trait), but the distinction appears to be virtually impossible to maintain as they write about their findings. As a result, these researchers report their calculated heritability estimates, but then often misconstrue them as meaning something about the strength-of-influence of genetic factors—Nature—on trait development. In a masterful treatment of this problem, Keller (2010) has considered both the causes and consequences of this sort of conceptual "slippage" (p. 34), which, she argues, has arisen from the fact that the word "heritable" has come to have more than one meaning. Without reciting her arguments, it might be enough to note here that although it seems like the heritability estimates generated from twin studies should tell us something out how *inheritable* various traits are, they actually cannot.

Because heritability statistics have been the subject of unrelenting criticism from philosophers, biologists, and psychologists for nearly four decades, it is unnecessary to recount here why they are widely recognized as being unable to address the kinds of Nature vs. Nurture questions Galton and his followers in behavioral genetics hoped they would. In virtual unanimity, theorists have come to question the value of heritability statistics, particularly in studies of human beings. Heritability, which is almost always the metric referenced by those attempting to argue that Nature or Nurture are more important in the development of a given a trait, is a statistic that is at worst meaningless and at best deceptive. Even leading behavioral geneticists now acknowledge that "heritability estimates are no longer important" (Johnson et al. 2009, p. 217).

A small army of scientists and philosophers of biology have identified a variety of misunderstandings that heritability statistics perpetuate. In an effort to protect unsuspecting readers from these common misinterpretations, I have pointed out in other publications (Moore 2002, 2006, 2008a, 2013) several things to keep in mind when one encounters these statistics. For instance:

- Heritability estimates tell us nothing about what causes an individual's traits (Johnson et al. 2009),
- Heritability estimates do not reflect the extent to which a trait is genetically determined and cannot be understood to reflect the *importance* of genes in the production of a person's traits,
- Heritability estimates are not measures of a trait's "openness" to environmental influence—they do not tell us how easily a trait can be affected by environmental factors (Lewontin 1974),
- Heritability estimates do not provide an accurate measurement of the likelihood that a trait will be "passed down" in a natural (i.e., not experimentally controlled) environment, so even 100 % heritable characteristics need not develop in the children of parents with that characteristic,
- Because some characteristics—for instance, the number of fingers present on normal human hands—are influenced by genetic factors that do *not* vary widely in human populations, these characteristics are not very heritable (Block 1995); no matter how counterintuitive it might seem, five fingers per human hand is not a heritable trait, given how behavioral geneticists define heritability,
- Heritability estimates reflect *environmental* variability, so the heritability of a trait in a population that develops in variable environments will be lower than the heritability of that same trait in a population that develops in less variable environments; thus, the heritability of a trait is not a characteristic of the trait at all, but is instead a characteristic of a studied population (Eisenberg 2004; Moore 2013).

As should be clear from this last point, heritability estimates cannot be generalized from the population that produced them to another population. Because this point has been misunderstood in the literature (Sesardic 2005), it warrants additional attention here. I have previously called attention to the fact that this caveat applies regardless of how similar two populations might appear; accordingly, I wrote "if alcoholism is [highly] heritable among Iowans, it need not be the case that it is [highly] heritable among Ohioans […] heritability estimates calculated for one population *do not apply* to another population" (Moore 2002, p. 47). Sesardic has argued that because I also believe genes and environments influence development symmetrically (i.e., they are always equally significant), it follows that "the non-generalizability of heritability implies the non-generalizability of environmental influences as well. Therefore, it would follow from Moore's pessimism about state-to-state inferences that if a new teaching strategy had good effects in schools in Ohio there would be no reason whatsoever to expect that the strategy would work in Iowa. This consequence is absurd…" (p. 80).

The absurdity here arises from Sesardic's misunderstanding of the central fact that *heritability estimates do not tell us anything about influences on trait development*; they tell us only how we can account for variation in a population. So, an environmental manipulation that *influences* the development of a scholastic competence in Ohio is likely also to influence the development of that scholastic competence in Iowa (just as a fictional genetic manipulation capable of influencing the development of a scholastic competence in Ohio is likely also to influence the development of that scholastic competence in Iowa). But because the *heritability* of a scholastic competence tells us nothing about what *influences the development* of that competence, it need not be the case that a study of the heritability of this competence in Ohioans would generate similar statistics as a study of the heritability of this competence in Iowans. If the factors that influence the development of a competence might not be equally variable for two different populations, the heritability of that competence in the two populations will differ, no matter how similar they (or their environments) might otherwise seem.

In spite of the fact that the heritability statistics generated by twin studies are unable to satisfactorily address questions about the relative importance of Nature and Nurture to the development of any of our traits, it remains the case that "twin studies [...] provide the bulk of the evidence for the widespread influence of genetics in behavioral traits" (Plomin et al. 2008, p. 78). Of course, the fact that genes have important effects on behavior in general is now apparent; because behavior is a product of a brain, and because a brain is *built* using genes that contribute to the brain's structure, chemistry, and functioning, anyone thinking about the relationship between Nature and Nurture should understand that when it comes to behavior, genes are always influential. But this insight does not rely on twin study data; as Johnson et al. (2009) note, "Once we accept that basically everything—not only schizophrenia and intelligence, but also marital status and television watching—is heritable [READ: associated with genetic factors], it becomes clear that specific estimates of heritability are not very important" (p. 220). Twin studies confirm the importance of genetic influences on behavior, but the heritability statistics they generate mislead many readers by suggesting that some characteristics are *more* influenced by genes than by environmental factors, or that some characteristics are *more* influenced by genes than are other characteristics. But Nature and Nurture always play essential roles in the development of all of our traits, so neither of these suggestions is accurate. Given this insight, why is it that some of our traits (e.g., the languages we speak) are *obviously* influenced by environmental factors whereas others (e.g., the structures of our faces) are not?

5 Overlooking Nurture's Effects

There are several reasons traits might appear to be unable to be influenced by environmental factors even when they can be. First, some of the factors that influence characteristics are present in *prenatal* environments, so we have little opportunity to directly witness their effects, which can be significant nonetheless. For example, there is evidence that a mother's diet can influence her infant's preferences for particular flavors (Mennella et al. 2001) or can influence the likelihood of her adult offspring being obese (Davenport and Cabrero 2009) or schizophrenic (Hoek et al. 1998). Likewise, the sounds that fetuses hear *in utero* can influence their behavioral characteristics once they are born (DeCasper and Spence 1986). Morphological characteristics that develop prenatally—a category that includes things like the bones in the face—also emerge as a result of interactions between genetic and non-genetic factors that occur *in utero* (e.g., see Hall 1988).

Second, some of the factors that might influence our characteristics are constant across human developmental environments, making it difficult to observe their influences. Because every human being grows up in an environment containing, for instance, oxygen and gravity, and *almost* every human being grows up in an environment containing, for instance, certain nutrients and communicative adults, it is impossible to casually observe the effects of such environmental factors. Nonetheless, such factors are likely to have important effects on the development of our traits, even if they cannot be invoked to explain *differences* among individuals. For example, although specific nutrients are known to influence human hair color (McKenzie et al. 2007), the effects of these environmental factors are not readily apparent to us because in many parts of the world the relevant nutrients are so plentiful that no one is malnourished in the specific ways that would reveal dietary influences on hair color. Likewise, the important role of gravity in the development of normal mammalian motor systems was undetectable until it was possible to study the effects on rats of developing as neonates in the microgravity environment present in the space shuttle's low-earth orbit (Walton et al. 2005; for further discussion of the importance of factors that *could* account for differences between individuals but that ordinarily do not because they ordinarily do not vary across individuals' developmental environments, see Griffiths and Tabery 2008).

Third, some of the factors that influence our characteristics are extremely subtle and might simply have escaped our notice. Studies of diverse species have now revealed a variety of effects of environmental stimuli on trait development, effects that bear a decidedly non-obvious relationship (Gottlieb 1991a) to the stimuli that produce them. For example, exposing chicks to their own toes influences their subsequent consumption of mealworms (Wallman 1979), exposing squirrel monkeys to either grasshoppers or crickets in their food influences their subsequent fear of snakes (Masataka 1993), and exposing mallard ducklings to their own embryonic vocalizations influences their subsequent preference for their mothers' assembly calls, even though the mothers' calls sound nothing at all like the embryos' vocalizations (Gottlieb 1991b). Considering how difficult it is to discover associations like these that seem entirely unpredictable, it is likely that non-obvious environmental contributors to development will ultimately be found to be a category that includes a large number of influential environmental factors that have yet to be recognized (for another good example, see King et al. 2005).

Finally, some of the factors that influence the development of our traits are not genes, but are nonetheless biological; steroid hormones are a good example. Biological chemicals like these do not fit behavioral geneticists' definition of

"Nature" (because they are not genes), but because they are produced within our bodies, they do not fit our intuitions about what should count as Nurture, either. Consider testosterone, a steroid hormone known to influence psychological characteristics as diverse as aggression and spatial cognition (see Archer 2006, or Mehta and Beer 2010, for references to the literature establishing the link between testosterone and aggression, and see Aleman et al. 2004, for evidence that experimentally administered testosterone affects visuospatial ability). Testosterone's effects on these characteristics means that *any* experience an individual has that influences testosterone levels could potentially influence their behavior. Importantly, this would be true regardless of whether or not the experience is one we would ordinarily associate with Nurture. So for example, when salivary testosterone levels are influenced by the experience of athletic competition (Edwards et al. 2006), we recognize this as an effect of Nurture (because some children experience more athletic competition than others). In contrast, when testosterone levels increase at the onset of puberty, similar effects on behavior can be expected even though experiencing the onset of puberty would ordinarily *not* be associated with Nurture. (It is for this reason that Gottlieb (1991a) suggested a broad and relational definition of experience that includes experiences other than those involving obvious learning). Should testosterone be considered an aspect of Nature or Nurture? The question makes little sense in light of what scientists now understand about how the molecules in our bodies are affected both by our genes and our experiences.

In summary, some of the environmental factors that influence development operate *in utero*, some are invariably present in human developmental environments, some do their work in extremely subtle ways, and we simply fail to recognize others as environmental factors at all (because even though they are *not* genetic and *can* be influenced by the external environment, they are located within a person's body). In each case, the influences of these factors are not easy to detect. As a result, casual observation sometimes suggests that we have some characteristics that are completely *un*influenced by Nurture. However, because genes only express their products in *contexts* and because their contexts influence what they do, the genome must be thought of as being reactive (Gilbert 2003), and non-genetic factors must be understood to always play a role in the development of our characteristics.

6 Genes in Contexts

It is in the discovery that genes do different things in different contexts that we can see most clearly how dichotomous thinking about Nature and Nurture must be erroneous. If a genome is associated with a characteristic in context A and that same genome is associated with a different characteristic in context B, it is clear that it makes no sense to think about either of the characteristics as being caused more by Nature than by Nurture or vice versa; the particular characteristic that develops depends critically on both the genes in question (Nature) and on the context in which those genes are being expressed (Nurture).

A good example of this type of environment-dependent phenotypic plasticity (West-Eberhard 2003) can be found in the development of the honeybee. Large numbers of honeybee larvae in a single colony can be genetically identical to one another, but a small number of these clones will develop into queens while the rest will become workers. Remarkably, workers are often half the size of queens, and unlike queens, they have sting barbs, short lifespans, and a behavioral repertoire required for food collection, among other major behavioral and morphological characteristics that distinguish them from queens (Carey 2011). The factor responsible for these differences is one even Galton recognized as Nurture: diet. While the larvae that become queens are maintained on a diet of royal jelly, their identical twin sisters that become workers are switched to a different "worker diet" after they turn 3 days old (Shuel and Dixon 1960). Therefore, what the genomes of these clones *do* depends on their nutritional context. But can we think of royal jelly as the factor that contains all of the information required for the construction of, for instance, mature ovaries, which are present only in queens? Of course not; critical information for the construction of ovaries is contained in the bees' genomes as well. The normal growth of ovaries in queens requires particular DNA *and* a particular developmental context, and this kind of collaborative construction of phenotypes during development is the rule among mammals as well.

Although theorists have thought of genes as providing information for trait construction at least since Francis Crick (1970) elucidated the "central dogma of molecular biology" in 1958, it is now clear that environments, too, provide information for trait construction (Lickliter and Berry 1990; Lickliter 2000). Thus, although the central dogma is still featured prominently in biology textbooks, its implication that DNA can be construed as single-handedly determining phenotypes is clearly wrong (Moore 2002). To the extent that textbooks represent genes as providing all of the information required for trait construction, they are masking what biologists currently understand about phenotypic development.

There are at least three different ways in which genes can be influenced by their contexts. First, genes can effectively be "turned on," "turned off," or rendered more or less active by chemical compounds that are normally involved in gene regulation. Because these compounds literally lie on top of genes, they are referred to as "epigenetic," and they have recently been the focus of an enormous amount of scientific attention (Bateson and Gluckman 2011; Carey 2011; Moore 2013; Uller, this volume). Although epigenetic phenomena have been observed since the early 1960s (e.g., Beutler et al. 1962), researchers have recently begun focusing on behavioral epigenetic phenomena, wherein specific *experiences* alter the activity of specific genes, thereby influencing subsequent behaviors. Among the most compelling findings in this domain have been those reported by Meaney (2010; Weaver et al. 2004). In this work, newborn rodents exposed to particular kinds of mothering grow up to be adults with particular ways of reacting to stressful situations. Meaney's lab has demonstrated that the parenting has its long-term effects by altering genetic activity in the offspring—not by changing the offspring's genes *per se*, but by epigenetically changing what those genes are *doing*. Although research on behavioral epigenetics in human populations is only now getting underway, several studies have already reported effects in people that are consistent with those observed in rodents (Beach et al. 2010; Borghol et al. 2012; McGowan et al. 2009; Oberlander et al. 2008), so there is good reason to believe that the experiences we have as we develop have significant effects on the activity of our genes. The implications of these findings for discussions about Nature and Nurture are so profound that one epigenetics researcher (Weaver 2007) subtitled his article on the epigenetic "programming" of offspring by their mothers' behaviors "Nature versus Nurture: Let's call the whole thing off."

Second, it has become clear that there is a particular class of genes that begin to function in neurons when they are activated by specific kinds of environmental stimulation. These genes are known as "immediate early genes," and they have been found to be able to respond to changes in light cycles in hamsters (Rusak et al. 1990) and in cats (Rosen et al. 1992), and to species-specific birdsongs in zebra finches and canaries (Mello et al. 1992). Primates like human beings have immediate early genes as well, and at least one of them has been found to be associated with various forms of learning (Okuno and Miyashita 1996) and memory (Davis et al. 2003). Again, the discovery of genes that are responsive to environmental stimulation reinforces the fact that it is an error to imagine that our bodies and environments are not in constant communication as they collaborate in the construction of our phenotypes.

Third, molecular biologists (e.g., Pan et al. 2008; Wang et al. 2008) now estimate that as many as 95 % of our genes undergo a process known as "alternative splicing," which enables a given gene to perform different functions in different contexts. For example, Amara et al. (1982) discovered that the gene that contributes to the production of the hormone calcitonin in the thyroid gland also contributes to the production of an entirely different product—a neuropeptide—when it is "alternatively spliced" in a different context (the hypothalamus). The fact that the same exact gene is capable of doing two entirely different things in different cellular contexts controverts the idea that genes operate independently of their environments. But if genes are *typically* capable of doing *many* different things as a function of how they are influenced by different contexts, the belief that characteristics can be determined exclusively—or even primarily—by genes would become increasingly untenable.

As it happens, alternative splicing does appear to work like this, rendering dubious the textbook notion that particular stretches of DNA are best thought of as "coding for" very specific products or as "controlling" very specific processes. For the purpose of illustration, imagine that a particular segment of genetic material contains information in the order ABCD. Given what molecular biologists now understand about alternative splicing, this segment of DNA could be spliced to yield a variety of different products, including products associated with other orders, such as ACD, BCD, AD, AC, DCBA, BDCA, DA, etc. (Noble 2006). It is as if a sentence that reads "Madison drove Terry to see the dog" could, in different contexts, mean "Terry drove Madison to see the dog," "Madison drove the dog to see Terry," or even "The dog drove Terry mad." It is not yet known for certain if this extreme flexibility characterizes most genes, but molecular biologists acknowledge that alternative splicing is "a universal feature of human genes" (Trafton 2008, p. 6, quoting Burge), so this kind of flexibility is certainly a possibility. Regardless, it has become clear that the idea that our genetic material contains a code that is capable of specifying particular predetermined phenotypic outcomes is false. In fact, genes typically behave as they do at least in part because of how they are effectively instructed to behave by the contexts in which they are operating. Simplistic notions of Nature and Nurture have no explanatory value in a system as complex as this one.

Given how extremely common alternative splicing is, it ought not be treated as a "special case" in biology curricula. Rather, by introducing students to multiple examples wherein different gene products—and consequently different processes and outcomes—are generated in different developmental contexts, such curricula could effectively emphasize that phenotype development is fundamentally a process involving the *co-action* of genetic and non-genetic factors. Such an approach would be an improvement over the still-popular approach that dogmatically emphasizes the one-way flow of information from DNA to phenotypes.

7 Rupturing Reaction Ranges

Because genetic activity is influenced by environmental factors, genes cannot determine the final forms of any of our characteristics independently of the contexts in which development is occurring. In the face of this conclusion, a common fallback position holds that genes can specify a *range* of possible phenotypes, and that the particular environment to which one is exposed dictates which phenotype within the range is the one that develops. In 1963, Gottesman put it this way: "A genotype determines an indefinite but circumscribed assortment of phenotypes, each of which corresponds to one [...] possible environment" (p. 254). Thus, this position effectively holds that what we inherit from our parents is a particular "potential" that may or may not be realized, depending on the experiences we have as we develop. But as intuitively appealing as this so-called "reaction range" idea is, the observed facts of development suggest that thinking about things in this way is not helpful and can actually be quite misleading.

As Platt and Sanislow (1988) explain, "empirical support for the reaction-range concept is questionable" (p. 254); instead, there appear to be no knowable limitations that constrain any particular genotype. This sounds like a radical claim, because it seems obvious that human beings cannot develop from an elephant genome, no matter what sort of environment we allow it to develop in! And in fact, genetic factors do constrain developmental outcomes. But because it is impossible for us to know the limits of any individual's potential, the mere existence of such unknowable constraints cannot have any practical implications for us.

Addressing this issue empirically, Lewontin (2000) discussed studies in which populations of genetically identical plants (*Achillea millefolium*) were allowed to develop in three different environments, namely at either 30, 1,400, or 3,050 m above sea level. Similar studies of *Drosophila melanogaster* examined how animals that had had large portions of their genomes cloned would respond when allowed to develop in a variety of different environments, namely at either 4, 21, or 26 °C. What is clear from all of these studies is that a single genotype, placed in a variety of different environments, can contribute to the development of a variety of different phenotypes. This finding on its own should not surprise anyone who has read this far into this chapter, but the *implication* of this finding is the surprising conclusion that genes cannot circumscribe phenotypes in any knowable way, rendering the range-of-reaction concept valueless. When faced with conclusive data in the mid-1950s that demonstrated that identical genomes react differently to different environments, Theodosius Dobzhansky—one of the key contributors to biology's modern synthesis of Darwinian evolution and Mendelian genetics—concluded that knowing what a particular genotype might be capable of would require empirically testing its development in *all* possible environments. Short of doing this impossibly comprehensive experiment, he wrote, "we can never be sure that any of these traits have reached the maximal development possible with a given genotype" (Dobzhansky 1955, p. 77). Thus, although the range of possible phenotypes associated with a genotype might be discoverable *in theory*, the fact remains that we can never know how a genotype might respond in some not-yet-tested environment; the limits of a genotype's reaction range cannot be known. And in case it was not obvious to readers why the range of all possible environments is infinite (and therefore untestable), Dobzhansky noted that "new environments are constantly produced. Invention of a new drug, a new diet, a new type of housing, a new educational system, a new political regime introduces new environments" (p. 75). As a result of this state of affairs, we can never confidently assert anything about genetic limits on an individual's developmental potential.

What is also clear from the kinds of studies presented by Lewontin (2000) is that different genotypes do not respond to different environments in similar ways. That is, it need not be the case that a genotype associated with the 'best' (or worst) phenotype in one environment is the same genotype associated with the 'best' (or worst) phenotype in a different environment. Instead, different genotypes have different environments that are optimal for them. Writing 16 years earlier, Lewontin addressed this issue directly using cloned corn plants as an example:

> … one genotype may grow better than a second at a low temperature, but more poorly at a high temperature […] modern corn hybrids are superior to those of fifty years ago when tested at high planting densities in somewhat poorer environments, while the older hybrids are superior at low planting densities and in enriched conditions. Plant breeding has then not selected for 'better' hybrids […] Thus genotype and environment interact in a way that makes the organism unpredictable from a knowledge of some average of effects of genotype or environment taken separately (Lewontin et al. 1984, pp. 268–269).

Because genotypes interact with their environments like this, we can never know *prior to performing the manipulation* how changing a person's environment might affect their development; manipulations that might have a desirable effect on one child cannot be guaranteed to have a desirable effect on a different child (or vice versa). Because a genotype associated with a "good" phenotype in one context could be associated with a "bad" phenotype in a different context, it is not possible to identify a particular genotype as generally "superior" or "inferior" to any other genotype. Given this reality, saying anything absolutely true about the "Nature" of anyone's genes is, for all intents and purposes, impossible. What a gene does depends on the environment in which it is operating. As West-Eberhard (2003) summed up the last several decades of thinking in this domain, "evolving organisms are universally responsive to the environment as well as to genes" (p. 3), so the discovery of this kind of developmental plasticity—wherein organisms develop in different ways in different contexts—should not surprise any of us, and educators should begin trying to stress for their students that genes are merely non-deterministic *contributors* to people's physical and psychological characteristics.

8 Influencing Traits

At the end of her 2010 book on Nature and Nurture, Evelyn Fox Keller argued that what "people want to know about" when they ask Nature/Nurture questions is really whether or not a given characteristic can be influenced by the circumstances in which a person develops. Although the answer to this question is now understood to be "yes" in all cases, this is not the final word on the issue. Many people assume that some traits can be *more* influenced by Nurture than can other traits, and further, that some traits can be more *easily* influenced by Nurture than can other traits. These claims seem intuitively reasonable given our experiences with living things, but they are not strictly true.

In many cases when it seems like we cannot influence the development of a trait (or cannot influence its development very much, or very easily), it is only because we do not understand *how* the trait develops. Because we understand that infants growing up around French-speaking adults will learn to speak French, we can manipulate the language a child learns by moving to France. In contrast, in the 1950s, before scientists understood the nature of the metabolic disorder called phenylketonuria (PKU), it appeared as if the development of the mental retardation typical of untreated children with PKU could not be similarly manipulated. These days, it is common to hear the claim that "a single gene is necessary and sufficient to cause [PKU]" (Plomin et al. 2008, p. 32), but although PKU can be understood in this way, our understanding of what this gene *does* permitted the discovery of a dietary manipulation that allows treated individuals to experience normal mental development even if they have the genetic abnormality associated with PKU. Prior to the implementation of this Nurture-based manipulation, the heritability of PKU was high—because human diets are virtually invariable in the extent to which they contain the amino acid associated with PKU, so the presence of PKU was associated with genetic variation only—but now that researchers understand something about the Nature-Nurture interactions that give rise to mental retardation in these cases, influencing outcomes for PKU patients is not particularly difficult. The same will be true of other conditions as we learn more about their development. Traits are likely to appear unchangeable when we do not yet understand how to change them.

It is no accident that the tools of quantitative behavioral geneticists (e.g., twin studies, adoption studies, heritability estimates) have left us with a confused understanding of this fact. In a textbook intended to be the definitive introduction to behavioral genetics, Plomin et al. (2008) note that "quantitative genetics, such as twin and adoption studies, depends on Mendel's laws of heredity but does not require knowledge of the biological basis of heredity" (p. 40). However, it is precisely an understanding of how genes mechanistically do what they do—in interaction with their contexts—that is required to comprehend how it is that highly heritable traits can nonetheless be easily and profoundly influenced by environmental factors.

Of course, just because all characteristics can theoretically be influenced by the contexts in which development occurs does not mean that a knowledgeable scientist could completely control the development of someone's phenotype, because some environmental manipulations are practically impossible to implement. If Keller is right that people who ask Nature-Nurture questions really want to know how easily a characteristic can be influenced by an environmental manipulation, it will not matter to them that the correct answer is "very easily, if you know how the characteristic develops"; such a person really wants to know how easy it might be to implement the manipulation. And because implementation is not always equally easy, all characteristics are not *in practice* equally easy to influence; after all, changing someone's diet, for instance, is currently easier than changing the gravitational field in which they develop!

Similarly, although it is true that all characteristics develop from gene-environment interactions, it matters very much *when* in development various things happen. So even if scientists were able to discover a hypothetical environmental manipulation that, when implemented in infancy, increases the IQ scores that treated babies achieve once they are adults, it could still be the case that after a certain point in development, that manipulation might have no effect on IQ at all. That is, just because it is true that Nurture has a role to play in the development of all of our characteristics does not mean that anything is possible at any given moment. To use another hypothetical example, even if psychologists fully understood the developmental origins of violent behavior, an adolescent's violent behavior could be intractable either because it is too late in her development to significantly influence those behaviors or because the environmental manipulation required to alter her behaviors is technically difficult to implement. As Keller put it, "perhaps we should rephrase the nature-nurture question, and ask, instead, how malleable is a given trait, at a specified developmental age?" (2010, p. 75).

To the extent that what matters to us are these kinds of questions, there is plenty of research still to be done, because scientists currently understand very little about how malleable particular traits are (although this is changing, as suggested by the publication of Bateson and Gluckman's 2011 book on developmental plasticity and robustness). But note that this understanding of Nature-Nurture interactions changes our focus from questions about whether or not particular traits are "innate"—or about how powerfully genetic versus environmental factors influence those traits—to questions about *how* and *when* traits develop. Such a change of focus is bound to help us as we grapple with individuals' and society's problems; in contrast to the correlational approach long used by quantitative behavioral geneticists, a developmental perspective encourages experimentation, and as such, it has the potential to reveal interventions that can actually be used in productive ways to influence developmental outcomes.

9 Conclusion

In the nineteenth and twentieth centuries, some scientists' ideas about Nature and Nurture were used to argue that certain people were inherently inferior to others; the belief that certain characteristics are determined by biology alone led in Germany to the systematic extermination of millions of people (Proctor 1988) and in the United States to large scale programs to sterilize individuals deemed socially "undesirable" (Kevles 1995). Ironically, the more we have learned about genetics in the past 50 years, the more we have come to understand that our characteristics are jointly determined by biological and environmental factors, that is, that all of our characteristics result from a unitary developmental process that relies on both "Nature" and "Nurture" for its functioning. Indeed, Bateson's characterization of the Nature-versus-Nurture debate as a "corpse" is appropriate, because it is clear now that Nature and Nurture are not oppositional influences on development; instead, they work collaboratively.

Although many theorists who read the academic literatures relevant to Nature and Nurture have understood for years that genes interact with their contexts to produce phenotypes, many high school students maintain misunderstandings about genes, for instance that genes operate deterministically (Shaw et al. 2008). It is likely that these misconceptions result from, or are perpetuated by, the content presented in introductory and advanced high school biology textbooks (Castéra et al. 2008; dos Santos et al. 2012; Gericke and Hagberg 2010). The idea that genes operate deterministically seems to be deeply ingrained in us, perhaps because Weismann proclaimed that "the germ-substance" (1894, p. 20) operates deterministically even before the world knew of Gregor Mendel, and 15 years before Johannsen (1911) had even coined the word "gene;" given this long history, it might not be surprising that some educators continue to teach genetics using Punnett squares and other tools that can be mistaken to support genetic determinism (see Jamieson and Radick, this volume). But because our conceptions about genes have such important consequences for all of us, it is important to find ways to teach genetics that convey how genes and environments operate collaboratively in the construction of phenotypes. An excellent way to ensure that this message is passed on to students would be to adopt a pedagogical approach that encourages study of the *emergence* of phenotypes in developmental time. To the extent that textbook writers and educators adopt such a developmental perspective, subsequent generations of students are likely to graduate from school understanding that DNA is merely one factor that contributes to the characteristics we observe in the living things around us.

As I was writing this chapter, the *New York Times* published an Opinion piece entitled "Sorry strivers: Talent matters" (Hambrick and Meinz 2011b), implying that people have some preordained level of competence—talent—that constrains what they can expect to achieve, whether in the arts, sciences, business, or sports, for example. As I indicated previously, it is certainly possible that some of us are in a developmental moment in which practice or striving might not have much influence on what we can achieve. In addition, there can be no doubt that scientists' understanding of how to improve people's performances in many domains is limited, so even if there are ways to improve people's skills, we might still be ignorant of those ways. But regardless of what is or is not possible for a given person to achieve from this moment forward, the idea that we are conceived with some quantity of competence that is predetermined by "Nature" is certainly false. "Talent," like all of our other characteristics, develops; it is not present in a fertilized egg any more than completely formed teeth are present in that same zygote. Thus, it is of as little value to talk about the extent to which "talent" contributes to a competence as it is to talk about the extent to which "Nature" contributes to a competence; what matters is how the competence *develops*. And it is only by studying the development of biological traits, psychological traits, and abilities—think eye color, IQ, or eye-hand coordination—that we can learn how to influence their emergence in individuals (in theory, either through genetic or environmental manipulations).

Hambrick and Meinz conclude their essay by noting pessimistically that "it would be nice if intellectual ability [...] were important for success only up to a point [...] But wishing doesn't make it so [...] Sometimes the story that science tells us isn't the story we want to hear." Intellectual ability *is* important, of course, but we ought not make the mistake of earlier generations and conclude that this ability is somehow unaffected by the experiences we have as we develop. Rather than studying the extent to which competence is influenced by factors we cannot yet control—for example, "working memory capacity" (Hambrick and Meinz 2011a)—we would be much better served by studying the *development* of such factors, so that we can learn how to helpfully influence their emergence. A focus on developmental processes—how they normally work and how we can influence them—rather than on questions about Nature and Nurture, will yield such insights in the future. In this case, the story science tells us is one we very well might want to hear.

Acknowledgments I am very grateful to both Kostas Kampourakis and Lisa Gannett for their helpful comments in response to earlier drafts of this chapter.

References

Aleman, A., E. Bronk, R.P.C. Kessels, H.P.F. Koppeschaar, and J. van Honk. 2004. A single administration of testosterone improves visuospatial ability in young women. *Psychoneuroendocrinology* 29: 612–617.

Amara, S.G., V. Jonas, M.B. Rosenfeld, E.S. Ong, and R.M. Evans. 1982. Alternative RNA processing in calcitonin gene expression generates mRNAs encoding different polypeptide products. *Nature* 298: 240–244.

Archer, J. 2006. Testosterone and human aggression: An evaluation of the challenge hypothesis. *Neuroscience and Biobehavioral Reviews* 30: 319–345.
Bateson, P. 2002. The corpse of a wearisome debate: Review of Steven Pinker's *The blank slate: The modern denial of human nature*. *Science*, 297: 2212–2213.
Bateson, P., and P. Gluckman. 2011. *Plasticity, robustness, development and evolution*. Cambridge: Cambridge University Press.
Beach, F. 1955. The descent of instinct. *Psychological Review* 62: 401–410.
Beach, S.R.H., G.H. Brody, A.A. Todorov, T.D. Gunter, and R.A. Philibert. 2010. Methylation at SLC6A4 is linked to family history of child abuse: An examination of the Iowa adoptee sample. *American Journal of Medical Genetics Part B* 153B: 710–713.
Beutler, E., M. Yeh, and V.F. Fairbanks. 1962. The normal human female as a mosaic of X-chromosome activity: Studies using the gene for G-6-PD deficiency as a marker. *Proceedings of the National Academy of Sciences of the United States of America* 48: 9–16.
Block, N. 1995. How heritability misleads about race. *Cognition* 56: 99–128.
Blumberg, M.S. 2005. *Basic instinct: The genesis of behavior*. New York: Thunder's Mouth Press.
Borghol, N., M. Suderman, W. McArdle, A. Racine, M. Hallett, M. Pembrey, C. Hertzman, C. Power, and M. Szyf. 2012. Associations with early-life socio-economic position in adult DNA methylation. *International Journal of Epidemiology* 41: 1–13. doi:10.1093/ije/dyr147.
Carey, N. 2011. *The epigenetics revolution*. London: Icon Books.
Castéra, J., M. Abrougui, O. Nisiforou, J. Turcinaviciene, T. Sarapuu, B. Agorram, F. Calado, and G. Carvalho. 2008. Genetic determinism in school textbooks: A comparative study conducted among sixteen countries. *Science Education International* 19: 163–184.
Crick, F.H.C. 1958. On protein synthesis. In *Symposia of the Society for Experimental Biology, XII*, The biological replication of macromolecules, 138–163. Cambridge: Cambridge University Press.
Crick, F. 1970. Central dogma of molecular biology. *Nature* 227: 561–563.
Davenport, M.H., and M.R. Cabrero. 2009. Maternal nutritional history predicts obesity in adult offspring independent of postnatal diet. *The Journal of Physiology* 587: 3423–3424.
Davis, S., B. Bozon, and S. Laroche. 2003. How necessary is the activation of the immediate early gene zif 268 in synaptic plasticity and learning? *Behavioural Brain Research* 142: 17–30.
Deater-Deckard, K., S.A. Petrill, L.A. Thompson, and L.S. DeThorne. 2006. A longitudinal behavioral genetic analysis of task persistence. *Developmental Science* 9: 498–504.
DeCasper, A.J., and M.J. Spence. 1986. Prenatal maternal speech influences newborns' perception of speech sound. *Infant Behavior & Development* 9: 133–150.
Dobzhansky, T. 1955. *Evolution, genetics, and man*. New York: Wiley.
dos Santos, V.C., L.M. Joaquim, and C.N. El-Hani. 2012. Hybrid deterministic views about genes in biology textbooks: A key problem in genetics teaching. *Science & Education* 21: 543–578. doi:10.1007/s11191-011-9348-1.
Edwards, D.A., K. Wetzel, and D.R. Wyner. 2006. Intercollegiate soccer: Saliva cortisol and testosterone are elevated during competition, and testosterone is related to status and social connectedness with teammates. *Physiology and Behavior* 87: 135–143.
Eisenberg, L. 2004. Social psychiatry and the human genome: Contextualising heritability. *The British Journal of Psychiatry* 184: 101–103.
Galton, F. 1874. *English men of science: Their nature and nurture*. London: Macmillan.
Galton, F. 1907. *Inquiries into human faculty and its development*. London/New York: J. M. Dent/E. P. Dutton (Originally published in 1883).
Gericke, N.M., and M. Hagberg. 2010. Conceptual variation in the depiction of gene function in upper secondary school textbooks. *Science & Education* 19: 963–994.
Gilbert, S.F. 2003. The reactive genome. In *Origination of organismal form: Beyond the gene in developmental and evolutionary biology*, ed. G.B. Müller and S.A. Newman, 87–101. Cambridge, MA: MIT Press.
Gottesman, I.I. 1963. Genetic aspects of intelligent behavior. In *The handbook of mental deficiency*, ed. N. Ellis, 253–296. New York: McGraw-Hill.
Gottlieb, G. 1991a. Experiential canalization of behavioral development: Theory. *Developmental Psychology* 27: 4–13.

Gottlieb, G. 1991b. Experiential canalization of behavioral development: Results. *Developmental Psychology* 27: 35–39.
Gottlieb, G. 1992. *Individual development and evolution: The genesis of novel behavior*. New York: Oxford University Press.
Gottlieb, G. 1997. *Synthesizing nature-nurture: Prenatal roots of instinctive behavior*. Mahwah: Lawrence Erlbaum Associates.
Gottlieb, G. 2007. Probabilistic epigenesis. *Developmental Science* 10: 1–11.
Griffiths, P.E., and R.D. Gray. 1994. Developmental systems and evolutionary explanation. *The Journal of Philosophy* XCI: 277–304.
Griffiths, P.E., and J. Tabery. 2008. Behavioral genetics and development: Historical and conceptual causes of controversy. *New Ideas in Psychology* 26: 332–352.
Hall, B.K. 1988. The embryonic development of bone. *American Scientist* 76: 174–181.
Hambrick, D.Z., and E.J. Meinz. 2011a. Limits on the predictive power of domain-specific experience and knowledge in skilled performance. *Current Directions in Psychological Science* 20: 275–279.
Hambrick, D.Z., and E.J. Meinz. 2011b. Sorry strivers: Talent matters. *The New York Times*. Retrieved from http://www.nytimes.com. Accessed 24 Mar 2013.
Harper, L.V. 2005. Epigenetic inheritance and the intergenerational transfer of experience. *Psychological Bulletin* 131: 340–360.
Harris, J.R. 2009. *The nurture assumption: Why children turn out the way they do*. New York: The Free Press.
Herrnstein, R.J., and C. Murray. 1994. *The Bell curve: Intelligence and class structure in American life*. New York: The Free Press.
Hoek, H.W., A.S. Brown, and E. Susser. 1998. The Dutch famine and schizophrenia spectrum disorders. *Social Psychiatry and Psychiatric Epidemiology* 33: 373–379.
Jablonka, E., and M.J. Lamb. 2005. *Evolution in four dimensions: Genetic, epigenetic, behavioral, and symbolic variation in the history of life*. Cambridge, MA: The MIT Press.
Johannsen, W. 1911. The genotype conception of heredity. *American Naturalist* 45: 129–159.
Johnson, W., E. Turkheimer, I.I. Gottesman, and T.J. Bouchard. 2009. Beyond heritability: Twin studies in behavioral research. *Current Directions in Psychological Science* 18: 217–220.
Johnston, T.D. 1987. The persistence of dichotomies in the study of behavioral development. *Developmental Review* 7: 149–182.
Johnston, T.D. 2010. Developmental systems theory. In *Oxford handbook of developmental behavioral neuroscience*, ed. M.S. Blumberg, J.H. Freeman, and S.R. Robinson, 12–29. New York: Oxford University Press.
Johnston, T.D., and L. Edwards. 2002. Genes, interactions, and the development of behavior. *Psychological Review* 109: 26–34.
Keller, E.F. 2010. *The mirage of a space between nature and nurture*. Durham: Duke University Press.
Kevles, D.J. 1995. *In the name of eugenics*. Cambridge, MA: Harvard University Press.
King, A.P., M.J. West, and M.H. Goldstein. 2005. Non-vocal shaping of avian song development: Parallels to human speech development. *Ethology* 111: 101–117.
Laland, K.N., J. Odling-Smee, and M.W. Feldman. 2001. Niche construction, ecological inheritance, and cycles of contingency in evolution. In *Cycles of contingency: Developmental systems and evolution*, ed. S. Oyama, P.E. Griffiths, and R.D. Gray, 117–126. Cambridge, MA: The MIT Press.
Lehrman, D.S. 1953. A critique of Konrad Lorenz's theory of instinctive behavior. *The Quarterly Review of Biology* 28: 337–363.
Lewkowicz, D.J. 2011. The biological implausibility of the nature–nurture dichotomy and what it means for the study of infancy. *Infancy* 16: 331–367.
Lewontin, R.C. 1974. Annotation: The analysis of variance and the analysis of causes. *American Journal of Human Genetics* 26: 400–411.
Lewontin, R.C. 1983. Gene, organism and environment. In *Evolution from molecules to men*, ed. D.S. Bendall, 273–285. Cambridge: Cambridge University Press.

Lewontin, R.C. 2000. *The triple helix: Gene, organism, and environment*. Cambridge, MA: Harvard University Press.
Lewontin, R.C., S. Rose, and L.J. Kamin. 1984. *Not in our genes*. New York: Pantheon.
Lickliter, R. 2000. An ecological approach to behavioral development: Insights from comparative psychology. *Ecological Psychology* 12: 319–334.
Lickliter, R. 2008. The growth of developmental thought: Implications for a new evolutionary psychology. *New Ideas in Psychology* 26: 353–369.
Lickliter, R., and T.D. Berry. 1990. The phylogeny fallacy: Developmental psychology's misapplication of evolutionary theory. *Developmental Review* 10: 348–364.
Lickliter, R., and H. Honeycutt. 2010. Rethinking epigenesis and evolution in light of developmental science. In *Oxford handbook of developmental behavioral neuroscience*, ed. M.S. Blumberg, J.H. Freeman, and S.R. Robinson, 30–47. New York: Oxford University Press.
Masataka, N. 1993. Effects of experience with live insects on the development of fear of snakes in squirrel monkeys, *Saimirisciureus*. *Animal Behaviour* 46: 741–746.
McGowan, P.O., A. Sasaki, A.C. D'Alessio, S. Dymov, B. Labonté, M. Szyf, ... and M.J. Meaney. 2009. Epigenetic regulation of the glucocorticoid receptor in human brain associates with childhood abuse. *Nature Neuroscience* 12: 342–348.
McKenzie, C.A., K. Wakamatsu, N.A. Hanchard, T. Forrester, and S. Ito. 2007. Childhood malnutrition is associated with a reduction in the total melanin content of scalp hair. *British Journal of Nutrition* 98: 159–164.
Meaney, M.J. 2010. Epigenetics and the biological definition of gene X environment interactions. *Child Development* 81: 41–79.
Mehta, P.H., and J. Beer. 2010. Neural mechanisms of the testosterone–aggression relation: The role of orbitofrontal cortex. *Journal of Cognitive Neuroscience* 22: 2357–2368.
Mello, C.V., D.S. Vicario, and D.F. Clayton. 1992. Song presentation induces gene expression in the songbird forebrain. *Proceedings of the National Academy of Sciences of the United States of America* 89: 6818–6822.
Mennella, J.A., C.P. Jagnow, and G.K. Beauchamp. 2001. Prenatal and postnatal flavor learning by human infants. *Pediatrics* 107(6): e88.
Moore, D.S. 2002. *The dependent gene: The fallacy of "nature vs. nurture."* New York: Times Books/Henry Holt & Co.
Moore, D.S. 2006. A very little bit of knowledge: Re-evaluating the meaning of the heritability of IQ. *Human Development* 49: 347–353.
Moore, D.S. 2008a. Espousing interactions and fielding reactions: Addressing laypeople's beliefs about genetic determinism. *Philosophical Psychology* 21: 331–348.
Moore, D.S. 2008b. Individuals and populations: How biology's theory and data have interfered with the integration of development and evolution. *New Ideas in Psychology* 26: 370–386.
Moore, D.S. 2011. Evelyn, Fox, Keller. 2010. The mirage of a space between nature and nurture. Durham: Duke University Press. ISBN: 978-0-8223-4731-6, 107 pages, price: $18.95. *Science & Education*. doi: 10.1007/s11191-011-9374-z
Moore, D.S. 2013. Behavioral genetics, genetics, & epigenetics. In *Oxford handbook of developmental psychology*, ed. P.D. Zelazo, 91–128. New York: Oxford University Press.
Niv, S., C. Tuvblad, A. Raine, P. Wang, and L.A. Baker. 2012. Heritability and longitudinal stability of impulsivity in adolescence. *Behavior Genetics* 42(3): 378–392. doi:10.1007/s10519-011-9518-6.
Noble, D. 2006. *The music of life: Biology beyond genes*. New York: Oxford University Press.
Oberlander, T.F., J. Weinberg, M. Papsdorf, R. Grunau, S. Misri, and A.M. Devlin. 2008. Prenatal exposure to maternal depression, neonatal methylation of human glucocorticoid receptor gene (NR3C1) and infant cortisol stress responses. *Epigenetics* 3: 97–106.
Okuno, H., and Y. Miyashita. 1996. Expression of the transcription factor zif268 in the temporal cortex of monkeys during visual paired associate learning. *European Journal of Neuroscience* 8: 2118–2128.
Oyama, S. 2000. *The ontogeny of information*. Durham: Duke University Press.

Pan, Q., O. Shai, L.J. Lee, B.J. Frey, and B.J. Blencowe. 2008. Deep surveying of alternative splicing complexity in the human transcriptome by high-throughput sequencing. *Nature Genetics* 40: 1413–1415.

Pinker, S. 2002. *The blank slate: The modern denial of human nature*. New York: Viking.

Platt, S.A., and C.A. Sanislow. 1988. Norm-of-reaction: Definition and misinterpretation of animal research. *Journal of Comparative Psychology* 102: 254–261.

Plomin, R. 1990. *Nature and nurture: An introduction to human behavioral genetics*. Belmont: Brooks/Cole.

Plomin, R. 1994. *Genetics and experience: The interplay between nature and nurture*. Thousand Oaks: Sage.

Plomin, R., J.C. DeFries, G.E. McClearn, and P. McGuffin. 2008. *Behavioral genetics*, 5th ed. New York: Worth.

Proctor, R. 1988. *Racial hygiene: Medicine under the Nazis*. Cambridge, MA: Harvard University Press.

Robert, J.S. 2004. *Embryology, epigenesis, and evolution: Taking development seriously*. New York: Cambridge University Press.

Rosen, K.M., M.A. McCormack, L. Villa Komaroff, and G.D. Mower. 1992. Brief visual experience induces immediate early gene expression in the cat visual cortex. *Proceedings of the National Academy of Sciences of the United States of America* 89: 5437–5441.

Rusak, B., H.A. Robertson, W. Wisden, and S.P. Hunt. 1990. Light pulses that shift rhythms induce gene expression in the suprachiasmatic nucleus. *Science* 248: 1237–1240.

Sesardic, N. 2005. *Making sense of heritability*. Cambridge: Cambridge University Press.

Shaw, K.R.M., K. Van Horne, H. Zhang, and J. Boughman. 2008. Essay contest reveals misconceptions of high school students in genetics contest. *Genetics* 178: 1157–1168.

Shuel, R.W., and S.E. Dixon. 1960. The early establishment of dimorphism in the female honeybee, *Apis mellifera,* L. *Insectes Sociaux* 7: 265–282.

Small, M.F. 2011. *Kids: How biology and culture shape the way we raise young children* [Kindle Edition]. New York: Anchor Books.

Stotz, K. 2006. With 'genes' like that, who needs an environment? Postgenomics's argument for the 'ontogeny of information'. *Philosophy of Science* 73: 905–917.

Thornhill, R., and C.T. Palmer. 2000. *A natural history of rape: Biological bases of sexual coercion*. Cambridge, MA: The MIT Press.

Trafton, A. 2008. Human genes sing different tunes in different tissues. *MIT TechTalk* 53(8): 6.

Wallman, J. 1979. A minimal visual restriction experiment: Preventing chicks from seeing their feet affects later responses to mealworms. *Developmental Psychobiology* 12: 391–397.

Walton, K.D., S. Harding, D. Anschel, Y.T. Harris, and R. Llinás. 2005. The effects of microgravity on the development of surface righting in rats. *The Journal of Physiology* 565: 593–608.

Wang, E.T., R. Sandberg, S. Luo, I. Khrebtukova, L. Zhang, C. Mayr, ... and C.B. Burge. 2008. Alternative isoform regulation in human tissue transcriptomes. *Nature* 456: 470–476.

Weaver, I.C.G. 2007. Epigenetic programming by maternal behavior and pharmacological intervention: Nature versus nurture: Let's call the whole thing off. *Epigenetics* 2: 22–28.

Weaver, I.C.G., N. Cervoni, F.A. Champagne, A.C. D'Alessio, S. Sharma, J.R. Seckl, ... and M.J. Meaney. 2004. Epigenetic programming by maternal behavior. *Nature Neuroscience* 7: 847–854.

Weismann, A. 1894. *The effect of external influences upon development*. London: Frowde.

West-Eberhard, M.J. 2003. *Developmental plasticity and evolution*. New York: Oxford University Press.

Wheeler, M. 2010. Obesity gene, carried by more than a third of the U.S. population, leads to brain tissue loss. *UCLA Newsroom*. Retrieved from http://newsroom.ucla.edu/portal/ucla/obesity-gene-carried-by-half-the-157028.aspx. Accessed 24 Mar 2013.

Yamagata, S., A. Suzuki, J. Ando, Y. Ono, N. Kijima, K. Yoshimura, et al. 2006. Is the genetic structure of human personality universal? A cross-cultural twin study from North America, Europe, and Asia. *Journal of Personality and Social Psychology* 90: 987–998.

Genomics and Society: Why "Discovery" Matters

Lisa Gannett

1 Introduction

In this post-Human Genome Project period, genomics has overtaken genetics. Research in genomics focuses on whole genomes rather than single genes, and has been facilitated by the development of technologies and statistical tools that permit the investigation of cellular interactions involving multiple sites across the genome.[1] Although research in genetics has always been considered to be of great practical value—with eugenics and agricultural breeding never far from sight—the social context in which genomics research is carried out is increasingly commercialized. Concerns have been raised about whether the ideal of value-free science, understood as the curiosity-driven pursuit of knowledge for its own sake, has been lost. Worries have been expressed that patent applications, consulting agreements, and reliance on industry grants are serving to compromise the objectivity of scientists, especially those working at publicly-funded universities, whose role it has been to engage in theoretical inquiry that protects society's long-term interest in the accumulation of fundamental knowledge (e.g., Brown 2008; Reiss 2010; Biddle 2012).

The chapter begins by considering ways in which philosophers of science have traditionally defended the objectivity of science against the intrusion of values from the social context within which science is practised, namely, by drawing three distinctions: between theory and practice (or basic and applied science), the context of discovery and context of justification, and facts and values. Regarding the distinction between theory and practice (or basic and applied science), the contention is that commercial applications follow downstream from theoretical developments in a

[1]This shows the inadequacy of descriptions of genes as distinct units on chromosomes in control of specific functions (see Burian and Kampourakis, this volume).

L. Gannett (✉)
Department of Philosophy, Saint Mary's University, Halifax, B3H 3C3, NS, Canada
e-mail: Lisa.Gannett@SMU.CA

K. Kampourakis (ed.), *The Philosophy of Biology: A Companion for Educators*, History, Philosophy and Theory of the Life Sciences 1, DOI 10.1007/978-94-007-6537-5_28,

science. Regarding the distinction between the context of discovery and context of justification, the contention is that while scientists might be motivated by profit in their choice of a research problem, logic and empirical data alone serve to justify the hypotheses they propose as solutions. Regarding the distinction between facts and values, the contention is that commercial interests and other values are subjective since unlike facts, which are objective, they are not given in sense experience. The chapter goes on to argue that these three defences of scientific objectivity and the value-free ideal in science cannot succeed in insulating research in genomics from the commercialized social context in which it is carried out. Instead, we find values all the way down. Theory is embedded in practice: prospective practical applications structure what theoretical knowledge is sought. Discovery matters: values embedded in the context of discovery inform the context of justification and become incorporated in the body of scientific knowledge. Facts and values are entangled: facts are not wholly empirical and detached from values. The chapter then presents a case study, specifically, the recent invention of the concept of biogeographical ancestry as a substitute for "race" in population genomics. The three traditional ways of construing the value-free ideal readily point to the value-ladenness of this research program, situated as it is in a commercial context and serving a range of practical aims, including the mapping of genes implicated in disease, forensic identification, and direct-to-consumer genealogy. But if, as the chapter argues, values go all the way down, the various defences of the value-free ideal no longer provide convincing bases for arbitrating between "good" and "bad" science. The chapter concludes by outlining some implications of these arguments for the status of epistemology in philosophy of science and for biology education.

2 The Ideal of Value-Free Science

Philosophers of science have traditionally defended the objectivity of science against the intrusion of values from the social context within which science is practised, namely, by drawing three distinctions: between theory and practice (or basic and applied science), the context of discovery and context of justification, and facts and values. This section traces the theory-practice distinction back to Aristotle in ancient times, Comte in the nineteenth century, and Polanyi and Cohen in the post-WWII period; the discovery-justification distinction to contributions made by Popper and Reichenbach in the 1930s; and the fact-value distinction to Hume in the eighteenth century and Poincaré early in the twentieth century.

2.1 Theory vs. Practice

In her chapter "Critical Silences in Scientific Discourse: Problems of Form and Re-Form," Evelyn Fox Keller (1992) uses contrasting metaphors of illumination to illustrate (and then critique) the value-free ideal of science as "pure", wholly theoretical, driven by curiosity alone, and separate from practical concerns. Keller begins with the view of "a distinguished physicist" that knowledge is "an expanding sphere of light in a background of darkness. Its only directionality is outward; it just grows, without direction, and without aim" (p. 81). By implication, this conception of science as "an undirected search for knowledge" (p. 77) holds that it is value-free. There is no preferred end, whether theoretical or practical, conscious or unconscious, that science is to serve. Presumably sustained merely by the diverse intellectual curiosities of individual scientists, knowledge grows directly outward, whilst maintaining spherical contours. This view of knowledge resembles the positivist ideal of value-free science, which rests on related distinctions between theory and practice, and basic and applied science.

The distinction between theory and practice goes back to Aristotle (1984). In *Nichomachean Ethics*, Aristotle distinguishes between three kinds of knowledge: productive knowledge (*techne*), practical knowledge (*praxis*), and contemplative knowledge (*theoria*). Productive knowledge is about making and relates to the fabrication of art and artifacts. Practical knowledge is about doing and includes ethics and politics. Contemplative knowledge is about thinking and concerns the search for "first principles". Aristotle privileges contemplative knowledge over productive and practical knowledge: its objects are necessary, constant, and eternal rather than contingent, changing, and temporary; it employs reason in search of truth for its own sake rather than for the sake of fulfilling specified ends. Aristotle argues that society should be organized in a way that allows a leisure class of male citizens to engage in contemplation as the highest form of human activity.

Historian of science Robert Proctor (1991) argues that the distinction between pure and applied science is a product of the nineteenth century (p. 68). For August Comte, as for Aristotle, different groups of people engaging in different sorts of activities are responsible for different forms of knowledge: scientists rely on speculation to arrive at the theoretical knowledge that is characteristic of "pure science"; production managers rely on action to arrive at the practical knowledge that is characteristic of technology, art, and industrial production; and engineers mediate between speculation and action, and theory and practice, in the production of "applied science". But in contrast to Aristotle's treatment of theory and practice as autonomous, Comte institutes an order that places theory epistemologically and temporally prior to practice. According to Comte, human intelligence has a "fundamental need" to understand nature, discern the causes of phenomena, and systematize facts. Theoretical knowledge provides the rational basis for human action because knowledge of nature's laws lends a predictive capacity that permits phenomena to be anticipated and modified by intervening in the causal chain. In Comte's memorable words: "science = foresight; foresight = action." However, theorizing must be carried out "totally removed from any practical consideration" (1974, p. 45). Only subsequently do nature's laws, discovered by scientists for knowledge's own sake, become tools in the hands of engineers, and brought to bear on applications in the service of industrialists' desires.

The nineteenth-century positivist distinction was taken up and reinvented in the period immediately following WWII as the distinction between basic and applied science, a distinction that remains familiar to us today. This development was motivated by social conditions of the time, somewhat different in Britain and the U.S.

In depression-era Britain during the 1930s, the idea of the social organization of science had become quite popular. This idea that science should be organized in ways that are conducive to fulfilling the interests of society as a whole, i.e., the working class and not just the elite, was promoted by the biophysicist J. D. Bernal and fellow Marxist and socialist scientists. The prospect was alarming to other scientists, however, among them the Hungarian-born chemist-turned-philosopher Michael Polanyi. At a December 1945 meeting on the planning of science held by a division of the British Association for the Advancement of Science, Polanyi's opening address defended the "social message of pure science". Polanyi blamed Marxism for the rise of fascism in Europe and destruction brought about by the war, and he called upon his fellow scientists to join the battle against Marxism by protecting "pure science" against the efforts of the social planners: "In the great struggle for our civilization science occupies a section on the front line. In the movement which is undermining the position of pure science I see one detachment of the forces assailing our whole civilization" (1951, p. 7). Scientists must "reassert that the essence of science is the love of knowledge and that the utility of knowledge does not concern us" (p. 6), thereby contributing to the preservation of truth as a value "more precious" than the "material welfare" that "applied science" serves (p. 6). To Polanyi's delight, he found that the popularity of the idea of the social organization of science had faded for his postwar audience.

In the U.S., the technological feats accomplished with the government's unprecedented support of science during the war so impressed members of Congress that they were open to maintaining federal funding for science when the war ended. The National Science Foundation was the result, though it did not come into existence until 1950 because of disagreements over its mandate and structure. Harvey Kilgore, a West Virginia senator, introduced legislation as early as 1942 to establish an agency under political control that would award grants and contracts for both basic and applied science in the social as well as natural sciences (Mazuzan 1994). Scientists welcomed the prospect of increased government funding for peacetime research, but they were wary of political control, tired of restrictions on scientific communication arising from government-imposed secrecy on the Manhattan and other wartime projects, and anxious to return to their labs—to scientific problems they found interesting, regardless of their potential applications. Engineer and science administrator Vannevar Bush, who headed the Office of Scientific Research and Development for the government during the war, in his 1945 report to President Roosevelt, *Science: The Endless Frontier*, argued that only "basic research" in medicine and the natural sciences, carried out at colleges, universities, and research institutes, should be supported, with scientists in control of funding decisions (Mazuzan 1994).

Historian of science I. B. Cohen wrote his 1948 book, *Science, Servant of Man: A Layman's Primer for the Age of Science*, as a philosophical defence of the postwar science policy advocated by Bush. Cohen draws on historical case studies to convince the U.S. public that investing in science would be money well spent.

Cohen adopts the positivist approach of Comte. He describes scientists as "dreamers" who, "activated by scientific curiosity and the desire to find the facts of nature and to master them", contribute to the advancement of civilization (p. 293). He characterizes science as proceeding as an undirected search for fundamental knowledge, focused entirely on discovery without regard for application. He contends that knowledge precedes application, often by significant stretches of time, and therefore that scientists are unable to foresee the uses to which their contributions will be put: "Readers of this book", Cohen writes, "should know full well that it is absolutely impossible in general to predict today what use will be made of fundamental knowledge tomorrow" (p. 294). Given its lack of direction and foresight, science is neutral concerning its prospective uses: "a double-edged sword" that can be used for both good and bad purposes in the military, industrial, clinical, and other contexts in which it is put to use. Only 3 years after Hiroshima and Nagasaki, Cohen was effectively absolving science and scientists of moral responsibility. For Cohen, as for Aristotle and Comte, science and its applications are in different hands: scientists may have discovered the physical theories behind the bombs, but engineers built them, and politicians decided to drop them. According to Cohen, "Science acts simply as a most efficient servant" (p. 288).

The distinction between theory and practice and its post-war formulation as a distinction between basic and applied science was also taken up by logical empiricism. Hans Reichenbach was among the logical empiricist émigrés from continental Europe who so strongly influenced the formation of philosophy of science as an academic discipline in North America. In his 1951 book, *The Rise of Scientific Philosophy*, Reichenbach claims, along Comtean lines, that "knowledge does not include any normative parts" (p. 277). Reichenbach distinguishes applied sciences like medicine and engineering from theoretical sciences like physics. He refers to "the urge to know, satisfied through studying scientific books or making scientific experiments" (p. 314), which motivates theoretical scientists. He distinguishes "functional" and "transcendental" conceptions of knowledge, and, in contrast to Polanyi, sides with the former "which regards knowledge as an instrument of prediction" (p. 252), rather than the latter which regards knowledge as a source of truth: "knowledge does not refer to another world, but portrays the things of this world so as to perform a function serving a purpose, the purpose of predicting the future" (p. 255). There is no mention of ends that might be served by such predictive capacities—the goal is prediction itself. As for applied science, Reichenbach adheres to the fact-value dichotomy and subjectivist or emotivist ethics that logical positivists and logical empiricists shared with their predecessors Hume and Poincaré (discussed below).

2.2 Context of Discovery vs. Context of Justification

Keller (1992) rejects the "distinguished physicist's" conception of science that rests on the positivist ideal of science as value-free since separated from practical concerns and technological applications. She contends that it is more accurate to regard scientific inquiry as "a great searchlight", with its beam directed by specific interests (p. 84), not "an expanding sphere of light in a background of darkness" (p. 81). Philosophers of science often use the distinction between the "context of discovery" and "context of justification" to reject this genre of science criticism.

Carl Hempel's (1966) account of scientific objectivity appeals to the distinction between the context of discovery, in which hypotheses and theories are proposed, and the context of justification, in which hypotheses and theories are accepted: "scientific objectivity is safeguarded by the principle that while hypotheses and theories may be freely invented and *proposed* in science, they can be *accepted* into the body of scientific knowledge only if they pass critical scrutiny, which includes in particular the checking of suitable test implications by careful observation or experiment" (p. 16). Thus, problem choice, fixed by Keller's "great searchlight", its beam directed by specific interests, belongs to the context of discovery, where it is admitted that values play a role in generating theories. Only the context of justification matters for objectivity, as it is here that logical relations alone are supposed to mediate between facts and theories, leaving no room for values. The examples Hempel uses to support the irrelevance of the origins of hypotheses and theories to their objectivity include the dream that inspired Kékulé's discovery of benzene's ring structure and the mysticism that inspired Kepler's planetary model. Hempel's choice of examples reflects the anti-psychologistic origins of the discovery-justification context distinction.

The expressions "context of discovery" and "context of justification" are introduced by Reichenbach in the early pages of his 1938 book, *Experience and Prediction*. However, the distinction Hempel makes between the proposal and acceptance of hypotheses and theories goes back earlier. John Stuart Mill, in criticizing William Whewell's views on induction, distinguishes between "Invention and Proof" as early as 1872 (Forster 2011, p. 102). Karl Popper, although critical of Mill's inductivism, in his defence of the hypothetico-deductive method in his 1934 *The Logic of Scientific Discovery*, incorporates the distinction between inventing or proposing theories, which he dubs the "psychology of knowledge", and proving or accepting theories, which he dubs the "logic of knowledge".

Popper's distinction between the psychology of knowledge and logic of knowledge is used to characterize two different aspects of "the work of the scientist": "putting forward and testing theories" (1985, p. 133). The logical concerns of philosophers have only to do with methods of testing theories. Questions about how theories originate are left to psychologists: "The initial stage, the act of conceiving or inventing a theory, seems to me neither to call for logical analysis nor to be susceptible of it. The question how it happens that a new idea occurs to a man—whether it be a musical theme, a dramatic conflict, or a scientific theory—may be of great interest to empirical psychology; but it is irrelevant to the logical analysis of scientific knowledge" (p. 133). Since the construction of theories involves "'an irrational element' or 'a creative intuition', in Bergson's sense" (p. 134), writes Popper, there can be no rational reconstruction of "the steps that have led the scientist to a discovery" (p. 134).

Reichenbach introduces the distinction between the context of discovery and context of justification, like Popper, as part of a demarcation project that establishes the objectivity of philosophy of science in contrast to other disciplines—for Reichenbach, sociology and psychology. Because Reichenbach considers knowledge to be a "sociological phenomenon" (1938, p. 3), he holds that epistemology is part of sociology as far as its descriptive task is concerned. In this descriptive task, epistemologists are interested only in "internal relations" that belong to the "content of knowledge" (p. 4): e.g., questions about the meanings of concepts, presuppositions of scientific method, and truth of sentences. Sociologists, in contrast, are interested in knowledge's "external relations": e.g., questions about the construction of technologies and class backgrounds of scientists. Although he initially suggests that epistemologists, like sociologists, describe knowledge "as it really is" (p. 3), Reichenbach retreats from this when he subsequently distinguishes between the approaches epistemology and psychology take to describing internal relations. Whereas psychologists describe how scientists actually think, epistemologists describe how scientists ought to think—an idealization that Reichenbach calls "rational reconstruction" (p. 5). Reichenbach introduces the distinction between the context of discovery and context of justification in order to delineate these separate concerns of psychologists and epistemologists. Psychologists are concerned with what happens in the heads of individual scientists when they come up with theories in the context of discovery. Epistemologists are concerned with making logical interconnections between thoughts explicit, as scientists might present their work to others, by rationally reconstructing theories in the context of justification. Epistemology's critical task, which assesses the logical validity of inferences, can then be executed. Discovery lacks the rational structure of justification: "the scientific genius has never felt bound to the narrow steps and prescribed courses of logical reasoning" (p. 5).

Thus, Popper's distinction between the psychology of knowledge and logic of knowledge and Reichenbach's distinction between the context of discovery and context of justification alike mark off philosophy of science's professional turf from psychology's. The distinctions are not identical, however. Popper's logic of knowledge involves the logical analysis of scientists' efforts to test theories, whereas Reichenbach's context of justification involves the logical analysis of scientists' theorizing as rationally reconstructed. Both distinctions, as well as the hypothetico-deductivism to which they are tied, have been influential in the philosophy of science, though Reichenbach's terminology won out. The discovery-justification distinction that emerged requires philosophers to attend only to the logical relations that obtain between observed facts and theories in the context of justification, what is supposed to be the rational and objective side of science. Anything supposed to be irrational or subjective can be left to psychologists. This places great confidence in the ability of rational decision procedures in the context of justification to guard against the intrusion into the content of science of any scientifically suspect ideas.

Reichenbach marks off philosophy of science's professional turf from sociology's by the distinction between internal and external relations, not the discovery-justification distinction. Writes Reichenbach: "A sociologist, for instance, might report that astronomers construct huge observatories containing telescopes in order to watch the stars, and in such a way the internal relation between telescopes and stars enters into a sociological description" (p. 4). Reichenbach assumes that external relations have no bearing on meaning, method, or evidence. And while the sociologist may be enriched by philosophical knowledge about internal relations, the philosopher has nothing to gain from sociological knowledge about external relations, notwithstanding the construal of epistemology's descriptive task as broadly speaking sociological. The discovery-justification distinction is introduced in order to separate the epistemologist's concerns with internal relations from the psychologist's—not the sociologist's whose concerns are with external relations. But among Reichenbach's fellow logical empiricists, the context of discovery was soon characterized in ways that expanded it to include "questions of the psychology and sociology of scientific discovery" (Nagel 1939, p. 226) and "the psycho-socio-historical aspects of science" (Feigl 1950, p. 187). Hence, the distinction between the context of discovery and context of justification came to demarcate epistemology not just from psychology but sociology and history, with philosophy of science continuing to claim rationality and objectivity for itself alone. Influences belonging to the social context in which science is practised were considered as irrational and subjective as Kékulé's dream, Kepler's mysticism, Popper's creative intuition, or Reichenbach's scientific genius.

2.3 *Facts vs. Values*

The distinction between facts and values represents facts as objective, based only in our sense impressions of a shared empirical world, and values as subjective, based in the interests, preferences, desires, or emotions of individuals. As mentioned already, in his conception of applied science, Reichenbach adheres to the fact-value dichotomy and a subjectivist ethics. Scientific attention to the ends of applied sciences can be directed only to the urges and habits that shape individual desire, and, like the context of discovery, this falls into the realm of the psychologist, not the epistemologist: "The decision for a goal is not an action comparable to the recognition of truth … the choice of the goal is not a logical act. It is the spontaneous affirmation of desires or volitions" (p. 314). Logical analysis can do no more than clarify relations between ends and means. Once subjective ends have been chosen, science can objectively determine which means will best bring about the desired ends. The fact-value distinction construed in this way by logical positivists and logical empiricists such as Reichenbach is a reflection of the influence on the development of that research tradition by eighteenth-century empiricist David Hume and nineteenth-century positivist Henri Poincaré.

It is in Hume's (1740) *A Treatise of Human Nature* that we find the prohibition against inferring an "ought" from an "is" for which he is well known, introduced almost as an afterthought:

> I cannot forbear adding to these reasonings an observation, which may, perhaps, be found of some importance. In every system of morality, which I have hitherto met with, I have always remark'd, that the author proceeds for some time in the ordinary way of reasoning,

> and establishes the being of a God, or makes observations concerning human affairs; when of a sudden I am surpriz'd to find, that instead of the usual copulations of propositions, *is*, and *is not*, I meet with no proposition that is not connected with an *ought*, or an *ought not*. This change is imperceptible; but is, however, of the last consequence. For as this *ought*, or *ought not*, expresses some new relation or affirmation, 'tis necessary that it shou'd be observ'd and explain'd; and at the same time that a reason shou'd be given, for what seems altogether inconceivable, how this new relation can be a deduction from others, which are entirely different from it. (p. 302)

Hume is not simply saying that sentences containing the word "ought" are not derivable from sentences containing the word "is"; he is making the theoretical argument that "ought" expresses a "relation or affirmation" that is "entirely different" than relations expressed by "is".

The relation or affirmation expressed by a moral "ought" or "ought not" concerns the "sentiments" that arise in us when we contemplate such "objects" as people's characters or actions. In Hume's classification of "perceptions" into "impressions" and "ideas", these sentiments count as impressions: the "agreeable" impression associated with virtue is a particular sort of pleasure, and the "uneasy" impression associated with vice is a particular sort of pain. "Morality, therefore," he writes, "is more properly felt than judg'd of" (p. 302). Judgements of truth and falsity enter only with the operation of reason on ideas, whether this involves the discovery of necessary relations in mathematics or contingent matters of fact in empirical science. Hume points out that because the moral sentiments arise in our subjective responses to objects, they are not candidates for truth and falsity and fall outside reason's purview and science itself, much like the secondary qualities: "Vice and virtue, therefore, may be compar'd to sounds, colours, heat and cold, which, according to modern philosophy, are not qualities in objects, but perceptions in the mind" (p. 301). "Morals excite passions, and produce or prevent actions," Hume tells us, and "[a]n active principle can never be founded on an inactive" (p. 294). Therefore, reason can do no more for morality than provide evidence of the existence of an object that "excites a passion" and discover the connection of causes and effects "so as to afford us means of exerting any passion" (p. 295).

In a 1913 essay, "Ethics and Science", Poincaré reiterates the logical fallacy introduced by Hume, but, anticipating the linguistic predilections of his logical positivist and logical empiricist successors, presents it as the "purely grammatical" impossibility of deriving imperatives from indicatives:

> If the premises of a syllogism are both in the indicative, the conclusion will also be in the indicative. For the conclusion to have been stated in the imperative, at least one of the premises must itself have been in the imperative. But scientific principles and geometrical postulates are and can be only in the indicative. Experimental truths are again in that same mood, and at the basis of the sciences, there is and there can be nothing else. That being given, the most subtle dialectician can juggle these principles as he may wish, combine them, and pile them up on one another. All that he will derive from this will be in the indicative. He will never obtain a proposition which will state: do this, or, do not do that; that is, a proposition which affirms or which contradicts morality. (p. 103)

Poincaré argues that, consequently, there can be neither a scientific morality nor an immoral science: he addresses the impossibility of a scientific morality to "the hopes" of those who want science "to place moral truths beyond all contestation as it has done for the theorems of mathematics and the laws stated by the physicists" (p. 102), and the impossibility of an immoral science to "the fears" of those who worry that scientific materialism will weaken or destroy morality. Cohen's (1948) formulation of postwar science policy similarly defends the value-freedom of science to those who would make science the "scapegoat" for wartime destruction: "Science does not, nay cannot, recognize of itself any distinction between right and wrong, and hence it cannot concern itself with moral issues. Science is never an *immoral* activity, but rather *amoral* – completely removed from the moral sphere" (p. 287).

Like Hume, Poincaré defends a subjectivist ethics forged by the sentiments, not reason. "The moral motor," he writes, "can only be something felt. It cannot be proved that we must have pity for the unfortunate; but let us be confronted with undeserved misery, a spectacle which is, alas! only too frequent, and we shall find ourselves aroused by a feeling of revolt" (p. 104). Also like Hume, and later Reichenbach, Poincaré points out science's contributions to morality is limited to providing the hypothetical reasoning that allows the actions demanded by moral imperatives to be fulfilled: "feeling provides us only with a general motive power for action. It will provide us with the major premise of our syllogism which, suitably to the occasion, will be in the imperative. Science, for its part, will provide us with the minor premise, which will be in the indicative, and will draw from it the conclusion, which can be in the imperative" (p. 105). Sometimes, Poincaré says, this hypothetical reasoning can play a role in resolving moral controversy by showing that the sentiments in apparent conflict can mutually satisfy their ends through a course of action. Hypothetical reasoning also underlies Cohen's defence of the value-neutrality of science and his denial of responsibility on scientists' parts—other than as citizens—for applications of their research: "Like most of the activities of man, science is a double-edged sword and can cut in either of two directions with equal facility [...] Science says only that if you follow a certain procedure x, then a certain result y will be sure to follow, and it will follow whether what you are doing is morally right or morally wrong [...] Science, the servant of man, will help you to attain whatever end you desire" (pp. 287–8).

Poincaré emphasizes more so than Hume the variability of the feelings upon which morality is based and the different degrees to which they are found in different people: "In some souls, some feelings predominate while in other souls other strings are always ready to vibrate" (p. 108). In doing so, Poincaré moves away from Hume's version of subjectivism that rests moral commonalities on the uniformity of human nature—recall his comparison of vice and virtue to secondary qualities like sounds, colours, heat and cold. In contrast to Hume and Poincaré, Reichenbach mentions not sentiments and feelings as the ends that motivate our actions but desires and volitions. In doing so, Reichenbach erases the distinction Hume draws between morality and self-interest: for Hume, moral sentiments or feelings are involved only when we consider people's characters and actions in a general sense, and not with respect to our particular interests. But as Hilary Putnam (2002) points out, imperatives driven by volitions cannot be rationally justified, unlike the categorical imperatives of Kant (p. 17). Ultimately, the influence of the logical positivists and logical empiricists on the development of philosophy of science in North America saw values relegated to the realm of subjectivity, as the idiosyncratic interests, preferences, desires, or feelings of individuals.[2]

3 Values All the Way Down

This section draws on criticisms of the value-free ideal in science by pragmatist, post-positivist, and feminist philosophers of science, including John Dewey, Hilary Putnam, Thomas Kuhn, Paul Feyerabend, Helen Longino, and Kathleen Okruhlik, to argue that it is values all the way down. The scientific production of knowledge is not insulated from the social, political, and economic contexts in which it occurs: theory is embedded in practice, the context of discovery matters to the context of justification, and facts and values are entangled.

3.1 Practice-Embedded Theory

Dewey's pragmatism rejects the mantle twentieth-century logical positivists and logical empiricists (and many of their critics) inherited from Comte. In "Philosophy's Search for the Immutable", Dewey challenges the Comtean construal of relations between theory and practice, knowledge and action. He warns readers not to be misled into believing that these are legitimate divisions just because people specialize in one or the other and wrangle over which is more important. According to Dewey, individual minds have a practical not epistemological function: "the ultimate ground of the quest for cognitive certainty is the need for security in the results of action [...] it is a strict truism that no one would care about *any* exclusively theoretical uncertainty or certainty" (1998 [1929], p. 108). By this, Dewey means that the ostensibly theoretical doubt that motivates scientific inquiry cannot arise unless some consequence bears on it. Without desires or purposes, one state of affairs matters no more than any other. On this view, Reichenbach's conception of "knowledge as an instrument of prediction" makes no sense unless embedded in a practical context in which greater value is attached to predicting some phenomena over others; nor does Cohen's characterization of scientists as "dreamers" motivated by directionless curiosity alone.

And yet, Reichenbach (1951) characterizes knowledge as a "sociological phenomenon" and scientists as engaged in an activity, with scientific activity, like any other activity in society, directed by goals: "In some sense, every human activity

[2] As Putnam (2002) argues, the analytic-synthetic distinction served to reinforce the dichotomizing of facts and values. For evaluative and normative claims to count as knowledge, they had to be either analytic or synthetic, and synthetic claims had to be facts based on direct sense experience.

serves the pursuit of a goal [...] In all such activities [...] there are moments where a choice is to be made; it is here that behavior exhibits valuation" (p. 313). Fellow logical positivist/logical empiricist Otto Neurath expresses a similar view: "The pursuit of sociology, of mathematics, of biology are activities like any other. Hence trends in scientific research are never socially neutral" (in Uebel 2000, p. S140). Reichenbach's reference to the evaluative dimension of scientists' choices of goals to pursue recalls an earlier discussion in *Experience and Prediction*. There, Reichenbach presents a scientist's decision about the purpose of inquiry (whether enjoyment, truth, or prediction) as an example of a volitional bifurcation. Reichenbach recognizes that not all decisions in science are governed by logical relations of truth and validity; some decisions are volitional. He distinguishes two kinds of volitional decisions: conventions and bifurcations. Conventions are volitional decisions that do not influence the content of knowledge because they involve logically equivalent conceptions, such as unit of length. Bifurcations are volitional decisions that do influence the content of knowledge. Reichenbach offers examples additional to scientists' decisions about the purpose of inquiry, which include scientists' choices about how to delimit (nonconventional) meanings for concepts and which (non-equivalent) language to use. Reichenbach characterizes volitional bifurcations as involving decisions that are not arbitrary but "of the greatest relevance" since they "will lead to consequences concerning the knowledge obtainable" (p. 147).

Notwithstanding this representation of volitional bifurcations as "of the greatest relevance", Reichenbach's examples do not sufficiently embrace the range of nonconventional decisions made by scientists and the influence they exert on the content of scientific knowledge. Consider the possible purposes of scientific inquiry he entertains: enjoyment (which he dismisses as being characteristic of play, not science), truth, and prediction. What about standardization of an experimental technique? What about treating disease? What about applying for patent rights? Keller (1992) aptly characterizes theories as "instruments" that serve as "vectors"; she writes: "scientific theories are tools for changing the world [...] Neither instruments themselves, nor the values, interests or efficacy associated with them are devoid of aim" (pp. 73–4). As we have seen, Keller argues that science is not "an undirected search for knowledge," metaphorically portrayed as "an expanding sphere of light in a background of darkness"; instead, scientific inquiry resembles "a great searchlight," with its beam directed by specific interests. The positivist ideal of science as value-free is mistaken, and while enjoyment, truth, and prediction may be among the purposes for which scientists engage in scientific activity, as Bernal points out in his 1939 *Social Function of Science*, science has a social function quite apart from the ways in which scientists conceive their work: just because scientists believe they are engaged in a search for truth or find science a noble or amusing pursuit does not mean that science is not directed to social and economic uses that satisfy the interests of those who finance it (p. 9).

Going beyond her alternative metaphor of scientific inquiry as a great searchlight, Keller suggests that practical concerns and technological applications do not just direct scientific inquiry in some directions rather than others but embed ostensibly theoretical fields of inquiry. In other words, the social interests and value preferences implicated in raising some questions but not others, isolating some phenomena but not others, and advancing some models but not others permeate the science we have. Eschewing ideals of not only value-free but value-neutral science, Keller says that genetics and nuclear physics contain within them "not only the possibility but the expectation" of eugenics and the bomb (p. 77). Should inquiry be directed to "the task of changing the world in different ways—perhaps, as some have hoped, giving us solar energy, rather than nuclear power; ecological rather than pathogenic medicine; better rearing rather than better breeding of our offspring" (p. 92), we would have a different science than we do. Dewey, with his pragmatist's conception of knowledge as instrumental, guided by interests and values, said something very similar in the aftermath of the August 1945 bombings of Hiroshima and Nagasaki by the U.S. Dewey (1981 [1945]) tells us that fears about the atomic bomb cannot be quelled without refusing power as a means of conflict resolution—that there cannot be an emancipatory science and technology without an emancipatory society, and that a science and technology that promotes human well-being must operate "within, not just outside of and against, the moral values and concerns of humanity" (p. 202).

3.2 "Discovery" Matters

The discovery-justification distinction has not escaped criticism from post-positivist philosophers of science such as Feyerabend and Kuhn and feminist philosophers of science such as Longino and Okruhlik.

Post-positivists argue that the context of discovery is not wholly irrational—some, even that there is a "logic of discovery"—and that the context of justification is not wholly rational. In *The Structure of Scientific Revolutions* (1962), Kuhn shows how paradigms serve as exemplars in periods of normal science: in the context of discovery, possible solutions to new puzzles are suggested by the ways in which similar puzzles have already been solved, and in the context of justification, what counts as an acceptable solution is determined by these existing solutions, not prescribed rules of rationality. In *Against Method*, first published in 1975, Feyerabend contends that the context of justification is no less anarchic than the context of discovery, with irrationality pervading both. Feyerabend claims that "*Copernicanism and other 'rational' views exist today only because reason was overruled at some time in their past*" (1988, p. 121). Such ideas survived "because prejudice, passion, conceit, errors, sheer pigheadedness, in short because all the elements that characterize the context of discovery, *opposed* the dictates of reason" (p. 121). Feyerabend examines the replacement of geocentrism by heliocentrism at the time of the scientific revolution and argues that empirical and rational support for Aristotelianism was undercut by rhetorical tricks on Galileo's part. Of course, for Kuhn, because of the incommensurability of paradigms in periods of revolutionary science, there is no pretence of rationality in theory choice. Kuhn (1977) attempts to respond to *Structure*'s critics by arguing that science's constitutive values preclude the charge that only "mob psychology" justifies theories on his account. Nevertheless, he continues to emphasize that the history of science shows us that no algorithmic decision-procedure governs theory choice in the context of justification. Rather, "every individual choice between competing theories depends on a mixture of objective and subjective factors" (p. 325).

As Longino (1990) points out, logical positivists and logical empiricists attribute the subjective and non-empirical elements of the context of discovery to aspects of individual psychologies: these "are treated as randomizing factors that promote novelty rather than as beliefs or attitudes that are systematically related to the culture, social structure, or socioeconomic interests of the context within which an individual scientist works" (p. 64). We see this in the anti-psychologistic origins of Popper's distinction between the psychology and logic of knowledge and Reichenbach's distinction between the contexts of discovery and justification. Postpositivist critics of these distinctions such as Kuhn and Feyerabend likewise assume that individual psychologies are responsible for the subjective and non-empirical elements that influence theory choice in the context of justification. But what about the social, cultural, and economic contexts that shape the ways in which science is practised: don't these matter as much or more than the prejudices, pig-headedness, or creative impulses of individual scientists?

In contrast to postpositivists such as Kuhn and Feyerabend, feminist philosophers of science have emphasized the contextual aspects of science, especially the influence of gender, race, and class biases given that science has historically been carried out, for the most part, by white, middle- or upper-class men. Feminist critics of the ideal of pure, value-free science have tended to focus their attention on the context of justification. In the article "Can There Be a Feminist Science?" (1987), Longino distinguishes between "constitutive" and "contextual" values. Constitutive values are those sanctioned by science's methodological norms—for instance, empirical adequacy, fruitfulness, simplicity, and scope. Contextual values are those personal, social, and cultural values deemed to be external rather than internal to the sciences. Longino considers different possible ways of making the argument that contextual as well as constitutive values operate in science. She writes: "One scholar is fond of inviting her audience to visit any science library and peruse the titles on the shelves. Observe how subservient to social and cultural interests are the inquiries represented by the book titles alone!" (p. 54). Because the mode of influence of contextual values in this case concerns what research gets done, and this belongs to the context of discovery where traditionalists freely admit the role of values, Longino opts instead to focus on the theory-ladenness of observation and underdetermination of theories by evidence that occur in the context of justification.[3] According to Longino (1987, 1990), because of the underdetermination of theory by evidence, there are no methodological rules that can provide any in-principle guarantee of excluding values from the context of justification, and this applies to contextual as well as constitutive values.

[3]The "theory-ladenness of observation" refers to the inability to describe observations in language that is independent of all theories; the "underdetermination of theories by evidence" refers to the inability to validate theories based solely on their logical relations to observational claims.

Longino's arguments resonate surprisingly well with some threads in logical positivism and logical empiricism that did not end up prevailing historically. As recent work in the history of the philosophy of science has shown, debates about the role of social and political values in theory choice took place among logical positivists and logical empiricists, beginning in Europe in the 1920s and continuing in North America through the 1940s and 1950s. This was particularly the case for the circle of philosophers meeting in Vienna long before Moritz Schlick's arrival in 1922, made up of Hans Hahn, Olga Hahn, Philipp Frank, and Otto Neurath. They were inspired by the mix of empiricism, instrumentalism, holism, and conventionalism found in Ernst Mach, Henri Poincaré, and Pierre Duhem. Neurath held that traditional epistemology would be replaced by the logic of science and behaviouristics (i.e., a theory of behaviour consistent with physicalism): the admission of behaviouristics reflected the need for historical and sociological explanations regarding the acceptance of theories given the underdetermination problem (Okruhlik 2005, p. 57). Frank believed that empirically equivalent theories might be judged according to their relative capacities for satisfying practical as well as cognitive purposes (Uebel 2000, pp. S141–S142). Thus, the underdetermination of Neurath, Frank, and Longino permits recognition not just of the influence exerted by social and political factors in the justification of theories but a "voluntarism" (Okruhlik 2005) whereby scientists may justifiably (without logical constraint) choose between two empirically adequate theories depending on which is most consonant with their values. Indeed, it has been suggested that Neurath and Frank welcomed sociological explanation as part of the context of discovery because they found it preferable to the "metaphysics" of individual choice that underdetermination would otherwise implicate (Howard 2003).

Claiming that "discovery" matters does not just recognize that social, cultural, and economic factors traditionally relegated to the context of discovery are also implicated in justification. Claiming that discovery matters also questions the autonomy of the two contexts and especially urges attention to be paid to the ways in which social, cultural, and economic factors specifically through their influence on problem choice and the generation of theories come to determine the content of knowledge—even though these aspects of science have been considered irrelevant to its content since part of discovery. And so, philosophers of science, even if their concerns are epistemological, must pay attention to discovery. Surprisingly, and somewhat ironically, Reichenbach provides support for this counsel.

Kuhn (1977) argues that the discovery-justification distinction is inadequate even as an idealization because of the lack of autonomy of the two contexts. It is the whiggishness and oversimplifications of "textbook science", he says, that lead us to ignore that scientists face equivocal evidence and good reasons to support either contender in choosing between theories, and the distinction between the contexts of discovery and justification leads us to ignore the influence that discovery bears on justification: "Considerations relevant to the context of discovery are [...] relevant to justification as well; scientists who share the concerns and sensibilities of the individual who discovers a new theory are ipso facto likely to appear proportionately frequently among that theory's first supporters" (p. 328). Okruhlik (1994) argues that there is no methodological guarantee in science that values belonging to the context of discovery will be purged in the context of justification. Theory evaluation is comparative: a test hypothesis is not held up to nature directly but in comparison with existing rival hypotheses. The decision procedure yields the theory that is epistemically superior but not necessarily true. And "if all these contenders have been affected by sociological factors, nothing in the appraisal machinery will completely 'purify' the successful theory" (p. 34). "Once you grant that social factors may influence the context of theory generation," writes Okruhlik, "then you *have* to admit that they may also influence the content of science" (p. 35). Okruhlik views this as a dimension of the underdetermination problem and not necessarily a threat to the discovery-justification distinction. Nonetheless, her account, like Kuhn's, raises questions about the autonomy of the two contexts by showing the interdependence of the activities occurring in each.

Recall that in *Experience and Prediction*, Reichenbach distinguishes between internal and external relations, discovery and justification, and truth and volition, with all three of these distinctions designed to circumscribe the domain of epistemology and protect the objectivity of knowledge. Reichenbach also assigns epistemology three tasks. Two of these tasks were mentioned in Sect. 2.2: the descriptive task involves coming up with a rational reconstruction, the analysis of which involves the critical task. Epistemology's third task is the advisory task: this task involves identifying those points at which volitional decisions occur and proposing alternatives. But Reichenbach is ambivalent about this role, likely because it stretches the limits of the "scientific philosophy". In effect, he swiftly dissolves it. Statements about what decisions actually take place are said to belong to epistemology's descriptive task because they involve an "object fact". Statements conveying the "logical fact" that these are decisions not statements, as well as the itemizing of decisions which may appear to be free or arbitrary but are not because they are entailed by the initial decision, are part of epistemology's critical task (p. 11). Reichenbach's example of an entailed decision is foregoing decimal system rules for addition once the initial decision to use the English system of measures is made. He uses the concept of entailed decisions to support a distinction between subjective and objective parts of scientific knowledge. Reichenbach contends that his conventionalist predecessors overestimated the number of volitional decisions and that, instead, most decisions are entailed, and therefore objective. He then minimizes problems posed by the subjectivity of the relatively few volitional decisions remaining by stating that as "basic decisions" from which subsequent decisions are entailed, many enjoy near-unanimous consent among scientists (p. 15)—perhaps a suggestion that these be reconceived as "objective" in the intersubjective sense.

However, Reichenbach also tells us that volitional bifurcations involve decisions that are "of the greatest relevance" since they affect the content of knowledge. Once again, Reichenbach's examples of basic decisions, which include calling things of the same kind by the same name and expecting science to provide methods for prediction, do not sufficiently embrace the range of possibilities. What about whether to apply for a research grant that requires industry partnership? What about choosing diseases to concentrate funding on? What about which variables to manipulate in an experimental setup? While these "volitional decisions" may enjoy widespread consent, scientists tend to agree on decisions that incorporate values which they share. Since scientists' shared values are shaped by the social, cultural, and economic features of the contexts in which they live and work, basic decisions are not "free" or "arbitrary" in any other than Reichenbach's limited sense. The decisions that basic decisions entail are not "objective" in the sense of being value-free or value-neutral either. Logical entailment guarantees that the evaluative content of basic decisions is carried through to subsequent decisions. In addition, entailed decisions are consequences of the logic that operates through society's institutional structures, and not just philosophers' formal systems. "Discovery" matters because features of the social, cultural, and economic contexts in which science is done have pervasive effects on the production of scientific knowledge.

3.3 Entanglement of Facts and Values

Philosophers of science have provided convincing critiques of the ideal of value-free science during recent decades. It is argued that values are implicated in the reporting of scientific findings (Dupré 2007); that there is no principled way to exclude the operation of "contextual" values from the context of justification given the underdetermination of theories by evidence (Longino 1987); that in policy-directed research, values are necessary for the weighing of potential risks where uncertainties exist and errors have consequences (Douglas 2000, 2007). These critiques are illustrated with a variety of case studies. John Dupré (2007) points out that some concepts are irreducibly normative (e.g., when evolutionary psychologists provide adaptationist explanations of rape, their studies of flies or ducks do not escape the meanings that attach to rape in human society) and that even numerical data are not necessarily value-neutral (e.g., when economists measure inflation by using the same scale for changes in prices of luxury goods and basic necessities, they ignore the differential impact of these). Helen Longino (1987) draws on neurobiological research on gender, hormones, and behaviour to argue that contextual values influence scientists' preferences for linear or interactionist models of causation, and whichever model they choose determines the relevance and interpretation of data. Heather Douglas (2000) appeals to studies on dioxins and cancer to show that because of the risk of error, "non-epistemic" values play a legitimate role in how scientists characterize and interpret data.

These critiques of the ideal of value-free science do not hinge on their authors' rejection of Hume's and Poincaré's dichotomous separation of facts and values. Longino allows that although scientific methodology does not guarantee freedom from contextual values, this does not mean that alternative explanatory models always have equivalent empirical support or that when they do, contextual rather than constitutive values decide between them. Dupré admits that there are paradigm cases of factual and evaluative statements, though he argues that much of our scientific language contains terms that are both factual and evaluative—especially concerning things that interest us. Douglas maintains that facts and values are logically distinct, and her account restricts the role of values in science to situations in which risks must be weighed. Both Douglas and Dupré accept modified versions of the distinction between basic and applied science: Douglas recognizes that non-epistemic values may be absent from research carried out "for pure curiosity" rather than "for use" (2007, p. 122), and Dupré allows that results in physics (or chemistry or mathematics) may be value-free because they address questions of "no immediate importance to us" except cognitively (p. 32).

In his 2002 book, *The Collapse of the Fact/Value Dichotomy*, Putnam, in contrast, argues that Hume's and Poincaré's dichotomous separation of facts and values and the associated inference that values are subjective, both of which so greatly influenced the logical positivists and logical empiricists, cannot be sustained. Insofar as we recognize that distinctions may apply and prove useful only in some contexts, it is valid to distinguish between facts and values. But nothing of metaphysical significance follows, and the dichotomous separation of facts and values is therefore illegitimate. Instead, we face a phenomenon Putnam calls "the entanglement of facts and values". According to Putnam, the entanglement of facts and values can be observed in several ways. While the role of epistemic values such as simplicity is widely accepted in science, Putnam points out that there is no external justification that can establish that these values are conducive to truth: any such investigation will inevitably be circular. In addition, thick ethical concepts such as "cruel" carry descriptive and normative connotations that are impossible to disentangle despite the efforts of noncognitivists to do so, whether by contending that the concepts are wholly descriptive or trying to factor them into descriptive and "attitudinal" components. Finally, since all perception involves concepts, when the relevant concepts are evaluative, values come to be properties of things: consider the connoisseur's characterization of a wine as "full bodied".

Putnam ventures a guess at why the dichotomization of facts as objective and values as subjective has prevailed for so long:

> There are a variety of reasons why we are tempted to draw a line between "facts" and "values"—and to draw it in such a way that "values" are put outside the realm of rational argument altogether. For one thing, it is much easier to say, "that's a value judgment," meaning, "that's just a matter of subjective preference," than to do what Socrates tried to teach us: to examine who we are and what our deepest convictions are and hold those convictions up to the searching test of reflective examination. (pp. 43–44)

Putnam contends that "warranted assertability" applies to statements of value no less than statements of fact. Appealing to the American pragmatist tradition, Putnam argues that the conditions conducive to inquiry generally—he mentions fallibilism, experimentalism, and democratization—hold also for value inquiry. Among the pragmatists, Putnam draws mostly on Dewey, finding inspiration in Dewey's contention that valuing something does not make it valuable: objectivity in value inquiry is achieved through submitting our valuations to reflection and criticism.

Indeed, as Putnam points out, Dewey maintained a logical distinction between facts and values, but rejected a metaphysical "fact/value dualism". Like the logical positivists and logical empiricists, Dewey (1998 [1929]) discarded a transcendental status for values: he criticized the western philosophical tradition's search for an unchanging, certain foundation for action in "the immutable", whether Platonic forms, reason, or God. But Dewey was also strongly opposed to subjectivism and emotivism in ethics. Dewey believed that objective value judgements and a scientific approach to ethics are possible. He argued that our values are made secure only as they become embodied in experience through our intelligently directed actions: "We should regard practice as the only means (other than accident) by which whatever is judged to be honorable, admirable, approvable can be kept in concrete experienceable existence" (p. 105). And why would we not prefer the concrete existence of such values to their transcendental being? According to Dewey (1920), "the experimental logic when carried into morals makes every quality that is judged to be good according as it contributes to amelioration of existing ills. And in so doing, it enforces the moral meaning of natural science [...] When physics, chemistry, biology, medicine, contribute to the detection of concrete human woes and to the development of plans for remedying them and relieving the human estate, they become moral; they become part of the apparatus of moral inquiry or science" (pp. 172–173).

4 Case Study: Biogeographical Ancestry and Race in Population Genomics

"Biogeographical ancestry" or "BGA" emerged as the product of the collaboration between biological anthropologist Mark Shriver's Pennsylvania State University laboratory and molecular biologist Tony Frudakis' now-defunct, Sarasota, Florida-based biotechnology start-up company, DNAPrintgenomics. The term "biogeographical ancestry" or "BGA" was introduced by members of Shriver's lab in a poster presentation at the meeting of the American Society of Human Genetics in 2000. The poster, titled "Genetic estimation of biogeographical ancestry", defines BGA as the biological, non-cultural component of "ethnicity": "Ethnicity is comprised of both biological and cultural components. Biogeographical ancestry (BGA) refers to the component of ethnicity that is biologically determined and can be estimated using genetic markers that have distinctive allele frequencies for the populations in question" (Pfaff et al. 2000).

BGA is both a concept and a technology. As a concept, BGA seeks to represent population genetic structure, i.e., the pattern by which genetic variability is distributed across the species. As a technology, proprietary DNA marker kits and computer algorithms are used to assign people to BGA groups based on their genetic makeup. The markers used, designated "ancestry-informative markers" or "AIMs", are single nucleotide polymorphisms (SNPs) that have been chosen on the basis of the degree to which their frequencies vary among population groups. BGA groups are delineated at multiple levels of resolution. The most comprehensive BGA groups, "East Asian", "Indo-European", "Native American", and "Sub-Saharan African", correspond with "race". At "a finer level" of "ethnicity within the

European group", we find "Mediterranean or Scandinavian", and at "a still finer level" of "groups of families within ethnic groups", we find the "O'Reillys" (Frudakis and Shriver 2003). Individuals belong to one or more BGA groups at a given level, i.e., as a matter of degree. This reflects BGA's development as a measure of "individual admixture" or "proportional ancestry" (Pfaff et al. 2000).

The Shriver lab and DNAPrint portray BGA as a means of construing race and ethnicity in appropriately scientific—"biological", "genetic", "natural", "heritable", or "objective"—ways. Consequently, its proponents hold that BGA eliminates the need for biomedical researchers to rely on racial and ethnic self-identification by their subjects that is based on social and political categories—a questionable approach in studies seeking to understand the genetic basis of complex diseases (Shields et al. 2005). I argue elsewhere that they do not succeed: BGA is itself a construction built upon race as race has been socially constructed in the European scientific and philosophical traditions, but especially in the United States (Gannett in preparation). In this chapter, I focus on the implications that BGA as a case study has for the ideal of value-free science. I show ways in which the ideal's supporting distinctions between theory and practice, the context of discovery and context of justification, and facts and values provide BGA's proponents with resources to promote the category's objectivity. However, an examination of the research context demonstrates that BGA-AIMs technology incorporates values that reflect its development in response to the post-HGP challenges of mapping genes for complex traits, forensic aims of predicting the physical appearance of unknown suspects from crime-scene DNA, and the promise of profits in pharmacogenomics.

4.1 BGA in Theory and Practice

From its inceptions, BGA's theoretical and practical aims have been inseparable. The authors of the 2000 poster attend to both sorts of aims in their forecast that as the number of available markers grows, "estimation of BGA may become a powerful tool for the elucidation of an individual's genetic and population history, as well as the identification of unknown samples in forensic cases" (Pfaff et al.). This reference to forensic cases is not merely speculative but related to the postdoctoral research Shriver had carried out just a few years previously. The research, which was funded by the National Institute of Justice, sought to identify markers "for estimating the ethnic affiliation of unknown suspects" using crime scene DNA. When Shriver published his results, he noted that it would be helpful to further develop the method "so that interethnic individuals, first- or second-generation hybrids of one or more populations, would be identified and classified appropriately" (Shriver et al. 1997, p. 963). BGA, as a measure of "individual admixture" or "proportional ancestry" (Pfaff et al. 2000), fulfills this goal.

Gene mapping is another practical aim. Since completion of the Human Genome Project, biomedical researchers have found that mapping genes involved in relatively common complex diseases like hypertension, asthma, cancer, and dementia has proved unexpectedly challenging. These conditions are referred to as "complex" diseases because multiple genetic, epigenetic, and environmental causal factors need to be identified and their separate roles and interaction effects unravelled. Some complex diseases occur with different frequencies in different "racial" and "ethnic" groups, and although they recognize the importance of epigenetic and environmental factors and discount genetic determinism, many researchers believe that genetic variants found at different frequencies in different groups are implicated. Shriver lab presentations and publications around the time of BGA's introduction explore the feasibility of "admixture mapping", which makes use of increased amounts of linkage disequilibrium (i.e., lack of random association among alleles of different genes) in "admixed" populations (i.e., populations to which long separated populations contributed in relatively recent evolutionary history) to isolate genes involved in complex traits. BGA is also promoted as a way to control for population structure when mapping genes using case–control association studies.

The commercial context in which the Shriver's lab research has been carried out was shaped by Shriver's collaboration with Frudakis, of DNAPrintgenomics, whose interests were in pharmacogenomics and DNA forensics. DNAPrint's report to the Securities Exchange Commission (SEC) for 2001 (filed April 9, 2002) provides information on several forensic products under development, including "Verity", a "classifier" panel of SNPs that "encode" physical traits such as skin colour, eye colour, and hair colour for "the inference of race from DNA", the "racial classes" being "Asian, African American, and Caucasian". Because DNAPrint's classifier of 64 "race-related SNPs" could be used to infer "major ancestral affiliation" only, Frudakis was interested in the "ancestral proportions" made possible by BGA. The consulting contract between DNAPrint and Shriver, which was signed in June 2002, recognizes Shriver as the "inventor" of the technique. The project outlined in the contract was to create a panel containing a "minimum and optimum set" of "Ancestry Informative Markers (AIMs)" that would support development of a "kit product that could be used to infer Ancestry Admixture Ratios in individual human beings" for "target ancestral groups". Shriver was to receive compensation as both a consultant and a member of DNAPrint's board of directors. And DNAPrint gained access to "DNA and data from over 3,000 reflectometry-qualified specimens of multiple ancestral backgrounds" and "exclusive and unlimited rights to exploit previous pigmentation results and analytical methods developed in Shriver's laboratory" (Editors 2002).

In a further commercial development, soon after its agreement with Shriver, DNAPrint added "recreational genomics" to its forensic and pharmacogenomic interests. The goal was to raise cash for the company, which had lost $5.46 million since late 1998. ANCESTRYbyDNA 2.0 was introduced as a direct-to-consumer (DTC) product in December 2002: it used a panel of 75 AIMs to genotype DNA obtained by mailed-in cheek swabs to provide the genealogically curious with estimates of their ancestral proportions from one or more of four "major" BGA groups: "East Asian", "European" (or "Indo-European"), "Native American", and "Sub-Saharan African". Also in December 2002, in furtherance of its pharmacogenomic interests, DNAPrint announced a whole-genome screening platform (ADMIXMAP) for mapping by admixture linkage disequilibrium (MALD) and admixture mapping

(AM), the latter a Bayesian refinement of the former developed by Paul McKeigue of University College London and University of Dublin in collaboration with Shriver. ANCESTRYbyDNA's forensic counterpart, DNAWitness 2.0, became available in May 2003: it used the same panel of AIMs to genotype DNA left at the crime scene to provide law enforcement officials with estimates of ancestral proportions from the same four major BGA groups for their unknown suspects. In August 2003, Frudakis and Shriver submitted a patent application covering all of these applications of the BGA-AIMs technology, claiming as inventions: (i) for forensics, DTC genealogy, and gene mapping, panels of 32, 71, and 331 AIMs that predict BGA; (ii) for DTC genealogy, personalized ancestral and genealogical maps; and (iii) for forensics, a database of digital photographs of individuals with similar ancestral proportions.

It is evident that the positivist ordering of theory and practice, with theory both epistemologically and temporally prior to practice, does not apply to the case of biogeographical ancestry. No doubt, it does not apply to much or even most of molecular genetics and genomics given the close ties between universities and industry, and governments' steadfast support of these ties—e.g., the 1980 Bayh-Dole Act in the U.S. allows the results of federally funded research to be patented; Genome Canada requires that scientists who apply for research grants obtain the promise of matching funds from industry.

The research carried out by the Shriver lab incorporates goals that range across the theoretical-practical spectrum. The same research report might respond to goals of both sorts, with support from the identical data set. For instance, studies that investigate the feasibility of using admixture mapping to identify genetic susceptibilities to disease in Hispanic and African American populations also theorize about the demographic history behind the sex-biased contributions to admixture and which population genetic models best explain admixture patterns (Bonilla et al. 2000; Parra et al. 2000). Similarly, although skin pigmentation in an African American population provides a model phenotype for admixture mapping for complex diseases, identifying such genes provides "the first step in understanding the molecular and evolutionary history of human pigmentation" (Norton et al. 2000; Pfaff et al. 2001). The same laboratory activities might be carried out to address both sorts of problems. For instance, Akey et al. (2002) use The SNP Consortium (TSC) database to compare allele frequencies in African American, Asian American, and European American population samples for the SNPs genotyped in all three groups. The study makes a theoretical contribution insofar as it identifies candidate genes as subject to natural selection because of Fsts (i.e., the "fixation index" or "F_{ST}" estimates the degree of genetic differentiation in subdivided populations) for SNPs that are unusually high or low, and the study makes a practical contribution insofar as it identifies SNPs with frequency differentials that qualify them as possible AIMs.

Although theoretical and practical aims coexist and mutually condition what research is carried out and how it is carried out, the traditional theory-practice dichotomy continues to play a legitimating role. In the collaboration between Shriver, an academic scientist based at a public university, and Frudakis, an industry scientist based at a biotechnology start-up company, we find that the legitimacy of technological and practical applications is justified by appealing to theoretical foundations: for example, the Frudakis-Shriver patent application validates the inventions claimed by citing research into the genetic basis of differences in skin pigmentation ostensibly carried out in order to better understand pigmentation as an evolutionary adaptation (Shriver et al. 2003; Bonilla et al. 2004). Similarly, DNAPrint's press release to promote DNAWitness 2.0's ability to predict phenotype from DNA left at the crime scene cites a Shriver lab publication in which skin pigmentation served as a model phenotype for admixture mapping using BGA-AIMS technology (Shriver et al. 2003). But developing a patentable invention that allows the racial profiling of unknown suspects based on crime-scene DNA was an aim embedded in the research program, and explains why skin pigmentation would be chosen at the outset as an evolutionary adaptation of interest and a complex trait for which genes might be mapped.

Of course, the theory-practice dichotomy is best poised to play a legitimating role that promotes the scientific validity of practical and commercial endeavours when research activities carried out in university laboratories and results published in peer-reviewed journals are portrayed in ways that highlight theoretical aspects and downplay associated applications. In *Biotech Week*'s September 2002 report of the R&D agreement between DNAPrint and Penn State, the jointly funded research is portrayed as using admixture mapping and genome screens to identify "the complex genetic determinants of variable human skin pigmentation, tanning/burn response, and melanoma risk" (Editors). Shriver is quoted as saying that he welcomes the "team effort" as "the most efficient means" of solving the "fascinating puzzle" of skin pigmentation already being investigated by his lab. No mention is made of the forensic "racial profiling" application already in the pipeline. Peer-reviewed articles published during 2003 by members of the Shriver lab draw attention to the "immense potential" of admixture mapping (Halder and Shriver 2003), and prominently display and offer definitions for "biogeographical ancestry" and "ancestry informative markers" in titles (Shriver et al. 2003), abstracts (Shriver et al. 2003), and keywords (Halder and Shriver 2003). Again, the focus is on biomedical applications, and forensics, DTC genealogy, and pending patents are not mentioned. Such a strategy reflects the interest of biotech companies in developing ties with academic scientists at prestigious universities because this provides scientific validation to their efforts and helps to attract investment. And when a scientist such as Shriver becomes a board member and receives shares in compensation for services and based on net sales of products, his aims are no longer separate from the company's.

4.2 Race Matters for BGA

In our 2004 paper "The ABO Blood Groups: Mapping the History and Geography of Genes in *Homo sapiens*", James Griesemer and I look at the inceptions of anthropological genetics in racial studies of blood group differences carried out during the first half of the twentieth century. Scientists believed that ABO and other blood group differences would support a more objective classification than traditional anthropological traits (e.g., skin colour, skull shape, etc.) because of their constant and hereditary nature. The Shriver lab and DNAPrint, as mentioned already, portray BGA as a means of assessing population structure by construing race and ethnicity in appropriately scientific—"biological", "genetic", "natural", "heritable", or "objective"—ways. These scientists make assumptions similar to those made by their predecessors. The ABO and other blood groups provided researchers during the first half of the twentieth century with traits upon which an objective classification might be based insofar as the underlying genetics was understood to be simple, with no more than a few alleles at a single locus controlling the trait. Likewise, the BGA researchers are confident that they are arriving at an objective classification because they are using DNA markers.

In both the historical blood group research and current BGA research, a priori groupings of people are necessary for comparisons to be made because the genetic differences involved are quantitative, not qualitative: populations differ in their characteristic frequencies and not in the presence or absence of either blood group alleles or the SNPs used as AIMs. Drawing a priori boundaries relies on categories of classification that resonate in the social milieu—social-political categories such as race, ethnicity, nationality, etc. The serologists assumed that their biological-anthropological categories of classification, which denote a posteriori groupings, since arrived at empirically, were objective. The Shriver lab and DNAPrint are similarly confident that empirical data alone validate their a posteriori classification scheme, the principle of which, in this case BGA, denotes a biological-anthropological, not a social-political category. However, the schema of a priori classification are not discharged in the course of arriving at the empirically validated schema of a posteriori classification.

For BGA and blood group researchers alike, a posteriori classifications are indebted to a priori classifications. AIMs are selected on the basis of the extent to which frequencies of candidate SNPs vary between designated populations, and populations are designated in ways that reflect the priorities and available resources of the U.S. based researchers. Shriver's initial efforts to identify markers (not SNPs but biallelic microsatellites) for forensic purpose targeted "African Americans", "European Americans", and "Hispanic Americans" because "together these constitute 95 % of U.S. residents" (Shriver et al. 1997, p. 963). An extensive effort to identify candidate SNPs made use of the SNP Consortium (TSC) database, which was established to facilitate pharmaceutical development in the U.S. and includes SNP allele frequencies for sampled "African Americans", "Asian Americans", and "European Americans" (Akey et al. 2002). Not surprisingly, the four major BGA groups are associated with those regions of the world that have contributed most to the formation of racial and ethnic identities in the U.S. For this reason, anthropologist Duana Fullwiley (2008) describes BGA-AIMs technology as a "tautological product" wholly embedded in "American racial taxa" (pp. 697–698).

And yet, drawing on Bruno Latour's (1999) theory of circulating reference, Griesemer and I argue that achieving objectivity depends on not discharging but holding onto the schema of a priori classification. Latour theorizes that scientists manage to represent the world with words through a series of conceptual and material displacements. We emphasize that each of these links along the chain of reference that is established involves a judgement, and that without the subjectivities associated with these judgements, objectivity, in Latour's sense of circulating reference, would be impossible to achieve. The judgements taken at all of these steps are not arbitrary, and nor is the availability of possible judgements unlimited: the range of judgements that can be made reflects the research context and its constitutive purposes. Since a range of judgements is available at each step, the judgements that are made along with the values they privilege must be assessed against the contrast class of judgements not made and values not privileged. In research contexts constituted by different purposes, when different judgements are made and different values are privileged, reference is put into circulation in different ways. The subjectivities of judgement make objectivity-as-circulating reference possible, but which judgements are made determine which chains of reference are established. This increases the probability that the purposes constitutive of the research context will be fulfilled since whichever chains of reference are established end up structuring the ways in which we explore, understand, and interact with "the world". Recall Keller's apt characterization of scientific theories as "instruments", "vectors", and "tools for changing the world" and her contention that genetics and nuclear physics contain within them "not only the possibility but the expectation" of eugenics and the bomb.

Latour's theory of circulating reference coupled with Griesemer's and my focus on the necessity of subjectivities of judgement for objectivity provide a basis for understanding the ways in which the "context of discovery" matters for the "context of justification", ways that are consonant with, but of greater materiality and depth than those suggested by the "theory-ladenness of observation" and "underdetermination of theory by evidence" described so well by post-positivist and feminist philosophers of science. Mapping the history and geography of genes in *Homo sapiens*, whether this involves the ABO blood groups or BGA, proceeds as a series of small steps, all of which require that choices be made, beginning with the assumption of an a priori group classification scheme and ending with the production of an a posteriori group classification scheme. The a priori classification of people on the basis of their social-political racial identities in the U.S. is just the first step/choice. Subsequent steps/choices reflect the various ways in which race is socially constructed, even just in the U.S. U.S. racial populations are considered "ancestral" groups such that DNA markers that occur at different frequencies in different groups are considered to be "informative" about ancestry, and ancestry is further specified to be "continental ancestry", with ancestors fixed to continents by defining what it is to be indigenous in terms of the so-called voyages of discovery. U.S. racial populations are also considered to be classes of individuals who differ in "pigmentation phenotype", i.e., in skin, eye, and hair colour, such that markers that occur at different frequencies in different classes are considered to be predictive of racial appearance. These steps/choices reflect purposes constitutive of the U.S. research context: controversy over the use of Office of Management and Budget (OMB) categories of race and ethnicity in biomedical research left ancestry as the preferred category; the potential market for DTC genealogy is overwhelmingly people who want to trace their family tree to ancestors who came to the U.S. from overseas; the uptake of the forensic product DNAWitness by law enforcement officials depends on whether racial profiling can be carried out in ways that conform to people's abilities to visualize race in stereotypic ways; the focus on African Americans and Hispanic Americans as admixed populations conforms to the one-drop and hypodescent rules that have prevailed in the U.S.,[4] while European Americans escape being considered an admixed population only because of the legacy of the White supremacist ideal of racial purity.

4.3 BGA's Entanglements

BGA proponents make several implicit appeals to the fact-value dichotomy. BGA is portrayed in entirely representational terms as an empirical measure of population structure. BGA-AIMs technology is portrayed as a way to assess population structure and determine race and ethnicity objectively on that basis, thus avoiding the subjectivities associated with reporting one's own or another's race or ethnicity. These appeals are challenged by Putnam's arguments for the entanglement of facts and values based on thick ethical concepts and concept-driven perception.

DNAPrint's FAQs for its genealogical customers defined BGA as "a simple and objective description of the Ancestral origins of a person, in terms of the major population groups". By demarcating separate spheres of objective facts and subjective values, the fact-value dichotomy supports assumptions that there is a (potentially) complete description of the world. The "User Manual" that customers received with their ANCESTRYbyDNA test results provided a historical interpretation of the four major BGA groups as "founder groups" that arose after migration out of Africa and gave rise to lineages that terminate in "present-day Europeans, Native Americans, Africans, and East Asians". A complete and accurate description of ancestral origins is considered possible, at least theoretically: confidence intervals attached to "mixture ratios" are explained to customers as reflecting the inability to "go back in time 200,000 years and keep track of every one of your ancestors" and the fact that not all variable sites of the genome are used as markers. It has been recommended that BGA-AIMs technology be used instead of self-reported race, ethnicity, and ancestry (e.g., in the U.S., according to the OMB categories of race and ethnicity) to control for population structure in biomedical research (NHGRI 2005). Using the McKeigue-Shriver approach of genotyping people in order to determine their "continental ancestry" proportions is considered to provide a suitably empirical assessment of population structure (Shields et al. 2005).

[4] The one-drop rule refers to persons being considered Black if they have any known Black ancestors (a single drop of "black blood"); the rule of hypodescent assigns persons of "mixed race" to the socio-economically subordinate group.

However, just as a thick ethical concept such as "cruel" carries descriptive and normative connotations that are impossible to disentangle, the labelling of genes as "disease genes" involves not just a description of their role in pathogenesis but a normative call to action. Normal and abnormal genes are identified on the basis of values that attach to differences among people and their designations as healthy or diseased, or desirable or undesirable states (Gannett 1998). When pharmaceutical profits are the aim, the complex traits that are likely to become the focus of the use of BGA-AIMs technology in mapping efforts are those that will be used by a significant proportion of the populace over a sustained period of time: DNAPrint's interest in variable responses to statin as a drug that lowers cholesterol levels is a good example. The use of BGA-AIMs technology in the mapping of complex traits, especially diseases that occur at different frequencies in different racial or ethnic groups, is part of the more widespread, post-HGP effort to use group genetic differences as a tool for gene mapping and a shortcut along the route to "personalized medicine". The assumption that BGA provides a wholly empirical representation of population structure for use in gene mapping ignores that conceptualizing population structure spatially as geographical stratification or temporally in terms of founder groups imposes discontinuities on continuities, and these discontinuities bring certain values with them: e.g., "continental ancestry" is not a value-free description of a natural boundary where land and sea meet—it embodies a social construction of race whereby the continents from which our ancestors come matters deeply to who we take ourselves and others to be and what our life prospects are.

The fact-value dichotomy assigns objective facts to a world that exists apart from and external to us and confines subjective values to our minds as expressions of feelings, emotions, interests, and preferences. Since sensory experiences provide us access to that world of facts, it is assumed that those experiences are best mediated by technologies that protect against idiosyncrasy. An abstract for a conference poster from the Shriver lab subscribes to this technology-mediated objectivity: "Variation in human pigmentation has been studied objectively using reflectance spectroscopy for over 50 years" (Norton et al. 2000). Several laboratories that piloted BGA-AIMs technology regarded BGA as a more accurate (objective) measure of race or ethnicity than the subject's (subjective) self-report. A 2002 study used ANCESTRY 1.0, its 31 AIMs proportioning BGA among "Native American, West African, and European", to confirm that a family that self-identified as "white" was indeed a family of "European descent" since the cardiac sodium channel gene SNP associated with cardiac arrhythmia and sudden death that was found in the family had been found previously only in African Americans (Chen et al.). BGA-AIMs technology is portrayed as correcting subjectivities associated with people's reports not just of their own race or ethnicity (e.g., as subjects in biomedical studies) but the race or ethnicity of others (e.g., eye-witness reports of criminal suspects).

DNAPrint promoted DNAWitness to law enforcement officials by characterizing the product as providing law enforcement with "a simple and objective description of your subject's ancestral origins" that eliminates the need to rely on "subjective" psychological profiles or eyewitness accounts. DNAPrint foresaw a future whereby, with a test kit in "every patrol car", the race of felony suspects could be immediately ascertained. The database of digital photographs that Frudakis and Shriver sought to patent and that were added as an upgrade to DNAWitness 2.5 in July 2004 were foreseen to allow "better physical profiling" when investigators are confronted with significant admixture and left wondering what race their suspect will appear to be: "By querying the database using a specific BGA admixture result, investigators can see for themselves what the range of variability is corresponding to that result for various features, such as skin shade, hair texture, nose shape, epicanthal eye folding etc." Frudakis introduces his 2008 book, *Molecular Photofitting*, hopefully predicting "that 20 years from now molecular photofitting will be standard practice, and we will be doing amazing things, such as using computers to provide most of the information on a person's drivers license from DNA left at a crime scene, even creating 'artist's renderings' from DNA 'eyewitness testimony'" (xiv).

However, this is not race but racialization. As Putnam points out, all perception involves concepts. To "see" race is not merely to observe differences in skin colour, in hair texture, in facial structure, etc. It is to perpetuate a way of seeing that is the basis for a system of social stratification that is the product of the European scientific and philosophical traditions since the Enlightenment. BGA-AIMs technology's allocation of proportionate ancestry among the four major groups, "East Asian", "(Indo-)European", "Native American", and "Sub-Saharan African", identifies the regions of the world that have contributed most to U.S. "racial" identities. The history of race and racism in the U.S. accompanies the technology as it moves between research laboratories, university laboratory and biotech company, corporate headquarters and law enforcement agencies, etc. To the extent that BGA-AIMs technology is able to facilitate the arrest of a suspect by predicting a racialized physical appearance on the basis of crime-scene DNA, this is not because BGA is a scientific portrayal of race that is wholly "objective"; rather, it is because race as it is socially constructed enlists physical and biological differences and invests these with sociocultural meanings, and these meanings continue to circulate and resonate. In its denial of the entanglement of facts and values and portrayal of facts as value-free, the fact-value dichotomy allows scientists to ignore the ethical questions that are implicated in their research—most obviously, in this case, whether BGA-AIMs technology and its accompanying research program are justified given that they help to perpetuate a system in which the racialization of crime and criminalization of race combine to make the description of a suspect's appearance seemingly unimaginable apart from race.

5 Conclusion

Philosophers often draw case studies from among the most successful episodes in the history of science: the Copernican revolution, Newtonian physics, Darwin's theory of evolution by natural selection. Even when these remarkable scientific achievements are superseded by others so that the superseded theories are no longer regarded as true (e.g., Newtonian physics by special relativity and quantum mechanics), this casts no aspersions on their objectivity. Cases drawn from contemporary science are more challenging, especially in a field like genomics that is rapidly changing, and even more especially for a research area like BGA that is politically charged. However, well-worked-out case studies from the history of science provide philosophers of science with epistemological tools that are used to assess the validity of knowledge claims associated with the more challenging cases. The ideal of value-free science is central to this normative approach insofar as it allows distinctions to be drawn between "good" (value-free) and "bad" (value-laden) science, "complete" or "mature" and "incomplete" or "immature" science, and science and "pseudoscience". With the loss of the value-free ideal, these distinctions are also lost.

The loss of the value-free ideal in science raises questions about the status of normative epistemology, the approach traditionally taken by philosophy of science. Logical positivists and logical empiricists adopted for philosophy the role of handmaiden to science, tasked with clarifying meanings and filling in missing steps in logical inference. Feminist and other post-positivist philosophers of science placed themselves as critics of science, their case studies showing repeated failures of numerous scientific disciplines and scientific methodologies more generally to fulfill the value-free ideal. Is philosophy of science left without a normative project, a successor project to logical positivism and logical empiricism's handmaiden or feminist and other post-positivist philosophies of science's critic? Some philosophers of science suggest that given the loss of the value-free ideal, the normative approach should be replaced by a descriptive approach that focuses on how scientists do reason and not how they ought to reason—an approach consistent with the direction taken by philosophy of science over the past several decades that has been called "the naturalistic turn" (Callebaut 1993). In its adoption of a descriptive, empirical approach, philosophy of science joins alongside other disciplines in science studies—history of science, sociology of science, anthropology of science, etc. These are disciplines that logical positivists and logical empiricists relegated to the context of discovery. But "discovery" matters. Theory is embedded in practice. The contexts of discovery and justification lack autonomy. Facts and values are entangled. Values matter all the way down. An epistemology for philosophers of science today needs to make room alongside facts, theories, and logical relations for values.

Objectivity reconceived may offer science a new ideal. It is not surprising that feminist philosophers of science have made such important contributions to thinking about objectivity: critiquing sexist and racist science in ways conducive to bringing about political change is made difficult once epistemological relativism is embraced. However, efforts to reconceive objectivity tend to focus on methodology and hypothesis testing and thus remain tied to the logical positivist and logical empiricist tradition (e.g., Longino 1990; Okruhlik 1994). I suggest that instead, we need to bring back the "object" in objectivity—this, in three senses. The first sense of "object" attends to how the object of knowledge is constructed in response to the particular aims embedded in the research program. The object is not nature's own but constituted from a context-dependent set of relations, belonging to the social as well as physical world, in a more Latourian than Lockean or Kantian fashion—as the concept of BGA shows. The second sense of "object" draws on scholarship from science and technology studies to focus on the materiality of scientific practice that constitutes the object of knowledge as the physical embodiment of the aims of the research program and facilitates its circulation from one site to another—e.g., BGA is not just concept but technology, accompanied by panels of AIMs, computer software, genealogical maps, digital photographs, etc. The third sense of "object" brings the purposes or ends of inquiry into consideration, and so demands that underlying values be an integral part of epistemological debate in science—e.g., this calls for scientists to reflect on the ethics of carrying out a research program in DNA forensics on racial profiling given the history of racism and racialization of crime in the U.S.

While it remains unclear what a successor project to logical positivism and logical empiricism's handmaiden and feminism and post-positivism's critic might encompass as a normative epistemology for philosophers of science, it is clear that scientists themselves must recognize that debates over values should be part-and-parcel of their research activities and not left to others, whether engineers or industrialists, bioethicists or citizens. Thomas Reydon et al. (2012) argue that the poor understanding of genetics and genomics by health care professionals and the general public presents a barrier for the potential of pharmacogenomics and personalized medicine to be realized. They urge scientists to accept that educating these groups is among their professional responsibilities, and correctly point out that public debate about the social and ethical implications of scientific and technological developments will be impoverished otherwise. But the duty of scientists to educate the public about the implications and limitations of knowledge is accepted by even the most stalwart defenders of the basic-applied distinction (e.g., Wolpert 1992, Ch. 8). Reydon et al. likewise assume this distinction in their presentation of two kinds of scientific literacy: literacy concerning facts of science versus literacy concerning the "socioethical implications" of science. I am arguing that we need to encourage ethical debate within scientific laboratories and science classrooms as well as in the public sphere.

Scientists can no longer appeal to the ideal of value-free science in order to insulate their research activities from wider debates in society over values. Consequently, in today's world, biology education cannot stick to laboratory methods, empirical facts, and theoretical models as if science occurs in a realm autonomous from society. Students need to grasp an understanding of science as a set of practices fully situated within society. Science educators might begin by drawing on materials from the history and philosophy of science that complicate the historically and philosophically naïve accounts of successful episodes in the history of science (e.g., the Copernican revolution, Newtonian physics, Darwin's theory of evolution, etc.) that appear in textbooks. But this is not enough. Students also need to be encouraged to grapple with what sorts of contributions they want to make as scientists and what sorts of expectations as citizens they should have of science. And this will ultimately depend on what kind of world they want to live in. As Keller (1992) writes, "Given our remarkable ingenuity, skill, and imagination, I have no doubt that, with sufficient interest, we could develop representations of natural phenomena adequate to the task of changing the world in different ways [...] We have proven that we are smart enough to learn what we need to know to get much of what we want; perhaps it's time we thought more about what we want" (p. 92).

Acknowledgements I am grateful to Kostas Kampourakis, not only for the invitation to contribute this chapter and the improvements brought about in it by his editorial feedback but especially for his encouragement and patience along the way. I would like to thank Kathleen Okruhlik and Anya Plutynski for their careful and insightful remarks, which also improved the text, as well as Jim Griesemer for his contributions to this work through our collaborative efforts. Material from this chapter was presented at too many conferences, workshops, and colloquia to name, but this in no way lessens my appreciation of those organizers, participants, and audience members. A chapter from Evelyn Fox Keller's *Secrets of Life, Secrets of Death* provided the impetus for this work, and I thank her for that.

References

Akey, Joshua M., Ge Zhang, Kun Zhang, Li Jin, and Mark D. Shriver. 2002. Interrogating a high-density SNP map for signatures of natural selection. *Genome Research* 12: 1805–1814.

Aristotle. 1984. Nicomachean ethics. In *The complete works of Aristotle: The revised Oxford translation*, vol. 2, ed. Jonathan Barnes. Princeton: Princeton University Press.

Bernal, J.D. 1939. *The social function of science*. London: Routledge.

Biddle, Justin B. 2012. Tragedy of the anticommons? Intellectual property and the sharing of scientific information. *Philosophy of Science* 79: 821–832.

Bonilla, C., E.J. Parra, C.L. Pfaff, K.G. Hiester, D.M. Sosnoski, S. Dios, F.O. Gulden, R.E. Ferrell, R.F. Hamman, and M.D. Shriver. 2000. *Native American admixture and its relationship to diabetes in a hispanic population of the Southwest*. American Society of Human Genetics annual meeting. http://www.ashg.org/genetics/abstracts/abs00/f194.htm

Bonilla, C., E.J. Parra, C.L. Pfaff, S. Dios, J.A. Marshall, R.F. Hamman, R.E. Ferrell, C.L. Hoggart, P.M. McKeigue, and M.D. Shriver. 2004. Admixture in the Hispanics of the San Luis Valley, Colorado, and its implications for complex trait gene mapping. *Annals of Human Genetics* 68: 139–153.

Brown, James R. 2008. The community of science®. In *The challenge of the social and the pressure of practice*, ed. Martin Carrier, Don Howard, and Janet Kourany, 189–216. Pittsburgh: University of Pittsburgh Press.

Callebaut, Werner. 1993. *Taking the naturalistic turn, or how real philosophy of science is done*. Chicago: University of Chicago Press.

Chen, S., M.K. Chung, D. Martin, R. Rozich, P.J. Tchou, and Q. Wang. 2002. SNP S1103Y in the cardiac sodium channel gene SCN5A is associated with cardiac arrhythmias and sudden death in a white family. *Journal of Medical Genetics* 39: 913–915.

Cohen, I.Bernard. 1948. *Science, servant of man: A Layman's primer for the age of science*. Boston: Little, Brown, & Co.

Comte, Auguste. 1974 [1830–1842]. Plan of this course, or general considerations on the hierarchy of the positive sciences. In *The essential Comte*, ed. Stanislav Andreski, 42–64. London: Croom Helm.

Dewey, John. 1920. *Reconstruction in philosophy*. New York: Henry Holt.

Dewey, John. 1981 [1945]. Dualism and the split atom: Science and morals in the atomic age. In *The later works of John Dewey, 1925–1953*, vol. 15, ed. Boydston Jo Ann, 199–203. Carbondale: Southern Illinois University Press.

Dewey, John. 1998 [1929]. Philosophy's search for the immutable. In *The essential Dewey, vol. 1: Pragmatism, education, democracy*, ed. Larry A. Hickman and Thomas M. Alexander, 102–112. Bloomington/Indianapolis: Indiana University Press.

DNAPrint/Annual Report for 2001, 9 April. 2002. http://www.secinfo.com/d1zWxj.3h.htm. Accessed 11 Dec 2012.

DNAPrint/Mark Shriver Consulting Contract, 12 June. 2002. http://www.secinfo.com/d1zWxj.328.d.htm#1stPage. Accessed 11 Dec 2012.

Douglas, Heather. 2000. Inductive risk and values in science. *Philosophy of Science* 67: 559–579.

Douglas, Heather. 2007. Rejecting the ideal of value-free science. In *Value-free science? Ideals and illusions*, ed. Harold Kincaid, John Dupré, and Alison Wylie, 120–139. Oxford/New York: Oxford University Press.
Dupré, John. 2007. Fact and value. In *Value-free science? Ideals and illusions*, ed. Harold Kincaid, John Dupré, and Alison Wylie, 27–41. Oxford/New York: Oxford University Press.
Editors. 2002. DNAPrint Genomics, Inc.: R&D Alliance Inked with Pennsylvania State University. Biotech Week. http://www.NewsRx.net
Feigl, Herbert. 1950. Logical reconstruction, realism and pure semiotic. *Philosophy of Science* 17: 186–195.
Feyerabend, Paul. 1988. *Against method*. London/New York: Verso.
Forster, Malcolm. 2011. The debate between Whewell and Mill on the nature of scientific induction. In *Handbook of the history of logic, vol. 10: Inductive logic*, ed. Dov M. Gabbay, Stephan Hartmann, and John Woods, 93–115. Amsterdam: Elsevier.
Frudakis, Tony N. 2008. *Molecular photofitting: Predicting ancestry and phenotype using DNA*. Boston: Elsevier/Academic Press.
Frudakis, Tony N., and Mark D. Shriver 2003. *Compositions and methods for inferring ancestry*. United States Patent Application 20040229231
Fullwiley, Duana. 2008. The biologistical construction of race: 'Admixture' technology and the new genetic medicine. *Social Studies of Science* 38: 695–735.
Gannett, Lisa. 1998. Genetic variation: difference, deviation, or deviance? Dissertation, The University of Western Ontario, London/Canada.
Gannett, Lisa. in preparation. Biogeographical ancestry and race.
Gannett, Lisa, and James R. Griesemer. 2004. The ABO blood groups: Mapping the history and geography of genes in Homo sapiens. In *Classical genetics research and its legacy: The mapping cultures of twentieth century genetics*, ed. Hans-Jörg Rheinberger and Jean-Paul Gaudillière, 119–172. London/New York: Routledge.
Halder, Indrani, and Mark D. Shriver. 2003. Measuring and using admixture to study the genetics of complex diseases. *Human Genomics* 1: 52–62.
Hempel, Carl. 1966. *Philosophy of natural science*. Englewood Cliffs: Prentice-Hall.
Howard, Don. 2003. Two left turns make a right: On the curious political career of North American philosophy of science at midcentury. In *Logical empiricism in North America*, ed. Gary L. Hardcastle and Alan W. Richardson, 25–93. Minneapolis: University of Minnesota Press.
Hume, David. 2000 [1740]. *A treatise of human nature*, ed. David Fate Norton and Mary J. Norton. Oxford: Oxford University Press.
Keller, Evelyn Fox. 1992. *"Critical silences in scientific discourse: Problems of form and re-form", in secrets of life, secrets of death: Essays on language*, 73–92. New York/London: Gender and Science/Routledge.
Kuhn, Thomas S. 1962. *The structure of scientific revolutions*. Chicago: University of Chicago Press.
Kuhn, Thomas S. 1977. *"Objectivity, value judgment, and theory choice", in the essential tension: Selected studies in scientific tradition and change*, 320–339. Chicago: University of Chicago Press.
Latour, Bruno. 1999. *Pandora's hope: Essays on the reality of science studies*. Cambridge, MA: Harvard University Press.
Longino, Helen E. 1987. Can there be a feminist science? *Hypatia* 2(3): 51–64.
Longino, Helen E. 1990. *Science as social knowledge: Values and objectivity in scientific inquiry*. Princeton: Princeton University Press.
Mazuzan, George T. 1994. *The National Science Foundation: A brief history*. http://www.nsf.gov/about/history/nsf50/nsf8816.jsp. Accessed 18 Oct 2012.
Nagel, Ernest. 1939. Probability and the theory of knowledge. *Philosophy of Science* 6: 212–253.
National Human Genome Research Institute's Race, Ethnicity, and Genetics Working Group. 2005. The use of racial, ethnic, and ancestral categories in human genetics research. *American Journal of Human Genetics* 77: 519–532.

Norton, H.L., C.L. Pfaff, E.J. Parra, N. Sylvester, D. Parrish-Gause, R.A. Kittles, and M.D. Shriver. 2000. *Skin pigmentation and individual admixture estimates*. American Society of Human Genetics annual meeting. http://www.ashg.org/genetics/abstracts/abs00/f1186.htm.

Okruhlik, Kathleen. 1994. Gender and the biological sciences. In *Biology & society: Reflections on methodology*, ed. Matthen Mohan and R.X. Ware, 21–42. Calgary: University of Calgary Press.

Okruhlik, Kathleen. 2005. Logical empiricism, feminism, and Neurath's auxiliary motive. *Hypatia* 19: 48–72.

Parra, E.J., C.L. Pfaff, K. Hiester, D. Sosnoski, Y. Jien, F. Gulden, P. McKeigue, and M.D. Shriver. 2000. *Pattern and dynamics of admixture in US African-American populations: A global perspective*. American Society of Human Genetics annual meeting. http://www.ashg.org/genetics/abstracts/abs00/f1196.htm

Pfaff, C.L., E.J. Parra, and M.D. Shriver. 2000. *Genetic estimation of biogeographical ancestry*. American Society of Human Genetics annual meeting. http://www.ashg.org/genetics/abstracts/abs00/f1195.htm

Pfaff, C.L., E.J. Parra, C. Bonilla, K. Hiester, P.M. McKeigue, M.I. Kamboh, R.G. Hutchinson, R.E. Ferrell, E. Boerwinkle, and M.D. Shriver. 2001. Population structure in admixed populations: Effect of admixture dynamics on the pattern of linkage disequilibrium. *American Journal of Human Genetics* 68: 198–207.

Poincaré, Henri. 1963 [1913]. Ethics and science. In *Mathematics and science: Last essays*. Trans. John W. Bolduc. New York: Dover.

Polanyi, Michael. 1951. *The logic of liberty: Reflections and rejoinders*. Chicago: University of Chicago Press.

Popper, Karl. 1985 [1934]. Scientific method. In *Popper selections*, ed. David Miller. Princeton: Princeton University Press.

Proctor, Robert N. 1991. *Value-free science? Purity and power in modern knowledge*. Cambridge, MA: Harvard University Press.

Putnam, Hilary. 2002. *The collapse of the fact/value dichotomy and other essays*. Cambridge, MA: Harvard University Press.

Reichenbach, Hans. 1938. *Experience and prediction: An analysis of the foundations and the structure of knowledge*. Chicago: University of Chicago Press.

Reichenbach, Hans. 1951. *The rise of scientific philosophy*. Berkeley: University of California Press.

Reiss, Julian. 2010. In favour of a Millian proposal to reform biomedical research. *Synthese* 177: 427–447.

Reydon, Thomas A.C., Kostas Kampourakis, and George P. Patrinos. 2012. Genetics, genomics and society: The responsibilities of scientists for science communication and education. *Personalized Medicine* 9: 633–643.

Shields, Alexandra E., Michael Fortun, Evelynn M. Hammonds, Patricia A. King, Caryn Lerman, Rayna Rapp, and Patrick F. Sullivan. 2005. The use of race variables in genetic studies of complex traits and the goal of reducing health disparities: A transdisciplinary perspective. *American Psychologist* 60: 77–103.

Shriver, Mark D., Michael W. Smith, Li Jin, Amy Marcini, Joshua M. Akey, Ranjan Deka, and Robert E. Ferrell. 1997. Ethnic-affiliation estimation by use of population-specific DNA markers. *American Journal of Human Genetics* 60: 957–964.

Shriver, Mark D., Esteban J. Parra, Sonia Dios, Carolina Bonilla, Heather Norton, Celina Jovel, Carrie Pfaff, Cecily Jones, Aisha Massac, Neil Cameron, Archie Baron, Tabitha Jackson, George Argyropoulos, Li Jin, Clive J. Hoggart, Paul M. McKeigue, and Rick A. Kittles. 2003. Skin pigmentation, biogeographical ancestry, and admixture mapping. *Human Genetics* 112: 387–399.

Uebel, Thomas E. 2000. Logical empiricism and the sociology of knowledge: The case of Neurath and Frank. *Philosophy of Science* 67 (Proceedings): S138–S150.

Wolpert, Lewis. 1992. *The unnatural nature of science*. Cambridge, MA: Harvard University Press.

Philosophical Issues in Human Pluripotent Stem Cell Research

Andrew W. Siegel

1 Introduction

Human pluripotent stem cell (HPSC) research offers much hope for alleviating the human suffering brought on by the ravages of disease and injury. HPSCs are characterized by their capacity for self-renewal and their ability to differentiate into all types of cells of the body. The main goal of HPSC research is to identify the mechanisms that govern cell differentiation and to turn HPSCs into specific cell types that can be used for treating debilitating and life-threatening diseases and injuries.

Despite the tremendous therapeutic promise of HPSC research, it has met with heated opposition. The principal HPSCs currently used in research are human embryonic stem cells (HESCs), the harvesting of which involves the destruction of the human embryo. HESCs are derived *in vitro* around the fifth day of the embryo's development (Thomson et al. 1998). A typical day-5 human embryo consists of about 150 cells, most of which comprise the trophoblast, which is the outermost layer of the blastocyst. HESCs are harvested from the inner cell mass of the blastocyst, which consists of 30–34 cells. The derivation of HESC cultures requires the removal of the trophoblast. This process of disaggregating the blastocyst's cells eliminates its potential for further development. Opponents of HESC research argue that the research is morally impermissible because it involves the unjust killing of innocent human beings.

Scientists recently succeeded in converting adult human skin cells into cells that appear to have the properties of HESCs by activating four genes in the adult cells (Takahashi et al. 2007; Yu et al. 2007). The reprogrammed cells – "induced pluripotent stem cells" (iPSCs) – could ultimately eliminate the need for HESCs.

A.W. Siegel (✉)
Berman Institute of Bioethics, Johns Hopkins University,
1809 Ashland Ave., 21205 Baltimore, MD, USA
e-mail: asiegel@jhu.edu

K. Kampourakis (ed.), *The Philosophy of Biology: A Companion for Educators*, History, Philosophy and Theory of the Life Sciences 1, DOI 10.1007/978-94-007-6537-5_29,

However, at present, the consensus in the scientific community is that both HESC and iPSC research should be pursued, as we do not yet know whether iPSCs have the same potential as HESCs or whether it is safe to transplant them into humans. Thus, the controversies around HESC research will continue, at least in the near-term.

While the main source of the controversy surrounding HPSC research lies in competing views about the value of human embryonic life, the scope of ethical issues in HPSC research is broader than the ethics of destroying human embryos. For example, there are ethical issues related to the creation of human/non-human chimeras and the derivation of human gametes from HPSCS that apply equally to iPSCs and HESCs. There is also the more general issue of how we should construct public policy in a pluralistic society in which there are opposing views about the ethics of the research. It is important that stem cell biology education extend beyond an inquiry into the biological properties of stem cells and further address the contested questions in metaphysics, ethics, and political philosophy that bear on the pursuit of research in the field (for biomedical research ethics more generally see Plutynski this volume). This chapter provides an overview of these issues.

2 The Ethics of Destroying Human Embryos for Research

The potential therapeutic benefits of HESC research provide strong grounds in favor of the research. If looked at strictly in terms of maximizing social utility, it's almost certainly the case that the potential health benefits from the research outweigh the loss of embryos involved and whatever suffering results from that loss for persons who want to protect embryos. However, most of those who oppose the research argue that the constraints against killing innocent persons to promote social utility apply to human embryos. Thus, as long as we accept non-consequentialist constraints on killing persons, those supporting HESC research must respond to the claim that those constraints apply to human embryos.

In its most basic form, the central argument supporting the claim that it is unethical to destroy human embryos goes as follows: It is morally impermissible to intentionally kill innocent human beings; the human embryo is an innocent human being; therefore it is morally impermissible to intentionally kill the human embryo. It is worth noting that this argument, if sound, would not suffice to show that all or even most HESC research is impermissible, since most investigators engaged in HESC research do not participate in the derivation of HESCs but instead use cell lines that researchers who performed the derivation have made available. To show that researchers who use but do not derive HESCs participate in an immoral activity, one would further need to establish their complicity in the destruction of embryos. We will consider this issue in Sect. 3. But for the moment, let us address the argument that it is unethical to destroy human embryos.

2.1 When Does a Human Being Begin to Exist?

A premise of the argument against killing embryos is that human embryos are human beings. The issue of when a human being begins to exist is, however, a contested one. The standard view of those who oppose HESC research is that a human being begins to exist with the emergence of the one-cell zygote at fertilization. At this stage, human embryos are said to be "whole living member[s] of the species Homo sapiens ... [which] possess the epigenetic primordia for self-directed growth into adulthood, with their determinateness and identity fully intact" (George and Gomez-Lobo 2002, p. 258). This view is sometimes challenged on the grounds that monozygotic twinning is possible until around days 14–15 of an embryo's development (Smith and Brogaard 2003). An individual who is an identical twin cannot be numerically identical to the one-cell zygote, since both twins bear the same relationship to the zygote, and numerical identity must satisfy transitivity. That is, if the zygote, A, divides into two genetically identical cell groups that give rise to identical twins B and C, B and C cannot be the same individual as A because they are not numerically identical with each other. This shows that not all persons can correctly assert that they began their life as a zygote. However, it does not follow that the zygote is not a human being, or that it has not individuated. This would follow only if one held that a condition of an entity's status as an individual human being is that it be impossible for it to cease to exist by dividing into two or more entities. But this seems implausible. Consider cases in which we imagine adult humans undergoing fission (for example, along the lines of Parfit's thought experiments, where each half of the brain is implanted into a different body) (Parfit 1984). The prospect of our going out of existence through fission does not pose a threat to our current status as distinct human persons. Likewise, one might argue, the fact that a zygote may divide does not create problems for the view that the zygote is a distinct human being.

There are, however, other grounds on which some have sought to reject that the early human embryo is a human being. According to one view, the cells that comprise the early embryo are a bundle of homogeneous cells that exist in the same membrane but do not form a human organism because the cells do not function in a coordinated way to regulate and preserve a single life (Smith and Brogaard 2003; McMahan 2002). While each of the cells is alive, they only become parts of a human organism when there is substantial cell differentiation and coordination, which occurs around day-16 after fertilization. Thus, on this account, disaggregating the cells of the 5-day embryo to derive HESCs does not entail the destruction of a human being.

This account is subject to dispute on empirical grounds. That there is some intercellular coordination in the zygote is revealed by the fact that the development of the early embryo requires that some cells become part of the trophoblast while others become part of the inner cell mass. Without some coordination between the cells, there would be nothing to prevent all cells from differentiating in the same direction (Damschen et al. 2006). The question remains, though, whether this degree of cellular interaction is sufficient to render the early human embryo a human being.

Just how much intercellular coordination must exist for a group of cells to constitute a human organism cannot be resolved by scientific facts about the embryo, but is instead an open metaphysical question (McMahan 2007a).

2.2 *The Moral Status of Human Embryos*

Suppose that the 5-day human embryo is a human being. On the standard argument against HESC research, membership in the species *Homo sapiens* confers on the embryo a right not to be killed. This view is grounded in the assumption that human beings have the same moral status (at least with respect to possessing this right) at all stages of their lives.

Some accept that the human embryo is a human being but argue that the human embryo does not have the moral status requisite for a right to life. There is reason to think that species membership is not the property that determines a being's moral status. We have all been presented with the relevant thought experiments, courtesy of Disney, Orwell, Kafka, and countless science fiction works. The results seem clear: we regard mice, pigs, insects, aliens, and so on, as having the moral status of persons in those possible worlds in which they exhibit the psychological and cognitive traits that we normally associate with mature human beings. This suggests that it is some higher-order mental capacity (or capacities) that grounds the right to life. While there is no consensus about the capacities that are necessary for the right to life, some of the capacities that have been proposed include reasoning, self-awareness, and agency (Kuhse and Singer 1992; Tooley 1983; Warren 1973).

The main difficulty for those who appeal to such mental capacities as the touchstone for the right to life is that early human infants lack these capacities, and do so to a greater degree than many of the nonhuman animals that most deem it acceptable to kill (Marquis 2002). This presents a challenge for those who hold that the non-consequentialist constraints on killing human children and adults apply to early human infants. Some reject that these constraints apply to infants, and allow that there may be circumstances where it is permissible to sacrifice infants for the greater good (McMahan 2007b). Others argue that, while infants do not have the intrinsic properties that ground a right to life, we should nonetheless treat them as if they have a right to life in order to promote love and concern towards them, as these attitudes have good consequences for the persons they will become (Benn 1973; Strong 1997).

Some claim that we can reconcile the ascription of a right to life to all humans with the view that higher order mental capacities ground the right to life by distinguishing between two senses of mental capacities: "immediately exercisable" capacities and "basic natural" capacities (George and Gomez-Lobo 2002, p. 260). According to this view, an individual's immediately exercisable capacity for higher mental functions is the actualization of basic natural capacities for higher mental functions that exist at the embryonic stage of life. Human embryos have a "rational nature," but that nature is not fully realized until individuals are able to exercise their capacity to reason. The difference between these types of capacity is said to be a difference between degrees of development along a continuum. There is merely a quantitative difference between the mental capacities of embryos, fetuses, infants, children, and adults (as well as among infants, children, and adults). And this difference, so the argument runs, cannot justify treating some of these individuals with moral respect while denying it to others.

Given that a human embryo cannot reason at all, the claim that it has a rational nature has struck some as tantamount to asserting that it has the potential to become an individual that can engage in reasoning (Sagan and Singer 2007). But an entity's having this potential does not logically entail that it has the same status as beings that have realized some or all of their potential (Feinberg 1986). Moreover, with the advent of cloning technologies, the range of entities that we can now identify as potential persons arguably creates problems for those who place great moral weight on the embryo's potential. A single somatic cell or HESC can in principle (though not yet in practice) develop into a mature human being under the right conditions – that is, where the cell's nucleus is transferred into an enucleated egg, the new egg is electrically stimulated to create an embryo, and the embryo is transferred to a woman's uterus and brought to term. If the basis for protecting embryos is that they have the potential to become reasoning beings, then, some argue, we have reason to ascribe a high moral status to the trillions of cells that share this potential and to assist as many of these cells as we reasonably can to realize their potential (Sagan and Singer 2007; Savulescu 1999). Because this is a stance that we can expect nearly everyone to reject, it's not clear that opponents of HESC research can effectively ground their position in the human embryo's potential.

One response to this line of argument has been to claim that embryos possess a kind of potential that somatic cells and HESCs lack. An embryo has potential in the sense of having an "active disposition" and "intrinsic power" to develop into a mature human being (Lee and George 2006). An embryo can mature on its own in the absence of interference with its development. A somatic cell, on the other hand, does not have the inherent capacity or disposition to grow into a mature human being. However, some question whether this distinction is viable, especially in the HESC research context. While it is true that somatic cells can realize their potential only with the assistance of outside interventions, an embryo's development also requires that numerous conditions external to it are satisfied. In the case of embryos that are naturally conceived, they must implant, receive nourishment, and avoid exposure to dangerous substances *in utero*. In the case of spare embryos created through in vitro fertilization – which are presently the source of HESCs for research – the embryos must be thawed and transferred to a willing woman's uterus. Given the role that external factors – including technological interventions – play in an embryo's realizing its potential, one can question whether there is a morally relevant distinction between an embryo's and somatic cell's potential and thus raise doubts about potentiality as a foundation for the right to life (Devolder and Harris 2007).

Some grant that human embryos lack the properties essential to a right to life, but hold that they possess an intrinsic value that calls for a measure of respect and places at least some moral constraints on their use: "The life of a single human organism commands respect and protection [...] no matter in what form or shape, because of the complex creative investment it represents and because of our wonder at the divine or evolutionary processes that produce new lives from old ones" (Dworkin 1992, p. 84). There are, however, divergent views about the level of respect embryos command and what limits exist on their use. Some opponents of HESC research hold that the treatment of human embryos as mere research tools always fails to manifest proper respect for them. Other opponents take a less absolutist view. Some, for example, deem embryos less valuable than more mature human beings but argue that the benefits of HESC research are too speculative to warrant the destruction of embryos, and that the benefits might, in any case, be achieved through the use of noncontroversial sources of stem cells (e.g., adult stem cells) (Holm 2003).

Many, if not most, who support the use of human embryos for HESC research would likely agree with opponents of the research that there are some circumstances where the use of human embryos would display a lack of appropriate respect for human life, for example, were they to be offered for consumption to contestants in a reality TV competition or destroyed for the production of cosmetics. But proponents of the research hold that the value of human embryos is not great enough to constrain the pursuit of research that may yield significant therapeutic benefits. Supporters of the research also frequently question whether most opponents of the research are consistent in their ascription of a high value to human embryos, as opponents generally display little concern about the fact that many embryos created for fertility treatment are discarded.

2.3 The Case of "Doomed Embryos"

When spare embryos exist after fertility treatment, the individuals for whom the embryos were created typically have the option of storing for them for future reproductive use, donating them to other infertile couples, donating them to research, or discarding them. Some argue that as long as the decision to donate embryos for research is made after the decision to discard them, it is morally permissible to use them in HESC research even if we assume that they have the moral status of persons. The claim takes two different forms. One is that it is morally permissible to kill an individual who is about to be killed by someone else where killing that individual will help others (Curzer 2004). The other is that researchers who derive HESCs from embryos that were slated for destruction do not cause their death. Instead, the decision to discard the embryos causes their death (Green 2002).

Both versions of the argument presume that the decision to discard spare embryos prior to the decision to donate them to research entails that donated embryos are doomed to destruction when researchers receive them. There are two arguments one might marshal against this presumption. First, one who wants to donate embryos to research might first elect to discard them only because doing so is a precondition for donating them. There could be cases in which one who chooses the discard option would have donated the embryos to other couples were the research donation option not available. The fact that a decision to discard embryos is made prior to the decision to donate the embryos thus does not establish that the embryos were doomed to destruction before the decision to donate them to research was made. Second, a researcher who receives embryos could choose to rescue them, whether by continuing to store them or by donating them to infertile couples. While this would violate the law, the fact that it is within a researcher's power to prevent the destruction of the embryos he or she receives poses problems for the claim that the decision to discard the embryos dooms them or causes their destruction.

3 The Ethics of Using Human Embryonic Stem Cells in Research

Assume for the sake of argument that it is morally impermissible to destroy human embryos. It does not follow that all research with HESCs is impermissible, as it is sometimes permissible to benefit from moral wrongs. For example, there is nothing *prima facie* objectionable about transplant surgeons and patients benefiting from the organs of murder and drunken driving victims (Robertson 1988). If there are conditions under which a researcher may use HESCs without being complicit in the destruction of embryos, then those who oppose the destruction of embryos could support research with HESCs under certain circumstances.

Researchers using HESCs are clearly implicated in the destruction of embryos where they derive the cells themselves or enlist others to derive the cells. However, most investigators who conduct research with HESCs obtain them from an existing pool of cell lines and play no role in their derivation. One view is that we cannot assign causal or moral responsibility to investigators for the destruction of embryos from which the HESCs they use are derived where their "research plans had no effect on whether the original immoral derivation occurred" (Robertson 1999). This view requires qualification. There may be cases in which HESCs are derived for the express purpose of making them widely available to HESC investigators. In such instances, it may be that no individual researcher's plans motivated the derivation of the cells. Nonetheless, one might argue that investigators who use these cells are complicit in the destruction of the embryos from which the cells were derived because they are participants in a research enterprise that creates a demand for HESCs. For these investigators to avoid the charge of complicity in the destruction of embryos, it must be the case that the researchers who derived the HESCs would have performed the derivation in the absence of external demand for the cells (Siegel 2004).

The issue about complicity goes beyond the question of an HESC researcher's role in the destruction of the particular human embryo(s) from which the cells he or she uses are derived. There is a further concern that research with existing HESCs will result in the future destruction of embryos: "[I]f this research leads to possible treatments, private investment in such efforts will increase greatly and the demand for many thousands of cell lines with different genetic profiles will be difficult to resist" (U.S. Conference of Catholic Bishops 2001). This objection faces two difficulties. First, it appears to be too sweeping: research with adult stem cells and non-human animal stem cells, as well as general research in genetics, embryology, and cell biology could be implicated, since all of this research might advance our understanding of HESCs and result in increased demand for them. Yet, no one, including those who oppose HESC research, argues that we should not support these areas of research. Second, the claim about future demand for HESCs is speculative. Indeed, current HESC research could ultimately reduce or eliminate demand for the cells by providing insights into cell biology that enable the use of alternative sources of cells (e.g., adult stem cells) (Siegel 2004).

While it might thus be possible for a researcher to use HESCs without being morally responsible for the destruction of human embryos, that does not end the inquiry into complicity. Some argue that agents can be complicit in wrongful acts for which they are not morally responsible. One such form of complicity arises from an association with wrongdoing that symbolizes acquiescence in the wrongdoing (Burtchaell 1989). The failure to take appropriate measures to distance oneself from moral wrongs may give rise to "metaphysical guilt," which produces a moral taint and for which shame is the appropriate response (May 1992). The following question thus arises: Assuming it is morally wrongful to destroy human embryos, are HESC researchers who are not morally responsible for the destruction of embryos complicit in the sense of symbolically aligning themselves with a wrongful act?

One response is that a researcher who benefits from the destruction of embryos need not sanction the act any more than the transplant surgeon who uses the organs of a murder or drunken driving victim sanctions the homicidal act (Curzer 2004). But this response is unlikely to be satisfactory to opponents of HESC research. There is arguably an important difference between the transplant case and HESC research insofar as the moral wrong associated with the latter (a) systematically devalues a particular class of human beings and (b) is largely socially accepted and legally permitted. Opponents of HESC research might suggest that the HESC research case is more analogous to the following kind of case: Imagine a society in which the practice of killing members of a particular racial or ethnic group is legally permitted and generally accepted. Suppose that biological materials obtained from these individuals subsequent to their deaths are made available for research uses. Could researchers use these materials while appropriately distancing themselves from the wrongful practice? Arguably, they could not. There is a heightened need to protest moral wrongs where those wrongs are socially and legally accepted. Attempts to benefit from the moral wrong in these circumstances may be incompatible with mounting a proper protest (Siegel 2003a).

But even if we assume that HESC researchers cannot avoid the taint of metaphysical guilt, it is not clear that researchers who bear no moral responsibility for the destruction of embryos are morally obligated not to use HESCs. One might argue that there is a prima facie duty to avoid moral taint, but that this duty may be overridden for the sake of a noble cause.

4 The Ethics of Creating Embryos for Stem Cell Research and Therapy

Most HESCs are derived from embryos that were created for infertility treatment but that were in excess of what the infertile individual(s) ultimately needed to achieve a pregnancy. The HESCs derived from these leftover embryos offer investigators a powerful tool for understanding the mechanisms controlling cell differentiation. However, there are scientific and therapeutic reasons not to rely entirely on leftover embryos. From a research standpoint, creating embryos through cloning technologies with cells that are known to have particular genetic mutations would allow researchers to study the underpinnings of genetic diseases *in vitro*. From a therapeutic standpoint, the HESCs obtained from leftover IVF embryos are not genetically diverse enough to address the problem of immune rejection by recipients of stem cell transplants. (Induced pluripotent stem cells may ultimately prove sufficient for these research and therapeutic ends, since the cells can (a) be selected for specific genetic mutations and (b) provide an exact genetic match for stem cell recipients.) At present, the best way to address the therapeutic problem is through the creation of a public stem cell bank that represents a genetically diverse pool of stem cell lines (Faden et al. 2003; Lott and Savulescu 2007). This kind of stem cell bank would require the creation of embryos from gamete donors who share the same HLA-types (i.e., similar versions of the genes that mediate immune recognition and rejection).

Each of these enterprises has its own set of ethical issues. In the case of research cloning, some raise concerns, for example, that the perfection of cloning techniques for research purposes will enable the pursuit of reproductive cloning, and that efforts to obtain the thousands of eggs required for the production of cloned embryos will result in the exploitation of women who provide the eggs (President's Council on Bioethics 2002; Norsigian 2005). With respect to stem cell banks, it is not practically possible to create a bank of HESCs that will provide a close immunological match for all recipients. This gives rise to the challenge of determining who will have biological access to stem cell therapies. We might construct the bank so that it provides matches for the greatest number of people in the population, gives everyone an equal chance of finding a match, or ensures that all ancestral/ethnic groups are fairly represented in the bank (Faden et al. 2003; Bok et al. 2004; Greene 2006).

There are, however, more general challenges to the creation of embryos for research and therapeutic purposes. Some argue that the creation of embryos for non-reproductive ends is morally problematic, regardless of whether they are created through cloning or in vitro fertilization. There are two related arguments that have been advanced to morally distinguish the creation of embryos for reproductive purposes from the creation of embryos for research and therapeutic purposes. First, each embryo created for procreative purposes is originally viewed as a potential child in the sense that each is a candidate for implantation and development into a mature human. In contrast, embryos created for research or therapies are viewed as mere tools from the outset (Annas et al. 1996; President's Council on Bioethics 2002). Second, while embryos created for research and therapy are produced with the intent to destroy them, the destruction of embryos created for reproduction is a foreseeable but unintended consequence of their creation (Fizpatrick 2003).

One response to the first argument has been to suggest that we could, under certain conditions, view all research embryos as potential children in the relevant sense. If all research embryos were included in a lottery in which some of them were donated to individuals for reproductive purposes, all research embryos would have a chance at developing into mature humans (Devolder 2005). Since those who oppose creating embryos for research would likely maintain their opposition in the research embryo lottery case, it is arguably irrelevant whether embryos are viewed as potential children when they are created. Of course, research embryos in the lottery case would be viewed as both potential children and potential research tools. But this is also true in the case of embryos created for reproductive purposes where patients are open to donating spare embryos to research.

As to the second argument, the distinction between intending and merely foreseeing harms is one to which many people attach moral significance. But even if one holds that this is a morally significant distinction, it is not clear that it is felicitous to characterize the destruction of spare embryos as an unintended but foreseeable side-effect of creating embryos for fertility treatment. Fertility clinics do not merely foresee that some embryos will be destroyed, as they choose to offer patients the option of discarding embryos and carry out the disposal of embryos when patients request it. Patients who elect that their embryos be discarded also do not merely foresee the embryos' destruction; their election of that option manifests their intention that the embryos be destroyed. There is thus reason to doubt that there is a moral distinction between creating embryos for research and creating them for reproductive purposes, at least given current fertility clinic practices.

5 Human/Non-human Chimeras Created with Human Pluripotent Stem Cells

Human cellular material has for decades been introduced into animals as a means of learning about human biological processes in disease and in developing and testing therapies. The transplantation of human cells into animals allows researchers to maintain the cells and study them *in vivo* when it is not morally or practically feasible to carry out the experiments in humans. The use of such chimeric animal models has been largely uncontroversial. But advances in stem cell technologies allow for the possibility of human/non-human chimeras that give some pause.

Consider the following: (1) It is in principle possible to graft HPSCs into animal embryos, with human cells making a substantial contribution to some cell types in a live-born animal. Human cells could theoretically contribute to the germ-line in the chimeric animal, so that if the animal were to breed it could produce a hybrid animal (i.e., one that has a full set of genes from each parent). (2) Grafting human neurons into a non-human primate brain would produce the best animal model for studying human neurodegenerative disorders, but populating a non-human primate brain with human cells could give rise to an animal with more human-like cognitive and behavioral capacities. These are some of the scenarios that give rise to concerns about the creation of human/non-human chimeras.

But what precisely is the moral foundation for these concerns? One view is that it is simply unnatural to mix human and non-human biological materials. Yet such an objection seems too broad, as it would encompass human/non-human chimeras that have long been used in research without controversy, such as mice with a single human gene or mice with human cancer cells. One could cite numerous other widely accepted forms of human/non-human admixtures, such as the use of pig heart valves in humans and human digestion of animal products. Thus, one who asserts this objection needs to distinguish putatively problematic instances of the unnatural from those that appear not to be troubling (Streiffer 2011).

One suggestion is that the chimeras that generate concern are ones in which the kind and quantity of human cellular material grafted into the animal creates a being that crosses species boundaries. Now, what constitutes a species is itself a contested a matter (see also Wilkins this volume). Biologists operate with various definitions of species – such as genetic isolation, reproductive isolation, common ancestry, or homeostatic property clusters – depending on the nature of inquiry they are engaged in (Robert and Baylis 2003, p. 3). Without a settled biological concept of species, we cannot definitively identify when species boundaries are crossed. At the same time, however, we do operate with social, moral, and legal constructs that assume sharp boundaries between humans and other animals. If a human/non-human chimera breaks down those boundaries, might we then expect social disruption and moral confusion?

Some bioethicists have suggested that such chimeras would be "threatening to the social order" because they would force us to "confront the possibility that humanness is neither necessary nor sufficient for personhood [i.e., full moral standing]." Confronting this possibility would, they argue, produce a state of moral confusion inasmuch as it would leave us with "no clear way of understanding our moral obligations to these beings," and compel us to "revisit some of our current patterns of behavior toward certain human and nonhuman animals" (Robert and Baylis 2003, p. 9).

To assess the merits of this claim, it will be helpful to recall some points that emerged in our discussion of the moral status of human embryos. First, there is already good reason to think that most of us reject that humanness is logically necessary for personhood, as we generally respond to fictional animals and aliens possessing human-like psychological and cognitive traits as moral persons. Second, many do not believe that being a human is sufficient to ground basic rights, as revealed by the common view that it is permissible to destroy embryos for research. On these views of moral status, a being's moral standing is a function of features (e.g., rationality and self-awareness) that are not logically tied to membership in a particular species. The existence of chimeras would not produce confusion about the principles that determine our obligations to them (Siegel 2003b). Rather, some chimeras could create novel questions for the application of those principles. For instance, there may be epistemic issues about the evidentiary bases for identifying when a being's moral status has changed – e.g., what behavioral and psychological tests would demonstrate increased capacities of the relevant kind.

But what about those who believe that being human is sufficient to confer full moral standing? Here, there is reason to anticipate moral confusion. Suppose we have a chimera that does not possess human-like psychological or cognitive capacities but is nevertheless comprised of a substantial amount of human cellular material. Is the amount of respect due to the chimera proportionate to the amount of human material it contains, or is respect for humanness an all or nothing matter? At what point, if any, does a chimera have enough human material to be treated as a human? Is it the amount of human material or the kind of human material that is more important? Those who view humanness as sufficient for full moral standing have not developed principles that would address these challenges, nor is it clear that there is any non-arbitrary basis for generating such principles. We should thus expect that certain human/non-human chimeras would leave some morally perplexed, and that this fact will heighten political controversy over the creation of chimeras.

6 Gametes Derived from Human Pluripotent Stem Cells

There are preliminary findings that suggest it might be possible to derive gametes from human pluripotent stem cells. Scientists have thus far succeeded in generating sperm and eggs from mouse ESCs and iPSCs and have produced offspring using these stem cell-derived gametes (Hayashi et al. 2011, 2012). While it may take many years before researchers successfully derive gametes from human PSCs, the research holds much promise for basic science and clinical applications. For example, the research could provide important insights into the fundamental processes of gamete biology and provide otherwise infertile individuals a means of creating genetically related children. At the same time, the research raises several controversial issues related to embryos, genetics, and assisted reproductive technologies (Matthews et al. 2009).

First, to establish that a particular technique for deriving human gametes from PSCs produces functional sperm and eggs, it is necessary to demonstrate that the cells can produce an embryo. This entails the creation of embryos through in vitro fertilization. Since it would not be safe to implant embryos created during the early stages of the research, the embryos would be destroyed or discarded. Thus, PSC-derived gamete research implicates all of the moral issues surrounding the creation and destruction of embryos for research.

Second, while the ability to generate PSC-derived eggs could reduce or eliminate the need for egg donors and thus help overcome concerns about exploitation of donors and the risks involved in egg retrieval, the prospect of being able to produce large numbers of eggs from PSCs would raise other concerns. As the capacity to identify disease and non-disease related alleles through preimplantation genetic diagnosis (PGD) expands, having large numbers of embryos would significantly increase the chances of finding an embryo that possesses most or all of the traits one wishes to select. This would be beneficial in preventing the birth of children with genetic diseases. But matters would become morally contentious if it could be used to select non-disease characteristics, such as sexual orientation, gender, eye color, and size. One common argument against using PGD in this way is that it could devalue the lives of those who do not exhibit the chosen characteristics. Another concern is that employing PGD to select for non-disease traits would fail to acknowledge the "giftedness of life" by treating children as "objects of our design or products of our will or instruments of our ambition" rather than accepting them as they are given to us (Sandel 2004, p. 56). Of course, one can question whether the selection of non-disease traits would in fact lead to devaluing other characteristics, and whether it would alter the nature of parental love. Nonetheless, the capacity to produce PSC-derived eggs would make these issues more pressing.

Third, the insights into the basic mechanisms of gamete biology this research potentially offers could greatly advance the prospects of germ-line genetic modification. As with genetic screening through PGD, germ-line genetic modification would be beneficial in preventing genetic diseases but controversial if used to engineer non-disease related traits.

Finally, in theory, it is possible to create eggs from PSCs derived from male cells and sperm from PSCs derived from female cells. Thus, a possible outcome of the research is that a same-sex couple could have a child that is genetically related to both partners. This would have the benefit of enhancing procreative liberty. But in countries like the United States, this method of reproduction would escalate hostilities in the culture wars. The forces that seek to deny same-sex marriage on the grounds that it violates the natural order of things would no doubt strongly oppose same-sex reproduction.

7 Stem Cell Research and Politics

Research into human pluripotent stem cells occurs in a social and political context where the moral controversies surrounding the research can significantly impact the regulation and funding of the research. How should we develop public policy in this contentious area of research? One approach would be to seek to advance one's own moral views through the political process while giving little or no weight to opposing views. The problem with this approach is that it creates instability, with policy subject to ever shifting political winds. It puts researchers in a position of perpetual uncertainty about whether they will be able to sustain their work. The alternative approach is to construct public policy through a "search for significant points of convergence between one's own understanding and those of citizens whose positions, taken in their more comprehensive forms, one must reject" (Gutmann and Thompson 1996, p. 85). The effort to build policy around areas of overlapping consensus among otherwise opposing views both manifests respect for the diversity of moral views that characterize pluralistic societies and provides a foundation for more stable policies.

The issue of government funding of stem cell research in the United States illustrates the challenges and possibilities related to formulating a public policy that accommodates competing moral viewpoints. In 2001, President Bush enacted a policy prohibiting federal funding of research on human embryonic stem cell lines created after the date the policy was announced. The idea behind the policy was to permit the research to go forward while immunizing the government against charges that its support of the research offers an incentive for the further destruction of embryos. The policy proved a significant impediment to stem cell science, as the cell lines that were eligible for funding under the policy were inadequate in number, quality, and diversity. In 2009, President Obama lifted the restriction against federal funding for the use of newly created HESCs. While supporters of HESC research welcomed this policy, it exposes the government to the charge that it is complicit in the destruction of human embryos, since funding work with new HESCs provides researchers with an incentive to harvest more lines. This policy fails to accommodate those who want to avoid government encouragement of the destruction of embryos and leaves the policy vulnerable to the vicissitudes of election cycles.

Ideally, we should attempt to develop a policy that would allow funding the use of HESCs irrespective of the time of their derivation without incentivizing the production of more lines. There are measures that, taken together, could accomplish this (Siegel 2004). First, there would need to be a prohibition on federal funding to researchers for work with HESCs that they derive. If researchers could receive federal funding only for the use of HESCs donated from non-federally funded researchers the prospect of federal funding would not serve as an incentive for a researcher to produce new cell lines. This limitation would still allow federally funded scientists access to new cell lines from the international research community and publicly funded researchers in states like California and Massachusetts. Second, it must be the case that the cells would have been harvested in the absence of demand for them from federally funded researchers. This measure is necessary to ensure that federal funding does not encourage researchers to derive HESCs for purposes of distributing them to other researchers. We can determine the purpose for which cell lines are derived through appropriate oversight and documentation requirements. If there is good documentation that a researcher derived hESCs for her own specific and legitimate uses and it further can be shown that she has actively used the lines in a manner consistent with the stated scientific rationale, that should be sufficient evidence that the lines were not derived for purposes of distributing them to other researchers.

While these measures place some constraints on federal funding of HESC research, they offer terms that those who otherwise morally oppose funding for research with new HESCs could accept. This suggests that we can sometimes find overlapping consensus between competing moral viewpoints when developing policy on HESC research. There will no doubt be issues where the opposing parties find themselves at an impasse, and where there is thus no stable resolution. Nonetheless, the search for convergence is essential to securing the political legitimacy of public policy in societies where there exists a plurality of reasonable but conflicting ethical, religious, and philosophical perspectives.

8 Stem Cell Research and Biology Education

When research in the biological sciences gives rise to moral and political controversy, it is important that biology students do not view the research in isolation from the social context in which it occurs. In the case of HPSC research, moral and political arguments inform individual choices about whether to participate in the research and social choices about what limits – if any – should be placed on the pursuit of the research. As such, there are a number of reasons why it is particularly valuable to introduce biology students to the philosophical debates around stem cell research.

First, we have seen that there are deep disagreements about such issues as when a human life begins, the moral status of human embryos, and the nature of complicity. The philosophical discussions of these issues reveal that there are sophisticated and reasonable arguments on both sides of the issues. It is important to engage the competing arguments because doing so helps one to critically assess and formulate one's beliefs and to cultivate the virtue of leading a reflective moral life. This engagement also helps to foster greater respect for opposing views and to enable one to offer well-reasoned responses to those who might challenge one's beliefs and practices.

Second, while biology students may generally be familiar with some of the ethical questions around human embryos as the sources of stem cells, little attention has been given to the fact that there may be uses of HPSCs – whether embryonic or non-embryonic in origin – that raise ethical concerns. As we saw earlier in the chapter, there are a number of ethical issues revolving around the creation of human/non-human chimeras and the production of human gametes from HPSCs. Even those who are liberal about HPSC research may have concerns if they consider the prospect of, e.g., the creation of human/non-human primate chimeric brains, the introduction of HPSCs into the germline of non-human animals, or the creation of human gametes from HPSCs for purposes of selecting non-disease traits in preimplantation genetic diagnosis.

Finally, the political dimensions of stem cell research suggest a role for biology education in highlighting connections between science and citizenship. In the area of HPSC research, researchers occupy the dual role of scientist and citizen. As scientists, they may have a crucial personal stake in the outcome of policy debates about their research. As citizens, they have both an interest in scientific progress and an interest in a political process that ensures a fair system of social cooperation. One way in which researchers can inform the political process is by articulating the nature and potential of the science. But they can go further. By attending to the nuances of the philosophical issues surrounding the research, researchers can also participate in formulating policy that manifests respect for pluralism. In introducing biology students to these issues, educators can thus offer the next generation of stem cell scientists tools that promote both science and justice.

Note: Sections 2–4 of this chapter were previously published in Siegel, A. "Ethics of Stem Cell Research", *The Stanford Encyclopedia of Philosophy (Fall 2008 Edition)*, Edward N. Zalta (ed.).

References

Annas, G., A. Caplan, and S. Elias. 1996. The politics of human-embryo research – Avoiding ethical gridlock. *The New England Journal of Medicine* 334: 1329–1332.

Benn, S.I. 1973. Abortion, infanticide, and respect for persons. In *The problem of abortion*, ed. Feinberg Joel, 92–103. Belmont: Wadsworth.

Bok, H., K.E. Schill, and R.R. Faden. 2004. Justice, ethnicity, and stem-cell banks. *Lancet* 364(9429): 118–121.

Burtchaell, J.T. 1989. The use of aborted fetal tissue in research: A rebuttal. *IRB: A Review of Human Subjects Research* 11(2): 9–12.

Curzer, H. 2004. *The Ethics of Embryonic Stem Cell Research* 29(5): 533–562.

Damschen, G., A. Gomez-Lobo, and D. Schonecker. 2006. Sixteen days? A reply to B. Smith and B. Brogaard on the beginning of human individuals. *The Journal of Medicine and Philosophy* 31: 165–175.

Devolder, K. 2005. Human embryonic stem cell research: Why the discarded-created distinction cannot be based on the potentiality argument. *Bioethics* 19(2): 167–186.

Devolder, K., and J. Harris. 2007. The ambiguity of the embryo: Ethical inconsistency in the human embryonic stem cell debate. *Metaphilosophy* 38(2–3): 153–169.

Dworkin, R. 1992. *Life's dominion*. New York: Vintage.

Faden, R.R., et al. 2003. Public stem cell banks: Considerations of justice in stem cell therapy. *The Hastings Center Report* 33: 13–27.

Feinberg, J. 1986. Abortion. In *Matters of life and death*, ed. T. Regan. New York: Random House.

Fizpatrick, W. 2003. Surplus embryos, nonreproductive cloning, and the intend/foresee distinction. *The Hastings Center Report* 33: 29–36.

George, R.P., and A. Gomez-Lobo. 2002. Statement of professor George (joined by Dr. Gómez Lobo). In *Human Cloning and Human Dignity: An Ethical Inquiry*, 258–266. Washington: Inform of the President's Council on Bioethics. http://bioethics.georgetown.edu/pcbe/reports/cloningreport/appendix.html#george.

Green, R. 2002. Benefiting from 'evil'; an incipient moral problem in human stem cell research. *Bioethics* 16(6): 544–556.

Greene, M. 2006. To restore faith and trust: Justice and biological access to cellular therapies. *The Hastings Center Report* 36(1): 57–63.

Gutmann, A., and D. Thompson. 1996. *Democracy and disagreement*. Cambridge, MA: Belknap Press of Harvard University Press.

Hayashi, K., et al. 2011. Reconstitution of the mouse germ cell specification pathway in culture by pluripotent stem cells. *Cell* 146(4): 519–532.

Hayashi, K., et al. 2012. Offspring from oocytes derived from in vitro primordial germ cell-like cells in mice. *Science* 338(6109): 971–975.

Holm, S. 2003. The ethical case against stem cell research. *Cambridge Quarterly of Healthcare Ethics* 12: 372–383.

Kuhse, H., and P. Singer. 1992. Individuals, human, and persons: The issue of moral status. In *Embryo experimentation: Ethical, legal, and social issues*, ed. P. Singer et al., 65–75. Cambridge: Cambridge University Press.

Lee, P., and R. George. 2006. Human embryo liberation: A reply to Peter Singer. *National Review Online*, January 25, 2006. http://old.nationalreview.com/comment/lee_george200601250829.asp.

Lott, J.P., and J. Savulescu. 2007. Towards a global human embryonic stem cell bank. *The American Journal of Bioethics* 7(8): 37–44.

Marquis, D. 2002. Stem cell research: The failure of bioethics. *Free Inquiry* 23(1): 40–44.

Matthews, D., et al. 2009. Pluripotent stem cell-derived gametes: Truth and (potential) consequences. *Cell Stem Cell* 5: 11–14.

May, L. 1992. *Sharing responsibility*. Chicago: University of Chicago Press.

McMahan, J. 2002. *The ethics of killing: Problems at the margins of life*. New York: Oxford University Press.

McMahan, J. 2007a. Killing embryos for stem cell research. *Metaphilosophy* 38(2–3): 170–189.
McMahan, J. 2007b. Infanticide. *Utilitas* 19: 131–159.
Norsigian, J. 2005. Risks to women in embryo cloning. *Boston Globe*, February 25, 2005, at A13.
Parfit, D. 1984. *Reasons and persons*. Oxford: Clarendon.
President's Council on Bioethics. 2002. Human cloning and human dignity: An ethical inquiry. http://bioethics.georgetown.edu/pcbe/reports/cloningreport/
Robert, J., and F. Baylis. 2003. Crossing species boundaries. *The American Journal of Bioethics* 3(3): 1–13.
Robertson, J. 1988. Fetal tissue transplant research is ethical. *IRB: A Review of Human Subjects Research* 6(10): 5–8.
Robertson, J. 1999. Ethics and policy in embryonic stem cell research. *Kennedy Institute of Ethics Journal* 2(9): 109–136.
Sagan, A., and P. Singer. 2007. The moral status of stem cells. *Metaphilosophy* 38(2–3): 264–284.
Sandel, M. 2004. The case against perfection. *Atlantic Monthly* 293(3): 51–62.
Savulescu, J. 1999. Should we clone human beings? *Journal of Medical Ethics* 25(2): 87–98.
Siegel, A. 2003a. Locating convergence: Ethics, public policy, and human stem cell research. In *The stem cell controversy*, ed. M. Ruse and C. Pynes. Amherst: Prometheus Books.
Siegel, A. 2003b. The moral insignificance of crossing species boundaries. *The American Journal of Bioethics* 3(3): 33–34.
Siegel, A. 2004. Temporal restrictions and the impasse on human embryonic stem cell research. *The Lancet* 364(9429): 215–218.
Smith, B., and B. Brogaard. 2003. Sixteen days. *The Journal of Medicine and Philosophy* 28: 45–78.
Streiffer, R. 2011. Human/non-human chimeras. In *The Stanford encyclopedia of philosophy*, Spring 2011 edn, ed. Edward N. Zalta. URL http://plato.stanford.edu/archives/spr2011/entries/chimeras/.
Strong, C. 1997. The moral status of preembryos, embryos, fetuses, and infants. *The Journal of Medicine and Philosophy* 22(5): 457–478.
Takahashi, K., et al. 2007. Induction of pluripotent stem cells from adult human fibroblasts by defined factors. *Cell* 131: 861–872.
Thomson, J.A., et al. 1998. Embryonic stem cell lines derived from human blastocysts. *Science* 282: 1145–1147.
Tooley, M. 1983. *Abortion and infanticide*. New York: Oxford University Press.
Warren, M.A. 1973. On the moral and legal status of abortion. *The Monist* 57: 43–61.
Yu, J., et al. 2007. Induced pluripotent stem cell lines derived from human somatic cells. *Science* 318: 1917–1920.

Ethics in Biomedical Research and Practice

Anya Plutynski

1 Introduction

This chapter will be divided into two sections. The first section will concern questions intrinsic to the conduct of biomedical research: What ethical guidelines should govern biomedical research, on human and nonhuman subjects? What are the responsibilities of researchers to their patients, and to the public at large? How can one avoid problems associated with research in vulnerable populations, and ensure fair subject selection? The second section will concern "extrinsic" questions of justice, and in particular, the allocation of biomedical research funds. The questions at issue here concern inappropriate influence of industry on the conduct of research, as well as concerns about distributive justice. For instance, which, among several drugs, treatments, or regimens, should be prioritized for research, who should have access to these drugs, and how can access be fair and just, given that many people in the third world can barely afford a minimum standard of living, let alone expensive drugs or treatments?

Addressing ethical questions in biomedical research requires a brief introduction to ethical theory. Ethics, or, normative ethical theory, is inquiry into the foundations and character of moral norms. Consider the moral norm, "thou shalt not kill." Why should we respect this norm? Different moral theories provide different answers to this question. For instance, according to a "deontological" moral theory, violating this moral norm would violate a duty, or, in Immanuel Kant's terms, the "categorical imperative": never act on a principle that you could not will as a universal law. Kant viewed this principle as morally equivalent to the principle that one should never treat persons as mere means, and only as ends in themselves. Clearly, one could not

A. Plutynski (✉)
Department of Philosophy, University of Utah, 215 S. Central Campus Dr.402 CTIHB, Salt Lake City 84112, UT, USA
e-mail: anya.plutynski@philosophy.utah.edu

K. Kampourakis (ed.), *The Philosophy of Biology: A Companion for Educators*, History, Philosophy and Theory of the Life Sciences 1, DOI 10.1007/978-94-007-6537-5_30,

will as a universal law that killing other persons (in general) is permissible. Different moral theories provide different rationales for such normative principles. For instance, an Utilitarian might argue that killing in self-defense is permissible, granted that this results in the greatest good for the greatest number. Utilitarians, in contrast to deontological moral theorists, argue that the end or purpose of an action may (sometimes) justify the means; or, as long as an action results in consequences that maximize the good (however this is measured, and this is a point of contention), the action is morally permitted, and may even be required. Moral philosophers debate the scope and interpretation of different moral theories, and competing interpretations of these theories have informed debates in applied ethics (see also Millstein, this volume), such as in biomedical research ethics.

Ethical theory is distinct from "metaethics," which is inquiry into the grounds or foundations of ethical knowledge, or the status of moral claims. For instance, metaethicists might ask the question: Does the claim, "Killing is wrong" express a truth? If so, how is this truth distinct from other kinds of truths? What gives us moral knowledge, and how is moral knowledge different from other kinds of knowledge? Can or should science help us answer moral questions? One issue that moral philosophers have recently taken up is whether and how the sciences – particularly biology, neuroscience, and psychology – bear on our understanding of moral norms and their justification. Moral psychologists investigate the determinants of moral judgments and behaviors. Such determinants might include information about patterns of behavior in response to environmental and social cues, neurological patterns or processes, or evolutionary or biological bases of different behavioral patterns. Results from such research might bear, for instance, on social policy that encourages more ethical behavior, or ameliorates poor conduct. A better understanding of human biology and psychological or biological bases of behavior may inform better social policies. However, one should always be cautious about using science to decide questions of policy; there is always a matter of ethical judgment in moving from "is" to "ought." People may be more likely to be organ donors, for instance, if organ donation is the "default" state. Yet, making this a social policy requires further ethical consideration of whether and why this particular default assignment violates personal autonomy. A Kantian might say that such a policy is not permissible, as a general rule. In contrast, a Utilitarian might regard it as permissible. One of the goals of reflection on ethical theories is to give students opportunities to develop the tools necessary to make and defend their own judgments with deliberative reasons.

Science educators can and should view their role as complimentary to that of philosophers in addressing questions in bioethics. For, there are a variety of ways in which biological science and bioethics are interrelated. First, understanding the biological science behind, e.g., stem cells, is crucial to addressing the ethical issues that arise in stem cell research. Knowing the differences in biological and research potential for embryonic versus somatic stem cells (and all the intermediates) is necessary for determining when and if harvest of the former is morally permissible (see Siegel, this volume). Philosophers of science and bioethicists might contribute to the biomedical sciences as well. Considering the foundations or rationale behind different statistical philosophies – e.g., the reasons why, or whether, randomized controlled trials are or should be considered the gold standard for evidence – could provide scientists with a better way to think about evidence-based medicine.

2 Biomedical Research Ethics: Intrinsic Issues

2.1 Research on Humans

Research ethics, traditionally conceived, concerns the conduct of research on human subjects. After the Nuremberg trials, which were military tribunals held after WWII, for the war crimes of Nazi leadership involved in the atrocities committed against both citizens and soldiers, research on human subjects has (ideally) been subject to international codes of ethics, such as the Nuremberg Code (1947) or the Declaration of Helsinki (1964). More recently, other national and international codes of ethics have been developed. For instance, one ethical framework for responsible conduct of research (1978), the "Common Rule," a U.S. Federal regulation, sets out the processes and requirements for approval of research on human subjects. In addition, there are international guidelines, such as those issued by the Council for International Organization of Medical Sciences (Bankowski and Levine 1992; Gallagher et. al. 2000). Each of these codes provides principles for ethical research on human subjects. Among such principles are requirements that subjects understand that they are being asked to participate in research, that they are fully informed of the nature of the research they are participating in, that they give consent to participate, and that they are able to end their participation. Also, research on human subjects must, according to the Nuremberg Code, produce on balance greater good for society, than any risk imposed on the subjects. In addition, research subjects must be chosen "equitably," vulnerable populations must be protected, and participants' privacy must be protected as far as possible. Arguably, none of these criteria were met in the Tuskegee syphilis study, a decades-long "natural experiment" to study the long-term effects of untreated syphilis on largely African Americans by the U.S. Public Health Service. The participants in this trial were systematically misled as to its aims, and many died after prolonged suffering. More recently, and in part as a result of these cases, several ethics codes emphasize procedural requirements, such as institutional ethical review and approval, as well as oversight, by an Institutional Review Board, or IRB, for any biomedical research. These standards create a variety of challenges concerning when and whether consent has occurred, how one weighs benefit and harm, or what constitutes "fair subject selection," or "adequate" review and/or oversight.

For instance, it's not always clear, at least in early-phase trials of new drugs or exceedingly toxic treatment regimes, whether and to what extent risks can be minimized to subjects. Clinical trials are attempts to systematically assess the safety and efficacy of new drugs, treatments or screening methods. A population is selected at random and broken into a "treatment" and "control" group, the latter of which may receive a "placebo" or sugar pill (without any pharmacogenetic properties), or may receive existing treatments, sometimes called the 'standard of care.' "Double blind" studies are where patients and researchers do not know to which group patients have been assigned. In early trials of cancer drugs, which often have debilitating side effects, as well as for "emergency" research on patients with critical or acute illnesses, or surgical innovations in desperate cases, ethical questions arise as to when and whether the benefits of such trials outweigh the risks (see, e.g., Agarwal and Emmanuel 2008). Early phase or "Phase 1" trials are attempts at determining the safety and efficacy of a novel drug or treatment. These are distinguished from latter stage trials, which test whether the new drug or treatment is more or less effective than the existing standard of care, and exactly how much. Especially for early phase trials, it is not always clear whether a patient fully understands the character of the trial. Some patients, particularly those who may have tried many previous treatments that all failed, may be eager to participate in a trial of a new drug or treatment, and have false hope as to its effectiveness. This false hope is an instance of what is called the "therapeutic misconception": the understanding by research subjects that they were asked to enroll in a trial because their physician thinks that there may be a therapeutic benefit to them through the research, and further, that their assignment to a study group is due to the physician's expectation that this is the right one for the patient. This is a conflation by patients of the goals of clinical medicine (to benefit individual patients) and the goals of the research (to answer a research question that may or may not result in benefit to the individual). In early phase trials, however, clinician-researchers are often only testing whether the drug or treatment has any effect whatsoever (though they usually have good reason to suspect it will). Social scientists have demonstrated that the therapeutic misconception is very common, particularly for patients and families in desperate circumstances. For instance, in the case of Jessie Gelsinger, the young man that died as a result of some of the first trials of genetic medicine, his father reported after Jessie's' death that he had no idea that the treatments had never been successful on human subjects before (Steinbrook 2008, p. 115). Yet, he had, at least by the standards of the institutional review board, given informed consent to participate in the trial. This sort of confusion and resulting situation – the Tier I trial of experimental treatments – inevitably creates an ethical dilemma: when, if ever, is it possible to get fully "autonomous" consent, particularly for patients who are subject to the perfectly reasonable (but perhaps false) hope that this treatment will save a life? What special ethical issues arise for underage patients and their parents in giving consent to participate? How can clinician-researchers communicate effectively about risk, given that many patients have a vague understanding of the role and tiered structure of clinical trials?

While bioethicists have offered various criteria for informed consent, it is sometimes difficult to know when these criteria have been met. CIOMS (the Council for International Organizations of Medical Sciences) developed a set of international guidelines for biomedical research on human subjects in 2002. These guidelines include, among others, a set of very specific and detailed requirements for informed consent. Subjects must be voluntarily participants, they must be free to withdraw, they should be made aware of the purpose of the research, the procedures involved, the research design, and the risks, harm, pain or discomfort likely to result from participation. Subjects must never be deceived or unduly influenced to participate. For instance, they must not be offered excessive compensation in exchange for participation, to avoid the risk of coercion. Design methods, such as placebo-controlled or double blind studies, must be explained in simple language. Provisions should be made, as far as possible, for privacy of subjects to be protected, and reasonable limits to confidentiality must be explained. Prior to a trial, the investigators should negotiate with the relevant stakeholders, including (if relevant), the host country, local health authorities, and infrastructure, to determine whether the research will be of benefit to the relevant community. All this must be subject to review of an institutional ethics review board. Finally, a signed form of evidence of informed consent must be obtained. These issues become particularly sensitive in "externally sponsored" research – i.e., when research is conducted in a host country by researchers from an external organization or pharmaceutical company. In these cases, particularly, the review of research protocol must be conduced by a host or local ethical review committee.

Faden and Beauchamp (1986, in Beauchamp et al. 2007) distinguish two senses of "informed consent," useful for approaching problems of participation in clinical trials. What they call "Sense 1" involves an "autonomous authorization" where patients have substantial understanding of the procedure or experiment, where they have agreed to participate absent any control or coercion, and when they intentionally authorize an intervention or procedure. "Sense 2," in contrast, is any "effective authorization" that follows the rules and procedures of some institution. They point out that one may obtain "informed consent" in Sense 1, without obtaining it in Sense 2, and vice versa. For instance, a patient may be fully informed, but not be legally permitted to participate, either because he/she is not of legal age to agree to participation in research, or because institutional procedures required for informed consent were not followed. In contrast, one may obtain official authorization from a subject, but patients may not be fully informed of the risks of the study. Protecting human subjects, Faden and Beuchamp argue, requires informed consent in both Sense 1 and Sense 2.

Entirely unique ethical questions arise for clinical trials in developing countries, where researchers have been known to conduct unsafe research on vulnerable populations. However, many such cases are difficult to assess. For instance, in Africa, a study tested a reduced dose of anti-HIV drugs in pregnant mothers, after it had already been shown that the full dose was effective in preventing infection or transmission of the disease from mothers to infants. This particular study has been a source of a great deal of controversy. Whether or not such trials should have been done, and whether fully informed consent was achieved in such cases is a matter of some controversy (Shah 2006). The aim of these trials, however, was to find interventions that the countries involved could afford. Success in this endeavor would yield a tremendous benefit, arguably justifying not providing subjects with the Western standard of care. Thus, some argue that in developing countries, where the locally available standard of care is often nothing, including patients in trails where less than the full standard of care provided in Western countries is morally permissible. However, in other cases, where researchers submitted subjects to risks that they may have avoided, exposing vulnerable individuals who may have had little alternative or inadequate opportunity for consent, such research violates several ethical codes of conduct.

This issue of what risks one may knowingly subject a research participant to, arises starkly in the context of placebo-controlled trials. In 1964, the Declaration of Helsinki required that placebo controlled trials should only be conducted when there was no available alternative current therapy. In 2004, Helsinki loosened the criteria for placebo controlled trials, such that they were permissible, where the use is "medically or scientifically necessary" to determine the efficacy or safety of a new drug or intervention, and only when withholding treatment for the condition investigated did not involve any serious harm to the patient. (Bankowski and Levine 1992; Gallagher et. al. 2000) does not permit placebo controlled trials except when there is no established effective intervention, when withholding treatment would at worst expose patients to temporary discomfort, or, when use of the intervention would not yield scientifically reliable results. This level of scrutiny is founded on a principle that is sometimes known as the requirement of "equipoise," or, the demand that researchers regard the treatment and control arm of a placebo-controlled trial as of equal benefit or risk.

What, however, is "scientifically" necessary? Must any new drug be established to be effective over and above a placebo, or only equally as effective as available treatments (called, "active control" or "equivalence" trial)? Miller and Emmanuel (2001) offer a "middle ground" between what they call "active control orthodoxy" (roughly, the view that placebo-controlled trials sacrifice the interests of patients in the service of science), and "placebo-control orthodoxy" (roughly, the view that only placebo controlled trials give adequate scientific results). According to them, both sides can agree that *some* placebo-controlled trials are clearly unethical (e.g., withholding life-saving drugs or treatment). However, there is legitimate disagreement about when it is *scientifically* necessary to have a placebo-arm of a clinical trial. Emmanuel and Miller argue that it is scientifically necessary to conduct a placebo trial only when high-placebo effect for some drug or treatment is likely, when conditions have a waxing and waning course, or frequent spontaneous remissions, or when the existing treatments are ineffective or only marginally effective, or, when a condition is so rare that it would be nearly impossible to get enough enrollment to do an equivalence trial. (An equivalence trial tests a new drug against an existing drug.) Finally, they argue that participants in any research trial should not be "substantially more likely to die, or have irreversible morbidity or harm" as a result of the study (Emmanuel et al. 2001, p. 917).

This requirement raises a number of concerns about new, life-saving treatments for deadly diseases. Treatments for cancer, for instance, often carry high risk of toxicity or morbidity to the patient. For instance, a common regimen for treatment of leukemia involves several rounds of chemotherapy, followed by bone marrow transplant, a procedure that itself carries risk of rejection. Unless cancer researchers in the 1950s were willing to put patients at serious risk, some of the present treatments would not be available. While many patients died as a result of toxicity from such trials, many more patients' lives have been saved; and of course, one can argue that since no effective alternative was available, such trials are morally permissible. This "consequentialist" rationale is often offered in such circumstances, where high-risk treatments carry potentially huge benefits, and when it's unclear whether trials of new treatment regimes must be halted. The flip side of this dilemma is when a drug or treatment regime appears overwhelmingly successful. If a new treatment is believed to be overwhelmingly more successful than the available alternative (whether the available alternative is nothing, or a relatively ineffective treatment), at what point is it necessary to halt the trial and offer the treatment to the placebo arm? Stopping rules for clinical trials emphasize that either benefit should be established, or risk shown to be so serious as to require halting – yet, sometimes the standards for establishing benefit, or harm, are contentious, or hinge upon statistical analysis of the design of the experiment itself (Stanev 2012). Establishing statistical significance for outcomes like reduced mortality is difficult, to say the least, when a disease is rare, or when its course is waxing and waning, and relapses are likely to occur.

The requirement of "clinical equipoise" has it that a clinical trial should not be conducted unless there is a lack of consensus in the professional community about whether one approach is better than another, in terms of efficacy and safety, either because there is a lack of good data or sufficient data to establish a significant difference between treatments. Some question whether such a requirement is ever met. Others wonder if there is some point that equipoise is overturned, and whether at that point, patients should be offered the more successful treatment in lieu of the treatment they were provided in the "control" or "active control" arm. While it is a great gift to future sufferers of disease and to science for patients to agree to participate in research, arguably, there is some point at which requiring them to continue to offer this gift comes at too great a cost.

When to end a trial is one side of a two-sided dilemma. The other side concerns access; sometimes when a drug is in early phase trials, only a limited population of subjects will have access. The matter of when and why selection of subjects is fair is a concern of ethical import. For instance, when a new drug might save lives, many patients could clamor for the opportunity to take the drug. How should such choices be made? When has access to a new trial been fair? Who should have the opportunity (or, bear the cost) of participating in a trial of a new drug? Patient activists have complained that restricting access is an injustice that must be addressed, particularly when a drug is shown to be effective. This issue first arose in the 1980s, when ACT Up, an AIDS activist group, argued that early HIV drugs should be made available sooner, to a wider population, even before phase 1 and 2 trials were completed. Similarly, many women took advantage of high dose chemotherapy regimes for later stage breast cancer in the 1980s, even before adequate clinical trials had been conducted to establish efficacy (Mukherjee 2010). Activists during the early trials of drugs for Her2-neu argued that more women with the Her-2 positive breast cancer should be permitted to participate in early trials. This raises a dilemma: is the cost (loss of life due to delayed access to some treatment or drug) worth the benefit (testing the drug's effectiveness and side effects) of completing all stages of early phase trials for life-saving drugs? At what point should drugs or treatments become available, if they should be, in early trials?

2.2 Research on Human Tissues, DNA, and Non-human Animal Subjects

Biomedical research is conducted not only on human beings, but also, on human tissue and biological materials. Both raise a host of ethical questions. With regard to biological materials, there are legitimate concerns about informed consent, as well as privacy and confidentiality. For instance, with respect to research on stem cells (see also Siegel, this volume), which are, (by and large) acquired from blastocysts generated via IVF, whose consent must be given for research? When an infertile couple undergoes IVF, how and when ought one to request consent for research using frozen (unused embryos)? Must consent be given before the embryos are created? Where there are donor eggs (or sperm), when, and how is consent required? There are also difficult questions attendant upon the use of genetic materials; when an individual is tested for a genetic disease, their genetic information may be used to generate future tests. Genetic testing and patenting draws upon genetic information from patients, but the profit for genetic tests goes to the holder of the patent, not the patient/subject. Courts have yet to agree that patients have a property interest in their DNA, but as genetic tests and genetic information become bigger business, this may change. Does broad consent to future uses of tissues or genetic material suffice for fair use and development of such tests, or, should the consent obtained be more specific with respect to the uses to which they may be put? And, is there an obligation to disclose results (particularly of genetic) research to subjects?

One case where such issues became contentious concerns patients who organized a foundation for Canavan disease, a relatively rare genetic disease. They argued that since they organized the foundation and hired a researcher, they should have an interest in any profits from a genetic test developed out of that research. However, the Supreme Court in Florida struck down their claims to all but unjust enrichment from the patent, requiring that any test using this genetic information not require a licensing fee (Mariner 2008). Patients and communities screened for such diseases should be fully informed of how this information will be used. In addition, it is important that the identity of patients or donors of genetic material, blood, tissue, or other biological material be protected.

Personalized medicine refers to health care that is tailored to individual patients. Recently, there has been a growing focus on the contribution that genetic information can provide to personalized care, and a corresponding investment in developing clinical tests for many "genetic diseases." Indeed, a significant percentage of funding for biomedical research goes toward the development of such tests, and discovery of genetic correlates for disease. There are, however, a variety of ethical issues that arise in the context of developing, providing and communicating the results of genetic testing (Reydon et al. 2012). First, there are many public misperceptions concerning the genetic causes of disease; belief in genetic "determinism," the assumption that there is a one-to-one relationship between the presence of a gene and disease, is fairly commonplace (Mills Shaw et al. 2008). The way that genetics is presented in introductory textbooks, and as it is understood and presented by elementary and high school instructors and in public forums, can be misleadingly deterministic (Gericke et al. 2012; Dougherty 2009; Moore 2008; Castera and Clement 2012). While there are a few clear cases where genetic testing can provide enormously helpful information to patients, for the most part, most diseases are not "Mendelian," but the product of complex genetic and environmental causes (see also Jamieson and Radick, this volume; Burian and Kampourakis, this volume; Moore, this volume). In other words, genetics contributes only in part to disease, and so one should be cautious about promoting the benefits of such tests (Sarkar 2012). Arguably, given the rather small relative genetic contribution for the diseases that cause the greatest mortality in the world, the rather generous investments that have been made in discovering and documenting genetic associations with disease may be better spent elsewhere (for further discussion, see below, on issues of distribution of funding in biomedical research).

There are, of course, some examples of very serious hereditary disease. PKU, phenylketonuria (an inability to metabolize phenylalanine), is sometimes understood as a genetic disease, since the disorder is due primarily to a specific mutation (though diet is an important component). Testing for this disease is done on all U.S. infants, and such testing is relatively uncontroversial, since it is easy to prevent development of the disease by altering diet, and failing to do so can result in delayed mental and social skills, hyperactivity, retardation or seizures. However, this disease can be detected with a simple blood test; it does not require "genetic" testing, per se, and symptoms can be prevented by managing diet. A stronger case for "genetic" disease might be Li Fraumeni syndrome. Families that have this syndrome share mutations in the TP53 or CHEK2 gene, which can lead to early onset of several cancers. If a patient possesses one or the other of these mutations, then early and more frequent screening for cancer is advisable. However, tests for serious hereditary disorders arguably benefit only a very few patients with a strong family history. Most genetic contributions to disease increase risk over and above background rates only a small amount. And so, while testing for genetic predispositions to disease may motivate healthy behavior and lead to health gains, it may not assist most patients, and could lead to unnecessary anxiety. Thus, clinicians and researchers ought to be very careful about promoting and communicating the information contained in such tests.

It is striking to consider that analogues to the above criteria for research on human subjects (informed consent, protection of vulnerable populations, risk outweighed by benefit) are starkly absent in the case of research on non-human animal subjects. While horrific experimental practices on animals (vivisection, or live dissection), are no longer in practice, animals are still used today to study the toxicity or carcinogenicity of various agents. The Environmental Protection Agency's regulatory levels of permissible lifetime exposure to toxins such as dioxin, mercury, or chlorine in our water, air, and food, are all largely the product of toxicological research on non-human animals. Such research may involve subjecting animals to quite high doses of toxic materials. Such work raises both ethical and epistemological questions. For instance, philosophers of science (Steel 2007; Elliott 2011) have explored whether and if so, under what circumstances, such extrapolation from non-human to human animals is warranted, as well as when and how toxicologists' assumptions about the shape of dose–response curves are problematic.

There is a range of views about when, if ever, research on non-human animals is permissible. Some (ALF, website; Regan 1983) argue that no research on animals is permissible. Others (Singer 1990; Frey 1980) contend that research on animals may be acceptable, but only for experiments whose benefits are expected to far outweigh the harms involved. The challenge comes with generating an objective measure of "harm." Degrazia (1999) argues that there is a sliding scale of moral consideration for animals, which gives moral weight to their interests in accordance with their cognitive, affective and social complexity. On this view, all sentient animals deserve moral protection, but some animals, particularly highly social animals with more complex cognitive lives, deserve greater protection. Almost all parties to the debate about research on non-human animals can agree that alternatives to use of animals in biomedical research should be used when possible. The challenge is to weigh the potential harm to non-human animals against the very important good of human health.

A variety of national and international organizations (including, e.g., the ICLAS (International Council for Laboratory and Animal Science)) have established guidelines for the respectful treatment of animals in research. Today, just as most institutions have IRBs for research with humans, most institutions have animal care oversight committees that review protocols that involve research with animals. Historically, there have been different (and sometimes, conflicting) guidelines in different nations for the care of animals used in research. The ICLAS guidelines were developed in part with the intention of making such conflicting standards consistent. Most national programs of oversight emphasize the three "R's": replacement, reduction, and refinement of animal use. That is, they all require that whenever possible, alternatives to animal research, reduced use, and use that involves more respectful and better care for animals is preferred. The ICLAS guidelines include the requirement that death and pain should be avoided whenever possible, that the earliest possible endpoint to an experiment should be used consistent with scientific objectives, and that animals should be monitored for behavioral, physiological, or clinical signs of pain, or distress. The working group also agreed that euthanasia should be conducted in a way that makes the animal's death as far as possible painless and distress free.

3 Issues Extrinsic to Research

3.1 *The Commercialization of Biomedical Research*

While biomedical research intrinsically raises a number of ethical questions, there are larger questions, which may be spoken of as "extrinsic" to research. There are (at least) three senses in which ethical violations or questions of justice can arise concerning the relationship of biomedical researchers to society at large. First, it is important for researchers to avoid conflicts of interest, or, undue influence from industry on the content or conduct of research. Second, more broadly, it is important for researchers to avoid wasteful use of research funds, particularly from public sources. One way to "waste" research funds is to choose only "safe," well-understood problems, to reproduce results or produce drugs or treatments that mimic or do not improve upon existing treatments. As a matter of research ethics, researchers should not only produce results, but also break new ground in research, and avoid simply producing more or less "redundant" results. Third, and finally, researchers should not be unduly influenced by commercial gain.

In part as a result of the Baye-Dole act, industry's share of total investment in biomedical research and development grew from approximately 32 % in 1980 to 62 % in 2000, while the federal government's share fell. Arguably, the commercialization of biomedical research has resulted in the larger portion of funds going towards research on the most profitable drugs and treatments, which in turn, reduces funding to test drugs that might benefit the few, or, the very poor. Developing drugs to treat the majority, the "worried well," or the diseases of the developed world is more profitable than developing drugs for diseases that affect the very few, or the very poor. Several philosophers have questioned whether this constitutes "well-ordered" science. (Reiss 2010; Reiss and Kitcher 2009; Pogge 2005; Stiglitz 2006a, b) While donations from charitable organizations like the Bill and Melinda Gates Foundation, for instance, have changed the landscape of biomedical research to some extent, there are still questions of justice and fairness concerning the organization of biomedical research, and provision of costly (patented) drugs to those in the developing world.

No less than 25 years ago, most biomedical research in the United States was conducted with public, rather than private, funds. The distribution of private versus public funding for biomedical research has been completely reversed today. Reiss (2010, pp. 431–432) explains:

> [...] profit-oriented companies have replaced academic and other non-profit organizations in conducting medical research such as clinical trials. For instance, in 1991, only 20 % of industry money for clinical trials went to commercial research units such as contract-research organizations (CROs) and site management organizations (SMOs); by 1998, the figure had risen to 60 % (Getz 1999). In 2003, CROs alone played a substantial role in 64 % of phase 1, 2, and 3 clinical studies, as compared with only 28 % in 1993 (Shuchman 2007)

Several have argued that this commercialization of biomedical research has led to undue influence, particularly by the pharmaceutical industry, on both the conduct and content of research (Bekelman et al. 2003). For example, industry sponsored research tends to have more favorable outcomes for tested drugs, compared to results from studies financed by independent organizations (Als-Nielsen et al. 2003, cf. Reiss 2010). There are, arguably, weakened standards for demonstrating effectiveness of drugs (Bodenheimer 2000), and suppressed or delayed publications of results that show side effects of drugs (Blumenthal et al. 1997; Biddle 2007). A striking example is the Vioxx debacle; Merck, the producer of Vioxx, apparently mischaracterized or under-reported possible cardiovascular side effects between 2000 and 2004.

Some might question whether such examples are rare, or exceptions to the rule. However, according to Sismondo (2007), it is not uncommon for:

> [...] medical journal articles [to be] researched and written by or on behalf of pharmaceutical companies, and then published under the name of academics who had played little role earlier in the research and writing process. In extreme cases, drug companies pay for trials by contract research organizations (CROs), analyze the data in-house, have professionals write manuscripts, ask academics to serve as authors of those manuscripts, and pay communication companies to shepherd them through publication in the best journals. The resulting articles affect the conclusions found in the medical literature, and are used in promoting drugs to doctors (Sismondo 2007, p. 286).

How common is such "ghost writing" and "ghost management" of research? Sismondo argues that it is quite common; in "comparing protocols and corresponding publications for industry-initiated trials approved by the Scientific-Ethical Committees for Copenhagen and Frederiksberg in 1994–1995" one study "found evidence of ghost authorship in 75 % of these publications"(Sismondo 2007). This "ghost writing" and "ghost management" of research, or the control by pharmaceutical companies on research conducted, analysis of data, and writing of articles, has transformed a good part of biomedical research into a marketing arm of pharmaceutical companies. Needless to say, this creates a host of questions about the scientific merit of such research, as well as the ethical strictures governing research.

Several researchers have questioned the innovativeness of drugs approved by the FDA. Angel's (2004) critical evaluation of the pharmaceutical industry argues that a significant proportion of drugs tested and approved are "me too" drugs – or, only purportedly "new" drugs that are in fact only slight modifications of existing substances designed to mimic existing available treatments that are as if not more effective. According to Reiss,

> 1,284 new drugs were approved by the FDA in the period 1990–2004. The FDA classified them as follows:
>
> - 289 (22.51 %) constitute a 'significant improvement compared to marketed products in the treatment, diagnosis, or prevention of a disease' (of these, 183 or 14.25 % of the total were 'new molecular entities' and 106 or 8.26 % of the total were improved variations on existing drugs such as new formulations, new combinations or new manufacturers);
> - 77.49 % of the approved drugs were *no better than existing drugs*. (http://www.fda.gov/cder/rdmt/pstable.htm. Cf. Reiss 2010)

The pharmaceutical industry is not singularly to blame for these compromises on quality biomedical research. Until relatively recently (Campbell et al. 2007), it was relatively common for clinicians and clinician-researchers to receive gifts or incentives from the pharmaceutical industry to either use or test various drugs, or recruit subjects for research. Campbell et al. (2007), reported that, even *after* the American Medical Association and the Pharmaceutical Industry adopted new codes of ethics to prevent this kind of conflict of interest, "most physicians (94 %) reported some type of relationship with the pharmaceutical industry, and most of these relationships involved receiving food in the workplace (83 %) or receiving drug samples (78 %). Over a third (35 %) received reimbursement for costs associated with professional meetings or continuing medical education, and more than one quarter (28 %) received payments for consulting, giving lectures, or enrolling patients in trials." In other words, despite efforts to prevent commercial interests influence on both which drugs are tested, or used, physicians and clinician-researchers continue to be subject to such influence.

3.2 From "Evidence-Based" to "Well-Ordered" Science

While these direct conflicts of interest are clear ethics violations, there are wider questions concerning the direction and content of research that deserve consideration. For instance, Groopman (2007) reports that common treatments or surgical procedures may fail to be tested for effectiveness against alternatives, in part because of a kind of "inertia" in clinical practice. If one has 'always' done X procedure or administered X drug, and one's peers all do the same, it's difficult to agree to participate in trials, or accept the conclusions of evidence-based trials that show such procedures or drugs ineffective, or less effective than previously believed. A famous example of this kind of inertia in medicine is the first trials of the Halstead radical mastectomy. Until as late as the mid-1970s, surgeons followed Halstead in assuming that "more was better" – or, that a more radical surgery was more likely to reduce chances of metastasis and death for breast cancer patients. It took a very determined surgeon, George Washington Crile, to convince practitioners to test this procedure against a less invasive surgery (Mukherjee 2010; Aronowitz 2007). Another instance of questioning the status quo has been the recent USPSTF's questioning of PSA screening regimes for prostate cancer. While it was relatively common in the 90s to offer this screening test to healthy men over 60, today, a "watch and wait" attitude is recommended, absent clear evidence of rising PSA levels and other risk factors for prostate cancer (Gardner 2011). More broadly, biomedical researchers should consider whether they have an ethical obligation to challenge the status quo, whether in the name of better basic science, and better treatment options. For instance, the idea that cancer is an epigenetic as well as genetic disease has only slowly gained acceptance. However, understanding cancer from a variety of perspectives – as a developmental, genetic, and epigenetic disease, arguably promotes better science and better medicine. For, intervening on epigenetic as well as genetic causes of disease may be a new avenue for treatment (Plutynski 2013).

Finally, it is no secret that there is an inequitable distribution of both health care services and the burden of disease across the world. Furthermore, of the close to 60 billion or so annually spent on medical research, 90 % of that is spent on diseases causing only 10 % of the global burden of disease. This inequitable distribution has been given a name: the 10/90 split (Benetar 2001). It is, arguably, a great injustice that most funds for biomedical research are spent on diseases that affect the wealthy few. There are obvious economic reasons for this; only the wealthy few can afford to spend $80 for a month's supply of a drug. Pharmaceutical companies argue that much of this money is reinvested in basic research; however, some contest this claim. Pharmaceutical companies have one of the highest profit margins of any industry, and a relatively high proportion of the budgets of pharmaceutical companies are devoted to marketing. Moreover, of money spent on basic research, some argue, the larger portion is spent on research for "me too" drugs – drugs for conditions that already have adequate treatments available (Angell 2004). Creating and marketing drugs that treat the "worried well" – such as anti-anxiety, anti-depressants – are more likely to garner profit than researching, for instance, drugs that might affect only a small number of those suffering from deadly diseases. Such diseases are sometimes called "neglected," or "orphan diseases" for this reason; such diseases are those that affect a few, or, diseases that affect only the very poor, and so drug development from such diseases do not have prospect of making high profits. In the developing world, a large number of individuals suffer from such orphan diseases: river blindness, typhoid, TB, etc.

In response to this serious concern, the Drugs for Neglected Diseases initiative was founded in 2003 by the non-profit groups Médicins sans Frontières, and several other international organizations, to develop new treatments for people suffering from neglected diseases, raise awareness through advocacy on research and development of drugs for neglected diseases; and strengthen existing research capacity in countries where neglected diseases are endemic. Several philosophers (Pogge 2005; Reiss and Kitcher 2009; Reiss 2010) have argued that as a matter of justice, priorities in biomedical researchers should be readjusted so as to give greater consideration to plights of the global poor. Pogge (2005) for instance, argues that cost of essential medicines to the global poor should be kept down by, effectively, making profits from pharmaceuticals proportional to their impact on the global disease burden. Pharmaceutical companies would still have patents and monopolies on their inventions, but innovation would be directed toward diseases that have the greatest impact. The cost of this scheme would be borne by the governments of advanced countries.

Similarly, Joseph Stiglitz, a Nobel-prize winning economist, proposed:

1. Introducing separate intellectual property regimes for different levels of development;
2. Provision of drugs at cost to developing countries;
3. Compelling innovating firms to provide licenses to (third-world) generics producers in the case of lifesaving drugs;
4. Creating a Medical Prize Fund (from public and philanthropic money). (cf. Reiss 2010, p. 439)

More radically, Brown (2008) and Reiss (2010) argue that the adverse effect of commercialization of biomedical research could be remedied by changing the intellectual property regime. While Brown suggests a whole-sale abandonment of the IP regime, ("IP" refers to Intellectual Property) Reiss argues for a more piecemeal, "adaptive management" approach, targeted at reducing the patent protections gradually, and reorganizing research gradually so as to reduce the adverse impact of commercial influence, by raising the standards for adequate tests of drugs, and changing the incentive structure to more significantly improve the health of the greatest number. Reiss calls his approach a "Millian" proposal, drawing upon Mill's moral commitment to the greatest good for the greatest number.

4 Conclusions

In this concluding section, we will address how and why biology students, future citizens and educators should reflect on ethical considerations in biomedical research. First, knowledge of not only science, but also, science's applications in biomedical research, and their ethical justifications, are important to consider. Citizens in a democracy are both impacted by, and play some role in implementing policy regulating such research. Second, undergraduate biology students are, of course, future clinicians and researchers. Thus, it is important that they are aware of the ethical proscriptions on biomedical research, and the history of experimentation on humans and non-human animals that led to these proscriptions. Clinician-researchers, in particular, need to consider how to balance their dual role of caretakers for patients, and scientists. Such considerations require an awareness of the variety of risks – including, but not limited to, risk of physical harm or co-morbidity, to patients, and their families. Future researchers must attend carefully to not only the physical harm and benefits of research, but also, the autonomy, integrity, and protection from abuse or coercion. This is important not only for the protection of overtly vulnerable populations, but any participant in a test of a new drug or treatment.

Second, as future researchers, they must consider their personal integrity, or, whether they have been unduly influenced by profit in the conduct of research. Further, biology students have the potential to change the course of future research. Deciding which questions to investigate requires thinking "outside" the box, or, challenging the status quo, and considering the larger context of their research (Kitcher 2001).

As biology students enter into their future careers, they should be encouraged to study not only ethical rules and guidelines, but also, the history of medicine. Understanding how and why the treatments that they now take for granted were hard won, has the potential to inspire students to challenge expectations and move biomedical research forward in their future careers. We will close with some exemplars of researchers who dared to try new treatments, or ask new research questions, and thereby changed the course of medicine.

Dame Cicely Saunders founded the first modern hospice, and is largely responsible for the development the discipline of palliative care. Saunders was a nurse and social worker in 1948, when a Polish Jewish émigré patient, David Tasma, left her funds to be "a window in your home." With these funds, Saunders devoted her life to developing better care for patients at the end of life. She eventually became a physician, and a pioneer researcher and activist on behalf of pain control and palliative care. She founded "hospice," a system care directed at attending to the physical, psychological and spiritual needs of dying patients, family and friends (Richmond 2005).

Sidney Farber, the "father" of modern chemotherapy, was a pathologist who conducted the first clinical trials of aminopterin, a folate antagonist, in children with acute lymphoblastic leukemia. He was the first to demonstrate that remission in this disease was possible; before this time, such patients were simply expected to die. Farber was also an activist on behalf of cancer research, and helped found the Jimmy Fund, which provided funds for much of the early research on cancer treatment (Mukherjee 2010).

Virginia Apgar, who received her M.D. in 1933, was responsible for the "Apgar" score, a test that is administered to every newborn in the U.S., and was an activist on behalf of vaccination, to prevent mother-to-child transmission of rubella, which can cause birth defects and miscarriages. Apgar also promoted the use of Rh testing, to prevent a disease in newborns that results from incompatibility between the mother and father's blood types (Changing the Face of Medicine, http://www.nlm.nih.gov/changingthefaceofmedicine/physicians/biography_12.html).

George Washington Crile, Jr. was a surgeon who questioned the orthodox surgery for women with breast cancer, the radical mastectomy, a disfiguring procedure, which involved removing not only the breast tissue, but also the surrounding muscle, tissue, and lymph nodes. He was able to demonstrate by clinical trials that this procedure was no more effective than a simple mastectomy, which removes only the breast, and for patients with less advanced cases, lumpectomy, or the removal of only a portion of the breast surrounding a small tumor. He was able to recruit activist patients to promote these clinical trials, including Babette Rosmund, a writer and activist on behalf of women with breast cancer (Mukherjee 2010).

We close with a quotation from Oreskes (1996, pp. 102–103) discussion of heroism and objectivity in science:

> The modem scientific enterprise has been historically characterized by two competing and to some extent contradictory images. Most prevalent is the image of a hyper-rational, dispassionate observer [...] His brain is his only essential organ. If we focus on this image, objectivity appears to be the central value of science [...] Objectivity is an epistemological ideal, to which scientific colleagues can reasonably be assumed to subscribe. But it is also an isolating ideal, one that is not generally shared by the world at large. Thus it can be an alienating ideal, which separates the scientist from others from whom he may want or need support.

One of the characteristics shared by many of the pioneering biomedical researchers mentioned above is concern for the welfare of others. It was not their heroism or narrow objectivity that made their success possible, but their subjective commitment to the value of human life and welfare.

References

Agarwal, M., and E. Emmanuel. 2008. Phase I oncology research. In *The Oxford textbook in clinical research ethics*, ed. E.J. Emmanuel et al. Oxford: Oxford University Press.

Als-Nielsen, B., W. Chen, C. Gluud, and L. Kjaergard. 2003. Association of funding and conclusion in randomized drug trials: A reflection of treatment effect of adverse events? *Journal of the American Medical Association* 290(7): 921–928.

Angell, M. 2004. *The truth about the drug companies*. New York: Random House.
Animal Liberation Front. Animal liberation frontline information service: *The A.L.F primer*. Website: www.animalliberationfront.com/ALFront/ALFPrime.htm Accessed 23 Mar 2012.
Aronowitz, R. 2007. *Unnatural history: Breast Cancer and American Society*. Cambridge: Cambridge University Press.
Bankowski, Z., and R.J. Levine. 1992. *Ethics and research on human subjects. International guidelines*. Geneva: Council for International Organizations of Medical Sciences (CIOMS).
Beauchamp, T., L. Walters, J. Kahn, and A. Mastroianni. 2007. *Contemporary issues in bioethics*, 7th ed. Belmont: Wadsworth Publishing.
Bekelman, J., Y. Li, and C. Gross. 2003. Scope and impact of financial conflicts of interest in biomedical research. *Journal of the American Medical Association* 289(4): 454–465.
Belmont Report. 1978. Excerpted in *Contemporary issues in bioethics*, 7th ed, ed. T.L. Beuchamp et al. Belmont: Wadsworth Publishing.
Benetar. 2001. cited in Benatar, S. and G. Brock. 2010. *Global health and global health ethics*. Cambridge: Cambridge University Press.
Biddle, J. 2007. Lessons from the Vioxx debacle: What the privatization of science can teach us about social epistemology. *Social Epistemology* 21: 21–39.
Blumenthal, D., E. Campbell, M. Anderson, N. Causino, and K. Louis. 1997. Withholding research results in academic life science. *Journal of the American Medical Association* 277(15): 1224–1228.
Bodenheimer, T. 2000. Uneasy alliance. *The New England Journal of Medicine* 342(20): 1539–1544.
Brown, J.R. 2008. The community of science! In *The challenge of the social and the pressure of practice: Science and values revisited*, ed. M. Carrier, D. Howard, and J. Kourany, 189–216. Pittsburgh: Pittsburgh University Press.
Campbell, E., R. Gruen, J. Mountford, L. Miller, P. Cleary, and D. Blumenthal. 2007. A national survey of physician-industry relationships. *The New England Journal of Medicine* 356(17): 1742–1750.
Castera, J., and P. Clement. 2012. Teachers conceptions about genetic determinism of human behavior: A survey in 23 countries. *Science & Education*. Online: doi:10.1007/s11191-012-9494-0
Changing the Face of Medicine: Celebrating America's Women Physicians. http://www.nlm.nih.gov/changingthefaceofmedicine/physicians/biography_12.html. Accessed 9 Apr 2012.
Declaration of Helsinki. 1964, 2004. Excerpted in *Contemporary issues in bioethics*, 7th ed, ed. T.L. Beauchamp et al. Belmont: Wadsworth Publishing.
Degrazia, D. 1999. The ethics of animal research: What are the prospects for agreement? *Cambridge Quarterly of Healthcare Ethics* 8: 22–34.
Dougherty, M.J. 2009. Closing the gap: Inverting genetics curriculum to ensure an informed public. *American Journal of Human Genetics* 85(1): 6–12.
Elliott, K. 2011. *Is a little pollution good for you? Incorporating social values in environmental research*. Oxford: Oxford University Press.
Faden, R.R., and T.L. Beauchamp. 1986. *A history and theory of informed consent*. New York: Oxford University Press.
Frey, G. 1980. *Interests and rights*. Oxford: Clarendon.
Gallagher, J., S. Gorovitz, and R.J. Levine. 2000. *Biomedical research ethics: Updating international guidelines: A consultation*. Geneva: Council for International Organizations of Medical Sciences (CIOMS).
Gericke, N., M. Hagberg, V. Carvalho Santos, L.M. Joaquim, and C. El-Hani. 2012. Conceptual variation or incoherence? Textbook discourse on genes in six countries. *Science & Education*. Online: doi:10.1007/s11191-012-9499-8.
Groopman, J. 2007. *How doctors think*. Boston: Houghton Mifflin Harcourt.
Harris, Gardner. 2011. U.S. panel says no to prostate screening for healthy men. *New York Times*. Retrieved 8 Oct 2011.
Kitcher, P. 2001. *Science, truth and democracy*. Oxford: Oxford University Press.
Mariner, W. 2008. Liability and compensation for injury of research subjects. In *The Oxford textbook of clinical research ethics*, ed. E.J. Emmanuel et al. Oxford: Oxford University Press.

Miller, F., and E. Emmanuel. 2001. The ethics of placebo-controlled trials – A middle ground. *New England Journal of Medicine* 345(12): 915–919.
Mills Shaw, K.R., K. Van Horne, H. Zhang, and J. Boughman. 2008. Essay contest reveals misconceptions of high school students in genetics content. *Genetics* 178(3): 1157–1168.
Moore, D.S. 2008. Espousing interactions and fielding reactions: Addressing laypeople's beliefs about genetic determinism. *Philosophy and Psychology* 21(3): 331–348.
Mukherjee, S. 2010. *The emperor of all Maladies; A biography of cancer*. New York: Simon and Shuster.
Nuremberg Code. 1947. Excerpted in *Contemporary issues in bioethics*, 7th ed, ed. T.L. Beauchamp et al. Belmont: Wadsworth Publishing.
Oreskes, N. 1996. Objectivity or heroism? On the invisibility of women in science. In *Osiris*, Science in the field (1996), vol. 11, 2nd series, 87–113. Chicago: University of Chicago Press.
Plutynski, A. 2013. Cancer and the goals of integration. *Studies in the history and philosophy of the biological and biomedical sciences*, ed. Alan Love and Ingo Brigandt, Special Issue. Online: doi:10.1016/j.shpsc.2013.03.019
Pogge, T. 2005. Human rights and global health: A research program. *Metaphilosophy* 36(1/2): 182–209.
Regan, T. 1983. *The case for animal rights*. Berkeley: Berkeley University Press.
Reiss, J. 2010. In favor of a Millian proposal for the reform of biomedical research. *Synthese* 177: 427–447.
Reiss, J., and P. Kitcher. 2009. Biomedical research, neglected diseases and well ordered science. *Theoria* 24(3): 263–282.
Reydon, T., K. Kampourakis, and G. Patrinos. 2012. Genetics, genomics and society: The responsibilities of scientists for science communication and education. *Personalized Medicine* 9(6): 633–643.
Richmond, C. 2005. Dame Cicely Saunders obituary. *British Medical Journal* 331: 238.
Sarkar, S. 2012. Complex genetic systems and disease. In *eLS*. Chichester: Wiley. doi: 10.1002/9780470015902.a0005887.pub2
Shah, S. 2006. *The body hunters: How the drug industry tests its products on the world's poorest patients*. New York: New Press.
Shuchman, M. 2007. Commercializing clinical trials—risks and benefits of the CRO boom. *New England Journal of Medicine* 357(14): 1365–1368.
Sismondo, S. 2007. Ghost management: How much of the medical literature is shaped behind the scenes by the pharmaceutical industry? *PLoS Medicine* 4(9): 1429–1433.
Singer, P. 1990. *Animal liberation*, 2nd ed. New York: New York Review of Books.
Stanev, R. 2012. Stopping rules and data monitoring in clinical trials. *The European Philosophy of Science Association Proceedings* 1: 375–386.
Steel, D. 2007. *Across the boundaries: Extrapolation in biology and social science*. Oxford: Oxford University Press.
Steinbrook, R. 2008. TheGelsinger case. In *The Oxford textbook of clinical research ethics*, ed. E.J. Emmanuel et al. Oxford: Oxford University Press.
Stiglitz, J. 2006a. *Making globalization work*. New York: Norton.
Stiglitz, J. 2006b. Scrooge and intellectual property rights: A medical prize fund could improve the financing of drug innovations. *British Medical Journal* 333: 1279–1280.

Environmental Ethics

Roberta L. Millstein

1 Introduction

Most biologists care about the organisms that they study: the individual organisms themselves, the particular species, or the ecosystems in which those organisms live. As professors or teachers, they are excited to share these passions with their students, and usually find a ready audience. Some even hope to share those passions with the wider public, especially because many species and ecosystems are threatened or endangered. And there are many large-scale human-caused phenomena that threaten organisms and their ecosystems: global climate change, habitat loss, water and air pollution, water shortages, invasive species, human overpopulation, increased extinction rates, etc. Many actions are proposed to deal with these threats: We should reduce our reliance on fossil fuels by providing new transportation and energy options. We should halt large-scale deforestation. We should assist local peoples so that they can afford to live in harmony with the organisms around them. We should improve agricultural methods to reduce impact on the environment. Some even call for widespread vegetarianism, or at least a reduction in the amount of meat consumed.

What is often left out of these discussions – what is often left out of the biology classroom – is *why* we should care about organisms and their ecosystems, and why we ought to consider taking these actions to protect them. Surely it isn't *just* because we find them personally interesting. We would not be considering, and enacting, such sweeping changes if that were all there were to it. And surely, even those who have not chosen to be biologists care about these organisms and ecosystems even if

R.L. Millstein (✉)
Department of Philosophy, University of California,
One Shields Avenue, 95616-8673 Davis, CA, USA
e-mail: RLMillstein@UCDavis.edu

K. Kampourakis (ed.), *The Philosophy of Biology: A Companion for Educators*, History, Philosophy and Theory of the Life Sciences 1, DOI 10.1007/978-94-007-6537-5_31,

they don't study them every day. As much as we sometimes think of humans as selfish, concerned only with money or with their own needs, widespread support for environmental organizations, environmental initiatives, and environmental reserves and public parks, suggests otherwise. But again, why should we care – why should students study these organisms and their ecosystems, and why should the general public act to protect them? Is it only because humans depend on non-human organisms and ecosystems? Or do the organisms and ecosystems matter ethically themselves?

The area of philosophy that studies these questions in depth is known as *environmental ethics*.[1] Although there are many facets to this field, one of the central ones has explored the question of which entities – only humans, all sentient organisms, all life, or ecosystems considered holistically – are deserving of moral consideration. More importantly, *why*? What characteristics of organisms and ecosystems contribute to or establish their moral status? There is an extensive literature discussing these questions. Of course, it is not expected that biology educators and their students will become environmental ethicists, although people with similar backgrounds can and have contributed productively to the field. Rather, my suggestion here is that biology educators and students explore the *reasons* for their beliefs as well as the beliefs of others in order to be more thoughtful about their own research and why it matters, and to be better able to communicate with others who may or may not share their beliefs.

In fact, there are at least three areas of environmental ethics that biology educators could profitably incorporate into their classes. The first is as already mentioned: an examination of who (or what) should be considered to be part of our moral community (i.e., the community to whom we owe direct duties), and why. The second area, related to the first, is the application of the different answers concerning the extent of the moral community to real environmental issues and problems. Students need to be aware of how the different answers concerning the moral community can imply conflicting answers for how we should act in certain cases and to think about ways to move toward conflict resolution. The third area in which environmental ethics can contribute is a more conceptual one, focusing on central concepts such as biodiversity, sustainability, species, and ecosystems. Exploring and evaluating various meanings of these terms will make students more reflective and thoughtful citizens and biologists, sensitive to the implications that different conceptual choices make. In what follows, I describe each of these areas in turn.

2 The Moral Community

In this section I canvass various answers to the question of who (or what) ought to be considered part of the moral community. The answers describe various increases to the moral community, beginning with humans only, expanding to include many

[1] In some ways, environmental ethics is a subdiscipline of ethics, and in other ways it is a subdiscipline of the philosophy of biology, but it might also simply be considered a discipline unto itself.

nonhuman animals, expanding still further to include all life, and finally, expanding to include entire ecosystems considered holistically. As with pretty much everything in philosophy, each of these views is subject to debate and disagreement. Thus, in what follows, I describe not only the views themselves and the reasons that support them, but also some major objections that have been given to those views. The idea is to give a sense of both strengths and weaknesses of each position.

2.1 *Humans Only*

Traditionally, the field of ethics considers only humans to be part of the moral community; in other words, it is *anthropocentric*. It concerns itself with issues concerning the behavior of humans towards other humans, considering questions such as "is it ever ethical to lie?" or "is there such a thing as just war?" Although a variety of ethical theories (and many variants of these theories) have been proposed, the two that have had the most influence in environmental ethics are utilitarian ethics and Kantian (deontological) ethics. I will describe each of these briefly; the reader should keep in mind that there is far more to each of these ethical theories (and ethics in general) than I will describe here (see also Plutynski, this volume).[2]

According to utilitarian ethics, actions are right (ethically justified) if and only if they would produce the greatest balance of happiness over unhappiness, taking into consideration all who would affected by the proposed action.[3] In one formulation, "happiness" is construed in terms of physical pleasure and "unhappiness" is construed in terms of physical pain; other formulations consider "higher" pleasures or other forms of happiness and well-being. In weighing the balance of potential happiness against that of potential unhappiness, all who have the capacity for these experiences and who would be affected by the proposed actions are considered equally (e.g., no favoritism is given for those who are of the same religion or nationality), although greater pains "weigh" more than lesser ones and greater happiness "weighs" more than lesser happiness. Furthermore, on the utilitarian view, the consequences of a proposed action are the *only* factor that goes into the determination of whether an action is right; other considerations about the type of action (e.g., whether lying is involved) are not relevant. Of course, if the action has not occurred yet, it can be difficult to know with any degree of certainty what the consequences of an action will be; the utilitarian thus makes the analysis given the best available information of likely consequences and their severity. If there is another action that would produce a greater balance of happiness over unhappiness, then *that* is the right action, according to the utilitarian.

The basic intuition behind utilitarianism is that, all things being equal, happiness is a good thing (and thus, it is good to bring more of it into the world) and unhappiness is a bad thing (and thus, it is good to try to reduce the amount of it in the world).

[2]For an accessible introduction to ethics, see Rachels and Rachels (2011) or Hinman (2012).

[3]Here I describe what has come to be known as "act utilitarianism." I will leave off discussion of other forms of utilitarianism, such as "rule utilitarianism."

There does not seem to be any *principled* way to say that one person's pain matters more than the equivalent pain experienced by another person, so all equivalent pains count equally.[4] Pain is pain. (As we will see below, *consistency of reasoning* is an important criterion in environmental ethics, and in ethics more generally). Furthermore, it seems as though the ability to feel pleasure and pain are *morally relevant characteristics*. If I stomp on a pen (which, of course, lacks the ability to feel pleasure and pain), I have not wronged the pen in the same way that I would wrong a person if I stomped on her foot. Even if I destroy the pen, I have not wronged *it*; at best, I have wronged other humans who might have made use of it, but that is a different matter.

In spite of its intuitive appeal, utilitarianism has been subjected to a number of well-known objections. For one, it seems to countenance the production of good consequences for the many at the expense of the few, as occurred with the infamous experimentation on Jews by the Nazis and the Tuskegee syphilis experiment on African Americans (see Plutynski, this volume). For another, even if one does have reasonable information about future consequences, it is difficult to know how to weigh them against one another, and different people can provide different reasonable weighings. Both of these objections (and others not mentioned) have been responded to in the literature, but other philosophers think that a Kantian approach to ethics, which lacks these problems, is superior.

Rather than take as its starting point the ability to feel pain and pleasure, a Kantian approach to ethics points out that beings such as humans that are rational and autonomous (i.e., can think, have a will of their own, and can make decisions and reason about them) are importantly different from *things*. Mere things (such as my pen) do not have a will to violate; there is thus nothing wrong in using them for whatever purposes we wish. However, Kant argues, we ought not to use rational and autonomous beings, with wills of their own, as mere things, as mere means to our ends. On a Kantian view, then, rationality and autonomy are morally relevant characteristics (seemingly a higher "bar" than the ability to feel pain and pleasure). Thus, Kant states that we ought to: "Act in such a way that you treat humanity, whether in your own person or in that of another, always at the same time as an end and never merely as a means" (Kant 1785, p. 43).[5] Kant does believe that we ought not to cause undue suffering to other animals, but this is because he believes that such actions would harm our abilities to act ethically towards other humans and not because of consideration towards the non-human animals themselves. Again, beings that are not rational and autonomous are mere things (and Kant included non-human animals in this category).

Perhaps foremost among the objections to Kantian ethics is its uncompromising nature. Always treating other humans as an end and never merely as a means has a

[4]Here I consider only the ethical weight of pain itself and not any other possible downstream consequences of that pain.

[5]This is known as the Categorical Imperative; there are other formulations, but again, I focus on the one that is most relevant for environmental ethics.

consequence that we should *never* lie (or violate the autonomy of other people in any way), regardless of circumstances. Few could live up to this, and even if we could – or could aspire to – it is not clear whether we *should.*

Setting aside the particular concerns with utilitarian ethics and Kantian ethics, some environmental ethicists challenge the implication that the moral community consists of humans only. After all, the members of many other species besides humans have the ability to feel pleasure and pain; on what grounds are they excluded from the moral community? Traditional (anthropocentric) utilitarians usually defend this by invoking *other* morally relevant criteria, such as the greater rationality of humans or their ability to engage in moral deliberations.[6] However, once one uses criteria such as greater rationality, greater autonomy, or ability to participate in moral deliberations, one has gone beyond the capacity of some humans, namely, those who are very young or those who suffer from a mental handicap or disorder. Indeed, it seems that the mental capacities of some non-human animals exceed those of some humans. This puts defenders of the human-only moral community in an uncomfortable position: they must (1) acknowledge that some humans do not fit their criteria and are thus not in the moral community and not deserving of direct moral consideration; (2) find some morally relevant characteristic that all humans have and all non-human animals lack (if one merely cites "being human," then one is merely showing a biased preference for one's species, i.e., being a "speciesist"); or (3) acknowledge that their account is based on a rank and indefensible inconsistency.

2.2 Animals

The inconsistency problem that plagues traditional anthropocentric ethics is often cited as a reason for recognizing that many (although perhaps not all – more on this in a moment) nonhuman animals ought to be considered as part of the moral community. The two most well-known proponents of this family of views are Peter Singer and Tom Regan; their views can be characterized as extensions to traditional utilitarianism and Kantian ethics, respectively. We will examine each of their views in turn.

Singer's animal-centered utilitarianism was in fact predicted by the founder of utilitarian ethics, Jeremy Bentham; in an oft-quoted passage, Bentham states:

> It may come one day to be recognized, that the number of the legs, the villosity of the skin, or the termination of the *os sacrum*, are reasons equally insufficient for abandoning a sensitive being to the same fate. What else is it that should trace the insuperable line? Is it the faculty of reason, or, perhaps, the faculty of discourse? But a full-grown horse or dog is beyond comparison a more rational, as well as a more conversable animal, than an infant of a day, or a week, or even a month, old. But suppose the case were otherwise, what would it avail? the question is not, Can they *reason?* nor, Can they *talk?* but, Can they *suffer?* (Bentham 1823: Chapter xvii, n. p. 122; emphasis in original).

---

[6]Or by making claims that humans feel greater pain and pleasure than other species – but those claims are difficult to substantiate, and do not seem to be true in all cases. It would be hard to show, for example, that one's pleasure in eating chicken outweighs the suffering of a chicken in a factory farm.

Singer, following Bentham's line of reasoning, argues that the *ability to suffer* is that which confers moral standing and that "the fact that [a being that suffers] is not a member of our own species cannot be a moral reason for failing to take its suffering into account" (Singer 1979, p. 194). To think otherwise, Singer asserts, would be "arbitrary" and "morally indefensible," analogous to the way that white slave owners who denied moral consideration to blacks were being arbitrary; the logic of racism and "speciesism" are the same (Singer 1979, p. 194). Singer clarifies that he does not think that all species are equal; some are more intelligent than others, some are stronger or better able to communicate, etc. However, he maintains that not all *humans* are equal on these grounds, either – and we would certainly not suggest that more intelligent humans be granted greater moral consideration than those who are less intelligent. Rather, Singer claims, the principle we accept, and ought to accept, is the *principle of equal consideration of interests*. On this view, equal interests count equally, regardless of the skin color of the human or the species of the animal. According to Singer, "the capacity for suffering and enjoyment is [...] sufficient for us to say that a being has interests–at an absolute minimum, an interest in not suffering"; this capacity, on Singer's view "is a *prerequisite for having interests at all*, a condition that must be satisfied before we can speak of interests in a meaningful way" (Singer 2001, pp. 7–8; emphasis in original). A rock, Singer asserts, is not sentient[7] and thus has no interests. Singer acknowledges that whereas some animal species (such as species of mammals and birds) are almost certainly sentient and many probably are (vertebrates), assertions that others are sentient are more dubious (insects, crustaceans, mollusks) or highly improbable (plants); such differences ought to be kept in mind when weighing potential interests (Singer 1979). For example, on Singer's account, the likely interests of a bird would outweigh the dubious interests of an insect. Finally, note that the principle of equal consideration of interests does not – by any means – dictate that the interests of nonhuman sentient animals outweigh those of humans! Rather, Singer claims only that the interests of all sentient beings who stand to be affected by a proposed action must be taken into consideration. And being taken into consideration means only that; it does not imply that all who are taken into consideration should ultimately be treated equally, since conclusions about treatment depend on the particulars of the case at hand (namely, the amounts of pain and pleasure that would be experienced by those who stand to be affected by the proposed action).

Singer suggests that his animal-centered utilitarianism would have numerous consequences for our dealings with our environment. Whereas an anthropocentric utilitarian might choose a less expensive yet more painful form of "pest" control, a Singer utilitarian might choose a less painful method (e.g., birth control for squirrels) even if it were more expensive. An anthropocentric utilitarian might be in favor of clearcutting a forest, whereas an animal-centered one would favor a more selective cutting (because of the negative impacts that clearcutting – a drastic change in

[7] I will use the term "sentience" (or "sentient") to refer to the capacity for suffering and enjoyment, although the reader should be aware that different authors use this term differently.

habitat – has on sentient species). And we might change the siting of human facilities (e.g., new power plants) to locations where they have less impact on other sentient species. However, Singer asserts, all else being equal, killing an endangered sentient animal is morally equivalent to killing a non-endangered one; the scales would tip in favor of the endangered species only if there were additional negative effects on *other* sentient species (including humans).[8]

Regan shares Singer's concerns about the inconsistency problems with traditional anthropocentric ethics. However, he rejects Singer's animal-centric utilitarianism on the grounds that utilitarianism (regardless of whose interests are included) is inadequate as a moral theory, in part because it countenances the production of good consequences for the many at the expense of the few, as mentioned above. One way to block this unpalatable implication, Regan claims, is by recognizing that it is not individuals' pains and pleasures that matter ethically, but rather, the individuals themselves. That is, he suggests that we ought to recognize that individuals have *inherent value*; furthermore, in order to avoid sexism, racism, discrimination on the basis of intelligence, etc., we need to recognize that all who have inherent value have it equally. Thus, all individuals with inherent value "have an equal right to be treated with respect, to be treated in ways that do not reduce them to the status of things, as if they existed as resources for others" (Regan 1985, p. 21). From this quotation, we can see the influence of Kantian ethics on Regan's thinking. However, unlike Kant, Regan assumes that all human beings have inherent value, regardless of their capacity; note that not all who are human are rational and autonomous and not all are capable of reasoning morally (e.g., infants and the mentally handicapped). Again, though, once you make that assumption, you must (in order to be consistent) recognize that all beings who share the same capacities as human beings also have inherent value; he states, "... the basic similarity is simply this: we are each of us the experiencing subject of a life, a conscious creature having an individual welfare that has importance to us whatever our usefulness to others" (Regan 1985, p. 22). This would seem to be a slightly more demanding criterion for inclusion in the moral community than that offered by Singer. However, Regan leaves open the question of whether entities that are not "experiencing subjects of a life" have inherent value as well, saying that "we do not need to know how many individuals have inherent value before we can know that some do" (Regan 1985, p. 23).

With regard to environmental issues, Regan would be opposed to *any* practice that treated experiencing subjects of a life as mere resources; each has a "fundamental right to be treated with respect." When it comes to animals in the wild, Regan urges us to "*let them be!*" (Regan 1983, p. 361; emphasis in original). This would imply, for example, halting the destruction of their natural habitat and increasing surveillance on poaching activities, with stiffer fines and longer prison sentences (Regan 1983). As for members of endangered species, Regan thinks we ought to protect

---

[8]The application of Singer's views and the other ethical views described in this section to various environmental challenges will be discussed in further detail in Sect. 3 below.

them, but only for the same reason that members of non-endangered species should be protected; members of endangered species should not receive *special* protection, on Regan's view (Regan 1983).

Some students may wonder if, by including non-human animals in the moral community, we are committed to saying predators act unethically when they kill other animals. This worry is understandable, but it encompasses a few misunderstandings. For one thing, most instances of predators killing other animals are required for the predators to sustain themselves; big cats, for example, arguably cannot survive on a vegetarian diet. And while there may be circumstances in which humans must kill other animals to survive, many of us are not currently in that situation. Many humans commit themselves to eating vegetarian or vegan diets and to avoid animal-derived products, products tested on animals, etc., so human survival (for many of us) does not seem to *require* the killing of other animals. If it did, that would likely change the ethical analysis of such killings. Perhaps more importantly, the worry about predators acting unethically confuses *moral agents* with *moral patients*. When it is recognized that an entity ought to be included in the moral community, it is as a moral patient, i.e., a being that is owed moral consideration or has moral rights. However, being a moral patient does not mean that one is a moral agent, i.e., a being that is held responsible for its actions. Consider, for example, human infants. They are uncontroversially members of the moral community, and yet they are moral patients without being moral agents; if a baby were to pick up a loaded gun left on the floor and shoot her sibling, we would not hold the baby morally responsible because she lacks the capacity (such as the ability to reason morally and to act on her reasonings) to be a moral agent. The worry about predators assumes that the criteria for inclusion in the moral community (i.e., for being a moral patient) are the same as the criteria for being held morally responsible for one's actions (i.e., for being moral agent), but the scenario of baby who accidently shoots her sibling shows that those two sets of criteria are arguably not the same. In the same way, non-human animals who lack the capacity for moral reasoning are not moral agents even if they are held to be moral patients.

Singer's and Regan's inclusion of (some) animals in the moral community is not without criticism. Mary Anne Warren (1983) argues that we must be more nuanced in our claims about which rights ought to be granted to which species, and that such rights must be based on the particular capacities of that species. For example, Warren suggests that members of species that seem to need and value freedom (humans, whales, migratory birds) have a greater right to liberty than members of species who do not. On a different tack, Gary Varner (2002) points out that humans can have biologically-based needs that make certain things in our best interests (such as getting enough ascorbic acid to avoid scurvy) even if we do not, or could not be expected to, consciously desire them. However, if biologically-based needs can give rise to interests, then it looks arbitrary to include only sentient beings (or only "experiencing subjects of a life") in our moral community, since plants (for example) have biologically based needs as well. Indeed, as Paul Taylor (1981) suggests, perhaps the criteria that we usually choose for determining entrance into the moral community (such as consciousness) are simply based on characteristics that are valuable for *us*, and are thus prejudiced and self-serving; for a plant, the ability to photosynthesize is far more valuable than consciousness.

2.3 All Life

Recall Singer's view that a being that lacks the capacity to experience suffering or enjoyment lacks interests. In his words, "If a being is not capable of suffering, or of experiencing enjoyment or happiness, there is nothing to be taken into account" (Singer 2001, p. 8). Such a claim practically challenges other philosophers to identify other characteristics that might be taken into account in determining which entities are part of the moral community, such as Varner's aforementioned criterion of "biologically-based needs" – a criterion that applies to all living beings, not just animals. More specifically, Taylor (1981) maintains that every organism has *a good of its own* which can be benefitted or harmed by our actions. Actions that preserve life and well-being, that keep an organism strong and healthy, are good for the organism, whereas actions that are detrimental to the life and well-being of an organism are bad for it. In other words, actions can be for or against the interests of an organism, where having "interests" does not require the entity to be conscious of those interests (or conscious at all). Views such as Varner's and Taylor's that include all living beings in the moral community have come to be known as *biocentric*.

Let's focus on Taylor's biocentric views in particular. Unlike Singer and Regan, Taylor does not choose a utilitarian or Kantian ethical orientation, but instead tries to incorporate aspects of both in his thinking. On Taylor's account, we adopt the moral attitude of *respect for nature* when we recognize that wild living things have inherent worth; inherent worth involves two general principles, and it is here that we can see both the utilitarian and Kantian influences, respectively. According to the first principle, the principle of moral consideration, every wild living being is a member of the Earth's community of life and is for that reason deserving of moral consideration–its good must be taken into account whenever it might be affected by our actions–although its good may need to be overridden to fulfill some other good. According to the second principle, the principle of intrinsic value, the realization of the good of any wild living being is something that is intrinsically valuable, meaning that "... its good is prima facie worthy of being preserved or promoted as an end in itself and for the sake of the entity whose good it is" (Taylor 1981, p. 201). Thus, "it must never be treated as if it were a mere object or thing whose entire value lies in being instrumental to the good of some other entity" (Taylor 1981, p. 201). In sum, to say that a wild living thing has inherent worth "is to say that its good is deserving of the concern and consideration of all moral agents, and that the realization of its good has intrinsic value, to be pursued as an end in itself and for the sake of the entity whose good it is" (Taylor 1981, p. 201).

The implications of biocentrism for environmental ethics are relatively clear. In deciding which actions to take, we must broaden the scope of the entities whose potential well-being or harm must be taken into account to include all living beings, whether sentient or not. If someone proposes building a movie theatre complex on the site where a species of endangered butterfly lives, the welfare of the butterfly must be factored into the decision. Or, if we find out that the production of ozone due to emissions from our automobiles is harming sequoias and other trees, we must consider whether we ought to change our transportation practices.

What is less clear is how we are to act in full accordance with a biocentric ethic: how to make such difficult decisions and whether it is possible to live in accordance with them. If we endorse Regan's animal-centered Kantianism, we can give up eating animals, give up animal experimentation, give up hunting and destruction of habitat (although all of this might demand quite a bit of effort on our part), but unfortunately our species cannot give up killing plants and continue to survive. All biocentrists acknowledge this, and all have developed ways of trying to balance competing interests (e.g., Taylor 1986; Varner 2002) but the challenge is to do so without continually defaulting to the interests of humans[9] or developing an ethic that humans are unable to live by. It is not clear that such a challenge can be met.

2.4 *Ecosystems*

All of the views discussed so far have focused on *individuals*, whether individual humans, individual sentient organisms, or individual living organisms. Perhaps that is why we encounter difficulties in sorting through ethical conflicts, especially for biocentrism. Would a holistic approach be preferable?

Aldo Leopold, who is usually understood to have a holistic approach, argues that ethics "has its origin in the tendency of interdependent individuals or groups to evolve modes of co-operation" (Leopold 1949, p. 201). Leopold hypothesizes that humanity's first ethics, where an ethic is "a limitation on freedom of action" or "a differentiation of social from anti-social conduct," dealt with relations between individual humans and later extended to relations between individuals and society. We accept restrictions on our actions within our society and co-operate with the fellow members of our society with whom we are interdependent. He then suggests that it is time that our ethics should be extended to the *land*, by which he means the community of "soils, waters, plants, and animals" (Leopold 1949, p. 204) with which we are also interdependent. That humans are interdependent with the land can be seen by looking at key points in history, which were actually the product of biotic interactions between people and land. For example, he contrasts the impact of grazing and plowing in Kentucky and the Southwest U.S.; the former led to useful

[9]Or, at least the most important interests of humans. Of course, determining what those are is a non-trivial matter and is bound to be controversial.

bluegrass whereas the latter led to the Dust Bowl,[10] "a progressive and mutual deterioration, not only of plants and soils, but of the animal community subsisting thereon" (Leopold 1949, p. 206). That humans are interdependent with the land can also be seen through an examination of what Leopold calls "the land pyramid":

> Plants absorb energy from the sun. This energy flows through a circuit called the biota, which may be represented by a pyramid consisting of layers. The bottom layer is the soil. A plant layer rests on the soil, an insect layer on the plants, a bird and rodent layer on the insects, and so on up through various animal groups to the apex layer, which consists of the larger carnivores (Leopold 1949, p. 215).

This energy flow is called a pyramid because the bottom layers are much more abundant than the top layers; prey tend to reproduce at a faster rate and to have more biomass overall than their predators. And humans, Leopold asserts, are at an intermediate layer with other omnivores. The land pyramid also captures lines of dependency (or "food chains"), such as "soil-corn-cow-farmer." As all organisms are part of many such lines of dependency, the pyramid is an illustration of the interdependence of the biotic community, a community whose "functioning depends on the co-operation and competition of its diverse parts" (Leopold 1949, p. 215).

Again, Leopold is suggesting that once we understand our interdependence with the land, we ought to extend our ethics to encompass it. The following two oft-quoted passages are taken to sum up Leopold's land ethic:

> In short, a land ethic changes the role of *Homo sapiens* from conqueror of the land-community to plain member and citizen of it. It implies respect for his fellow-members, and also respect for the community as such (Leopold 1949, p. 204).

> A thing is right when it tends to preserve the integrity, stability, and beauty of the biotic community. It is wrong when it tends otherwise (Leopold 1949, pp. 224–225).

These quotations – especially the second quotation – have been understood as claiming that the primary ethical value adheres to ecosystems, considered holistically, rather than the individuals that compose ecosystems; this view has come to be known as *ecocentrism* (J. Baird Callicott 1987 is probably the philosopher who has been most influential in defending this interpretation of Leopold). Consider, for example, an ecosystem in which deer were plentiful and predators absent. From a biocentric point of view, this might (albeit temporarily) be a positive state of affairs, given the number of flourishing deer; however, such an ecosystem would lack integrity (the predators are missing) as well as stability (eventually, the growing deer population would overgraze the area and starve to death). Thus, from an ecocentric point of view, this would not be a desirable state of affairs, i.e., it would be a state of affairs that we ought to avoid bringing about.

Much discussion has gone into trying to understand what Leopold means by "integrity," "stability," and "beauty." However, Leopold gives us at least a reasonable idea of

[10]The Dust Bowl was a phenomenon that occurred in the 1930s in the south-central plains of the United States. After decades of poor farming practices that destroyed the grasses protecting the soil, an extended drought hit the region, leading to massive dust storms and consequent loss of topsoil. Not much grew in the area for about a decade.

what he means by the first two of these. In discussing integrity, Leopold describes wildflowers and songbirds of Wisconsin, many of which cannot be put to economic use but which are nonetheless "entitled to continuance" as members of the biotic community (Leopold 1949, p. 210). From this we can infer that a loss of continuance (i.e., an extinction) is a loss of integrity. With respect to stability, Leopold notes that:

> When a change occurs in one part of the circuit, many other parts must adjust themselves to it [...] Evolutionary changes, however, are usually slow and local. Man's invention of tools has enabled him to make changes of unprecedented violence, rapidity, and scope (Leopold 1949, p. 217).

From this we can infer that by "stability" Leopold did not mean "unchanging"; rather, in saying that we ought to act so as to preserve stability, he was suggesting that we ought to reduce the scale and the speed of the changes that we make to be more similar to ones induced by non-human forces. That is, we should only make changes that the rest of the biotic community can adjust to. Surely, the creation of the Dust Bowl is an example of an instance in where we did not do so.

Again, I think the implications of ecocentrism – of including ecosystems in the moral community – for environmental ethics are relatively straightforward. In preserving integrity and stability, we need to act to protect[11] endangered species. Furthermore, we need to put special emphasis on preserving the interactions between species, such as predator–prey or pollinator-pollinated. It might turn out that certain species are more central for such relationships; thus, Leopold might urge us to especially protect such "keystone species." Finally, biodiversity is often taken to be a factor that contributes to stability, so an ecocentrist would generally seek to preserve biodiversity.

One criticism that has been made against Leopold is that ecology no longer relies on stability models; in response, alternative ecocentrisms have been developed that do not rely on preserving stability (e.g., Hettinger and Throop 1999).[12] Another criticism is that Leopold's ecocentrism amounts to "environmental fascism" because humans are mere members of the biotic team, with the same moral standing as any other member (Regan 1983). Regan suggests that ecocentrism has the implication that if

> the situation we faced was either to kill a rare wildflower or a (plentiful) human being, and if the wildflower, as a 'team member,' would contribute more to the 'integrity, stability, and beauty of the biotic community' than the human, then presumably we would not be doing wrong if we killed the human and saved the wildflower (Regan 1983, p. 362).

[11] Here one might worry about what ecocentrism is committed to with respect to "natural" (as opposed to human-caused) extinctions. By my reading, the cause of the extinction is not relevant, but rather, whether the extinction threatens the ecosystem of which the organism is a part. And here, I think that Leopold's message is that we should generally err on the side of caution and assume that the organism in question is important to the ecosystem, given our ignorance of the interdependencies between organisms. Also, given widespread human-caused global changes such as global warming, it would be hard to say that humans did not have a hand in any given extinction.

[12] Although I think the death of stability models in ecology can be exaggerated, it is certainly worth thinking about how we ought to behave towards ecosystems that are not naturally stable. I leave the empirical question of what percentage of ecosystems are best characterized in terms of stability models and what percentage are best characterized in terms of instability models to ecologists.

In other words, ecocentrism seems to allow individuals – sentient individuals – to be sacrificed for the good of the whole.

2.5 Reflections on the Composition of the Moral Community

I have presented what I take to be the major ethical positions in environmental ethics; they vary in the entities included in the moral community (humans only, some animals, all life, ecosystems). It should be acknowledged that my presentation is neither complete in its depth nor its breadth (i.e., there are variations on each of these positions and there are more positions). In each case, I briefly presented the reasons supporting the position, while pointing out major criticisms. I should also acknowledge that responses to those criticisms can be found in the literature; perhaps these are successful and perhaps they are not. The question remains, then – where do we go from here? In particular, what do we teach our biology students, many of whom may not be interested in becoming environmental ethicists but who will be facing difficult environmental questions as biologists or as citizens?

We could simply choose one and defend it against the criticisms. Alternatively, we could present them all and let the students decide. However, I would like to present a third alternative, one that is inspired by a defense of holism/ecocentrism. Don Marietta (1999) suggests that no real ecocentrist is as extreme as Regan implies; note, for example, that Leopold says that his land ethic "*implies respect for his fellow-members*" as well as "respect for the community as such" (quoted above; emphasis added). If we were meant to respect the fellow members of our biotic community, then it would not seem as though killing off a human to save a wildflower would be justified. Perhaps instead we would seek out ways for both to co-exist. Marietta also points out that any ethical theory which reduces the value of a person to only *one* aspect of their life – their function in the ecosystem, *or* that they are alive, *or* their self-awareness, *or* their rationality, *or* their ability to feel pain – is leaving out characteristics that are morally significant. To choose one of these over the other, Marietta suggests, would be to have an incomplete moral account, one that simply dismisses the important moral insights gained over generations of human relationships.[13] Thus, Marietta argues, we have many kinds of duties: to each other, to the community, to future generations, to non-human animals, to all living beings, and yes, to the environment as a whole.

[13] Indeed, what does ecocentrism tell us about the ethics of lying? What does Kantian ethics tell us about the ethics of siting a power plant in the habitat of an endangered species versus a non-endangered one? What does utilitarian ethics tell us about an invasive plant species that is replacing the native ones (if, for example, it turned out that the effects on sentient creatures were negligible)? It seems as though there are questions on which each ethical theory will simply be silent and thus fail to provide any insight or guide for our behavior.

One way of adopting Marietta's position, then, is to see that each of the ethical positions described above has merit. Each picks out important, ethically relevant characteristics, and thus, none should be discarded in favor of one of the others. On this view, all living creatures as well as ecosystems considering holistically ought to be considered part of the moral community, which is not to say that our human rationality won't *at times* (but not all the time) be what is most morally significant. Of course, as Marietta acknowledges, "[t]rying to take into account everything that is morally relevant forces us to face complexity, conflicts, and confusions, but there is no justifiable alternative to hard study and the making of hard choices" (Marietta 1999, p. 244). In other words, we *can* make it easier on ourselves by choosing only one morally relevant characteristic, but we cannot *defensibly* do so.

In the next section, I illustrate how the different ethical theories (and the morally relevant characteristics that they invoke) can be used as a lens to illuminate different aspects of an environmental issue, pointing out the sorts of conflicts that can arise and making general suggestions for how one might solve them.

3 Applications to Environmental Issues

As discussed at the outset of this chapter, our society faces innumerable environmental challenges that raise a wide range of ethical issues. Thus, my remarks here will of necessity be illustrative rather than comprehensive.

I like to think of the different ethical theories as "lenses" that help to highlight different aspects of a case. Utilizing a range of theories to understand particular environmental issues generates a series of considerations that can help bring relevant aspects of a case to the fore. A Kantian lens forces one to consider whether any humans or non-humans are merely being used as instruments to bring about some other purpose. This would be the case if, for example, one was considering siting a power plant in a neighborhood where the locals were not consulted or given incomplete information, or were not adequately compensated for their harms. A utilitarian lens forces one to exhaustively consider all sentient beings (human and non-human) who will be affected by an action, to try to estimate to what extent they are affected and whether those effects are positive or negative, and the likelihood of those consequences actually occurring. A biocentrist lens reminds us to take into account organisms other than the sentient ones that tend to draw our attention first. Finally, an ecocentrist lens lets one literally see the forest instead of just the trees, focusing on identifying and preserving the interdependencies between the species and thinking in terms of factors that would promote the longer term stability of the ecosystem.

Of course, for any given case some considerations may turn out to be more relevant than others. The point is simply that by analyzing through the lens of each ethical theory, one may turn up aspects of a case that might not otherwise have been obvious.

In making this kind of analysis, it is important to realize that there are typical ways in which the different ethical theories will tend to conflict. Consider again the case of siting a power plant. Suppose a site has been identified. Suppose the quickest, cheapest way to get the plant built is by informing and involving the local citizenry in the most cursory way possible. Suppose that the plant will have negative health effects on the local citizens, but that the plant will provide cheap energy to thousands of customers, most of whom live far enough away to avoid the negative health consequences. In such case, a utilitarian might argue that the benefits of building such a plant outweigh the harms. However, a Kantian ethicist would suggest that the local citizenry is being used as a means to obtain profit for corporations and cheap energy for customers. They would insist that the locals be properly informed of the health risks and that they have a say in whether the plant gets built, or at least that an acceptable means of compensation is worked out. (Here I consider only the anthropocentric forms of these theories; including other sentient animals might change the analysis).

Another classic sort of conflict is between an animal-centered perspective and an ecocentric one (Sagoff 1999). Consider, for example, the wild Hawaiian pig, a hybrid of Asian and European pigs that was brought to the Hawaiian Islands.[14] The Hawaiian Islands are known for their extraordinary biodiversity, with species that are found nowhere else, but the native plant species evolved in the absence of hooved animals and the pigs are very destructive to them. The feral pigs are also harmful to native birds by creating wallows that avian malaria-carrying mosquitoes reproduce in and by eating nestlings of birds that nest on the ground. Finally, their digging leads to erosion and siltation of streams and reefs. In other words, the pigs are wreaking havoc on Hawaiian ecosystems and causing the extinction of species; an ecocentrist would thus tend to seek the removal of the pigs. The Nature Conservancy sought to control the pig population using snares, but People for the Ethical Treatment of Animals (better known as PETA) objected that the snares were cruel, with pigs often dying of starvation rather than asphyxiation. Some Hawaiians do not like the snares, either; hunting dogs can also get caught in them, and some Hawaiians like to hunt the feral pigs (thus, the goal has been to control rather than eradicate). From an animal-centered point of view, then, the pigs are suffering great harms only for doing what pigs do naturally. So, while the snares may be justifiable from an ecocentrist point of view, they are not from an animal-centered one. (Note that an animal-centered utilitarian and an animal-centered Kantian would agree on this point; the pigs are both suffering and being used as a means to preserve ecosystems, with their basic rights to life being taken away).

If we are not to choose one ethical theory over another in advance (as Marietta urges, the *particular details* of each case matter), how should we handle such conflicts? The immediate answer is fairly obvious – try to satisfy each theory and each set of competing values to the greatest extent possible – and yet much ground can be gained by following it.

In the power plant case, a thorough utilitarian analysis would seek out the *best* balance of happiness over unhappiness, not just a situation where happiness outweighs unhappiness. Thus, the utilitarian should consider: are there other sites

[14]My discussion of this case relies heavily on "Case 25" of Patrick Derr and Edward McNamara's 2003 *Case Studies in Environmental Ethics*, a book I recommend highly as a source of diverse case studies to prompt discussion and thought.

where fewer people would be harmed? What sorts of compensation can be offered to locals – money? Jobs? Health care? Relocation? Will fully informing them of the potential risks and allowing them to participate in solutions raise their level of happiness and thus overall happiness for all affected? There are, of course, costs to each of these courses of action, but they might still be able to produce a reasonable utilitarian balance while going some way toward allaying Kantian concerns.

The Hawaiian feral pig situation is even more challenging, especially if we are seeking not to eradicate the pig from the Islands entirely and not to harm any other species in the process. Still, however, it seems worth exploring other ways to control the damage that pigs cause to the ecosystem in ways that cause less suffering: other sorts of traps, or a pig-specific virus, perhaps? An alternative possibility is to set aside pig-free areas and areas where pigs are allowed. I don't claim to have the answers to this challenging problem; I suggest only that we try to move beyond the pigs vs. ecosystems framing of it. It is interesting to compare the Hawaiian situation to the reintroduction of wolves to Yellowstone National Park, an act that was meant to help restore balance to the ecosystem (and it has), but of course, was at the expense of deer and other wolf-prey. My point here is not that we should introduce a pig predator into Hawaii (that might cause other problems!) but to note that it is the extreme suffering of the pigs that is creating the ecocentric/animal-centric conflict. In situations where the suffering of the animals is less (or at least, normal for the species), the conflict is less,[15] and that provides a guide to our actions.

Another conflict between ethical theories is illustrated by U.S. President Bill Clinton's establishment of the Grand Staircase-Escalante (GSE) National Monument in southern Utah in 1996.[16] President Clinton's action pre-empted plans to open up a large-scale coal mining operation in the area, thus angering local residents who had been looking forward to jobs in the coal industry. The GSE was established because of the diversity of habitats (from desert to coniferous forest) and the diversity of species within those habitats. There are also significant archaeological and paleontological sites in the area. However, after the designation of the GSE, many noted that the previously infrequent human visits to the area that had preserved the different ecosystems and species had now become frequent ones, putting those very ecosystems and species at risk. So, did President Clinton do the right thing?

The worry about frequent visitation is easily dealt with; many parks have passed policies to limit human impact by limiting the number of visitors and their activities or limiting the types and numbers of vehicles that can be used in the park, etc. However, a conflict still remains; from an anthropocentric point of view, there is the loss of future jobs and the loss of local autonomy (and so the decision seems to have been wrong one), whereas from an ecocentric – and perhaps biocentric and

[15]However, an animal-centric Kantian ethic implies that killing the pigs in order to preserve the ecosystem would never be justified; non-lethal means of controlling the pigs would have to be found. Disseminating birth control to the pigs might be one such means.

[16]My discussion of this case relies on "Case 16" in Derr and McNamara (2003).

animal-centric – point of view there is the preservation of ecosystems and species (and so the decision seems to have been the right one). So, can this conflict be resolved, and if so, how?

Were we able to go back in time and re-do this decision, we could certainly make an effort to involve the local residents in the decision (although some would argue that these sorts of areas are for all people to enjoy, not just the local residents). More than that, though, we could again seek to maximize different values. Here, more detail is of the essence.[17] On the human side of the equation: How badly do the local people need jobs? What sorts of jobs are created by tourism, and how do they compare monetarily? How do they compare in terms of length of employment or in quality of life for the employee? Can energy needs be met in ways other than coal, ways that might also be job-producing (and perhaps safer)? How much enjoyment will having the area set aside as a park bring to its visitors? On the ecosystem side of the equation: How *much* diversity is in this area; how does it compare to other areas? Are there unique species? Endangered species? Unique habitats? Endangered habitats? To what extent would these be threatened by the coal mining?

The answers to these questions matter. If it turns out that the tourism jobs are sufficiently comparable to the mining jobs and that this is a "biodiversity hotspot," then it seems as though President Clinton made the right decision. However, if the mining jobs are far superior, if locals desperately needed them, and if the area was not all that biodiverse after all, then it seems as though President Clinton made the wrong decision. However, my point is not to try to settle this case; rather, it is to say that, as hard as it is to balance competing values, asking and answering more detailed questions about the effects on all the organisms and ecosystems involved can go a long way toward making the best solution easier to see. (Getting people to agree to it and getting it implemented are even more challenging – but those actions lead us outside the scope of ethics).

4 Conceptual Issues

Much of the discussion above relies on contended biological concepts: species, ecosystems, biodiversity, life, etc. (see also Justus, this volume). Other conceptual questions may arise as well; for example, the question, "What is sustainability?" In discussing environmental ethics with their students, biology educators should be aware of these controversies and how they affect our ethical analyses. Once again, these are large issues, and I will only scratch the surface.

Consider the concept "ecosystem." What is an ecosystem? Can one ecosystem be cleanly delineated from another? Or is it only that some ecosystems are able to be delineating clearly (e.g., on an island or in a pond) and others are "clear enough" (i.e., with real but fuzzy boundaries)? Or is there no privileged way of delineating

---

[17] Again, as Marietta emphasizes, such decisions should not be made in the abstract.

ecosystems? It is evident that the answers to these questions affect how we understand ecocentrism – in other words, they affect what we take ourselves to be trying to preserve – and yet there has been much disagreement about the proper answers (Odenbaugh 2007). One sort of problematic case is illustrated by the Mauhoun river basin in Burkina Faso.[18] This area is inhabited by two species of tsetse flies, *Glossina tachinoides* Westwood and *Glossina palpalis gambiensis* Vanderplank. The region appears to be patchy, habitat-wise, with some areas preserved as reserves while other areas have undergone agricultural development. However, when we look at the behavior of the two species of flies, we find that one forms a panmictic breeding population whereas the other only breeds within the reserves. So, is it one ecosystem or several? Or, is the concept of ecosystem species-relative? [19]

Or consider the concept of "species" (see Wilkins, this volume). The most common one among biologists (or at least those biologists who study animals[20]) is the biological species concept. As articulated by Ernst Mayr, this is the view that "Species are groups of interbreeding natural populations that are reproductively isolated from other such groups" (Mayr 1996, p. 264). However, by one count, there are 26 concepts of species in the literature (Wilkins 2008); the question "what is the concept of 'species'?" (or even "are there multiple legitimate concepts of 'species'?") is hotly contended. So, when we say that we want to preserve species, what is it that we are trying to preserve? Consider, for example, the California Tiger Salamander (CTS) and the Barred Tiger Salamander (BTS).[21] The CTS and the BTS had been allopatric for ~5 million years, with the CTS in California and the BTS in Texas, but because of their value as bait, in the 1940s and 1950s bait dealers from the Salinas Valley imported thousands of BTS larvae into California. Since then, the BTS and the CTS have been producing viable hybrids. The BTS and the CTS have distinctive phenotypes and genetic characters. Are they two species or one? The CTS have been listed as "endangered" under the Endangered Species Act, but are they really a distinct species? What about the hybrids? Again, how we answer these conceptual questions affects how we understand the ethical issues.

Biodiversity is another thorny conceptual issue; it is often spelled out in terms of number of species. Even if we can settle the question of what species are, however, it's not at all clear that sheer number of species is really the issue. Some species are quite different from one another, whereas others are quite similar. Is it that we want to preserve great numbers of species, regardless of how similar they are? Or do we want to preserve the greatest number of different species? (And how should that be characterized?) Or, should we consider preserving diverse subspecies? Diverse genomes? Diverse habitats? In other words, is species even the right level of the biological hierarchy at which to preserve diversity? (For discussion of the concept of biodiversity, see, e.g., Norton 1994; Callicott et al. 1999; Gaston 2004).

---

[18] My discussion of this case relies on Peck (2009).

[19] As Peck (2009) argues.

[20] Biologists who study microbial life, for example, do *not* tend to endorse the biological species concept (see Bourrat et al. this volume); for one, some microbes do not reproduce sexually and so the biological species concept seems inapplicable to them.

[21] My discussion here relies on that of Fitzpatrick and Shaffer (2004, 2007).

The reader may have noticed that a subtle shift occurred in the last paragraph from the conceptual to the normative. That is, we can seek to ask the question, "what is biodiversity?" but that question very quickly becomes "what should we preserve?" Thus, the normative and the conceptual questions are intertwined. Note that this is to some extent true of the species and ecosystem concepts as well; although those conceptual debates sometimes occur outside of the normative realm, once they are in the normative realm it is hard to keep the normative issues separated from the conceptual ones. In other words, if we are asking "what is the concept 'species'?" with an eye to preserving endangered species, it seems as though we are partly asking what it is we would like to preserve: Distinctive genetic and phenotypic characteristics? Distinctive past evolutionary histories? Distinctive future evolutionary histories? Or something else? Similar considerations apply to ecosystems; are we preserving inter-species relationships, particular combinations of species, particular "ecosystem services," or something else?

Some concepts are even more value-infused – sustainability, for example. Sustainability is the buzzword of the day, and yet it is highly ambiguous. Are we trying to sustain only human activities, or are non-human activities included (Callicott and Mumford 1997)? Are we trying to sustain "business as usual," or might sustainability involve sacrifice? If agricultural sustainability is the issue, are we trying to sustain certain types of processes or a certain agricultural yield, and then, statically or dynamically (Blatz 1992)? Should sustainability be linked to carrying capacity or ecological footprint (Vanderheiden 2008)? Should we be aiming for land health rather than sustainability (Newton and Freyfogle 2005)?

Being aware of the controversies over concepts like these and how these controversies affect ethical discussions can help biology educators get students to think critically about important environmental issues of our time.

5 Conclusion

I have argued that biology students can benefit from: (1) understanding different ethical theories concerning the environment as well as the reasons offered in support of and against them, in order to better understand their own views as well as the views of others; (2) seeing how different ethical theories can be used as lenses to help understand different aspects of concrete cases and to work towards solutions that maximize different values; and (3) recognizing that key concepts are often themselves controversial as well as value laden, encouraging them to explore environmental issues in all of their complexity. I thus encourage biology educators to incorporate these issues in their classes.

Acknowledgments Thanks to multiple classes of environmental ethics undergraduates at CSU East Bay and UC Davis, for helping to shape my view of what the interesting and important issues in environmental ethics are. Thanks also to my 2012 graduate seminar at UC Davis for an enlightening quarter discussing the concepts of "biodiversity" and "sustainability." Finally, thanks to Kostas Kampourakis and James Justus for extremely helpful referee comments.

References

Bentham, J. 1823. *An introduction to the principles of morals and legislation*. Oxford: Clarendon Press.

Blatz, C.V. 1992. The very idea of sustainability. *Agriculture and Human Values* 9(4): 12–28.

Callicott, J.B. 1987. *The conceptual foundations of the land ethic. In companion to a Sand County Almanac*. Madison: University of Wisconsin Press.

Callicott, J.B., and K. Mumford. 1997. Ecological sustainability as a conservation concept. *Conservation Biology* 11(1): 32–40.

Callicott, J.B., L.B. Crowder, and K. Mumford. 1999. Current normative concepts in conservation. *Conservation Biology* 13(1): 22–35.

Derr, P.G., and E.M. McNamara. 2003. *Case studies in environmental ethics*. Lanham: Rowman & Littlefield.

Fitzpatrick, B.M., and H.B. Shaffer. 2004. Environment-dependent admixture dynamics in a Tiger Salamander hybrid zone. *Evolution* 58(6): 1282–1293.

Fitzpatrick, B.M., and H.B. Shaffer. 2007. Introduction history and habitat variation explain the landscape genetics of hybrid Tiger Salamanders. *Ecological Applications* 17(2): 598–608. doi:10.1890/06–0369.

Gaston, K.J. 2004. What is biodiversity? In *Biodiversity: An introduction*, ed. K.J. Gaston and J.I. Spicer, 1–18. Oxford: Blackwell.

Hettinger, N., and B. Throop. 1999. Refocusing ecocentrism: De-emphasizing stability and defending wildness. *Environmental Ethics* 21: 3–21.

Hinman, L. 2012. *Ethics: A pluralistic approach to moral theory*, 5th ed. Belmont: Wadsworth.

Immanuel, K. [1785] 1993. *Grounding for the metaphysics of morals*, 3rd ed. Trans. J.W. Ellington. Indianapolis: Hackett.

Leopold, A. 1949. *The land ethic. In a Sand County Almanac: And sketches here and there*. Oxford: Oxford University Press.

Marietta Jr., D.E. 1999. Environmental holism and individuals. In *Environmental ethics: Concepts, policies, theories*, ed. J. DesJardins, 238–246. Mountain View: Mayfield.

Mayr, E. 1996. What is a species, and what is not? *Philosophy of Science* 63: 262–277.

Newton, J.L., and E.T. Freyfogle. 2005. Sustainability: A dissent. *Conservation Biology* 19(1): 23–32.

Norton, B.G. 1994. On what we should save: The role of culture in determining conservation targets. *Systematics and Conservation Evaluation* 50: 23–29.

Odenbaugh, J. 2007. Seeing the forest and the trees: Realism about communities and ecosystems. *Philosophy of Science* 74: 628–641.

Peck, S.L. 2009. Whose boundary? An individual species perspectival approach to borders. *Biological Theory* 4(3): 274–279.

Rachels, J., and S. Rachels. 2011. *The elements of moral philosophy*, 7th ed. New York: McGraw-Hill.

Regan, T. 1983. *How to worry about endangered species. In the case for animal rights*. Berkeley: University of California Press.

Regan, T. 1985. The case for animal rights. In *In defense of animals*, ed. P. Singer, 13–26. Oxford: Blackwell.

Sagoff, M. 1999. Animal liberation and environmental ethics: Bad marriage, quick divorce. In *Environmental ethics: Concepts, policies, theories*, ed. J. DesJardins, 317–322. Mountain View: Mayfield.

Singer, P. 1979. Not for humans only: The place of nonhumans in environmental issues. In *Ethics and problems of the 21st century*, ed. K.E. Goodpaster and K.M. Sayre. Notre Dame: University of Notre Dame Press.

Singer, P. 2001. *Animal liberation*. New York: Ecco Press.

Taylor, P. 1981. The ethics of respect for nature. *Environmental Ethics* 3(3): 197–218.

Taylor, P. 1986. *Respect for nature: A theory of environmental ethics*. Princeton: Princeton University Press.
Vanderheiden, S. 2008. Two conceptions of sustainability. *Political Studies* 56: 435–455.
Varner, G. 2002. Biocentric individualism. In *Environmental ethics: What really matters, what really works*, ed. D. Schmidtz and E. Willott, 108–120. Oxford: Oxford University Press.
Warren, M.A. 1983. The rights of the nonhuman world. In *Environmental philosophy: A collection of readings*, ed. R. Eliot and A. Gare, 109–134. St. Lucia: University of Queensland Press.
Wilkins, J. 2008. Species, kinds, and evolution. *Reports of the National Center for Science Education* 26(4): 36–45.

Glossary

a posteriori A concept, idea, judgment, or knowledge obtained only after empirical investigation.

a priori A concept, idea, judgment, or knowledge held before empirical investigation.

adaptation, ahistorical definition of A biological trait that confers an advantage to its possessors which consequently survive and reproduce better than others in a particular environment.

adaptation, historical definition of A biological trait which plays a role in the ecology of an organism lineage as a result of a history of selection for this trait because it has played the particular role.

adaptation, process definition of The evolutionary process by which populations become adapted to their environment.

adaptationism A family of views about the power, prevalence, and importance of natural selection. Particular adaptationist views involve different empirical, methodological, or philosophical commitments.

adaptationism, empirical The view that most biological traits are adaptations. This is a claim about the world that is testable.

adaptationism, explanatory The view that evolution by natural selection occupies a special place in our scientific world view and changes our image of humanity and human nature. This philosophical view may or may not have consequences for scientific practice.

adaptationism, methodological The view that looking for adaptations first is a good investigative strategy for evolutionary biology. This is a claim about how best to do evolutionary science.

adaptedness An assessment of fit between some biological trait of an organism and the current environment that this organism inhabits. Degrees of adaptedness often translate into a measure of evolutionary fitness.

analogy A relation of similarity between traits in two or more organisms with different body-plans. A pre-Darwinian concept and term.

K. Kampourakis (ed.), *The Philosophy of Biology: A Companion for Educators*, History, Philosophy and Theory of the Life Sciences 1, DOI 10.1007/978-94-007-6537-5,

ancestry informative marker (AIM) DNA variants (usually single nucleotide polymorphisms) whose frequencies vary sufficiently between populations so that they can be predictive of membership in those populations.

animism The perception of non-living objects as alive.

anomaly A phenomenon that is not readily explainable by established theories.

apomorphy A trait in a set of homologous traits which is derived (apomorphic), i.e. in an innovative condition with respect to the condition in a reference ancestor.

argument from design Traditional argument purporting that organisms are outcomes of intentional design and therefore evince the existence of God as the source of the design.

artifact Any object intentionally created by humans which exhibits properties designed to serve a particular purpose.

artificial kind A category of entities that exist as a result of human actions or interests, or conventions.

astrobiology A field of science concerned with life other than familiar life, especially on planets or moons other than Earth.

behavioral genetics The study of how genetic factors influence behavioral characteristics. In contrast to *molecular* behavioral geneticists, who use the tools of molecular biology to study how specific DNA segments influence behavior, *quantitative* behavioral geneticists use the tools of population genetics to study how behavioral and genomic variation are related.

belief Although there are many theories as to the precise nature of belief, it is commonly held that belief is a mental state in which one accepts a particular proposition to be true.

biocentric ethics Ethical theories that hold that all individual living beings are part of the moral community.

bioethics The study of ethical questions in medicine and in biomedical research.

biogeographical ancestry (BGA) Representation of population structure at various hierarchical levels that assigns individuals' proportionate ancestry in one or more populations based on their genomic makeup.

bioinformatics The science concerned with collecting, categorizing, managing, storing, processing, retrieving, disseminating, mining, and querying biomedical data and information appropriately and efficiently by computational means.

bioinformation (biological information) Information in the biological realm. Information implies a relationship between: (1) a message which may be any event, linguistic, or otherwise; (2) a system of reference which the message informs the receiver about; and (3) a receiver. Bioinformation occurs when there are biological entities—nucleic acids, cell cytoplasm, proteins, antibodies, neurons, sensory organs, organisms, or even ecosystems—involved as such in the informational relationship.

biological advantage An ability that increases an organism's potential to stay alive, including its chances to survive and reproduce. Such abilities are involved in the performance of a biological role, and they are advantageous precisely because they enable the organism to perform this role better than certain alternatives.

Biometricians A group of late-nineteenth/early-twentieth century biologists who were united by their belief that the understanding of inheritance depended on statistical analysis of the full range of biological data in randomly combining populations. This belief entailed a commitment to continuous variation as the material for natural selection, and blending inheritance. The biometrical research programme was based largely on observations of natural populations in the field (including analysis of human traits). The most prominent promoters of biometry were Francis Galton, Karl Pearson and W.F.R. Weldon.

blastocyst An early embryo of approximately 150 cells produced by cell divisions following fertilization. The blastocyst is a spherical cell mass consisting of an outer layer of cells (the trophoblast) and a cluster of cells in the interior (the inner cell mass).

causal account of explanation Any account of explanation according to which an explanation of a phenomenon is a description of the state of affairs of which the phenomenon is an effect. Many causal accounts involve identification of that which is necessary and sufficient for an effect's occurrence.

cell line Cells that are grown and maintained in culture for research or therapy.

ceteris paribus clauses Provisos meaning "All other things being equal" that are attached to certain claims.

chimera An entity that is comprised of whole cells originating from different organisms.

chromosomal theory of the gene This theory was fairly well established with the publication of *The Mechanism of Mendelian Heredity* by T. H. Morgan and his coworkers in 1915. According to this theory, the term 'gene' referred to a segment of a chromosome which, when activated or deactivated, performed a certain function or had a characteristic effect.

circadian rhythms Endogenously controlled oscillations of approximately 24 h in many physiological processes (e.g., basic metabolism and body temperature) and behaviors (e.g., locomotion and cognitive performance).

cladistics (phylogenetic systematics) The classification of organisms according to their evolutionary history as reconstructed by their shared traits.

clinical research Research conducted in a clinical setting – e.g., on the effectiveness of some drug, screening or treatment regimen.

community A set of populations of different species that interact in *some* way and to *some* degree. Other features biological communities are often claimed to possess are clear spatial boundaries, internal regulation, and a determinate successional trajectory following perturbation. Communities, unlike ecosystems, do not encompass the abiotic environment.

complementarity model The idea that science and religion provide complementary narratives about the same reality.

condition, necessary A property, characteristic or feature that an entity must have in order to be a member of a given category.

condition, sufficient A property, characteristic or feature that is enough to guarantee that an entity is a member of a given category.

conflict model The idea that science and religion are intrinsically in conflict and always have been.

constraints Any properties or processes that limit or facilitate evolutionary change by biasing what is or is not possible. They can be interpreted in terms of adaptation (e.g., constraints prevent optimal phenotypes from evolving) or development (e.g., constraints facilitate specific types of variation for natural selection to act upon).

context of discovery The invention of theories; because of the role played by psychological, social, and historical factors, not considered of interest to philosophers.

context of justification The process by which hypotheses and theories are accepted. According to this, logic and empirical data alone serve to justify the hypotheses and theories proposed.

corroboration Corroboration refers to a body of evidence providing further support for a proposition or hypothesis that has been supported by some earlier evidence. For example, if an hypothesis is supported by some evidence and then additional evidence in support of the hypothesis is discovered, this new evidence is said to corroborate the hypothesis.

counterfactual conditionals If-then statements concerning what would have happened under certain hypothetical circumstances that did not actually take place.

creation The ontological belief that everything that exists does so due ultimately to God as the ground of all being and existence.

creationism The belief that God creates by a series of miracles to bring about the universe in general and this planet and humankind in particular. Young Earth creationism perceives the world to have been created in six literal days of 24 h sometime within the last 10,000 years. Old Earth creationism accepts the scientific account of the age of the Earth but believes that the creation of life and of living creatures occurred by a series of miraculous interventions.

deduction One of two fundamental types of reasoning (the other being induction). Particularly used in logic and mathematics, though useful in science as well. Deductive reasoning begins with one or more premises which — if true and properly logically linked to each other — yield a conclusion that must also be true. For instance, IF all men are mortal; and IF Socrates is a man; THEN it follows that Socrates is mortal.

Deductive-Nomological (D-N) account of explanation The D-N account of explanation requires that in an explanation, whatever is doing the explaining (explanans) logically implies the phenomenon to be explained (explanandum)—that is, the explanandum results from the explanans via a sound deductive argument. Further, at least one premise of the argument must be a natural law.

deism the idea that God creates the physical laws needed for the universe to exist and function, but otherwise does not interact with the universe.

design teleology A mode of teleological explanation which suggests that a feature exists for some purpose because it is intentionally designed to fulfill it.

Teleological explanations based on design are appropriate for artifacts but not for organisms or non-living natural objects (e.g. stones, clouds).

design, intentional Blueprint, as in the attribution of the adaptations of organisms to the Creator.

design, natural Pattern or arrangement arising by natural processes, as in the adaptations of organisms by natural selection.

development The processes of growth, change, and transformation that organisms undergo in their life cycle, such as from a fertilized egg to a sexually mature adult. It is also called ontogeny (adjective: ontogenetic). Development includes changes that entail increasing order, via the differentiation and integration of specialized parts.

developmental plasticity Modifiability of the phenotype during development. 'Developmental plasticity' is typically considered synonymous with 'phenotypic plasticity'.

DNA information Since the middle of last century, DNA has been often identified as the informational molecule par excellence. It has become commonplace to say that DNA "encodes," "contains," or "stores" information; even that it "transmits" or "conveys" hereditary information from one generation to another. What these expressions really mean is that DNA plays an important role in certain bioinformational relationships, by usually playing the role of a message, i.e. as a small factor of great specificity in relation to a given function and displaying a high potential for variability. DNA possesses precisely these characteristics in relation to reproduction and metabolism. However, this does not force us to identify bioinformation simply with a property of DNA but, instead, as a complex relation in which DNA has an important role.

domains Superkingdoms of life – the most fundamental division of living entities into the three groups of Archaea, Bacteria and Eukarya.

dominance In classical genetics, a quality inherent in one of a pair of alleles in a diploid organism, the phenotype of which is manifest in the heterozygote.

drift An indiscriminate sampling process that typically produces a pattern of random variability.

ecocentrist ethics Ethical theories (particularly that of Aldo Leopold, or inspired by Aldo Leopold) that hold that ecosystems, considered holistically, are part of the moral community.

ecological stability The tendency of an ecological system to remain the same. The tendency takes three characteristic forms. With respect to biological communities, more stable communities are less changed by perturbations, more rapidly return to a reference state or dynamic following perturbation, and are able to sustain stronger perturbations than less stable ones. These different mechanisms by which systems tend to remain the same capture the sense in which there may be a balance of nature.

embryo In humans, the organism that develops from the time of fertilization until the end of the eighth week of gestation, at which point it is called a fetus.

environmental ethics An area of philosophy concerned with the systematic study of right and wrong behavior toward non-human entities as well as humans.

Among other things, environmental ethics pays particular attention to the composition of the moral community.

epigenetics This term captures all processes of phenotype organisation above the level of DNA sequence. Epigenetics is now often more narrowly defined as the set of (molecular) mechanisms involved in regulating gene activity (often specifically during development rather than in adulthood).

epistemology Epistemology is the branch of philosophy that addresses questions having to do with the nature of knowledge and rational belief.

essence The necessary properties of a thing that make it the kind of thing it is. In logic, essence is the set of necessary criteria of a general notion or term. Essence is often contrasted to "accidental" or contingent properties.

essentialism The idea that classes or kinds must have jointly necessary and severally sufficient conditions. That is, a general term like "animal" must be definable in terms of properties that only animals jointly have.

essentialism, biological (taxic) The claim that biological kinds or taxa must have necessary shared properties or traits that no other kind or taxon does.

essentialism, scientific The claim that natural kinds must have modally necessary shared properties that nothing else does.

ethical theory A systematic and overarching account of right and wrong behavior. Examples include Kantian ethics, utilitarian ethics, ecocentrism, and biocentrism.

ethics An area of philosophy concerned with the systematic study of right and wrong behavior.

eugenics The scientific improvement of the human species by encouraging reproduction of individuals or groups with "desirable" characteristics and discouraging that of individuals or groups considered "unfit".

eukaryote A macro- or microorganism that is not in Archaea or Bacteria, possessing well defined cellular compartments, such as the nucleus.

evidence The word "evidence" has many different meanings in ordinary language. However, when philosophers and scientists speak of evidence they generally have in mind the sort of thing that can make a belief rational. This can be things such as experimental data, mathematical proofs, perceptual experiences, memory, and so on.

evil, problem of How to account for the existence of physical and moral evil in the world if the world's features are the outcome of God's direct creation ; also known as the theodicy problem.

evo-devo A constellation of biological disciplines that investigate the evolution of development (how patterns and processes of ontogeny vary and change over time) and the developmental basis of evolution (how ontogenetic processes causally impact the evolution of organismal traits).

evolution In biology, the process of change that over eons of time accounts for the origin and diversity of species and for the adaptations of organisms.

evolutionary game theory Imported and modified from the field of economics, evolutionary game theory studies the behavior of populations of agents who repeatedly engage in strategic interactions. Changes in the populations are typically understood as resulting from differences in birth and death rates (differential fitness) resulting from different strategies.

evolutionary morality The idea that morality evolves as an adaptation and/or that evolution indicates what moral code humans should adopt.

evolvability The capacity or disposition to evolve, usually ascribed to a group of organisms (e.g., a population or lineage). Evolvability is often described as depending on other properties, such as modularity (e.g., increased modularity leads to increased evolvability).

exaptation An adaptive trait which originally evolved for reasons unrelated to its current biological role.

experimental sciences Sciences marked by the lack (or relatively lower importance) of historical components affecting the phenomena under study. In physics and chemistry, for instance, researchers can experiment on particles or molecules without having to be concerned about the past history of such objects, since they behave the same regardless of such history.

explanandum In an explanation, that which is to be explained. Usually, the target of explanation is some specific phenomenon, a group of seemingly related phenomena, or a law-like regularity.

explanans In an explanation, that which explains a phenomenon (the explanandum).

explanation In common use, an explanation is a set of statements or an account that makes something clear. There are competing notions of explanation in philosophy of science, but they all assume that there is something that distinguishes scientific explanation from mere description. For example, scientific explanations provide us with an understanding of why particular events or types of events occur. Paradigm explanations in science are 'why' explanations, e.g.: "the reason why dinosaurs went extinct is because a large asteroid impacted the Earth at the end of the cretaceous period."

explanatory pluralism The view that there is no objectively privileged level of explanation. Rather, higher-level explanations (e.g. as formulated in biology or psychology) are expected to remain valuable, even if we possess complete lower-level explanations (e.g. in physics or chemistry).

falsification A theory is falsifiable if and only if an observational consequence can be derived from the theory. The idea is that if such a consequence can be derived, then an observation designed to determine whether the consequence obtains provides a genuine test of the theory, and if the test fails then the theory must be false.

fetus In humans, the developing organism from about eight weeks after fertilization until birth.

fitness A measure of evolutionary success, often broken down into two components: viability (survival) and fecundity (reproduction).

function A role in an organization. The role of a component in the organization of a system whose very existence is an organizational problem might be called an 'essential' function. The functions of the parts and activities of artifacts in the organization of their ability to meet our expectations (such as the keyboard's function to enter text) are 'artifact functions'. The functions of the parts and activities of organisms in enabling their continued existence are 'biological functions' or 'biological roles'.

functional explanations In biology, functional explanations answer the question why certain organisms have a certain trait rather than some specific alternatives. Such explanations point out that the trait to be explained is advantageous to those organisms because some of their other traits are functionally dependent on that trait, and so it is these traits that explain its presence. Functionally dependent means that the ability to maintain the living state of an organism with the traits doing the explaining would diminish if the trait to be explained were replaced by an alternative, whereas replacing the trait to be explained would not make much difference or have a negative effect if the organism lacked the traits doing the explaining.

fusion model The attempt to fuse scientific and religious ideas into a single discourse.

gene composition and localization Refers to what genes are made of (e.g. DNA) and where they are located within the cell (e.g. nucleus).

gene concepts, definite These are specific concepts of the gene, which are committal, at least to some degree, about the structure or location of genes. What is typically required is a mixed mode of identification in terms of both structure and function. Thus, one must specify the substrates out of which genes are built and the structures that deserve to be identified as genes in order to individuate genes among the factors contributing to the relevant functional state. For this class of gene concepts, the choice of a phenotype is crucial in determining what counts as a gene; when the phenotype is an amino acid sequence, genes will be individuated differently than when the phenotype is something like the suppression of the expression of certain other genes.

gene concepts, referentially indefinite causal (or functional) These gene concepts are based on very open-ended indefinite, functional descriptions consisting of two parts. The first part specifies a difference in the phenotype of the organism bearing a gene or gene variant (i.e., allele) – e.g., tall vs. short – whereas the second requires a pattern of transmission of the factor(s) responsible for the change. A schematic formulation of a referentially indefinite functional gene concept is the following: a gene for trait x is any stably inherited factor that causes an organism (or certain cells of the organism), given the rest of what it has in common with conspecifics, to have the potential for manifesting x, where x will (or can be made to) appear under the appropriate developmental plus environmental circumstances.

gene function Refers to the contributions/consequences of genes to traits affected by them (e.g. be implicated in the synthesis of a particular peptide).

gene networks Gene regulatory networks represent all the regulatory interactions existing between a number (which can be a large one) of genes. In such models, genes are represented as nodes and regulatory interactions as links between the nodes. Various mathematical and computational methods are used to model the dynamics of these networks.

gene structure Refers to which features of genes (e.g. base sequence) correlate with the phenotypes they produced (e.g. a particular kind of protein), or how they "store" and transmit some kind of information.

gene-centric perspective, on evolution The notion that evolutionary insights can emerge from viewing genes (particular DNA sequences) as units of selection by virtue of their different rates of proliferation within and among organismal genomes.

genetic accommodation Gene frequency change due to selection on variation in the regulation, form, or side effect of novel traits in the subpopulation of individuals that express the trait.

genetic annotation Reading *annotation* here as "commentary" or "explanation," the methods and technologies used to identify the locations of genes (as well as the coding regions in a genome) and determine specifically what those genes do.

genetic code Nucleotides are composed of a sugar (deoxyribose in DNA, ribose in RNA), a phosphate group, and one of four different nitrogen-containing bases, namely, adenine, guanine, cytosine, and thymine in DNA (uracil replaces thymine in RNA). These four bases are like a four-letter alphabet, and triplets of bases form three-letter words or *codons* that comprise the "information" which identifies an amino acid or signals a function. The specific correspondence between codons and amino-acids is the genetic code.

genetic material Any nucleic acid with the propensity to be inherited and to interact with other cellular components as a source of sequence information, eventually affecting or being implicated in cellular processes with local or extended impact.

genetic sequencing The methods and technologies used to determine the specific order of the bases in a molecule of RNA (adenine, guanine, cytosine, and uracil) or DNA (adenine, guanine, cytosine, and thymine).

genomics Biological research that focuses on whole genomes, i.e. the base sequence of the genetic material of organisms.

heredity The transmission of material from ancestors to descendants.

heritability (in the broad sense) A statistic that specifies the proportion of variation in a characteristic across a population, which can be accounted for by variation in that population's genes. (Note: Although the words sound similar, *heritability* need not accurately reflect how "*inheritable*" a characteristic is).

heterochrony The different times or different speeds with which the different parts of the body are formed during the development of the two organisms under comparison.

historical sciences Sciences marked by a significant historical component affecting the phenomena they study, meaning that the characteristics of an object of study (e.g., a living organism) depend on the past history of similar types of objects (e.g., their line of descent). The emphasis is on observation-based, rather than experiment-based hypothesis testing. They include evolutionary biology, geology and astronomy, among others.

homology A relation of sameness between two or more traits in two or more organisms, or within the same organism, usually evaluated in an evolutionary context. A complex concept, so that the precise meaning is often given through specific adjectivation (e.g., serial homology). For the historical concept of homology, homologous traits in a set of organisms are those that derive from the same trait in the most recent common ancestor of those organisms.

homoplasy A relation of similarity between two traits in two or more organisms that do not derive from the same trait in the most recent common ancestor of those organisms, but due to independent evolution (e.g. convergence).

human embryonic stem cells Undifferentiated human cells that are derived from the inner cell mass of developing blastocysts and that are (a) self-renewing, (b) pluripotent, and (c) capable of indefinitely dividing without differentiating in culture.

human pluripotent stem cells Human cells that have the capacity to differentiate into all tissues of an organism, but are not able to form embryonic components of the trophoblast and placenta and so are not alone capable of sustaining the full development of an organism.

in utero In the uterus.

in vitro Literally, "in glass"; in a test tube, culture dish, or other artificial environment.

in vivo Within a living organism.

induced pluripotent stem cells Stem cells created by converting adult human skin cells into cells that are pluripotent and self-renewing.

induction One of two fundamental types of reasoning (the other being deduction). Particularly used in science, though also at the basis of much commonsense inferences. Philosophers recognize various types of induction, but essentially the approach consists in making generalizations from particular instances. For example, from the observation that all known living organisms are characterized by some type of information-carrying molecule (DNA or RNA), one can induce that all terrestrial (and perhaps extraterrestrial) organisms do too.

information entropy (Shannon entropy) A measure of uncertainty, usually expressed in bits, whose mathematical formula is $H(S)=-K \sum_i P(s_i) \cdot \log P(s_i)$, where H is the entropy of a source, S is a source (that is, a discrete random variable), s_i is one of the possible values of S, $P(s_i)$ is the probability of s_i, and K is a positive constant. In more intuitive terms, information entropy enables us to estimate the amount of uncertainty reduced on average by each symbol produced by a given source.

informational molecule Any molecule capable of participating in a bioinformational relationship, either as a message, receiver, or a reference system such as a fragment of DNA, a neurotransmitter, an antigen, and a protein. When speaking specifically about genetic or hereditary information, it is very usual to ascribe the role of message to DNA or RNA, the role of reference to the proteins, and the role of receiver to the molecules of the cytoplasm, such as those that make part of a ribosome. This ascription is not arbitrary, but it is worth noting that the same molecule may be involved in different informational relationships with different roles. For instance, an mRNA molecule may be seen as the reference of a DNA fragment, but it can be also seen as a message regarding a protein.

inheritance The reception by offspring of material from parents. Inheritance was traditionally understood to involve property, but was later invoked to also explain how offspring have characteristics like their parents' characteristics. Many theorists today recognize that characteristics cannot be inherited; instead, they are *built* by offspring during development, using genetic and non-genetic materials inherited from parents.

inner cell mass A cluster of cells attached to the inner wall of the blastocyst. Embryonic stem cells are isolated and cultured from cells that form the inner cell mass. In development, the inner cell mass gives rise to the organs and tissues of the organism.

intelligent design The idea that some intelligent, supernatural agent has influenced the history of organismal life on Earth. Intelligent design proponents claim that it is possible to infer the past action of an intelligent designer from some features of extant organisms (e.g., cellular structures, genetic information). The label 'intelligent design' was created and promoted by American creationists to hide the religious aspects of creationism. It was developed as part of the attempt to integrate anti-evolution views into US public high schools, where, due to the constitutional separation of state and church, no religious views may be taught. There is no intelligent design theory that would predict and explain a variety of biological phenomena. Instead, intelligent design textbooks contain (unsound) arguments against evolutionary theory.

interdisciplinarity It involves efforts to combine different fields or disciplines in order to address a problem or a family of problems. Interdisciplinarity arises from the recognition that a particular discipline is not able to analyze and explore important aspects of a problem.

irreducible complexity A subsystem of an organism (e.g., a molecular mechanism) consisting of parts that interact so as to fulfill a function, is irreducibly complex when the removal of any part leads to the system no longer performing the function. Intelligent Design proponents consider e.g. the vertebrate eye or the bacterial flagellum as irreducibly complex systems which cannot have evolved by gradual evolution based on natural selection; they consider such systems as evidence for intelligent design since they believe they could only have been crafted *ex nihilo* for their current roles by an intelligent agent or Creator God.

Kantian ethics Ethical theories inspired by the work of Immanuel Kant, emphasizing (particularly in environmental contexts) that members of the moral community ought to be treated as ends in themselves rather than mere things, tools, or instruments.

knowledge One knows a proposition when: (1) the proposition is true, (2) one believes the proposition, (3) one's belief in the proposition is based on sufficiently strong evidence, and (4) one satisfies whatever condition is required to handle the Gettier problem.

laws Facts (such as Newton's laws of motion and gravitation, according to classical physics) that play certain special roles in connection with scientific explanations, natural necessity, counterfactual conditionals, and inductive reasoning.

levels of selection Different levels (such as genes, cells, individuals, or kinship groups) at which natural selection can operate in a biological hierarchy.

life That which distinguishes animate objects from inanimate objects. A precise characterization is not available.

logical necessity The sense in which truths that hold as a matter of logic alone could not have been false. Truths that lack logical necessity are "contingent."

macrobe A non-microscopic organism.

mechanism A structure performing a function in virtue of its component parts, component operations, and their organization. The orchestrated functioning of the mechanism is responsible for one or more phenomena.

mechanism decomposition Identifying the parts that constitute a mechanism and the activities or operations these perform.

mechanism recomposition Determining how the parts are organized and their operations orchestrated in the generation of the phenomenon.

mechanistic account of explanation Especially relevant in the biological sciences, this account views explanations as representations of causal mechanisms, viz., physical structures that perform some function in virtue of their constitutive parts, organization, and operation. A mechanistic explanation is the explanation of a phenomenon by identifying the working parts of the responsible mechanism, i.e. the parts that perform the various operations that go into producing the phenomenon.

Mendelians A group – often seen in opposition to the biometricians – espousing and promoting the work of Gregor Mendel as the best way to elucidate inheritance. The Mendelians believed in particulate inheritance as exemplified by Mendel's results. Contra the biometricians, the Mendelian research programme centered on selective breeding of experimental subjects in the laboratory. William Bateson is generally considered to be the leading Mendelian in Britain in the early twentieth century.

Mendelism The study of inheritance based on the premise of discrete, heritable "particles", as suggested by Mendel's "re-discovered" results in 1900.

microbe A microscopic organism or virus.

missing links Taxa intermediate between major groups of organisms which have not yet been discovered.

model-based science An approach to understanding complex real-world systems through the use of simpler hypothetical systems that resemble their target real-world systems in some relevant respects.

modularity The property of being a module or distinguishable unit (e.g., a segment), behaving in a quasi-autonomous fashion. Modularity allows for evolutionary change to occur in one trait without detrimentally affecting another trait or the entire organism.

molecular clock The hypothesis that the rate of molecular evolution is approximately constant for each different type of molecule.

molecular evolution The study of the evolutionary patterns and processes of biological macromolecules.

molecular systematics The use of molecular data to infer evolutionary relationships.

moral community The entities that are deserving of moral consideration; the entities to which an ethical theory applies. Different ethical theories offer different criteria for inclusion in the moral community, resulting in different candidates for the correct moral community.

natural kind A class of entities in nature that exist independently of human cognition or ideas, actions or interests. Natural kinds are usually held to require essences.

natural selection An evolutionary process that occurs when heritable variation in features or traits of organisms in a population produces difference in reproductive success. The result is the differential reproduction of organisms with different traits.

natural selection, creative view of Natural selection, properly conceived, evolves genuinely functional goal-directed adaptive traits only by working over much time and many generations on small variants in traits that first arise independently of the utility they subsequently acquire as they move toward fixation. This was Charles Darwin's view.

natural selection, eliminative view of In this view, natural selection is either favoring or eliminating organisms whose traits are or are not adapted *from the outset*. Selection is thus conceived as an eliminative force that discriminates among whole organisms rather than the slightly variant traits they bear and that adaptations are nothing but retained accidents. This is the view of natural selection summarized in the phrase "survival of the fittest", a phrase coined by Herbert Spencer and adopted by Charles Darwin.

natural theology Systematic arguments purporting to demonstrate the existence and attributes of God based on the features of the natural world.

naturalism, metaphysical Asserts that the material phenomena studied by science are all that exist. This includes the agency of persons and social phenomena, based on the idea that the reasoning and actions of humans are material (as opposed to immaterial or supernatural) phenomena, studied and explained by the cognitive and social sciences. Creationists and intelligent design proponents have claimed that a commitment to methodological naturalism entails metaphysical naturalism. This is false, for while endorsing methodological naturalism, scientists with religious beliefs reject metaphysical naturalism insofar as they believe in the presence of a non-material deity.

naturalism, methodological A claim about the aims and methods of science which asserts that science studies natural (as opposed to supernatural) phenomena only, supports its claims with empirically accessible evidence and explains by appeal to material causes. Methodological naturalism does not maintain that there are no supernatural phenomena (such as the presence of a deity), it merely asserts that science does not and cannot study the supernatural. For this reason, methodological naturalism has to be distinguished from metaphysical naturalism. Creationists and intelligent design proponents reject methodological naturalism (and thus the scientific approach), as they intend to infer the supernatural from empirical observations.

nature and nurture Traditionally seen as independent factors that contribute to the development of biological and psychological characteristics. Now understood to be two poorly defined but mutually interdependent factors that contribute to development.

neutral theory of molecular evolution A theory of molecular evolution that claims that the majority of observed changes in biological macromolecules

(DNA, RNA, and proteins) are neutral or nearly neutral. That is the behavior of these observed changes is dictated by random drift, rather than selection. The neutral theory combines both drift and selection, since selection is presumed to operate on a number of molecular changes.

niche There are two senses of the 'niche' concept. One focuses on the causal role a species plays in the overall community dynamics, a species' "way of making a living." The other focuses on distinct portions of the abiotic environment in which species persist and reproduce. The latter makes sense of the idea there are "vacant niches" into which species can successfully colonize or invade.

NOMA model Stands for 'Non-Overlapping Magisteria', a phrase popularised by the late Stephen Jay Gould to suggest that science and religion address different questions and are therefore located in quite separate compartments.

nominalism The view that only individual things exist, and no universal kinds do. The name is from the Latin for "name", as nominalists hold that general classes of things are just names, usually used for convenience. Methodological nominalism – aims at describing how a thing behaves, and especially, whether there are any regularities in its behavior.

nomological principles See "laws".

non-genetic (epigenetic) inheritance Refers to the many different mechanisms in addition to the transfer of DNA by which the parental phenotypes (or more remote ancestors) affect the development of their offspring. This definition is the same as a recent definition of 'parental effect'. Non-genetic inheritance includes cellular epigenetic inheritance, which is the transmission from mother to daughter cell of variation in the molecular epigenetic regulation of gene expression, but can also involve other mechanisms, including behavioral interactions between parents and offspring.

non-living natural object Any natural object other than organisms (e.g. clouds, rocks).

novelty (evolutionary) A trait which has no obvious homology with any other trait in another organism or the same organism, and whose origin cannot be easily traced back to a modification of a body structure already existing in the ancestral lineage leading to that organism. In other words, an evolutionarily new trait that is a qualitative departure from the ancestral condition (e.g., avian feathers), which is sometimes defined as neither homologous to a trait in an ancestral taxon nor serially homologous to any trait of the species. Often invoked in evo-devo to focus investigation on the developmental origin of variation rather than adaptation by natural selection.

organization A system is organized for a certain property or ability if the latter is critically dependent not only on the system's material composition, but also on the arrangement of its components and on the order and timing of their activity. The notion of 'organization' can be oblivious to both the way in which the organization came into being and the way in which it is maintained (if it is at all maintained), and so does not assume the existence of an organizer in any sense of that term.

origin of life The transition from a nonliving suite of chemicals to a living system that occurred on Earth approximately 4 billion years ago and gave rise to all known life.

orthology A relation between homologous genes that are present in different organisms and have evolved from a common ancestral gene by means of speciation events.

overdetermination In philosophy of science, a term applied to situations in which a subset of information is sufficient to predict or explain a particular outcome. For instance, a relatively small number of geological clues have been enough to establish that a large asteroid hit the Earth 65 million years ago, probably contributing to the extinction of dinosaurs and other species. It is the same phenomenon that makes it possible to identify the culprit of a crime even though there are no eyewitnesses or actual footage of the crime itself.

paralogy A relation between homologous genes that are present in the same organism or in different organisms and have evolved from a common ancestral gene by gene duplication.

phenetics Also called "numerical taxonomy". The classification of organisms by the closeness or not of their similarities, arbitrarily selected.

phenotypic accommodation Refers to the capacity of organisms for mutual adjustment of different parts during development to produce a functional phenotype even when perturbed by genetic or environmental input.

philosophy The word *philosophy* comes from two Greek words: *philos* deriving from *philein* "love," and *sophos* meaning "wisdom." Love here means something like an intense desire for something, while wisdom is arguably a kind of knowledge gained from experience, whether this is practical experience (gained from living life with all of its ups and downs) or theoretical experience (gained from understanding, evaluating, critiquing, and synthesizing ideas, positions, and concepts). Ever the theoretician, the philosopher has always been the person who not only desires to look deeper into some claim, idea, argument, event, or state of affairs by questioning assumptions and challenging status quo thinking, but also attempts to explain and systematize aspects of reality as it is perceived.

philosophy of biology A sub-discipline of philosophy, the concern of which is the meta-leveled attempt on the part of philosophers, biologists, and other thinkers to understand, evaluate, and critique the methods, foundations, history, and logical structure of biology in relation to other sciences, disciplines, and life endeavors so as to better clarify the nature and purpose of biological science and its practices.

plasticity The property of being flexible during development as a function of internal or external perturbations during the life cycle of an organism, such as the ability to produce different traits (behavioral or morphological) depending on environmental cues (e.g., diet or population density). Plasticity refers to the variety of developmental processes or traits that can be generated viably from a single genotype or genome.

pleiotropy The phenomenon whereby one gene has a causal impact on multiple phenotypic traits. Pleiotropy can affect evolutionary change by constraining what genetic changes are possible or encouraging properties such as modularity (e.g., modules as discrete sets of pleiotropic interactions).

plesiomorphy A trait in a set of homologous traits which is primitive (plesiomorphic), i.e. in the same condition with respect to the condition in a reference ancestor.

prokaryote A microorganism that belongs to Archaea or Bacteria; possessing cellular structures that are less obviously compartmentalized than in cells of non-prokaryotes.

proximate causes Proximate causes are immediate, mechanical influences on the phenotype that explain how internal (e.g. hormonal) and external (e.g. temperature, day length) factors combine to elicit or generate a specific character.

rate constancy in evolution The idea that the rate of observed substitutions in a protein or nucleotide sequence is approximately constant with regard to time. Rate constancy is usually explained as the result of the action of random drift on neutral or nearly neutral sites.

rate variability in evolution The idea that the rate of observed substitutions in a protein or nucleotide sequence varies over time. Rate variability is expected if the sequence is selected since the variability will reflect variations in the environment.

recessiveness In classical genetics, a quality inherent in one of a pair of alleles in a diploid organism, the phenotype of which is not manifest in the heterozygote.

reductionism Reductionism can mean many different things, depending whether one talks about ontology, scientific theories or methodology and explanatory strategies. In science it often implies that higher-level phenomena (e.g., as studied in biology or psychology) must be explained by more fundamental, lower-level laws or processes (e.g., in molecular biology, or even fundamental physics).

reductionism methodological The approach used in science to take something to pieces in order to understand its components and determine how they function.

reductionism ontological The idea that a system is 'nothing but' its components and that there is no more to say once the components have been analysed and described.

research ethics Ethical standards for the responsible conduct of research.

resilience The rate a system returns to a reference state or dynamic following a perturbation. Lake communities that return to a reference state or dynamic quickly after an incident of thermal pollution, for instance, are more resilient than those with slower return rates following similar incidents.

resistance Inverse of the magnitude a system changes relative to a reference state or dynamic following a perturbation. Resistance is often characterized as the tendency of a system to remain the same when perturbed. A biological community that can withstand severe drought with little change, for instance, is more resistant than one modified dramatically.

saltative evolution Evolution by sudden or abrupt and discontinuous changes ("leap" = saltus in Latin).

selection teleology A mode of teleological explanation which suggests that a feature exists in a population because it is being selected for its beneficial consequences to its bearers. Teleological explanations based on natural selection are appropriate for organisms but not for artifacts or for non-living natural objects (e.g. stones, clouds).

selfish gene Any stretch of DNA sequence that displays self-perpetuating evolutionary behavior without apparent benefit to the organism.

sexual genome The full suite of genetic material within each cell of a sexual species.

small probability arguments A common type of argument raised by creationists against evolution which asserts that the origination of complex biological features (e.g., anatomical structures, cells, or genetic information) by means of Darwinian evolution is too unlikely to be credible. All such small probability arguments against evolution are fallacious, as it is easy to generate events with a probability smaller than any bound (e.g., the outcome of repeated tosses of a coin). Therefore, an event (e.g., the origin of complex biological structures) having a small probability does not imply the improbability of the theory assigning this probability (e.g., Darwinian evolution).

symbiosis A broad term that covers parasitic, mutualist and commensal interactions between biological entities; interactions may be obligatory or facultative; endosymbiosis refers to symbioses that take place within cells (as opposed to between cells or organisms).

symplesiomorphy A plesiomorphy shared by members of a taxon.

synapomorphy An apomorphy shared by members of a taxon.

synthesis, Modern The standard theoretical framework in evolutionary biology, a synthesis of the original Darwinism and of Mendelism, based on the modern theory of population genetics. Achieved from the 1920s through the 1940s.

synthesis, Extended A proposed extension of the Modern Synthesis to take into account a broader range of biological phenomena (e.g., phenotypic plasticity), to incorporate new disciplines (e.g., evo-devo, genomics), and to factor in new concepts (e.g., evolvability).

systematics The scientific practice of classifying objects, usually biological organisms, by the relations between them. It is similar to but not identical with taxonomy.

systemic approach (or systems thinking) A general methodology holding that a system cannot be explained by decomposing it in its parts and then looking at their properties in isolation. Components must be understood in the context of the whole system. Systemic approaches focus on circular rather than linear causality.

systems biology A group of experimental, analytical and modeling approaches, developed to explain how biological properties emerge through complex interactions. Largely based on integration of data from functional genomics methods, systems biology focuses on dynamical modeling and analysis of large networks (e.g. gene regulatory networks or metabolic networks).

taxon A natural (usually biological) group.

taxonomy In biology the discipline of identifying and describing species and subspecific kinds. It is the basis on which systematics is undertaken.

teleology (teleological explanation) A mode of explanation in which some property, process or entity is explained by appealing to a particular result or consequence that it brings about. There are two distinct types of teleological explanations: teleological explanations based on design and teleological explanations based on natural selection.

theism the idea of a personal creator God who interacts with his creation and answers prayer.

theistic evolution The idea that God works through the evolutionary process to bring about his will for the Earth and its biodiversity in general, and for humankind in particular.

theodicy Vindification of the justice of God in establishing a world in which evil and suffering exist.

tolerance The range and intensity of perturbations a system can sustain and still return to a reference state or dynamic. A grassland that can withstand a severe drought and eventually return to its prior character, for example, is more tolerant than one that is altered irrevocably.

transposable element Any of a class of DNA sequences that can move from one chromosomal site to another, often replicatively.

triadic view of information A concept of information that implies a relationship between: (1) a message which may be any event, linguistic, or otherwise; (2) a system of reference which the message informs the receiver about; and (3) a receiver.

trophoblast The blastocyst's outer layer of cells. It is responsible for implantation and develops into the placenta.

ultimate causes Ultimate causes are used in historical explanations (e.g., natural selection) that explain why an organism has one character rather than another.

unification account of explanation The unification account states that explanation consists in unifying diverse and distinct sets of facts by connecting them to more general patterns and principles.

utilitarian ethics Ethical theories inspired by the work of Jeremy Bentham and John Stuart Mill, emphasizing that actions are right insofar as they produce the greatest balance of happiness over unhappiness for all who stand to be affected by the action under consideration.

variation/variability, genetic/environmental Refers to the differences detectable among the members of a population, either in their genetic material or in the environments in which they developed.

volitional bifurcation Decision not necessitated by logic but made freely by a scientist, which influences the content of knowledge.

well ordered science A term referring to how and whether research (particularly in biomedicine, but not necessarily exclusively in these fields) asks the right questions in the right ways, where this means that such questions and their answers benefit the greatest number.

xenology A relation between homologous genes that are present in different organisms because of interspecies (horizontal) transfer of genetic material.

The manufacturer's authorised representative in the EU is Springer Nature Customer Service Centre GmbH, Europaplatz 3, 69115 Heidelberg, Germany. If you have any concerns regarding our products, please contact ProductSafety@springernature.com

Printed and bound by CPI Group (UK) Ltd, Croydon, CR0 4YY
30/09/2026
02225510-0019